D0164048

economic BOTANY

Beryl Brintnall Simpson

The University of Texas at Austin

Molly Conner Ogorzaly

Zilker Botanical Garden

economic BOTANY

Plants in our world

third edition

Boston Burr Ridge, IL Dubuque, IA Madison, WI New York San Francisco St. Louis
Bangkok Bogotá Caracas Kuala Lumpur Lisbon London Madrid Mexico City
Milan Montreal New Delhi Santiago Seoul Singapore Sydney Taipei Toronto

McGraw-Hill Higher Education

A Division of The ***McGraw-Hill*** *Companies*

ECONOMIC BOTANY: PLANTS IN OUR WORLD, THIRD EDITION

Published by McGraw-Hill, a business unit of The McGraw-Hill Companies, Inc., 1221 Avenue of the Americas, New York, NY 10020.

Some ancillaries, including electronic and print components, may not be available to customers outside the United States.

 This book is printed on recycled, acid-free paper containing 10% postconsumer waste.

5 6 7 8 9 0 QPD/QPD 0 9 8 7 6 5

ISBN 0–07–290938–2

Publisher: *Michael D. Lange*
Senior sponsoring editor: *Margaret J. Kemp*
Senior developmental editor: *Kathleen R. Loewenberg*
Marketing managers: *Michelle Watnick/Heather K. Wagner*
Project manager: *Jill R. Peter*
Production supervisor: *Kara Kudronowicz*
Design manager: *Stuart D. Patterson*
Cover/interior designer: *Jamie A. O'Neal*
Cover image: *Super Stock*
Senior photo research coordinator: *Lori Hancock*
Compositor: *Precision Graphics*
Typeface: *10/12 Times Roman*
Printer: *Quebecor World Dubuque Inc.*

Library of Congress Cataloging-in-Publication Data

Simpson, Beryl Brintnall.
Economic botany : plants in our world / Beryl Brintnall Simpson, Molly Conner Ogorzaly. — 3rd ed.
p. cm.
Includes bibliographical references (p.) and index.
ISBN 0–07–290938–2
1. Botany, Economic–United States. 2. Botany, Economic. I. Conner-Ogorzaly, Molly. II. Title.

SB108.U5 S56 2001
581.6—dc21 00–049549
CIP

www.mhhe.com

To our families, friends,
and colleagues

Brief Contents

Contents

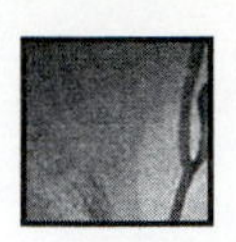

Preface

In the introduction to the second edition we remarked on the drastic changes that had occurred in the eight years since the first edition. This time, with only four years between editions, the changes are as startling. During these four years, the global population has passed six billion. While the predicted increases by 2050 are now lower than forecast in 1996, the number of people that will need to be fed, clothed, and housed will continue to rise sharply in developing countries for at least the next decade. New developments such as genetic engineering hold promise for improving crop yields to meet these needs.

Four years ago virtually no genetically engineered crops were grown. Today over 50% of the U.S. soybean crop is planted with genetically engineered seed. Genetic engineering has had mixed successes and some failures. Insects have evolved resistance to the toxins engineered into crops as rapidly as they evolved resistance to synthetic pesticides.

Since 1996, the use of new and phylogenetic methods has continued to provide data that allow us to determine the evolutionary relationships of flowering plants and to ascertain more accurately relatives of food crops. Rapid gains in biotechnology are allowing us to produce improved crops and exploit plants as factories for non-plant compounds. Research showing the antioxidant benefits of natural foods has caused a growing interest in plants as sources of bioactive substances. The herbal or dietary supplement industry has exploded during the last five years with natural plant products showing continued promise not only as medicines, but also as insecticides, and sources for new fibers. In this edition, we have included many of the exciting recent discoveries and tried to foresee where breakthroughs may be made in the future.

The organization of this edition follows that of the first two, but as can readily be seen, this new edition differs in size, format, and much of the content. The change in size and format will make the book more affordable and easier to read. We have excluded the color photos present in the second edition to reduce costs and because images are now so readily available on the World Wide Web. We have also included many new photos that are up to date and more inclusive of other parts of the world. Finally, this edition contains more boxes that highlight interesting ancillary material about the plants included in different chapters.

In the text we provide detailed coverage of the major uses of plants in the world today. Rather than being encyclopedic, our concentration is on species of major economic importance in the Western World. We have tried to provide a balanced treatment of the plants discussed by including aspects of history, morphology, chemistry, and modern usage.

We have kept the mixture of the historical and new illustrations in order to provide a sense of history but we have made significant changes in response to our readers' and reviewers' suggestions. Dark photographs have been replaced, new drawings have been substituted, and references have been updated and expanded. With these references and the current easy access to computer databases, any student should easily be able to research further a topic in economic botany. Technical data such as production figures have been brought up to date.

In the first two editions, we provided two introductory chapters to provide background information about general biology and botany. We have combined those chapters into a single introductory chapter that omits much of the basic chemistry and genetics of the earlier editions. This change reduced the size of the book and provides a more focused introduction to economic botany. Chapter 2 reviews current ideas on the adoption of an agricultural way of life by people in different parts of the world. Every year, new archeological finds add some new piece of information to this fascinating part of our history.

The body of the text, Chapters 3 to 17, deals with important angiosperm and gymnosperm crop species. We have tried to include as many new findings as possible throughout these chapters. In the first five of these chapters, we group food plants by the parts of the plants (fruits, leaves, stems, and roots) that are harvested for food. Legumes and grains are treated individually because of their importance in human nutrition. In the discussions of the plant organs that provide food, we emphasize their primary ecological functions and the ways humans have exploited them. An example is fruit pulp, which is important in nature for seed dispersal by animals. The sweet pulp serves as a food that "pays off" such dispersers; another is starch stored in tubers or roots that serves as a carbohydrate reservoir for some plant species. Such natural accumulations of carbohydrates were independently discovered by human beings in many parts of the world and incorporated into the repertoire of food plants.

In Chapters 8 to 16, we present products that are primarily extracted from plant parts. Included are substances such as volatile oils, alkaloids, latexes, and fibers. These products are grouped according to their use: spices and perfumes, textile fibers, bioactive compounds, etc. For each group of plant products we explore the natural occurrences, chemistries, and functions within the plants in which they occur. Because of the economic importance of bioactive substances, we have devoted separate chapters to medicinally important plant products, psychoactive drugs, caffeine-containing plants, and alcoholic beverages. Chapter 11 now includes information on the most commonly used herbal remedies, and Chapter 14 has recent findings on how alcohol may affect the body.

Chapter 17 concludes our survey plant uses of flowering by explaining the history and current importance of ornamental plants. For many people today, the primary contact they will have with growing plants will be ornamentals, either landscape or houseplants. Gardening has become the most popular hobby in America.

We have expanded Chapter 18 on algae by including new information on bioactive algae that produce newly discovered toxins and research on the medical potential of algal compounds. Chapter 19, Uses of Plants in the Future, has been revised to look back at what became of the predictions we made in 1986 and 1994 and to look forward to what the future might bring.

Many people have helped with advice, criticisms, photographs, and their time. In particular we thank Maureen Bonness, Don Cheney, Dennis Cornejo, Garry Fox, Greta Fryxell, Paul Horny, Jack Neff, Don Tindall, B. L. Turner, and C. Todzia for photographs. Tina Dooley and The University of Texas Health Sciences Center provided materials and help with photographs as did Central Market of Austin. Jack Neff, Bob Ogorzaly, Jonathan Simpson, Gina Ogorzaly, and Dairne McLoughlin kindly read new parts of the manuscript and made helpful suggestions. Many individuals, companies, and organizations sent us photographs to replace or supplement those in the earlier editions. These individuals are acknowledged under the photos provided. As always, our families—Jack, Jonathan, and Meghan, Bob, Mae Song, and William gave us moral support. We also appreciate the help and encouragement of our editors: Michael Lange, Marge Kemp, and Kathy Loewenberg.

We are particularly indebted to the reviewers who carefully read one or even several chapters. They provided us with an enormous amount of helpful advice on subjects ranging from pedagogy to chemistry. These reviewers include: Harvey Ballard, Jr. (Ohio University), Peter Bernhardt (Missouri Botanical Garden), James Caponetti (University of Tennessee–Knoxville), David Czarnecki (Loras College), W. Hardy Eshbaugh (Miami University, Ohio), Lee Jahnke (University of New Hampshire), Lawrence Kaplan (University of Massachusetts), Robert Ornduff (University of California, Berkeley), Bruce Parfitt (University of Michigan–Flint), Steve Smith (University of Arizona), Donald Sutton (California State University at Fullerton, Mission Viejo), John Utley (University of New Orleans), Robert Waaland (University of Washington), and Garrison Wilkes (University of Massachusetts, Boston).

We also thank the many University of Texas classes in Botany 351 who passed along suggestions and corrections during the years that we have used the book since its first printing.

Beryl Brintnall Simpson
Molly Conner Ogorzaly

1

Plants and Their Manipulation by People

Our association with plants predates our human condition. Fossil teeth of early hominids show us that the ancestors of *Homo sapiens* were omnivorous and probably consumed, and used as tools, a wide variety of plants. Excavations of cave dwellings occupied over 300,000 years ago have revealed that Neanderthal man, the extinct species closest to modern humans, gathered walnuts, hazelnuts, pine nuts, rose hips, and roasted Chinese hackberry seeds. Such archaeological findings have shown that the history of humans' association with plants does not begin at a certain point when humans "discovered" that they could eat or use plants. Rather, these finds indicate that the human ability to manipulate plants became increasingly sophisticated over time.

Although a human appreciation of plants is probably innate, the ability to exploit plants so successfully is in large part a result of our unique capacity to transmit knowledge culturally. As a consequence, generations were able to learn from previous ones, and knowledge accrued over time. Plants were tried and discarded or added to the repertoire of those already used. Because different kinds of plants were available in different parts of the world, various peoples built up their own inventories of useful plants.

It has been estimated that there are about 300,000 edible plant species. Of these, humans have historically eaten about 2500 with some regularity, but only 150 have entered into modern world commerce. Once people began to use some species preferentially over others or to sow the seeds of selected individuals, they began to alter the plants they used. Over time, people were able to generate crops that produced greater quantities of nutritious substances or were less bitter than their wild progenitors. Wild sources were ultimately abandoned in favor of species that humans were able to modify into particularly productive or pleasing crops. This trend has led to our present situation, in which only about 20 species, all highly modified by humans, are of major economic importance (see Fig. 19.7). Chapter 19 discusses some of the ways in which we are now trying to cope with this dangerous dependency.

Early humans would have chosen plants that appealed to their senses of color, odor, or taste. In some cases, this appeal reflected natural plant adaptations for fruit dispersal by animals. In others, humans learned that certain plant organs contained substances that could be used for food. By observing naturally occurring fires, humans learned to exploit plants as sources of fuel. Using natural tree and vine structures as models, humans began to use plants to fashion their own crude dwellings and ropes. In all instances, humans capitalized on products that are naturally present in plants. Some of these products are used by plants for metabolism and growth (carbohydrates), to attract animal pollinators or seed dispersers (colors, perfumes, and sugar-rich pulp), or to repel herbivores (toxic compounds). Still others provide suppleness or strength to the plant body (cellulose).

Human beings have also been able to invent novel uses for many plant products. Seed hairs from cotton, functional in nature as a dispersal mechanism, turn into cloth when gathered and processed by humans. Gums exuded by plants to ward off infection become thickening agents for articles as diverse as chocolate milk and mining operations. Chemicals that naturally deter plant predators are converted by people into substances that regulate human fertility, and wood that provides towering trees with structural support is reduced to pulp for paper on which to record human knowledge.

We depend on plants because they provide compounds essential for our existence. Plants produce **primary compounds** such as sugars and proteins that are used in a plant's basic physiology and form in part the foundation of our food web. Plants also produce an array of chemicals known as **secondary compounds** because they are not usually integral to basic metabolism but function to attract animals as well as to help the plant avoid infection, parasitism, and predation. Secondary compounds that are effective against parasites and predators may be toxic to people as well. As we shall see repeatedly in our discussions of food and beverage plants, human use of plants for food has often been correlated with the loss or reduction of poisonous compounds. Our effort to balance exploitation of some plant compounds while avoiding others is an underlying theme throughout this book.

To appreciate both the original and current uses of plants, one must understand the nature of the compounds themselves, the ways in which they are produced and sequestered, and the structure of the plants producing them. In this first chapter we provide a general introduction to the structure of flowering plants. We also outline the reproductive systems of flowering plants to serve as background for a discussion of the changes in economically valuable plants that can be produced by human selection. In chapters dealing with specific plant products we highlight explicit kinds of plant compounds, their chemistry, natural functions, and human uses.

What Are Plants?

To most people, plants are green, insensitive, rooted organisms that normally produce flowers and fruits (Fig. 1.1). Although these characteristics are true for the plants we normally encounter, they do not adequately define plants or convey an idea of the variation that exists among them. Historically, botany (plant biology) included almost all organisms that were not animals. Consequently, botany traditionally encompassed bacteria, fungi, algae, and the land plants. Today, it is recognized that a division of the living world into a plant and an animal kingdom does not reflect the evolutionary history, or **phylogeny,** of the organisms that now exist on the earth. Various attempts have been made in recent times to produce classifications that are more realistic in terms of relationships of organisms, but none has proved to be completely adequate. Since 1990, biologists have made great strides in figuring out the evolutionary history of life on the earth and the relationships among green plants and between green plants and animals (Fig. 1.2).

A discussion of what makes green plants special should start with the cell, the fundamental structural unit of life. Green plants have cells that are enclosed in a tough cell wall made primarily of cellulose (Fig. 1.3). Inside this wall is the plasma membrane, which limits movements of substances into and out of the cell. Animal cells also have plasma membranes, but they are not surrounded by a cellulose wall. Within the plasma membrane is the cytoplasm, which contains the organelles, small recognizable bodies that have specific functions, and the nucleus. From a human point of view, the most important organelles in green plants are the chloroplasts. Chloroplasts, the nucleus, and the mitochondria (where respiration takes place) all contain DNA. The DNA present in these organelles carries the genes for most of the processes carried out within them. Chloroplasts are green because they contain chlorophylls, pigments that are of primary importance in the absorption of light energy from the sun. It is within the chloroplasts that the process of photosynthesis occurs.

Photosynthesis consists of the chemical breakdown of water molecules with the release of oxygen gas and the incorporation of carbon into glucose. Solar energy fuels this fixation, or capture, of inorganic carbon dioxide into molecules that can be metabolized by living organisms. Green plants, a few bacteria, blue-green algae, and a few other kinds of single-celled creatures are the only organisms that can naturally convert radiant energy into chemical energy. In doing so, they synthesize an organic carbon-containing sugar molecule, that yields chemical energy when degraded, or "burned." During the initial step in the fixation of carbon, energy from the sun is used to split the hydrogen atoms from the water molecules, releasing oxygen. This reaction, known as the light reaction (Fig. 1.4), yields chemical energy and hydrogen atoms that enter into the synthesis of glucose (Fig. 1.5), a six-carbon sugar, during a second, "dark," reaction.

The process during which glucose is broken down is called **respiration.** During this process, oxygen is consumed, water and carbon dioxide are released, and energy is produced. Both plants and animals respire, but plants fortunately photosynthesize more glucose than they respire,

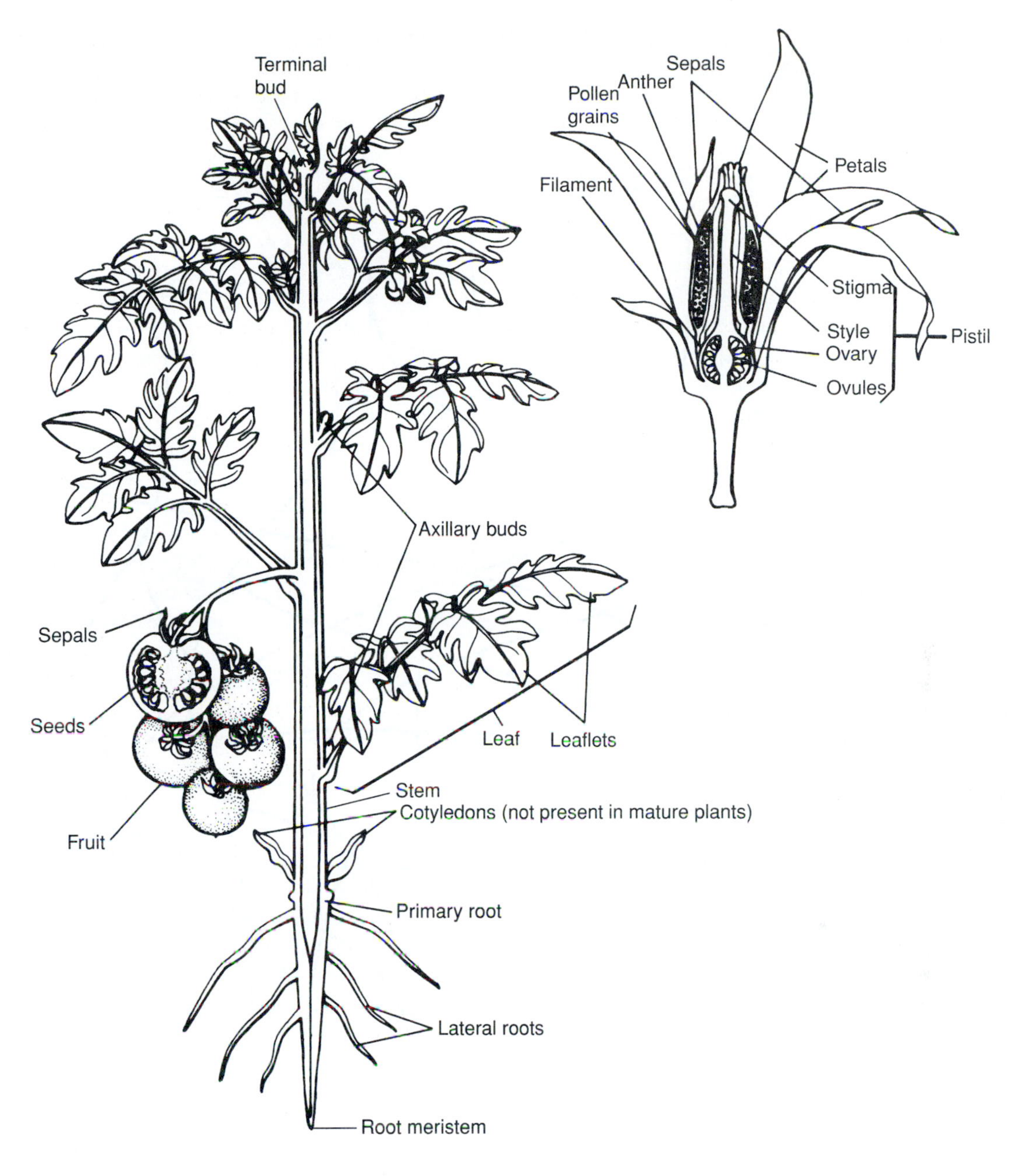

FIGURE 1.1
A tomato plant shows the basic characteristics of a flowering plant: the underground root, a shoot system consisting of stems and leaves, and the reproductive structures borne on the shoots. Axillary buds arise in the axils of leaves and produce lateral branches. The terminal bud at the apex of the plant is responsible for upward growth.

allowing them to grow and store carbohydrates. Animals can therefore consume portions of a plant's stored photosynthesized material. Removal of portions of the aerial parts of a green plant usually does not destroy the ability of a plant to carry out its vital processes.

Glucose, the primary product of photosynthesis, is most commonly found in a cyclic form with an oxygen atom forming one apex of the ring (Fig. 1.5). This molecule, in addition to being the fundamental fuel for all organisms, is a primary building block for two other extremely important polysaccharides that humans obtain from plants, starch and cellulose.

There are many different kinds of plants, ranging from single-celled green algae to eucalyptus trees. **Angiosperms,** or flowering plants, are those we most commonly encounter and are the group of primary importance for human beings. Consequently, most of this book is concerned with flowering plants of major economic

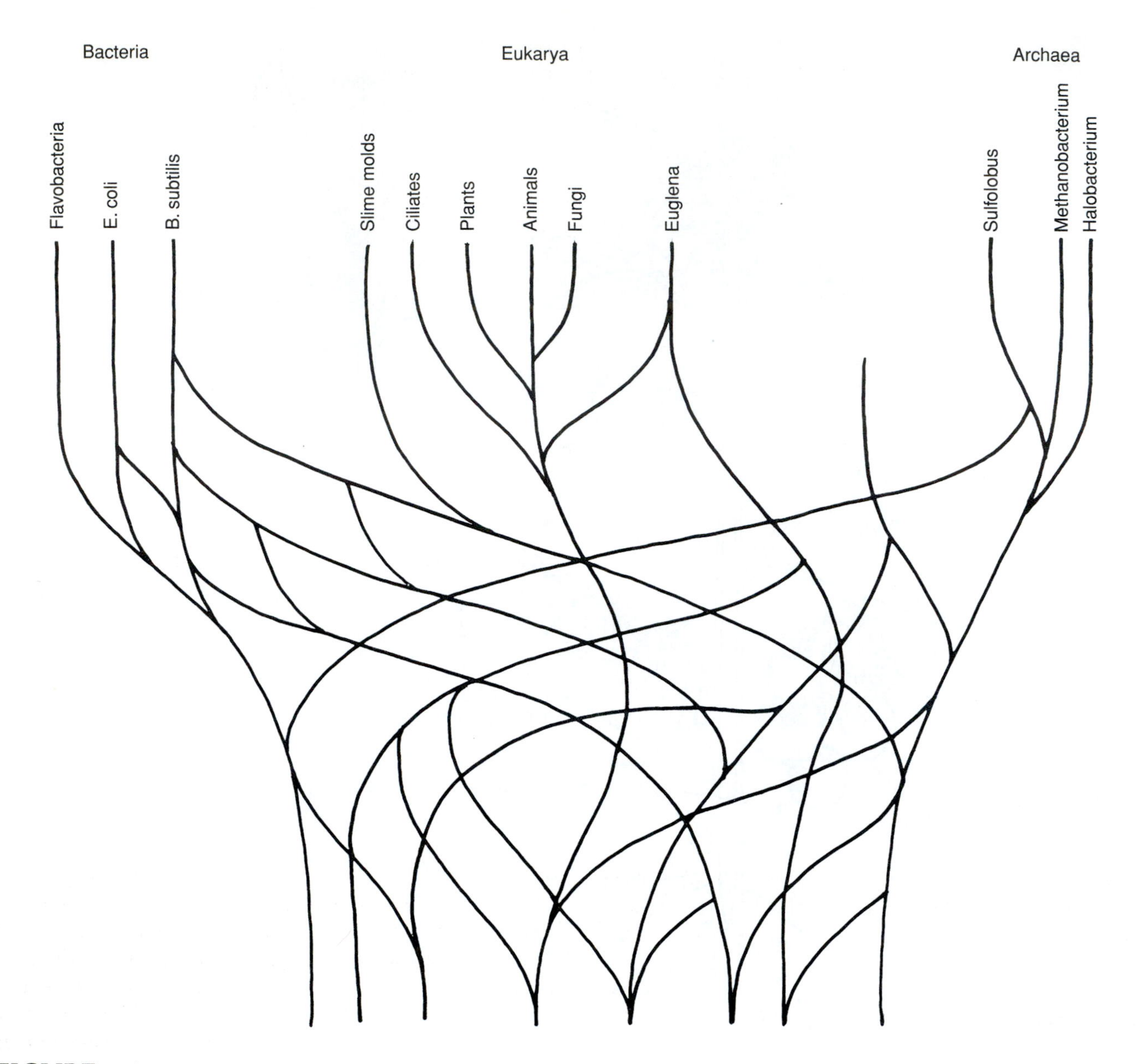

FIGURE 1.2
Despite intensive efforts by scientists using molecular methods, the root of the tree of life remains elusive and it is unclear which of the three main branches (Bacteria, Eukaryota [organisms possessing cells with membrane bound nuclei] and Archaea) is basal to the other two. As indicated by the illustration, recent evidence suggests that there was more extensive lateral exchange of genetic information within and between these lineages than previously thought. The unlabeled branch indicates an extinction.

importance. **Gymnosperms,** plants that lack flowers but produce seeds, also yield many important chemical compounds (Chapters 8 and 10) as well as wood (Chapter 16). We therefore include them sporadically throughout this book. In Chapter 18 we will also discuss algae, a heterogeneous assemblage of organisms that reproduces by a diverse array of mechanisms.

As indicated by their name, the flowering plants are unique in their production of flowers. The structure of the flower itself and the processes involved in the formation of seeds are what set this group apart from all other plants. Except as ornamentals, flowers themselves are not economically significant plant organs. However, since flowers are important in taxonomy and necessary in plant-breeding programs, we include in this chapter a detailed discussion of them and the process of reproduction. Before discussing plant reproductive structures, we briefly describe the other major plant organs: roots, stems, and leaves. For each of these organs, we emphasize their primary functions in the life of most plants and point out their various modifications for natural secondary functions. In some cases, natural modifications have altered organs in ways that particularly suit

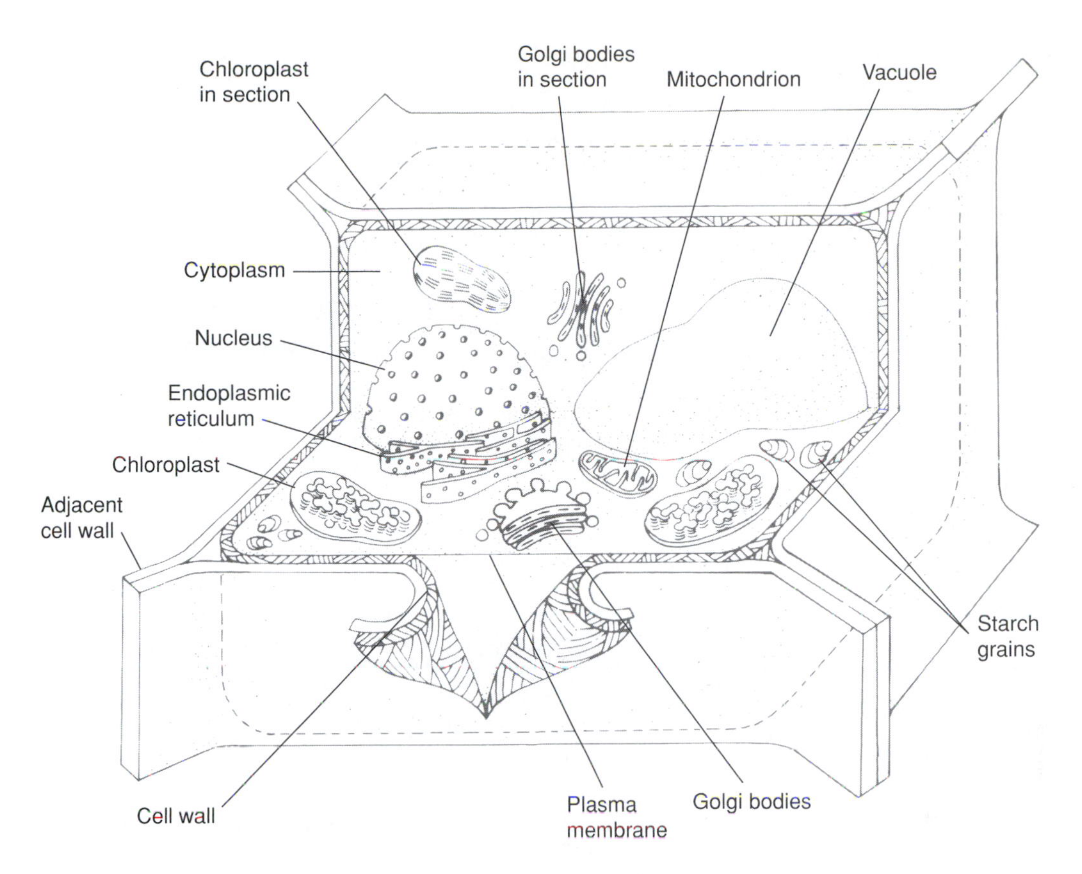

FIGURE 1.3
A diagram of a basic plant cell.

them for human use. Humans have often intensified such modifications to enhance the desirable qualities of the plants they have selected.

Vegetative Structures

Roots

The first organ to emerge from most germinating seeds is the root (Fig. 1.6). The roots of most plants perform the vital function of absorbing both water and nutrients from the soil. Nutrients can be simple minerals (iron, phosphorus, calcium, etc.) or more complex organic substances such as vitamins, hormones, or amino acids that have leached out of decaying matter. In addition to absorption, the roots of angiosperms normally anchor plants in the soil so that the process of absorption is not interrupted and the plant can remain in a position to carry out photosynthesis effectively.

The root system of most species consists of a dominant central taproot (Fig. 1.7) from which radiate lateral, or secondary, roots. The taproot is developed from the **radicle,** or embryonic root, and is the largest and dominant root. Secondary roots can also produce lateral roots, and they can branch as the root system expands. Water from the soil is generally absorbed only by root hairs that are borne along a short region behind the growing tip of a root (Fig. 1.8). Water and nutrients move from the root hairs into the central portion of the root (Chapter 7) and eventually up the stem through a conductive tissue known as the xylem. Since the root grows at its tip, it continuously moves into new areas of the soil that may provide the plant with additional nutrients. The absorption of nutrients is usually facilitated by the symbiotic association of roots with mycorrhizae, nonparasitic fungi that facilitate the uptake of nutrients in exchange for carbohydrates.

Not all plants have roots like the typical kind just described. Variations on this theme are numerous and reflect the kind of plant, the environment in which it grows, and its life history. For example, if we pull up a bunch of grass and examine its root mass, it becomes obvious that there is no taproot. The roots form a shallow, very branched, fibrous system (Fig. 1.7). Fibrous roots are formed when the radicle dies and new, approximately equally sized, roots emerge from tissue at its base. Because these roots are produced by the stem rather than root tissue, they are called **adventitious roots.** If we look at the roots of many species of tall trees growing in wet, tropical environments, we find that the roots begin to spread out from the tree trunk several feet above the ground and form a system of buttresses. These buttress roots provide stability for the shallowly rooted trees. Another kind of supportive roots,

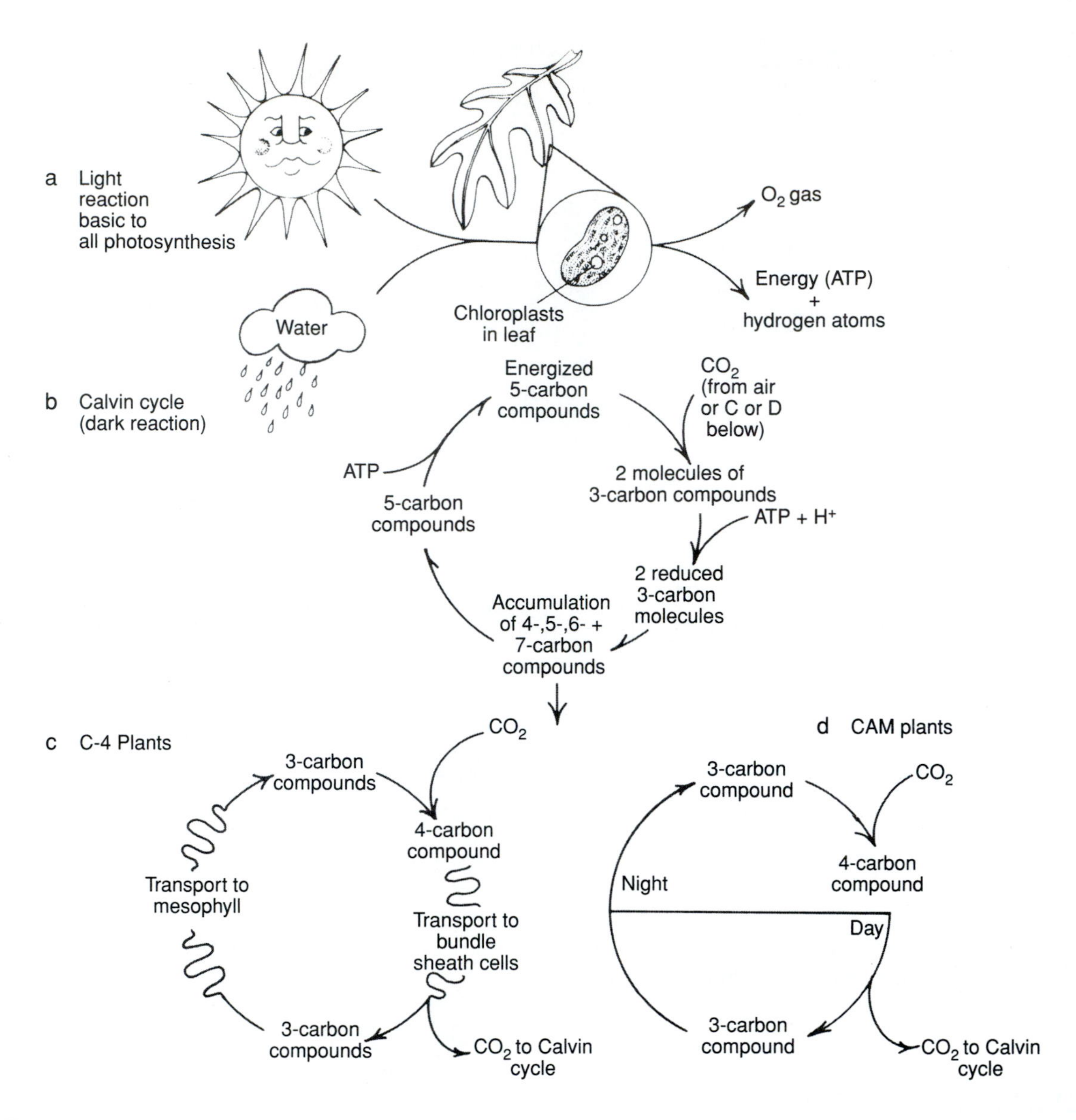

FIGURE 1.4
A schematic diagram illustrating the three ways in which plants carry out photosynthesis. (a) In all cases there is an initial reaction that splits water using energy from the sun. The products of this reaction are energy in the form of ATP, hydrogen ions, and oxygen gas. (b) In C-3 plants that follow the Calvin-Benson pathway, all the photosynthetic reactions occur in the leaf mesophyll cells. (c) In C-4 plants carbon from carbon dioxide is fixed (inserted into an organic molecule) to form a 4-carbon molecule, oxaloacetic acid, in the leaf mesophyll (see Fig. 1.13). A second 4-carbon molecule, malic acid, produced by the reduction of the oxaloacetic acid is then transported to layers of cells around the vascular bundles (bundle sheaths) where the reactions of the Calvin cycle (b) occur. Once hexose (any 6-carbon sugar) is formed, a 3-carbon compound (pyruvic acid) is transported back to the mesophyll. A third photosynthetic system known as crassulacean acid metabolism (CAM) is a modified C-4 pathway. It differs from the C-4 process in that all reactions occur within a single cell but carbon fixation and sugar synthesis are separated in time. (d) Incorporation of carbon occurs at night, and the Calvin cycle portion of the process occurs during the day. Since this strategy permits the plants to keep their stomates closed during the day, it is considered to be an adaptation to desertic conditions. CAM photosynthesis is found in many succulents.

known as prop roots (Fig. 1.9), emerge from the stems of a few species of annual monocotyledons such as corn. Prop roots, like fibrous roots, are adventitious roots.

In some situations, roots have been modified into structures that absorb water not from the soil but from other plants. Organs modified for parasitic absorption of water and nutrients are called **haustoria.** Still another modification of the typical root is common in biennial species, or species that germinate, grow, reproduce, and die over a 2-year period. Biennials spend the first year producing a leafy plant. As winter approaches, the plant transports sugars to the taproot, which becomes swollen with stored products. If the root is not harvested, the plant will use the stored energy during the second year of growth. This root modification, correlated in nature with the life history of the plant, provides human beings with an easily harvested, abundant supply of food.

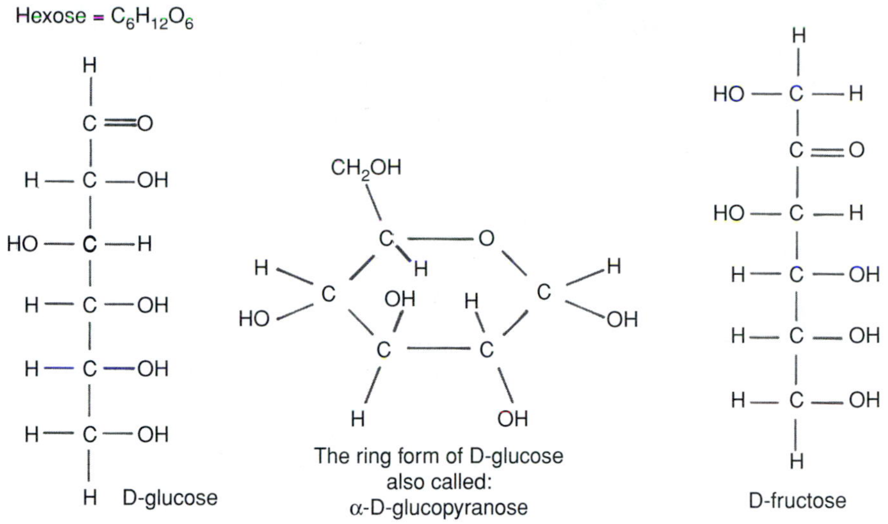

Hexose = $C_6H_{12}O_6$
D-glucose
CH_2OH
The ring form of D-glucose
also called:
α-D-glucopyranose
D-fructose

FIGURE 1.5
All these forms, or isomers, of hexose have the same molecular formula, $C_6H_{12}O_6$, but they differ in the ways in which the atoms are attached. By using structural formulas such as these rather than simple molecular formulas, scientists can communicate the relative placements of the various atoms.

Coleoptile
Radicle
Corn seed
Root
hairs
Radicle
Bean seed
Cotyledons

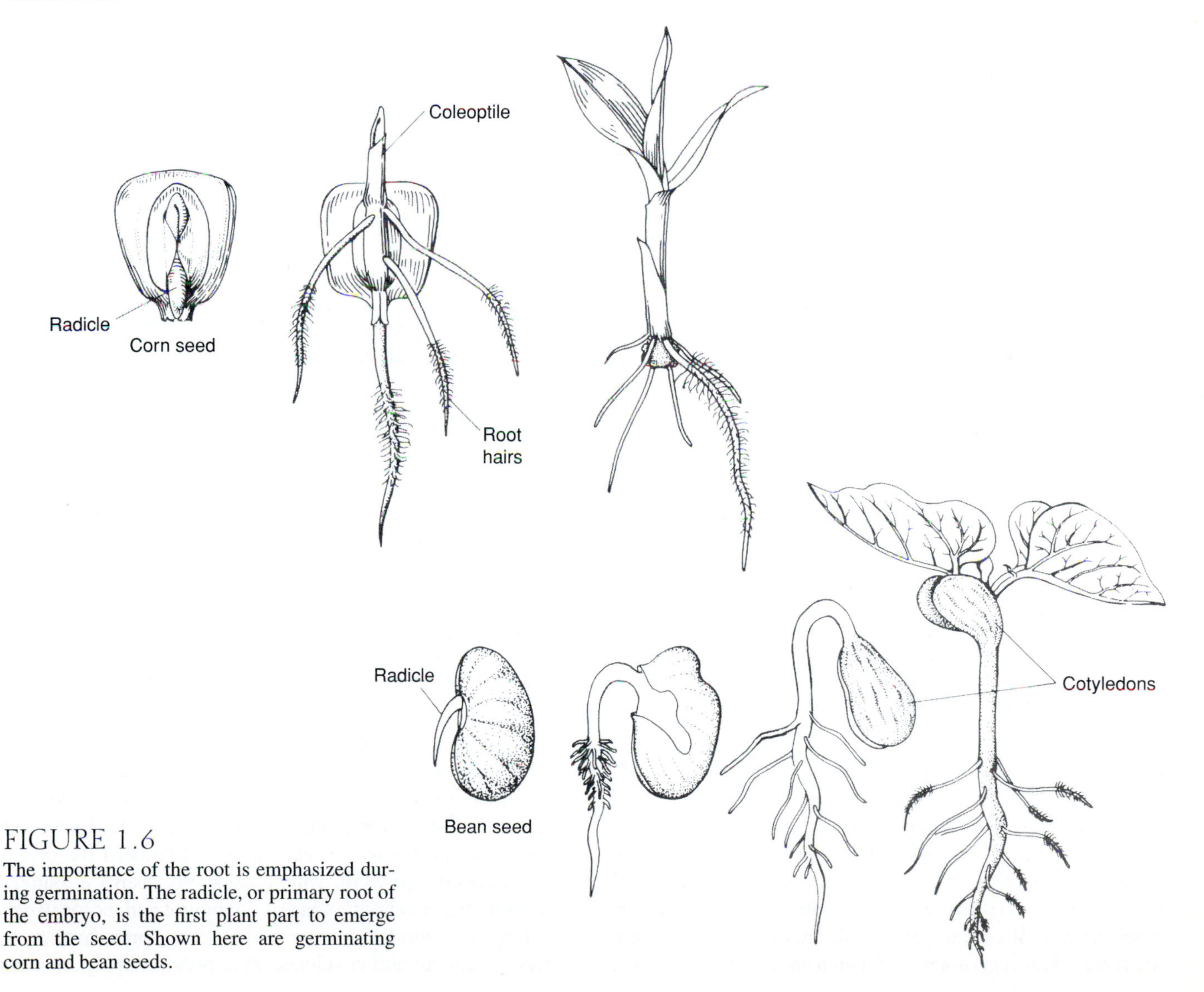

FIGURE 1.6
The importance of the root is emphasized during germination. The radicle, or primary root of the embryo, is the first plant part to emerge from the seed. Shown here are germinating corn and bean seeds.

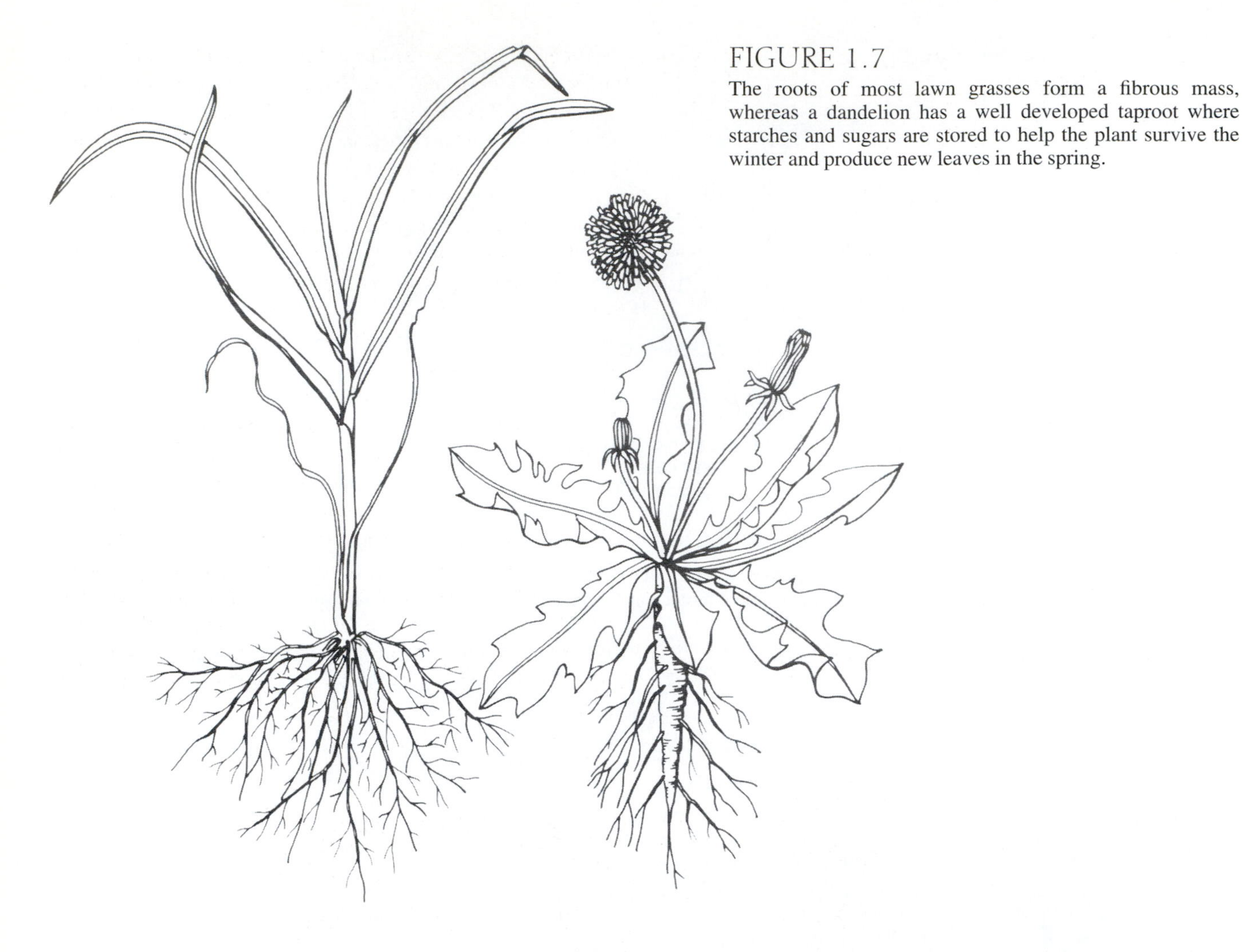

FIGURE 1.7
The roots of most lawn grasses form a fibrous mass, whereas a dandelion has a well developed taproot where starches and sugars are stored to help the plant survive the winter and produce new leaves in the spring.

Most of our temperate "root" crops (Chapter 7) have been selected from naturally occurring storage taproots of biennial plants. Some perennials, plants that live and flower over several years, also produce fleshy storage roots. In many instances, such plants are native to arid regions where dry periods cause the dieback of photosynthesizing tissues.

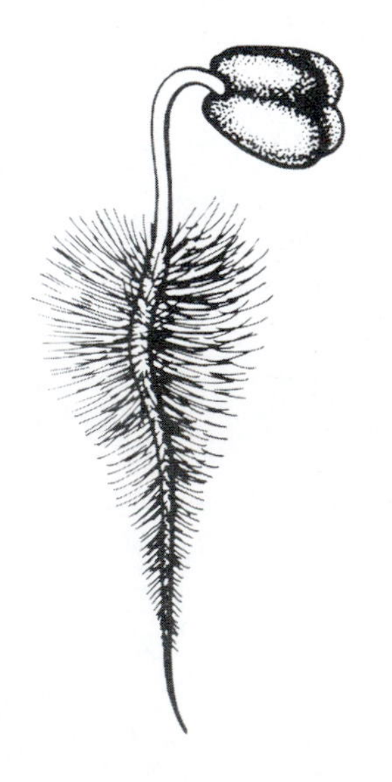

FIGURE 1.8
A radish seedling root covered with root hairs.

The Shoot System

The shoot system of a plant consists of the stems and leaves. Since these two organs are usually readily identifiable, we discuss each separately.

Stems Plant stems provide a framework, or supportive structure, that bears branches or leaves or both in such a way that phytosynthetic surfaces are maximally exposed to the sun. They also display flowers to pollinators and fruits to dispersal agents and house the conductive system (Fig. 1.10) consisting of **xylem** tissues, which conduct water and dissolved nutrients from the roots to other parts of the plant, and the **phloem,** which transports products synthesized in the leaves and stems throughout the plant (or stored in the roots during the winter). Stems are generally more or less cylindrical and enveloped by a protective covering. In

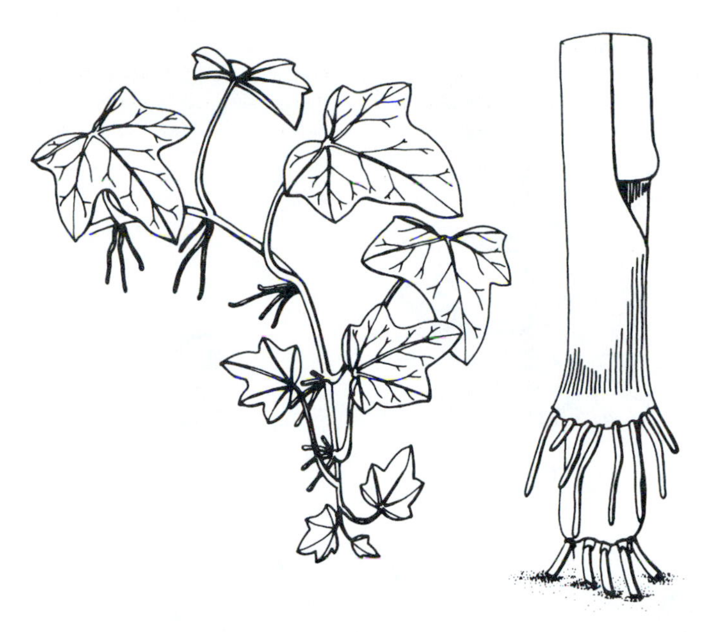

FIGURE 1.9
The aerial roots that develop along the stems of ivy are considered adventitious, as arc thc prop roots that develop on corn.

herbaceous plants, which have little or no wood, the covering is a relatively thin epidermis and the xylem and phloem occur in bundles, with the xylem closer to the center of the stem and the phloem closer to the epidermis. Woody plants, perennials with permanent stems that increase in girth by adding layers of wood, have a protective bark (Chapter 16) and a continuous ring of phloem under the bark and a mass of xylem forming most of the tissue in the central part of the stem. Additional phloem and xylem cells are produced by a layer of cells called the vascular **cambium,** which lies between them. Cambium cells retain the ability to divide and can thus continue to provide additional xylem, phloem, and other cells as the stem ages and enlarges.

Despite the basic functions common to all stems, many different or accessory functions have evolved (Fig. 1.11). In desert regions, we can find leafless species because the elimination of leaves greatly decreases the surface area of a plant and, consequently, its water loss. However, the lack of leaves presents something of a problem since photosynthesis normally occurs in leaves. Leafless species have gotten around this problem by shifting photosynthesis to the stems. As a result, stems of such species are green. In cacti, the most diverse group of leafless plants in the New World, the stems are also swollen and store water.

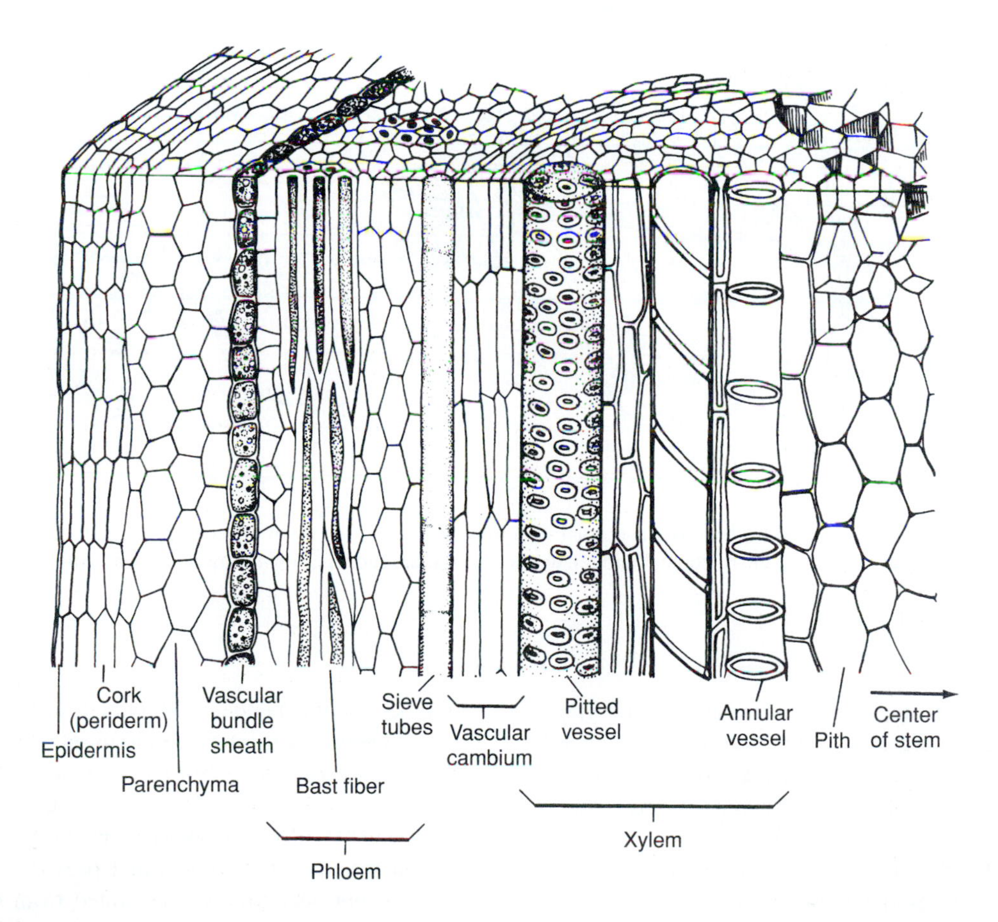

FIGURE 1.10
Longitudinal section of a dicot stem showing the outer epidermal cells, the location of fibers that strengthen the stem, and the vascular cambial cells that produce the xylem and phloem cells.

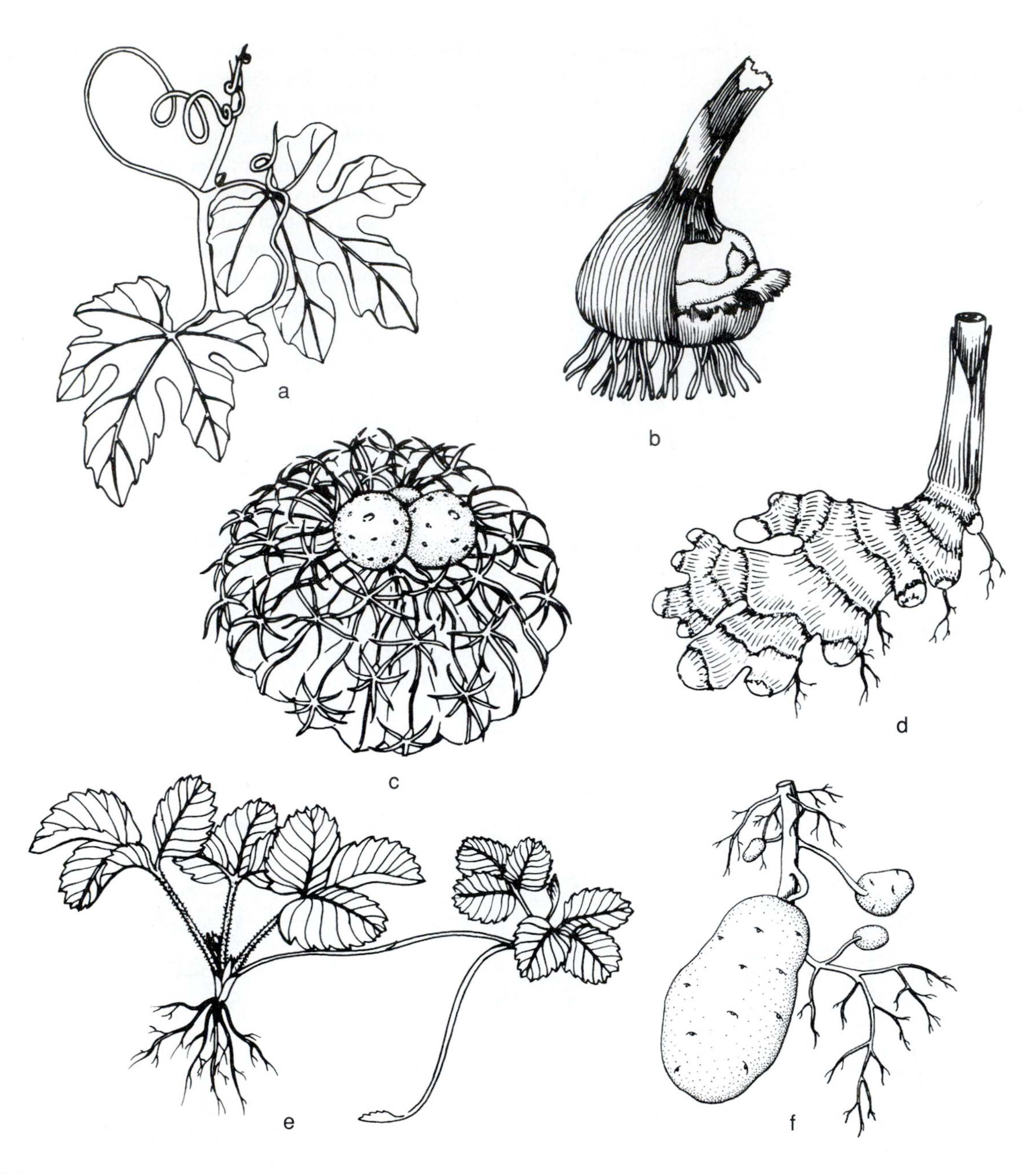

FIGURE 1.11
Natural selection has led to numerous modifications of the normal stem form into (a) the tendril of a grape, (b) the corm of a crocus, (c) the whorl of glochids at the base of cactus spines, (d) the rhizome of ginger, (e) the stolons, or "runners," on a strawberry plant, and (f) the tuber of a potato.

A different stem modification occurs in vines, such as grapes (Fig. 1.11a) in which some stems have become leafless tendrils that aid in holding the clambering plants fast to almost any support encountered. A contrasting modification of stems is that of rigid, pointed stem tips that function as thorns.

Various kinds of stem modification are important for asexual reproduction. Some plants produce runners (Fig. 1.11e), stems that creep along the surface of the soil and eventually root at some distance from the parent plant. Once the original connection disintegrates, the daughter plant becomes an independent individual. Asexual reproduction can also occur by means of underground stems. These structures often look more like roots than stems but, unlike most true roots, they are capable of producing lateral shoots. Several kinds of underground stems are fleshy because they contain stored carbohydrates that provide energy for shoot development once they are separated from the parent plant. Many of our important vegetables such as potatoes are modified storage stems (Chapter 7). Spherical underground storage stems are called **tubers** (Fig. 1.11f), and horizontal, swollen underground stems are called **rhizomes** (Fig. 1.11d). Vertically compressed underground stems encased in dry, scale-like leaves are **corms** (Fig. 1.11b). Humans take advantage of the carbohydrates sequestered in underground stems by harvesting them when storage is complete but before they begin to initiate new shoots. Many common starchy vegetables such as potatoes (Fig. 1.11f) or true yams are underground stems that functioned in the natural ecology of the ancestors of our crops as asexual reproductive units.

Leaves Leaves come in a variety of shapes and sizes, but most are flat, green, and borne along stems (Fig. 1.12). By

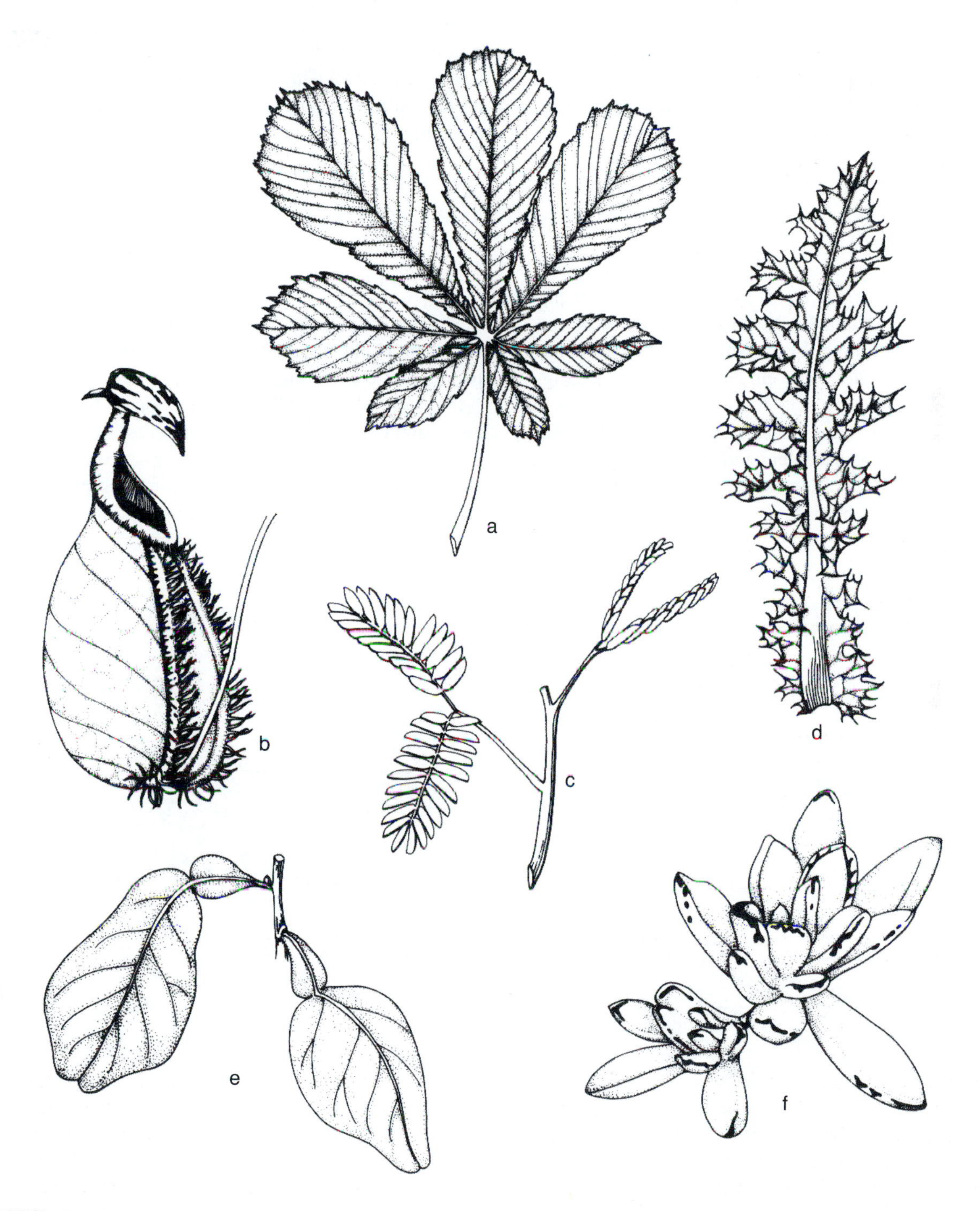

FIGURE 1.12

Leaves exhibit a wide range of variation related to the environmental conditions in which they occur. (a) A palmately compound leaf of a horse chestnut. (b) A *Nepenthes* leaf modified into a pitcher for capturing insects. (c) The pinnately compound leaves of the sensitive plant close in response to a touch. (d) Thistle leaves have stiff, tips that deter predation. (e) Citrus leaves have flattened petioles and contain packets of volatile oils. (f) The fleshy leaves of many succulents store water in arid climates.

exposing flat surfaces to the sun, a plant maximizes both the surface area exposed to sunlight and the rate of gaseous exchange necessary for photosynthesis and respiration. The movement of gases into and out of leaves occurs through small openings called **stomates.** Stomates can be found on both surfaces of a leaf, allowing gases to move across the protective layers of the upper and lower epidermis and the cuticle. In a "typical" leaf, a middle layer, or **mesophyll,** composed of elongate **palisade** parenchyma cells and widely spaced **spongy parenchyma** cells, is sandwiched between the epidermal layers (Fig. 1.13). Photosynthesis occurs primarily in the cells of the palisade layer, which contain large numbers of chloroplasts. Bundles of xylem and phloem branch through the parenchyma, transporting water and minerals to the photosynthesizing cells and carrying away synthesized products.

Although most leaves serve only as sites of photosynthesis, additional or alternative leaf functions occur in many kinds of plants (Fig. 1.12). In the common poinsettia, the bright red leaves resembling flower petals attract pollinating insects to the inconspicuous flowers they surround (see Fig. 17.39). Where browsing by animals is a threat, leaves bear spines (Fig. 1.12d) or have reduced blades and become rigid. Leaves can also be succulent and store water, particularly in dry environments. Like stems, leaves can be modified into tendrils that help hold a climbing plant to a substrate (Fig. 1.14a). In some species such as onions, which have leaves that emerge directly from the ground, leaf bases have been modified for carbohydrate storage underground. As winter or a dry period approaches, photosynthesized material is transported to the leaf bases. When frost or aridity kills back the exposed portions of the leaves, the fleshy leaf bases remain protected. The swollen collections of leaf bases surrounding the short underground stem forms a **bulb** (Fig. 1.14b) that provides energy for sprouting the next season. Like storage taproots, rhizomes, tubers, and corms, bulbs are commonly used by people because they constitute natural, concentrated sources of food.

Reproductive Structures

Flowers

For many people, the most aesthetically appealing parts of flowering plants are the flowers themselves. Flowers attract people and other animals such as insects, birds, bats (Fig. 1.15), and rodents with a combination of form (Fig. 1.16), color, and odor. All of this attractiveness is, however, just a part of the real business of a flower—sexual reproduction. Unlike most vertebrate animals, flowering plants generally produce both "male" and "female" reproductive structures

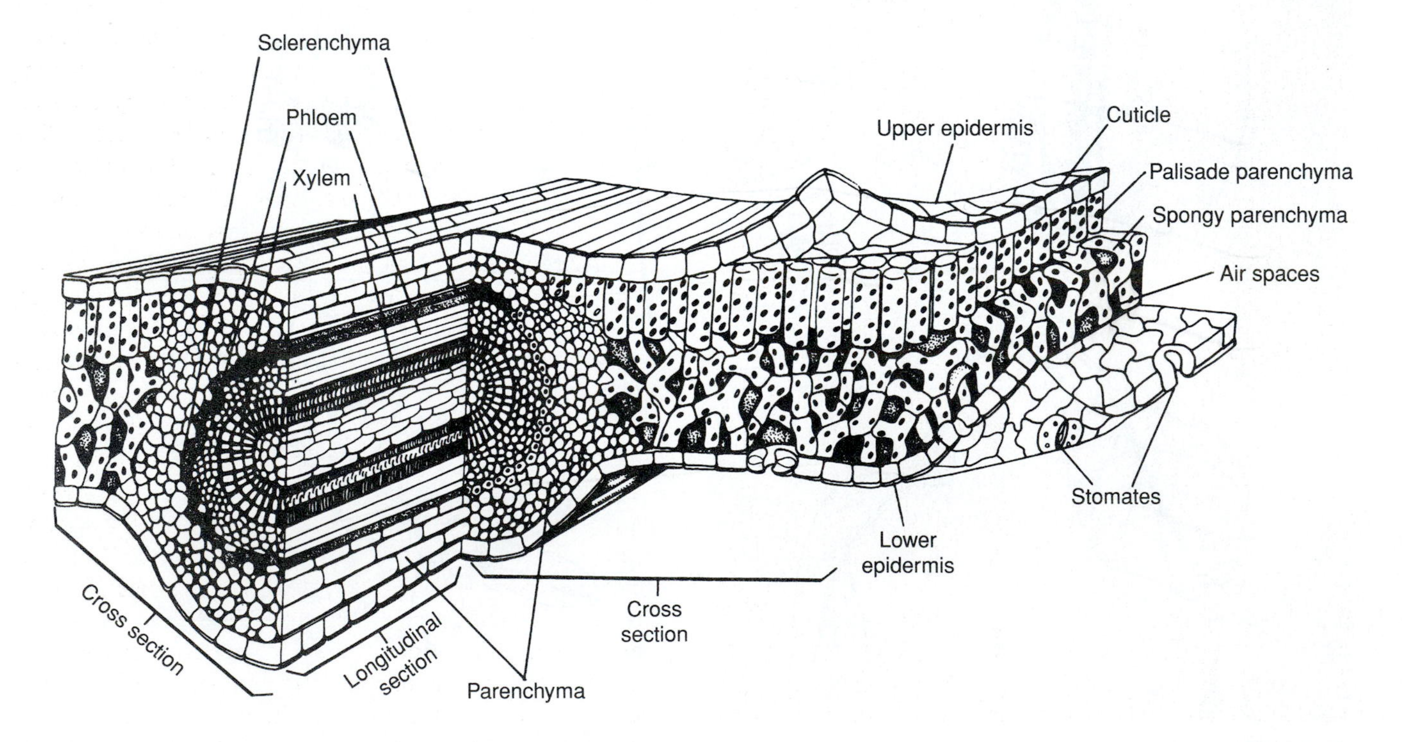

FIGURE 1.13
This leaf section shows how various cells are organized to promote photosynthesis. Gas exchange is regulated by the opening and closing of the stomates in the epidermal layer. The elongated cells of the palisade layer contain the chloroplasts where carbon dioxide enters the photosynthetic process and leads to the production of sugar (see Fig. 1.4). Vascular traces throughout the leaf deliver water and minerals and carry away synthesized sugars.

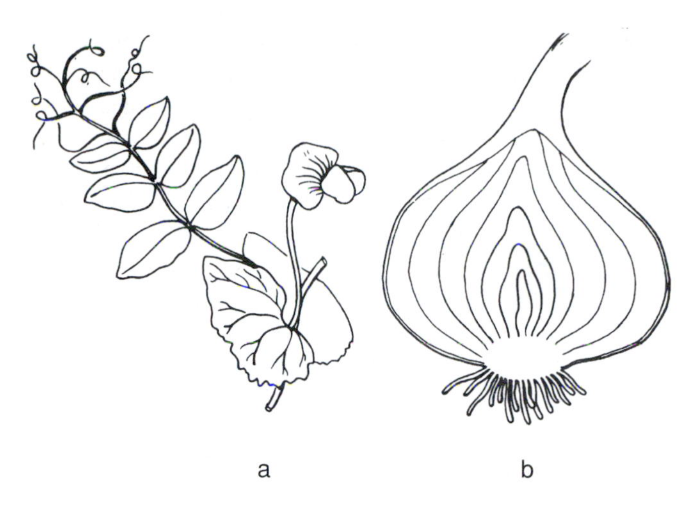

FIGURE 1.14
(a) In peas, leaves have been modified into tendrils. (b) The bulbs of onions consist of overlapping fleshy leaf bases that store nutrients and persist underground during the winter.

FIGURE 1.15
Even mammals can serve as pollinators. This San Pedro cactus is pollinated primarily by night flying bats.
(Photo by M. Tuttle, Bat Conservation International.)

on the same individual and most frequently in the same flower. A flower with both sexes is considered to be **perfect** and flowers with only one functional sex are **imperfect.** A plant that produces separate male and female flowers on the same plant is **monoecious** (*"one house;"* Fig. 1.17). Species that have individual plants that bear only male or only female flowers are **dioecious,** a term indicating that there are two kinds (*"two houses"*) of plants.

Flowers normally have an outer ring of green sepals (collectively, the **calyx**) and an inner ring of colored petals (collectively, the **corolla;** Fig. 1.18). In the center of the flower is a variable number of **stamens** and a **pistil;** stamens produce the male and the pistils, the female **gametophytes,** or sex-bearing organs. The stamens have a stalk, or **filament,** supporting the **anthers,** where pollen is produced. Pollen grains constitute the fully formed male gametophytes inside which male gametes are produced. Every pistil has a stigmatic region where pollen is received. In most plants, the stigma is a relatively small surface capping an elongate style that rises from the top of the ovary. The pistil commonly has a swollen base (ovary) that contains the ovules where the female gametes are produced. Each **ovule** consists of the female **gametophyte** and **nucellular** tissue enclosed by one or two **integuments** (Fig. 1.19). The region not covered by the integuments is the **micropyle.** It is through this small opening that the male gametes enter the ovule.

Sexual reproduction always involves a division that halves the number of chromosomes. This reduction division, known as meiosis, produces haploid (n) sex cells. Sex cells thus contain one half the diploid ($2n$) complement of chromosomes present in somatic, or body, cells. In flowering plants, the diploid individual is called the **sporophyte** because it produces the haploid **spores** (Fig. 1.20). The spores produce **gametophytes** that house the haploid sex cells (**gametes**) that will fuse during fertilization into a diploid **zygote** cell that gives rise to a new sporophyte.

Although meiosis is the starting point in the process of both male and female sex cell formation, final gamete formation follows different pathways in male and female plant organs. Formation of male gametes is a relatively straightforward process. As the anthers develop, microspore mother cells inside the mass of dividing tissue begin to undergo meiosis (Fig. 1.20). Each of the haploid cells (spores) produced by meiosis will produce a pollen grain, or male gametophyte. After the meiotic division, a specialized pollen wall begins to form around each cell. While the wall is forming, the cell within each pollen grain enlarges to its final size and then divides mitotically (Fig. 1.21). Cell membranes, but not cell walls, are produced between the daughter cells within the pollen grain. One of the daughter cells becomes the **vegetative,** or tube, **cell,** and the other becomes the **generative cell.** The generative cell will subsequently divide to form two male gametes. Pollen can be shed in the two-cell stage before the generative cell divides, or it can be shed with three cells if the generative cell has already undergone a mitotic division.

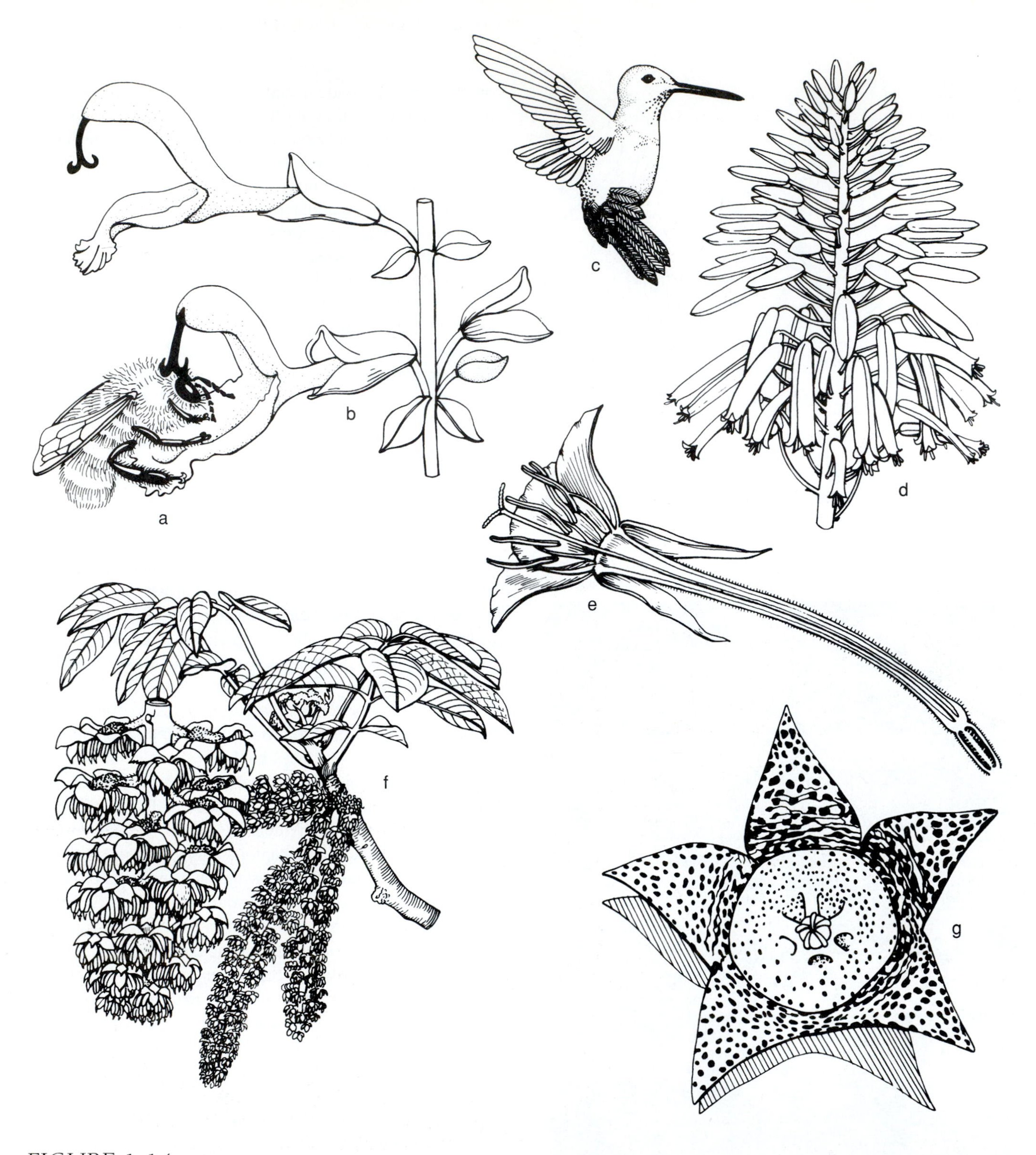

FIGURE 1.16

(a) Bees pollinate about a third of our crop plants. (b) This *Salvia* is typical of bumblebee flowers in that it has a landing platform. When the bee visits a flower to obtain nectar, the anthers touch her head and back to deposit pollen. When she flies to another flower, pollen from the previously visited flower is transferred to the new stigma. (c) Hummingbirds are the most important bird pollinators in the New World. Different kinds of birds serve as pollinators in the Old World. (d) Aloe, a native of South Africa, is pollinated by sunbirds in its natural habitat but by hummingbirds in warm areas of the Americas where it has been introduced as an ornamental. The long corolla tubes and copious nectar found in aloes are characteristic of bird-pollinated flowers. (e) The evening primrose is hawkmoth-pollinated and exhibits several characteristics associated with pollination by nocturnal moths. These include opening at night, a sweet scent, a white corolla tube, and copious nectar. (f) Wind-pollinated flowers such as those of the walnut are usually small, green, odorless, and nectarless. To facilitate pollen dispersal, the male flowers are borne in long catkins with the stamens well exposed. The female flowers have exposed feathery stigmas that easily collect wind-borne pollen. (g) *Stapelia* flowers lie close to the ground and are mottled maroon and cream-colored. This floral color, combined with a fetid odor, attracts carrion flies that mistakenly visit the flowers in search of rotting meat. While visiting a flower, a fly will pick up pollen and possibly also deposit some from another flower on the stigma while she lays eggs.

FIGURE 1.17
Wax melon flowers, like those of all members of the squash family, are unisexual, but the plants are monoecious.

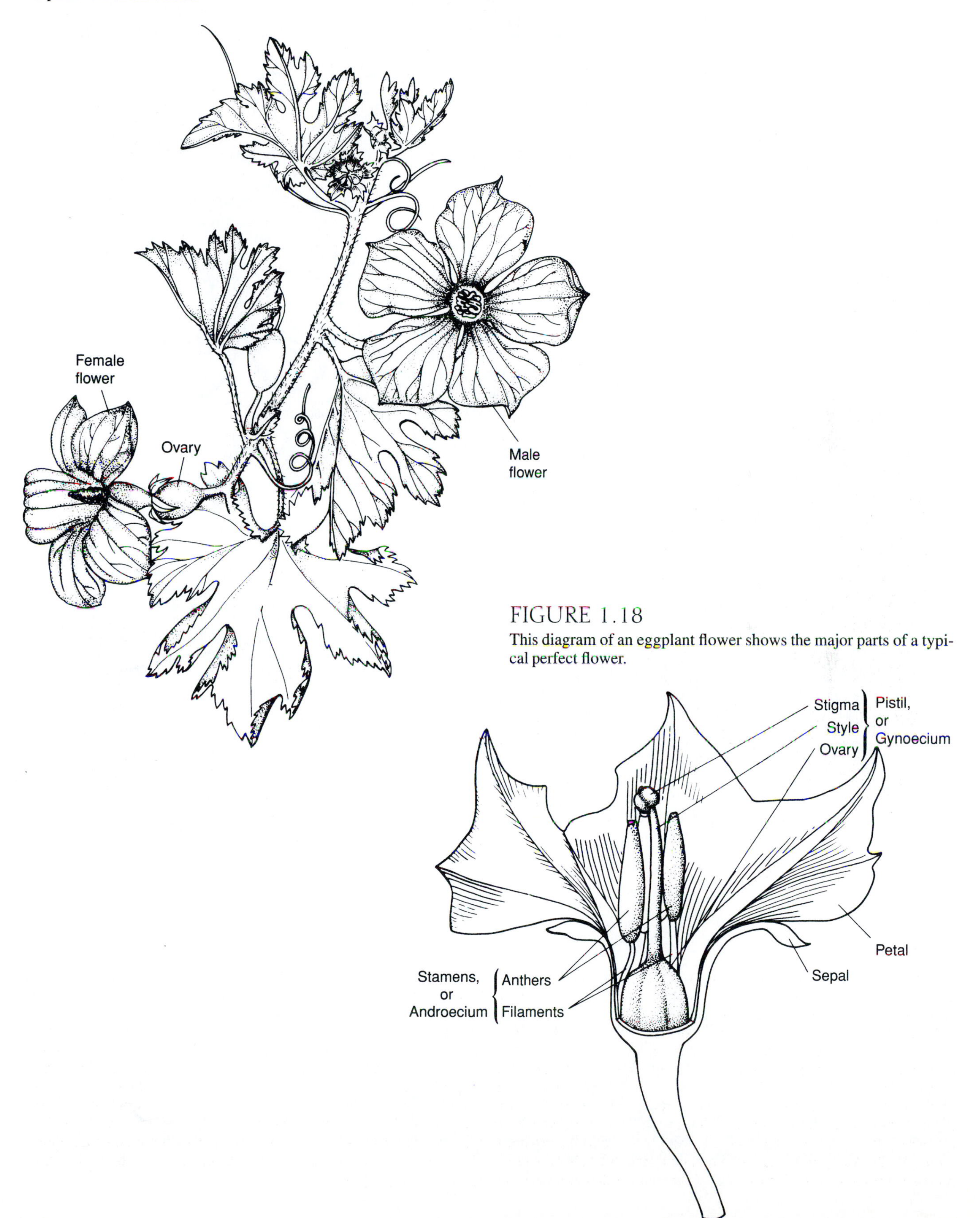

FIGURE 1.18
This diagram of an eggplant flower shows the major parts of a typical perfect flower.

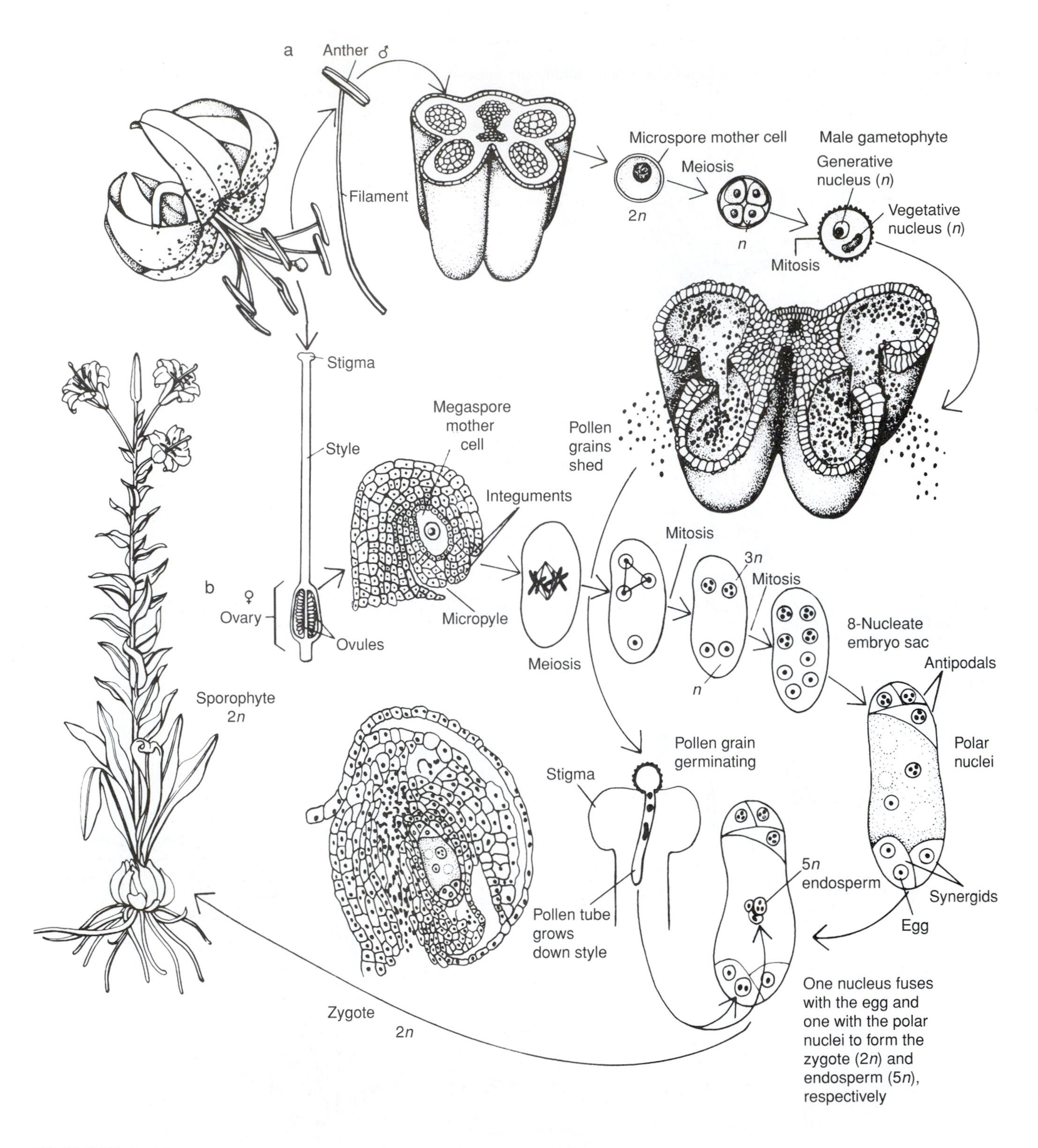

FIGURE 1.19
As exemplified by this lily flower, meiosis in the anthers (a) and ovary (b) produces haploid sex cells, or gametes. Fertilization involves the fusion of a haploid nucleus from the pollen grain with the haploid nucleus of the egg in the ovule and an additional fusion of the second male gamete and two polar nuclei in the ovule. This process, called double fertilization, produces a zygote from the first fertilization and a food tissue known as endosperm from the second. The zygote is diploid, and the endosperm is polyploid. In this example using a lily, the embryo sac is formed from all four of the original products of meiosis. Three of the original haploid nuclei fuse into a triploid nucleus. Both this triploid nucleus and the haploid nucleus undergo two successive mitoses, yielding eight nuclei in all. The egg nucleus and its synergids are haploid. The antipodals are triploid. The polar body is formed by a triploid nucleus and a haploid nucleus. After double fertilization, the endosperm tissue is pentaploid since it has been formed from the fusion of five nuclei.

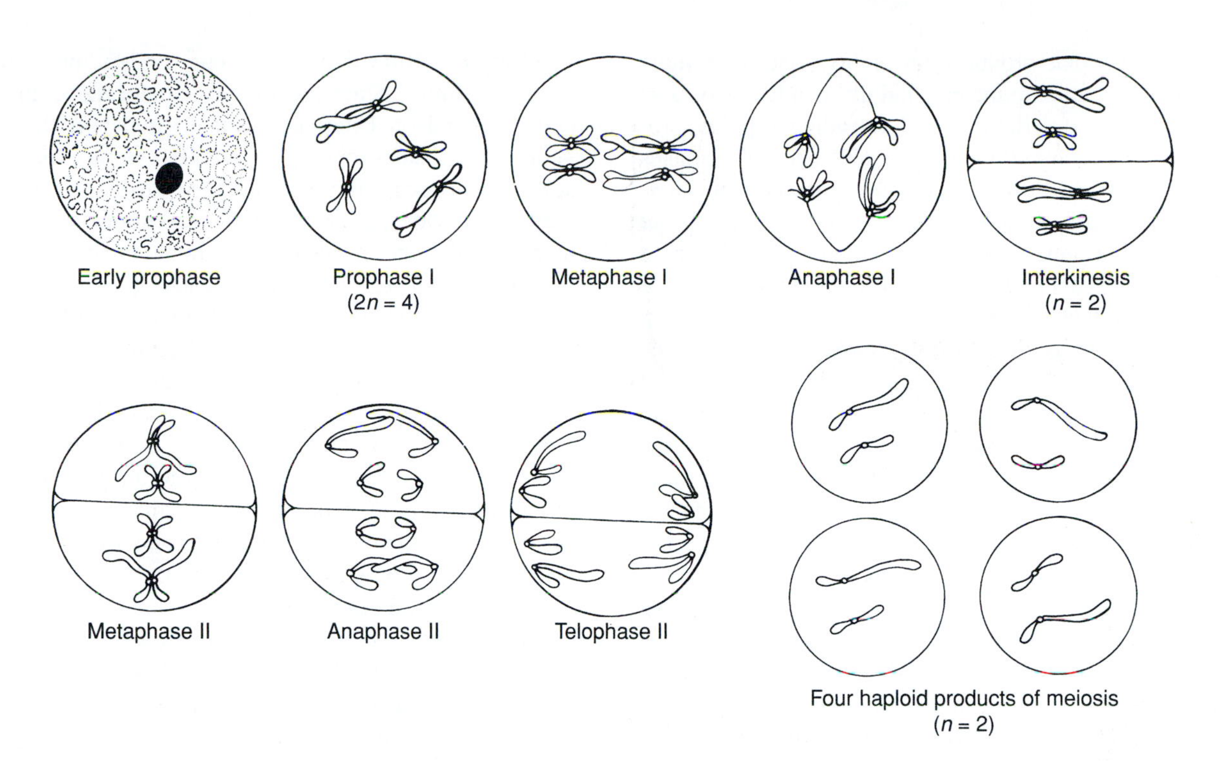

FIGURE 1.20
The process of meiosis, shown diagrammatically, produces two divisions of a diploid ($2n$) cell. The first is a reduction division yielding two haploid (n) cells. There is no change in the number of copies of each chromosome during the second division.

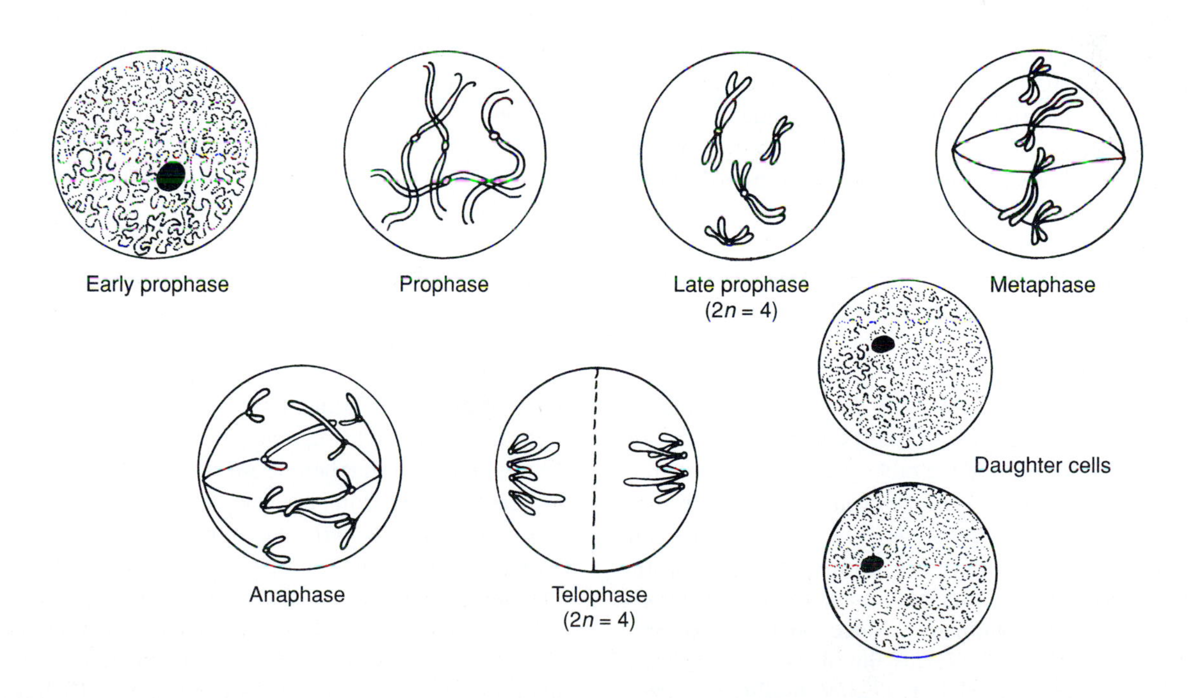

FIGURE 1.21
Mitosis is a single division yielding two daughter cells having the same number of copies of every chromosome as the parent cell.

The process that produces an embryo sac, or **female gametophyte,** containing the egg and its attendant apparatus is more intricate and variable than that leading to pollen production. Within each rudimentary ovule, one cell initiates female gamete production. This cell undergoes a meiotic division, but the resulting four haploid cells rarely all participate in embryo sac formation. The most common type of embryo sac formation appears to involve the disintegration of three of the original products of meiosis. The remaining cell then enlarges by absorbing and crushing surrounding cells. The nucleus of this cell then divides three times to form eight nuclei. The nuclei are roughly positioned so that three are located opposite the micropyle, three at the micropylar end, and two in the center of the grossly enlarged cell. The central nucleus near the micropyle is the **egg,** flanked by two nuclei. Opposite the egg are the three other cells. The two nuclei in the center of the embryo sac are the **polar nuclei.** In some of these types of embryo sacs, only two, or none, of the original nuclei resulting from meiosis degenerate. An example of one of these rarer forms of embryo sac development is shown in Figure 1.19. In this case, one of the polar nuclei has three sets of chromosomes; the other has only one. Different kinds of embryo sacs also differ in the final number of polar nuclei.

The second step in sexual reproduction after gamete production is fertilization, the fusion of male and female gametes to form a diploid **zygote.** The cell divisions of the zygote produce the embryo.

A pattern of haploid cell formation by meiosis followed by the fusion of two haploid gametes into a diploid zygote is common to all sexually reproducing organisms. Angiosperms have a second process that occurs during sexual recombination. As we saw above, the pollen grain contains two or three cells when it is shed. If only two are present when the pollen is dispersed, the third is produced later by a mitotic division. The embryo sac, or female gametophyte, as explained above, also contains more than one cell. When pollen lands on a compatible stigma, it begins to grow down through the style toward an embryo sac. It is during the period of pollen tube growth that the second male gamete is produced if it was not present earlier. The vegetative cell in the tube directs the growth of the pollen tube, but it does not fuse with any of the nuclei in the female gametophyte. Once a pollen tube penetrates an embryo sac through the micropyle, the tip ruptures, releasing the two male gametes. One of the male gametes fuses with the haploid egg cell to form the diploid zygote, and the other gamete fuses with the polar nuclei to form a cell with two or more sets of chromosomes. Repeated mitotic divisions (Fig. 1.19) of this cell produce a nutritive tissue, the **endosperm,** which is absorbed by the embryo either before it completes development or at the time of germination. Since the process of sexual reproduction in angiosperms involves two separate fusions of male haploid nuclei, it is called **double fertilization.** Double fertilization resulting in endosperm tissue is a unique feature of the angiosperms.

Before fertilization can occur, pollen must reach a receptive stigma. Since most plants cannot move, they generally depend on external agents to transport the pollen. Animal pollinators (chiefly insects) use flower form, color, and odor to locate flowers, but they visit flowers because they derive some reward from them. Without such rewards, there would be no incentive for animals to continue to visit flowers. Floral rewards are usually foodstuffs: nectar, pollen, oils, or even flower parts. In a few special cases, chemicals used in insect mating behavior or in nest construction are collected from the flowers. A few species of plants even deceive insects by mimicking a food source or a female insect of the same species. For example, some flowers, such as *Stapelia* (Fig. 1.16g), emit an odor resembling that of rotting meat, causing female carrion flies to land on them, crawl around (and become dusted with pollen), and even lay eggs. The flies then go off in search of more "rotting meat," which often turns out to be another flower. In the process, pollen gets moved from one flower to another.

Flowers that smell like rotting meat may be repellent to humans, but many of the odors that attract birds and bees are often pleasant to humans as well. Compounds that produce floral scents have therefore played an important role in the perfume industry (Chapter 8).

Flowers that use wind or water to move pollen from one flower to another are usually green, odorless, and devoid of any reward (Fig. 1.16f). The fact that production of attractant structures or chemicals was unnecessary led to selection for other investments of these plants' energy. Wind- and water-pollinated plants tend to produce large amounts of pollen and have large stigmatic surfaces that can intercept the passively moving pollen. Anyone who suffers from pollen-induced allergies can attest to the large quantities of pollen being shed into the air during "allergy" season.

Fruits

Every sexually reproducing flowering plant bears some type of fruit that serves the purpose of protecting the seeds and dispersing them to areas where they can germinate. The structure of a fruit often gives a clue as to the mechanism of dispersal. Fruits dispersed by the wind are dry and light in weight and frequently have wings or tufts of hairlike appendages that help them float in the air (Fig. 1.22b). Water-dispersed fruits are buoyant and encased in protective waterproof coverings. Animal-dispersed fruits show variable characteristics depending upon the way in which they are transported. In relying on animals for seed dispersal, the plant must use the animal while preventing the seeds from being destroyed. Seeds carried externally by mammals (including humans) or birds usually have barbs, hooks (Fig. 1.22f, g), or sticky outer coatings that promote adherence of the fruits to skin, fur, or feathers of animals

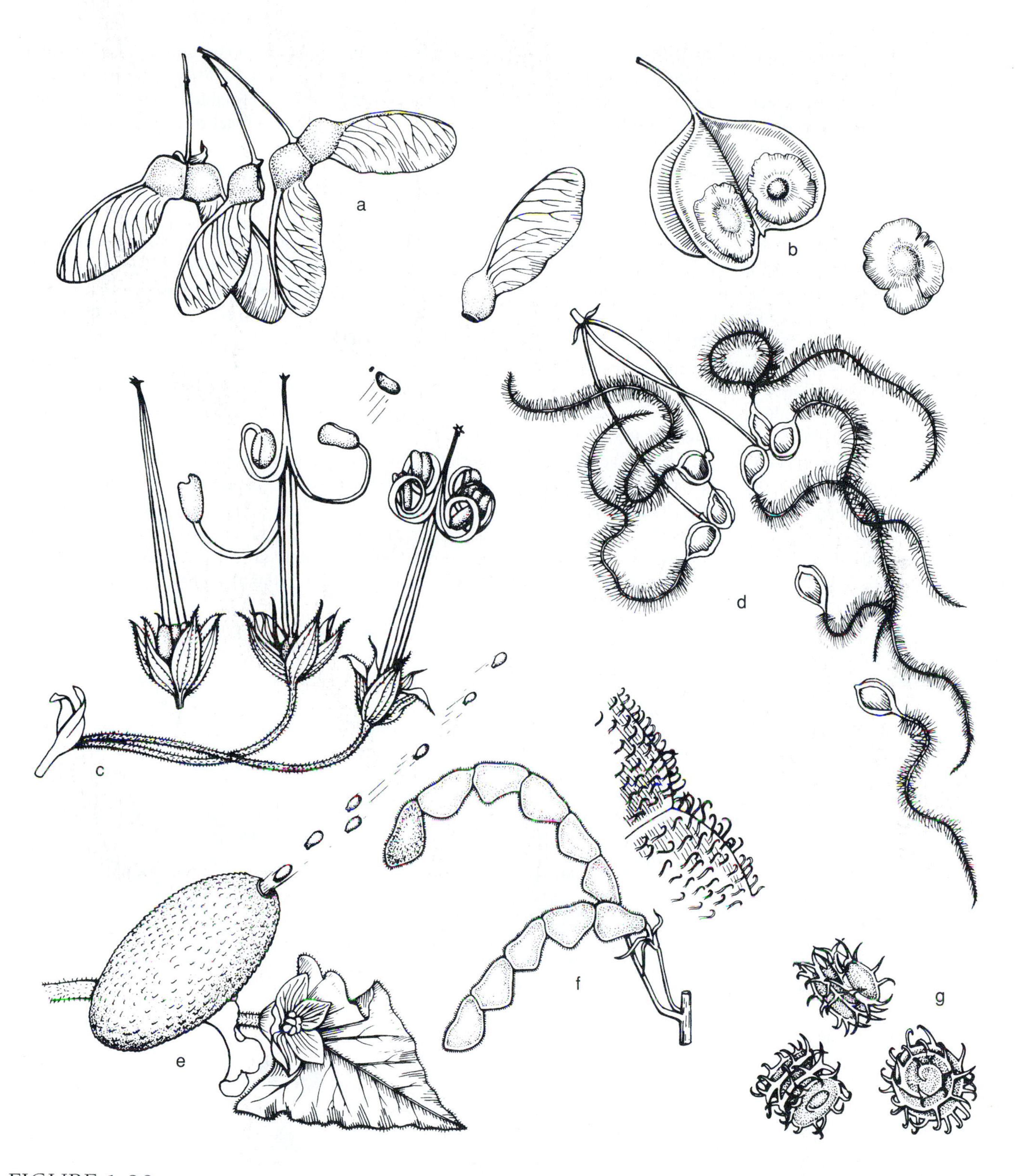

FIGURE 1.22
Wind-dispersed fruits like (a) the samaras of maple trees, (b) the winged seeds that are released from *Dioscorea* capsules, and (d) the fruits of *Clematis* all have morphological adaptations that foster movement in the air. The seeds of many geraniums (c) are catapulted as the follicles dry and dehisce, while the squirting cucumber (e) gets its common name from its unusual method of shooting its seeds. The seeds of *Hedysarum canadense* (f) and alfalfa (g) are equipped with barbs or hooks that help them adhere to passing animals that serve as agents of dispersal.

that brush against them. Eventually, the fruits fall off, possibly in a place where the seeds they contain can germinate. Some fruits forcibly eject the seeds away from the parent plant (Fig. 1.22c, e), whereas those of other species simply drop to the ground to be passively moved by rain or wind. In Chapters 3–6 we discuss in detail fruits used by people.

Plants also actively attract dispersing animals by producing edible fruits. Having edible fruits benefits the plant, however, only if animals digest the pulp of the fruit but not the seed (or seeds) it surrounds. Protection against seed destruction by fruit-eating animals is achieved in several ways. One way is to produce tiny seeds with a seed coat impervious to digestive juices. Animals usually consume the entire fruits of species with such seeds, but the seeds pass through the digestive tract unharmed and are deposited with the feces. An alternative method of protecting the seed is to produce medium-sized, or very large, seeds that are too tough to be crushed by the teeth or beak of fruit-feeding animals. Animals often carry off such fruits, but they eat only the fleshy outer portions and discard the seeds.

Fleshy, animal-dispersed fruits are a natural food source for humans, although within the array of potentially edible fruits there are many that are too bitter or unpleasant for the human palate. All the items sold in grocery stores as "fruits," and many of our "vegetables," have been selected from naturally occurring animal-dispersed, fleshy fruits (Fig. 1.23).

Seeds

Within fruits there are usually one or more seeds that contain an embryo, variable amounts of stored food, and a protective coat derived from maternal tissues. An angiosperm embryo consists of one or two seed leaves (**cotyledons**), a region that will develop into the shoot system, and the radicle, which can ultimately grow into a taproot or wither away to be replaced by adventitious roots. In some species, the cotyledons absorb all the stored food by the time the seed is shed (Fig. 1.24). The two halves of a peanut are good examples of cotyledons replete with stored food. The tiny precursors of the first true leaves can be seen in the notch of one of the two halves. In monocotyledons, the other major group of angiosperms, there is only one (*mono*) seed leaf. In grasses, an especially important group of monocotyledons for humans, food for the developing embryo is stored as a copi-

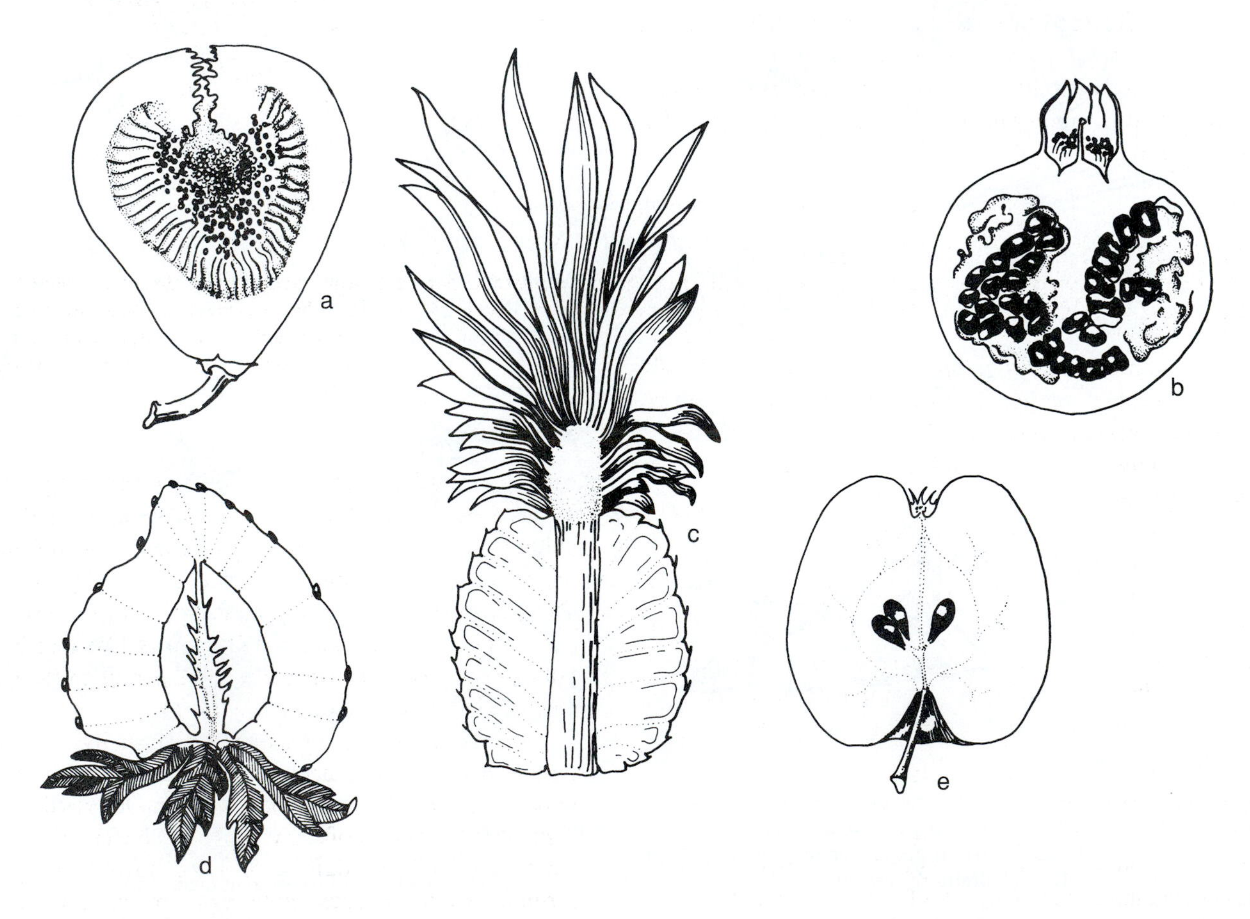

FIGURE 1.23
Animal-dispersed fleshy fruits are preadapted to serve as human foods. Some common fleshy fruit crops are the (a) fig, (b) pomegranate, (c) pineapple, (d) strawberry, and (e) apple.

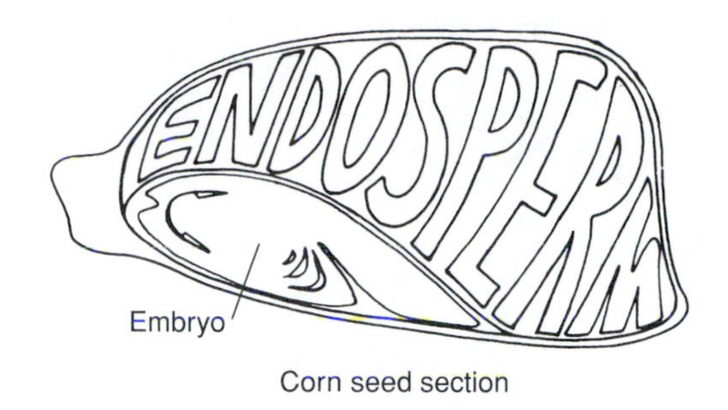

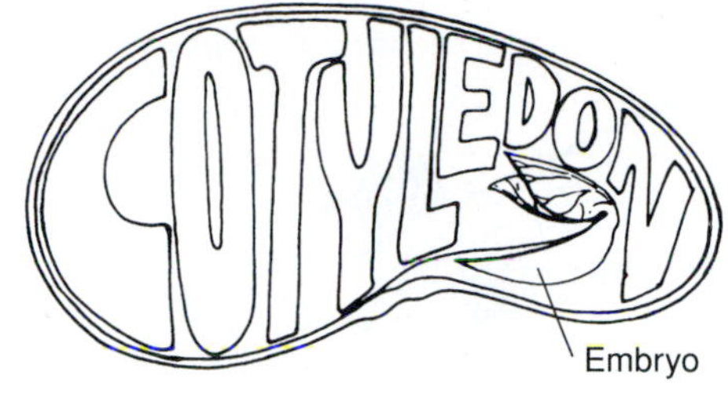

FIGURE 1.24
Seeds must contain reserves of stored food that provide energy for germination and seedling establishment. In a grain of corn, the endosperm tissue provides nutritive material for the embryo. In a common bean, the food reserves are in the cotyledons, or seed leaves.

ous mass of starchy endosperm (Fig. 1.24). In other monocotyledons, such as the coconut, the food supply consists primarily of fats and oils.

Seeds have several features that make them good sources of food for humans. First, they tend to be highly caloric because they contain comparatively large amounts of fats or starches that provide energy to the germinating embryo. Some, such as beans and peanuts, also contain significant amounts of protein. Second, seeds are often produced in large quantities, particularly in annual plants. Third, seeds of many species can be harvested readily. In perennials, the harvesting of seeds does not disturb either the roots or the photosynthetic apparatus of the plant that produced them.

Traditional Methods of Plant Manipulation

We have outlined the common pattern of sexual reproduction in plants, but compared with vertebrate animals, plants are veritable sexual experimentalists, reflected in their wide array of mating and reproductive systems. Farmers have long recognized and taken advantage of the alternatives to sexual crossing between distinct individuals that are common among plants. Similarly, variability in features such as chromosome number and the ability to regenerate new individual plants from fragments of tissue have led to the developments of novel crop varieties. These reproductive alternatives have helped humans manipulate plants in ways that led to our modern, highly selected crop species.

Inbreeding

The most common breeding system in angiosperms is **outcrossing,** the fusion of gametes from different individuals, although most angiosperm species have flowers with both male and female sex organs. The presence of both male and female gametophytes within the same flower or on the same plant raises the possibility of **inbreeding,** or **self-fertilization** (Fig. 1.25). The term *inbreeding* is used in agriculture not only for the fusion of gametes within a flower (or within an individual) but also for the fusion of gametes from separate, but genetically very similar, plants. Self-fertilization can occur only if a plant is **self-compatible:** that is if the pollen of that plant is capable of germinating on the stigmas and fertilizing the eggs of its own flowers.

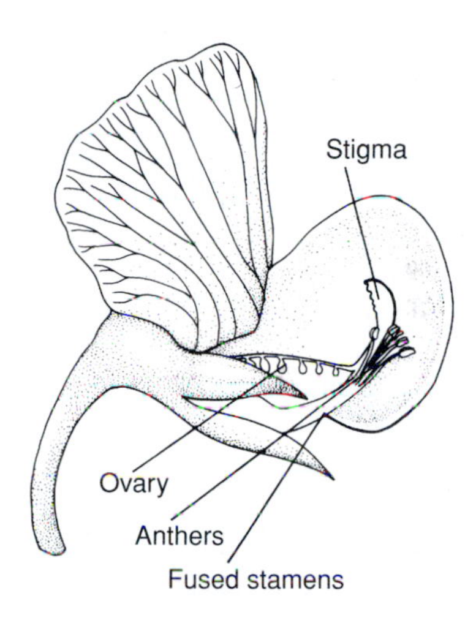

FIGURE 1.25
Common beans naturally outcross, but the pea shown here has been selected for self-pollination. The lack of an incompatibility reaction, the proximity of the anthers and stigma, and the synchronous shedding of pollen and receptivity of the stigma contribute to successful self-fertilization.

In nature, most species appear normally to outcross, despite being self-compatible. To ensure outcrossing, plants in nature employ a number of strategies. Outcrossing can be assured if the pollen of a given plant is incapable of germinating on its own stigmas, growing down the styles, or forming viable zygotes with eggs of its own flowers. Flowers that are unable successfully to self-fertilize are called **self-incompatible.** If fertilization can occur within a plant, the plant is considered to be self-compatible.

FIGURE 1.26
Outcrossing is fostered in wild oats by protogyny (a). The stigmas appear and are receptive before the anthers of the same flowers emerge and release their pollen. In *Fuchsia,* outcrossing is fostered by protandry (b). The stamens in a given flower mature and release their pollen before the stigma of that flower is ready to accept pollen.

Outcrossing can be fostered in both self-incompatible and self-compatible flowers if pollen is released before or after the stigma of the same flower is receptive or if the anthers and the stigmas are spatially separated (Fig. 1.26). Outcrossing, even at low levels, ensures variability in the offspring of an individual.

Why is variability selected for in natural populations? Outcrossing leads to genetic diversity among the offspring of a plant. All evidence points to the fact that, in the long run, organisms in nature that produce variable offspring will ultimately leave more survivors than those that produce uniform offspring. The greater success of a variable array of offspring is apparently related to the fact that the environment is not uniform, in either space or time. Since there is no way to determine in what kind of germination site a seed will land, or what climatic and biological factors it will face during its lifetime, natural selection can not lead to a single guaranteed successful **genotype.**

Although outcrossing is the norm, annual herbs and many weeds characteristically self-pollinate and produce seeds that are largely derived from the fusion of gametes produced within a single flower. Because such plants are mostly independent of external agents for successful pollination and fertilization, seed set tends to be high. In terms of the genetic consequences of self-fertilization, there is a reduction in variability in the offspring of an inbred individ-

FIGURE 1.27

Producing seed by manually crossing flowers of plants with known, but different, genotypes may take more effort than simply allowing natural pollination to occur, but the resulting hybrid plants can exhibit some qualities superior to those of their parents. In this 1880 advertisement, the hybrid vigor of the Mikado tomato was promised to make it superior to the old varieties.

(Courtesy of the Smithsonian Institution.)

ual because all of the gametes involved in recombination contain a limited array of alleles.

From the human point of view, the fact that flowers of the same plant (or genetically very similar individuals) can produce viable seeds is often a desirable character because it permits the rapid production of genetically identical, or **homozygous,** individuals. Agriculturists favor such homozygous, or inbred, lines of crops because they produce uniform stands of selected genotypes of known quality or lead to synchronous maturity, or both. Inbred lines are also used as parental stocks for the production of hybrid seed. The interest in hybrid seeds is related to the fact that plants resulting from the crossing of two inbred lines are often larger and produce larger seed crops than either homozygous parent, a phenomenon known as **hybrid vigor** or **heterosis** (Fig. 1.27). If the seeds of the crops grown from hybrid seed are used for further planting they will produce a heterogeneous array of offspring because of meiosis and genetic recombination. Consequently, new hybrid seed, sold by companies that produce seed solely for planting agricultural crops, must be purchased each year. Figure 1.28 shows how inbred lines are used to produce hybrid corn.

The ability of some plants to self-fertilize is of particular significance to humans for another reason. In many cases, a crop is grown for its seed or fruit. Especially in large fields planted with a single crop, it is often difficult to achieve successful pollination of all the flowers if they are completely dependent on insects or wind for pollen transfer. Humans have, therefore, often selected for mutations that facilitate self-pollination. Figure 1.29 diagrams the postulated changes in the flowers of cultivated tomatoes brought about by human selection from the natural outcrossing system to the self-pollinating system now found in all commonly cultivated varieties. Other crops, such as wheat, peas, and grapes, are all predominantly self-pollinating in nature. If selection for self-fertilization is not possible, effective pollinators must often be artificially provided during flowering periods. Commercial beekeepers often rent hives to farmers with orchards, cotton fields, or alfalfa grown for seed during flowering times in order to ensure a high level of pollination.

Asexual Reproduction

Asexual, or vegetative, reproduction is another method of propagation that has often been useful for human manipulation of plants. When plants reproduce asexually, the "daughter" plants are genetically identical to the "parent" because no meiosis and recombination has occurred. All the resultant plants thus formed can be considered to form a single **clone.** The word *clone* means an identical replica, in this case genetically identical individuals. One kind of asexual reproduction involves the production of seeds directly from diploid maternal cells. No male-female gamete fusion is involved so the seeds are identical genetically to the mother. These seeds can be stored and planted exactly like sexually produced seeds.

Vegetative propagation refers to the use of pieces of leaves, stems, rhizomes, or tubers to regenerate new individuals. Because vegetative propagation employs pieces of mature tissue to start plants, it is often more efficient than planting seeds. Young seedlings are more vulnerable to fungal attack than are shoots from rhizomes or tubers. Plants also tend to mature faster if grown from asexual parts than if grown from seed because the time delay involved in seedling establishment is circumvented. Both of these forms of asexual reproduction, even more than inbreeding, ensure crop uniformity that is advantageous in many agricultural operations.

Grafting is one artificial method of asexual reproduction that ensures the perpetuation of the genotype of the plant from which the graft was taken and that produces a

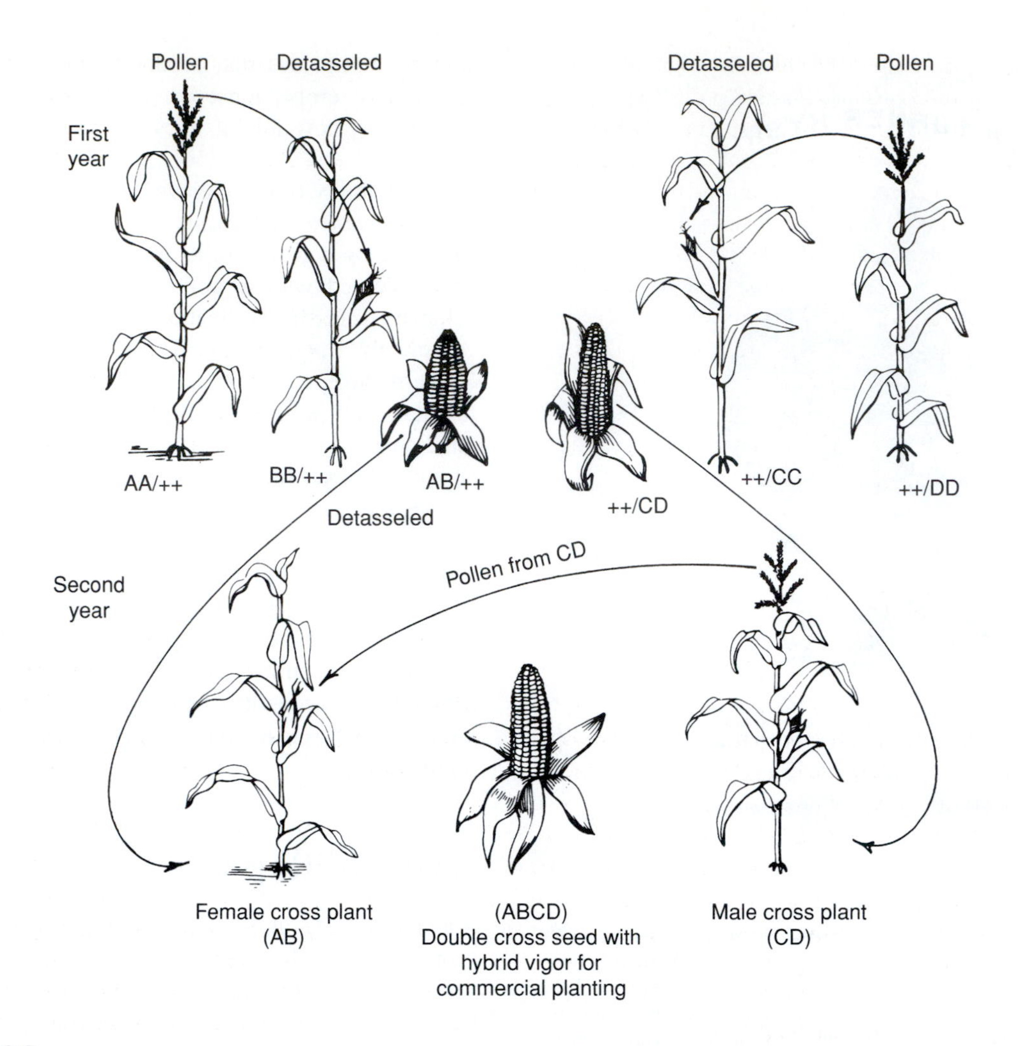

FIGURE 1.28
The production of the hybrid corn from inbred, homozygous lines. The cross shown is called a double cross because the breeder is interested in a hybrid that is heterozygotic for two genes, one having alleles A and B and the second, alleles C and D. The breeder's plan is to use homozygous lines AA and BB for gene 1 and CC and DD for gene 2. The ++ refers to undesirable alleles of gene 2 in AA and BB plants and of gene 1 in plants CC and DD. Once the initial cross has yielded a heterozygotic individual for allele 1 (AB/++) and a heterozygotic individual for allele 2 (++/CD), the breeder can simply cross these two plants for the desired combination of ABCD.
(After a USDA illustration.)

FIGURE 1.29
Outcrossing was fostered in the wild tomato by the special spatial of the anthers and the stigma. Pollen shaken from the pores of the anthers did not reach the longer style of the same flower. Humans have selected for a shorter and shorter style until now the stigma is below the level of anthers and pollen shaken from the anthers will naturally fall on the stigma. In this way, we are assured a high level of fruit set in this self-compatible species.

"plant" as mature as a sapling several years old. In the process of grafting, a branch or a bud of a desirable woody tree or shrub is joined to a rootstock or stem of another individual (Fig. 1.30). Care is taken to assure that the living cambium layers of the scion (branch or bud) and the stock (rootstock or stem) are aligned so that the actively dividing tissues come into contact and grow together. The cut tissues are protected from desiccation until the graft takes hold. If the graft is successful, the crop-bearing part of the plant will express the genes of the plant from which the scion was taken. A variety of stocks can be used depending on local soil and climate characteristics. Selection of a stock is therefore determined by its cold tolerance, disease resistance, and so on in a particular environment.

Polyploidy

Large-scale changes in chromosome number have played a major role in plant evolution and in the history of domesticated crops. Chromosome number increases can occur if homologous chromosomes fail to separate at the first meiotic division, leading to a "gamete" with an unreduced diploid, instead of a haploid, chromosome number. If these diploid cells fuse with gamete cells that have the normal haploid number, zygotes having three times the haploid number of chromosomes will be formed. If a gamete cell with a double number of chromosomes fuses with another unreduced diploid gamete, the zygote will have four times the number of chromosomes of the normal haploid cells. Multiples of the haploid number (n) greater than 2 ($2n$) are called **polyploids.** The kind of polyploidy reflects the number of times the haploid number is repeated. A cell with three times the haploid number is a triploid ($3n$); a cell with four times the haploid number is a tetraploid ($4n$); and one with six times the haploid is a hexaploid ($6n$). Table 1.1 lists some cultivated crops that are believed to be of polyploid origin, their chromosome numbers, and presumed ploidy level.

Polyploidy is relatively common in plants and represents a way in which hybrid individuals with unlike chromosome sets can become reproductively successful. Normally, if a plant is formed from two gametes with dissimilar chromosomes, it will have difficulty in producing functional gametes because there will not be a true partner for each chromosome at meiosis. In other words, although there may be an even number of chromosomes, a given chromosome will not recognize, and pair with, another chromosome. If, however, each gamete contains the diploid number of chromosomes and two such gametes fuse to form a zygote, the zygote will have two diploid sets of chromosomes. As a result, every chromosome will have at least one partner chromosome (homolog), and regular meiosis can occur in the organism with the $4n$ number of chromosomes. Polyploids can also be formed by a doubling of the hybrid chromosome complement after the zygote is formed. Figure 1.31 shows how this process works. A polyploid that has dissimilar diploid sets of chromosomes such as that shown in Figure 1.31 and present in such species as the rutabaga is called an **allopolyploid.** If all of the diploid sets of chromosomes in the polyploid of an organism are the same (as in the snapdragon), the plant is called an **autopolyploid.**

Polyploid plants are usually larger and in many cases have larger fruits or seed crops than their diploid parents. These are traits that humans may find desirable, and farmers have perpetuated them without realizing, until the advent of modern cytology, that they were due to an increase in chromosome number. Now that we know how polyploidy acts, and how to induce it, we can experimentally produce polyploids of many crop species. In subsequent chapters, we point out important crop plants (such as bread wheat, Chapter 5) that are polyploids. In many cases, the original increase in chromosome number occurred as a chance accident thousands of years ago but produced a type of plant that humans perpetuated over others. In other cases, modern

TABLE 1.1 Some Major Crop Species of Presumed Polyploid Origin

COMMON NAME	SCIENTIFIC NAME	FAMILY	APPARENT BASE NUMBER	CHROMOSOME NUMBER	PLOIDY LEVEL
Coffee	*Coffea arabica*	Rubiaceae	$x = 11$	$2n = 44$	Tetraploid
Cotton	*Gossypium hirsutum*	Malvaceae	$x = 13$	$2n = 52$	Tetraploid
Potato	*Solanum tuberosum*	Solanaceae	$x = 12$	$2n = 48$	Tetraploid
Strawberry	*Fragaria xananassa*	Rosaceae	$x = 7$	$2n = 56$	Octaploid
Sugar cane	*Saccharum officinarum*	Poaceae	$x = 10$	$2n = 80$	Octaploid
Tobacco	*Nicotiana tabacum*	Solanaceae	$x = 12$	$2n = 48$	Tetraploid
Wheat, Bread	*Triticum aestivum*	Poaceae	$x = 7$	$2n = 42$	Hexaploid
Wheat, durum	*Triticum turgidum*	Poaceae	$x = 7$	$2n = 28$	Tetraploid

Source: Adapted from N. W. Simmonds. 1976. *Evolution of Crop Plants.* New York, Longman.

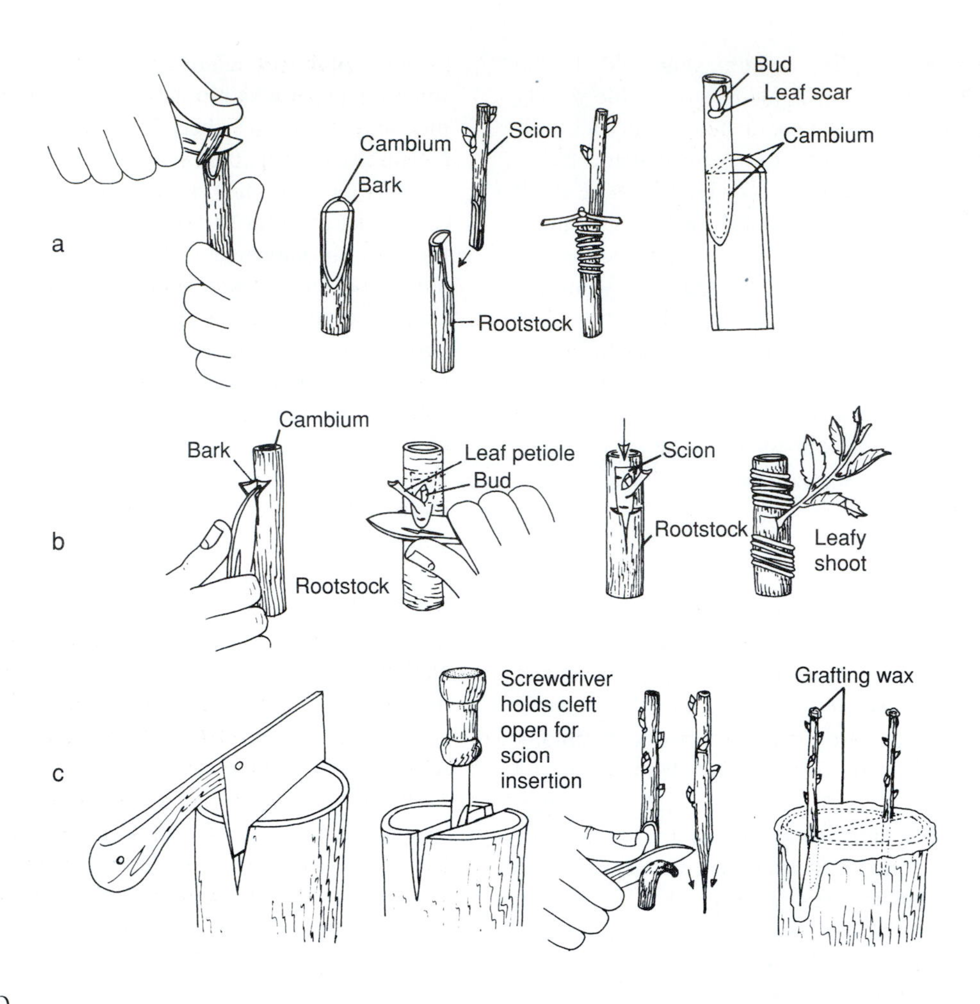

FIGURE 1.30
The essence of grafting is the positioning of the cambium layers of the scion (bud or stem) and the stock (rootstock) on top of one another so that they will grow together. Once the graft has taken, the rootstock will provide the nutrients to a shoot system that resembles the plant from which the scion was taken. (a) Splice graft. (b) Bud graft. An indication that the graft has been successful is the abscission of the leaf. (c) Cleft graft. After fusion occurs, the two scions are assessed and the less vigorous is cut, leaving only one scion.

breeding programs have incorporated the artificial production of polyploids into their methods for the development of new varieties.

Variation and Selection in Flowering Plants

Since the time humans first began to tend useful plants, they have been inadvertently and consciously selecting for the qualities they desired. In either case, selection initially took place without any real understanding of why various methods were effective. Now we understand why and how selection works, and we have been able to manipulate it to our advantage.

Throughout this book, we frequently make references to changes brought about by humans in cultivated plants, including the development of new varieties with high yields, increased nutritional value, improved appearance, and pest resistance. Changes in the genetic makeup of plants by humans is called **artificial selection** to distinguish it from **natural selection,** the major driving force of evolution in wild populations. Table 1.2 shows the requirements and consequences of evolution by natural selection. Under the action of natural selection, evolution will occur because of the differences in the numbers of fertile offspring individuals leave. This differential reproduction is known as relative fitness. For selection to work, there must be genetically inherited variation present in populations on which selection can act. Within this spectrum of variability, natural agents, or humans, carry out selection by differentially allowing certain individuals to reproduce.

How does genetic variation arise in the first place? As far as we know, the only way in which an entirely new allele is produced is by mutation, an alteration of the chemistry or reading of the DNA strand along which that gene lies. This alteration must be of sufficient magnitude to lead to the production of a different protein. Sometimes, mutations create alleles with such strange end products that the cells containing

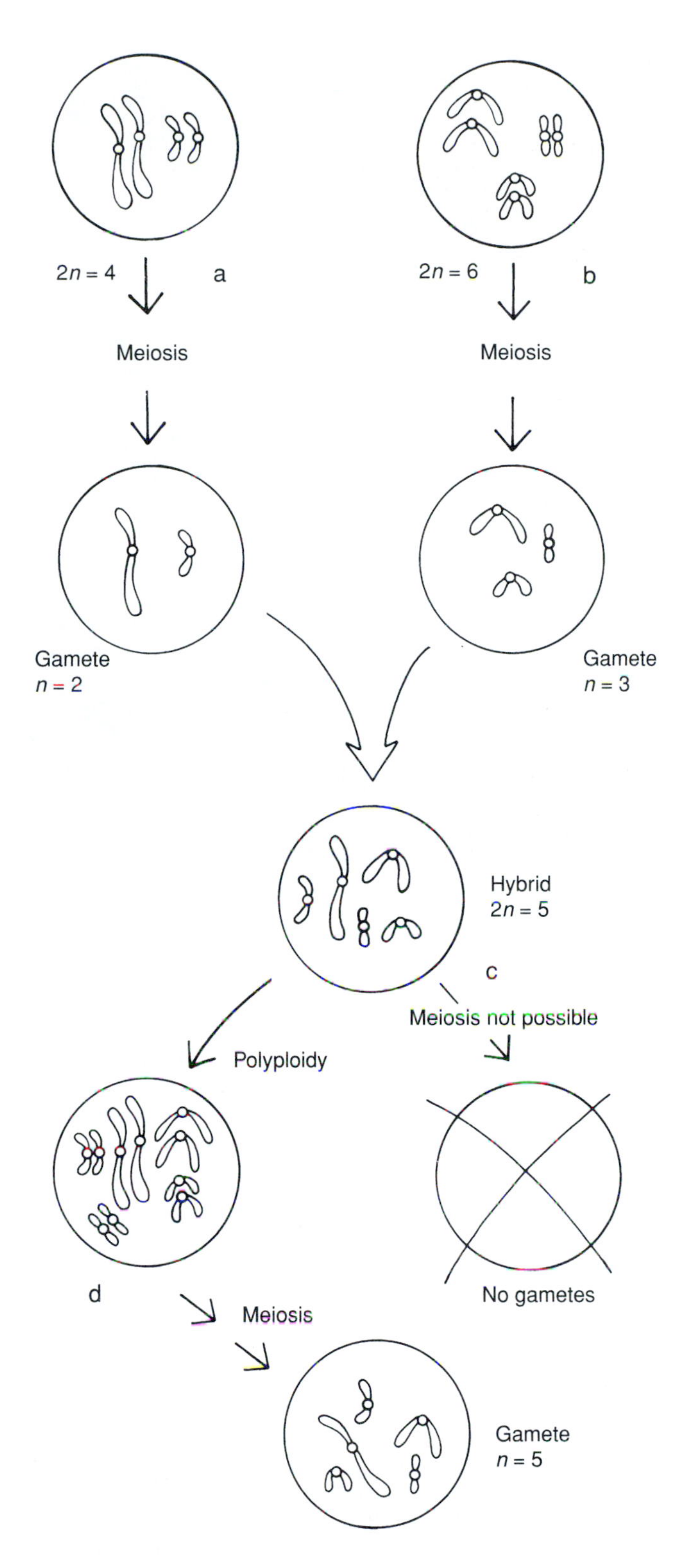

FIGURE 1.31

When gametes from two genetically dissimilar individuals (a and b) fuse to form a zygote (c), the adult plant that results may not be able to form gametes because during meiosis the chromosomes cannot pair properly and thus cannot undergo a reduction division. If a doubling of the chromosome number occurs in the zygote (d), a tetraploid hybrid will result. Meiosis can occur in the adult that results from this mating because each chromosome will have a homologous partner with which to pair.

them cannot function properly and die or do not reproduce. In some cases, however, a new allele may be neutral, or may even produce an effect superior (in a given situation) to that of the existing alleles. If such a mutation occurs in cells that are involved in reproduction, it can be passed on and subsequently spread within a population.

Humans have taken advantage of the fact that out of the many millions of mutations that occur, a few produce changes in the organisms possessing them that are "superior" from the human point of view to the previously existing types. People can then perpetuate the new types. Humans can also subject reproductive tissues to **mutagens,** mutation-creating substances, such as x rays or chemicals that affect the DNA in order to induce mutations that they hope will produce desirable changes. They can then screen the resultant tissues, or progeny of the exposed parents, for possible superior plants.

Mutations that produce a new allele will not necessarily spread to other individuals in the population. The only way in which new alleles will move into new combinations in nature is by periodic reshuffling of chromosomes or their parts. This shuffling occurs every time meiosis and sexual recombination is completed. Since there are thousands of genes in each organism, and usually numerous alleles of any one gene normally found among the individuals of a population, sexual recombination can produce almost infinite combinations of the various alleles found in a population.

Geographic Variation in Plants

Although humans strive for uniformity in crop plants once a desired type has been developed, genetic variability is generally the rule in nature. As we have stressed, the entire process of sexual recombination is a mechanism for the continual reshuffling of genes that provides a variable array of individuals. In nature, changes in gene frequencies are often caused by either living agents (such as predators) or nonliving factors (such as soil type) in the environment. Since selective factors change across geography, we find that populations of the same organism have diverse mixtures of genotypes in different parts of its range. Such variation is called **geographic variation.** Eventually, however, there is a limit to the variation in physiological tolerances or competitive ability of a particular kind of organism. For this reason, every kind of organism has a natural geographic limit.

One would expect that the range of environments in which humans could grow particular crops would normally be restricted by these environmental constraints. Often, however, humans have been able to extend the range of a crop by either producing an artificial environment (for example, that in a greenhouse) or growing it as an annual during a favorable

TABLE 1.2 Conditions and Consequences of Natural Selection

Natural selection can be defined as a process that will occur if a population has:

1. *variation* among individuals in some attribute or trait,
2. a consistent relationship between that trait and a *difference in fitness,* and
3. *inheritance* for the trait between parents and their offspring.

The process will result in a change in the frequency of the trait that will differ:

1. among age classes or life history stages beyond that expected from development, and
2. from what would be expected from genetic recombination.

Source: Modified from J. A. Endler. 1986. *Natural Selection in the Wild.* Princeton, N. J., Princeton University Press.

season. A common example is the tomato, a native of American tropical regions that cannot survive freezing temperatures. People can grow tomatoes in the northern United States because they place them in gardens after the danger of frost is past and harvest them before winter again sets in. New plantings must be made every year because both seeds and adult plants are killed by very cold conditions.

People can also extend the range of a crop plant by taking it from the area of the world in which it is native to another area with a similar type of climate. Under natural circumstances, the geographic range of these plants was restricted not only by environmental factors but also by physical barriers such as oceans or mountain ranges. In many cases, we have transported a plant in which we are interested over such natural boundaries. These large-scale, often intercontinental, movements provide a side benefit because the natural pests of the crop are often left behind. The successful establishment of rubber plantations in Java and vanilla orchards in Africa are examples of important crop plants that are now grown most extensively in areas other than those of their native home. Occasionally, however, introduced plants become invaders and spread rapidly because they are not held in check by their natural predators, parasites, and competitors (Fig. 1.32).

FIGURE 1.32

Kudzu, a member of the legume family, was introduced into the United States as cattle forage. However, it quickly spread and has become a rampant weed in the southeast part of the country, clambering over telephone wires, houses, barns, and other structures. As it spreads, it chokes out other plants as is shown here in a summer and winter scene in Georgia. Kudzu has established itself so thoroughly in the absence of its natural enemies that communities have given up on controlling it and have been looking for uses for the voracious weed.

(Photos courtesy of J. Anthony.)

Beyond Traditional Methods

Even if we were to preserve all of the variation currently present in our important crops, there would be limits to the changes that we could bring about by conventional selection practices. Selection acts by allowing differential reproduction of individuals with certain traits. Traits "chosen" by way of selection have to be among those present in the genotypes of individuals in a population. In some cases, a species simply does not have the traits we would like it to have. In other cases, the traits might have once been present but have inadvertently been lost. Some crops do not lend themselves to breeding programs. Crops such as bananas and potatoes are usually vegetatively propagated since they have extremely high ovule or seed abortion rates. Finally, breeding programs can generally be carried out only within species or between closely related species. A gene that produces a particularly effective insect repellent in a milkweed cannot be transferred to tomatoes with traditional methods.

Since 1990 there has been an explosion in the possibilities for modifying crop species using various techniques of biotechnology ranging from protoplast manipulation to genetic engineering. Plants that produce their own pesticides that are nontoxic to humans and fruits that ripen to their peak flavor without progressing to the mushy stage are

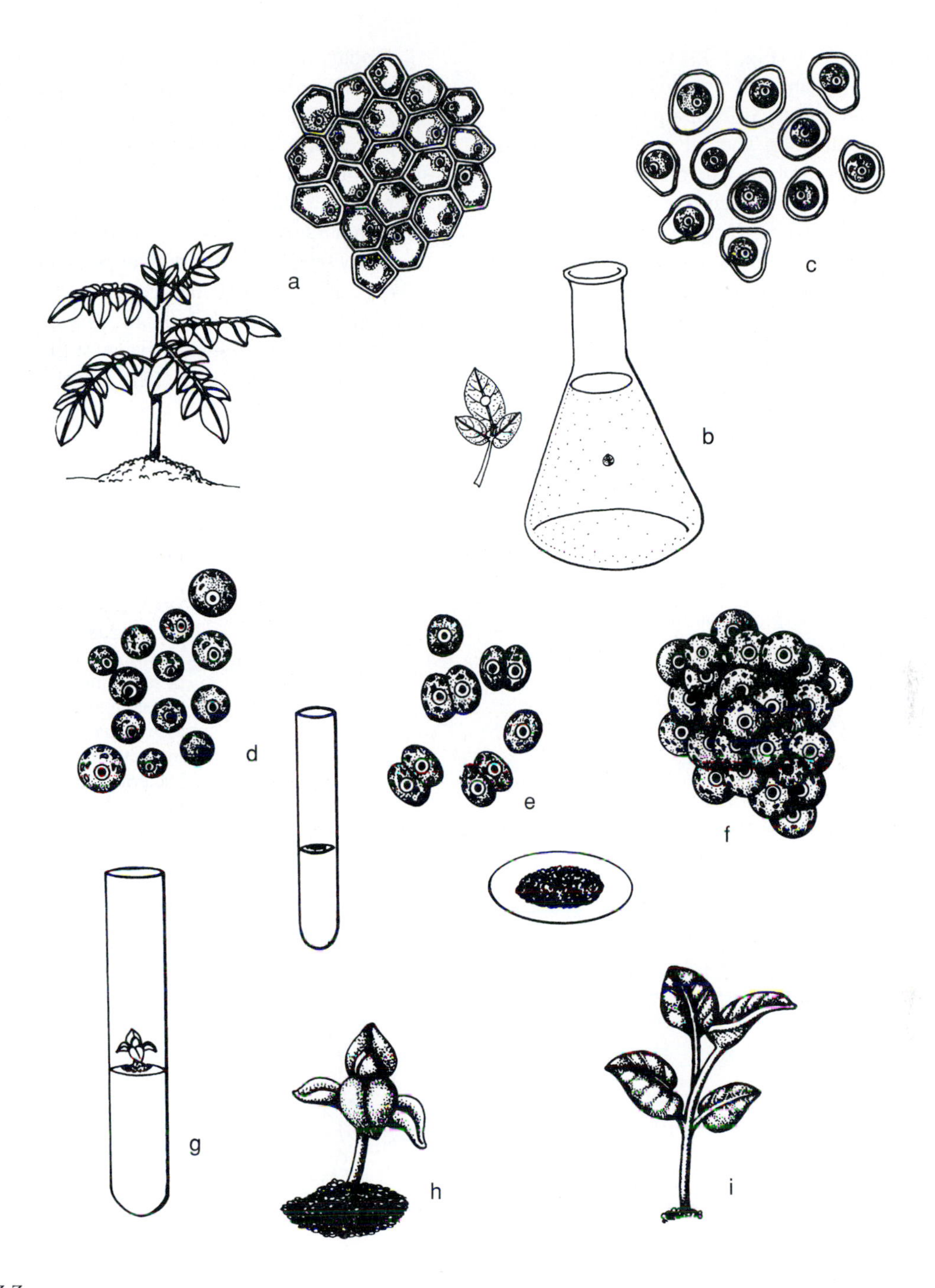

FIGURE 1.33
Protoplasts, plant cells with their cell walls removed, can be used in a variety of ways for biotechnology. They can be tested individually for a particular trait and then used to generate new plants. In this figure, a piece of leaf tissue is removed from a plant (a) and placed in a solution (b) containing enzymes that separate the cells (c) and dissolve the cell walls (d), yielding the protoplasts. Protoplasts of different varieties or species can be fused (e) to form hybrid cells that divide to form an amorphous callus tissue (f). Callus cells placed in proper conditions (g) can produce a new hybrid plant (h,i).

no longer impossible dreams. Genetically engineered plants that possess many of these properties have already been produced and are widely grown in the United States (see Box 4.2). These novel innovations are possible because we are no longer restricted to traditional methods of crossing plants with different traits as the means of introducing desirable characters into crops. We can now insert genes from one plant species into another or even from animals or bacteria into plants. We are not yet, however, at a stage where we can create completely new genes. Functional genes have to be present in an organism that can be duplicated and then inserted into species in which they can be expressed.

Protoplast manipulation (Fig. 1.33) was one of the first techniques developed in plant biotechnology. In this procedure, pieces of tissue are removed from a plant and the cells are separated from one another. The cell walls are then dissolved, leaving simple protoplasts (membrane-bound cells). These protoplasts can then be assayed like bacterial cells for traits such as disease resistance. Once a resistant type has been identified, the cells are put in a solution that causes them to form clumps of **callus** (amorphous tissue). Properly treated, these clusters of tissue can produce entire plants that are genetically and cytoplasmically identical to the original plant.

Today, new techniques in genetic engineering actually allow us to alter the genetic makeup of individuals. One kind of genetic engineering employs **plasmids,** tiny circular pieces of bacterial DNA, as carriers of genes from one species to another. The process involves locating a gene of interest on a chromosome of a donor species. The DNA of the donor chromosome is cut on either side of the gene using **restriction endonucleases** (such as *Eco*RI in Fig. 1.34), enzymes that recognize particular sequences of nucleotides on the DNA chain. The ends where the DNA was cut are termed "sticky" because they will tend to relink with complementary ends. The circular DNA of a plasmid

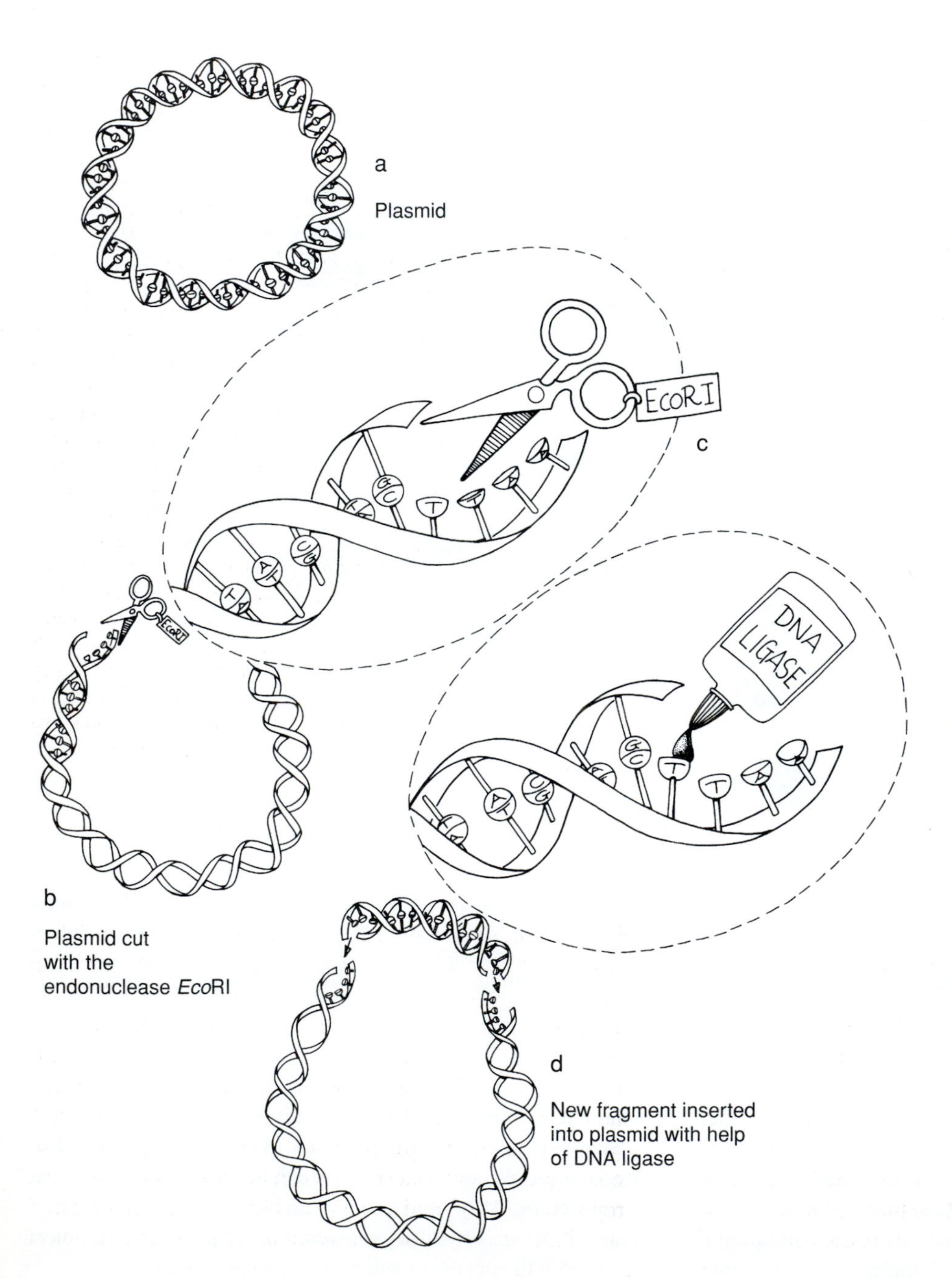

FIGURE 1.34

Endonucleases are enzymes that cut DNA at precise positions along the DNA chain. The positions where the cut is made are determined by a particular string of nucleotides. Endonucleases occur naturally in bacteria and are named for the bacteria from which they are obtained. *Eco*RI, shown here, is the first endonuclease obtained from the bacterium *Escherichia coli.* It recognizes the DNA nucleotide string GAATT (G = guanine; A = adenine; T = thymine) and cuts the string between the G and the A, leaving "sticky" ends that easily attach to complementary sequences of nucleotides. (a) An intact plasmid (a small circular piece of double-stranded DNA found in bacteria). (b) *Eco*RI cuts the plasmid at its precise location. (c) A fragment containing a gene of interest that has been cut from a chromosome using *Eco*RI is added to a solution of plasmids cut with *Eco*RI, and ligase is added. Ligase is an enzyme that promotes the bonding of the DNA backbone to form a continuous complementary DNA strand once the nucleotides have been aligned. (d) In some cases the DNA fragments with the gene of interest attaches to the two sticky ends of the plasmid DNA, forming a hybrid plasmid-gene molecule. In other cases the plasmid simply reforms. For genetic engineering, the hybrid molecules are used.

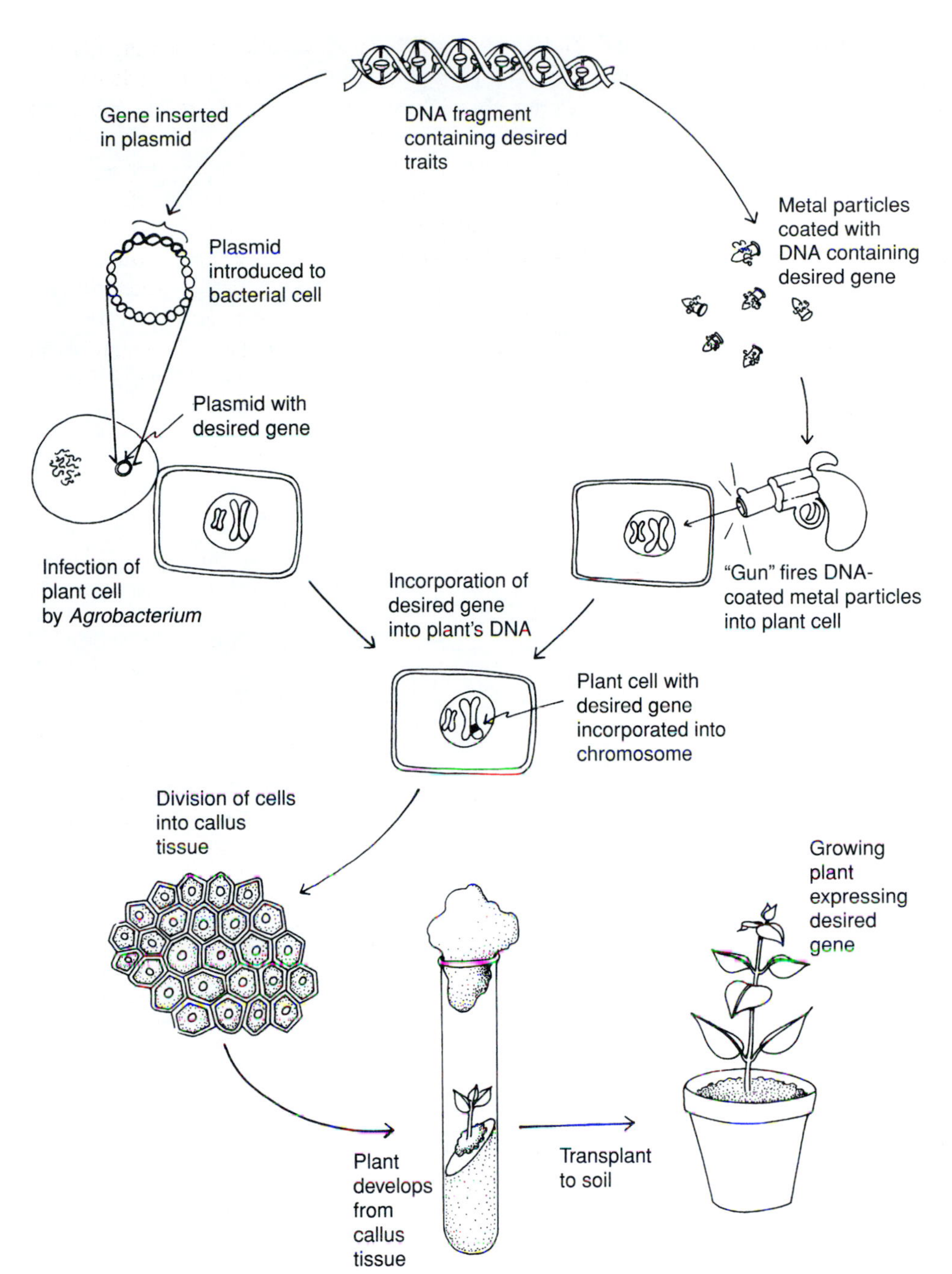

FIGURE 1.35

Genes can be moved from one species into another using genetic engineering. The most common process in plants involves the use of a bacterium, *Agrobacterium tumefaciens,* and its commonly associated plasmid. This bacterium infects numerous dicotyledonous plants, including many crop species, and causes galls on the tips of the plants. The disease is caused by the insertion of plasmid DNA into the host DNA after infection. The host subsequently expresses the plasmid DNA, resulting in abnormal growth of the plant crown. Botanists have made use of the ability of the plasmid to insert its DNA into the host chromosomes. A gene of interest is cut from another plant, animal, or bacterium, using an endonuclease, and is inserted into the plasmid. The disease-causing properties of the plasmid are disabled, and the plasmid is introduced into *Agrobacterium* cells. The *Agrobacterium* is used to infect plant cells. In some cases the gene of interest will be inserted into the plant's nuclear DNA by the plasmid and will be expressed. The cells that have successfully incorporated into the new DNA can then be used to generate entire plants that express the novel gene. Genes can also be inserted into cells by literally shooting DNA fragments into cells. The DNA fragments are made to adhere to minuscule metallic "bullets" that are shot into cells. By accident, the fragment of DNA sometimes gets incorporated into the cell's nuclear DNA. Cells can be screened to determine if they have the new piece of DNA. If they do, they are cultured like those transformed using *A. tumefaciens.*

is then cut in a single spot using the same enzyme. The DNA gene fragment from the donor species and the open plasmid are then mixed. Some of the opened plasmids reform into circles, but, in some cases, the ends of the gene fragment align with the plasmid ends and the fragment is incorporated into the plasmid. The plasmid-gene complex can then be introduced into the cells of an infectious bacterium (Fig. 1.35).

The most common bacterium used in genetic engineering is *Agrobacterium tumifaciens.* This bacterium often possesses a plasmid called the ti plasmid that causes crown gall disease in many plants. Because *Agrobacterium* can infect plants (whereas the ti plasmid alone cannot), the bacterium is used as a means of infecting the plant with the ti plasmid. This plasmid is particularly important because it can transfer part of its genetic material into the chromosomes of the host plant. Molecular biologists take advantage of this ability and piggyback the gene of interest on the ti plasmid after disabling the gene that produces the crown gall disease. The hope is that the introduced gene will be incorporated into the crop species' DNA and begin to code for the original desired trait. Other methods of incorporating new genetic material include shooting genetic DNA sequences adhering to microscopic pieces of metal into cells (Fig. 1.35, Fig. 19.11). Pieces of DNA can also enter cells if the cells are subjected to an electrical current that produces pores in the membranes of protoplasts. In both cases, some of the DNA will be incorporated by random processes into the nuclear DNA of some of the cells.

The technology of plasmid transfer is well advanced, and genes have been successfully transferred into several different higher plants (see Table 19.2). Unfortunately, major problems with using plasmid transfer to alter characters arise when dealing with traits that are controlled by more than one gene. The transfer of several genes, most of which may occur on different chromosomes, appears to be impossibly difficult. Moreover, as the number of genes per trait increases, the probability that all will function successfully in a foreign background decreases. As we discuss different crops such as tomatoes and canola, we will indicate where successful gene transfers have been made and what newly engineered crops are being tested for possible use as commercial varieties.

The Nature of Plant Species

So far, we have referred primarily to "kinds of organisms" and have talked about reproduction among members within such groups. What we have essentially been discussing are species. People seem to have an intuitive feel for the grouping of organisms into species (e.g., crab grass, Bermuda grass, giraffes, horses, donkeys), but scientists disagree even among themselves about the definition of "a species." Most zoologists define a species as a set of individuals or populations that are capable of interbreeding among themselves in nature but that are reproductively isolated from other such individuals or populations. Botanists have difficulties with this definition because of the variability of plant reproductive systems and the common occurrence of polyploidy in plants. The production of polyploids creates individuals usually unable to produce fertile offspring when crossed with the diploid plants that gave rise to them because the offspring have an odd number of chromosomes ($3n$, $5n$, etc.) and thus cannot produce gametes because they cannot undergo successful meiosis. Since the "parents" and their offspring are reproductively isolated, they would be considered different species using the common zoological definition of biological species.

A second problem with the use of reproductive barriers as the criterion for defining a plant species is the ease with which plants hybridize, making the phrase "reproductively isolated" difficult to apply. Consequently, botanists tend to consider as a species a group of populations that were derived from a single ancestor and that can be distinguished morphologically from other groups of populations. Botanists have thus often also tended to give names to cultivated species such as lettuce if they have diverged morphologically from their wild progenitors. However, there has recently been a trend toward considering both the cultivated plant and its wild ancestor as components of a single species in order to indicate the recency of the divergence and the fact that the morphological differences are a result of human manipulation.

FIGURE 1.36

Linnaeus after his return from Lapland, wearing the traditional Lap costume.

(Reproduced from *Medicine and the Artist [Ars Medica]* by permission, Philadelphia Museum of Art, Carl Zigrosser, Dover Publications, Inc. 1970.)

The Naming of Plants

Our current system of naming plants is governed by an international system of rules called the *International Code of Botanical Nomenclature,* which specifies how correct names are to be determined for various groups of plants. This scheme uses a system of hierarchical ranks (species, genus, family, order, etc.) that are useful for information transfer, but it tends to imply an equivalency of the ranks between unrelated groups. The code does not specify the biological basis for what should be called a species, genus, or family. It merely gives instructions for how to apply a particular name once a decision about the rank of a group of organisms has been made and a proper description has been provided. The most commonly used rank is the species. Every species is given a Latin name consisting of two parts, forming a binomial (*bi* for two and *nomial* referring to name). This system, known as the **binomial system of nomenclature,** was developed by Carl Linnaeus (Fig. 1.36) in Sweden and was first used consistently by him in 1753.

The first part of the binomial is the genus name and the second, the species name. Both parts of the name are needed to identify a particular species, and both names are always written in italics. The genus name, but not the species name, begins with a capital letter. Every species of plant (and animals under the provisions of a different code) has only one correct name. Despite language, colloquial, or common names, the scientific name is the same all over the world. Thus, scientists are able precisely to refer to a particular species even if they do not speak the same language. Often, the Latin name given to a species describes one of its attributes. Table 1.3 gives some Latin descriptive adjectives frequently used as specific names of cultivated plants.

Since the generic name is part of a species' designation, we might ask, What is a genus? Unfortunately, there is no real definition of a genus. It is a category that contains a collection of species that share a common ancestor and therefore one or more defining characters. Likewise, at a higher level of grouping, genera that are believed to have originated from a common ancestor and that share several characters are clustered together in a plant family. As in the case of species, the names of families are in Latin. Family names begin with a capital letter and normally end with the letters -aceae, but they are not written in italics. For example, Ros**aceae** is the name for the family containing the rose genus, *Rosa.* The code allows the use of some old names not ending with "-aceae" for several important plant families. Table 1.4 gives the alternative names allowed for families of economic importance that we discuss in this book. Plant families are often easy to recognize because of some trait that is common to almost all members. For example, the bean family includes many genera with many species, but all members have legumes, or "beans," for fruit. The dandelion family contains hundreds of genera and thousands of species, but all have a similar type of flowering structure, a **head,** consisting of flowers crowded together at the top of a stem and surrounded by leafy structures (Fig. 1.37). Throughout this book we will use species names and indicate the families to which they belong to help in ordering them into evolutionary groups. We shall often use the family as a convenient way to group species of economic importance.

TABLE 1.3 Common Latin Names Used for Species and Their English Meanings

LATIN	ENGLISH
aestivus	summer
alba	white
ambiguus	doubtful
americanus	from America
annuus	annual
arabicus	from Arabia
arboreus	treelike
argenteus	silvery
arvensis	from cultivated fields
asiaticus	from Asia
biennis	biennial
blandus	agreeable
bulbiferus	bulb-bearing
caeruleus	blue
communis	common
edulis	edible
erectus	upright
esculentus	edible
foetidus	bad-smelling
glabellus	smooth
japonicus	from Japan
luteus	yellow
maritimus	from (near) the sea
minor	small
odoratus	fragrant
officinalis	official, medicinal
oleraceus	from a kitchen garden
perennis	perennial
sativus	cultivated
silvestris	from the woods
tinctorius	useful in dyeing
usitatissimus	very useful
vulgaris	common

Source: Abstracted from L. H. Bailey. 1933. *How Plants Get Their Names.* New York, Macmillan.

Note: Species names that are adjectives must agree in gender with their generic names. Those given here are masculine adjectives.

TABLE 1.4 Alternative Names of Important Angiosperm Families

TRADITIONAL NAME*	RECOMMENDED NAME
Monocotyledons	
Palmae	Arecaceae
Graminae	Poaceae
Dicotyledons	
Leguminosae	Fabaceae
Umbelliferae	Apiaceae
Cruciferae	Brassicaceae
Labiatae	Lamiaceae
Compositae	Asteraceae

*Traditional names frequently came from some characteristic aspect of the members of the family. The name Umbelliferae is derived from the shape of the flowering stalk. Leguminosae comes from the type of fruit, a legume, found in all the species of the family. The recommended name is based on the name of a genus in the family, not on a character.

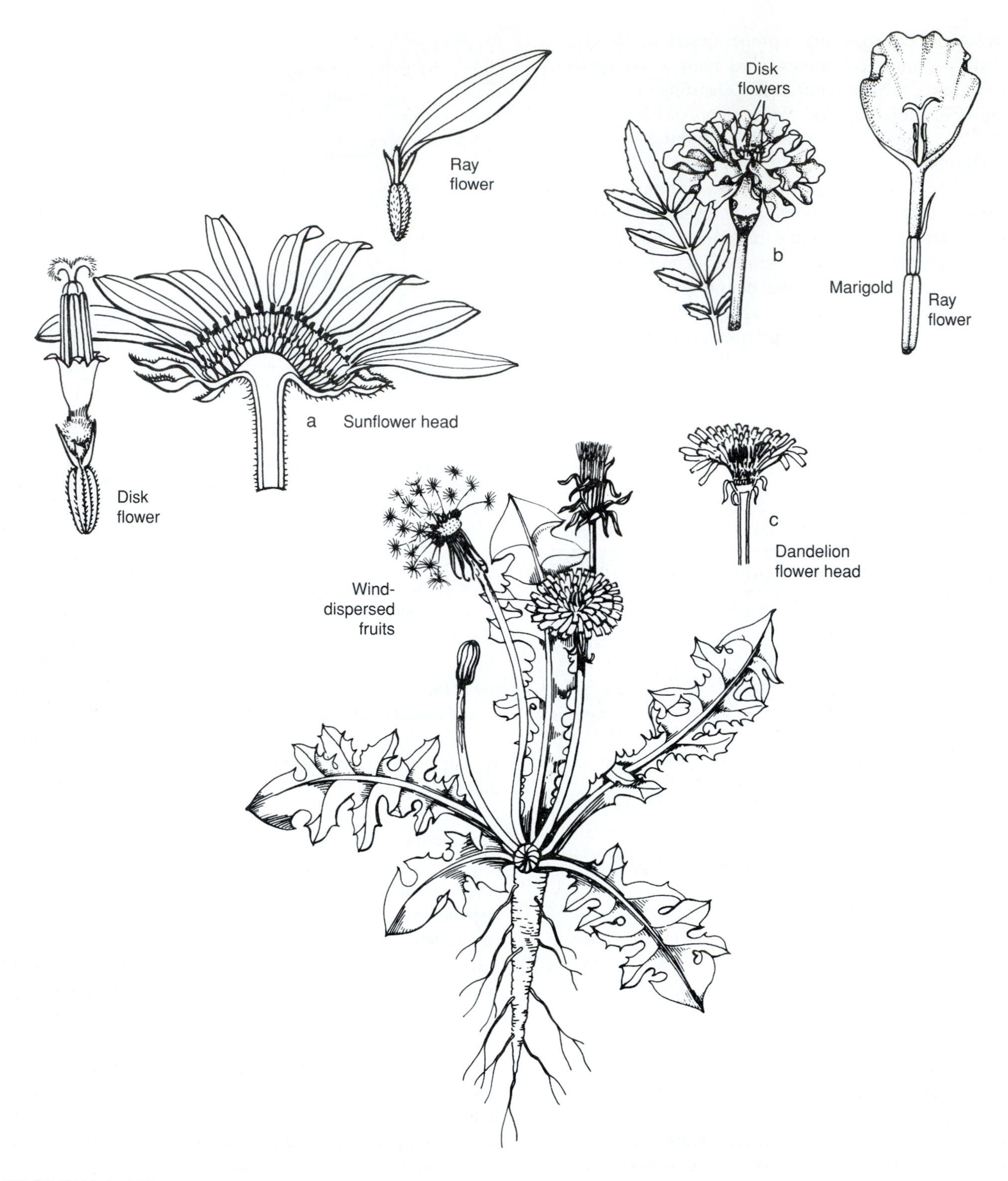

FIGURE 1.37
The sunflower head (a) consists of a ring of female ray flowers around the outside and numerous perfect disk flowers in the center. The marigold (b) has fewer disk flowers and more ray flowers. A dandelion head (c) is composed of perfect flowers, each with a long, strap-shaped corolla reminiscent of the rays in the sunflower.

TABLE 1.5 Categories Used in Classifying Economically Important Higher Plants

CATEGORY	EXAMPLE	COMMON NAME
Example from gymnosperms		
Kingdom	Phyta	Vascular plants
Division	Coniferophyta	Conifers
Class	Pinidae	
Order	Pinales	
Family	Pinaceae	Pine family
Genus	*Pinus*	Pine
Species	*P. palustris*	Longleaf pine
Example from monocotyledons		
Kingdom	Phyta	Vascular plants
Division	Anthophyta	
Class	Liliopsida	Monocotyledons*
Order	Cyperales	
Family	Poaceae	Grass family
Genus	*Zea*	Corn and teosinte
Species	*Z. mays*	Corn
Example from dicotyledons		
Kingdom	Phyta	
Division	Anthophyta	
Class	Magnoliposida	Dicotyledons*
Order	Fabales	
Family	Fabaceae	Bean family
Genus	*Phaseolus*	Beans
Species	*P. vulgaris*	Common bean

Source: Based on H. C. Bold, C. J. Alexopoulos, and T. Delevoryas. 1980. *Morphology of Plants,* 4th ed. New York, Harper and Row; and on A. Cronquist. 1981. *An Integrated System of Classification of Flowering Plants.* New York, Columbia University Press.

*The older conceptions of the monocotyledons and the dicotyledons have been somewhat modified in the past few years because the dicotyledons are not a natural group. Evolutionarily, the water lilies, historically placed in the monocotyledons, belong with the dicotyledons and the monocotyledons share an ancestor with some dicotyledons. Nevertheless, the terms are still widely used because the plants so designated are easily recognized morpholgically (see Table 1.6).

Sometimes, species have geographic populations that are distinctive enough to merit a name. Under the rules of the *International Code of Botanical Nomenclature,* such populations could be classified as subspecies or varieties. In practice very distinctive populations set apart geographically from others in the species tend to be called subspecies. Less distinctive populations are often treated as varieties. Table 1.5 gives the various categories most frequently used in classifying plants and provides examples of each.

Unfortunately, the word **variety** used for crops or horticultural plants has a different meaning than it does for natural populations. Varieties of cultivated plants are distinctive types selected by humans. Under the *International Code of Botanical Nomenclature,* varieties with Latin names are evolutionary entities differentiated from others by natural selection. The florist's or farmer's variety is a human construct analogous to breeds of dogs or cattle. In Chapter 17, we describe the way in which horticultural varieties are named. We also refer to cultivars in various parts of our discussions. Cultivars are forms of a species selected for by humans. They are the botanical equivalent of breeds in animals.

Determining the Relationships of Plants

Biologists generally try to group organisms in such a way that the classification reflects the evolutionary history, or phylogeny, of the organisms involved. Physical characteristics (flower structure, fruit type, leaf arrangement and architecture) are used to aid in decisions about the relatedness of different flowering plant species. Families of plants are defined by one or more of such shared features. Yet, determining relationships unambiguously is not an easy task because of convergence and divergence of characters during evolution. In the case of convergence, we may erroneously judge two species to be related because of a similarity that is due to factors other than a shared evolutionary history. For example, agave plants (see Fig. 15.22) are often confused with cacti by nonbotanists because they are fleshy and spiny, appear leafless, and grow in deserts. The physical features are the result of natural selection for water conservation and do not reflect a close evolutionary relationship. The true

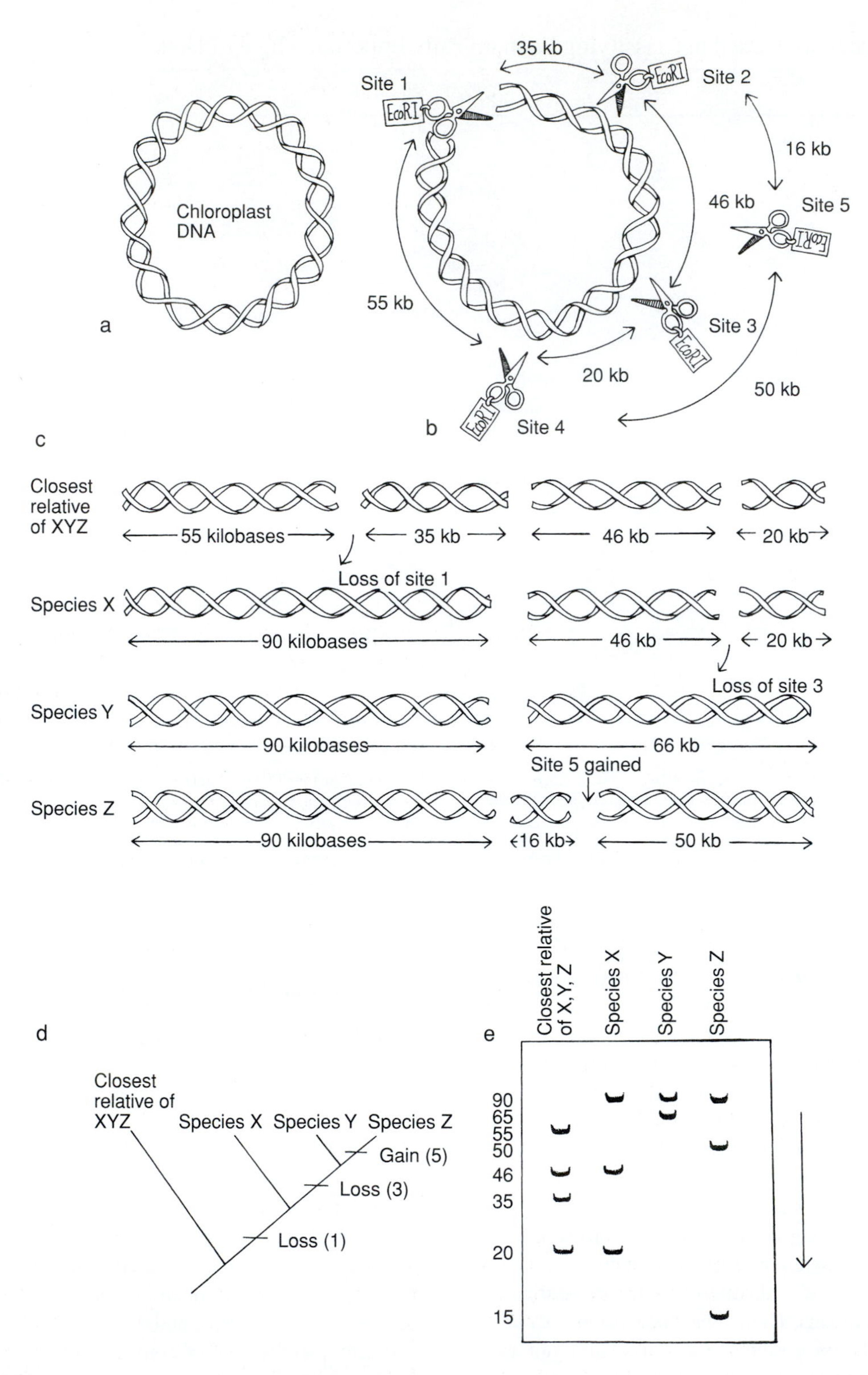

FIGURE 1.38

The pattern of gains and losses in sites (restriction sites) recognized by endonucleases in chloroplast DNA can be used to help piece together the relationship of plants. (a) In the chloroplast, DNA forms double-stranded circular molecules. The chloroplast DNAs of the flowering plants labeled X, Y, Z, and their closest relative W can be mixed with the enzyme EcoR1 that cuts the DNA in every position where there is a nucleotide string of GAATT. The chloroplast DNA of species W is cut in four places (b) leading to the pattern of fragments shown on the top line of (c). The fragments are then separated by size on a starch gel through which an electrical current flows. The gel is blotted to a filter and the fragments are determined, and "maps" are made of the locations of the restriction sites, 1, 2, and 3. Species X, Y, and Z have lost site 1. Consequently, they have (or had) only three fragments. Species Y and Z have lost site three and therefore have only two fragments. However, species Z has gained a new site 5, and therefore has secondarily produced three fragments (c and e). The gain or loss of a site can be determined by adding the length of the fragments (55 + 35 = 90, 46 = 20 + 66, 66 = 16 +50) and following the pattern of change starting with the pattern found in the relative (d). Once the pattern of gains and losses is determined, a hypothesis of relationships can be made.

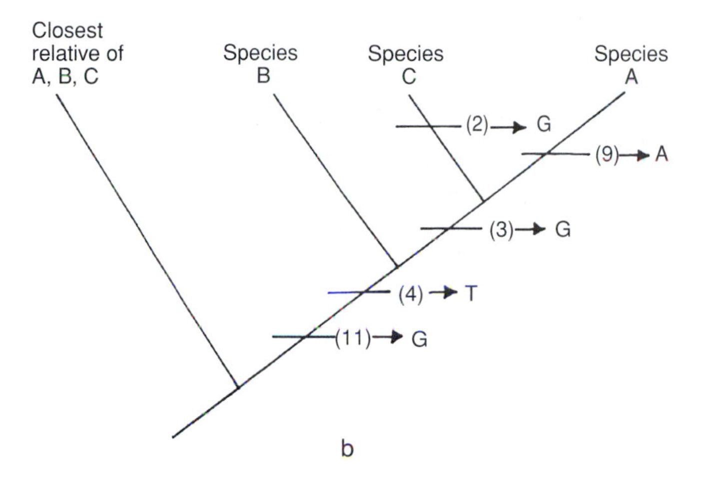

FIGURE 1.39

Genes or parts of a gene can be sequenced for different species, and the species can be assessed for changes in the sequence of nucleotides. (a) Here, as an example a DNA fragment of 11 base pairs has been sequenced for species A, B, and C and their closest relative. The sequences show that species A, B, and C share a change from their relative in position 11, where the nucleotide C has changed to G, and in position 4, where a T has changed to an A. Species A and C share a change from C to G in position 3. Species A has the unique feature of a change from T to A in position 9. (b) These changes can be used to propose the relationships shown in a branching diagram.

relationships of these groups are determined by characters less influenced by environmental factors. Such characters can include floral features, fruit structure, or genetic sequences. In the case of divergence, unequal evolution can cause a species to look very different from its closest relative because selection under different circumstances can cause changes in many characters in one descendant of a common ancestor and not the other. As a result, one member of a related pair of species might physically resemble distantly related species in more characters than it does its closest relative. In some cases, observed similarity of species would result in an erroneous classification.

Since 1990, molecular methods have been increasingly employed to help infer evolutionary relationships. In general, two methods have been used. The first compares patterns of DNA fragments generated by using the same kinds of **restriction enzymes** used in genetic engineering. This method uses the identities of fragments generated when DNA of related species is cut by the same enzyme to determine the number and positions of sites recognized by the restriction enzyme. Figure 1.38 shows how this procedure can be used to deduce the relationships among three hypothetical species of plants. The second technique involves determining the actual nucleotide sequence of particular genes or other sections of DNA and determining the amount of similarity between them (Fig. 1.39). For plants, comparison of DNA restriction fragment patterns is often done using chloroplast DNA, although nuclear DNA has also been used. Chloroplast DNA has proved to be especially useful for restriction fragment pattern analysis because the amount of DNA in these organelles is much smaller than that in the nucleus and the number of fragments produced is correspondingly manageable. There are also many copies of each chloroplast DNA molecule in a green plant cell, and the DNA in these molecules seems to change at a slow enough rate to allow reliable comparisons between species. Using molecular methods, researchers have been able to add additional characters to the traditional morphological characters and increase our confidence in determining the relatives of several economically important plant species.

An example of how molecular methods have both confirmed and challenged traditional classifications of the flowering plants can be seen by comparing the most commonly used system of angiosperm classification (the Cronquist system) with that suggested from data amassed from many different laboratories using the DNA sequence of five genes. Cronquist,

TABLE 1.6 Characters Traditionally Used to Characterize Dicotyledons and Monocotyledons

DICOTYLEDONS	MONOCOTYLEDONS
Two cotyledons	One cotyledon
Leaves with veins forming a network	Leaves with parallel major veins
Floral parts in sets of 5 or 4	Floral parts in sets of 3
Pollen with 3 openings (slits, pores, or both)	Pollen with no openings or 1 opening
Mature root system with a primary (tap) root, adventitious roots, or both	Mature root system completely adventitious
Vascular bundles in a ring around the pith	Vascular bundles scattered

Source: Modified from A. Cronquist. 1988. *The Evolution and Classification of Flowering Plants.* New York, New York Botanical Garden.

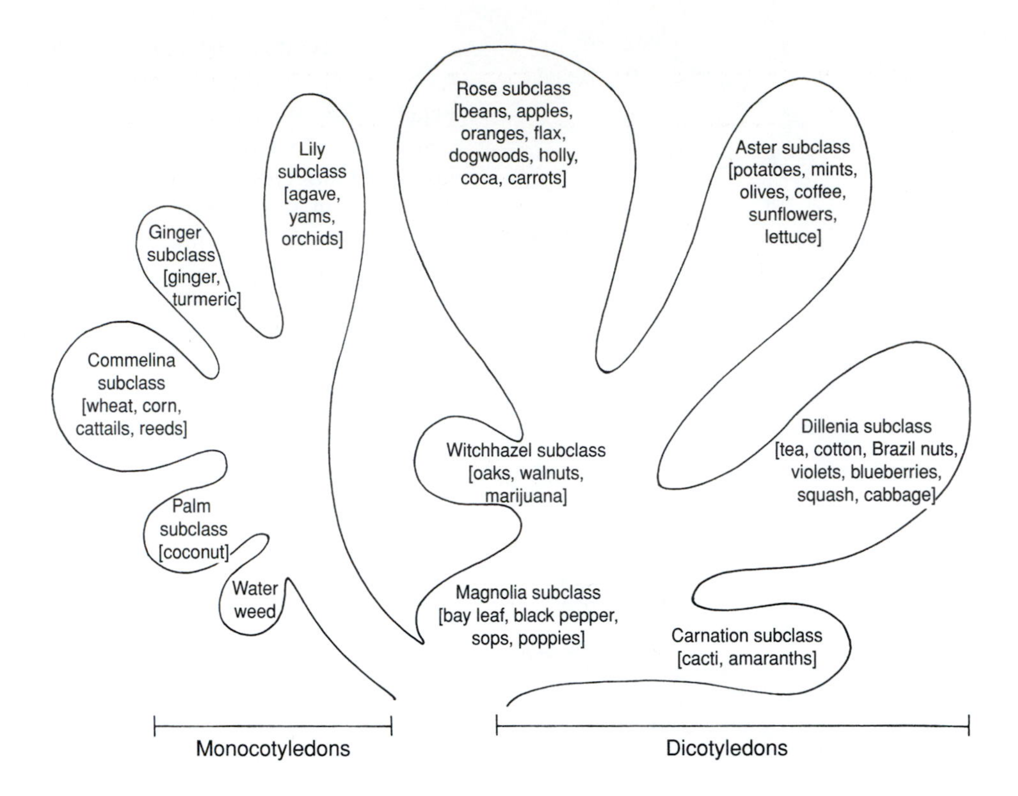

FIGURE 1.40

The Cronquist system of angiosperm systematics is the most commonly used in the United States. Cronquist based his classification on morphology, chromosome numbers, and chemistry. This diagram shows his concept of how the major groups of flowering plants are related to one another. The size of the protruding arms is somewhat proportional to the number of species in each group. As did most taxonomists before 1995, Cronquist considered the monocots and the dicots as distinctive, independent lineages.

as did almost all plant taxonomists before him, recognized two major divisions of angiosperms: the Monocotyledonae and the Dicotyledonae. Plants belonging to the first group have one fleshy seed leaf (cotyledon), whereas dicotyledonous plants have two seed leaves. These features are generally correlated with several other characters (Table 1.6).

Within the monocotyledons there are five major groups, one of which contains the palms, another the grasses, and the others ginger, the lilies, and the water weeds (Fig. 1.40). The dicots have traditionally been divided into a number of major groups, each named after an important family it contains. As can be seen in Figure 1.40, a basal group in the Cronquist system contains the magnolias, and more advanced groups contain (among other plants) oaks, violets, and asters. Diagrams such as Figure 1.40 are often drawn in this way to indicate the relationships of the groups depicted. Groups placed close to one another are believed to have shared a common ancestor more recently than groups separated from one another. In this figure we have listed some of the important plants in each of these groups. Many of these will be discussed in more detail in later chapters.

The pattern of relationships reconstructed using the five-gene data set (Fig. 1.41) does not agree with the Cronquist scheme presented above. For example, monocots form a natural group, but they appear to have arisen within the magnolia group of the dicotyledons. Consequently, dicotyledons as a whole and the cluster containing magnolia, considered by many to be the most primitive in the angiosperms, are not natural groups. Figure 1.41 shows the general phylogenetic relationships of many economically important plant groups based on the five-gene study. As we discuss various crop species, we will point out instances in which molecular studies have added to our understanding of their origins.

We have outlined in this chapter the structure of plants, their reproductive systems, and the use of classical and molecular methods for crop improvement and for determining phylogeny. After an examination in the next chapter of the origins of agriculture, we elaborate on all of these aspects in economically important plants starting first with

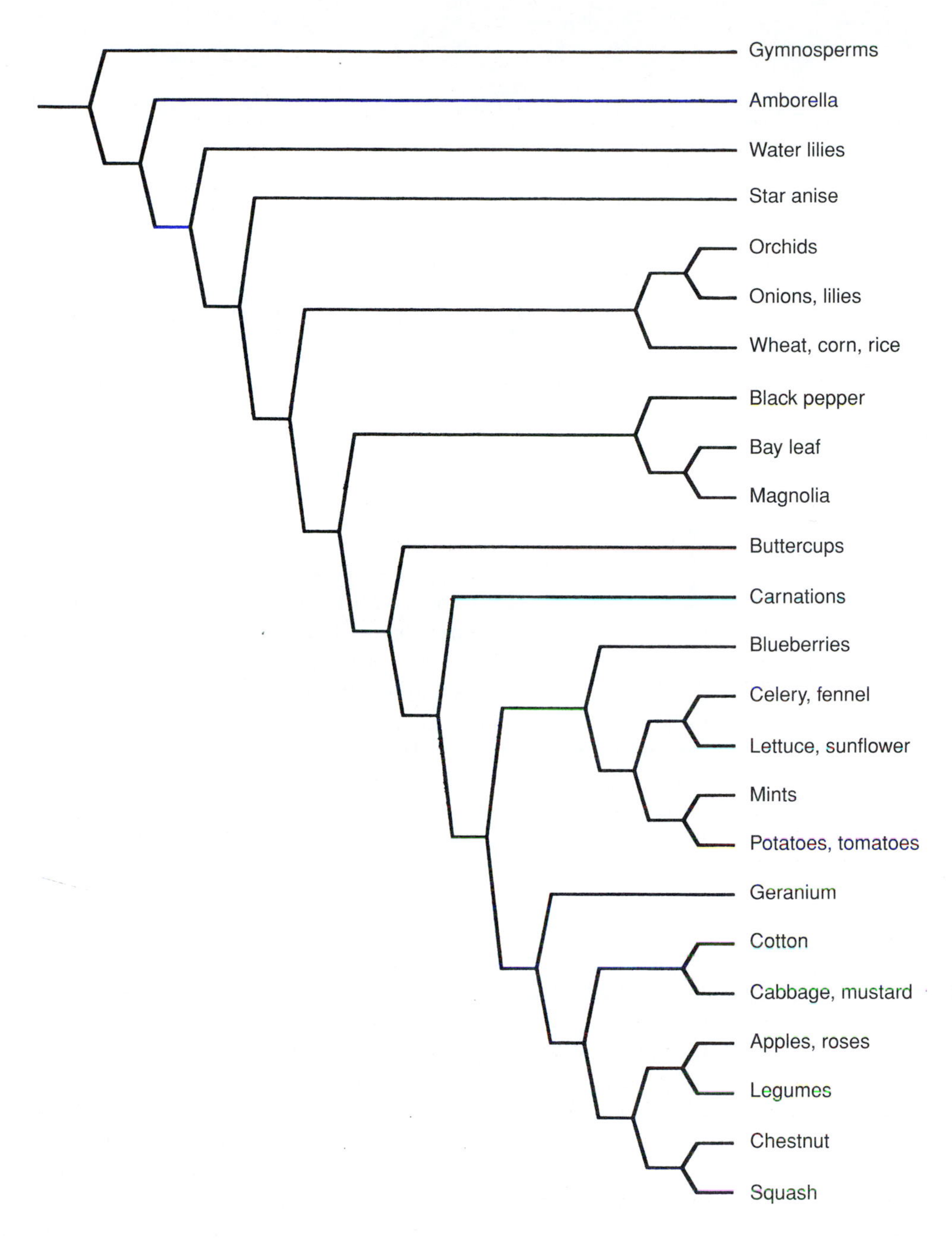

FIGURE 1.41
Molecular data have altered many of our ideas about the relationships among flowering plants. A 1999 study of 560 species used DNA sequences of five different genes to reconstruct the overall phylogeny of the angiosperms. *Amborella* is now believed to have shared an ancestor with all of the rest of the extant flowering plants. Monocots are a natural group that shares an ancestor with some of the dicotyledons. The dicotyledons are not a natural group because the monocots are embedded within them.

products derived from entire plant organs. These include fruits, seeds, stems, and roots—our primary foodstuffs. The second part of the book treats plant products that are extracted from various plant parts, often different organs from a diverse array of species. In these chapters we concentrate on the chemistry of the products, how we procure them, and the ingenious ways in which we have learned to process them to meet our needs.

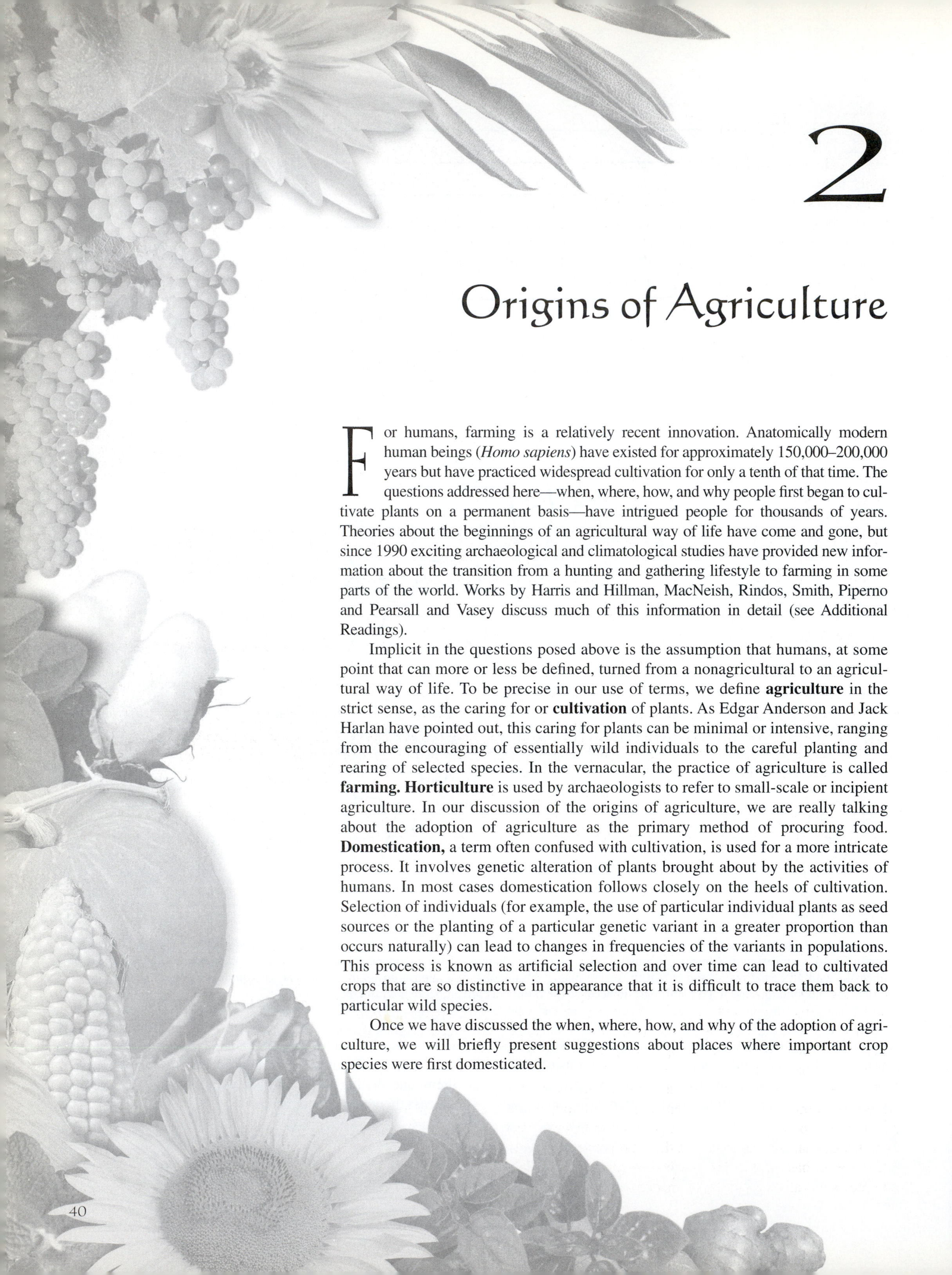

2

Origins of Agriculture

For humans, farming is a relatively recent innovation. Anatomically modern human beings (*Homo sapiens*) have existed for approximately 150,000–200,000 years but have practiced widespread cultivation for only a tenth of that time. The questions addressed here—when, where, how, and why people first began to cultivate plants on a permanent basis—have intrigued people for thousands of years. Theories about the beginnings of an agricultural way of life have come and gone, but since 1990 exciting archaeological and climatological studies have provided new information about the transition from a hunting and gathering lifestyle to farming in some parts of the world. Works by Harris and Hillman, MacNeish, Rindos, Smith, Piperno and Pearsall and Vasey discuss much of this information in detail (see Additional Readings).

Implicit in the questions posed above is the assumption that humans, at some point that can more or less be defined, turned from a nonagricultural to an agricultural way of life. To be precise in our use of terms, we define **agriculture** in the strict sense, as the caring for or **cultivation** of plants. As Edgar Anderson and Jack Harlan have pointed out, this caring for plants can be minimal or intensive, ranging from the encouraging of essentially wild individuals to the careful planting and rearing of selected species. In the vernacular, the practice of agriculture is called **farming. Horticulture** is used by archaeologists to refer to small-scale or incipient agriculture. In our discussion of the origins of agriculture, we are really talking about the adoption of agriculture as the primary method of procuring food. **Domestication,** a term often confused with cultivation, is used for a more intricate process. It involves genetic alteration of plants brought about by the activities of humans. In most cases domestication follows closely on the heels of cultivation. Selection of individuals (for example, the use of particular individual plants as seed sources or the planting of a particular genetic variant in a greater proportion than occurs naturally) can lead to changes in frequencies of the variants in populations. This process is known as artificial selection and over time can lead to cultivated crops that are so distinctive in appearance that it is difficult to trace them back to particular wild species.

Once we have discussed the when, where, how, and why of the adoption of agriculture, we will briefly present suggestions about places where important crop species were first domesticated.

A Time Frame

Let us begin with the least controversial question: When did people first consciously begin to grow plants on a large-scale basis, or, when did humans first become farmers rather than hunter-gatherers? Because there are methods for dating traces of prehistoric agriculture and because most scientists are willing to accept the oldest date until an even older one is conclusively established, this is the easiest to answer.

What are the ways of finding and dating evidence of early agriculture? The first step is to find fossil human encampments or resting places. Evidence that agriculture was practiced includes abundant remains of plants that are known to have been cultivated or of tools used for preparing soil, cultivating, or harvesting. Human skeletons can provide information about the kinds of plants eaten and the ways in which food was prepared. Grasses, for example, have a ratio of two stable carbon isotopes (^{12}C and ^{13}C) different from that of most other plants eaten by people. Changes in the ratio of these isotopes in human skeletons over time from grass ratios to that found in grains can indicate a shift to grains as a primary source of food. Similarly, the consumption of large amounts of grains can be documented by examining the kind and amount of wear on the teeth of archaeological skeletons.

An important source of data is fossilized plant material in archaeological digs that differs significantly in quantity from wild plants in nearby areas. The plant and animal remains used for such analyses are generally found in garbage dumps, strewn around ancient hearths, or as pollen in pots or on pot fragments around these sites. Changes in plant characters such as seed size are considered evidence that humans altered, or domesticated, the plants involved. Pollen grains are particularly useful for identifying plant remains because they are exceedingly resistant to decomposition and because many have characteristic sculpting (Fig. 2.1a) that allows botanists to determine the species or genus from which they came. Charred seeds and plant parts can also often be identified by comparing gross or cellular features with those of modern counterparts. **Phytoliths** (Fig. 2.1b), silica particles found inside the cells of many plants, can also be used to identify plant remains.

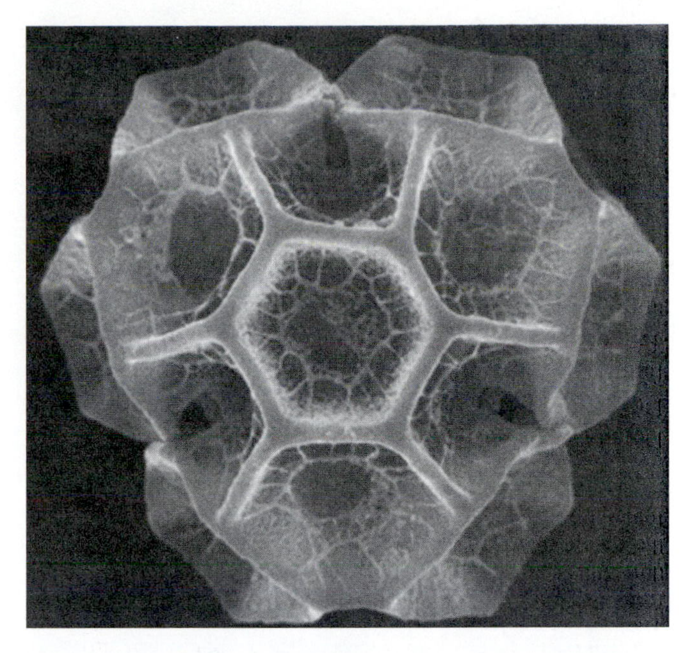

a

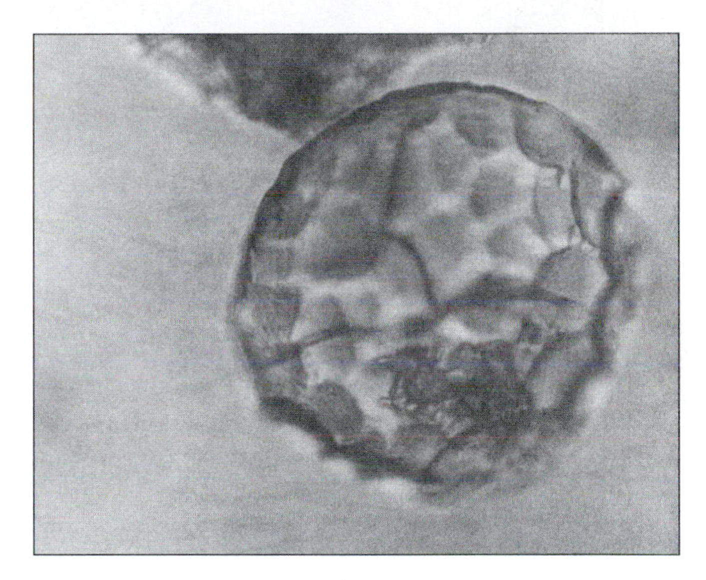

b

FIGURE 2.1

Fossilized pollen and phytoliths are used to identify plant remains in archaeological digs because they often have characteristic morphologies. Shown here (a) is a pollen grain of *Barnadesia,* a member of the sunflower family and a phytolith (b) from the rind of a squash (*Cucurbita* spp.). The pollen grain is magnified 1733 times, the phytolith 160 times.

(Pollen courtesy of Zaiming Zhao and the phytolith Dolores Piperno.)

Dating is usually done by what is called the carbon-14 (written ^{14}C) method (Fig. 2.2), which makes use of the natural occurrence of a radioactive carbon isotope that decays at a known rate. All living organisms are made up primarily of carbon atoms. The usual form of carbon (^{12}C) has a molecular weight of 12, but several other isotopes with atoms containing different numbers of neutrons also occur. One of them, ^{14}C, has a molecular weight of 14. This isotope is formed continuously in the atmosphere by the interaction of cosmic rays and nitrogen. The process of ^{14}C formation and disintegration has been taking place for so long that the ratio between ^{14}C and ^{12}C in the air has more or less reached an equilibrium. When an organism is alive, it constantly incorporates carbon into its body. In the case of plants, most of the carbon incorporated, or "fixed," is taken into the plant in the form of carbon dioxide (CO_2). Since ^{14}C and ^{12}C occur in a constant ratio in the atmosphere, they are incorporated into plant tissues (with some selectivity) in that ratio. Once a plant dies or is collected

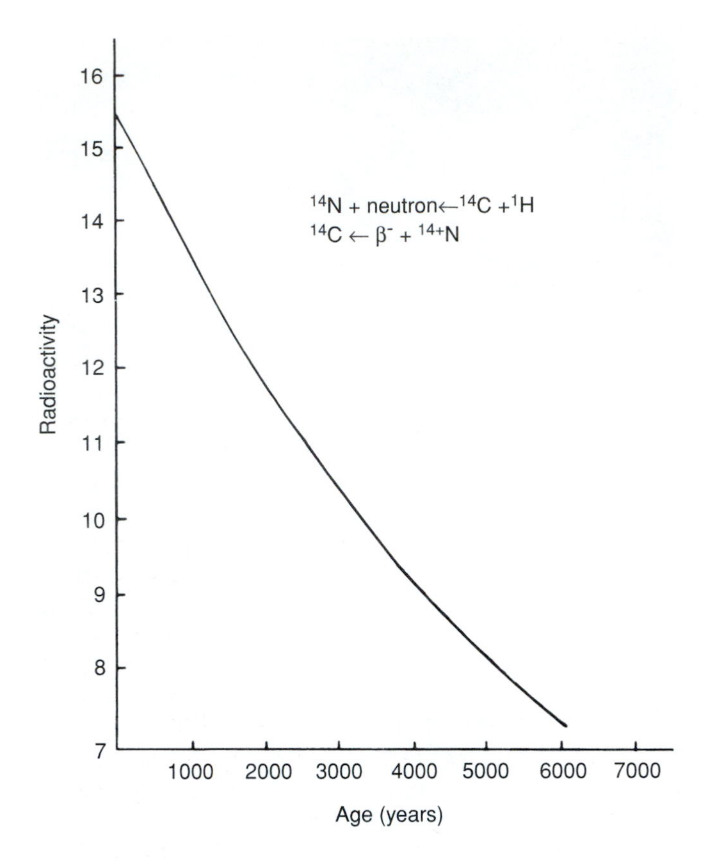

FIGURE 2.2

The decay curve for carbon-14 clearly shows that the half-life of the isotype is between 5000 and 6000 years (actually 5730 ± 40). The radioactivity is measured in disintegrations per minute per gram of carbon.

(Adapted from W. F. Libby. 1955. *Radiocarbon Dating.* Chicago: University of Chicago Press.)

by humans, the incorporation ceases. From that time onward, there is only decay of the ^{14}C into nitrogen while the amount of ^{12}C stays constant. When one measures the ratios of ^{12}C and ^{14}C in a piece of organic fossil material, a formula can be used to calculate the time since the death of the organism. Carbon-14 slowly degenerates back into nitrogen with a half-life of about 5700 years. This degeneration time provides reliable dates for ages between 5000 and 50,000 years ago, a time span adequate for dating most archaeological remains. This method can be used even for cooked or burned organic material since neither heating nor charring alters the ratio of ^{12}C to ^{14}C.

Although the ^{14}C technique sounds straightforward, accurate dating requires meticulous care and the use of uncontaminated material. In addition, it requires a comparatively large amount of material, so in many cases archaeologists have had to use pieces of wood or bone associated with the fossilized objects of interest to provide a date. When one uses material other than plants parts themselves, there is always a chance that a date will be incorrect. Weathering and other disturbances can lead to the mixing of material of unequal ages, so a precise date is often difficult to determine.

A newer method of dating is accelerator mass spectrometry (AMS), in which molecules of carbon compounds are accelerated and smashed into detectors allowing an analysis of their proportions of ^{12}C and ^{14}C carbon atoms. This method uses such small quantities of material that samples of the actual fossils rather than the other carbon sources associated with them can be used. Since 1990 several dates for domesticated plants have been changed (in all cases they have become more recent) once this method of dating was applied to samples of the crop material itself.

What have the data from such studies revealed? First, all dates before 15,000 years ago indicate that the plant and animal material used by humans was wild, indicating that humans were still hunters and gatherers. Second, the earliest dates firmly ascribed to evidence of agriculture (as ascertained by the finding of domesticated crops) are after 12,000 years ago (Box 2.1). Finally, on the basis of evidence from numerous archaeological sites, it is also undisputed that by 2000 years ago most major civilizations were sustained by agriculture. It was during the period from 10,000 years ago to 2000 years ago that people's entire way of life and the course of civilization changed in most of the world. Once city-states became dependent on their own production of food, they virtually never reversed the process by returning to gathering as a means of sustenance.

Was There a Cradle of Agriculture?

Because the earliest dates indicating the large-scale practice of agriculture come from archaeological sites around the eastern edge of the Mediterranean (Fig. 2.3), this region has been labeled the cradle of agriculture, or the Fertile Crescent. To people today this looks like an unlikely place to begin cultivating plants since the terrain is hilly and rocky, the soil comparatively poor, and the climate arid.

Although this area has yielded the oldest dates for the practice of agriculture, cultivation seems to have begun at about the same time in many other parts of the world. In Asia, a 1999 report based on phytolith data concluded that rice was being cultivated 10,000–11,000 years ago in the Yangtze River valley. In 1998 a seed of a domesticated squash found in Oaxaca, Mexico, was dated as 10,000 years old. In Ecuador, phytoliths of a different domesticated squash also indicate domestication by 10,000 years ago. Cultivation probably began at about the same time or only slightly later in northeastern Africa, but the scanty data available indicate only that agriculture was not adopted south of the Sahara until

Box 2.1

Not So Revolutionary After All

The phrase Neolithic Revolution refers to the dramatic change in the human condition following the adoption of agriculture. Early evidence suggested that agriculture arose suddenly and was the basis upon which civilization rested. How it developed was not originally debated because it was considered to be an inevitable stage in human cultural development. For many years, the standard dogma prevailed that humans must have settled down before they could develop agriculture. It was also generally thought that hunting-gathering followed a technological progression to herding and then to farming and that agriculture was independently originated as a way of life in only a few major centers of origin. Since 1970, numerous archaeological studies have proved those concepts to be simplistic. Studies by Piperno and her colleagues have shown that in the neotropics, people who were semisedentary swidden (slash-and-burn) farmers practiced small-scale agriculture or horticulture as early as 10,000 years ago. For thousands of years, people in these areas apparently lived in family groups that grew crops to supplement what they could gather or hunt. They tended to remain in one place for a limited amount of time, much as many Central and South American indigenous people did before the end of the nineteenth century. Phytoliths and other plant remains of domesticated squashes and corn from the eastern United States, Ecuador, Panama, and Mexico suggest that crops may have been domesticated in areas other than the classic "centers" of the Mexican plateau and the Andean region of Peru.

In the Old World, studies to date indicate that plants were domesticated before animals. Although plant remains from regions such as the Mideast and eastern China still provide the oldest dates for highly developed agricultural societies, incipient agriculture may have started before 13,000 years ago. Moreover, some ancient agricultural settlements such as Çatalhöyück in Turkey (6000 B.C.) appear to have lacked any form of central government or stratified society once thought to be necessary for the maintenance of large farm-based communities.

Where do these findings leave us in terms of the origins of agriculture? First, it appears that we must think of the adoption of agriculture as a continuum from cultivating a few plants (horticulture) that supplement hunting and gathering to complete dependence on agriculture as a means of food procurement. In the past, researchers have focused on the latter, in part because of interest in the major civilizations that followed after the switch to agriculture. The supposed revolution was simply the culmination of thousands of years of varying intensities of cultivation. In terms of the concept of "centers of origin" of agriculture, what we may have designated as centers may simply have been the earliest successful, dominant civilizations that relied on the cultivation of crops collected from a large surrounding geographic region.

One fact, however, remains clear. The switch to an agricultural lifestyle was always correlated with population sizes too large to survive by foraging. If a community eventually reached a size that could not be maintained even by primitive agriculture, it moved or died out. In regions where agricultural production could be increased, population sizes began to grow at unprecedented rates. With a 1999 world population size of 6 billion and a projected population of 10 billion in 2050, we now need to worry if the entire world as an agricultural ecosystem can support such a global population size.

much later. Still, with dates from the eastern Mediterranean, Southeast Asia, and Central America being rather close, the question becomes whether agriculture began in only one place and the knowledge of its practice spread around the world from that source or whether it was adopted independently in several widely separated areas. It is indisputable that diffusion played an important role in the spread of agriculture in parts of the world such as Europe. However, the majority of botanists, anthropologists, and archaeologists believe that agriculture was also independently adopted in many places.

Of Myths and Men

How did humans learn to cultivate plants? Both oral history and written history are replete with speculations about how people first came to know about agriculture. In the Old World, ancient cultures around the Mediterranean favored goddesses as the purveyors of the secrets of agriculture. In ancient Egypt two of the five children of Geb and Nun (the second couple created by Ra), Isis and Osiris, married each other and ruled Egypt when Geb abdicated his throne. The couple ruled well and are credited with forbidding humans to eat one another (implying a practice of cannibalism in pre-Egyptian civilizations?) and with teaching people many skills. Osiris taught humans about growing grain and making beer. Isis (Fig. 2.4) developed the practice of embalming, a skill she had to learn, according to legend, when she restored the mutilated body of Osiris to life.

Many people remember the Greek myth of Demeter's daughter, Persephone, who was abducted by Pluto, the god of the underworld. Persephone was repulsed by Pluto and vowed that she would not eat until he released her.

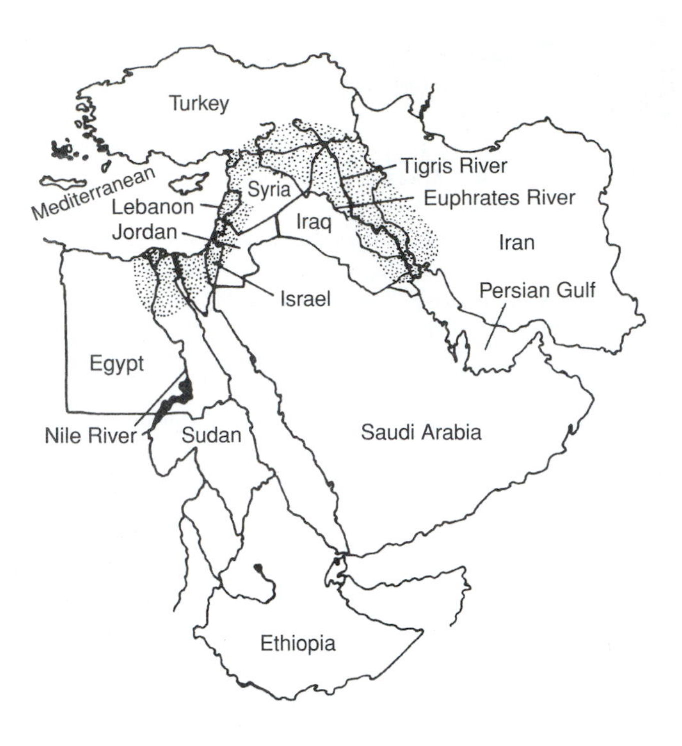

FIGURE 2.3

The Fertile Crescent (shaded), a curved area of land extending from the Nile Valley along the Mediterranean to northern Syria and down the Tigris and Euphrates valleys toward the Persian Gulf, is the region in which the earliest records of agriculture have been found.

Demeter roamed the earth, searching for Persephone. During those months she met many mortals, including the family of Metanira, who befriended her. Metanira's son, Triptolemus, was dying, and Demeter, to express her appreciation for the kindness shown to her, tried to make the boy immortal. Metanira, who did not realize that Demeter was a goddess, snatched her son from the ashes where Demeter had placed him as part of the transformation ceremony. Although the boy's chance of becoming immortal was thwarted, Demeter nevertheless provided him with the knowledge of plowing and cultivating so that he could teach other men (Fig. 2.5). As she continued her search, Demeter found evidence that made her believe that her daughter had been drowned, and she cursed the earth so that animals died and plants would not grow. When she finally learned the truth, Demeter pleaded with Zeus to restore her daughter to her. Zeus agreed to do so only if Persephone had eaten nothing during her stay in the underworld. Unfortunately, Persephone had given in and accepted a pomegranate offered to her and had eaten the pulp from one or more seeds. As a result of her weakness, she was forced to marry Pluto because she had not been strong enough to carry out her vow, but she had to live with him for only a third of the year. This story is part of the Greek agricultural saga that explains the origin of agriculture. By linking the times when Persephone was below the ground to the 4 months of weather inhospitable for growing crops and her yearly return to the 8-month growing season, the myth also provided a rationale for the cycle of agriculture in the Near East.

In Chinese mythology, Shen Nung, the apparently fictitious second emperor of China (Fig. 2.6), taught humans to use the hoe and the plow. Fire was his symbol, and the Chinese legend also attributes to him the knowledge of how to clear and burn forests. As in other forested regions, fire was used to help clear the forest for agriculture, and burning the debris produced ash that served as a short-term fertilizer.

In Central America the Aztec god Quetzalcoatl (Fig. 2.7) appears in many myths. In one legend he disguised himself as a black ant and carried a grain of corn from

FIGURE 2.4

The Egyptian goddess Isis is shown kneeling with her winged arms protectively outstretched. Redrawn from a wall painting in the Tomb of Seti, approximately 1373 to 1202 B.C., Thebes.

FIGURE 2.5
Demeter, the Greek goddess of the earth, and her daughter, Persephone, offer a libation to Triptolemus before he sets out to introduce agriculture to the world. Redrawn from a red-figured vase in the British Museum.

Tonacatepel to Tomoanchán and gave it to humans for cultivation. To the south, along the western highlands of the central Andes, the great Inca empire developed its own gods and myths. For the people of this empire, the knowledge of how to sow seed came directly from Father Sun, who sent a son and a daughter to teach people to revere him, build houses, and plant seeds.

Almost all native peoples developed their own explanations of how humans learned about the purposeful cultivation of plants. The central idea of most of these legends was that the knowledge of agriculture was a "gift" from the gods. An interesting exception to this notion is found in the book of Genesis (Fig. 2.8). In contrast to the myths of other cultures, the Judeo-Christian account states that God told Adam and Eve after they had eaten the forbidden fruit, "In the sweat of thy face shalt thou eat bread, till thou return to the ground." As punishment, "the Lord God sent them from the Garden of Eden to till the ground from whence they were taken." Obviously, in this case having to practice agriculture was considered a burden incurred by misbehaving rather than a gift or a blessing.

In more recent times authors have discarded mystical or religious explanations from the origin of agriculture and sometimes have replaced them with a "Eureka! I've discovered it" theory. Edgar Anderson, a longtime student of the interactions between plants and people, developed this idea in his famous "dump heap hypothesis." He thought the knowledge of cultivation came about because

FIGURE 2.6
Shen Nung was reputed to be an emperor of China in about 3000 B.C. and is credited with showing people how to cultivate plants and teaching people about the potential uses of different herbs.

FIGURE 2.7
The Aztec god Quetzalcoatl in his manifestation as a plumed serpent, immortalized in stone at Chichen Itza.

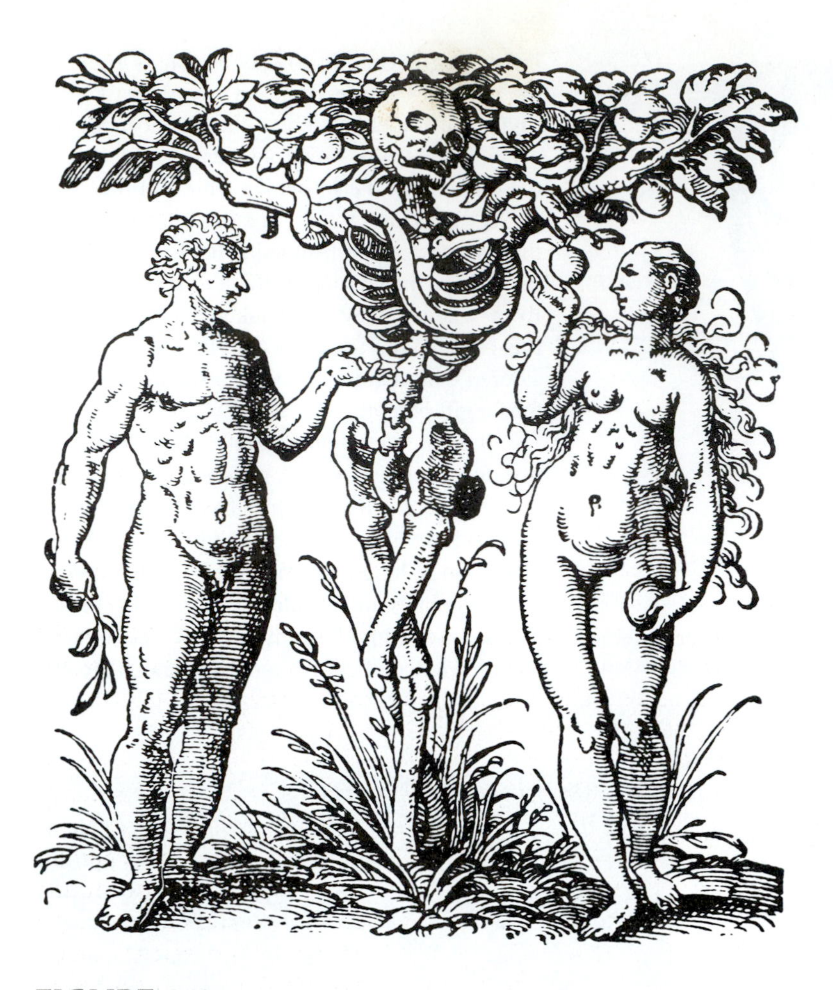

FIGURE 2.8
In the biblical story of the origin of agriculture, Adam and Eve were driven from the Garden of Eden as punishment for eating the forbidden apple and from that time onward had to earn their sustenance by practicing agriculture.
(Woodcut by Jost Amman, printed by Peter Fabricius in 1587, Frankfurt.)

FIGURE 2.9
The Abyssinian in this frieze (884–859 B.C.) is holding a structure that appears to be a male date inflorescence. Because of this and similar drawings, it has been suggested that humans understood since very early times that pollination is required to ensure seed set in many crops.

of the human practice of dumping refuse in a particular spot near a cluster of dwellings. Humans might have thrown on the dump uneaten seeds of grains, squashes, and other plants gathered from the wild. Since the wild relatives of many cultivated plants naturally occur in open and often disturbed habitats, the seeds would have readily germinated. Soon the dump heap would have become a thriving, if untidy, vegetable garden. An enterprising individual eventually would have seen that she or he could purposefully create a dump heap and ensure a supply of edible plant products close to home.

Another approach to the idea that humans had to figure out how to farm was put forth by Carl Sauer. Sauer envisioned people as living in settlements in areas with abundant wild food and water and a large array of plants amenable to domestication. In such villages, people had enough leisure time to conceive of and implement ways of improving life. They consciously invented farming in order to ensure a more permanent supply of plant food nearby. In Sauer's thinking, people must have become sedentary before they adopted a farming existence.

In contrast to the discovery or invention theories, there is now a considerable body of evidence indicating that modern nonagricultural people have the knowledge necessary to grow crops but choose not to do so until something "pushes them over the edge." By extrapolation, many botanists think that people also knew how to grow plants before the adoption of agriculture (Fig. 2.9). Perhaps preagricultural curers or shamans even grew a few plants for medicinal uses. The question of how people learned to farm becomes irrelevant if nonagricultural people possessed the knowledge to become farmers. The important question then becomes why populations in different parts of the world shifted independently and completely to an agricultural way of life. One might therefore ask, What tipped the balance?

Why Farm?

Perhaps the most interesting and widely debated question about the origin of agriculture is why humans, within a short period of time and in different places, suddenly adopted an agrarian way of life. One might think that the answer is relatively simple: the adoption of agriculture was part of the progress of civilization from primitive cultures to modern technological society. This kind of human determinism was a common notion in the late nineteenth and early twentieth centuries. Primitive humans were conceived of as living a deprived life. The progress of civilization and the improvement of the human lot were viewed as a series of revolutions each of which led to a higher stage in cultural evolution.

However, since 1960 these ideas have changed completely. Recent advances in archaeology, better dating of fossil material, and in-depth studies of modern nonagricultural peoples have indicated the extent to which they understand, exploit, and even manage their plant landscapes. In contrast to early ideas, modern studies have suggested that traditional people are not nutritionally deprived and that primitive agriculture does not necessarily improve the human lot. In its initial stages, agriculture was a more difficult way of making a living than traditional hunting and gathering, according to many modern scholars. These scientists propose that people turned to agriculture only when forced to do so.

One of the most famous studies of a modern nonagricultural tribe was carried out by Lee and Devore on the !Kung bushmen, pygmies who live in the Kalahari Desert in southern Africa. These researchers found that this group of hunter-gatherers did not lead a life of hardship or suffer from malnutrition, contrary to the prevalent notion at the time. In fact, the !Kung were selective in the plants that they used and received an adequate diet from a combination of collected plants and a few items of game. The !Kung recognized about 105 species of edible plants but normally used only the 14 they most preferred. Their diet usually consisted of an average of 96 g (2.5 oz) of protein and 2355 calories a day. Modern estimates suggest that for a 70-kg (154-lb) moderately active man, 60 g (2.1 oz) of protein and about 2300 calories per day constitute an adequate diet. In view of the fact that the !Kung are pygmies, their diet would appear to be very good. If one uses another method of comparison—the amount of work expended to acquire this amount of protein and this caloric intake—one again finds that the !Kung did not have to work harder than did primitive agriculturists. The !Kung averaged about 2.5 days per week per person for food acquisition, or between 400 and 1000 hours per person per year. In primitive agricultural societies 1000 person-hours per year is the lower limit necessary for subsistence. Obviously, a switch to agriculture does not necessarily provide a life with more leisure time, as had previously been supposed.

Lee and his associates found that when asked about plants the !Kung were very knowledgeable about many aspects of plant growth and ecology. If asked why they did not grow plants purposefully, the !Kung tended to answer, Why should we? Why should they indeed? Perhaps one might say the !Kung and other modern nonagricultural people live in highly productive areas with a constant natural source of food, but since many of these groups live in the Kalahari Desert, this is not the case.

As a result of studies like those of Lee and Devore, most scholars today conceive of agricultural practices as neither a sudden technological breakthrough nor an inevitable part of people's progress to higher levels of civilization. In contrast, they view the adoption of an agricultural way of life as a logical step for people who know how to grow plants to take when it becomes necessary or the most practical course of action.

Several theories have been proposed to explain why a change to farming might have become necessary or why the balance shifted in its favor. These include population pressure, environmental change, and coevolution between plants and the people who became increasingly dependent on them. These hypotheses are not necessarily mutually exclusive, however. An adverse change in climate could reduce the natural resources available for gathering and hence the number of people a region could support. The effect of the climatic change would have the same effect as an actual increase in the number of people by effectively reducing the resources available per person.

Climatic change has been one of the first and most cogent factors suggested for "forcing" a shift to farming. About 50 years ago it was proposed that after the retreat of the last major Pleistocene glaciation in Europe (approximately 12,000–10,000 years ago), the climate around the eastern Mediterranean became drier, forcing people in the Near East to live crowded together around water sources. The theory states that humans confined to such oases had to practice agriculture to obtain enough food for survival. When studies in the 1970s suggested that the Near East became wetter rather than drier at that time, the hypothesis seemed untenable. However, modern studies with better-dated material indicate that there was indeed an intensification of aridity in this region beginning about 13,500 years ago. Climatic change is therefore a probable factor in the increased dependence on farming, at least in the Near East.

Another, not mutually exclusive, explanation for the abandonment of hunting and gathering is that human populations simply outgrew the resources available for hunting and gathering. Presumably, humans started out gathering wild plants such as grains, storing any excess that they accumulated. Once they had built up a sizable store, they would have had to stay in one place to use and protect the hoard. Food surpluses would have spurred local population increases until populations were too large to exist simply by continuing to gather wild plants.

By incorporating the effect plants had on humans into a model of coevolution, one researcher, David Rindos, has expanded on the idea that humans promoted the evolution of the plants they used. He posits that artificial selection made domesticated plants more and more dependent on humans for survival. Humans in turn became increasingly dependent on domesticated plants as populations grew. In groups that practiced agriculture, population numbers rose, making those groups stronger than bands of hunters and gatherers. These settlements were more fit in the sense that they overpowered small groups of people and hence left more of their own offspring. This process would be a continuing one, with settlements expanding into cities, the most powerful of which overcame others. Contributing to the success of certain cities would be improved varieties and species of crops selected by people. Rindos conceives of this process as coevolutionary, with plants and humans mutually affecting each other's Darwinian fitness. Although this process probably did occur, it was partially a cultural one and cannot be considered coevolutionary in a biological sense because there is no evidence that the human traits specifically involved with farming are genetically based. Hence, they are not hereditary and therefore could not have evolved by natural selection (see Table 1.2) imposed on humans by the plants used in agriculture. There is good evidence, however, for the evolution of domesticated plants imposed by humans via artificial selection.

Implicit in the climatic change hypothesis and explicit in the model of the anthropologist Mark Cohen is the notion that stress resulting from population pressure was the main factor that caused humans to resort to agriculture as a primary support system. There are problems with this view, primarily the fact that there is no direct evidence that population levels rose to a critical point at the time of the switch to agriculture. All researchers have is the circumstantial evidence that the human diet changed in composition at about the time of the emergence of agriculture in the Near East. Analyses of the contents of archaeological refuse heaps suggest a change from a diet of plants and large vertebrates to one of plants mixed with less preferred foods such as land snails and even insects.

One might consequently ask if human population sizes always increase. The !Kung and similar tribes were able to maintain constant population sizes before extensive contact with white people. It has not been possible to determine why they were able to maintain stable population sizes, although recent evidence indicates that pregnancy rates are kept low by extremely frequent nursing that suppresses ovulation. If ancient people were also able to maintain stable population sizes, population pressure would not have been present. Obviously, it will be very difficult ever to know with certainty if populations 10,000 years ago were too dense in certain areas to be able to live as hunter-gatherers (Box 2.1).

As we have indicated, there is no consensus about why humans became farmers. There is no reason to assume that there is only one reason or that the same reason for turning to agriculture prevailed in different areas. People and populations under the same conditions may or may not respond in similar ways. Richard MacNeish has elaborated on the concept of different routes to agriculture in a three-part hypothesis that proposes a number of different pathways for the development of farming across cultures. He suggests that certain conditions were necessary and sufficient in one region that developed an agricultural lifestyle but were not of primary importance in other areas. Nevertheless, there seem to be prerequisites common to all situations in which agriculture was independently initiated. These include the presence of plants that could be domesticated and could calorically support humans, the knowledge (acquired over time) of how to grow and manage crops, and a decrease in the ability of the landscape to support humans with hunting-gathering lifestyles.

Origins of Particular Crops

Although there are few definite answers to the questions of when, where, how, and why agriculture was adopted as a way of life, there is better evidence for the possible times and sites of the domestications of many specific crops (Fig. 2.10). People seldom think about the fact that the varieties of foods they commonly eat have been drawn from numerous and widely separated parts of the world (Figs. 2.11, 2.12, and 2.13), but even typical Italian food such as tomato sauce, chicken cacciatore, and pizza would not have existed before the discovery of America because there were no tomatoes, red and green peppers, and zucchini in Italy. Italians obviously had a diet quite different 500 years ago from that which they enjoy today. Similarly, apple pie and green peas seem typically American, yet both were unknown in the New World until Europeans introduced them.

It is fascinating to trace the origins of cultivated plants even though some have become so modified over years of domestication that finding their ancestral homes is similar to solving a mystery. As people become more aware of the need for maintaining diversity in cultivated crops (Chapter 19), they realize that such studies are not simply an intriguing pastime but a necessity.

Alphonse de Candolle, a Swiss, was the first person to undertake a serious study of crop plants and their origins. In 1882 he published a book titled *Origins of Cultivated Plants,* in which he tried to determine the original homes of the cultivated species by combining information gathered from studies of geography, linguistics, archaeology, and written history with his own extraordinary knowledge of plants. His notions about how wild plants were transformed into cultivated crops seem quaint in view of modern knowledge of genetics and selection, but many of his conclusions about the places of origin of domesticates have been shown to be correct.

One of the most comprehensive treatments of crop plant origins came from a Russian about 70 years later. Nikolay Vavilov tried to find centers of origin (gene centers) of cultivated species. As a basis for determining the location of those centers, Vavilov made two assumptions: that the areas where wild relatives of cultivated species can now be found are the most likely sites of original domestication and that these centers should be areas in which one can find large amounts of natural variation in crop plants. From his research, Vavilov delimited six, soon expanded to eight, major centers of domestication in the world. Later, he and his collaborators increased the list further, but the eight centers generally used are: the Chinese center, the Central

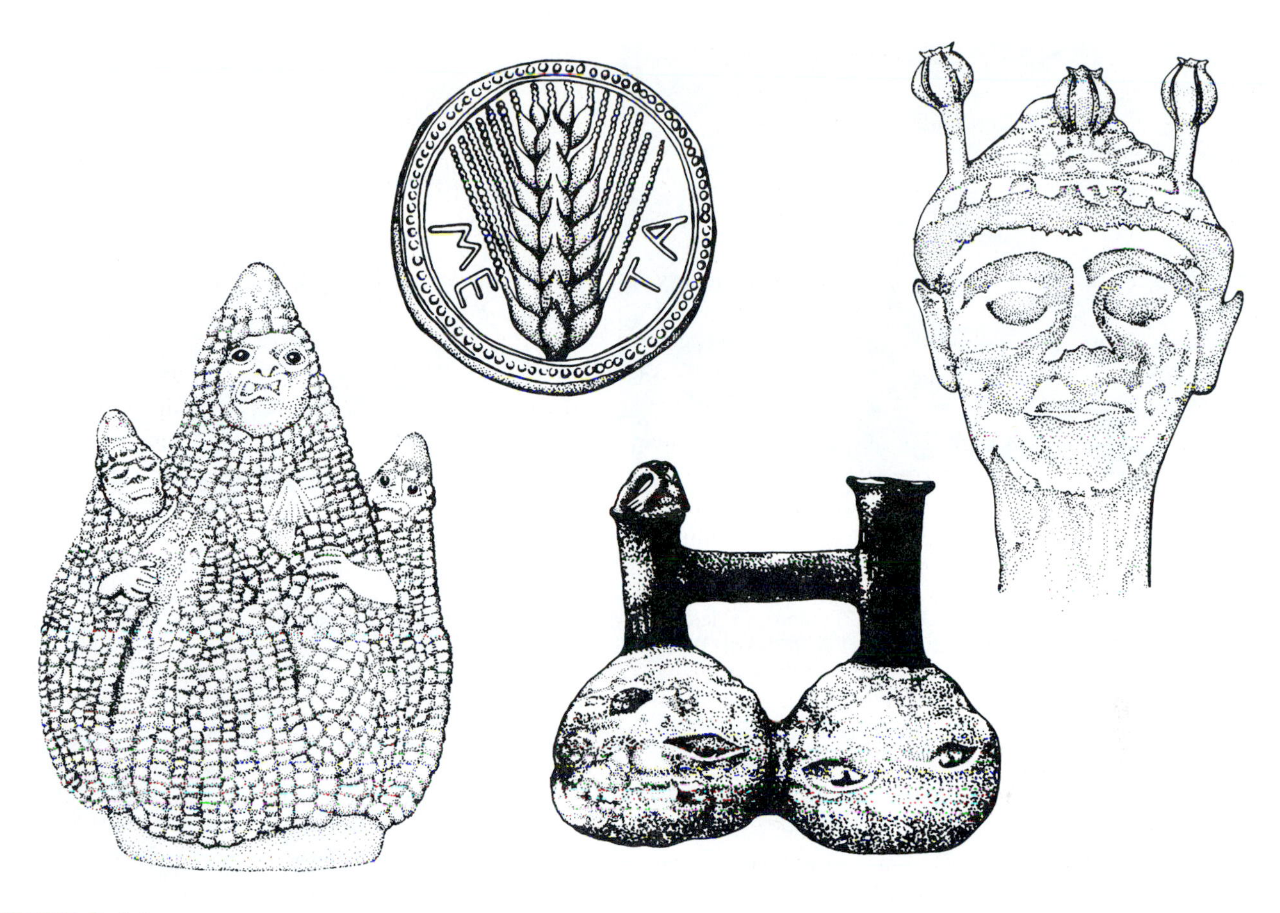

FIGURE 2.10

Archaeological evidence helps in reconstructing the history of particular crop plants. (Upper left) Barley, an important item of trade in ancient times, depicted on a Greek coin. (Upper right) Poppy capsules bedecking the head of a stone sculpture from Crete, approximately 1400 B.C. (Lower left) A vase portraying three corn gods left by the Mochica people, who grew corn under irrigations from the first century through the third century A.D. in Peru. (Lower right) A chimu potato pot redrawn from an artifact in the British Museum.

FIGURE 2.11

Plants first domesticated in the Mediterranean or Near Eastern region. (1) sage, (2) thyme, (3) oregano, (4) parsley, (5) mint, (6) wheat, (7) barley, (8) oats, (9) fig, (10) lentils, (11) pomegranate, (12) olives, (13) leeks, (14) rosemary, (15) lettuce, (16) artichoke.

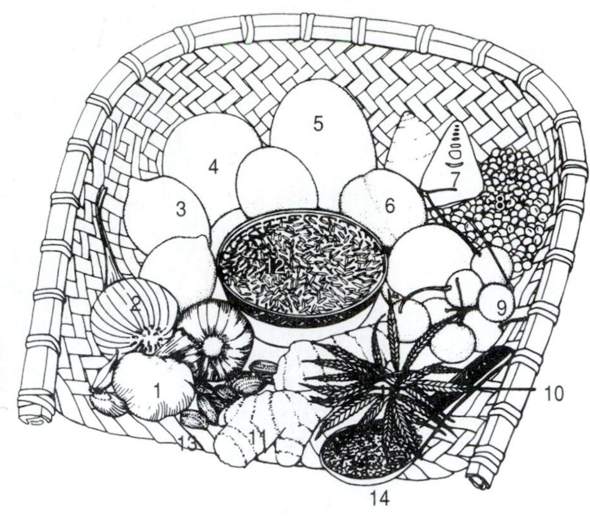

FIGURE 2.12

Plants first domesticated in Asia. (1) Garlic, (2) onions, (3) lemons, (4) oranges, (5) mangoes, (6) peach, (7) bamboo shoots, (8) soybeans, (9) cherries, (10) hemp, (11) ginger, (12) rice, (13) almonds, (14) tea.

FIGURE 2.13

Plants first domesticated in the New World. (1) Jerusalem artichoke, (2) pinto beans, (3) tomatoes, (4) pepper, (5) vanilla, (6) papaya, (7) avocado, (8) wild rice, (9) potatoes, (10) green beans, (11) peanuts, (12) corn, (13) jicama, (14) pineapple.

Asian center, the Middle Eastern center, the Indian center, the Mediterranean center, the Abyssinian center, the Mexico–Central American center, and the central Andean center. In the original formulation of six centers, the first two were considered the Southeast Asian center and the next two were combined as the Southwest Asian center. Table 2.1 lists some of the crops that Vavilov proposed had originated in these centers.

In light of modern knowledge of genetics and data from archaeological findings, many of Vavilov's conclusions seem naive. Current reconstructions of the process of domestication of a particular crop involve the compilation of information from cytology, ecology, plant systematics, archaeology, and ethnology. Careful assessment of the probable areas of origin of many important crops has indicated that domestication occurred over a much more diffuse region than Vavilov supposed. Particularly in Africa and South America, there do not seem to have been restricted regions in which a large number of crops were first brought into cultivation.

Jack Harlan has synthesized the modern data and developed a concept of centers and noncenters. Simply stated, noncenters are areas so large that designating them as centers is meaningless. A case in point is Africa. Although a variety of crops were first domesticated in Africa, they were brought into cultivation in different parts of that continent.

TABLE 2.1 Vavilov's Eight Centers of Origin and Some of the Crops Proposed to Have been Domesticated in Each

CENTER	COMMON NAME	SCIENTIFIC NAME	FAMILY
Chinese	Buckwheat	*Fagopyrum esculentum*	Polygonaceae
	Hemp	*Cannabis sativa*	Cannabaceae
	Mulberry	*Morus alba*	Moraceae
	Orange, sweet	*Citrus sinensis*	Rutaceae
	Peach	*Prunus persica*	Rosaceae
	Poppy	*Papaver somniferum*	Papaveraceae
	Soybean	*Glycine max*	Fabaceae
	Tea	*Camellia sinensis*	Theaceae
	Tung	*Aleurites* spp.	Euphorbiaceae
Central Asian	Apple	*Malus domestica*	Rosaceae
	Carrot	*Daucus carota*	Apiaceae
	Grape	*Vitis vinifera*	Vitaceae
	Onion	*Allium cepa*	Liliacea
	Pea	*Pisum sativum*	Fabaceae
	Pear	*Pyrus communis*	Rosaceae
	Radish	*Raphanus sativus*	Brassicaceae
	Spinach	*Spinacia oleraceae*	Polygonaceae
Middle Eastern, Near Eastern	Alfalfa	*Medicago sativa*	Fabaceae
	Fig	*Ficus carica*	Moraceae
	Flax	*Linum usitatissimum*	Linaceae
	Hazelnut	*Corylus avellana*	Betulaceae
	Lentil	*Lens culinaris*	Fabaceae
	Melon	*Cucumis melo*	Cucurbitaceae
	Quince	*Cydonia oblonga*	Rosaceae
	Rye	*Secale cereale*	Poaceae
	Wheat, einkorn	*Triticum monococcum*	Poaceae
	Wheat, emmer	*T. turgidum* var. *durum*	Poaceae
Indian, including Indo-Malayan	Banana	*Musa* spp.	Musaceae
	Breadfruit	*Artocarpus altilis*	Moraceae
	Chick-pea	*Cicer arietinum*	Fabaceae
	Citron	*Citrus medica*	Rutaceae
	Coconut	*Cocos nucifera*	Arecaceae
	Mango	*Mangifera indica*	Anacardiaceae
	Pepper, black	*Piper nigrum*	Piperaceae
	Rice	*Oryza sativa*	Poaceae
	Safflower	*Carthamus tinctorius*	Asteraceae
	Sesame	*Sesamum indicum*	Pedaliaceae
	Sugarcane	*Saccharum officinarum*	Poaceae
	Yam	*Dioscorea alata*	Dioscoriaceae
Mediterranean	Asparagus	*Asparagus officinalis*	Liliaceae
	Beet	*Beta vulgaris*	Chenopodiaceae
	Cabbage	*Brassica oleracea*	Brassicaceae
	Carob	*Ceratonia siliqua*	Fabaceae
	Lavender	*Lavendula angustifolia*	Lamiaceae
	Leek	*Allium ampeloprasum*	Alliaceae
	Lettuce	*Lactuca sativa*	Asteraceae
	Olive	*Olea europea*	Oleaceae
Abyssinian	Barley	*Hordeum vulgare*	Poaceae
	Castor bean	*Ricinus communis*	Euphorbiaceae
	Coffee	*Coffea arabica*	Rubiaceae
	Millet, African	*Eleusine coracana*	Poaceae
	pearl	*Pennisetum glaucum*	Poaceae
	Okra	*Abelmoschus esculentus*	Malvaceae
	Sorghum	*Sorghum bicolor*	Poaceae

TABLE 2.1 *Continued*

CENTER	COMMON NAME	SCIENTIFIC NAME	FAMILY
Mexico-Central American	Avocado	*Persea americana*	Lauraceae
	Bean, common	*Phaseolus vulgaris*	Fabaceae
	Cacao	*Theobroma cacao*	Sterculiaceae
	Corn	*Zea mays*	Poaceae
	Cotton	*Gossypium hirsutum*	Malvaceae
	Potato, sweet	*Ipomoea batatas*	Convolvulaceae
	Pepper, red	*Capsicum* spp.	Solanaceae
	Squash, winter	*Cucurbita moschata*	Cucurbitaceae
Central Andean (including most of South America)	Manioc	*Manihot esculentum*	Euphorbiaceae
	Peanut	*Arachis hypogea*	Fabaceae
	Pineapple	*Ananas* spp.	Bromeliaceae
	Potato, white	*Solanum tuberosum*	Solanaceae
	Rubber	*Hevea brasiliensis*	Euphorbiaceae
	Squash	*Cucurbita maxima*	Cucurbitaceae
	Tobacco	*Nicotiana tabacum*	Solanaceae
	Tomato	*Solanum esculentum*	Solanaceae

Source: Selected species with names updated from N. I. Vavilov, 1951. *The Origin, Variation, and Breeding of Cultivated Plants. Chronica Botanica 13.* Waltham, Mass., New York, Ronald Press.

Note: Although many of Vavilov's determinations about centers of domestication have withstood the test of time, recent evidence has suggested that several of the crops listed in this table arose and/or were domesticated in regions different from those he proposed.

Designating the entire African continent as a center does not convey any information beyond the fact that a plant was probably native to Africa and was first used by humans there. Designating Africa as a noncenter is a way of indicating the diffuse nature of the places in which various crops were domesticated. Harlan considered the Near East, Asia (north China), and Mesoamerica as centers of agricultural origins. In his interpretation, these centers are areas in which agriculture independently arose, and each was associated with a large region over which the domestication of a few crops occurred. Nevertheless, although Harlan's views are probably realistic considering the evidence for the independent origins of agriculture and recent data about the areas of the first cultivation of many important crop plants, workers still refer to Vavilov's centers when discussing economically important plants. In this text, however, we will assess each plant independently rather than trying to force it into a system of centers of domestication.

Fruits and Nuts of Temperate Regions

When it comes to fruits, botanical terminology bears little relationship to everyday speech. Nutrition charts group edible plants into categories of fruits, vegetables, grains, and nuts, an artificial but generally understood classification. A typical American grocery shopper would define fruits as sweet, aromatic plant products generally eaten as dessert or a first course. This shopper would probably lump together as vegetables plant parts eaten as entrées or as side dishes and typically seasoned with salt and pepper. The very origins of the words reflect these ideas. The term *fruit* comes from the Latin *fruor,* meaning "to enjoy," while *vegetable* has its roots in the Latin *vegetare,* "to enliven or animate." Presumably, it has long been appreciated that fruits in the vernacular sense were eaten mostly for pleasure whereas vegetables were necessary for life. However, to a botanist the word *vegetable* refers to plant matter in general, whereas fruits are specific plant organs. Botanically, green beans and cucumbers are fruits. Tomatoes and eggplants are both fruits classified as berries. Strawberries and blackberries are also fruits, but botanists do not consider them berries (Table 3.1).

Because all the foods we commonly call fruits and nuts, and many of the starches and everyday vegetables we eat, are fruits from a botanical point of view, we discuss fruits in detail in this and the next three chapters. Here we define fruits, outline their classification, and explore dessert fruits and nuts from **temperate** regions, areas that typically experience winter freezes. In Chapter 4 we cover dessert fruits and nuts of warm regions where freezes rarely occur. In Chapters 5 and 6 we discuss two particularly important groups of fruit, grains and legumes.

What Is a Fruit?

What, then, are fruits? How are they formed, and what kinds are there? A fruit is a mature, or ripened, ovary of a flower along with its contents and any adhering accessory structures. A fruit usually includes seeds resulting from the maturation of the ovules after successful fertilization of the egg and polar nuclei (Fig. 1.19). However, both seed and fruit maturation can occur without fertilization. Seed production without fertilization is a kind of a sexual reproduction known as **agamospermy** (without sperm). A seed that is produced agamospermously contains genetic and cytoplasmic

TABLE 3.1 Fruit Types

TYPE	EXPLANATION	EXAMPLE
SIMPLE	Fruit from a flower with only a single ovary; ovary can be simple or compound	
Dry fruits	Pericarp dry when mature	
Indehiscent	Pericarp does not split at maturity	
Achene	1-seeded, pericarp thin, seed coat separate from ovary wall	Sunflower
Grain (caryopsis)	1-seeded, pericarp fused with ovary wall	Corn, rice, wheat
Nut	1-seeded with hard or bony pericarp wholly or partially surrounded by a husk of bracts	Hazelnut, walnut
Dehiscent	Pericarp splits to release the seeds	
Follicle	Podlike fruit that splits along the ventral side	Milkweed
Legume	Pods that split along 2 opposite edges	Beans, peas, soybeans
Silique	Pericarp splits along the juncture of 2 carpels	Mustard
Capsule	Pericarp opens with pores or slits	Opium poppy
Schizocarp	Fruit splits into 2 to several achenelike segments with free seeds inside	Dill "seed"
Fleshy fruits	Part or all of the pericarp fleshy at maturity	
Berry	1- to several-seeded, mesocarp fleshy, endocarp soft	Grape, tomato
Drupe	1- to 2-seeded, mesocarp fleshy, endocarp usually hard	Peach, plum
Hesperidium	Special kind of berry with leathery rind and oil glands dotting the surface	Citrus fruits (lime, orange)
Pepo	A berrylike fruit with a hard rind of combined accessory tissues and pericarp	Melons, squash
Pome	A berrylike fruit with most of the flesh derived from a floral cup and receptacle	Apple
AGGREGATE	Fruits formed from several separate ovaries within a single flower	Raspberries
MULTIPLE	Formed by the fusion or collective shedding of fruits of numerous independent flowers	Fig, pineapple
AGAMOSPERMOUS	Fruits with seeds derived from maternal tissues only, no fertilization occurs	Dandelion
PARTHENOCARPIC	Fruits formed without fertilization; can have agamospermous seeds or be seedless	Banana, navel orange

material derived solely from maternal tissue. Although pollen sometimes triggers the development of such seeds, it contributes nothing to them. A fruit that matures without seed formation is called **parthenocarpic.** In some cases agriculturists have selected for seedless fruits by vegetatively propagating parthenocarpic mutants, such as navel oranges, that spontaneously mature fruits without fertilization. Hormones can also be used to trigger fruit production without fertilization, as is commonly done with seedless grapes. Seedless herbaceous crops such as watermelons can also be produced by making crosses that yield sterile triploid plants.

A fruit is classified by the number of ovaries that produced it (one to several), the position of the ovary relative to the apparent bases of the petals, whether the fruit is fleshy or dry at maturity, and the way the seeds are shed. Table 3.1 provides a basic outline of the common fruit types discussed throughout this book. Table 3.2 contains a listing of the plants discussed in this chapter.

An ovary can be simple or compound. The notion of simple and compound ovaries involves the concept of the carpel, which comes from a hypothesis called the **foliar theory.** According to this theory, the ovary was derived from a leaflike structure on which ovules were borne (Fig. 3.1). During the course of evolution this leafy structure was supposed to have folded, producing the closed ovary characteristic of the angiosperms. A **carpel** is a female reproductive structure derived from a single, folded, leaflike unit. A **simple ovary** is derived from one such structure and therefore has one carpel. A **compound ovary** is produced by the fusion of two or more of these leafy structures (Fig. 3.2). The number of carpels in a compound ovary is therefore equal to the number of ancestral leaflike structures fused into the pistil. This explanation of the carpel's origin is not universally accepted, but virtually all descriptions of flowering plants use the number of carpels in the ovary as one of the diagnostic characters of a species.

A fruit derived from one ovary (single simple or compound) is a **simple fruit.** The nature of the fruit wall depends on whether the ovary is superior or inferior. A **superior ovary** is one borne above the insertion of the sepals and petals. An **inferior ovary** is positioned below the attachment of the other floral parts (sepals, petals, etc.). If a fruit results from the maturation of a simple, superior ovary, it will have one or more seeds surrounded by a **pericarp,** or fruit wall, made up of three layers; the endocarp, mesocarp, and exocarp. The last three terms refer to the inner (*endo-*), middle (*meso-*), and outer (*exo-*) layers of the ovary wall (Fig. 3.3). The pericarp can develop in various ways, leading to the assortment of fruit types that characterize different species of flowering plants (Table 3.1). For example, the pericarp can

TABLE 3.2 Plants Discussed in Chapter 3

CROP	SPECIES	FAMILY	NATIVE REGION
Almond	*Prunus dulcis*	Rosaceae	Central Asia
Apple	*Malus domestica*	Rosaceae	Western Asia
Apricot	*Prunus armeniaca*	Rosaceae	Asia
Blackberries	*Rubus* spp.	Rosaceae	North temperate regions
Blueberries	*Vaccinium*	Ericaceae	
High-bush	*V. corymbosum*		North America
Low-bush	*V. angustifolium*		North America
Rabbit eye	*V. ashei*		North America
Cherries	*Prunus*	Rosaceae	
Sour	*P. cerasus*		Near East
Sweet	*P. avium*		Near East
Chestnut	*Castanea*	Fagaceae	
American	*C. dentata*		North America
Chinese	*C. mollissima*		China
European	*C. sativa*		Eurasia
Japanese	*C. crenata*		Asia
Cranberries	*Vaccinium oxycoccos*	Ericaceae	North America
	V. macrocarpon		North America
Grapes	*Vitis*	Vitaceae	
Fox	*V. labrusca*		North America
Wine, table	*V. vinifera*		Eurasia
Hazelnut (filbert)	*Corylus avellana*	Corylaceae	Eurasia
Kiwi fruit	*Actinidia deliciosa*	Actinidiaceae	Asia
Olive	*Olea europaea*	Oleaceae	Mediterranean
Peach	*Prunus persica*	Rosaceae	Asia
Pear	*Pyrus communis*	Rosaceae	Asia
	P. pyrifolia		
	P. bretschneiderii		
Pecan	*Carya illinoinensis*	Juglandaceae	North America
Pistachio	*Pistacia vera*	Anacardiaceae	South-central Asia and Mediterranean
Plums	*Prunus*	Rosaceae	
American	*P. americana*		North America
Cherry	*P. cerasifera*		Western and Central Asia
European	*P. domestica*		Europe
Japanese	*P. salicina*		Japan
Quince	*Cydonia oblonga*	Rosaceae	Eurasia
Raspberries	*Rubus*	Rosaceae	
Black	*R. occidentalis*		North America
Red	*R. idaeus*		Eurasia and North America
Strawberry	*Fragaria*	Rosaceae	
Chilean	*F. chiloensis*		Western North and South America
Common	*F. ananassa*		Hybrid
Virginia	*F. virginiana*		North America
Walnut	*Juglans*	Juglandaceae	
Black	*J. nigra*		North America
Persian	*J. regia*		Eurasia

remain thin and become dry during fruit maturation. The resultant dry fruits can be shed as entire units, or they can split open to release their seeds. In the first case, the fruits are considered to be dry and **indehiscent** (nonsplitting), and in the second case, dry and **dehiscent** (splitting). Maple or sunflower fruits are indehiscent dry fruits, whereas geranium fruits are dehiscent dry fruits (see Fig. 1.22). Dry fruits, or their seeds, are generally dispersed by wind, water, gravity, or animals to which they have become attached. A **grain,** or **caryopsis,** is a particularly important kind of dry indehiscent fruit in which the seed coat is fused to the ovary wall (see Fig. 5.4). **Legumes** are dry, dehiscent fruits from a single carpel that split along the two opposite margins of the pod (see Fig. 6.1).

In many cases, the pericarp becomes fleshy as the fruit matures (Table 3.1). The maturation of a single superior ovary within a flower into a fruit with a fleshy pericarp produces a berry. Berries can contain one or several seeds, but

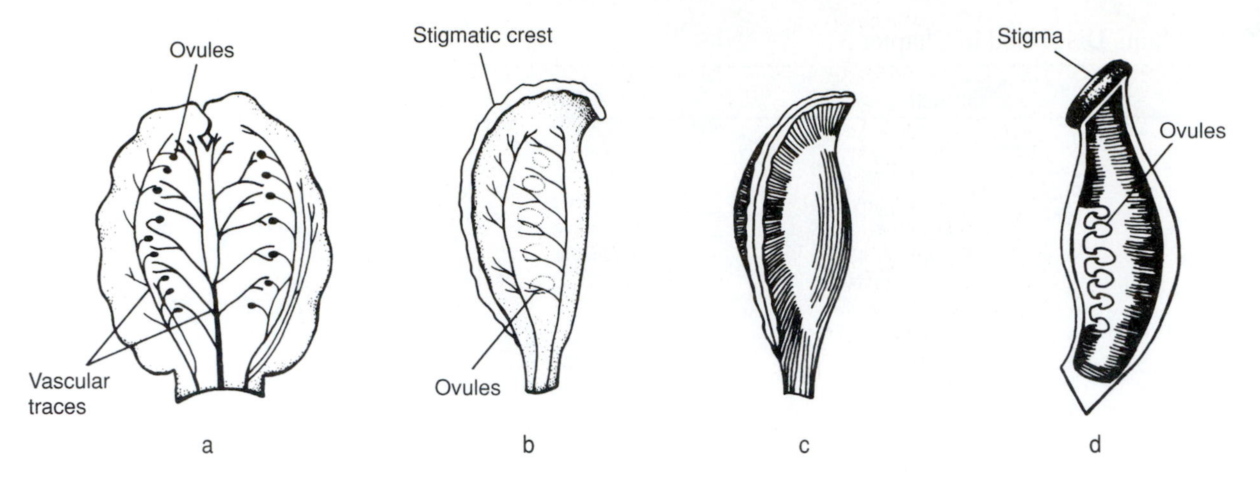

FIGURE 3.1
(a) It is postulated that the carpel was derived from the folding of a leaflike structure on which the ovules were borne. (b–d) The modern carpel of *Drimys winteri* is believed to resemble this early folded leaflike structure. The vascular traces inside the *Drimys* carpel are shown in b; c presents an exterior view, and d a cutaway view.

the seeds are never enclosed by a hard endocarp. Berrylike fruits with the seed or seeds enclosed in a hard, stony endocarp are **drupes** (Fig. 3.3d). Citrus fruits are special kinds of berries called **hesperidiums** (see Fig. 4.1) because they have a leathery rind dotted with packets of oil and specialized fleshy hairs projecting from the inner wall of the endocarp.

If there are numerous simple superior ovaries borne within a single flower, each ovary matures to form part of an aggregate fruit (Table 3.1). Strawberries, raspberries, and blackberries are aggregate fruits and not true berries (Fig. 3.4a).

The pericarps of inferior ovaries are fused with the bases of the floral parts surrounding them. The apparent fruit wall therefore consists not only of the ovary wall, but also of the extra, or **accessory** (nonovary), parts. The fleshy portion can result from an enlargement of any one (or all) of the pericarp layers or even by a swelling of accessory tissues (Fig. 3.4b). Fruits produced by inferior ovaries are technically accessory fruits, but they are usually called by the names of the simple fruits they resemble. Cranberries and blueberries are examples of accessory fruits generally referred to as berries. Bananas are also accessory fruits. Many plant families have distinctive accessory fruits that are given names. For example, fruits of the squash family are called **pepos** (from the Latin for pumpkin) because of their distinctive hard rind. Fruits of some members of the rose family, such as the apple, are formed when the floral cup *and* the receptacle around the ovary swell to form the majority of the fruit pulp. These fruits are known as **pomes** (Fig. 3.4b).

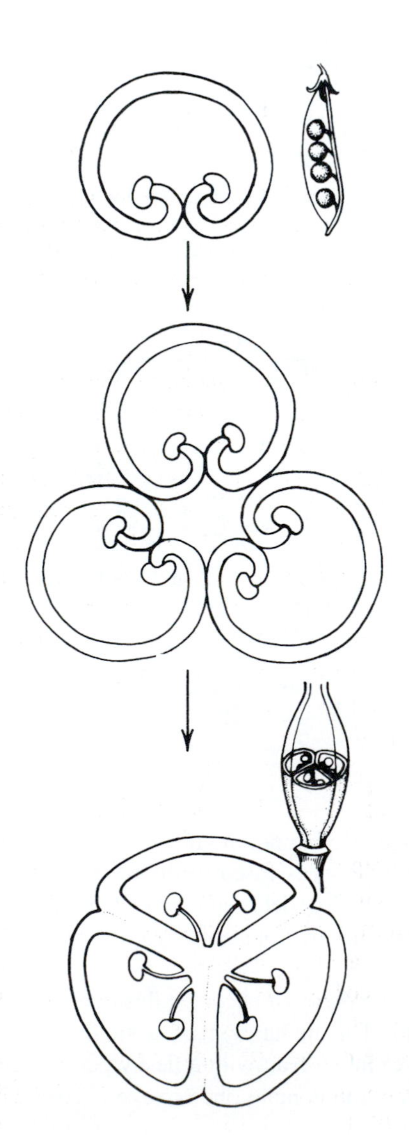

FIGURE 3.2
The fusion of several carpels is presumed to have led to a compound ovary. The pea pod (top) is an example of a simple ovary that could have been derived from the folding of a single leaflike structure. The fusion of several such structures (or carpels) presumably led to a compound ovary (bottom) as is found in tomatoes or peppers.

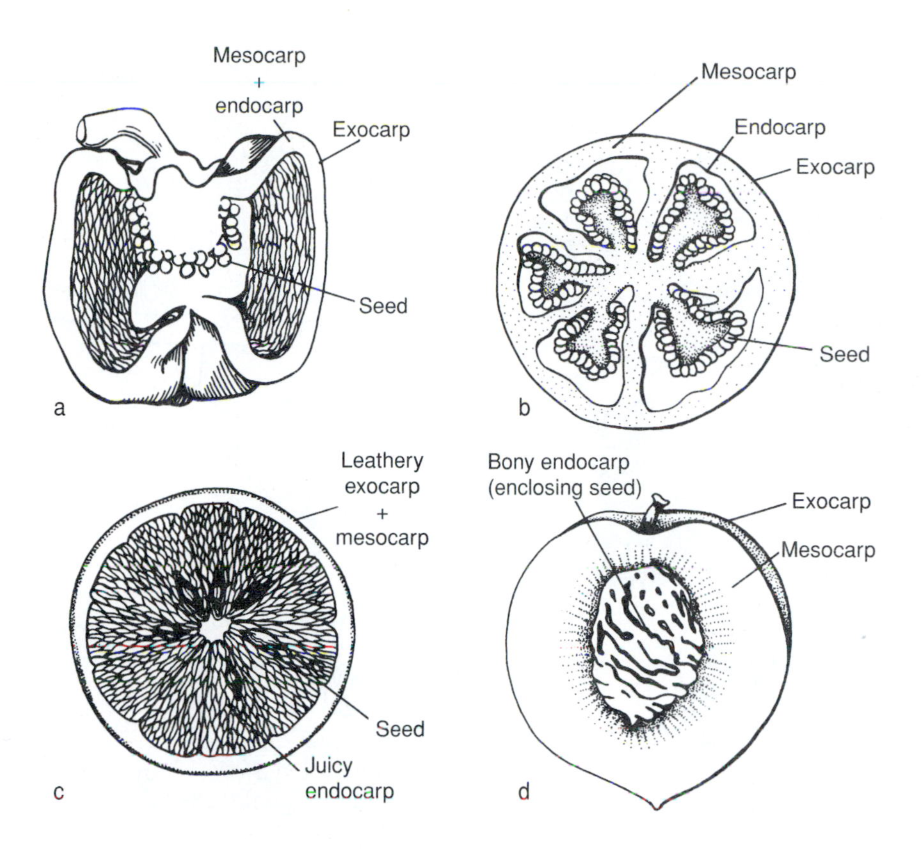

FIGURE 3.3
The exocarp, mesocarp, and endocarp layers of the pericarp can develop differently, leading to a variety of fruit types. (a) A slightly expanded mesocarp gives rise to the fleshy rind of the pepper. The central core is composed of the placental tissues on which the seeds are borne. (b) A tomato also has a relatively fleshy mesocarp and an elaboration of the seed-bearing (placental) tissues that produces the "jelly" inside the fruit. (c) The fleshy part of an orange is made up of swollen hairs borne on the inside of the endocarp. (d) A greatly expanded mesocarp produces the sweet flesh of a peach.

Finally, a few, mostly tropical, fruits are formed by the tight packing of fruits from many separate flowers. Fruits such as pineapples, figs, and breadfruits constructed in this way are **multiple** fruits (see Fig. 4.25). In a sense, multiple fruits are false fruits since each is composed of many fruits. Each segment of a pineapple or each tiny achene inside a fig is a complete fruit.

Fleshy fruits have evolved to attract animals that will eat them but spit out, drop, or pass the seeds unharmed away from the mother plant. Since the succulent fruit tissue is not a source of nutrition for the embryo or for the later development of the seedling, a plant puts as few resources as possible into fruit pulp. Since sugars are the immediate products of photosynthesis, they are less "expensive" for a plant than are more complex compounds that require further synthesis or contain limiting nutrients such as nitrogen. As a result, fleshy fruit parts tend to consist primarily of flavored sugar solutions held in a cellulose matrix. With the exception of a few fruits, such as the avocado and the olive, fleshy fruits are watery, low in calories, and good sources of water-soluble vitamins.

In contrast to the pulpy parts of fleshy fruits, the seeds contain nutrients that are used by the embryo for emergence and establishment. Seeds therefore possess appreciable quantities of high-energy material such as fats, oils, and starches. In nature, any animal that destructively consumes seeds is a predator that reduces a plant's reproductive potential. Consequently, many plants have seeds that are protected by hard shells, seed coats, or toxic compounds. Humans have learned to overcome the various devices for seed protection, taking advantage of the nutrition packed into seeds. Tables 3.3 and 3.4 highlight the nutritive differences between fruit pulps and seeds by comparing the caloric values of fleshy fruit crops with nuts, an important group of foods derived from seeds.

Apples and Their Relatives

Apples are the most important temperate fruit tree crop in the world. Apples and their relatives plums, peaches, and strawberries all come from the rose family (Rosaceae). These fruits differ from one another, however, in their structures and are consequently considered to belong to different subgroups of the family.

The Rosaceae are divided by most botanists into four groups, or subfamilies: the spirea subfamily (Spiraeoideae), the apple subfamily (Maloideae), the plum subfamily

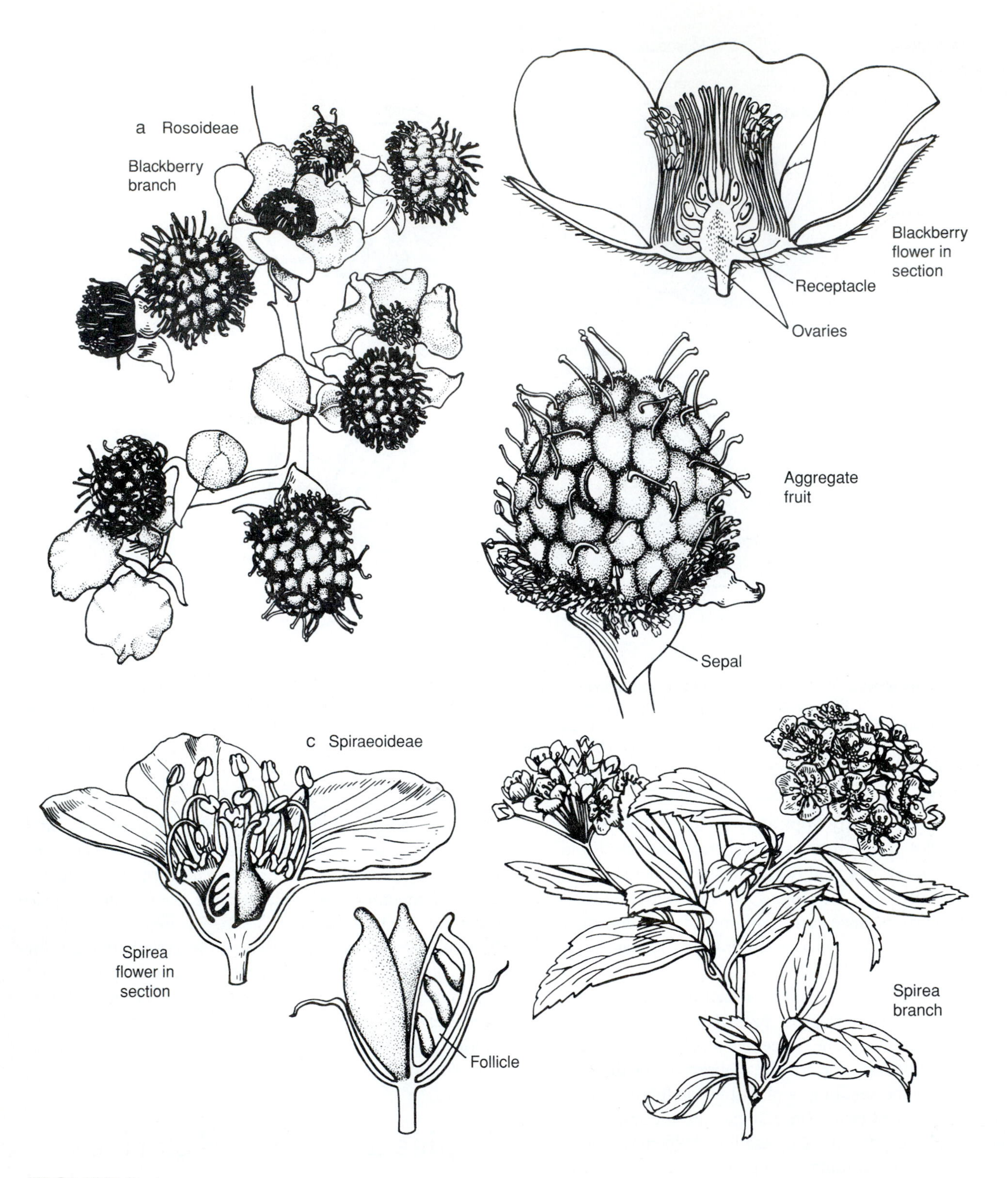

FIGURE 3.4

The Rosaceae are divided into several subfamilies on the basis of flower and fruit characters. These include (a) the Rosoideae, which has fruits formed primarily from simple ovaries grouped into aggregate fruits such as blackberries and raspberries; (b) the Maloideae, with fruits such as the apple and pear, that develop in part from a floral cup surrounding the compound ovary, producing an accessory fruit known as a pome; (c) the Spiraeoideae, with capsulate or follicular fruits such as those of the ornamental *Spiraea;* and (d) the Amygdaloideae, with fruits formed from simple superior ovaries developing into drupes such as cherries, peaches, and plums.

(*Spiraea* and blackberry cross section after Baillon; the rest drawn from nature.)

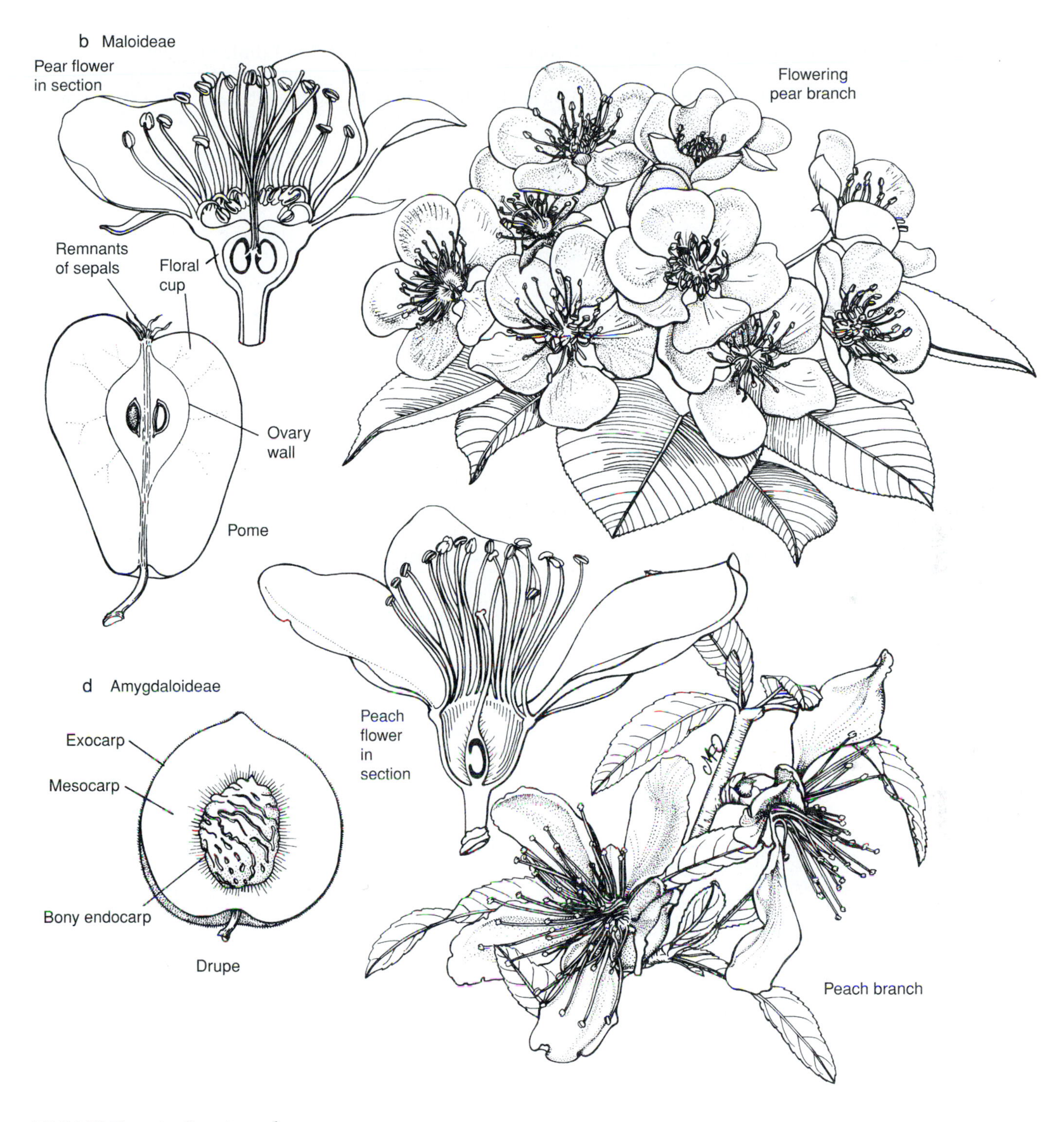

FIGURE 3.4 *Continued*

(Amygdaloideae), and the rose subfamily (Rosoideae). The last three are important sources of edible fruits (Fig. 3.4). The Maloideae produce pome fruits (Table 3.1). The Amygdaloideae contain species that produce drupes as fruits, and plants in the Rosoideae produce aggregate fruits. Figure 3.4 illustrates the structural differences in the fruits of these four subfamilies.

Apples, Pears, and Quinces

The apple subfamily contains two important genera of fruit plants, *Malus* (apple) and *Pyrus* (pear), as well as the less well known *Cydonia* (quince). Apples (Fig. 3.5) account for nearly 60 percent of the world's temperate fruit tree production

TABLE 3.3 Nutritional Composition of Selected Temperate Fruits, Based on 100-g Edible Portion

FRUIT	WATER, %	CALORIES	PROTEIN, g	FAT, g	CARBOHYDRATE, g	VITAMIN A, µg	VITAMIN C, mg
Apple	88	35	0.3	0.1	9	17	14
Apricot	87	31	0.9	0.1	7	405	6
Blackberry	85	25	0.9	0.2	5	80	15
Blueberry	83	62	0.7	0.5	15	17	14
Cherry, red	83	48	0.9	0.1	11	25	11
Cranberry	88	46	0.4	0.7	11	4	11
Grape	82	60	0.4	0.1	15	17	3
Olive, green	77	103	0.9	11.0	—	180	—
Peach	89	33	1.0	0.1	8	58	31
Pear	84	40	0.3	0.1	10	18	6
Plum	84	36	0.6	0.1	9	295	4
Raspberry, red	87	25	1.4	0.3	5	6	32
Stawberry	90	27	0.8	0.1	6	8	77

Source: R. A. McChance and E. M. Widdowson. *The Composition of Foods,* 5th ed. 1992. Royal Chemical Society, Bungay, Suffolk, England, Clay.

TABLE 3.4 Nutritional Composition of Selected Temperate Nuts, Based on 100-g Edible Portion

NUT	WATER, %	CALORIES	PROTEIN, g	FAT, g	CARBOHYDRATE, g	VITAMIN A, µg	VITAMIN C, mg
Almond	4	612	21	56	7	0	—
Chestnut	52	170	2	3	37	—	—
Hazelnut (filbert)	5	650	14	63	6	—	—
Pecan	4	689	9	70	6	50	0
Pistachio	1	331	10	30	5	71	0
Walnut	3	688	15	68	3	0	0

Source: R. A. McChance and E. M. Widdowson. *The Composition of Foods,* 5th ed. 1992. Royal Chemical Society, Bungay, Suffolk, England, Clay.

(Table 3.5). Cultivated species are usually lumped under the species *Malus* X*domesticus* (or *Malus* X*asiatica*). The genus is thought to have arisen in Asia but had spread to North America long before humans arrived on the continent. Nevertheless, despite the expression "as American as apple pie," our cultivated edible apples come from the Old World. Apples have been eaten in Eurasia since prehistoric times, and selection must have been practiced very early, since the Romans are known to have had at least 22 apple varieties. By A.D. 1670 enough varieties had been selected to allow the grand duke of Tuscany to give a banquet during which he served 56 varieties of apples. Apples were apparently first planted in North America in the 1620s on Bunker Hill, where the famous Revolutionary War battle is said to have occurred. Europeans were routinely grafting apples before settling

FIGURE 3.5

Improvements in shipping and techniques for delaying fruit ripening, such as storage under high carbon dioxide concentrations allow an increasing array of apple varieties to be offered year round in U.S. markets. Here, we have 20 apple varieties for sale at Central Market, Austin, Texas. Left to right from the top: Pink lady, Empire, Red granny, Fuji; Crabapple, Golden delicious, Cameo, Jonagold; Crispin, Jonathan, Cortland, Mutsu; Idared, Winesap, Braeburn, Red Delicious; Jonamac, New York, Rome, Gala.

Box 3.1

Johnny Appleseed: An American Tale

Every American has heard of Johnny Appleseed, who trudged ragged and alone across the wilderness in the early part of the nineteenth century, sowing apple seeds. The very nature of the man's life contained the elements around which legends are built. The most romantic of these tales recounts that John Chapman, the son of a clergyman, fell in love with the daughter of one of his father's enemies. His suit was summarily rejected by his sweetheart's parents, who moved westward, taking their daughter with them. Chapman dreamed up the idea of carrying seeds to pioneers while he searched for his love. When he found her, he supposedly learned that she had died only the previous year of fever and a broken heart. Chapman subsequently became Johnny Appleseed in earnest as he traveled westward in a ragged cloak with a pot on his head, planting trees as a purely humanitarian act.

The notion of Johnny Appleseed's unrequited love dates from the early part of this century and the writings of Reverend Newell Dwight Hillis in *The Quest of John Chapman.* Unfortunately, almost none of this story is true. Chapman's mother died when he was 2 years old, and his father was a farmer and carpenter who served in the Revolutionary War. When he reached maturity, Chapman chose the nursery business as his profession but harbored a lust for travel. At the age of 18 he hiked west to central New York State, where he joined the Swedenborgian (New Church) movement and began serious missionary activities. While preaching across the still wild Midwest, Chapman lectured on apples and planted orchards, choosing sites with a practiced eye. Many of these orchards later became settlements and eventually cities. Planting apples might seem like a frivolous activity when one is pushing into unsettled territory, but apples were a mainstay of early settlers. The fruits were pressed for juice, relished fresh, or fermented into a mildly alcoholic drink that could be kept over the winter. Apples stored well or could be dried or cooked down into jamlike apple butter. Chapman practiced a form of selection by using the seeds of particular trees. Nevertheless, his trees produced a motley array of apples compared with those of cultivated orchards, which consisted of uniform hybrids maintained by grafting. However, there was a method to his madness. Seeds were easy to carry and germinated readily. Although they produced trees that bore comparatively poor apples, they would be bearing fruit when settlers arrived. Moreover, they provided hardy stocks already in place for subsequent grafting.

Throughout his life Chapman was known as a kind man, a vegetarian who told delightful tales and shared his meager food and seeds with anyone regardless of ability to pay. He made friends with the Indians and learned several native languages. His appearance with bare feet, cut-off britches, a coat consisting of a sack with a hole cut for his head, and a pot topping it all off has been used as an indication of his eccentricity. However, cut-off britches, bare feet, and the use of old sacks or "towlinen" for clothing were not uncommon among pioneers. The pot was only one of many things he used to cover his head. The nickname Johnny Appleseed plus the image of an individual who toiled selflessly for the good of his fellow pioneers provided the basis of an American folk tale. For over a century, poets, chroniclers, and even cartoon moviemakers have embellished the legend. Johnny Appleseed's botanical and religious missionary work finally ended in 1842 with his death of pneumonia near Fort Wayne, Indiana.

North America, and orchards in the colonies were also established from grafted trees. However, as pioneers such as Johnny Appleseed (Box 3.1) pushed westward, trees were often planted from seed; these seed plantings eventually resulted in the increase in variation in apples in America relative to those in Europe. This variability allowed North America to take the lead when horticulturalists began to breed new varieties. The Red and Yellow Delicious, developed in the United States, tend to dominate in world trade, but there has been a resurgence in the diversity of apple varieties available in supermarkets. Today we can choose from American-developed Baldwin, Cortland, Granny Smith, Jonathan, McIntosh, and Rome, among others as well as imported and domestic Fuji and Royal Gala.

Apples normally have an ovary with five carpels, but one or more often contain no seeds. The fleshy part of the apple that we eat is a floral cup that expands as the fruit develops after fertilization (Fig. 3.4b). The actual ovary wall lies inside the core line.

Apples grown today are either diploids ($2n$ = 34) or triploids ($2n$ = 51) resulting from crosses of diploid ovules and haploid pollen. Because the initial production of these triploid hybrids is time-consuming and costly, superior varieties that have been developed are perpetuated by grafting. Grafting also permits the use of a variety of rootstocks that have different adaptations for hardiness or pest resistance and are therefore better suited to a variety of localities. Rootstocks are produced by vegetative propagation using cuttings.

The apple has been involved in folklore since ancient times. In Greek mythology, one of Hercules's tasks was to obtain the golden apples of the Hesperides. The apples were well guarded because they were supposed to impart immortality. The generic name of the apple, *Malus,* comes from the Latin word *malus,* or "bad," referring to Eve's fateful nibble

TABLE 3.5 1999 World Production of Fruits and Nuts of Temperate Regions

COMMON NAME	TOP 5 COUNTRIES	1000 METRIC TONS	TOP CONTINENTS	1000 METRIC TONS	WORLD 1000 METRIC TONS
Almonds	United States	627	North and Central America	747	1,575
	Spain	272			
	Iran	112	Europe	417	
	Italy	88	Asia	366	
	Morocco	65	Africa	152	
			Oceania	12	
Apples	China	19,490	Asia	29,529	56,971
	United States	4,791	Europe	15,764	
	Germany	2,026	North and Central America	5,720	
	Italy	2,115			
	France	1,954	South America	2,727	
			Africa	1,482	
Apricots	Turkey	538	Asia	1,412	2,674
	Spain	150	Europe	743	
	Italy	142	Africa	328	
	Pakistan	190	North and Central America	124	
	France	186			
			South America	62	
Chestnuts	China	117	Asia	337	521
	Korea, Rep.	109	Europe	148	
	Turkey	70	South America	34	
	Italy	69	—		
	Japan	33	—		
Grapes	Italy	9,208	Europe	29,682	58,727
	France	6,800	Asia	13,230	
	United States	5,946	North and Central America	6,648	
	Spain	4,818			
	Turkey	3,650	South America	5,220	
			Africa	2,954	
Hazelnuts (filberts)	Turkey	580	Asia	612	786
	Italy	117	Europe	139	
	United States	34	North and Central America	34	
	China	12			
	Spain	10	—		
			—		
Olives	Spain	4,500	Europe	10,276	14,737
	Italy	3,373	Asia	2,340	
	Greece	2,068	Africa	1,870	
	Turkey	1,550	South America	126	
	Tunisia	950	North and Central America	124	
Peaches and nectarines	China	3,000	Asia	4,587	11,326
	Italy	1,429	Europe	3,845	
	United States	1,340	North and Central America	1,518	
	Greece	600			
	Spain	912	South America	792	
			Africa	478	
Pears	China	7,920	Asia	9,725	15,613
	Italy	931	Europe	3,467	
	United States	855	North and Central America	906	
	Spain	603			
	Argentina	540	South America	811	
			Africa	494	

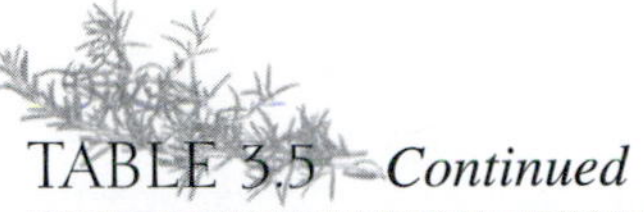

TABLE 3.5 Continued

COMMON NAME	TOP 5 COUNTRIES	1000 METRIC TONS	TOP CONTINENTS	1000 METRIC TONS	WORLD 1000 METRIC TONS
Pistachios	Iran	313	Asia	420	476
	United States	50	North and Central America	50	
	Turkey	40			
	Syria	34	Europe	6	
	China	30			
Plums	China	2,707	Asia	3,647	
	United States	680	Europe	2,500	7,306
	Germany	316	North and Central America	765	
	Yugoslavia, Fed. Rep.	460			
	Romania	319	South America	253	
			Africa	114	
Raspberries	Russian Fed.	95	Europe	298	355
	Yugoslavia, Fed. Rep.	56	North and Central America	54	
	Poland	54			
	Germany	32		—	
	Hungary	20		—	
Strawberries	United States	785	Europe	1,207	
	Spain	367	North and Central America	924	2,796
	Japan	198			
	Poland	171	Asia	526	
	Korea, Rep.	155	Africa	54	
			South America	63	
Walnuts	China	275	Asia	604	
	United States	254	North and Central America	274	1,184
	Iran	145			
	Turkey	120	Europe	268	
	France	23	South America	25	

Source: FAOSTAT Database Gateway. On line at http//:apps.fao.org/ (under "All databases" then Crops, Primary).

in the garden of Eden. There is some dispute, however, about the identity of the golden apples of mythology or the fruit in the garden of Eden because some authorities maintain that apples did not grow in the Mideast or around the Mediterranean Sea when those episodes were supposed to have occurred. According to this view, a late climatic change allowed the movement of wild apples into this region. To these workers, apricots are a more likely candidate than apples to be the fruit mentioned in Mideastern stories. Apricots are called golden apples today in the Mideast and were the most abundant fruit in Palestine (next to figs) in early historical times. Nevertheless, most authorities believe that apples are native to the Mideast and see no reason to doubt that they figured prominently in early legends.

There are more recent tales involving fruits that are unmistakably apples. William Tell, a Swiss hero, is said to have been forced to shoot an apple from his son's head because he refused to pay homage to Austrian conquerors. It is rumored that Newton discovered gravity when an apple fell on his head. In the United States, legends of Johnny Appleseed, based on the life of John Chapman (Box 3.1), recount the migration of the colorful arborist from Massachusetts as he trekked westward planting his beloved fruit trees.

Although less popular than apples in the United States, pears are the second largest temperate fruit tree crop worldwide (Table 3.5). There are many species of pear that appear to hybridize freely and to have been involved in the production of the most widely cultivated species, *Pyrus communis.* The genus *Pyrus* originated in Asia, and cultivated species were introduced into North America only after it was settled by Europeans. In Eurasia pears have been cultivated since at least Classical Grecian times. Selection in pears has been primarily for different fruit colors, flavors, and shapes. Propagation is generally done by means of grafting, using rootstocks grown from seed. The slightly gritty texture of pears is caused by the presence of **stone cells,**—short, more or less square or ovoid cells with thick cell walls. These hard cells are grouped throughout the flesh of the fruit. In Asia, *Pyrus pyrifolia* and *P. bretschneiderii* are important species of pear. The first of these is the Asian pear now commonly sold in American supermarkets.

The third economically important fruit of the apple group is the quince (*Cydonia oblonga*) (Fig. 3.6). Quinces are sometimes used as rootstocks for pears and, like pears, are more popular in Europe than in the United States. Quince fruits tend to have a hard, aromatic flesh with

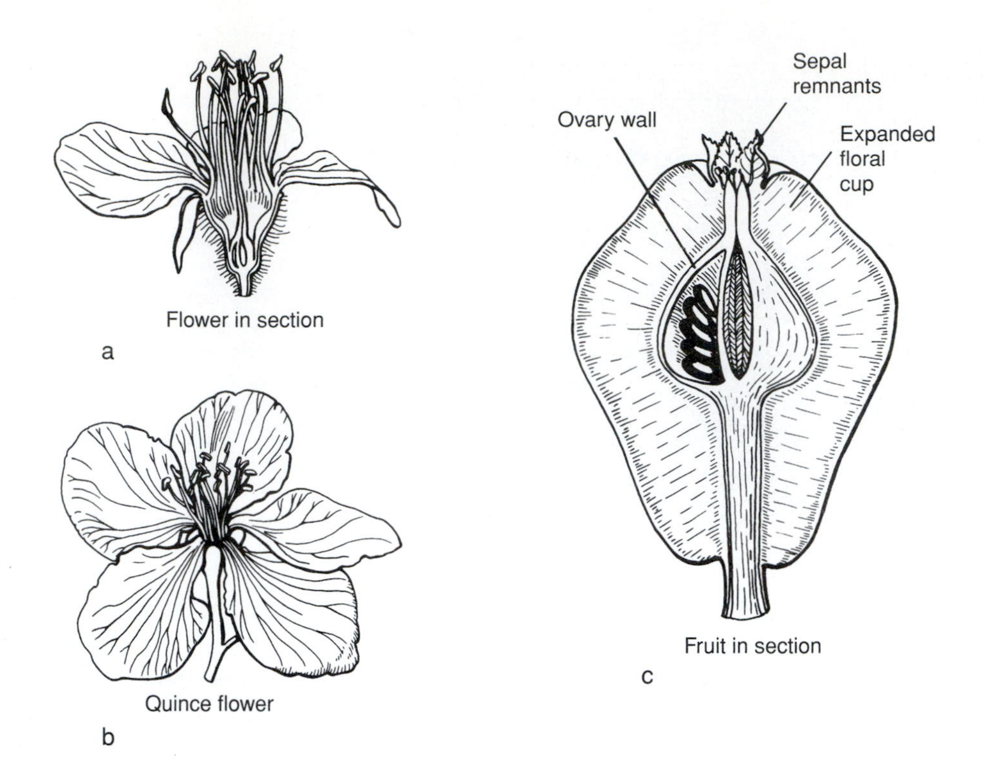

FIGURE 3.6

Quince trees can reach 4 to 6 m (15 to 20 ft) in height and are among the first plants to bloom in the spring. The ovaries embedded in a floral cup (a) of lovely pink and white flowers (b) develop into yellow-skinned pome fruits similar to pears in form and size (c). Unlike pears, however, quinces contain up to 20 seeds in each carpel and flesh that is too hard and tart to eat raw. Fortunately, the pulp softens and turns pink and delicious when cooked with sugar.

(After Baillon.)

noticeably more stone cells than occur in pears. Consequently, most quinces are made into preserves and conserves.

Plums and Other Stone Fruits

Within the plum group (Amygdaloideae) there are several important fruit species, all belonging to the genus *Prunus.* These are plums (primarily *P. domestica*), cherries (*P. avium* and *P. cerasus*), peaches (*P. persica*), and apricots (*P. armeniaca*). In all these fruits the part of the fruit eaten is the fleshy mesocarp (Fig. 3.4d). The exocarp is the "skin" of the fruit. The endocarp is modified into the "pit," a hard stone-like casing that provides protection for the seed.

The center of origin of all the various fruits in this group seems to have been central or western China. There are native species of plums and cherries in the New World, but peaches and apricots did not occur in North or South America before the arrival of Europeans.

Plums, peaches, and apricots have been cultivated for 2000 years or more. Plums in particular are an important European crop (Table 3.5), with hundreds of developed varieties. Plums come from a variety of species that were domesticated in different regions of the Northern Hemisphere: the common European plum, *Prunus domestica,* in Europe; the cherry plum, *P. cerasifera,* in western and central Asia; the Japanese plum, *P. salicina,* in China; and the American plum, *P. americana,* in North America. Although the European plum is the dominant crop species, it is the most recently derived, probably from mixtures of the cherry plum and other wild plums. Selection in plums has been for uniformity in fruit production. Consequently, plums are grafted onto vegetatively propagated rootstocks. In plum orchards in the United States today, trees that mature the same kind of plum but at different times of the summer are planted in separate rows. This arrangement allows a prolonged harvest without the need to alter any procedures or machinery. Plums are cultivated in the United States in every state except Alaska, but the primary area of production is along the West Coast.

Prunes are particular varieties of plums or plums that have been dried. Despite the common modern association of prunes with laxative effects, their production dates back to times before modern methods of preservation because, once dried, they can be kept for long periods without spoiling.

As there are for plums, there are New World cherries, but cultivated cherries are essentially all derived from Old World species. Their cultivation in the United States (Fig. 3.7) began with the early settlers, but the most important year for the U.S. cherry industry was 1847, when two brothers named Lewelling settled in Oregon after an ox car trip from Iowa with

FIGURE 3.7
Cherries are often harvested mechanically by shaking trees and catching the falling fruits on a tarpaulin placed around the base of the tree.
(Photo courtesy of Cherry Marketing Institute.)

several hundred cherry seedlings. One of the brothers eventually selected the bing cherry (named after their Chinese housekeeper), the most important cultivar in the United States today, and the Lambert, the second most important U.S. variety.

Cherries have a special meaning for Americans because of their association with young George Washington. The famous story of Washington's chopping down the cherry tree and confessing the deed to his father is, unfortunately, fictitious. Parson Weems created the story for the fifth edition (1806) of his book *The Life and Memorable Actions of George Washington.* Weems seems to have been more concerned with providing a good example of honesty rather than accuracy. Today, cherry trees cultivated for their flowers rather than their fruits adorn the Tidal Basin of Washington, D.C. These trees, a gift from the Japanese government, signal the advent of spring with a display that draws visitors from all over the country.

Cherries are grouped according to taste into sweet and sour types. Sweet cherries are usually considered to belong to *Prunus avium* (a diploid; $2n = 16$), and sour types to *P. cerasus* (a tetraploid; $2n = 32$). Within each of these groups, cherries are classified on the basis of the color and firmness of their flesh.

Maraschino cherries are made from sweet cherries, usually the Royal Anne variety, that have been bleached, deseeded, and soaked in a sugar solution to which red food coloring and flavoring have been added. The production of maraschino cherries dates back several hundred years to Italy, where a cordial, maraschino (named after the marasca cherry most commonly used in its preparation), was produced. The French later developed the sweet type of maraschino that has now replaced the original bitter Italian liqueur.

Peaches (*Prunus persica*) are the third most important fruit crop in the United States after oranges and apples. They are native to China, where they have been cultivated for thousands of years and where they symbolize long life. Spaniards introduced peaches to Mexico in the late 1500s. Since that time their cultivation in the New World has spread, and many selections have been made that allow their growth in a variety of climates. The United States is a leading peach producer (Table 3.5). Peaches are an important crop in California, Arkansas, Texas, Georgia (the Peach State), and New Jersey.

Peaches fall into two major classes, freestones and clings. Freestones are peaches that have pits that separate easily from the flesh. The mesocarp of cling peaches adheres tenaciously to the pit. Like apples and plums, peaches are generally propagated by grafting once a superior cultivar has been selected or discovered. Nectarines, an increasingly popular fruit, are peaches without furry exocarps. They are all descended from a mutant that was noted by a nurseryman and subsequently propagated by grafting.

Apricots (*Prunus armeniaca*) originated in China and were brought to Greece from Persia by Alexander the Great and to the New World by Spaniards, who established orchards in California in 1770. In ancient times, they figured prominently in folklore. In Persia apricots were called "seeds of the sun." Chinese legends ascribed prophetic powers to them. Confucius is said to have worked out his philosophy under an apricot tree, and the indiscretions of the duchess of Malfi in John Webster's English blood tragedy were uncovered when she ate apricots and claimed that they were bitter, a sign that she was pregnant. This device worked in Webster's play because of a common belief at the time that apricots taste extremely bitter to pregnant women and could therefore be used to "test" for pregnancy. Apricots have also been associated with healthiness. In Hunza, a small kingdom in the Himalayas, apricots are an important part of the diet. The astounding long life and robust health of the people in this tiny country have often been attributed to apricots. Pluots are a man-made fruit created by crossing a plum with an apricot.

A Rose Is a Rose, But Berries Are Not Always Berries

The last important subfamily of the Rosaceae is the rose group (Rosoideae) named after its most beloved horticultural member. It contains the principal false berries of the fruit trade; strawberries, raspberries, and blackberries. Although all these fruits contain many separate mature ovaries from a single flower, their fleshy portions have different origins. Most of a strawberry (Fig. 3.8) consists of the swollen convex receptacle on which the ovaries were borne. When a strawberry is picked, the entire fleshy receptacle still surrounded by the green sepals is pulled from the stem. The individual fruits are tiny brown achenes that dot the surface of the luscious red portion. Blackberries and raspberries consist of a conglomeration of individual swollen ovaries.

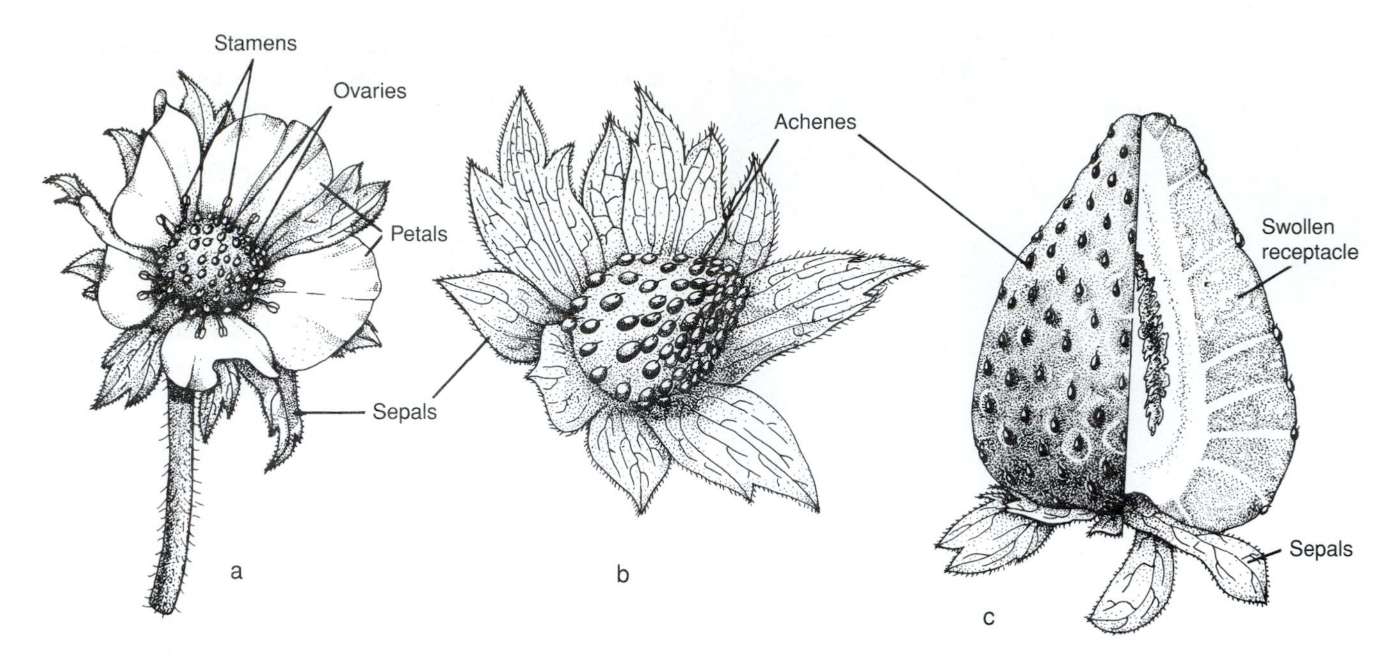

FIGURE 3.8
The many separate ovaries (a) within a strawberry flower become tiny brown achene fruits as they mature (b). (c) These fruits dot the surface of the red, fleshy part of the fruit produced by the swelling of the receptacle on which the ovaries were borne. A strawberry is an aggregate fruit, not a berry, because it is formed from many separate carpels within a single flower.

Each juicy sphere formed by a single carpel adheres to those next to it, forming a cap of succulent beads. The receptacles of these fruits are not swollen and remain on the plant when the cap is pulled free. These berries are consequently hollow. Each bump on a raspberry or blackberry is an individual fleshy drupe with a seed inside.

Strawberries (*Fragaria* X*ananassa*) are one of the most beautiful fruits in the rose family. Their delicate yet distinctive flavor, their intense color, shape, and odor, and their high contents of vitamins A and C make them an increasingly popular fruit in American markets (Table 3.5). The generic name *Fragaria* is derived from the Latin *frago* (to be fragrant) because of the fruit's delightful scent. The Anglo-Saxon name *strawberry* probably refers to the plant's habit of sending out runners, or "strewing" itself across the ground. Strawberries occur wild in Eurasia and the Americas. The fruits of all species are edible and have been eaten by native peoples of both the Old and New Worlds. Romans relished strawberries, and for peasants during the Middle Ages they were one of the few delicacies available. European strawberries tend to be small, however, and few berries are produced on each plant. As North and South America were explored, new species of strawberries were found and sent to Europe. One such species was the Virginia strawberry, *F. virginiana,* a North American species. Other strawberries were found in Canada, and in 1646 an exceptional species, *F. chiloensis,* was imported from Chile. A spontaneous cross between the Virginia and the Chilean species growing together in a European garden about 1750 produced our most commonly cultivated species. Today strawberry cultivars have been developed that yield more fruit per hectare per time than any other temperate dessert fruit.

In addition to strawberries, the rose subfamily contains blackberries and raspberries. Many people simply lump the two together as "brambles," but the fruits actually come from a number of species of the genus *Rubus* that occur natively in North America, Europe, and Asia. The principal raspberries grown commercially are *R. idaeus* (red), a widespread north temperate species, *R. occidentalis* (black), a North American species, and their hybrids. Black raspberries are rarely marketed as fresh fruit, but they are grown in the Pacific Northwest as a source of purple dye. This dye, which is used to label meats with the familiar USDA ratings of prime, choice, and so on, is completely edible since it is nothing more than raspberry juice. Unlike raspberries which pull free from the conical receptacle when picked, blackberry drupelets adhere to the receptacle, leading to fruits with a solid center. Domesticated blackberries are tetraploids, hexaploids, or octaploids resulting from various crosses between species from Europe, western North America, and eastern North America.

A cross between an octaploid California blackberry and a European raspberry first found in the California garden of Joshua Logan gave rise to the loganberry, a popular fruit in the western United States, where it is used for pies and jams. Boysenberries are hybrids between the same two species but at a higher ploidy level.

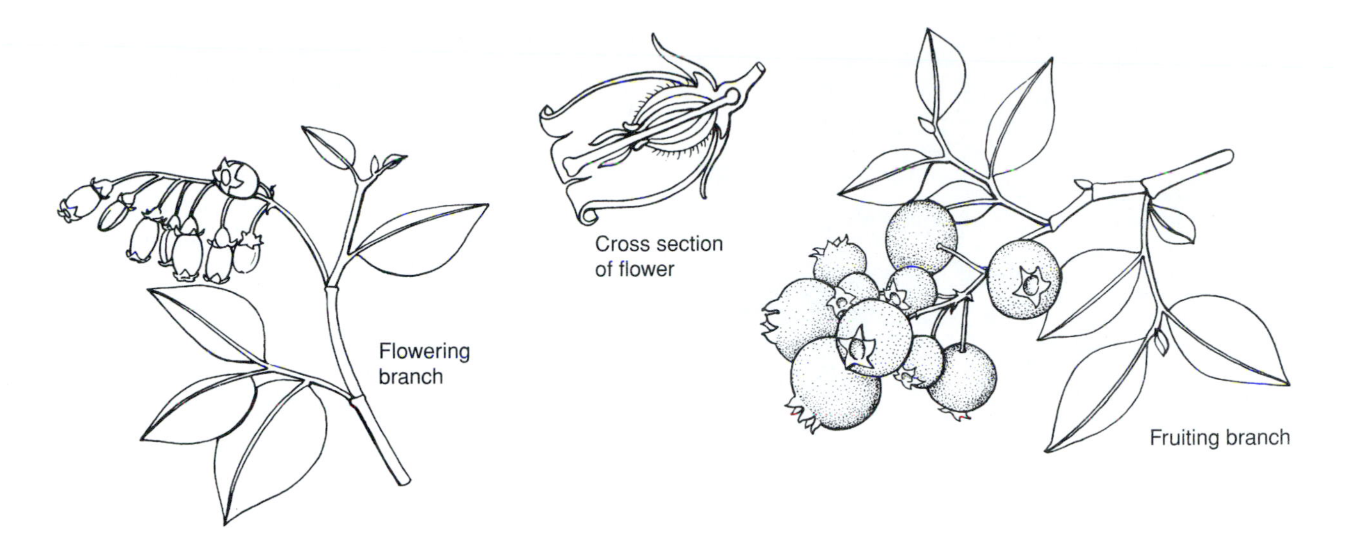

FIGURE 3.9
Blueberry plants have small urn-shaped flowers characteristic of many members of the heath family. Because the ovary is inferior, blueberries are actually accessory berries.

Blueberries and Cranberries

Although blueberries and cranberries are not important on a worldwide basis, both are popular in the United States. Blueberry pie is almost as American as apple pie, and cranberries are an indispensable part of a Thanksgiving meal. Both berries belong to the Ericaceae, or heath family, and both were domesticated within the last 200 years. The family is probably best known for its ornamental shrubs: azaleas and rhododendrons. Like these ornamentals, cranberries and blueberries thrive best in acid soil. Blueberry cultivation is limited not only by its soil preferences but also because the berries are difficult to pick. Three species, high-bush blueberries (*Vaccinium corymbosum*), low-bush blueberries (*V. angustifolium*), and rabbiteye blueberries (*V. ashei*) yield commercial fruit in the United States (Fig. 3.9). All of these species are North American, with the first two found in the northern United States and Canada and the last native to the warm regions of the U.S. Southeast.

Cranberries are usually placed in the same genus as blueberries, although some botanists separate them as the genus *Oxycoccus.* Two species, *Vaccinium macrocarpon,* of North America, and *V. oxycoccos,* a widespread north temperate species, are cultivated. Cranberry plants are creeping shrubs that are grown in boggy areas that can be periodically flooded. Flooding prevents winter desiccation, protects plants from freezing, helps regulate fruiting times, and facilitates harvesting. The ripe berries are shaken loose from the plants and are scooped up by large machinery as they float to the surface of the water (Fig. 3.10). The first efforts to manage cranberries in the United States consisted of the early nineteenth century practice of building dikes around marshes where they grew. They have been grown in large-scale operations only during the last 50 years; acreage planted to cranberries increased from 100 hectares (247 acres) in 1930 to over 40,000 hectares (98,800 acres) in 1995.

FIGURE 3.10
The wet method of harvesting cranberries.
(Photo courtesy of Ocean Spray Cranberries, Inc., Plymouth, Massachusetts.)

Grapes

One of the fruits that has most inspired people, yet equally often has led to their downfall is the grape (Fig. 3.11). Both effects are, however, caused by the liquid product of grapes, not the fruit itself. We discuss wine and its derivatives in detail in Chapter 14. Undoubtedly, the eating of grapes preceded the

FIGURE 3.11
Cultivation of table grapes in Syria.
(FAO photo by R. Faidutti.)

discovery of their potential for producing a fermented beverage. The most widely cultivated species of grapes (Fig. 3.11) is *Vitis vinifera* (Vitaceae), a woody perennial vine native to middle Asia. Cultivation of the vines was under way by at least 8000 B.C. Before that time, nomads would mark trees on which superior vines grew so that they could relocate them as they wandered across the Mideast. The Greeks began to raise grapes during the first century B.C.; viticulture later moved westward into Italy, France, Germany, and Spain. Many of the table, or eating, varieties of *V. vinifera* produce poor wine, although some, such as the muscats, produce both an interesting wine and a fine eating grape. There are at least 175 table grape cultivars. Thompson seedless, Perlette, and an increasing array of rose, red, and black seedless grapes are among those developed in the United States. Seedless grapes are particularly popular table grapes because they are easy to eat, but because they lack seeds they must be propagated by grafting.

The most important native New World grape is *V. labrusca,* the fox grape. Early settlers in the northeastern part of the United States selected the Concord and Catawba grapes from this species. Concord grapes are now the most important American grape for juice, jams, and specialty wines. This grape initially became popular because of its robust flavor. For grape juice to be prevented from fermenting, it must be pasteurized or processed by ultrafiltration. Pasteurization was formerly the most common method of preventing fermentation, but it altered the delicate taste of *V. vinifera* juice. The hardy flavor of Concord grapes was able to survive the effects of pasteurization. Consequently, it became the traditional juice grape. In Europe, where ultrafiltration is common, *V. vinifera* is used for juice as well as wine making.

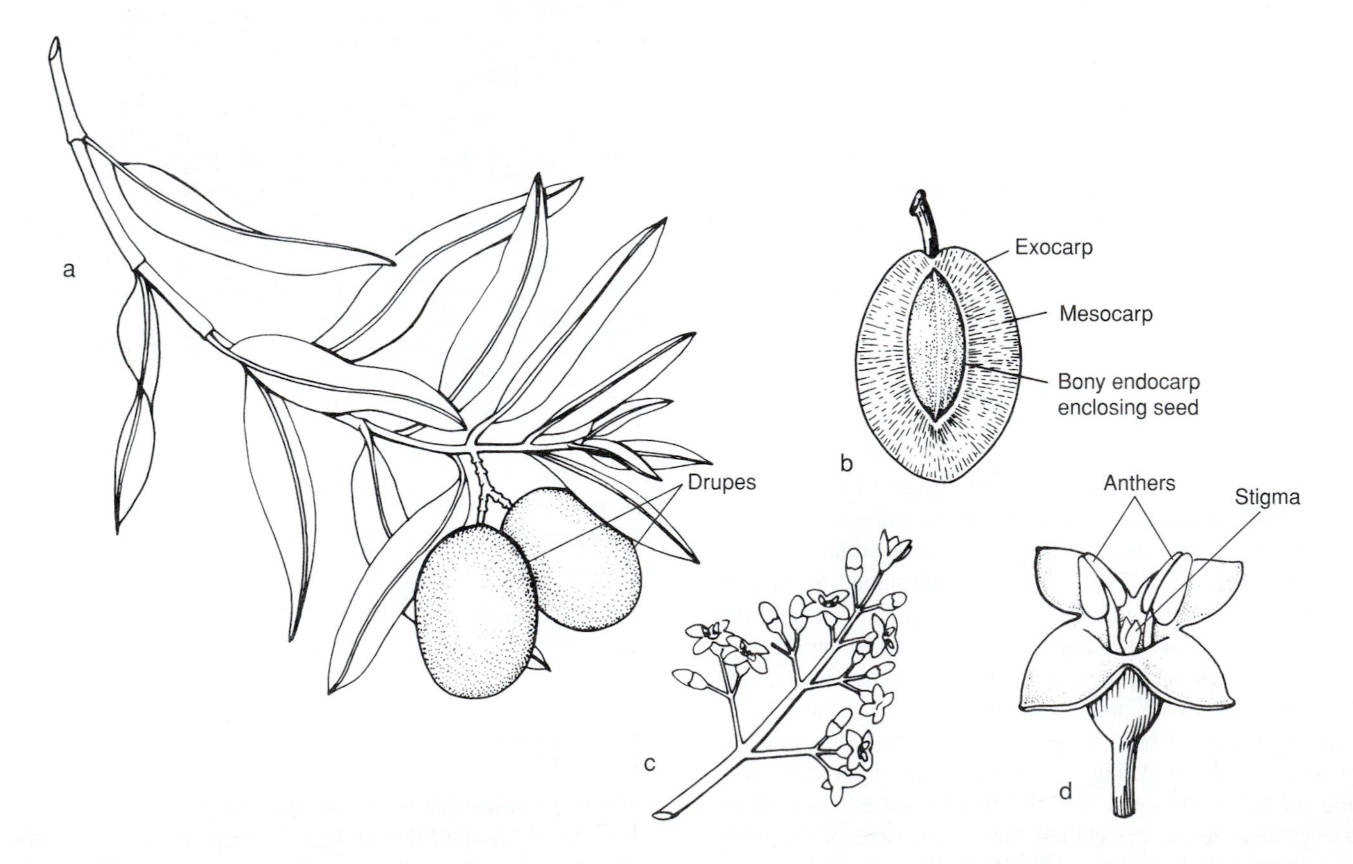

FIGURE 3.12
Olives are drupes formed from flowers borne on trees with gray-green foliage. (a) branch, (b) longitudinal section of a fruit (c) cluster of flowers, (d) individual flower.

Box 3.2

Not Just a Martini Decoration

Our increasing awareness of the health benefits of olive oil along with our growing appetites for gourmet foods have spawned a burgeoning assortment of olives on our grocery shelves and in restaurants. These delectable fruits now come in an incredible array of sizes, colors, and tastes (Fig. 3.13). Olives do not start out as succulent fruits, however, because they contain oleuropein, a water-soluble phenolic compound that imparts an extremely bitter and unpleasant taste to whole fruits unless it is neutralized or flushed from the flesh. Because oleuropein is water-soluble it is not a problem when olives are pressed or extracted for oil. Various methods, collectively referred to as "curing" can remove or break down oleuropein. Olives can actually be cured simply by cracking and soaking them in fresh changes of water each day for at least 25 days. This method is simple but labor-intensive and time-consuming, so most olives are usually cured using an alkali or salt solution or even sugar.

The most common types of olive seen in the United States are the black and green varieties that are produced in California, where an alkaline method of leaching out the oleuropein was developed. The processing of both kinds of olive depends on lye, sodium hydroxide, which impregnates the flesh and dissolves the phenolic compounds. For black olives, the fruits are allowed to ripen and then are pitted before being placed in the lye solution. During this time, air is bubbled through the solution, oxidizing the fruits and turning them black. After the curing is complete, all traces of the lye are washed away and a small amount of iron salt (ferrous gluconate) is added to retain the color. These olives are usually packed in a mild saltwater solution in cans and then heat-sterilized. California green olives undergo the same treatment except without the oxidation or iron salt. They are usually sold in glass jars and often are stuffed with pimientos or almonds.

In Spain, green olives are fermented for 4 to 6 weeks in an acid solution and then in an 8 percent salt solution. They are usually sold in salt brine. Sicilian olives are also green and cured in a salt solution to which lactic acid is added. This treatment makes them crisp as well as salty. Most other olives are purple, brown, or black. Greek olives are allowed to ripen on the tree and then dry-cured with salt. They are often rubbed with olive oil and sold in bulk, but they can also be packed in vinegar. Italian Gaeta olives are produced the same way (although they can be brine-cured) but end up a black, wrinkled, mild-flavored olive. Southern Italian varieties are sometimes oven-dried to yield intensively flavored fruits. Kalamata olives, which are cut in half and salt brine–cured, are purple-black and usually packed in vinegar. Nicoise olives from France are cured in a variety of ways but yield the characteristic small, brown-green very flavorful olives that are a necessary component of a dish prepared à la Niçoise. The Chinese usually dry-cure olives with salt, sugar, or honey. Sweet, preserved olives are eaten like candy, whereas those covered with salt are used to make a soup fed to people with a sore throat.

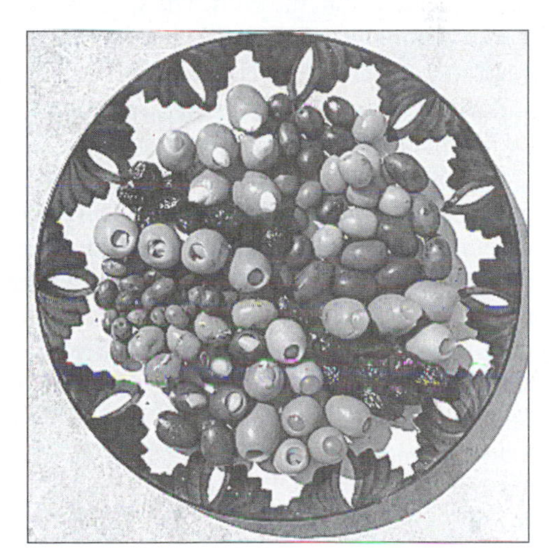

FIGURE 3.13
Fruits of olive trees are harvested at different stages and processed in different ways, which results in a variety of types. Shown clockwise from the top are almond stuffed, gaeta, anchovy stuffed, kalamata, onion stuffed, thasos, pimento and habanero stuffed, garlic, cracked green, nicoise, mushroom stuffed, and moroccan oil cured olives.

Raisins are grapes that have been carefully dried. Drying is sometimes carried out in a sulfur dioxide atmosphere to prevent fungal and bacterial growth. Varieties of grapes used for raisin production have been selected for a soft texture, reduced stickiness, and pleasing flavor. Common grape varieties that produce good raisins are the Sultana, Black Corinth, and Muscat of Alexandria. In the United States, almost all raisins are produced in California.

Olives

Often considered merely a flavoring or decorative agent on pizzas or in Greek salads, olives (Fig. 3.12) have been an extremely important fruit and source of oil (Chapter 9) for Mediterranean people for over 5000 years. Olives, *Olea europaea* (Oleaceae), are native to the region bordering the Mediterranean Sea, where they still grow wild. Olive pits dated to 3000 B.C. have been found in excavated settlements in various parts of Israel. Fresh olives are extremely bitter because they contain oleuropein, a compound that must be broken down or altered to make olives palatable. In ancient times olives were processed by drying, salting, or pickling. These procedures removed some but not all of the bitterness. In about 1900 a process was developed in California (where olives had been cultivated since their introduction in 1769) that produced the first truly tender, nonbitter olives. The process entails treating ripe olives with sodium hydroxide to hydrolyze the bitter oleuropein. Today, a wide array of olives cured in a variety of ways are available (Box 3.2). Despite the increased popularity of olives, only about 1 or 2 percent of the world's crop is eaten as fruit. Most olives are pressed for oil (Chapter 9).

Kiwi Fruit

Kiwis (*Actinidia deliciosa,* Actinidiaceae) are a comparatively new addition to U.S. supermarkets. Until 1930, kiwis were simply collected from the wild. The species is native to Asia and was originally called a Chinese gooseberry. The fruit was not a commercial success until it was introduced into New Zealand, where it was aggressively marketed. In 1959, New Zealand held a contest to rename the fruit in an effort to bolster sluggish sales. Under the new name of kiwi fruit, supplied by the lucky contest winner, production has increased dramatically. Kiwis are dioecious vines (Fig. 3.14) that are now grown in orchards, with male vines interplanted with female, fruit-bearing vines. The pulp of the ovoid, fuzzy brown fruits is delicate in flavor and blends well with other foods. In addition, slices of the fruit with their translucent, green flesh surrounding a narrow ring of tiny black seeds are striking and produce attractive desserts.

FIGURE 3.14
Because kiwi vines are dioecious, both male and female plants are needed for fruit production. Male (right) and female (left) blossoms of a kiwi vine.
(Photo by George Lepp, Corbis.)

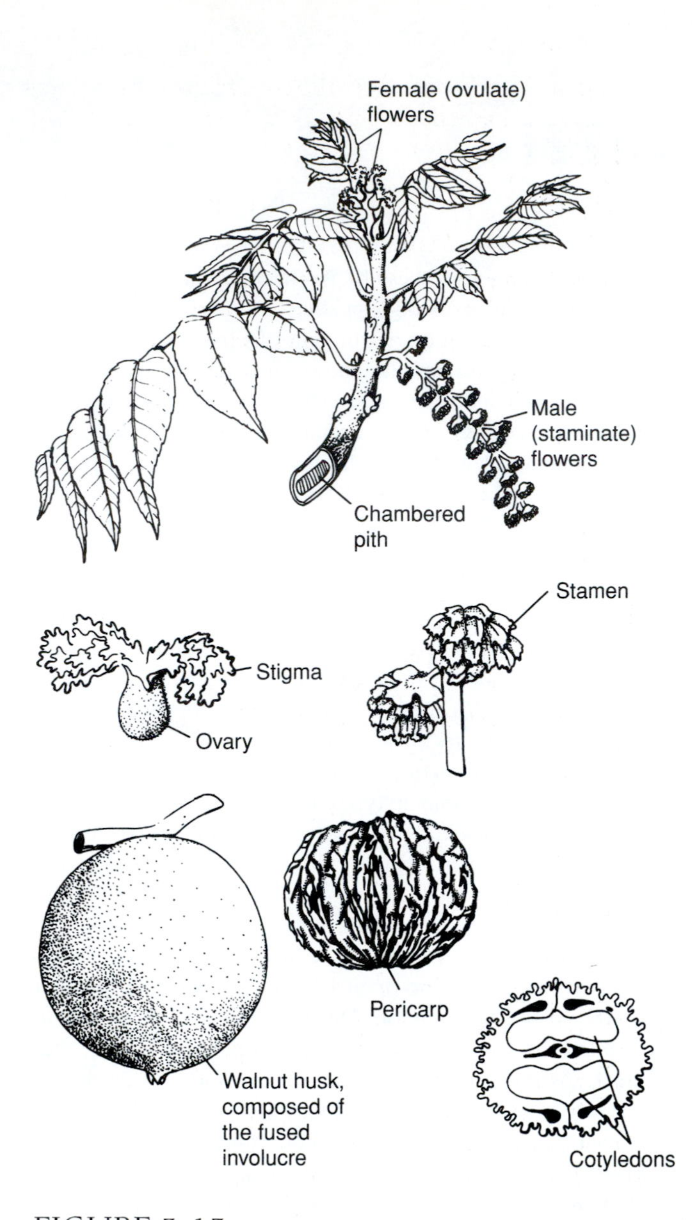

FIGURE 3.15
Walnuts are produced by large monoecious trees that can reach over 30 m in height at maturity. Up to 40 male flowers are borne in catkins that dangle in the wind, releasing their pollen before the leaves expand in the spring. The female flowers occur singly or in small groups. The ovaries, which are surrounded by a tough husk derived from the sepals, mature into leathery green fruits. Once the husk dries and splits, the true nut with its hard pericarp is exposed and shed.

Temperate Nuts

We commonly call any oily seed that we eat raw or toasted a "nut" even though botanically it may not be a true nut. Strictly speaking, nuts are a particular kind of dry fruit that has a single seed, a hard pericarp, and a husklike involucre partially enclosing the fruit. Chestnuts, filberts (hazelnuts), pecans, and walnuts are nuts both botanically and in the vernacular. Peanuts and almonds, however, are not. Peanuts are seeds borne in a pod of a species in the bean family, and almonds are encased in the endocarp of a drupe like a plum. The edible portion of all nuts, regardless of their origin, is the embryo, which has greatly enlarged embryonic leaves (cotyledons) that have absorbed the entire endosperm during development. Results of a study of 86,000 female nurses published in 1998 revealed that women who ate more than 5 ounces of nuts a day had one-third fewer heart attacks than those who rarely ate nuts. Apparently, the benefits accrued regardless of the kind of nut (or whether the "nut" was even a true nut in the cases of peanuts and almonds). It is unclear, however, whether the positive effects were due to some common quality of nuts or to a difference in eating habits and lifestyles of the women who ate or did not eat nuts. Table 3.4 shows the nutritional composition of the major seeds of temperate regions eaten as nuts, and Table 3.2 lists their primary areas of domestication. Chestnuts, filberts, pecans, and walnuts are borne on trees that are native components of the deciduous forests of Eurasia and North America. All of these trees are monoecious and have small, wind-pollinated flowers.

Pecans and walnuts belong to the Juglandaceae, or walnut family. Despite its name, the Persian walnut, *Juglans regia,* is native to southeastern Europe and Asia. The black walnut, *J. nigra* (Figs. 3.15 and 3.16), a North American native, is a relatively unimportant commercial species because it has a strong taste and an exceedingly hard shell. Its primary use is in ice cream and candy. Persian walnuts were blanched, pulverized, and soaked in water in Europe in the eighteenth century to provide a milk substitute that was a staple in many households. Although walnut fruits were used to nourish humans, the trees were used to ward off insects. Walnut trees are notorious for inhibiting the growth of other plants surrounding them. This type of inhibition, known as **allelopathy,** is caused by chemicals in the leaves that are leached out by rain and soak into the ground around the trunk. Apparently, this property led farmers to believe that the trees had insecticidal properties and to plant them near barns to keep flies away from the animals. Unfortunately, the effect of walnuts on other plants does not extend to insects. In the United States, production of the introduced Persian walnut is centered in California.

Pecans and their hickory nut relatives and butternuts belong to the genus *Carya.* The most important commercial species of pecan is *Carya illinoinensis* (Fig. 3.16), native to Mexico and the American Southwest. Pecans are now extensively grown in the southeastern part of the United States, where they have become a prominent part of the regional cuisine. Selection from the wild of thin-shelled forms combined with the development of successful grafting techniques led to the production of disease-resistant, easy-shelling varieties. Selection has also improved disease resistance and uniformity of fruit production. Despite its relatively recent introduction into trade, the pecan has over 300 named varieties. Annual production in the United States exceeds 52,000 tons.

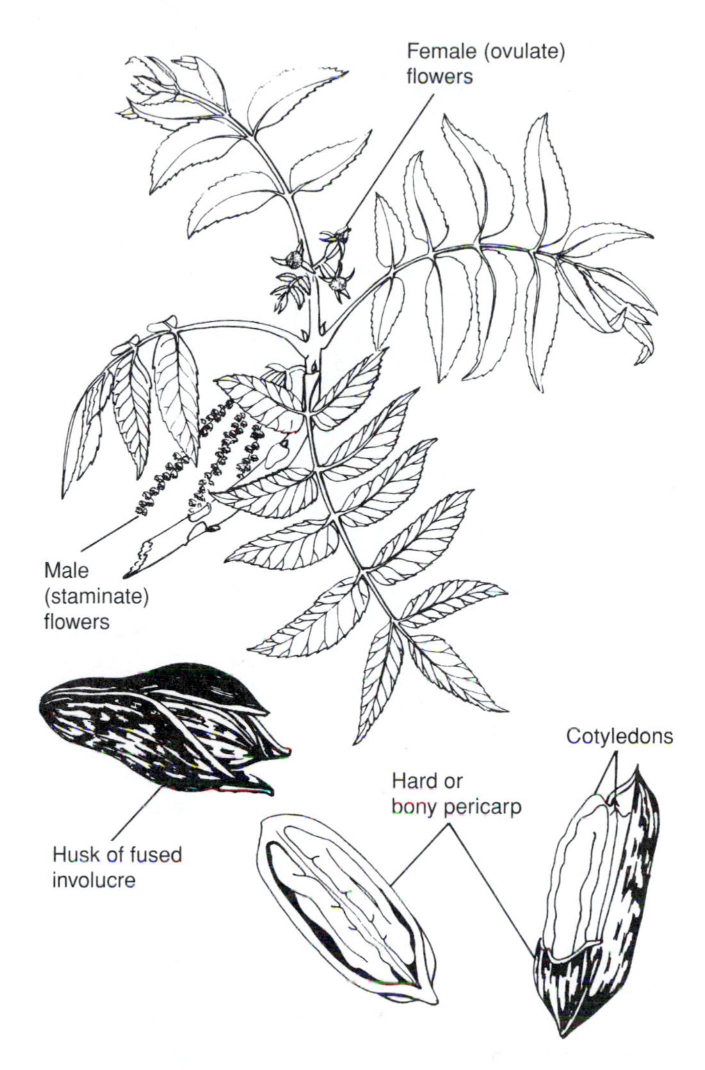

FIGURE 3.16

Pecan trees, like walnut trees, are large, monoecious, and deciduous. They have gray, furrowed bark and produce a handsome spreading crown. Male flowers are produced on branched, hanging stalks, and female flowers come in spikes that are clustered into groups of 2 to 10. The husks surrounding the mature ovaries split at maturity along four lines to release the fruits encased in the pericarp "shell."

In addition to the pecan, North America once produced a significant quantity of chestnuts. Before 1900, roasted American chestnuts, *Castanea dentata* (Fagaceae), were common in the northeastern United States. In addition to the American chestnut, there are three other commercially important species, the European chestnut (*C. sativa*), the Japanese species (*C. crenata*), and the Chinese chestnut (*C. mollissima*). In about 1890 the chestnut blight, caused by a fungus of the genus *Cryphonectria,* arrived in the United States from Asia. The disease presumably spread to the American species from nursery plants imported from the Orient. Between 1904, the time of its first reported occurrence in the United States, and 1950 the disease had killed or infected virtually all of the U.S. fruit-bearing *C. dentata* trees. The Japanese and European chestnuts, more resistant to the fungus than the American species, are now commercially grown in California.

Chestnuts are borne three to a cluster (Fig. 3.17). Each nut is produced by a single flower, but three flowers are grouped together in a compact inflorescence surrounded by spiny, fleshy bracts. When the nuts are mature, the three bracts spread apart, revealing the chestnuts. The nuts are usually picked before they are completely mature but when the bracts are beginning to separate. Ripening is completed after harvesting. Unlike most nuts, chestnuts are usually eaten without appreciable drying. In Europe they are considered one of the most versatile and delicious nuts and are used in candies and pastries, are pureed as a vegetable, and are eaten as a plain or roasted nut.

Hazelnuts, also known as filberts or cob nuts (Fig. 3.18), occur natively across the north temperate zone, but only the European species, *Corylus avellana* (Corylaceae), is extensively cultivated as a crop. The origin of the common names for this nut and its purported supernatural qualities are much

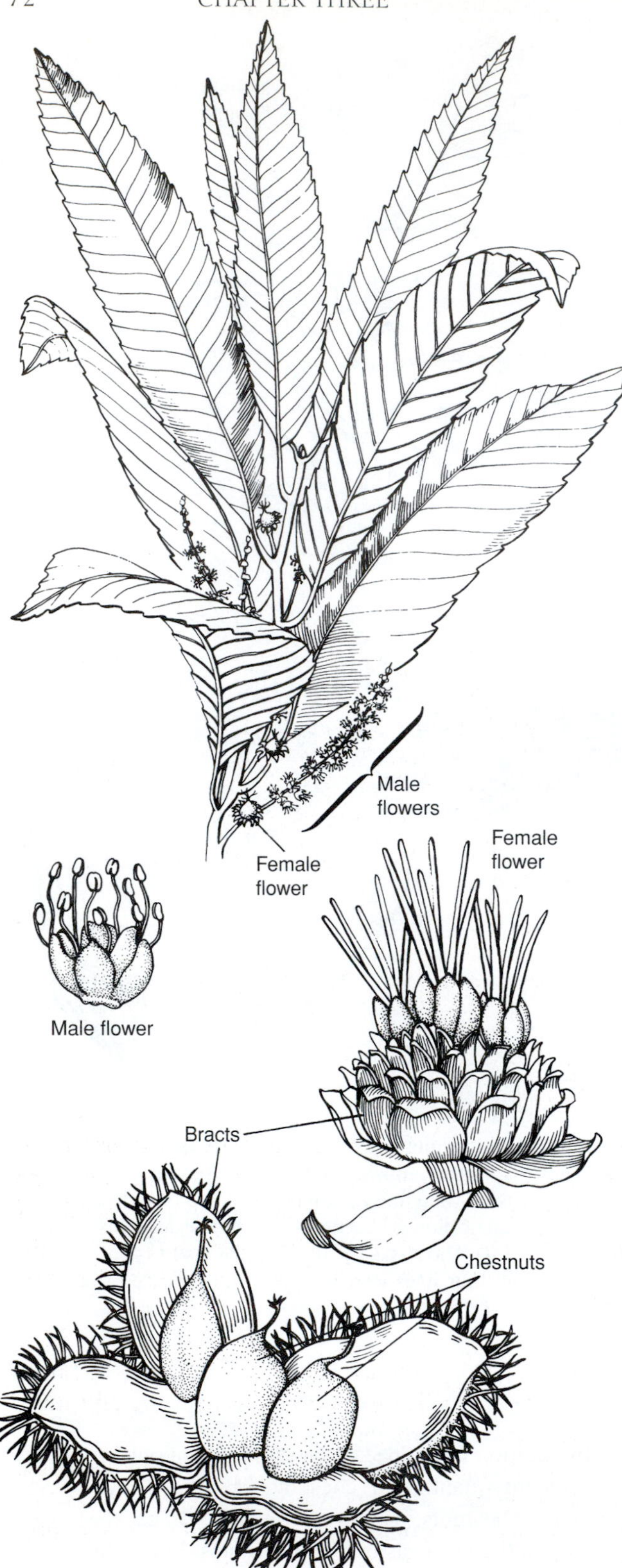

FIGURE 3.17
Chestnuts are borne on tall forest trees with coarsely toothed leaves. Male flowers occur along most of the length of the 10- to 30-cm-long flowering stalks. The female flowers, borne at the bases of the same stalks, occur in groups of three. At maturity, the spiny bracts surrounding each group of female flowers split to release the nuts.

(After Baillon.)

FIGURE 3.18
Filberts, or hazelnuts, are enclosed in distinctive leafy bracts, unlike the leathery husks of walnuts and pecans.

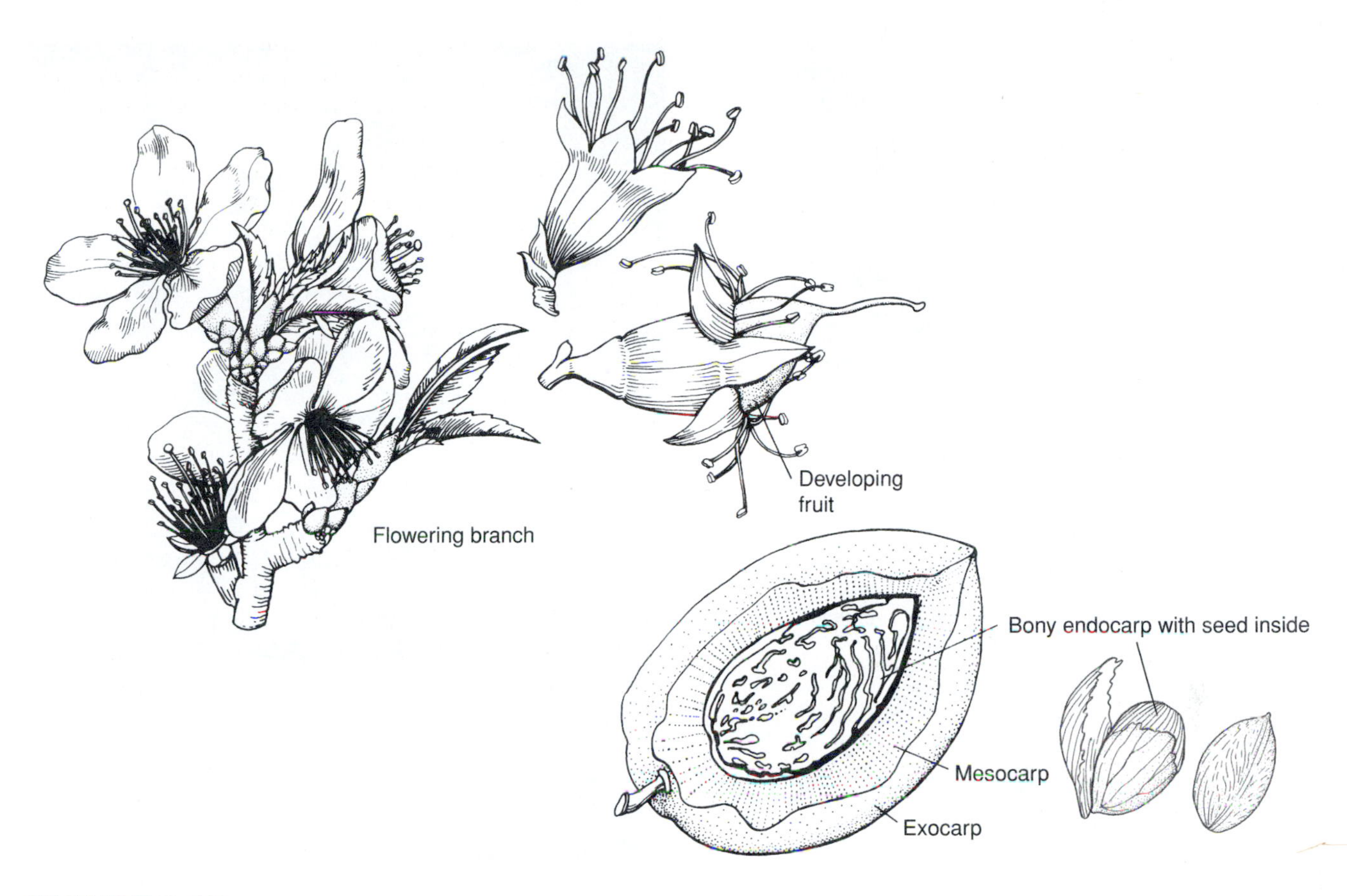

FIGURE 3.19

The almond has the same kind of lovely flowers as a peach or cherry tree, but the mesocarp of the almond fruit is tough and leathery rather than soft and juicy. Almonds are not considered true nuts because only the endocarp, not the entire ovary wall, is hard and bony. The edible part of the almond is the seed.

debated. Various authors ascribe the name to St. Philibert because the nuts ripen in Europe around St. Philibert's Day. Others say that the nut was named for King Philibert of France. Still others claim that the name refers to "full beard," a descriptive term of the bract that accompanies the nut. The origins of the names hazelnut and cob nut are just as obscure. In ancient times filberts were considered to have special powers that allowed divination. Divining rods are still made from hazelnut wood. In Greece the nuts symbolized fertility, and in other parts of Europe the filbert tree was believed to be effective against lightning and witches.

Filberts are much more popular in Europe than in the United States. In Italy and France, the nuts are ground into a fine powder that is used to flavor ice cream and candies. Hazelnut paste is also used much like peanut butter.

Among the temperate nuts, almonds (Fig. 3.19) are unique because they belong to a family (Rosaceae) noted for its showy insect-pollinated flowers and fleshy fruits. Almonds, *Prunus dulcis,* belong to the same genus as plums, apricots, and peaches. In contrast to their juicy relatives, almonds have a mesocarp that expands little as the fruit matures. Removal of the leathery exocarp and mesocarp leaves the familiar almond inside the "shell" (endocarp). Like apricot and peach seeds, almond seeds contain amygdalin, a cyanide-producing compound that gives them the characteristic "almond" flavor and odor. The seeds of peaches and apricots are not eaten because they are too bitter for most people's taste and contain dangerously high levels of amygdalin.

Almonds are native to central Asia, where they have been cultivated for thousands of years. In the United States almonds are cultivated in California, where the climate is similar to that of the almond's native region.

Monks brought almonds to North America during the Spanish mission period. Large-scale production in the United States began only after 1900. California now produces 70 percent of the world's production.

Slightly more exotic than the common dessert nuts is the pistachio, which has become an important crop in California. The distinctive, slightly resinous flavor of the pistachio, *Pistacia vera* (Anacardiaceae) (Figs. 3.20 and 3.21), has endeared it to many who love the yellow-green kernels salted or incorporated into ice cream, cakes, or nougat candies. Pistachios are small dioecious trees native to the eastern Mediterranean region and central Asia, where they have been cultivated for over 3000 years. Female trees

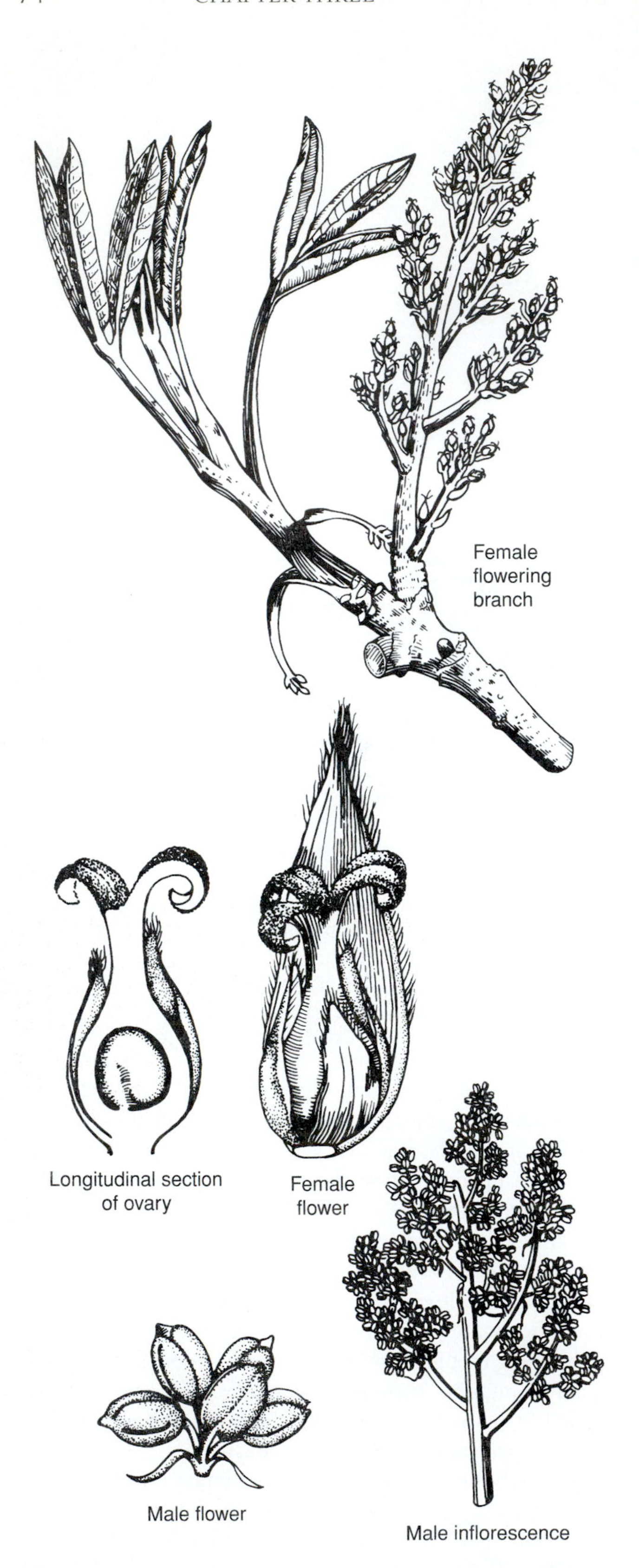

FIGURE 3.20

Pistachio nuts are produced only on the female trees of the dioecious species. Both male and female flowers lack petals. Each female flower contains a single ovule that develops into a fruit containing the familiar pistachio nut.

(After Baillon.)

FIGURE 3.21

After fertilization, female pistachio flowers develop into peach-colored clusters of pistachio nuts.

(Photo courtesy of California Pistachio Commission.)

produce highly reduced flowers in axillary inflorescences. The fruits are small ovoid drupes. What we consider the nut is the seed enclosed by the endocarp "shell," with the surrounding pulp removed before marketing.

The practice of dyeing pistachio nuts vivid red was begun by an American entrepreneur from the East Coast who wanted to hide the blemishes on the shells of imperfect nuts. The practice has been perpetuated, especially in the eastern United States, where consumers have become accustomed to red pistachios, but naturally colored nuts are more commonly seen today.

Fruits and Nuts of Warm Regions

The words *tropical fruits* evoke images of displays of color and exotic, heady flavors. Common fruits such as tomatoes, green peppers, and squash don't fit these images, yet they are all native to tropical regions and none can survive winters with hard freezes. However, some of them are native to regions with a wet-dry seasonality or mountains with cool daytime temperatures. Using human ingenuity, we have circumvented the inability of tomatoes, melons, and peppers to withstand freezes by planting them as annual crops that germinate, flower, and produce harvestable fruit during the spring or summer. Most tropical fruits, however, come from woody perennials that cannot germinate and produce fruits in a few months. Growth of such species is therefore restricted to areas that do not experience freezes. The tropics have many more kinds of edible fruits than do temperate regions, yet temperate fruits dominate our markets and constitute an appreciable proportion of the commercially available fruits in tropical regions as well. One reason for the limited production of many tropical fruits is their perishability. As a result, most are gathered from the wild or cultivated on a limited scale and consumed locally. The few exceptions include bananas, pineapples, and avocados, all of which became readily available in the nineteenth century when methods of shipping them without spoilage were developed. Luckily, the selection of tropical fruits offered in this country is expanding. We begin this chapter with well-known fruits native to warm regions and then treat the less familiar fruits found on the grocer's shelves (Table 4.1).

Citrus Fruits

Orange juice is an integral part of the "all-American breakfast." However, in the early twentieth century oranges were considered a luxury in northern markets and were prescribed as cold or consumption remedies by physicians. Today, orange trees produce more fruit than any other perennial fruit tree crop in the United States (compare Tables 3.5 and 4.2). Oranges and their relatives—limes, lemons, citrons, grapefruits, and tangerines—all belong to the same genus, *Citrus* (Rutaceae). Kumquats belong to the closely related genus *Fortunella.* All these citrus species grow on trees with shiny evergreen leaves that typically have expanded and flattened petioles. The first species are believed to have arisen in southeast Asia, but cultivated citrus fruits

TABLE 4.1 Plants Discussed in Chapter 4

CROP	SPECIES	FAMILY	NATIVE REGION
Avocado	*Persea americana*	Lauraceae	Mexico
Banana	*Musa Xparadisiaca*	Musaceae	Hybrid
	M. acuminata (*AA progenitor*)		Southeast Asia
	M. balbisiana (*BB progenitor*)		
Bitter melon	*Momordica charantia*	Cucurbitaceae	Asia
Bottle gourd	*Lagenaria siceraria*	Cucurbitaceae	Africa, Western Asia, South America
Brazil nut	*Bertholletia excelsa*	Lecythidaceae	South America
Breadfruit	*Artocarpus altilis*	Moraceae	Polynesia
Carambola	*Averrhoa carambola*	Oxalidaceae	Asia
Cashew	*Anacardium occidentale*	Anacardiaceae	South America (Brazil)
Chayote	*Sechium edule*	Cucurbitaceae	Mexico
Cherimoya	*Annona cherimoya*	Annonaceae	South America
Citron	*Citrus medica*	Rutaceae	Asia
Coconut	*Cocos nucifera*	Arecaceae	Indo-Pacific region
Cucumber	*Cucumis sativus*	Cucurbitaceae	South-central Asia
Date	*Phoenix dactylifera*	Arecaceae	Mediterranean
Durian	*Durio zibethinus*	Annonaceae	South America
Eggplant	*Solanum melongena*	Solanaceae	Asia
Fig	*Ficus carica*	Moraceae	Mediterranean
Grapefruit	*Citrus paradisi*	Rutaceae	Hybrid (*C. maxima* X *C. sinensis*)
Guava	*Psidium guajava*	Myrtaceae	Central America
Kumquat	*Fortunella spp.*	Rutaceae	China
Lemon	*Citrus limon*	Rutaceae	Hybrid (trihybrid origin involving *C. medica, C. maxima,* and another genus)
Lime	*Citrus aurantifolia*	Rutaceae	Hybrid (trihybrid similar parents to lemon)
Litchi	*Litchi chinesis*	Sapindaceae	China
Luffa	*Luffa acutangula*	Cucurbitaceae	Asia
	L. cylindrica	Cucurbitaceae	Asia
Macadamia nut	*Macadamia integrifolia*	Proteaceae	Australia
Malay apple	*Syzygium malaccense*	Myrtaceae	Southeast Asia
Mandarin	*Citrus reticulata*	Rutaceae	China
Mango	*Mangifera indica*	Anacardiaceae	Southeast Asia
Melon (cantaloupe, honeydew, etc.)	*Cucumis melo*	Cucurbitaceae	Africa, Asia
Okra	*Abelmoschus esculentus*	Malvaceae	Asia
Orange	*Citrus* spp.	Rutaceae	
Bitter	*C. aurantinum*		Hybrid (*C. reticulata* X *C. maxima*)
Sweet	*C. sinensis*		Hybrid (*C. reticulata* X *C. maxima*)
Papaya	*Carica papaya*	Caricaceae	Central America
Passion fruit	*Passiflora edulis*	Passifloraceae	South America
Pepper, sweet, hot	*Capsicum annuum*	Solanaceae	Mexico
Pineapple	*Ananas comosus*	Bromeliaceae	South America
Pomegranate	*Punica granatum*	Punicaceae	Asia
Pummelo	*Citrus maxima*	Rutaceae	Southeast Asia
Rambutan	*Nephelium lappaceum*	Sapindaceae	Malaysia
Rose apple	*Syzygium jambos*	Myrtaceae	Southeast Asia
Sapodilla	*Manilkara zapota*	Sapotaceae	Mexico
Soursop	*Annona muricata*	Annonaceae	Central and South America
Sapote	*Pouteria* spp.	Sapotaceae	American tropics
Mammey	*P. sapota*		
Green	*P. viridis*		
Squash	*Cucurbita* spp.		
	C. maxima	Cucurbitaceae	South America
	C. argyrosperma		S. Mexico
	C. moschata		Mexico or S. America
	C. pepo		Mexico, Eastern United States
Sweetsop, or custard apple	*Annona squamosa*	Annonaceae	West Indies, South America
Tangerine	*Citrus reticulata*	Rutaceae	China
Tomato	*Solanum esculentum*	Solanaceae	Western South America
cherry	*S. esculentum* var. *cerasiforme*		Central America
Watermelon	*Citrullus lanatus*	Cucurbitaceae	Central Africa
Wax gourd	*Benincasa hispida*	Cucurbitaceae	Southern China
Zucchini	*Cucurbita pepo*	Cucurbitaceae	Mexico

TABLE 4.2 1999 World Production of Fruits and Nuts of Warm Regions

COMMON NAME	TOP COUNTRIES	1000 METRIC TONS	TOP CONTINENTS	1000 METRIC TONS	WORLD 1000 METRIC TONS
Avocados	Mexico	906	North and	1,349	2,259
	United States	130	Central America		
	Dominican Republic	155	South America	385	
	Indonesia	114	Asia	253	
	Brazil	75	Africa	179	
			Europe	64	
Bananas	India	10,200	Asia	25,680	57,383
	Brazil	5,591	South America	14,039	
	Ecuador	4,563	North and	8,086	
	China	3,984	Central America		
	Philippines	3,549	Africa	7,189	
			Oceania	945	
Cashew nuts	India	430	Asia	638	1,179
	Nigeria	152	Africa	411	
	Brazil	124	South America	124	
	Tanzania	106	North and	7	
			Central America		
Chilles and green peppers	China	7,513	Asia	10,715	18,024
	Mexico	1,469	Europe	2,705	
	Turkey	1,390	Africa	2,077	
	Spain	908	North and	2,191	
	Nigeria	783	Central America		
			South America	301	
Coconuts	Indonesia	14,710	Asia	41,798	48,568
	Philippines	11,000	Oceania	1,931	
	India	10,000	North and	1,885	
	Sri Lanka	1,999	Central America		
	Thailand	1,430	Africa	1,873	
			South America	1,087	
Cucumbers and gherkins	China	15,912	Asia	22,315	28,551
	Turkey	1,4000	Europe	4,002	
	Iran	1,302	North and	1,750	
	United States	1,077	Central America		
	Japan	800	Africa	393	
			South America	73	
Dates	Iran	918	Asia	3,224	5,049
	Egypt	850	Africa	1,794	
	Iraq	600	North and	22	
	Saudi Arabia	600		Central America	
	Pakistan	540	—		
			—		
Eggplants	China	11,026	Asia	19,543	21,166
	Turkey	850	Africa	799	
	Japan	490	Europe	709	
	Egypt	560	North and	106	
	Italy	361	Central America		
			South America	6	
Grapefruit and pommelo	United States	2,286	North and		
	Cuba	420	Central America	3,060	4,892
	Israel	335	Asia	1,027	
	China	260	South America	417	
	Argentina	216	Africa	355	
			Europe	50	
Lemons and limes	Mexico	1,211	Asia	3,060	9,477
	Argentina	1,020	North and	2,317	
	India	1,000	Central America		
	Iran	891	South America	1,982	
	United States	741	Europe	1,503	
			Africa	518	

TABLE 4.2 *Continued*

COMMON NAME	TOP 5 COUNTRIES	1000 METRIC TONS	TOP CONTINENTS	1000 METRIC TONS	WORLD 1000 METRIC TONS
Mangoes	India	12,000	Asia	18,449	23,815
	Mexico	1,504	Africa	2,189	
	Thailand	1,250	North and Central America	2,063	
	Mexico	1,120			
	Philippines	950	South America	1,068	
			Oceania	44	
Melons*	China	5,262	Asia	10,234	16,190
	Turkey	1,800	Europe	2,502	
	Iran	1,215	North and Central America	2,045	
	Spain	943			
	United States	680	Africa	1,039	
			South America	239	
Oranges	Brazil	20,552	South America	23,552	63,098
	United States	12,401	North and Central America	17,142	
	Mexico	3,329			
	China	2,257	Asia	11,564	
	Spain	2,448	Europe	5,459	
			Africa	4,865	
Papayas	Brazil	1,700	South America	2,127	5,084
	Thailand	542	Asia	1,286	
	India	450	North and Central America	604	
	Indonesia	336			
	Mexico	498	Africa	284	
			Oceania	18	
Pineapples	Thailand	1,734	Asia	6,220	12,385
	Brazil	1,640	South America	2,433	
	Philippines	1,495	Africa	2,143	
	India	1,100	North and Central America	1,441	
	China	925			
			Oceania	145	
Squashes, pumpkins, and gourds	China	3,325	Asia	9,517	14,904
	India	3,300	Europe	2,237	
	Egypt	620	Africa	1,440	
	Iran	500	South America	855	
	Argentina	430	North and Central America	592	
Tangerines, mandarins, clementines, and satsma	China	5,955	Asia	10,321	15,971
	Spain	1,750	Europe	2,434	
	Japan	1,176	South America	1,362	
	Iran	726	Africa	1,070	
	Brazil	680	North and Central America	706	
Tomatoes	China	17,919	Asia	41,246	91,664
	United States	9,940	Europe	19,789	
	Turkey	6,600	North and Central America	13,424	
	Italy	5,369			
	India	5,300	Africa	10,800	
			South America	5,990	
Watermelons	China	27,300	Asia	39,914	50,940
	Turkey	3,925	Europe	3,880	
	Iran	2,472	Africa	3,240	
	United States	1,673	North and Central America	2,300	
	Egypt	1,500	South America	1,494	

Source: FAOSTAT Database Gateway. Online at http://apps.fao.org/ (under "All databases" then Crops, Primary). For some crops harvested late in 1999, 1998 figures are given.
*Data from *FAO Production Yearbook for 1997,* Rome, FAO.

are thought to be derived from species native to India, China, and Myanmar (Burma).

All citrus species have the same fruit type, a **hesperidium,** which is basically a berry with a leathery skin formed by the combined exocarp and mesocarp (Fig. 4.1). The skin, like the leaves (see Fig. 1.12e) and other vegetative parts of the plant, is covered with pockets filled with aromatic oils. These oil-filled cavities are characteristic of the citrus family and are responsible for the aroma given off when citrus peels are bruised, grated, or twisted. The endocarp of citrus fruits bears highly

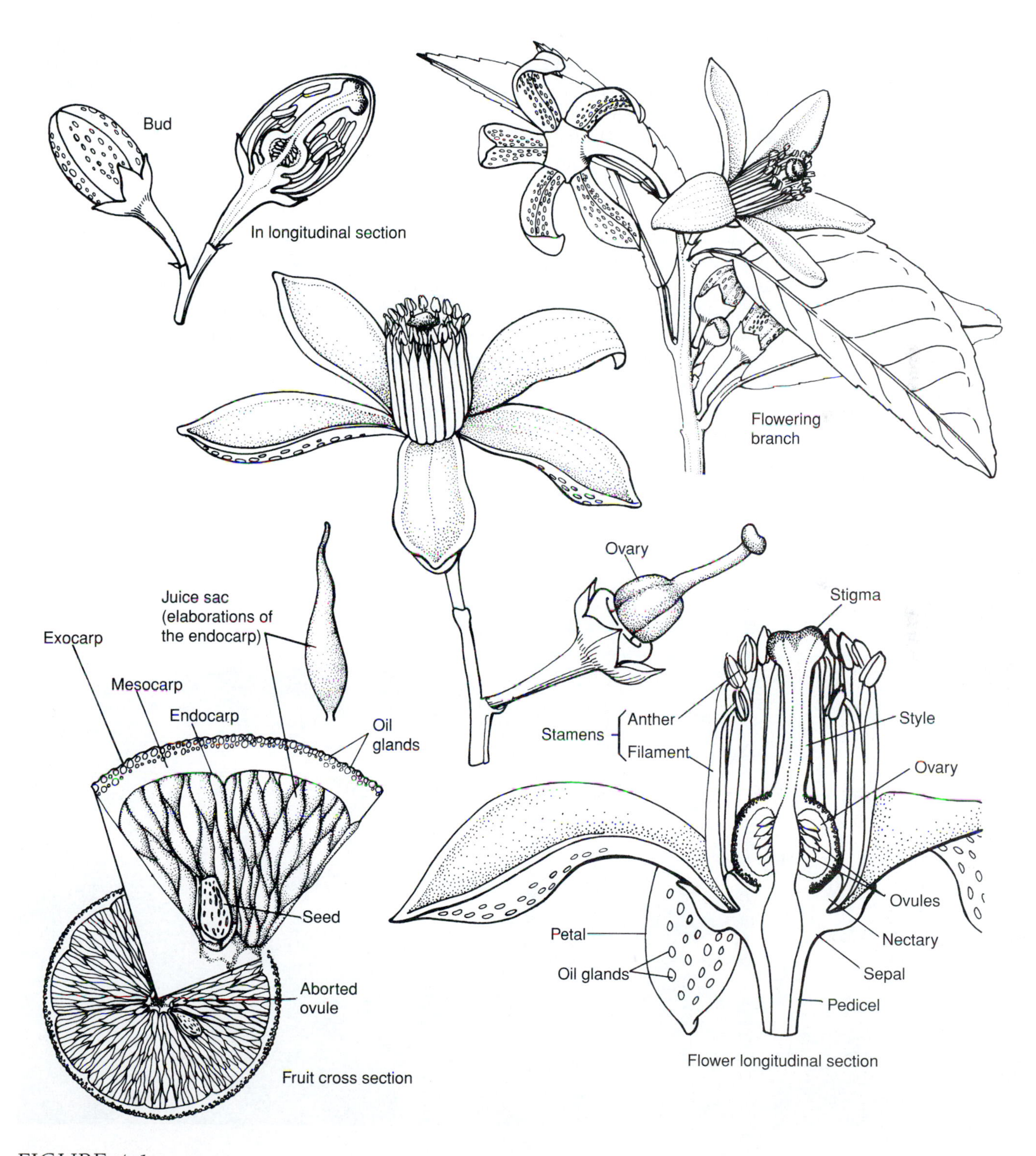

FIGURE 4.1
A branch, bud, flower, and fruit of a sweet orange, showing development from the initiation of the bud to the production of the mature fruit, which is called a hesperidium.

modified fleshy hairs that fill the cavities of the ovary and surround the seeds. These hairs, or sacs, are the parts of citrus fruits that we eat or squeeze for juice. They contain a watery solution of sugars and acids. In some species the acid concentrations decrease with age and produce fruits that we eat without the addition of sugar.

The cultivated *Citrus* species cannot tolerate severe freezing temperatures. Most do best where there is good soil moisture but abundant sun and reasonably dry air. In the past, most citrus fruits were grown from seed, but with the rise in popularity of certain cultivars, many of which are partially or completely sterile, bud grafting has become the most common method of propagation. Bitter oranges, pummelos, and citrons are still frequently grown from seed.

The sweet orange (*Citrus sinensis*) (Fig. 4.1) is now the most widely grown fruit in the world, but, with the wild ancestors gone, we can only try to reconstruct its origin. Some authors believe that the orange is derived from an unidentified or extinct Chinese species, but most now think that it resulted from selection of a hybrid between a tangerine and a pummelo. Oranges were considered by some to be the "golden apples" of Greek mythology that the goddess of fertility gave to Hera when she married Zeus. Hera planted the seeds in the garden of Hesperides, hence the designation *hesperidium* for the citrus fruit type. Oranges were transported along caravan routes from the Orient to the Persian empire. The Moors brought them to Spain and used them medicinally and in religious services. The Spanish and Portuguese later introduced them into their New World territories in the 1500s.

The English word for the color orange is derived from the Arabic name of the fruit, but fully ripened oranges are not necessarily orange. In areas where temperatures never get cool (e.g., Thailand) oranges remain mottled yellow and green to maturity. Cool temperatures thus contribute to the orange color because they promote the formation of carotene, an orange pigment, and the breakdown of chlorophyll. If temperatures in an area fluctuate while the oranges are ripening, the fruits can change from orange to green and vice versa. Merchants, realizing the public's desire for "orange" oranges, even though it makes no difference in their taste, spray batches of green or blotchy oranges with ethylene to induce a color change.

Orange juice, like lime juice, is effective in preventing scurvy, a disease resulting from a deficiency of vitamin C. The spread of oranges followed the paths of seafaring explorers who wanted to ensure supplies of the fruit along their routes. Nevertheless, up through the eighteenth and nineteenth centuries, sweet oranges were a delicacy reserved for the affluent. When it was discovered that oranges could be grown in temperate climates if they were protected from freezes, wealthy individuals began to grow them in glasshouses. The possession of such "orangeries" became a status symbol (Fig. 4.2). Oranges were first grown

FIGURE 4.2

In the summer, plants taken from an orangerie are set out in a formal arrangement in the gardens of Versailles near Paris. Oranges were so prized in northern Europe that the wealthy built glasshouses to protect the trees and ensure a winter supply of the fruit.

FIGURE 4.3
An Agricultural Research Service (ARS) engineer and the program administrator for the Florida Department of Citrus examine oranges harvested by a newly developed canopy shaker.
(Photo by Keith Weller, courtesy of ARS.)

in Florida in 1565, but it was not until the United States took possession of the peninsula in 1820 that sweet oranges became an important U.S. commodity (Fig. 4.3)

There are three main classes of hybrid sweet oranges: bloods, normals, and navels. Bloods have streaks and patches of red in the pulp that make them unattractive to many Americans, although they are popular in Europe. The most commonly grown normal type is the Valencia orange. Modern Valencia cultivars were introduced into California in 1876 and into Florida in 1870 by an Englishman who had become familiar with the variety in the Azores. Valencia oranges are now the most important variety grown in Florida. They produce a richly flavored juice that sets the standard by which other orange juices are judged. Because we have come to expect orange juice to taste a certain way, commercial orange juice is often blended to ensure consistency in flavor and texture. In many cases the juice is even filtered, freeze-dried, and reconstituted with controlled amounts of citrus peel and pulp.

Large-scale navel orange production is a recent phenomenon, but a type of navel orange was known in Europe at least 350 years ago (Fig. 4.4), and similar seedless oranges were either purposely propagated or arose spontaneously in most orange-growing regions. The "Washington" navel, or modern commercial variety, is derived from a mutation that occurred in Brazil before 1820. Navel oranges are formed because the flowers produce an abnormal small ovary on top of the regular ovary. When the fruit matures, the second ovary becomes the button, or navel. The original ovary produces the sweet fleshy pulp that we eat. Sterility is correlated with the aberrant production of the navel. The pollen produced by navel orange flowers is nonfunctional, and the ovules almost never produce seeds. New plants must be produced vegetatively, usually by grafting. Navel orange production in Florida began in 1835, and in California it started 3 years later. Navels are now the leading orange variety grown in California. Unlike Valencias, navel oranges are primarily eating oranges because they lack seeds, are easy to peel, and separate easily into segments.

Despite the modern preeminence of sweet oranges in the New World, they were not the first citrus grown in the Americas. Three others—the lemon, *Citrus limon;* the bitter orange, *C. aurantium;* and the citron, *C. medica*—arrived almost 100 years earlier. Columbus brought seeds of these three species to the New World on his second voyage in 1493 and appears to have established them successfully on the island of Hispaniola. Bitter oranges with a hybrid origin

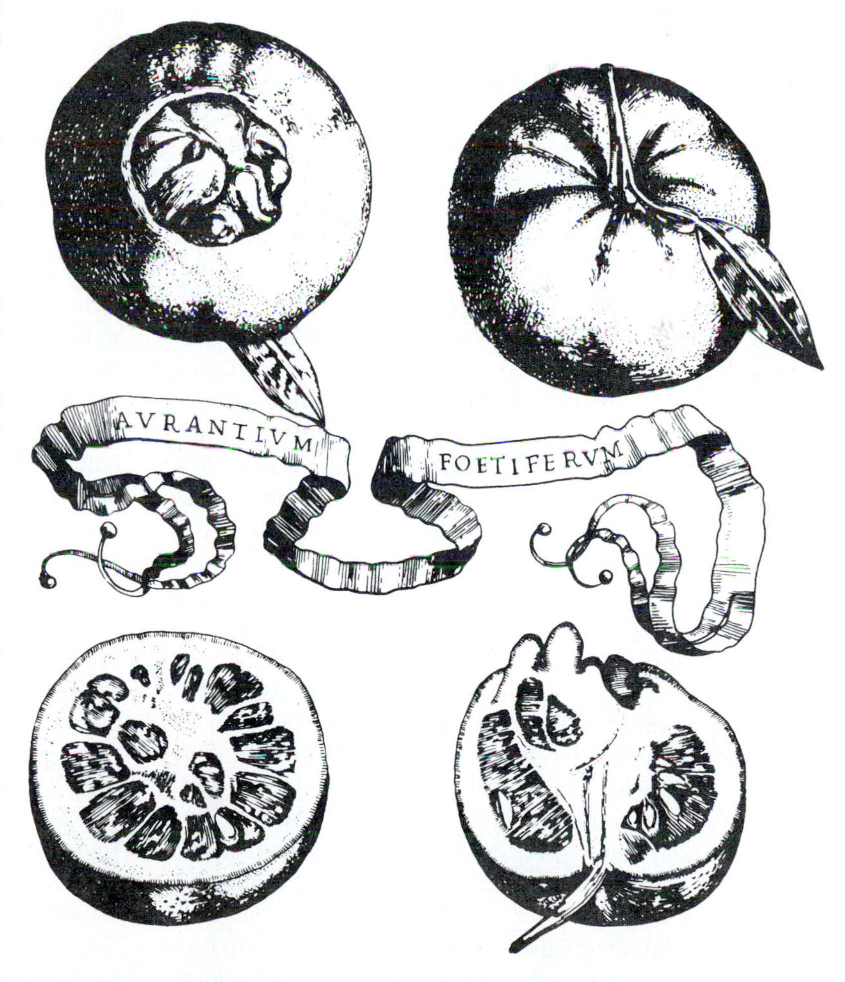

FIGURE 4.4
The original of this drawing of a navel orange, probably a bitter orange, is used as evidence the navel forms of citrus were known in Europe at least 350 years ago.
(Redrawn from a woodcut by Ferrari in *Hesperides,* published in 1646.)

similar to that of sweet oranges are now used almost exclusively for making preserves, marmalade, and orange liqueurs. Their fragrance is extracted as neroli oil used in the perfume trade. The pulp of citrons is rarely eaten, but the peel is often candied and used in confections such as fruitcakes. Although native to southeastern Asia, citrons were grown in Mesopotamia in 4000 B.C. and were known to the ancient Greeks. Linnaeus used Pliny's name, *citron,* as the basis for the generic name of all the "citrus" fruits.

Like citrons and bitter oranges, lemons are practically never eaten alone, but their juice is used for flavoring everything from beverages to meat dishes. Lemons also appear to be of hybrid origin, with the citron, the pummelo, and a species of *Microcitrus* the probable parents. Of all of the citrus fruits, lemons are the most exploited for secondary purposes. The oils in lemon rinds are extracted for perfumes, cleaning products, and deodorants. The inner portion of the rind is a valuable source of pectin (see Fig. 10.10). Lemons are usually picked before they are fully mature and "cured" for 4 months. During this time they become yellow, the skin becomes thinner, and the flesh juicier.

Morphologically, the most distinctive member of the citrus group is *C. aurantifolia,* known as the Mexican or West Indian lime. The lime apparently arose from the same three-way cross as the lemon (Table 4.1). The Arabs were using limes by the tenth century A.D. and introduced them into Europe in the twelfth or thirteenth century. Women in the Renaissance French court carried limes with them for cosmetic purposes, to whiten their skin and redden their lips. Lime cultivation in the New World began in the seventeenth century in Spanish and Portuguese territories.

When the correlation between the consumption of fresh fruits or fruit juices and the absence of scurvy was realized, British sailors were ordered by the admiralty to drink a daily ration of lime juice. This juice kept especially well on long voyages and appeared to be effective in small doses. We now know that it is the high vitamin C content of limes that prevents scurvy, but without knowing why, the British saved thousands of lives with their kegs of lime juice. To this day English sailors are known as "limeys."

Grapefruits (*Citrus paradisi*) originated as a result of a spontaneous cross between a pummelo or shaddock (*C. maxima*) and a sweet orange (*C. sinensis*) that occurred before 1750 in the West Indies, probably on Barbados. Thus, the grapefruit can be considered a New World product and constitutes a "species" that is only a few hundred years old. One of its ancestors, the pummelo, is little known in the Americas today but is relished in its native Thailand. Pummelos were apparently brought to the West Indies by the English or Dutch.

Commercial production of grapefruits began in the United States in 1880. Grapefruits with pink flesh arose as a bud mutation in a Florida plantation and were propagated by asexual grafting. The "ruby reds" (with deep red-pink flesh) were developed in McAllen, Texas, in 1929. The flavonoids in grapefruit have been shown to prevent the breakdown of certain drugs taken for heart disorders, leading to dangerously high levels of the drugs.

The last important citrus fruit, the tangerine, Mandarin or clemintine orange, *Citrus reticulata,* was originally cultivated in China and is still an important fruit in the Far East. Recent numerical and chemical studies indicate that this species is a biological species, not a natural or artificial hybrid. Cultivation in the United States began in 1850, but the fruit has never enjoyed the popularity of oranges or grapefruits. The lack of acceptance is perhaps due to the short fruiting season in the United States, which prevents the fruit from reaching the same sweet quality found in Indian and Asian fruit.

Squashes and Their Relatives

In both the Old World and the New World, members of the squash family (Cucurbitaceae) figured in the early history of humankind. Corn, beans, and squash, known as the "three sisters," were the mainstays of early agricultural peoples of the southwestern United States (see Fig. 5.27), Mexico, and Peru. For native New World peoples, squashes provide not only edible flesh but also seeds relatively rich in the essential sulfur-containing amino acids. The blossoms are collected in some areas and are used in soups and sauces. The shelled, toasted seeds are sold as *pepitas.* Fossilized edible squash remains from Mexico have been dated to about 10,000 B.C. Undoubtedly, wild squashes were collected and eaten long before they were cultivated. In Africa, where melons but not squash are native, hunter-gatherer tribes such as the !Kung in the Kalahari Desert gather wild watermelons. Other melons figure in the earliest Egyptian writings.

As was implied above, squashes are all native to the New World, and most melons are indigenous to Africa and Asia. Both melons and squashes are "squashes" in that they belong to the squash family. All cultivated members of the family are derived from species that are vines in the wild. Most are annuals, and the domesticated species discussed here are all monoecious. Because these plants are annuals, they are planted each year from seed. All members of this family have fruits called **pepos,** produced from inferior ovaries that have a hard rind composed of tissues of the ovary wall fused with the lower parts of the calyx and corolla (Fig. 4.5). Because of their hard rinds, many squashes and melons store well if not bruised or cut.

The naming of true squashes (members of the genus *Cucurbita*) (Fig. 4.6) has been in a state of hopeless confusion for years, in part because the common names of the edible squashes bear little relationship to the species from which they come (Table 4.3). There are five widely cultivated species, four of which—*Cucurbita pepo, C. moschata, C. maxima,* and *C. argyrosperma*—are commonly encountered in American supermarkets. All these squashes are known only as domesticated species and all are diploids ($2n = 40$) and were probably first cultivated for their seeds. Wild cucurbits

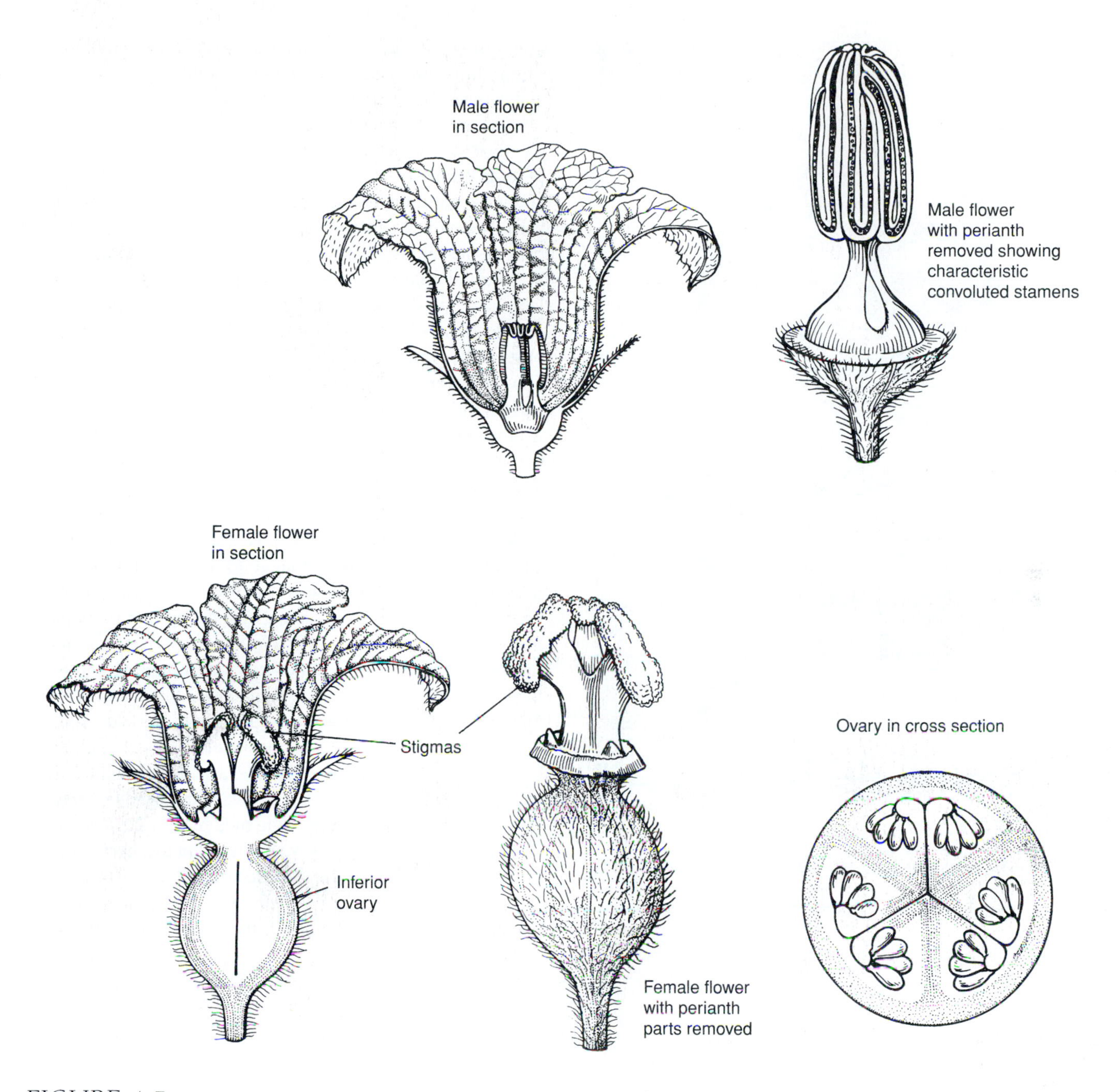

FIGURE 4.5

Squash fruits are formed from inferior ovaries with the rind of the fruit a composite structure of the ovary wall and lower parts of the calyx and corolla.

(After Baillon.)

usually have a thin rind and bitter flesh. Table 4.3 lists some of the distinguishing characters of the four species and the common names of products from each one. *Cucurbita pepo,* perhaps independently domesticated in Mexico and the eastern United States, is now the most widely cultivated and versatile species of squash. Fossil remains believed to belong to this species have been uncovered in Mexico and have been dated to be 10,000 years old. Phytoliths of a squash, probably from coastal Ecuador, have been dated to 10,000 B.P. Since the earliest of these fossils are from Mexico, this species also appears to have first been domesticated in Central America and subsequently taken to South America. Although *C. pepo* and *C. moschata* are naturally trailing vines, there has been selection for bush types in both species that take up less space in fields and are comparatively easy to tend and harvest.

As is indicated in Table 4.3, several *Cucurbita* species are sold as "winter squashes." Several species are called pumpkins, with the result that one is never sure what is

FIGURE 4.6
Winter squash varieties include (left to right from the top) banana, red kuri, hubbard; sweet dumpling; delicata, white pumpkin, acorn; turban, carnival acorn; kabocha, orangetti and spaghetti.

going to be carved or spooned out of a can of pumpkin pie filling. Immature fruits of *C. pepo* are sold as zucchini, vegetable marrow, or summer squash.

Whereas the New World cucurbits provide important staple food crops, mature melons of the Old World members provide primarily dessert fruits. A few, such as the cucumber and luffa, are eaten while immature, before the seeds become hard. Most of the commercially exploited Old World members of the family tend to be mild-flavored or sweet. These include the common melons and the watermelon. Modern, large-scale production of all these melons is possible because of selection beginning in 1911 for resistance to fungal and other pathogenic infections. Watermelons (*Citrullus lanatus*), now grown extensively in California and Texas, are African natives that were appreciated by Europeans in ancient times. By the eleventh century A.D., the Chinese were also growing them. Watermelons contain between 87 and 92 percent water and have a high sugar content compared with other melons, yet the pulp is acidic. Because the acid curdles milk, watermelon is not used as a flavoring for ice cream. Melons, *Cucumis melo*, are also thought to be of African origin because wild forms occur there. However, the earliest references to melons are from China, suggesting that domestication of some forms occurred first in Southeast Asia. The major melon groups today, the cantaloupes and honeydews, have their origin in western Asia. References to melons do not appear in Egyptian or Greek writings, but they do occur in texts produced at the end of the Roman empire. Selection eventually led to a wide array of varieties, including cantaloupe and Persian, musk, Cranshaw, and honeydew melons (Box 4.1). These sweet, refreshing melons may not seem very similar to cucumbers, but they are placed in the same species. However, cucumbers (*Cucumis sativus*) (Fig. 4.7) are only rather distantly related to their African cousins. Although it is unclear where cucumbers arose, they were domesticated in

TABLE 4.3 Commonly Used Species of Squashes and Ways to Distinguish Them

SPECIES	COMMON NAMES	PEDICEL AND FLESH	SEEDS
C. pepo	Summer squash, zucchini, winter squash, acorn squash, crookneck, spaghetti squash, pumpkin, ornamental squashes	Pedicel hard, irregularly shaped and 5-angled at base, thin, deeply furrowed, often prickly; flesh fibrous, most notable in the spaghetti squashes	Beige, with smooth-edged margins, blunt ends
C. moschata	Winter squash, butternut squash, pumpkin	Pedicel moderately hard, smoothly 5-lobed at the base; flesh moderately fibrous	Light-colored with beige scalloped margins, blunt ends
C. argyrosperma	Winter squash, hubbard squash, turban squash, pumpkin, green striped cushow	Pedicel hard, with five rounded ridges at the base; flesh moderately fibrous	Slightly scalloped margins, pointed ends
C. maxima	Winter squash, winter marrow, pumpkin	Pedicel soft, spongy, thick very slightly furrowed; flesh fine-textured and used in most canned "pumpkin pie" fillings	Tan with very narrow, smooth margins, blunt ends

Source: Adapted from M. Yamaguchi. 1983. *World Vegetables*. Westport, Conn., AVI.

Box 4.1

How to Choose a Perfect Melon

It can be difficult for a novice to choose a good melon from the produce section of the supermarket. A melon that will ripen should have no stub of the stem still attached because a melon picked too soon may get soft but will never become sweet. Melons develop all the sugar they will ever contain while still on the vine. Home gardeners can pick their melons when they are "fully ripe" or when they "slip" from the vine without effort. For commercial sales, melons can be picked slightly before they are fully ripe, but they should still be at the "full slip" stage (detaching readily flush with the melon). Several other characters to look for are slight give when pressed (especially at the end opposite the stem), a sweet aroma, and a sloshing sound when the melon is shaken, which indicates that the seeds are loose within the cavity. Muskmelons or cantaloupes should have a netted pattern on the outer surface that does not rub off easily. Watermelons are a bit more difficult to judge. The common practice of thumping a watermelon to see if it produces a hollow sound is not necessarily a good test because it can lead to the purchase of an overripe melon.

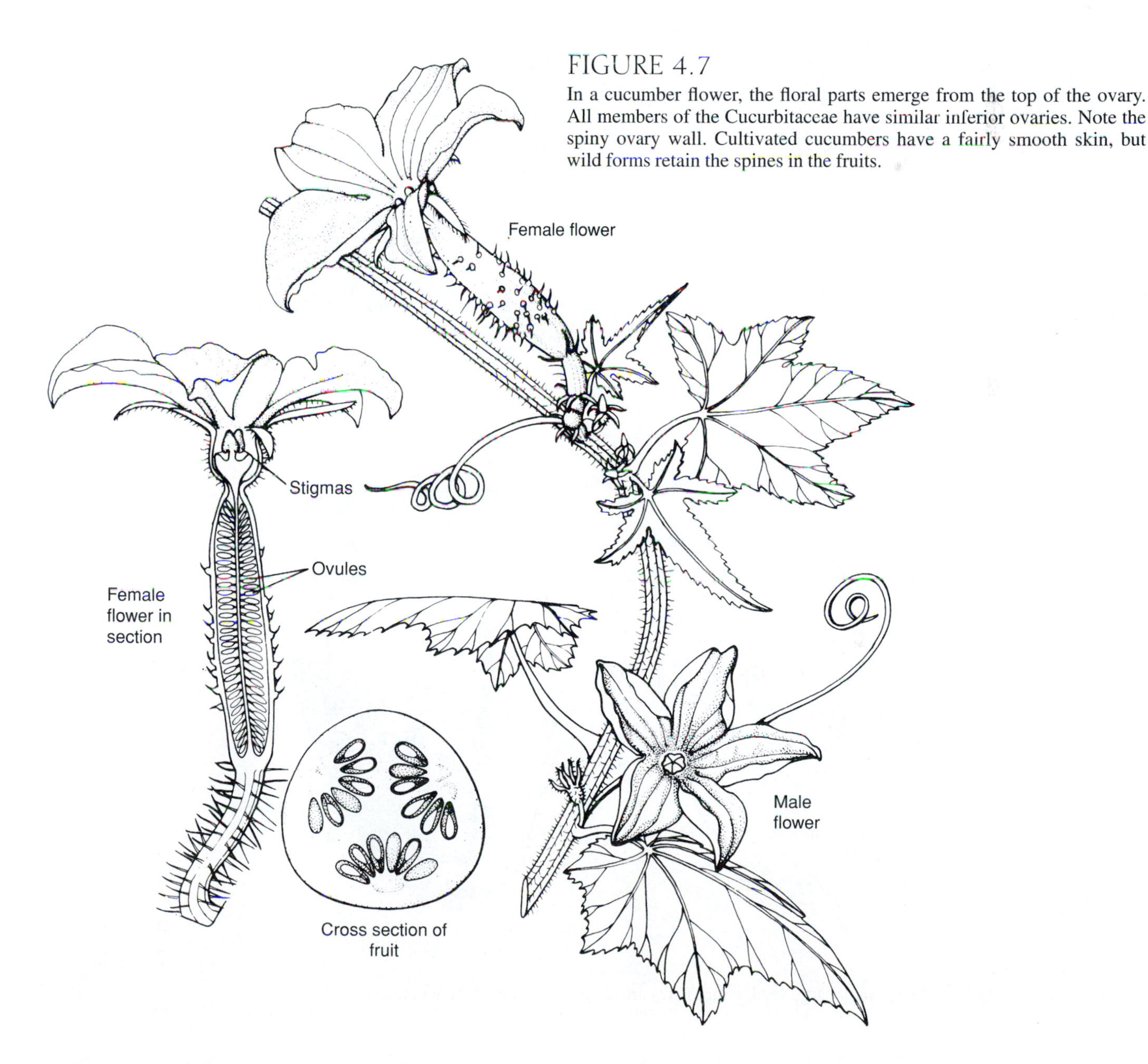

FIGURE 4.7
In a cucumber flower, the floral parts emerge from the top of the ovary. All members of the Cucurbitaccae have similar inferior ovaries. Note the spiny ovary wall. Cultivated cucumbers have a fairly smooth skin, but wild forms retain the spines in the fruits.

either central or western Asia. The long, narrow fruits might be considered nature's version of the modern Thermos since they have often been carried on caravan journeys as a source of water. Egyptian slaves were reportedly provided with rations of leeks and garlic as protection against sunstroke and with cucumbers as a source of water. High water content and a bland flavor were prized qualities in the arid regions of Asia Minor, and travelers introduced cucumbers throughout the Mediterranean region. Special varieties of cucumbers are grown for use in making pickles. Pickling, which consists of immersing cucumbers in a brine or vinegar solution, is an ancient method of preserving perishable fruits and vegetables.

Several other members of the Cucurbitaceae are important in Latin America. The white-flowered calabash, or bottle gourd, *Lagenaria siceraria,* has continued to baffle biologists interested in the origins of domesticated plants because it appears to have been cultivated in Ecuador and Peru over 7000 years ago and in Egypt over 3000 years ago. How the species managed to be transported across either the Pacific or the Atlantic Ocean before European contact has been the subject of many debates. Most workers now consider the disjunctive distribution to be a natural one, attributable to the rafting of fruits across an ocean. Archaeological collections from Peru show a striking diversity of gourd shapes, several of which had apparently been selected by Indians for specific functions. Gourds of a certain shape were used as *chicha* (corn beer) (Chapter 14) cups, others served as water vessels, and still others served as bowls.

The chayote (mirliton or chowchow; *Sechium edule*) (Fig. 4.8) was also domesticated in pre-Columbian times in Mexico. The pear-shaped green, occasionally spiny fruits are increasingly seen in U.S. supermarkets. They can be boiled, fried,

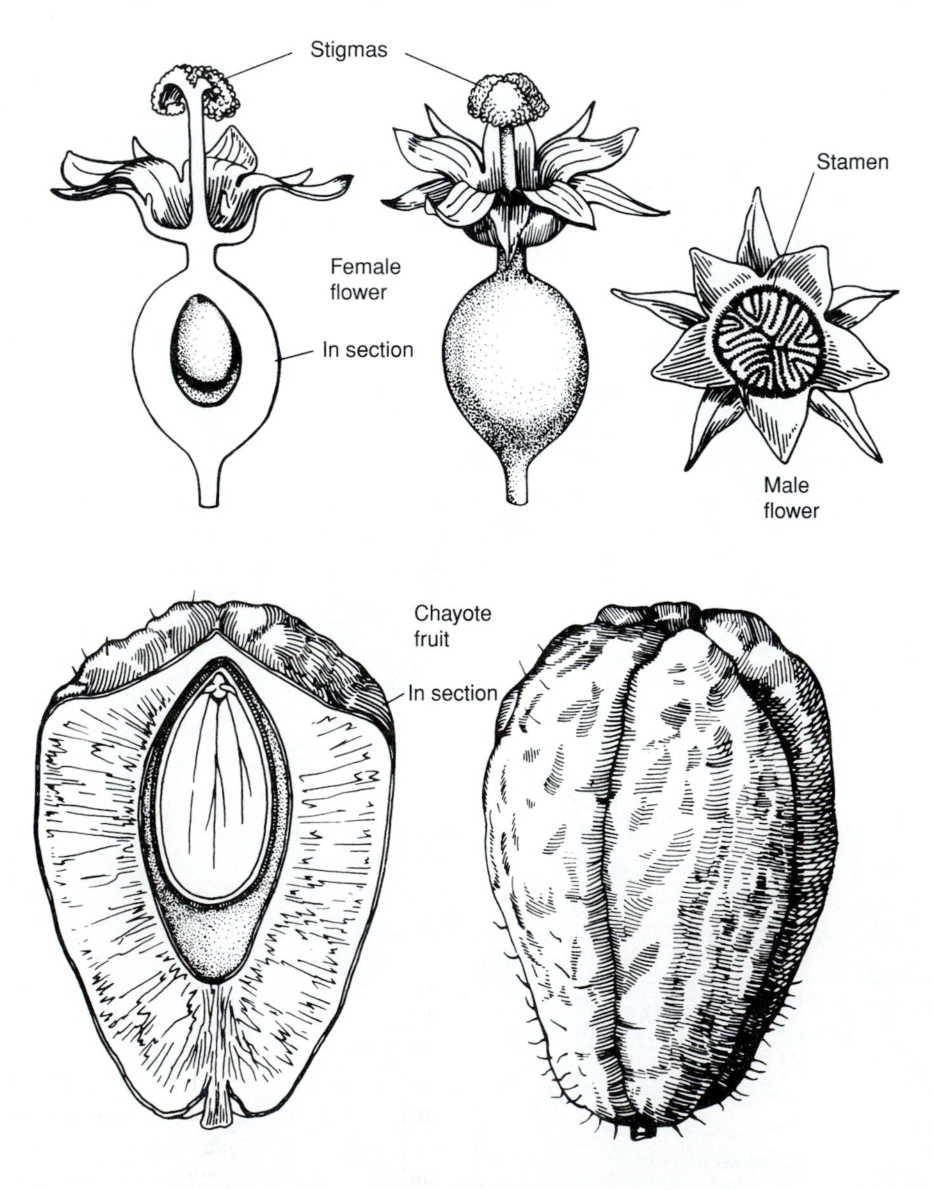

FIGURE 4.8
Because each chayote contains a single large seed, entire fruits are planted to establish new vines.
(After Baillon.)

FIGURE 4.9
The luffa gourd is eaten when young and tender, but as the fruits mature, they become tough and fibrous. The mature fruits can be picked or allowed to remain on the vine until after the first frost. The soft portions of the fruits are removed from the fibers by retting. The network of rough tissue that remains serves as a body scrubber.

stuffed, or eaten in salads. Unlike most other members of the squash family, which have numerous seeds embedded in the endocarp tissues, chayotes have only a single large seed.

Other cucurbitaceous genera—luffa (primarily *Luffa acutangula* and *L. cylindrica*) (Fig. 4.9), bitter melon (*Momordica*), and wax gourd (*Benincasa;* see Fig. 1.17)—are commonly used in Asia. Luffa fruits are eaten only when they are very young. Old fruits are allowed to rot partially, and the soft, disintegrating cells are washed away. The remaining network of fibrous tissue yields a luffa "sponge" used for scrubbing.

Tomatoes, Peppers, and Eggplants

It is hard to think of Italian food without evoking visions of fragrant pots of tomato sauce, yet tomatoes are American and their inclusion in Old World cuisines came only after the discovery of the New World. In fact, it is hard to imagine that tomatoes (*Solanum esculentum*), today the most important fruit vegetable crop in the United States, were slow to be accepted in Europe because of their similarity to poisonous *Solanum* species.

There is now little doubt that the ancestors of modern cultivated tomatoes are natives of the west coast of South America. The three red-fruited species of the genus all occur in Peru or on the Galápagos Islands. However, the cherry tomato (*Solanum esculentum* var. *cerasiforme*), which appears to be the direct ancestor of domesticated forms (Fig. 4.10), grows wild in Central America as well as in Peru. Archaeological data suggest that domestication occurred in Mexico, not South America. The Mayans called the fruit *xtomatl* or *tomatl,* corrupted by the Spanish into *tomate.* The English substituted an *o* for the *e.* Once the colorful fruits were brought back to Europe, the Spanish and the Italians were the first to accept them. Elsewhere in Europe and in the British colonies acceptance was much slower because of persistent misconceptions. Some believed that tomatoes had aphrodisiac properties. The French called tomatoes *pommes d'amour* (love apples), but this was apparently a misinterpretation of the Italian name *pomo d'oro* (golden apples) or a variant of *pomi dei Moro,* a name that referred to the use of the fruit into Europe by the Moors.

Tomatoes were initially thought to be poisonous because many European members of their family (Solanaceae) have bitter fruits containing toxic or hallucinogenic compounds (Chapter 12). The German common name, "wolf peach" reflected the belief that the fruits could be used in mystic ceremonies to evoke werewolves. Linnaeus formalized this belief by naming the species *Lycopersicon esculentum,* Latin for "juicy wolf peach." Recent evidence from studies of chloroplast DNA restriction fragments (Chapter 1) has shown that tomatoes are a nontuberous member of the genus *Solanum.* The correct name is consequently *Solanum esculentum,* but old habits are hard to change.

Tomatoes were taken to Europe from Mexico by the Spanish, but they were introduced to northeastern North America by the British, where they were grown primarily as ornamental plants. Until 1800 the fruits were relegated to a role as a pustule remover. According to an old farm journal, doubts about the fruit's edibility were dramatically removed in 1820, when Colonel Robert Gibbon Johnson announced that at noon on September 26 he would eat a bushel of the dreaded fruits. Two thousand people thought him mad and turned out to witness the event. To their astonishment, he survived. In the ensuing 20 years tomatoes became a relatively popular fruit, but the spectacular surge in tomato cultivation came after 1920. Most of this increased productivity was due to the rise of processed forms of the fruits such as tomato juice, canned tomatoes, tomato paste, catsup, and various salsas.

Wild tomatoes are outbreeding perennial herbs with small red berries that have two carpels (Fig. 4.10k, l). During the process of domestication there has been selection for self-pollination (see Fig. 1.29) to ensure high fruit set. Even so, tomato flowers have to be vibrated to shake the pollen from

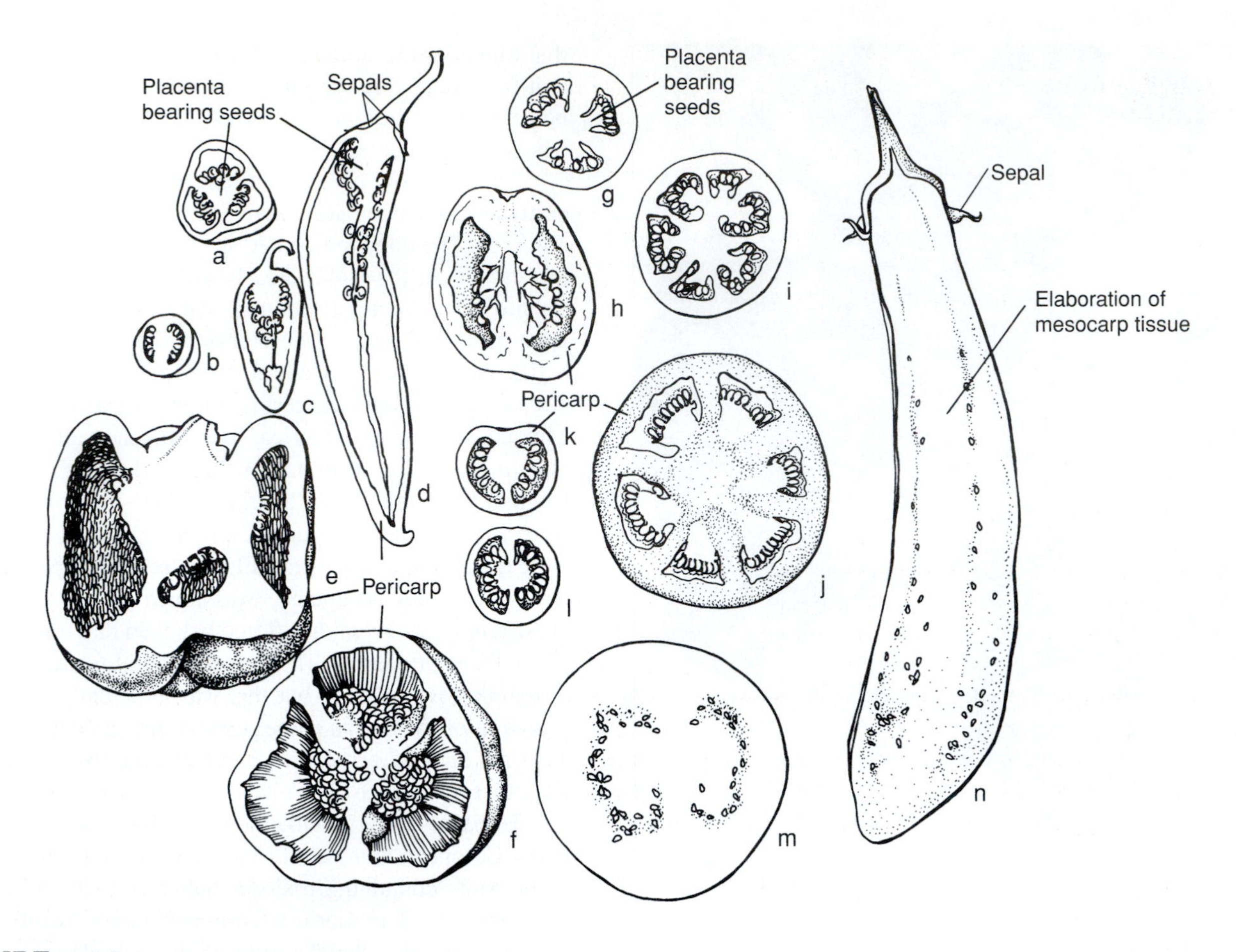

FIGURE 4.10

Cross and longitudinal sections of fruits of the Solanaceae (nightshade family). (a–d) hot peppers; (e–f) sweet bell peppers: (g–h) paste (plum) tomatoes; (i, j) salad tomatoes; (k, l) cherry tomatoes; (m, n) Japanese eggplant. All are berries. The **placenta** is the tissue on which the seeds are borne. In the tomato, the placenta produces a gelatinous substance that surrounds the seeds with the characteristic "jelly." The flesh of the eggplant is an elaboration of the mesocarp. There appear to be no seeds in the eggplant because the fruit is used in an immature stage.

the tubular anthers (Fig. 4.11). In open fields wind currents are sufficient to dislodge the pollen, but in greenhouses the plants must be artificially shaken. Selection has also produced a variety of colors and shapes of tomatoes, but until the tools of biotechnology became available, researchers were unable to overcome the problem of tough or mealy, flavorless commercially grown tomatoes (Box 4.2).

Like tomatoes, sweet peppers, *Capsicum annuum* (Solanaceae) (Figs. 4.10 and 4.11), are native to the New World and were probably first domesticated (as hot peppers) in Mexico. Archaeological sites at Tehuacán, Mexico, have yielded pepper seeds dated to be almost 7000 years old. Early peppers all seem to have been pungent types (Chapter 8); selection for sweet varieties must have occurred later. Most of the peppers produced in the United States today are sweet, or bell, varieties (Fig. 4.12). The green peppers that are most commonly seen are the immature stage of peppers bred so as to remain in this stage.

Another commonly encountered edible fruit of the Solanaceae is the eggplant (*Solanum melongena)* (Figs. 4.10 and 4.11). India or perhaps southern China is thought to be its native home. Sometime in the fifteenth century eggplant cultivation spread to Europe and later to the New World. Two other species of eggplant constitute important fruits in Africa, where they were domesticated, but *S. melongena* is the only species now used on a worldwide basis. Like zucchini, eggplants are picked when immature; their immaturity accounts for the apparent absence of seeds in the fruit pulp. Characteristics of eggplants that have tended to suppress their popularity in the United States are the browning of the flesh once the fruits are peeled or skinned and a tendency toward bitterness. Although the name eggplant may seem inappropriate for the large, ovoid black-purple fruits that are now marketed, varieties common a few hundred years ago had small white-skinned fruits that resembled birds' eggs.

Coconuts

There is a South Seas proverb, "He who plants a coconut tree plants food and drink, vessels and clothing, a habitation for himself, and a heritage for his children." Because of the

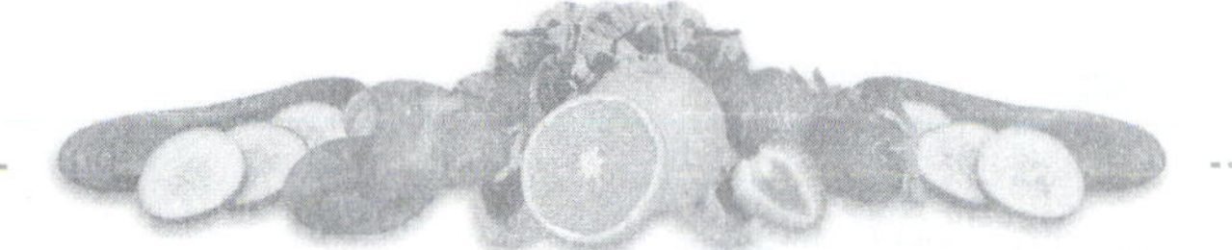

Box 4.2

How to Make a Better Tomato

Anyone who has tasted vine-ripened tomatoes knows how different they taste from the dull red, cardboard-like varieties sold in supermarkets most of the year. The reasons for the difference in taste between the home-grown fruits and supermarket offerings lie in the fact that tomatoes ripened to perfection on the vine start to disintegrate so rapidly that they cannot be stored and shipped successfully. Tomatoes to be sold as fresh grocery store produce are picked before they are mature or when they are only beginning to turn colorful and then are ripened at the time of selling. Fruits that have been picked green are tough because of a lack of the proper ethylene-generated maturation, or they are mushy because the intercellular matrix deteriorates. They also lack the sugar that accumulates very rapidly at the peak of ripening when tomatoes are left on the vine. In the early 1990s two firms, the Cetus Corporation and Calgene, found ways to allow tomatoes to mature but not overripen.

Tomatoes belong to a group of fruits that undergo an extremely rapid terminal burst of ripening. During this period there is an accelerating production of ethylene, a plant hormone that is responsible for the series of events that lead to the final color change, softening, and flavor production characteristic of natural ripening. During this period, enzymes are produced that begin to degrade the fruit wall. Scientists at the Cetus Corporation isolated the strand of messenger RNA that codes for the compound that just precedes ethylene in the biochemical pathway leading to ethylene. They next determined the nucleotide sequence of the DNA that produced the RNA and then made a complementary, or **antisense,** version. They subsequently inserted the antisense sequence into a coding strand of the nuclear genome. An antisense sequence has the complementary nucleotides substituted for each of the original nucleotides in the DNA chain. It thus produces a strand of RNA that is complementary to, and thus binds to, the normal RNA. The antisense RNA blocks expression of the RNA leading to the ethylene precursor. Consequently, ethylene is not produced and the final burst of ripening does not occur. Later treatment with propylene reverses this effect and allows the final ripening. The fruits produced are similar in flavor and texture to those ripened on the vine.

Calgene tried a different approach. Scientists from this company isolated the gene that codes for the enzyme that degrades pectin and allows softening to occur once a fruit has ripened. They determined the DNA sequence of the gene for this enzyme and produced an antisense version that they introduced into the nucleus of tomato plants. Overripening, or deterioration, does not occur because production of the degrading enzyme is blocked. Plants with copies of the antisense gene can consequently be allowed to ripen fruits on the vine. In May 1994, after extensive tests, the Flavr savr™ tomato, as the Calgene product is called, was approved for sale, heralding the beginning of genetically engineered foods.

versatility reflected in this adage, coconuts have earned the designation of the greatest provider in the tropics. Coconuts, *Cocos nucifera* (Arecaceae), can yield oil, fiber, food, building material, and drink. Since we discuss oil production in Chapter 9 and fiber in Chapter 15, we mention here only the history of the coconut and its use as a primary food. Discussing its origin is not, however, an easy task. Controversy has raged for years about the coconut's native home, because the fruits were present on the Pacific coasts of South America, Southeast Asia, and Polynesia before Europeans reached the New World. Various geographers have suggested that ancient voyagers crossed the Pacific before 1492, but evidence currently supports the hypothesis that coconuts are native to the Indo-Pacific region and dispersed across the ocean passively by means of currents. The species was perhaps first domesticated in Malaysia. Wild types that have a thicker husk and less liquid inside when immature do not occur in southern Africa or South America. Coconuts are now planted wherever the climate supports their growth.

Coconut palms are monocotyledons with a trunk composed of sheathing leaf bases. Plants begin to flower and fruit several years after germination and establishment. The species is monoecious, with numerous male flowers and a few female flowers on each inflorescence. The fruit is formed from a flower with three carpels (each represented by an "eye" of the coconut), only one of which develops. The mature fruit contains one seed, but the embryo itself is small and is located near the stem end. Initially, the endosperm is a liquid containing free nuclei. This liquid, called coconut water or coconut milk, is drunk from green coconuts in many tropical countries. As the endosperm matures, cell walls form around the nuclei, and the endosperm solidifies into an oil-rich layer of coconut "meat" inside the seed coat. It is the solidified endosperm that we eat in pieces or grated in cookies, cakes, and pies. The coconut "milk" used in many Asian recipes is made by soaking grated coconut meat in water and squeezing out the oil-rich liquid.

If mature coconuts are not used or harmed, the embryo can germinate within the coconut because there is no dormancy period. The germinating seedling eventually extends the tip of the cotyledon through one of the eyes. The base of the embryo swells into an absorbing organ that eventually fills the entire cavity of the coconut as it digests the endosperm (Fig. 4.13). The swollen organ, called a

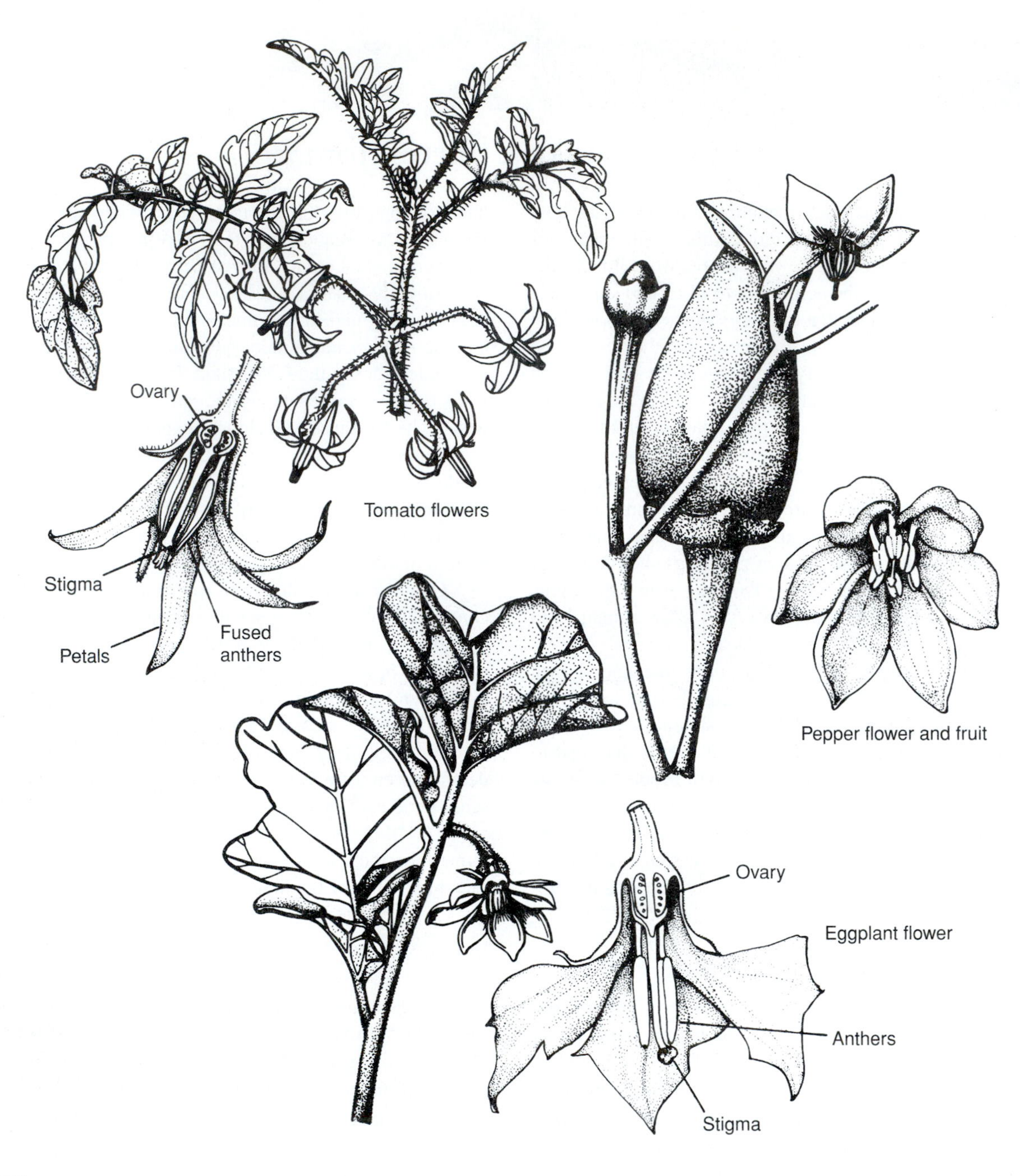

FIGURE 4.11
Flowers of the Solanaceae typically have five fused petals with five or fewer stamens attached to the corolla. The anthers form a characteristic cone above the star-shaped petals. The ovary is superior and generally has two carpels.

coconut apple, can also be eaten. In addition to providing coconut meat, or copra, for eating or oil production (Chapter 9), the stalk can be fed to animals. The inflorescence stalk can also be tapped, and the exuded sap can be used as a basis for fermentation. Occasionally, the growing tip of a coconut palm is removed and eaten as a vegetable, but since this results in the death of a valuable tree, it is rarely done. Hearts of palm are therefore usually obtained from other palm species.

Dates

The fruits of the date palm, *Phoenix dactylifera* (Arecaceae) (Fig. 4.14), were an important food in biblical times. Fruits of wild relatives of the modern commercial date were presumably gathered for thousands of years by tribes wandering across arid regions from India to the western Mediterranean, but by about 4500 B.C., dates had been domesticated. The Muslims believed that the date palm was

FIGURE 4.12
Children of migrant workers harvesting bell peppers in a North Carolina field.
(Photo by S. Rotner, courtesy of the United Nations.)

FIGURE 4.13
Coconuts are harvested from trees that can reach up to 24 m (80 ft) in height. Once the embryo has germinated, it produces an absorbing organ that enlarges as it consumes the rich endosperm. This organ, known as the "coconut apple," is sometimes eaten.
(Germinating coconut modified after W. H. Brown. 1935. *The Plant Kingdom.* Boston, Ginn.)

made from the dust left after God made Adam. Certainly, dates would have seemed like a divine gift to desert travelers because they indicated the presence of water, which meant food and relief from the parching heat. Dates are also highly nutritious, containing 60 to 70 percent sugar and 5 percent protein. It is thus not surprising that the date palm was considered the "tree of life" by the Bedouin people.

Date palms are dioecious, but as early as 2300 B.C., agriculturists had learned to hang a male inflorescence in a female tree to enhance pollination (Fig. 2.9). Modern machines now blow pollen across the female flowers when the stigmas are receptive.

Pineapples

Pineapples, *Ananas comosus* (Bromeliaceae), are indigenous to the New World and were widely cultivated by native people by the middle of the fifteenth century. Columbus described pineapple fruits and noted during his second voyage that they resembled pine cones, a similarity that led to the common English name. Native Americans considered the pineapple a symbol of hospitality and used the sweet juice for making an alcoholic beverage and a poultice and as a component of arrow poisons. We have retained the association between the pineapple and hospitality by incorporating pineapples into motifs used over doorways or on statuary lining driveways.

The spiny yellow-brown items we find in the store are multiple fruits formed by the fusion of the flower stalk, bracts, and ovaries of 100 to 200 independent flowers (Fig. 4.15). The edible portions are bracts and corollas that become fleshy. Modern cultivars are parthenocarpic

FIGURE 4.14
Dates are wrapped to protect the developing fruits from the sun and predation by insects and birds.
(Photo courtesy of California Date Commission.)

(develop fruit spontaneously without fertilization), although seeds can be obtained by carefully pollinating the flowers. Fruit initiation and maturation are carefully controlled by spraying the plants with hormones. Most cultivated varieties are seedless. Consequently, commercial plantings are made by inserting either the leafy portion that is borne above the fruiting stalk (Fig. 4.16), or suckers arising from axillary buds, into the soil.

Pineapples were spread around the world in the 1500s by Portuguese, Dutch, and Spanish traders. They were introduced into Hawaii in the early nineteenth century, but it was only after a young Yankee entrepreneur, J. D. Dole, encouraged the natives to grow the plants that pineapple production became an important source of income in Hawaii. Today tourism has supplanted farming as the primary industry, and the center of pineapple cultivation has shifted to the Far East and Africa. Pineapples are easily canned or pressed for juice, but they contain a proteolytic enzyme, bromelain, that prevents their use in desserts containing gelatin or milk. Because of this protein-degrading enzyme, pineapples have been used as a meat tenderizer.

Bananas

Bananas (Fig. 4.17) are today so readily available in the United States and such an important food in some Latin American countries that it is easy to forget their recency in the New World. Bananas and their relatives in the genus *Musa* (Musaceae) are all native to eastern Asia and Australia. Various species of the genus have been used for fiber (Chapter 15), food, and ornamentation in these areas since prehistoric times. It was even proposed by Sauer that agriculture first arose in Southeast Asia, with the banana among the first plants cultivated. Although few people now support Sauer's idea, there is no doubt about the importance of wild bananas in the lives of preagricultural and early agrarian peoples in Southeast Asia.

Domestication and eventual commercial production of the modern edible banana involved hybridization and polyploidy of two diploid species, *Musa acuminata* and *M. balbisiana.* The first of these species provides a chromosome set called the "A genome," and the second provides a chromosome set labeled the "B genome." Modern cultivated polyploids contain various dosages of these sets of chromosomes and can therefore have combinations ranging from the diploid AA and BB sets to triploids with AAA, AAB, ABB, and BBB mixtures to various tetraploid blends. The common 'Gros Michele' variety usually seen in American groceries has an AAA combination, and the large 'Horn Plantain' has an AAB mixture. All edible bananas, regardless of their polyploid level are now lumped under the name *Musa Xparadisiaca.* The transformation of wild bananas, which have a thick peel, numerous large seeds, and small amounts of pulp surrounding the seeds, involves two changes. One is sterility, and the second is spontaneous development of the fruit (parthenocarpy) without fertilization.

From their native southwestern Pacific home, bananas spread to India by 600 B.C. They were probably introduced independently into Africa and eastward across the Pacific. In 1522 bananas were taken from the western coast of Africa to the Canary Islands as food for slaves being sent across the Atlantic. The first clones to occur in the New World were planted in the West Indies in the seventeenth century. Included in these early clones were at least one type of sweet banana and a starchy variety that tastes best when cooked before eating. Whereas bananas are usually simply eaten fresh or fried in the New World, they are used for beer and are steamed, boiled, dried, or roasted in their native region of Asia. In fact, starchy bananas are used in Asia in ways similar to the ways potatoes are used in the United States.

The story of the banana's rise to prominence as a fruit in U.S. markets is intimately linked to the history of the United Fruit Company, which was formed in 1899 by the merger of

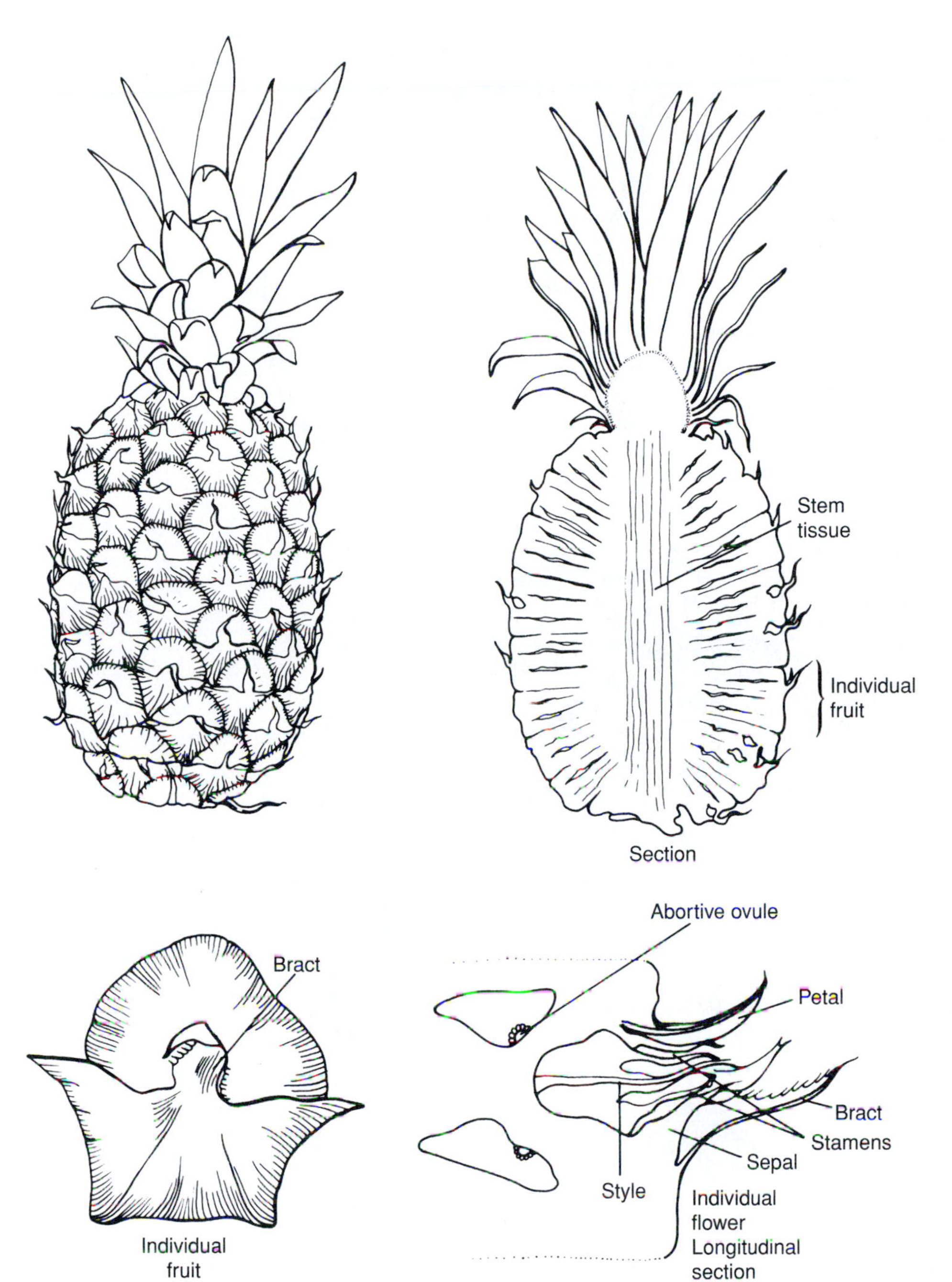

FIGURE 4.15

A pineapple fruit is a multiple fruit composed of 100 to 200 ovaries of separate flowers fused together and to the flowering stem. The remnants of the floral parts produce the prickles on the knobbly surface.

FIGURE 4.16

A pineapple's spirally arranged sword-shaped leaves, seen in these plants being harvested, is typical of bromeliads.

(Photo by Banoun/Caracciolo, FAO.)

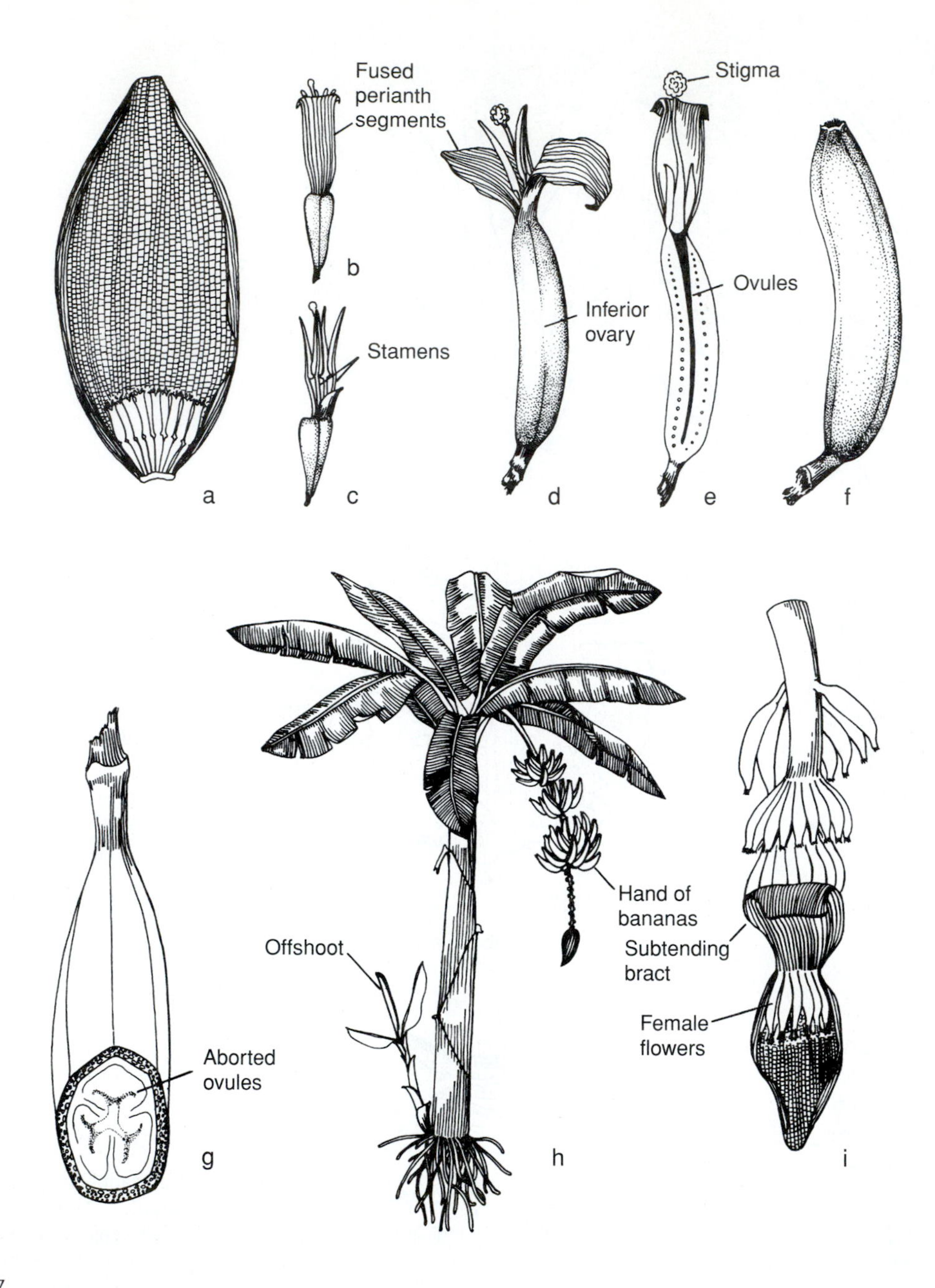

FIGURE 4.17

The banana is considered the world's largest herb. (a) A cluster of male flowers with its bract. (b) Male flower with the perianth attached. (c) Male flower with the perianth removed. (d) Female flower. (e) Female flower in longitudinal section. (f) The fruit, an accessory berry formed from the inferior ovary. (g) A fruit cut open to show the aborted seeds in the sweet pulp composed of placental tissue. (h) A plant in flower with an offshoot that can produce a new plant when the flowering plant is cut down. (i) Female flowers clustered in a bract.

two smaller companies. These companies had established shipping routes from Central America to ports such as New York, but the original method of transportation was by sailing schooner. Since such vessels were at the mercy of the winds, trips often took so long that shipments rotted before they arrived at their destination. In 1888, steamships began to replace the schooner, and only a few years later, in 1901, refrigerated ships began to make trips to Europe and the United States. The combination of fast, reliable steamships and refrigeration led to an explosion in the production of bananas, and the United Fruit Company began to expand into Panama (1903), Guatemala (1907), El Salvador, and Costa Rica (1920s). In the 1930s and 1940s the company extended its operations into Colombia and Ecuador. For several decades United Fruit essentially owned the land and controlled the workers on huge acreages in several Central and South American countries, causing them to be referred to as "banana republics." By World War II, in the face of public pressure and political reorganization within those countries, the influence of the United Fruit Company began to wane. Still, the importance of banana cultivation in many tropical American countries is reflected by the fact that bananas now constitute the major source of carbohydrate for millions of their people.

Even with a good shipping system, cargoes of bananas were often ruined because carbon dioxide and ethylene produced by bruised and ripe bananas accelerated ripening and caused rotting of entire shipments. Experiments eventually showed that if bananas are kept at 10°C (56°F) and 90 percent humidity, ripening is delayed. Conversely, ethylene can be used at a later time to induce ripening of green bananas. Once the physiology of banana ripening was understood, shippers were able to prevent large-scale spoilage and even ship green bananas that they could ripen at will when they reached their destination.

Bananas are planted using other corms, pieces of corms, suckers or any of numerous vegetative parts. Adult plants are essentially giant herbs, although they are called trees and their cylindrical shoots are commonly considered trunks (Fig. 4.17). Lateral stems form at the base of the main shoot, producing suckers, most of which are removed in plantations. Plants start to flower 2 to 6 months after initial shoot elongation. The flowers are borne on long, pendant inflorescences with the female flowers at the base of the bisexual stalk and male flowers at the apex. Female flowers are produced in clusters, each consisting of two whorls of flowers protected by a large bract. As a whorl of 12 to 20 female flowers becomes mature, the subtending bract peels back. Each cluster of female flowers produces a bunch, or "hand," of bananas when it matures. In many regions immature bananas are covered with plastic bags to prevent damage from soil, sun, dust, birds, and insects.

Bananas are often cut by hand by teams of harvesters consisting of cutters and backers. After checking that a stem is ready for harvesting, the cutter cuts the inflorescence stalk while a backer stands ready to receive the heavy mass of fruit on a thick pad covering his or her shoulder. In this way, fruit damage is minimized. Care must be taken to avoid bruising or wounding the bananas, because such injuries cause the fruits to release ethylene. For shipment, only perfect bananas are used. They are picked green, washed, dipped in pesticide, and packed.

After harvesting the fruit, the cutter decapitates the plant and cuts up the leaves. Banana plants can be replanted each year, or a sucker (offshoot) of each of the previous year's plants can be allowed to develop. If the initial planting was in rows, a sucker from the same side of each of the plants can be left to develop. The following year, the rows simply shift a few centimeters. This practice of allowing natural vegetative reproduction to produce subsequent plants can continue over a 5- to 25-year period. Bananas can yield impressive amounts of carbohydrate per unit area, with yields similar to those of potatoes (28 to 68 tons per hectare, or 11 to 27 tons per acre).

Figs

The fig genus, *Ficus,* contains over 800 species. Many species of the genus, such as the "rubber plants" commonly sold in pots, are cultivated, but only *Ficus carica* produces a commercially important fruit. This fig, a native of the Mediterranean region, is mentioned in Egyptian documents (2750 B.C.) (Fig. 4.18) and is frequently referred to in the Bible. Adam and Eve were supposed to have sewn together fig leaves to cover themselves once they had learned to be ashamed of their naked bodies. To the Greeks, an edible fig was a *sykon,* a designation that led to the modern term *syconium* for the fruit, which is actually an inflorescence enclosed inside receptacular tissue.

Wild figs are dispersed by birds or bats that eat the fruits and later defecate the tiny seeds. People generally use cuttings for propagation of cultivated varieties. A few figs are either self- pollinating or parthenocarpic. Common figs such as Kadota and Mission figs can produce fruits without pollinators, but many wild figs and most of the edible varieties have a very elaborate pollination system involving tiny wasps that mate inside the fruits (Fig. 4.19). Newly emerged and mated females carrying pollen from the male flowers of their natal figs fly to immature figs and enter them. While laying eggs in modified female flowers of the immature figs, females pollinate normal, fertile female flowers. Smyrna figs, historically the most popular edible figs, contain only female flowers in their syconia. To produce fruits, the flowers must be pollinated with pollen from male flowers in syconia on different trees. These pollen-donor trees are

FIGURE 4.18
Figs and baboons.
(Redrawn from an Egyptian frieze of the twelfth century B.C.)

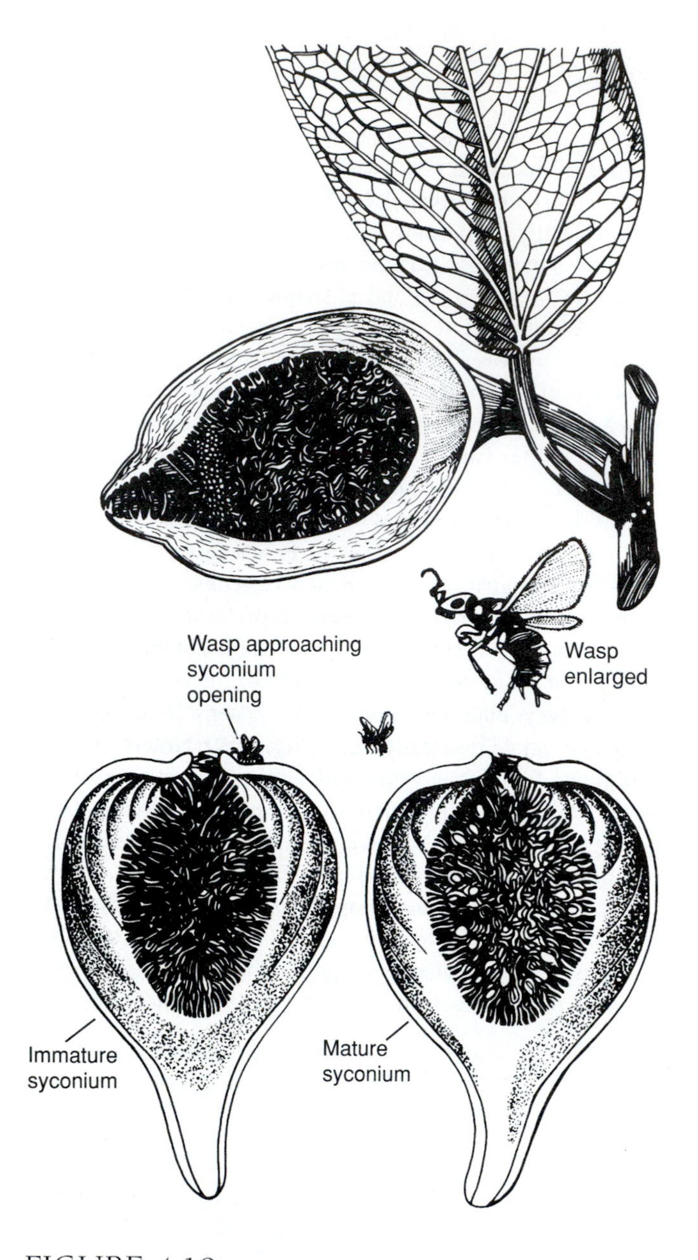

FIGURE 4.19

The pollination of most figs is an intricate process involving species-specific wasps that lay their eggs in the ovaries of short-styled female flowers within the enclosed inflorescences, or syconia. When the young wasps emerge, they mate before escaping from the syconium where they were born and carry some pollen from male flowers in this syconium to long-styled female flowers in the syconium where they will lay their eggs.

(After von Marilaun.)

called caprifigs. In the Mediterranean region Smyrna figs and caprifigs naturally grow together. When Smyrna figs were brought to the New World, the pollination system was not fully appreciated and the inedible caprifigs were left behind. Naturally, the Smyrna trees failed to produce fruits. In the 1880s, caprifig trees and (by 1900) the wasps were imported, leading to successful American fig cultivation.

Avocados

Avocados, or alligator pears (*Persea americana,* Lauraceae) (Fig. 4.20), have a controversial history and seem to defy our theories about animal-dispersed fleshy fruits. Fleshy fruits are generally low in calories and consist primarily of water and sugar or, in the case of bananas and breadfruits, starch as well. Avocados, in contrast, have a mesocarp that is extremely rich in oil like that of olives. Up to 30 percent of the pulp of cultivated varieties (on a dry-weight basis) can be oil. As a result, avocados have the highest energy-containing fruit pulp (from 2000 to 2800 calories per kilogram) known. Moreover, the seed is not protected in any way from the sharp teeth or digestive juices of animals that feed on the fruit. It has been suggested that extinct large animals were the original dispersers of avocados, but the postulation of a large animal does not explain how a seed without a hard endocarp is protected in its journey from the mouth through the digestive system of such an animal. The natural dispersal of avocados thus remains a mystery.

Avocados are known only as a cultivated species, yet they appear in some of the oldest archaeological deposits in southern Mexico (7000 B.C.). It is possible that avocados were independently domesticated three times in the Americas, giving rise to what have traditionally been considered the West Indian, Guatemalan, and Mexican (Drymifolia) varieties (Fig. 4.20). The Mexican variety has a small (about 250 g, or 8 oz) fruit with a thin smooth skin, a seed loose within the cavity, and a 30 percent oil content. Guatemalan fruits weigh between 500 and 1000g (1 to 2 lb) and have a thick warty skin. These two varieties are grown most commonly in California. The West Indian types grown in Florida are about the same size, with a thick but smooth

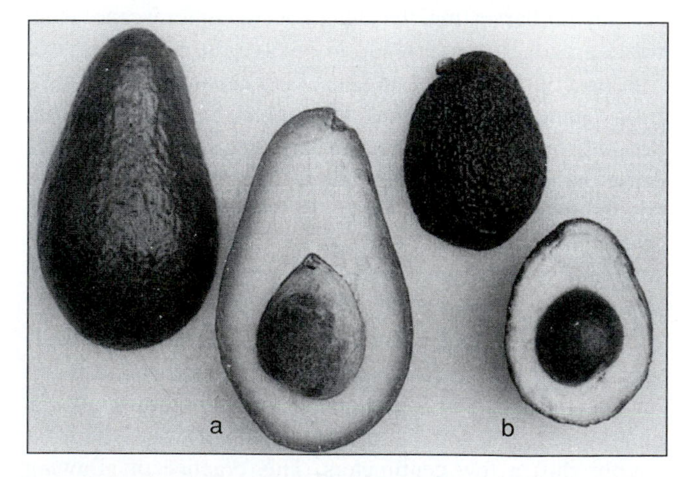

FIGURE 4.20

In American markets, we can find (a) smooth-skinned West Indian avocados grown in Florida and (b) small, warty avocados, which are hybrids between the Mexican and Guatemalan varieties, commonly grown in California.

skin and a mesocarp with only 8 to 10 percent oil. Recent molecular studies have indicated that the three varieties share a common ancestor and that the Guatemalan variety has been produced by hybridization of two other species of *Persea.*

Avocados were first planted as a commercial venture in the United States in Florida in 1893, but acceptance was slow. It is often said that the recent popularity of the avocado is a direct result of a calculated publicity campaign. Having failed to combat historical adverse feelings about the fruit, growers (on the advice of an advertising agent) began loudly to deny that the fruits had any aphrodisiac properties. In the 1920s few other claims would have been able to guarantee an immediate demand. The campaign was a success, and the popularity of avocados has been increasing ever since. The relationship between avocados and sex had, in fact, been claimed by native Americans. The word *avocado* comes from the Aztec word *ahuacacuahatl,* meaning "testicle tree," which refers either to its stimulating properties or to the appearance of the fruits, which commonly hang in pairs. Avocados will not ripen on the tree, so they are picked when they reach a specific gravity of about 0.96. They are then allowed to rest for at least 25 hours and finally are exposed to ethylene to complete ripening.

Mangoes

Among all the truly tropical fruits, mangoes, *Mangifera indica* (Anacardiaceae), are probably the most popular on a worldwide basis. They are particularly appreciated in their native region of Southeast Asia, but they have now been spread throughout the tropics. In times of emergency, the cotyledons of the seed can be ground into a flour substitute and the leaves of the trees can be used for cattle food. The fruit is so revered that leaves are hung above doorways as a symbol of goodness in southeastern Asia, and the shape of the mango is a sacred symbol in India.

Mangoes are naturally large trees, reaching over 30 m (98.25 ft) in height. When they are planted and irrigated in relatively dry areas, the trees are much smaller and the plantations tend to resemble orchards. The flowers of mangoes are small and are borne in long clusters. Usually only one flower on each cluster develops a fruit. In full fruit, a mango tree looks as if someone had tied fruits on the branches because the fruits hang on long, bare stalks (Fig. 4.21). The usually stringy pulp of a mango is the mesocarp that surrounds a stony endocarp enclosing the large seed.

Mangoes belong to the same family as poison ivy and, like many members of that family, produce a latex that is irritating to some individuals. Fortunately, latex production is restricted to the leathery skin, so peeling the fruit generally eliminates the source of irritation.

Mangoes were introduced to Brazil by the Portuguese in the early 1700s and were brought from South America to the West Indies in 1742. They now constitute an important seasonal food source for the poor populaces of Jamaica and Haiti. Until recently few were exported, but they are now commonly found in many U.S. supermarkets. Commercial operations usually consist of grafted, dwarf trees. The fruits are picked green to avoid bird predation and later are ripened by exposure to ethylene.

Okra

Although the native region of okra, *Abelmoschus esculentus* (Malvaceae), is still uncertain, it is thought to be either southwestern or south-central Asia. The domesticated species must have spread very early because okra was used in ancient Egypt; it spread from there to the Far East and Europe. Okra first reached the New World in the 1600s, and in the 1700s it was adopted as a crop by French immigrants in Louisiana. They referred to the vegetable by its African name, *gumbo,* and used okra with such frequency that it became an integral part of Cajun cooking.

Okra is a member of the Malvaceae, or cotton family, with hibiscus-like flowers that produce capsules resembling elongated cotton bolls (Fig. 4.22). Many people find the mucilaginous coating on okra seeds objectionable, but the gum acts as a thickening agent in soups, stews, and gumbos. Okra has maintained a place in American agriculture because it can tolerate even the hottest southern months and still produce a substantial fruit crop at the end of the long growing season.

Pomegranates

Pomegranates, *Punica granatum* (Punicaceae) (Fig. 4.23) are subtropical fruits native to the region from Iran to the Himalayas. Their importance in the ancient world was as much symbolic and artistic as nutritional. The Hebrews embellished their buildings with pomegranate motifs, and the globular fruits with their numerous juicy red seeds (see Fig. 1.23) eventually became a symbol of fertility and abundance. This symbolism was carried to eastern Asia, where pomegranates were offered to wedding guests, who were expected to throw them onto the floor of the honeymoon suite so forcefully that they shattered, scattering their seeds. This practice was believed to ensure fertility and a large number of offspring for the newlyweds. The broadcasting of pomegranate seeds has another, more sobering connotation. The French word for the fruit, *grenade,* is applied to a hand-thrown weapon used extensively since the seventeenth century that scatters not seeds but deadly metal fragments.

The Moors took pomegranates with them when they conquered Spain in A.D. 800. They used the seeds for food and the rinds of the fruit as a source of tannin for leather. The Spanish introduced the fruits into the New World many centuries later. Pomegranates are now grown from the

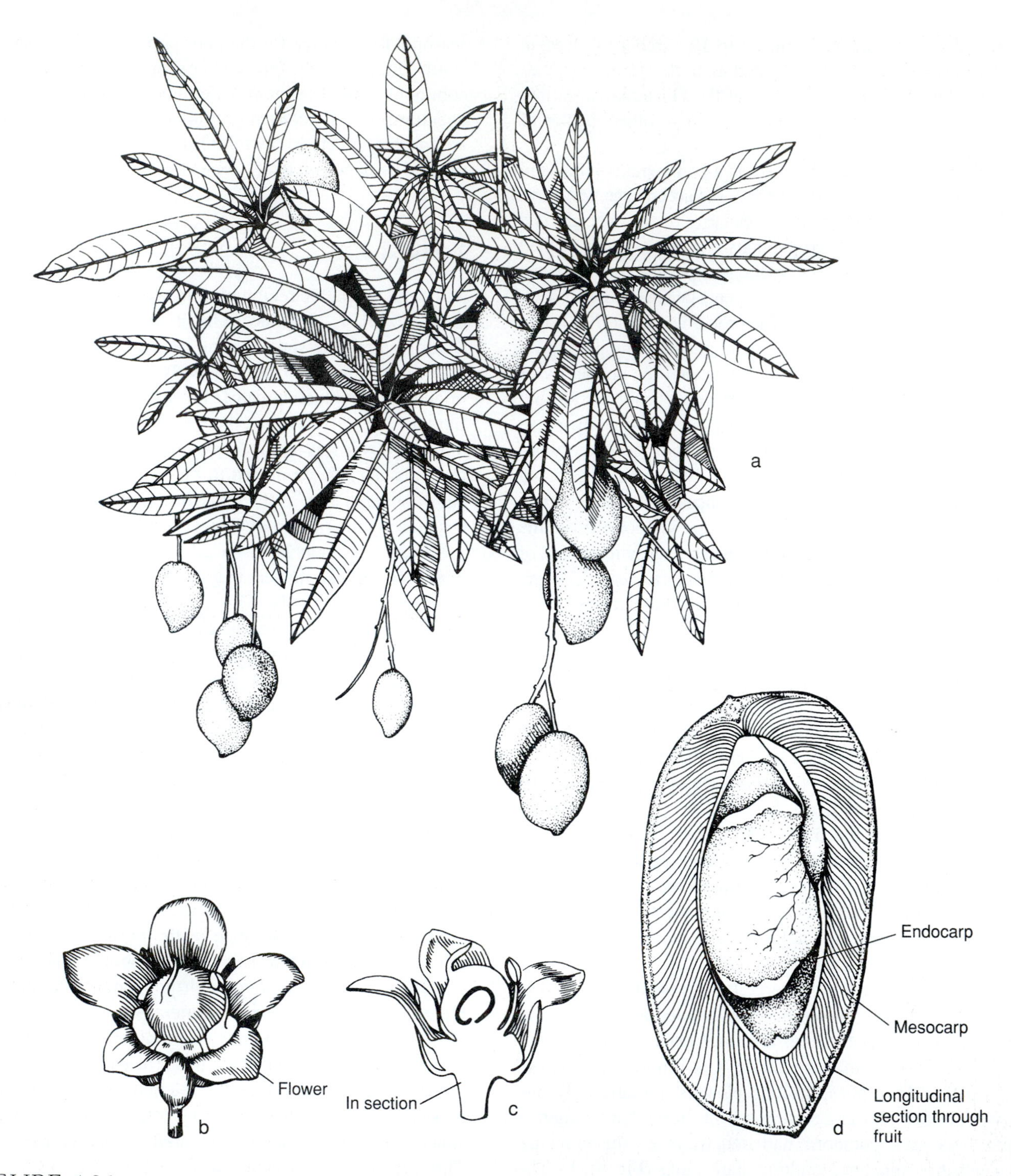

FIGURE 4.21
(a) Mangoes dangle from trees on long stalks. (b) A single mango flower shown (c) in longitudinal section. (d) The fruit in longitudinal section shows the fleshy mesocarp and the hard endocarp.

southern United States to northern Chile and Argentina. In many areas the trees are planted primarily as ornamentals because their flowers as well as their fruits are attractive (Fig. 4.23). In addition to being eaten fresh, the seeds can be pressed to yield a juice. If sweetened with sugar, the juice becomes grenadine syrup.

Papayas

Papayas (*Carica papaya,* Caricaceae) are smooth-textured, musky-flavored fruits native to Central America but little known in the United States until recently. They are called paw paws in Britain, Australia, and the West Indies. In parts

FIGURE 4.22
The appearance of okra flowers betrays the species' kinship with *Hibiscus* and cotton. Because of its shape, the Greeks named the fruit *okra*, meaning "lady's fingers."

FIGURE 4.23
A flower and bud of the pomegranate. Pomegranate flowers are large and attractive. The fruits are also striking when they are ripe and crack open to reveal the deep red seeds embedded in the pale cream placenta.

of Central and South America they find their way into practically every meal during the fruiting season. The demand in the United States for papayas has not traditionally been for the fresh fruit but for papain and chymopapain, proteolytic enzymes extracted from the latex exuded when the skin of the unripe fruit is scored. Papain is so effective in breaking down proteins that it forms the basis of commercial meat tenderizers. If is also used medicinally to aid digestion and in chewing gum and cosmetics.

Papaya plants are short-lived, soft-wooded trees that have a characteristic umbrella-shaped growth form. The flowers are borne along the stems in leaf axils. By the time the fruits are mature, the leaves have been shed, leaving clusters of the ovoid fruits emerging directly from the stem. Inside each fruit is a mass of slimy (each 0.5 cm in diameter) black seeds (Fig. 4.24). For latex tapping, green fruits are used, beginning when they are 10 cm (3.9 in.) in diameter. The skin is scored diagonally in such a way that only the outer part of the exocarp is cut. The latex that oozes out is collected the next day. A single fruit can be tapped several times.

A Sample of More Exotic Tropical Fruits

Although many tropical fruits can be tasted fresh only where they are grown, an increasing array appear in specialty food stores or even supermarkets. Within their native ranges some are extremely important foods for local people during the fruiting season. Among these is the breadfruit, which has a history as novel as the fruit itself (Box 4.3).

The sops, species of the genus *Annona* (Annonaceae), are American natives. Several, such as the sweetsop, or custard apple (*A. squamosa*), the soursop (*A. muricata*) (Fig. 4.25), and the cherimoya (*A. cherimoya*), are eaten as fruits or used for juice throughout the neotropics. The flowers of *Annona* species have numerous pistils, which fuse to form irregular round to oval compound fruits that have a characteristic segmented appearance on the outside. The sweetsop is the most widely consumed of the annonas because selection has led to consistently high quality in the fruits and to seedlessness. The species also tolerates a wider range of climates than do its relatives. It has been introduced to Florida and is grown there on a very limited scale. To some people the cherimoya is the finest of all of the sops, but its growth is restricted to montane tropical regions between elevations of about 1400 and 2000 m (4585 to 6550 ft).

The carambola, or star fruit (*Averrhoa carambola*, Oxalidaceae), is a tart tropical fruit. The tartness is caused by the presence of calcium oxalate crystals in the flesh, which dissolve in the saliva when the fruit is eaten, forming oxalic acid. The fruits are yellow, 8 to 15 cm long, and star-shaped, reflecting the five-carpellate fruit characteristic of the oxalis family (Fig. 4.28). Native to Indonesia, star fruits are now planted throughout the tropics.

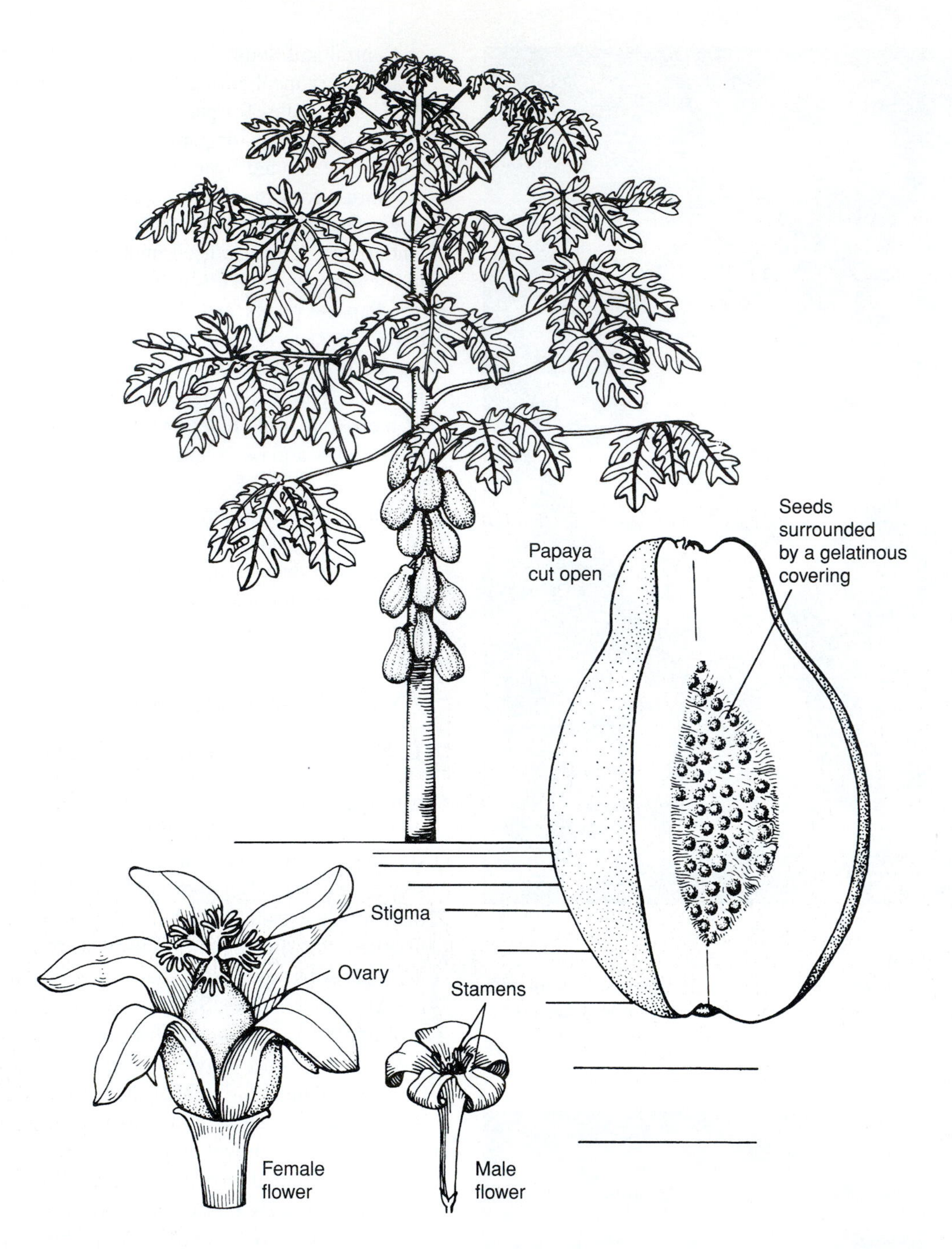

FIGURE 4.24
Papaya flowers are cauliforous, with the flowers and fruits emerging directly from the plant's stem.

FIGURE 4.25
Soursops are aggregate fruits formed by the fusion of the fruits of many flowers. They are among the tropical fruits increasingly seen in North American grocery stores.

(Photo courtesy of USDA.)

Box 4.3

Breadfruit

Early European explorers were amazed when they first saw breadfruits, describing the plants as producing masses of "bread" (hence the English name). A native of Polynesia, breadfruit, *Artocarpus altilis* (Moraceae), has been cultivated since ancient times. Breadfruits are tall (to 20 m) trees that have compound unisexual inflorescences. Female inflorescences develop into a round, multiple fruit about 10 to 30 cm in diameter (Fig. 4.26). The moist edible pulp is formed by abortive

FIGURE 4.26

The breadfruit, like the pineapple, is a multiple fruit, 10 to 30 cm in diameter. Cooked, the flesh resembles potatoes in flavor and texture. (a) Flower and flowering branch (After Baillon.) (b) Longitudinal section of a portion of a female flower. (c) Longitudinal section of part of a fruit.

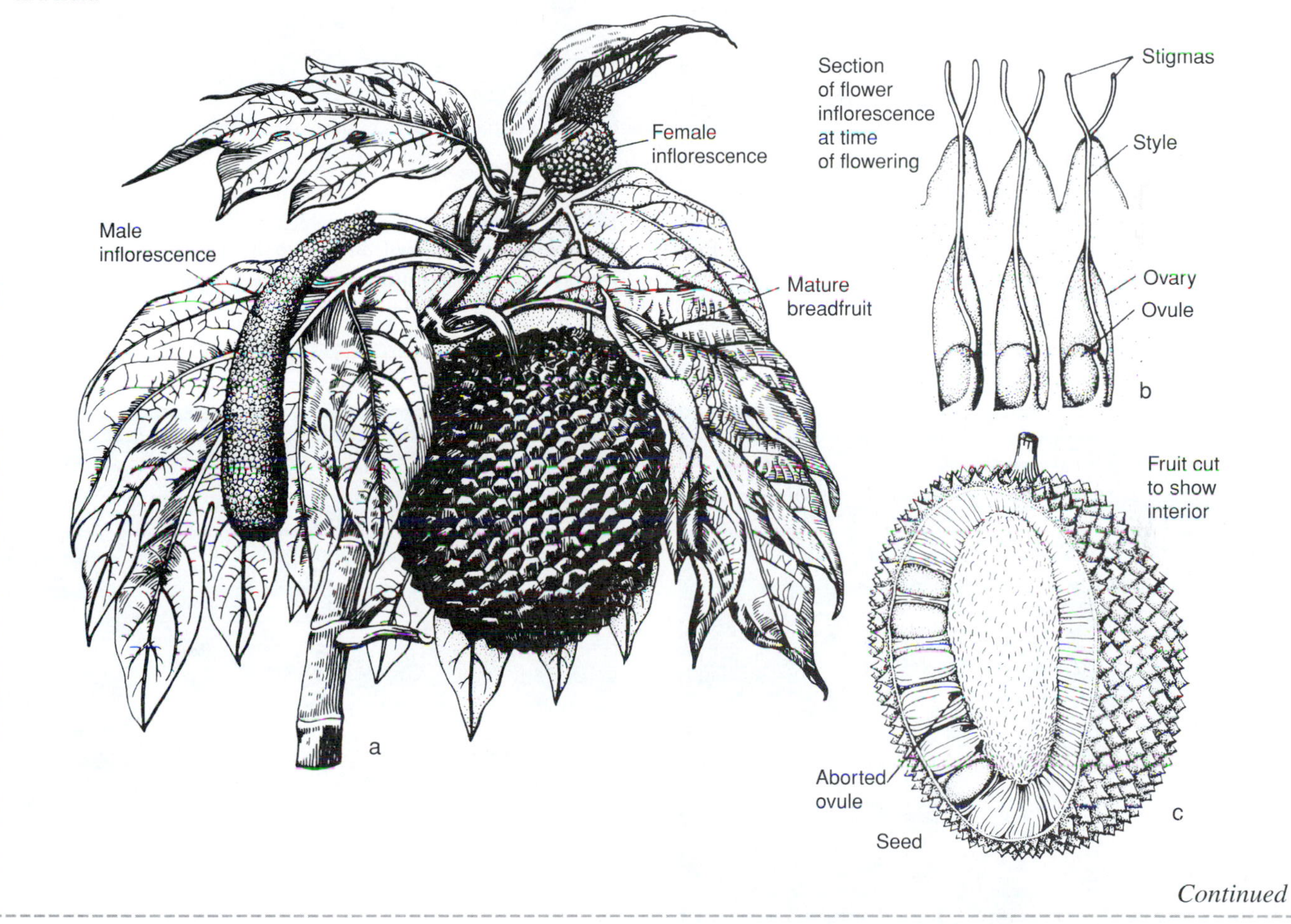

Continued

Most Americans encounter litchis as canned fruits accompanying fortune cookies at the end of a Chinese meal. The so-called nuts, *Litchi chinensis* (Sapindaceae), are natives of China. They are cultivated primarily in southern China and, to a limited extent, in subtropical areas of Florida. Litchis and their Malay relatives rambutans (*Nephelium lappaceum,* also Sapindaceae) are produced on trees 9 to 30 m tall. The edible portions of the warty and spiny fruits (Fig. 4.29) are the succulent, smooth-textured arils that cover the seeds. The arils are commonly sold in cans in the United States.

Passion fruits have become familiar to most Americans through advertisements for various fruit punches. To many people they seem to represent the essence of tropical fruits. It was the flowers, however, and not the fruits that originally attracted Europeans to the vines (Fig. 4.30). The intricate and showy flower parts were considered by religious immigrants to represent the instruments of Christ's passion. The Christian imagery of the flowers is detailed in Figure 4.30. Plants were consequently taken to Europe, where they were grown as curiosities in greenhouses. Although the fruits had been eaten by Native Americans long before European contact, passion

Box 4.3 *Cont.*

flowers. The few seeds that are produced are not normally eaten but can be used in times of food shortage. An average fruit usually contains about 20 percent carbohydrates and when cooked is similar in taste and consistency to a white potato.

In an attempt to capitalize on this bounteous (no pun intended) and potentially cheap natural source of carbohydrates, Britain sent Captain Bligh (Fig. 4.27) and *H. M. S. Bounty* to Tahiti to collect breadfruits for introduction into the New World colonies. In 1787 Captain Bligh headed out from England with a ship equipped to be a floating greenhouse. It took Bligh and his men 6 months to collect seedling breadfruits and prepare them for the long journey to the West Indies. During their stay in Tahiti, where they collected the plants, the crew began to savor the easy island life and compliant women. Soon after setting sail for the West Indies, the crew rebelled and put Bligh and 18 loyal seamen adrift in a longboat with a small cache of rations (Fig. 4.27). Bligh survived the mutiny, crossing 3618 nautical miles of the Pacific Ocean to Timor. Undaunted, he soon returned to Tahiti in another ship, *H. M. S. Providence,* and during his second voyage successfully established the trees in Jamaica. Today they grow throughout the humid tropics.

FIGURE 4.27
H.M.S. *Bounty,* the ship on which the hapless Captain Bligh attempted to bring breadfruit trees to the West Indies.
(Photo courtesy of National Maritime Museum, London.)

FIGURE 4.28
The carambola develops from a five-carpellate ovary in such a way that the fruit is star- shaped in cross section.
(Photo by Scott Bauer, courtesy of Agricultural Research Service, USDA.)

FIGURE 4.29
Bright red, warty litchis hang in clusters ready to be harvested.
(Photo by Barry Fitzgerald, courtesy of Agricultural Research Service, USDA.)

fruit cultivation began only a few hundred years ago. Of the 55 or so edible species, *Passiflora edulis* (Passifloraceae) is now grown on the largest scale. This species, a native of Brazil, was introduced to Hawaii in 1810. Since the species grows best in tropical montane habitats at an elevation of about 2000 m, Hawaii provides excellent growing conditions. In northwestern South America, *P. mollissima* is commonly grown.

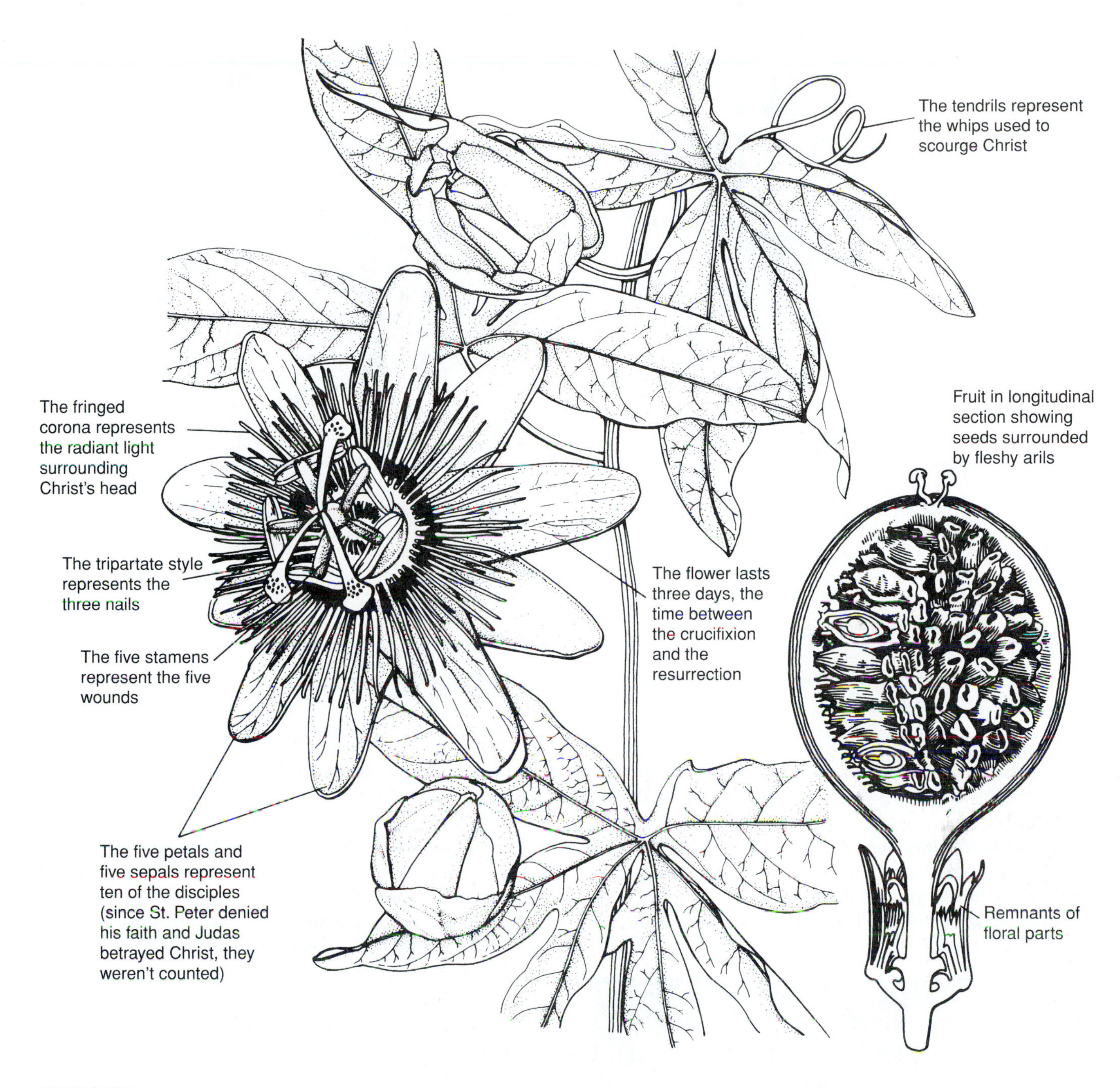

FIGURE 4.30

The exotic floral morphology of the passion fruit led to its association with the crucifixion of Christ.

(Section of fruit after Baillon; vine and flower from nature.)

The parts of passion fruits that are eaten or pressed for juice are the arils surrounding the seeds. Each fruit has a tough pericarp surrounding numerous seeds, all of which are embedded in fleshy, aromatic red-yellow arils. Because the plants are vines, plantations of passion fruits resemble vineyards. The perennial plants are evenly spaced, and the branches are trained along wires strung a meter or so above the ground.

Malay apples, *Syzygium malaccense* (Myrtaceae), and rose apples (*S. jambos*) are popular West Indian fruits. Both species, introduced from southeastern Asia, have a tart, fresh, crisp flavor. The fruits are pear-shaped and are either red or purple on the outside.

Another myrtaceous fruit is the guava (*Psidium guajava*) (Fig. 4.31), which has spread from its native home in Central America to the Old World tropics. The pear-shaped fruits of the guava are yellow when ripe and contain more vitamin C than do most citrus fruits. Because guavas have a distinctly pungent taste when eaten raw, they are most commonly stewed or processed into jams, jellies, or pastes.

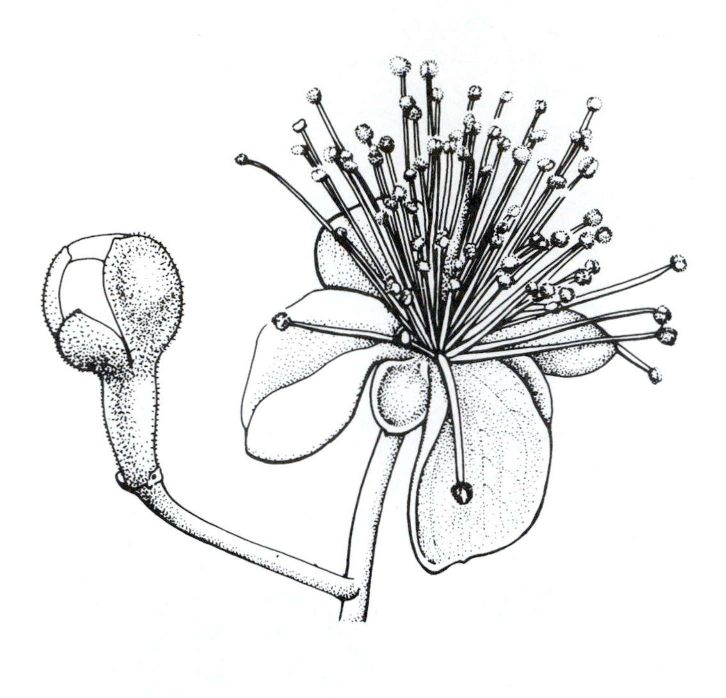

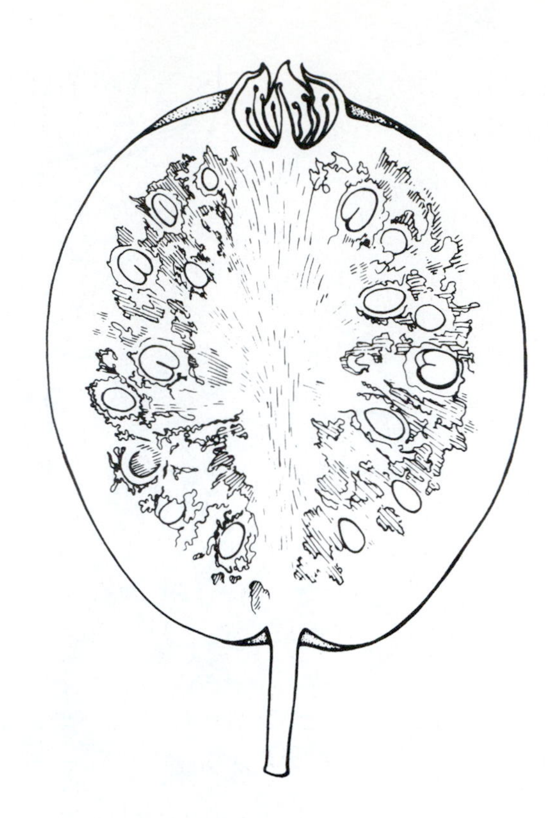

FIGURE 4.31
Because they contain many small seeds and tend to be bitter, guavas are usually processed into jams and jellies.
(Flower from nature; section after Baillon.)

Sapote is a name given to an assemblage of species, the most common of which belong to the Sapotaceae. Two of these, the mammey sapote (*Pouteria sapota*) and the green sapote (*P. viridis*), are native American species that produce fruits about 5 to 20 cm (2 to 8 in.) long with one to four seeds. The edible portion is the firm reddish mesocarp surrounding the seed clusters. These two species are related to the more common sapodilla (*Manilkara zapota*), the fruit of the tree from which we obtain latex for chewing gum (Chapter 10).

FIGURE 4.32
The combination of a repulsive odor and exquisite flavor have led to the controversial reputation of durians.
(Photo by Todd Barkman.)

Durians (*Durio zibethinus*) (Fig. 4.32), members of the Bombacaceae, or baobab family, are known for tasting like heaven and smelling like hell. The fruits have an odor so offensive that several countries (e.g., Malaysia) have banned the fruits in public places such as airplanes and hotel rooms. Despite their odor, durians are considered the "king of all fruits" in many parts of Southeast Asia.

The source of the odor, which has been described as a mixture of strong cheese, rotting onions, and turpentine, appears to be indole compounds that have been implicated in causing some physiological reactions such as an increase in heart rate. Individual fruits are football-sized and so spiny that they need to be handled with care. Inside the tough pericarp are numerous seeds, each enclosed in a yellow or pinkish aril. The flesh itself is sweet and custardlike in consistency.

Tropical Nuts

Although it seems that most of our common nuts are produced by trees of temperate regions (Chapter 3), several important nuts come from warm subtropical or tropical areas. One of the most expensive and, to many people, the most delicious of tropical nuts is the macadamia

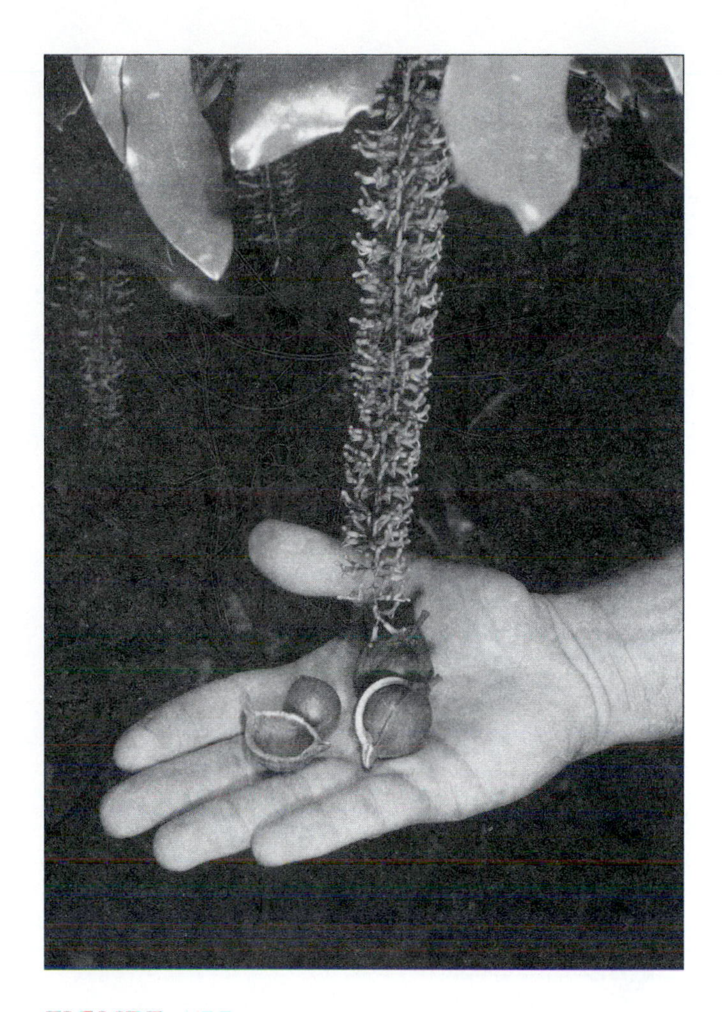

FIGURE 4.33
Macadamia nuts and flowers.
(Corbis photo by George Lepp.)

(*Macadamia integrifolia*) (Fig. 4.33). Macadamias are members of the Proteaceae, a family known primarily for its exotic inflorescences (see Fig. 17.29). Although native to Australia, macadamias are now produced primarily in Hawaii, where they have been cultivated as ornamentals since the 1800s. Hawaii provides well-drained soils and abundant rainfall, which are ideal for macadamia growth.

Commercial macadamia fruit production was initially hindered by the sporadic fruiting of macadamia and the extremely hard seed coats that made neat shelling of the nuts almost impossible. The problem of freeing macadamia nuts successfully from their shells was eased somewhat when it was discovered that drying the fruits before cracking facilitates the operation. Nevertheless, a hefty pressure of 300 lb per square inch must still be exerted to crack open the shells. Although macadamia nut production is still limited by climatic conditions, asynchronous fruiting patterns, and tenacious shells, Hawaii now enjoys a 6-month (July through January) season during which orchards yield an abundant harvest of the sweet, buttery nuts.

The cashew, another popular nut in the United States, is native to forests of northeastern South America. Cashews (*Anacardium occidentale,* Anacardiaceae) are related to poison ivy and mangoes and, like them, contain toxic compounds in various parts of the plant. The latex in the hard pericarp coat, for example, is toxic, but the seed itself is edible. Cashews are produced on curious fruits. The "nut" is the embryo borne inside a hard shell. Once the fruit reaches its full size, but while it is still immature, the receptacle begins to swell. Eventually the receptacle becomes large, fleshy, and red (Fig. 4.34). Presumably, the fleshy portions attract animals that disperse the seeds. Because of the irritating compounds in the fruit wall, South American natives generally

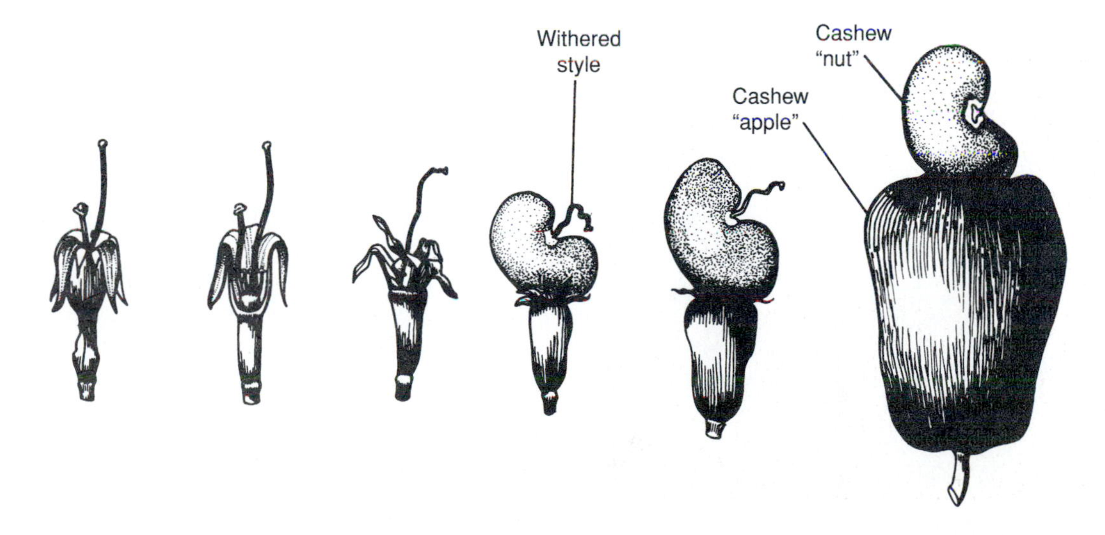

FIGURE 4.34
The development of a cashew. One of the stamens surrounding the ovary is characteristically larger than the others. The cashew "nut" develops from the fertilized ovule inside the ovary, and the "cashew apple" develops from the stem and receptacle.
(Modified after W. H. Brown. 1935. *The Plant Kingdom.* Boston, Ginn.)

eat only the fleshy portion. In Brazil, the fleshy fruit, called the apple, is occasionally crushed and fermented into a wine called *kaju.*

Brazil nuts (*Bertholletia excelsa* and other species of the Lecythidaceae) are also native to the South American tropics and are borne in a peculiar fashion. The flowers have a striking appearance, and the fruits are large woody spheres weighing 1 to 2.5 kg (2 to 5 lb) each (Fig. 4.35). Within the massive fruits are 12 to 24 wedge-shaped seeds arranged in rings. Each seed is encased in a stony seed coat. Each tall rain forest tree can produce over 300 capsules, called pods. Needless to say, standing under a Brazil nut tree when the fruits fall can be dangerous, and pickers are often injured during the harvest season. Almost all the world's supply of the nuts is collected from wild trees, a rare occurrence in food production.

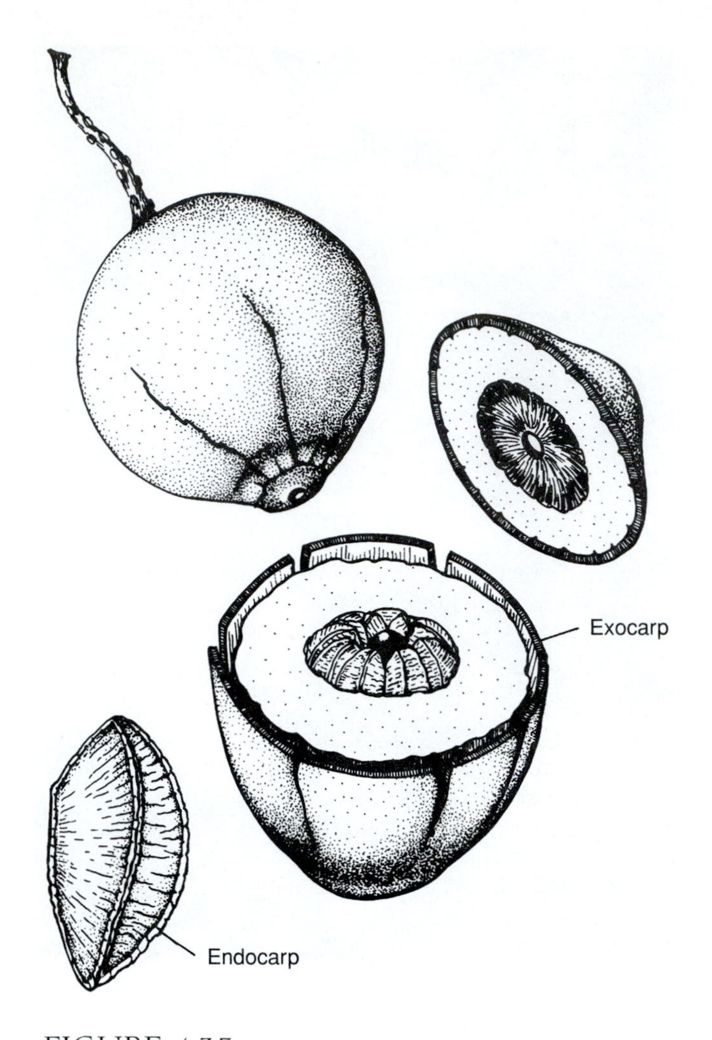

FIGURE 4.35
Brazil nuts are the seeds of species of large tropical forest trees. Between 12 and 20 seeds develop within the 1- to 2.5-kg fruits. The natives in many regions of Brazil call the fruits "monkey pods" because the inquisitive primates apparently get their hands stuck inside the hard exocarp when they reach inside fruits that have been gnawed open by rodents.

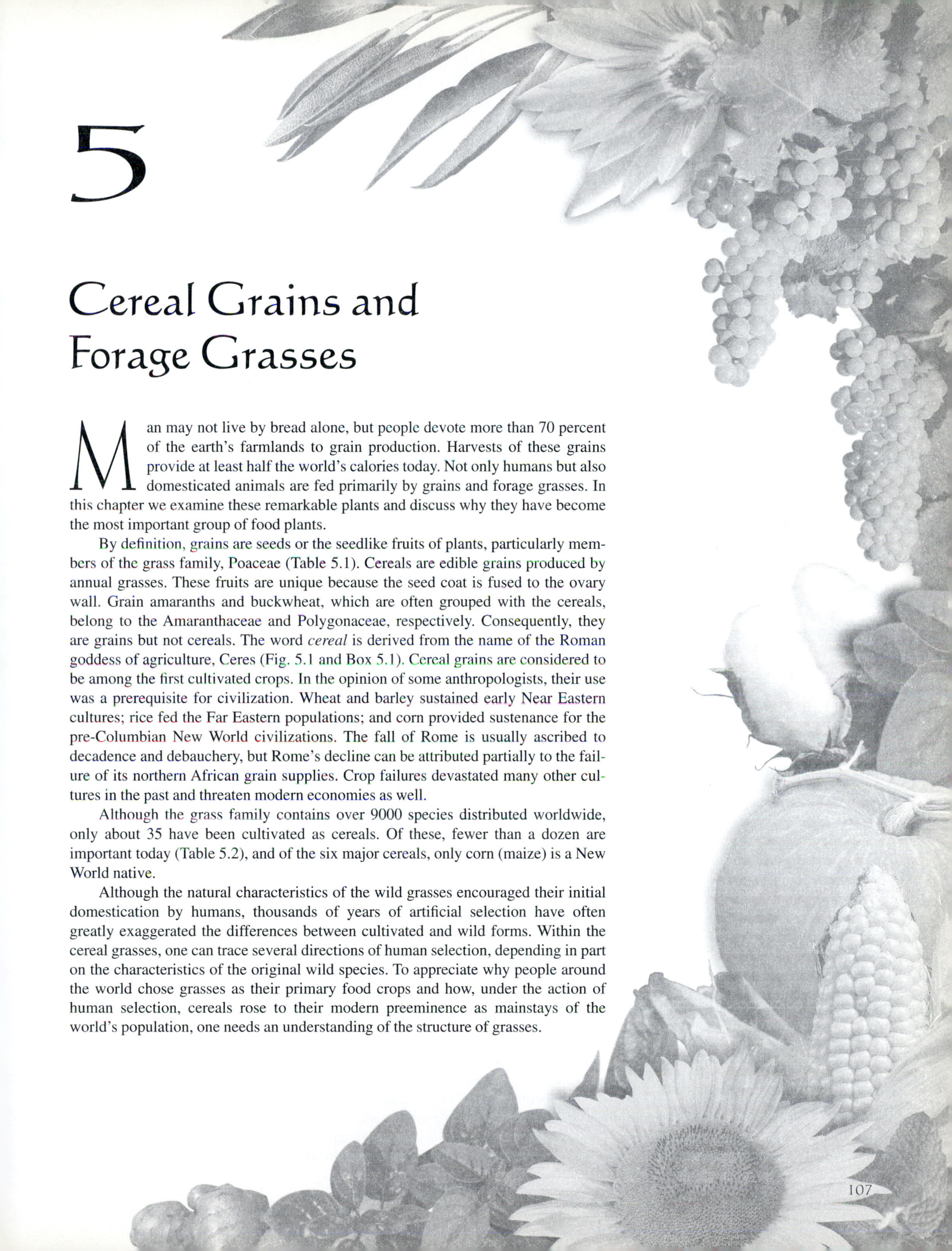

5

Cereal Grains and Forage Grasses

Man may not live by bread alone, but people devote more than 70 percent of the earth's farmlands to grain production. Harvests of these grains provide at least half the world's calories today. Not only humans but also domesticated animals are fed primarily by grains and forage grasses. In this chapter we examine these remarkable plants and discuss why they have become the most important group of food plants.

By definition, grains are seeds or the seedlike fruits of plants, particularly members of the grass family, Poaceae (Table 5.1). Cereals are edible grains produced by annual grasses. These fruits are unique because the seed coat is fused to the ovary wall. Grain amaranths and buckwheat, which are often grouped with the cereals, belong to the Amaranthaceae and Polygonaceae, respectively. Consequently, they are grains but not cereals. The word *cereal* is derived from the name of the Roman goddess of agriculture, Ceres (Fig. 5.1 and Box 5.1). Cereal grains are considered to be among the first cultivated crops. In the opinion of some anthropologists, their use was a prerequisite for civilization. Wheat and barley sustained early Near Eastern cultures; rice fed the Far Eastern populations; and corn provided sustenance for the pre-Columbian New World civilizations. The fall of Rome is usually ascribed to decadence and debauchery, but Rome's decline can be attributed partially to the failure of its northern African grain supplies. Crop failures devastated many other cultures in the past and threaten modern economies as well.

Although the grass family contains over 9000 species distributed worldwide, only about 35 have been cultivated as cereals. Of these, fewer than a dozen are important today (Table 5.2), and of the six major cereals, only corn (maize) is a New World native.

Although the natural characteristics of the wild grasses encouraged their initial domestication by humans, thousands of years of artificial selection have often greatly exaggerated the differences between cultivated and wild forms. Within the cereal grasses, one can trace several directions of human selection, depending in part on the characteristics of the original wild species. To appreciate why people around the world chose grasses as their primary food crops and how, under the action of human selection, cereals rose to their modern preeminence as mainstays of the world's population, one needs an understanding of the structure of grasses.

TABLE 5.1 Grains [Poaceae (Graminae)] Discussed in Chapter 5*

COMMON NAME	SPECIES	NATIVE REGION
Barley	*Hordeum vulgare*	Southwestern Asia
Corn	*Zea mays*	Mexico
Goat grass	*Triticum speltoides*	Southwestern Asia
	T. longissima	Southwestern Asia
Millet		
Finger	*Eleusine coracana*	Africa
Pearl	*Pennisetum glaucum*	Africa
Oats	*Avena sativa*	Europe
Rice		
African	*Oryza glaberrima*	Africa
Common	*O. sativa*	Southeastern Asia
Texas wild	*Zizania texana*	North America
Wild	*Z. palustris*	North America
Rye	*Secale cereale*	Near East
Sorghum	*Sorghum bicolor*	Tropical Africa
Teosinte	*Zea mays* subsp. *mexicana*	Mexico
	Z. mays subsp. *parviglumis*	
Tripsacum spp.	*Tripsacum* spp.	Tropical America
Triticale	× *Triticosecale*	Human-made
Wheat		
Bread	*Triticum aestivum*	Southwestern Asia
Einkorn	*T. monococcum*	Southwestern Asia
Emmer	*T. turgidum* var. *dioccum*	Near East

*Most names mentioned in Tables 5.4 and 5.5 are not included here.

The Grass Plant

Grasses are ubiquitous. In nature, the matted roots of grasses hold down soils from the Arctic tundra to prairies, and tropical savannas to football fields, while planted grasses carpet manicured suburban lawns. However, perhaps because of their green, inconspicuous nature, one hardly seems to notice grasses, even though they are constantly underfoot. Once examined closely, however, this group of monocotyledons proves to be elegant in design and different in many respects from the "generalized" plant outlined in Chapter 1.

Starting at the roots, one finds that grasses have a fibrous root system (Fig. 5.2) and lack a dominant taproot. Grasses also have a pattern of growth different from that of most other kinds of plant (Fig. 5.2). Some grasses have one or a few major stems called **culms** that produce sequential branches as they grow. Others branch only at the base, producing more of less erect stems called **tillers** (see Fig. 5.5). Grass leaves consist of a lower part, the sheath, which encircles the stem, and an upper flattened divergent part called the **blade.** Where the base of the sheath attaches to the stem (the **node**), the stem is solid. The section between two nodes, or the internode, is hollow.

Vegetative reproduction is common in perennial grasses. Many species have **rhizomes,** underground creeping stems that grow outward as the grass clump ages, forming clones. Still others produce horizontal branches on the surface of the soil (**stolons**) (Figs. 5.2 and 1.11e), often rooting at the nodes as they go. This kind of vegetative growth permits the rapid spread of grasses that are inserted into lawns in the form of starter plugs.

The floral structures of grasses are even more highly specialized than the stems and leaves. Most grass flowers contain both male and female parts. Corn and wild rice, which have unisexual flowers borne on different parts of the plant, are notable exceptions. Grass flowers are highly reduced and are borne in compound inflorescences consisting of few to many **spikelets** (Figs. 5.2 and 5.3). Each spikelet is a flowering branch with one to several florets. A **floret** (Fig. 5.2) is composed of two bracts that enclose a minute flower. The brace closest to the floret is the **palea;** the next one is the **lemma.** The tip of the lemma can extend beyond the body of the bract, forming a long structure called an **awn** that aids in dispersal. Florets typically have a superior ovary topped by two feathery stigmas, three stamens (sometimes six) on long filaments, and two microscopic, scale-like remnants of the petals that swell at flowering to force open the bracts and expose the stamens and stigma.

At the base of each spikelet (or cluster of florets) are additional bracts called **glumes** (Fig. 5.15). Once fertilized, the single ovary matures into a fruit in which the seed coat and the ovary wall (pericarp) are fused into a complex structure known in lay terms as the **bran.** The different components of the bran

Box 5.1

From Cereals to Cereal

If they were asked to define *cereal,* chances are that most Americans would say a breakfast food eaten in a bowl with milk. The rise of breakfast cereal to this prominent position in the minds of millions of people is a testament to the power of American big business. The story of the Kellogg Company is a classic rags-to-riches saga of a hardworking, enterprising, religious young man who became one of the country's most powerful corporate magnates.

To many people the name Kellogg is synonymous with Battle Creek, Michigan, and the company's founder, William Keith (W. K.) Kellogg. However, W. K. was not the only Kellogg involved in the development of the breakfast cereal industry. Without the initial inspiration of his brother, Dr. John Harvey Kellogg, W. K. would never have launched the company at all. The Kellogg brothers were raised as Seventh-Day Adventists, a sect that advocates a healthy diet that prohibits alcohol, coffee, tea, and meat. Harvey Kellogg grew up to become a physician, returning to Battle Creek to work in the Adventists' sanitarium, where he eventually assumed the directorship. In this position, he was able to implement his ideas on sensible eating habits.

The timing for his dietary recommendations could not have been better, since by the end of the nineteenth century the majority of the U.S. population had moved from rural to urban areas. Despite that demographic shift, the diets of many of those city dwellers still consisted of the heavy, high-caloric meals that had been necessary when working on a farm. City dwellers on such country diets complained about indigestion and tended to be overweight. Kellogg prescribed low-fat diets consisting of grains, nuts, and fruit. Harvey began to experiment with natural foods to overcome what those diets lacked in flavor. He invented (or reinvented) peanut butter, granola, and a caramel cereal beverage to take the place of coffee.

While Harvey was advancing his career in Battle Creek, W. K. was trying and failing at a number of occupations. Recently married and unemployed, he asked his brother for a job. Harvey appointed him steward of the sanitarium in 1880, and for 15 years the two brothers labored together to expand the business and develop new dishes. After inventing shredded wheat, they tried flattening dry, whole grains between metal rollers to produce a product that was both light and tasty. All their attempts failed until one day W. K. spied some damp whole wheat grains that had begun to mold. Instead of tossing them out, he decided to run them through the rollers. To his surprise, each grain came off the metal cylinders as an irregular, flat flake. When baked, the flakes were light and crispy. By accident, the brothers had discovered the process, now known as "tempering," by which grains absorb enough moisture to become pliant. The Kelloggs began to produce this new cereal (minus the mold), which they called Granose. They served it to patients at the sanitarium, who became so fond of the flakes that they began to order boxes sent to their homes after they were discharged. By 1900 the Kelloggs were producing over 100,000 pounds of flakes a year. Eager entrepreneurs began to copy the process, and soon numerous companies were making wheat flakes. Corn flakes were born in 1902, when the brothers developed a process for removing the bran and germ from corn kernels and then tempering and flaking them.

Four years later W. K. decided to strike out on his own, borrowing money and buying a majority of shares in the manufacturing company. By 1909 W. K. was producing a million boxes of corn flakes annually. Other companies tried to copy his shredded wheat and corn flakes, but W. K., a master of advertising, developed the corn flakes girl, the cereal box with his name in script, and the slogan "Only the real product bears this signature." Later, Kellogg introduced a cereal box that was fun to read and, soon thereafter, the idea of prizes and games inside the packages.

Most imitators of Kellogg's products soon failed. C. W. Post was one who did not. Post had been a patient at the sanitarium and knew all about the development of the toasted flakes. Although he left without being cured, he soon abandoned his wheelchair and plunged into business. Instant Postum was his imitation of Kellogg's caramel coffee substitute, Post Toasties were his rejoinder to Corn Flakes, and Grape Nuts were his counterpart to Granose. Post today is one of the main subsidiaries of General Mills, but it has always trailed behind its Kellogg model. Almost 100 years after its founding, Kellogg still controls about 40 percent of the U.S. breakfast cereal market and 80 percent of the international trade. The company claims that it has remained strong because it has followed its founder's original intention of providing nourishing, grain-based foods that fit the lifestyles of Americans. Modern concerns about the health risks of traditional bacon and eggs breakfast and recognition of the benefits of fiber echo Harvey Kellogg's original reasons for inventing breakfast cereals and account for the recent surge in their popularity.

TABLE 5.2 1999 World Production of Grains

COMMON NAME	TOP 5 COUNTRIES	1000 METRIC TONS	TOP CONTINENTS	1000 METRIC TONS	WORLD 1000 METRIC TONS
Barley	Germany	13,342	Europe	80,146	133,567
	Russian Fed.	11,250	North and Central America	19,601	
	Canada	12,968	Asia	23,094	
	France	9,542	Oceania	5,247	
	United Kingdom	6,730	Africa	4,147	
Maize	United States	242,254	North and Central America	271,894	604,930
	China	129,297	Asia	169,442	
	Brazil	32,503	Europe	71,154	
	Mexico	18,492	South America	52,486	
	France	15,160	Africa	39,453	
Millets (combined)	India	10,300	Asia	14,973	28,621
	Nigeria	5,457	Africa	12,566	
	China	3,800	Europe	793	
	Niger	1,850	North and Central America	200	
	Burkina Faso	972	South America	57	
Oats	Russian Fed.	5,315	Europe	16,551	33,900
	Canada	3,656	North and Central America	5,903	
	United States	2,126	Oceania	1,877	
	Poland	1,501	Asia	1,356	
	Germany	1,347	Australia	1,222	
	Australia	1,222	South America	958	
Rice	China	200,000	Asia	533,264	588,523
	India	127,600	South America	20,995	
	Indonesia	49,500	Africa	17,811	
	Bangladesh	29,856	North and Central America	11,790	
	Vietnam	28,116	Europe	3,234	
Rye	Poland	5,599	Europe	18,377	29,212
	Russian Fed.	3,752	Asia	1,166	
	Germany	4,319	North and Central America	569	
	Ukraine	1,200	South America	61	
	Belarus	1,000	Africa	30	
Sorghum	United States	15,151	North and Central America	21,940	68,074
	India	11,000	Africa	21,070	
	Nigeria	8,443	Asia	17,853	
	China	5,857	South America	4,936	
	Mexico	6,296	Oceania	1,667	
Wheat	China	144,400	Asia	261,702	563,649
	United States	62,812	Europe	175,000	
	India	71,007	North and Central America	91,917	
	France	37,050	Oceania	15,172	
	Russian Fed.	31,000	Africa	15,563	
			South America	14,589	

Source: FAOSTAT Database Gateway. On line at http//:apps.fao.org/ (under "All databases" then "Crops, Primary").

FIGURE 5.1
Because of the importance of grains in international trade, an art deco interpretation of Ceres, the goddess of agriculture, was placed on the top of the Chicago Board of Trade building.
(Courtesy of Chicago Board of Trade.)

can be distinguished only microscopically (Fig. 5.4). Inside the bran is a layer of cells known as the **aleurone layer.** The cells of this layer are typically rich in protein and fats. The aleurone layer plays a key role in the germination of grain seeds because it secretes the enzymes that break down the endosperm starch into sugars that are absorbed by the germinating embryo (the **germ**).

In contrast to legumes (Chapter 6), which store food in the cotyledons, grasses store energy for germination and seedling establishment in the endosperm (see Fig. 1.24). This tissue contains predominantly starch, although it can also contain protein and traces of fat. The embryo itself contains high concentrations of proteins, fats, and vitamins. Thus, when eaten whole, grains contain carbohydrates, proteins, fats, and some vitamins (Table 5.3). Grains can have the bran, the aleurone layer, and the germ removed, leaving the starchy endosperm, which sometimes contains small amounts of protein. The process of removing the layers surrounding the endosperm is called **polishing** or **pearling.**

A high fruit set and abundant, highly caloric endosperm make grains particularly attractive as crops. Many grasses that grow in seasonally dry climates are self-compatible (able to self-fertilize) annuals that put all their energy into fruits that survive the dry season in the soil. Humans are interested in the highest possible yields per unit area of cultivation, and grains are among the species with the maximum fruit production per unit plant weight. Modern yields greatly exceed those of wild grains because people have made use of the natural characteristics of grass species to select for greater yields.

Directions of Selection in Grains

Humans have modified the habit of cereal grasses in two ways to increase the total amount of grain that can be collected easily at one time. First, in grasses that ancestrally had a tillering habit, people have selected for erect growth and synchrony of tiller formation. Since each tiller is terminated by an inflorescence (Fig. 5.5), tillers that begin growth at the same time tend to mature seeds at the same time. By replacing sporadic tiller formation with simultaneous tiller initiation, early farmers ensured that a single harvesting of grains such as wheat, rye, and barley would collect almost all the seeds produced.

In grains that have branched stems, the lateral branches as well as the main stem usually bear one or more inflorescences. Since the plant grows upward, lateral branches are produced sequentially from the base to the tip, preventing the lateral shoots and the main shoot from maturing fruits at the same time. In tropical cereal crops with robust branching culms, such as sorghum, there has been a selective trend toward the elimination of branching and the production of a strong central stem with all of the fruit production concentrated in one or a few inflorescences. Modern cultivars of corn, sorghum, and pearl millet all have large, single unbranched stems, whereas their ancestors or early cultivars were branched.

In more recent times there has also been selection for short plant stature. A major problem with many grains, particularly when they are grown in areas much wetter than their native habitats, is **lodging.** This refers to the matting together of tall stalks that have been bent over by wind or rain. The matted stalks cannot disentangle and return to an upright position. Once lodging has occurred, the stalks generally rot or become impossible to harvest with combines. The ancestral home of many cereal grains of the Near East is seasonally dry. When humans tried to grow these grains in wet areas, lodging became a serious problem. Intensive selection programs for increased stalk strength and for

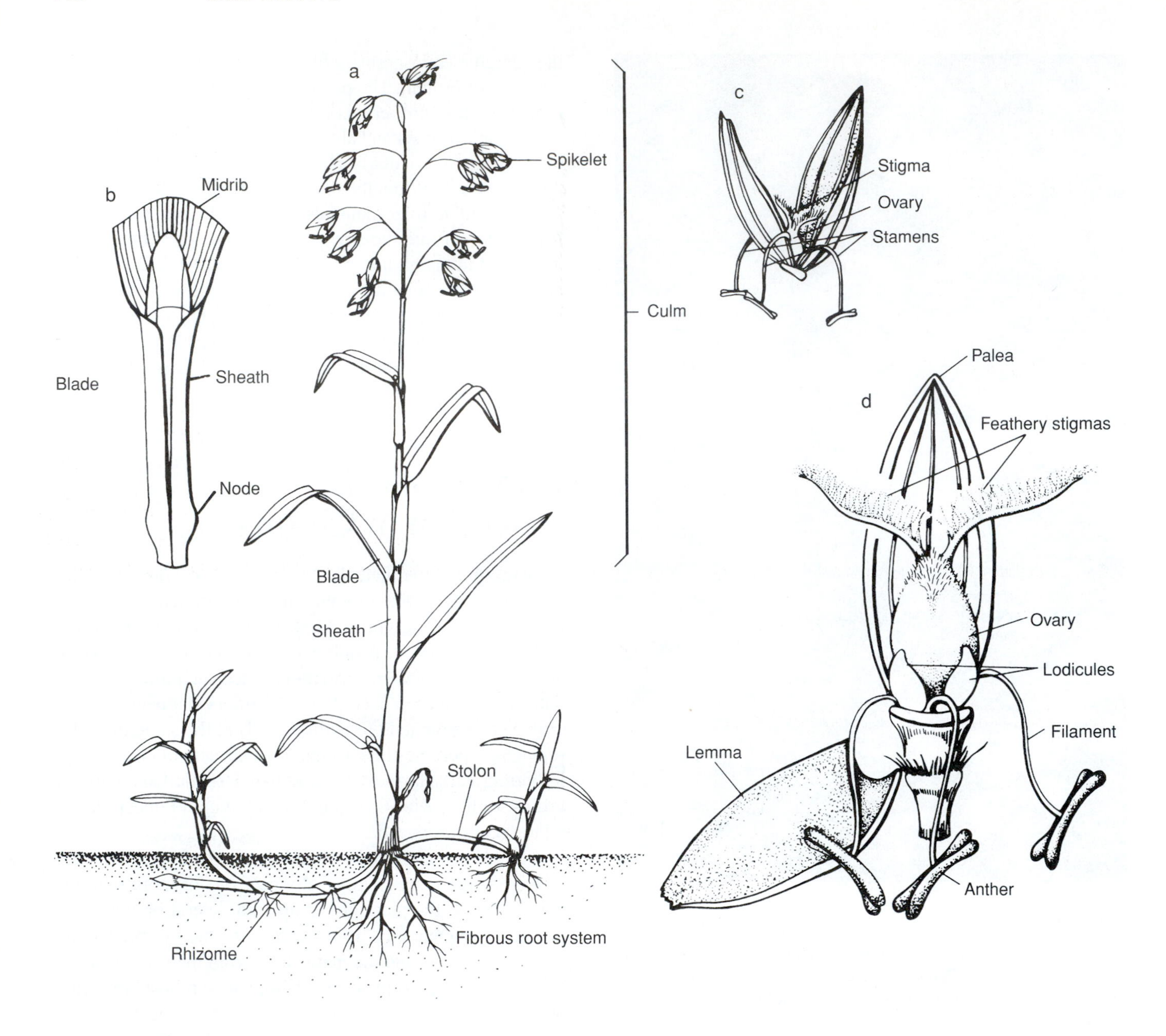

FIGURE 5.2

A generalized grass plant with perfect flowers, a main culm, and the ability to reproduce by means of stolons or rhizomes. (a) entire plant; (b) close-up of a sheathing leaf base; (c) floret enclosed in the palea and lemma; (d) an opened floret with the lemma removed, showing reduced petals (lodicules), three anthers, and two feathery stigmas.

dwarf varieties of wheat, rice, and even sorghum have allowed a dramatic increase in the cultivation of these cereals, primarily in the tropics.

Under the influence of human selection, the inflorescences of the cultivated cereals have undergone even more drastic changes that have the vegetative structures. One of the initial changes in the cultivated cereals was the retention of the grains on an inflorescence once they matured. All flowering plants face the problem of dispersing their seeds from the rooted parent to areas suitable for germination. In many grasses the inflorescences become brittle at maturity and shatter when jostled by wind, pressure from passing animals, or neighboring plants. In this process, which is called **shattering,** the spikelets are flung away from the parent. From the human point of view, shattering is an undesirable character because the seeds scatter in all directions when harvesting is attempted. Any mutation or series of mutations that prevented shattering would have been selected because seeds of the nonshattering type would have been collected preferentially and subsequently used as seed sources for the next year's planting. Since nonshattering is a genetically controlled, single-gene character, nonshattering plants became the predominant type used in cultivation after a short time.

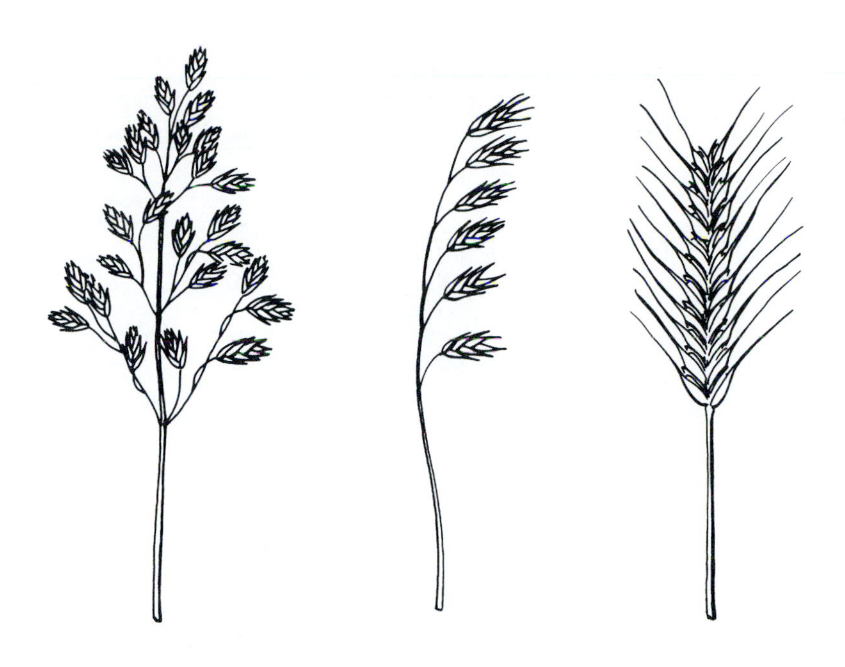

FIGURE 5.3
Grasses usually have compound inflorescences with spikelets arranged in a panicle (left), raceme (middle), or spike (right).

Whether or not the trait was intentionally selected for by humans, all the major cereal crops now have nonshattering inflorescences.

Another major change in the morphology of Old World grains such as barley, rye, and oats that occurred during their domestication is related to the ease of fruit separation from the inflorescences once the stalks have been harvested. Grains can separate from the stalks below the attachment of the bracts, or the fruits can fall free from the enclosing lemmas, paleas, and glumes (bracts or chaff). Fruits of primitive and wild grasses remain permanently enclosed in the floret or entire spikelet, making them difficult to use as food. These are known as hulled grains. The activity of removing the fruits from the bracts is known as **threshing** (Fig. 5.6). Threshing was formerly done by hand and involved beating the stalks. Hand threshing is an arduous job, and any mutation that lessened the task would have been seized upon by humans. Domesticated cereals have grains in which the bases of the palea, lemma,

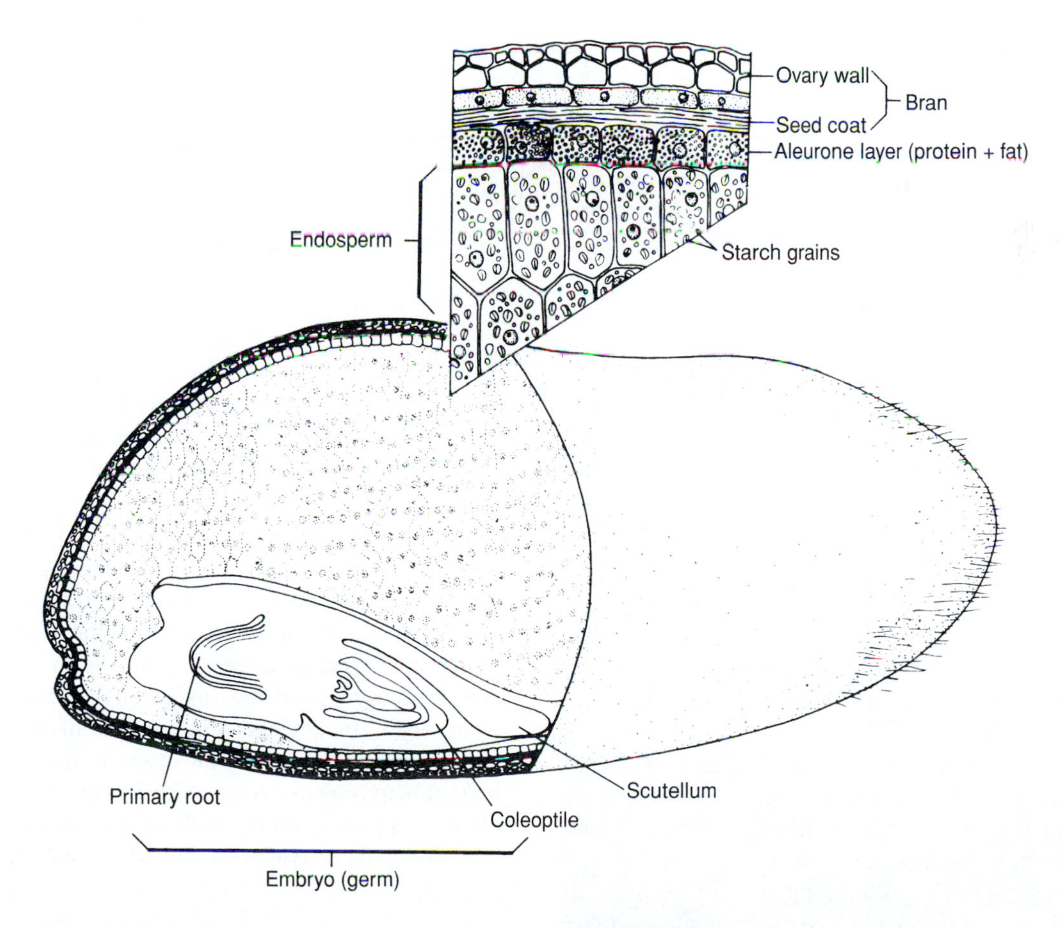

FIGURE 5.4
Cross section of a generalized grain. Note that the ovary wall and the seed coat are distinguished here even though they are fused and are not discernible to the naked eye.

TABLE 5.3 Nutritional Composition of Grains, Based on a 100-g Edible Portion

GRAIN	WATER, %	CALORIES	PROTEIN, g	FAT, g	TOTAL, g
Barley, pearled	11.1	349	8.2	1.0	78.8
Corn, field	13.8	348	8.9	3.9	72.2
Oats, rolled	8.3	390	14.2	7.4	68.2
Rice, brown	12.0	360	7.5	1.9	77.4
Sorghum	11.0	332	11.0	3.3	73.0
Wheat, whole grain					
Durum	13.0	332	12.7	2.5	70.1
Hard red winter	12.5	330	12.3	1.8	71.7

Source: Data from *Composition of Foods*. Agricultural Handbook No. 8. Rev. 1963. Washington, D.C., USDA.

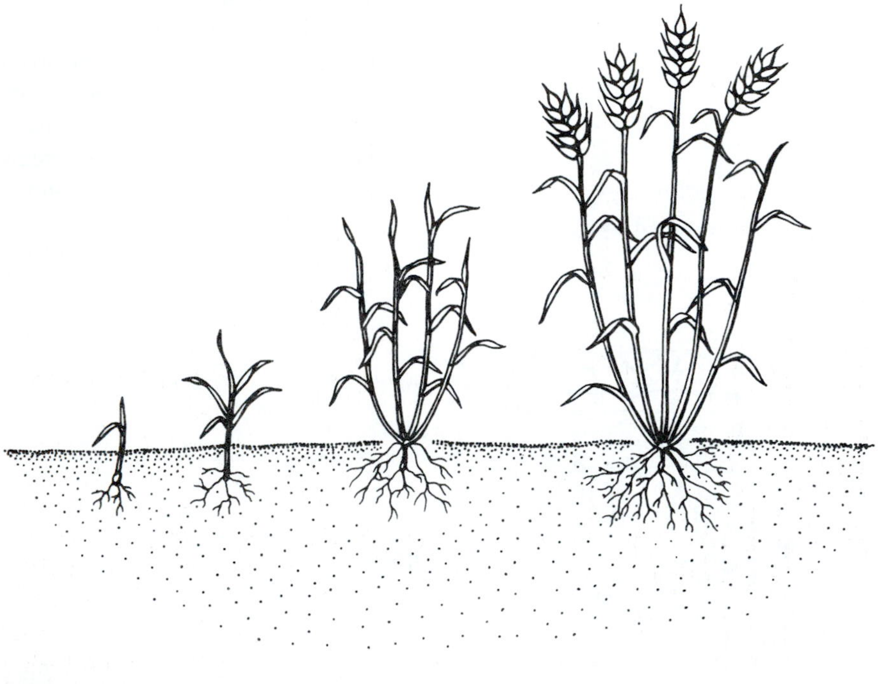

FIGURE 5.5
In tillering grains that have been selected for simultaneous filler formation, all the inflorescences tend to mature at the same time. Each of the separate stems arising from the base of the plant is a tiller.

and glumes break easily, allowing easy removal of the fruits. The appearance of these "free-threshing" or naked types in the archaeological record documents early selection for mutations that facilitated the separation of the fruits from the chaff. Free-threshing types solved the problem of developing a grain that would not shatter before harvest but that would separate easily from the enclosing bracts once it was harvested. Hulled versus free-threshing is a simple Mendelian trait.

Light fragments of broken chaff that are mixed with the threshed grain are eliminated by sifting and **winnowing** by hand or on a sieve, an operation that involves throwing the grain-chaff mixture into the air (Fig. 5.7). The lightweight chaff fragments are blown away by the wind, while the heavier grain falls back down onto the ground or sieve; seeds smaller than the grain fall through the sieve. In industrialized countries, modern machines called **combines** thresh and winnow grains as they are harvested in the field, but threshing and winnowing by hand are still practiced in many parts of the world.

Separation of the fruit wall from the seeds is even more difficult than separation of the grains from the inflorescence and chaff because of the fusion of the seed coat with the ovary wall. In ancient times this fruit wall (bran) was removed by rubbing the grains (wheat, rice, or occasionally barley) between two stones with enough pressure to abrade off the unwanted layers without crushing the seeds (Fig. 5.8). Today these grains are hulled, dehusked, and polished

FIGURE 5.6
Mother with baby threshing wheat in Nigeria.
(Photo courtesy of Corbis.)

FIGURE 5.7
This Egyptian wall frieze from about 1450 B.C. shows workers carrying the grain in baskets (lower left) to the threshing floor, where it is spread out (lower right) for cattle to trample (upper right) to separate the seeds from the husks. The grain is cleaned by subsequent winnowing (upper left) in shallow trays. Note the irregular sequence of the figures.

FIGURE 5.8
A Peruvian girl grinding grain on a stone floor.
(Photo by Phil Schermeister, Corbis.)

(pearled) to various degrees by machines. The complete removal of the fused fruit and seed coat (which generally also pulls away the embryo) is very common for wheat and rice and leads to the production of white flour and white rice. Removal of the germ can serve a practical purpose because degermed grains have a comparatively long shelf life. Grains that are not degermed tend to turn rancid over time because the oils in the embryo become oxidized, particularly if the grains are milled or crushed. **Groats** are grains, often broken, from which the fruit wall has been removed.

Additional modifications of grass inflorescences during domestication include an increase in individual grain size, restoration of fertility to sterile florets, a reduction in awns, selection for rapid and synchronous germination of seeds, changes in the kinds of storage products in the endosperm, and increased tiller production. When we discuss each major grain crop, we will point out the particularly important kinds of artificial selection to which it has been subjected.

Major Grain Crops

Barley

On the basis of archaeological evidence, barley (*Hordeum vulgare*) appears to have been one of the first cereals domesticated. Archaeological digs in the Fertile Crescent (primarily in Syria and Iraq) have uncovered charred domesticated barley kernels about 10,000 years old. Wild forms of barley are all two-rowed (Fig. 5.9). In the Fertile Crescent, the oldest undisputedly cultivated forms are also two-rowed. All species of barley have clusters of three spikelets joined at the base to form a unit on each side of the inflorescence. In wild and primitive domesticates only one of each cluster of

FIGURE 5.9
Two- and six-row barley.

three spikelets has a fertile floret. Consequently, wild varieties have only two grains per node, forming a two-rowed fruiting stalk. If all the spikelets develop a fertile floret, the fruiting stalk bears six rows of grain. During domestication, human selection led to the establishment of fertility in all three spikelets, so that by 6000 B.C. six-row forms appear in the archaeological record. Like nonshattering, restoration of fertility is determined by a single gene.

Initially, barley appears to have been ground and made into pastes. However, since raw starch, even mixed with water, is relatively unappetizing, the grain was probably first toasted by being heated on stones. The pastes might also have been boiled into porridges over coals in stone-lined pits. By Egyptian times barley had become such an integral part of the culture that it had its own hieroglyphic symbol. Similarly, the ancient Greeks depended primarily on barley and greatly expanded its uses. In addition to pastes, they produced variously flavored baked barley breads. They also learned to soak the grains in water before drying and grinding to make them more digestible. This process was the forerunner of malting (Chapter 14). The ground meal, once it was remixed with water, was highly susceptible to fermentation. Baking and beer making thus became part of the same operation. Barley flour was also commonly mixed with ground flaxseed and baked on griddles. Among the common items in the diet of ancient Egyptians and Greeks, therefore, were barley pastes, porridges, and breads complemented by figs, olives, and goat's milk and cheese.

Barley reigned as the king of the cereals until about four millennia ago, when it was supplanted by tetraploid wheats. More than half the barley grown in the United States today is used as feed for livestock, and about a quarter of the harvest is used in the brewing of beer and whiskey (Chapter 14). Modern selection has been directed toward synchronous tiller production, short stems, and tolerance to cold and saline soils. Selection has been for better feed quality and more protein in grains grown for feeding livestock. For grain used in malting, there has been selection for uniformity of endosperm composition.

Wheat

Wheat is now considered the "staff of life." Its initial cultivation was probably synchronous with or slightly later than that of barley (but perhaps later than rye), since the wild wheats that were taken into cultivation are native to the same regions as wild barley. Wheat was originally less popular than barley, and the grain on the staff of Demeter, the Greek goddess of agriculture, was barley, not wheat. By the time the Book of Genesis was written, wheat had become the dominant cereal in the eastern Mediterranean area and was the grain from which Adam and Eve were forced to make their own bread.

Wild and early domesticates of wheat were diploid ($2n = 14$) and have been classified as *Triticum monococcum.* At some point during the early cultivation of this species natural mutations occurred that suppressed shattering. These mutants were incorporated and changed the cultivated form into a type of wheat known as **einkorn,** "one grain" (Fig. 5.10a). Today these wheats are cultivated only in a few relic areas. By 6000 B.C. and independently of human selection, an einkorn wheat and another species thought by some to be a wild goat grass, *T. speltoides,* but by others to be *T. longissima* or *T. searsii* hybridized. It is difficult to determine the species that hybridized with einkorn wheat because the chromosomes contributed by this ancestral species appear to have undergone many changes since the initial cross. The hybrids of the diploid species were presumably sterile, but when a doubling of the chromosomes occurred, a fertile tetraploid (*T. turgidum,* $2n = 28$) was formed. One variety of wheat, **emmer** (*T. turgidum* var. *dioccum*) (Fig. 5.10e), underwent a mutation that caused the bases of the bracts to collapse at maturity. This change made the separation of the fruit from the chaff relatively easy and resulted in the free-threshing type of wheat known as **durum** wheat (*T. turgidum* var. *durum*) (Fig. 5.10b). Perhaps even more important than the free-threshing nature of these grains was the change that polyploidy made in their ability to produce raised bread.

FIGURE 5.10
Wheat varieties. (a) Einkorn (*Triticum monococcum*, $n = 7$). (b) Durum (*T. turgidum* var. *durum*, $n = 14$). (c) Polish (*T. polonicum*, $n = 14$). (d) Poulard (*T. turgidum*). (e) Emmer (*T. turgidum* var. *dicoccum*). (f) Club (*T. aestivum* var. *aestivum*, $n = 21$). (g) Spelt (*T. aestivum* var. *spelta*).

Tetraploid wheat contains appreciable quantities of two proteins (gliadin and glutenin) that are known as **gluten.** Because of these proteins, flour from bread wheat becomes elastic when it is mixed with water and kneaded. If yeast is added to the mixture, the yeast cells carry out their normal process of fermentation, using the small amount of natural sugars in the flour. The yeast cells release carbon dioxide as a product of fermentation. This gas becomes trapped in the spongy protein mass, and the dough rises (Fig. 5.11). Baking "sets" the dough by drying the starch and denaturing the protein. While the dough is baking, the gas expands, forming many larger bubbles. The dough sets around the bubbles, solidifying the spongy bread matrix before the gas escapes. Cooking also kills the yeast cells.

Before selection for free-threshing wheats, wheat was commonly heated to facilitate threshing. Flour from heat-treated grain was useless for making raised bread because the heat denatured the proteins it contained and prevented the formation of the elastic spongy mass so important for raised bread. Wheat naturally has more gluten than do other grains and is the only grain capable of forming an airy dough if the proteins remain in their natural state. However, until free-threshing types appeared, the advantages of gluten were obscured. The formation of the tetraploid emmer wheat produced a free-threshing grain with an especially large amount of gluten. These features combined to produce wheat grains that were easily separated from the bracts (without heat) and capable of producing soft, raised bread. Durum wheats are known today as hard, or macaroni,

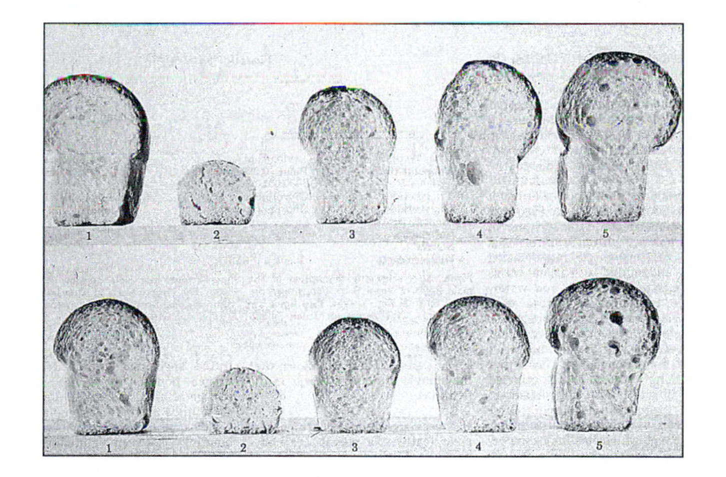

FIGURE 5.11
The differences in loaf sizes in this picture show the effect of gluten in producing an airy loaf of bread. In both rows the loaf on the far left is made from an average flour containing 13 percent gluten (protein). Next to the end loaf on the left are gluten-free loaves and, continuing to the right, loaves made with gluten-free flour to which 10 percent, 13 percent (as in the left loaf), and 16 percent protein has been added, respectively. In the top row are breads baked with flour containing normal amounts of fat. The breads in the bottom row are made with defatted flours. The differences in loaf heights (a measure of lightness) is obviously due to the amount of protein, not to the possession of fat or starch.
(Photo courtesy of Y. Pomeranz and K. F. Finney.)

wheats and are grown primarily in areas of low rainfall such as the Mediterranean region, central Russia, and the northern Great Plains. These wheats are now used primarily for pasta and noodle making because another event in the history of wheat produced a grain superior even to durum wheat for making bread.

Shortly after the appearance of agriculture, a hexaploid species arose that had six sets of chromosomes ($2n = 42$), four from a tetraploid emmer wheat and two from a diploid species ($2n = 14$) generally assumed to have been *T. tauschii* (*Aegilops squarrosa* in the older literature). Modern molecular studies of the DNA fragments of the chloroplasts (see Fig. 1.38) of wheat have even shown that emmer wheat was the female parent in this cross. The hexaploids that are also free-threshing are known botanically as *T. aestivum.* Hexaploid wheat has an endosperm that is especially high in protein and surpasses other wheats in bread making. There are now over 20,000 cultivars of bread wheat, including the hard red and club wheats (Fig. 5.10f). Spelt wheat (Fig. 5.10g) is a hulled variety of *T. aestivum.*

Wheat has not always been the major crop that it currently is in North America (Figs. 5.12 and 5.13). The Spanish brought the grain to the New World in 1520, and English colonists also tried to grow it in their early villages. Almost all these early agricultural attempts failed. Mennonite immigrants who settled in western Pennsylvania were one of the few groups to be successful with wheat in the eastern United States. As settlers moved westward, they carried seeds with them, but before the 1870s wheat growing in the Midwest was plagued with problems similar to those experienced in the East. Periods of drought, strong wind, and various pests took a heavy toll on wheat crops. In the 1870s immigrants began to pour out of central Russia, hoping to dodge compulsory military service or find religious freedom. Like many other groups of immigrants, they settled in regions climatically similar to the ones they had left behind, such as the Great Basin, Oklahoma, and parts of Texas. These areas have a strongly seasonal continental climate, with dry conditions during the time when wheat matures, and an annual rainfall of only 32 to 87 cm (13 to 35 in.). Most of the wheat brought by immigrants from Russia was "Turkey" wheat, although it is debated whether the variety originally came from Turkey or from the nearby Crimea. Because of its ability to grow well in dry regions with deep sandy soils, this particular variety made wheat farming feasible in the United States, and it subsequently led to the development of successful American varieties.

FIGURE 5.12

An agricultural engineer examining a grain sample collected from the grain flow sensor on his combine.

(Photo by Bruce Fritz, courtesy of Agricultural Research Service, USDA.)

FIGURE 5.13

Wheat harvest at the Great Plains Research Station in Colorado.

(Photo by Scott Bauer, courtesy of Agricultural Research Service, USDA.)

Two later developments contributed to the success of wheat growing in the United States. One development helped alleviate pest problems, and the other improved the quality of flour, the principal product obtained from wheat. Although selection improved many aspects of cultivated wheat, it failed until recent times to produce disease-resistant varieties. A major group of wheat pests are the wheat **rusts,** a group of fungi many of which infest and ruin cereal fruiting stalks. Beginning in 1907, the U.S. Department of Agriculture (USDA) and the Minnesota Agricultural Experiment Station began looking for ways to combat black stem rust caused by the fungus *Puccinia garminis.* Chemical pesticides proved not to be feasible, and destruction of the alternate host of the fungus, the introduced European barberry, *Berberis vulgaris* (Berberidaceae), did not significantly reduce infestation. Breeders finally learned that resistance could be conferred through a change in a single gene. Enormous breeding programs were begun and still continue. Agricultural agencies now collect and maintain stocks of mutants of this gene that can be used to breed resistance into current varieties.

Just as resistance to rust is conferred by a single wheat gene, the ability of the rust to overcome the resistance appears to be just as simply controlled. Consequently, rust mutations that permit the rust to infect previously immune wheat varieties continually arise, necessitating the incorporation of a different form of the resistance gene into the crop. Our understanding of the genetics of rust immunity and our ability to introduce alleles for resistance quickly into seed stocks have allowed wheat to become one of the major crops in the United States.

Until recently, modern genetic engineering had little impact on wheat and other cereals because the bacterium (*Agrobacterium tumifaciens*) used to introduce genes into dicotyledons does not infect monocots. By firing projectiles (see Fig. 1.35) coated with plasmids that contain desired genes into wheat embryos or callus tissue, researchers had their first success in introducing herbicide-resistant genes into wheat. With constantly improving technology, other innovations should soon follow.

The second important event that encouraged wheat production was the development of a modern industrial flour mill with steel rollers that replaced traditional millstones. Successive layers of coarse to fine mesh silk cloth allowed the rapid sifting of a fine white flour. With improved industrial milling facilities, white flour, traditionally associated with the rich, became readily available to all. The germ and the bran are now often sold as animal feed.

In addition to its widespread appeal, white flour has a longer shelf life than does brown (whole wheat) flour because it lacks the embryo oils that tend to become rancid over time. Graham flour, which contains the germ and the endosperm but not the bran, also becomes rancid quickly. Nevertheless, white flour is much less nutritious than whole wheat flour. It is lower in protein and fat and lacks the fiber of the bran and the vitamins of whole wheat flour. Most white flour sold in the United States is "enriched" with a mixture of vitamin B_1 (thiamine), niacin, and iron. Whole wheat or partially refined flour and specialty flours are now used increasingly in breads as consumers become aware of their need for the fiber, vitamins, and minerals afforded by whole grains. Wheat "berries" (soaked whole grains), sprouted wheat, cracked wheat, and wheat germ are also frequently added to the specialty breads.

Rye

It was previously thought that rye (*Secale cereale*), in contrast to wheat and barley, which were both deliberately brought into cultivation, began its association with people as a weed in wheat and barley fields. Instead of fighting the weed, people simply adopted it as a cultigen in cool regions. Consequently, rye is commonly referred to as a **secondary crop.** However, new findings of domesticated rye from Syria dated by accelerator mass spectrometry to be 13,000 years old suggest that it may have predated other grains by almost 3000 years. Nevertheless, it never assumed the importance of barley and wheat in the Near East. Rye, like barley and wheat, is native to southwestern Asia. Initial changes in rye paralleled those of the primary crops with which it grew. Beginning about 1800 B.C. its cultivation spread across Europe. By the eighteenth century, it was a major grain in Russia.

Rye is a hardy grain that is capable of germinating at temperatures as low as 1°C (34°F) and maturing when temperatures are as low as 12°C (55°F). Its roots can reach 2 m (over 6.5 ft) in length, allowing it to grow in relatively dry habitats. Rye has been called the "poor person's wheat" because it has historically been grown in northern Europe in areas where wheat does not grow well. German, Czech, Polish, and northern European black breads are a reflection of the widespread cultivation of rye in those regions. Unlike the protein in wheat, the protein in rye is useless for making a loaf of raised bread. Consequently, even black bread usually contains wheat as well as rye flour. In the United States "rye" breads are made with half wheat flour and half rye flour. Rye flour contains comparatively large amounts of the important amino acid lysine, but rye is cultivated less today than are most of the major cereals (except triticale) discussed in this chapter. Besides its use as a bread flour, rye is consumed as forage and feed, planted for erosion control, and fermented in the production of rye whiskey and Dutch gin. Because of its use as a forage crop, some rye varieties have been selected for winter growth and digestibility by livestock.

Triticale

Wheat and rye are classified in different genera but share a distant common ancestor. This kinship is reflected by the fact that they can be crossed (although with some difficulty) to produce fertile offspring. Although this potential for hybridization has been known for over 100 years, only since the late 1960s have plant breeders taken advantage of it to produce a fertile hybrid crop. The result, triticale, or *Triticosecale,* has been hailed as

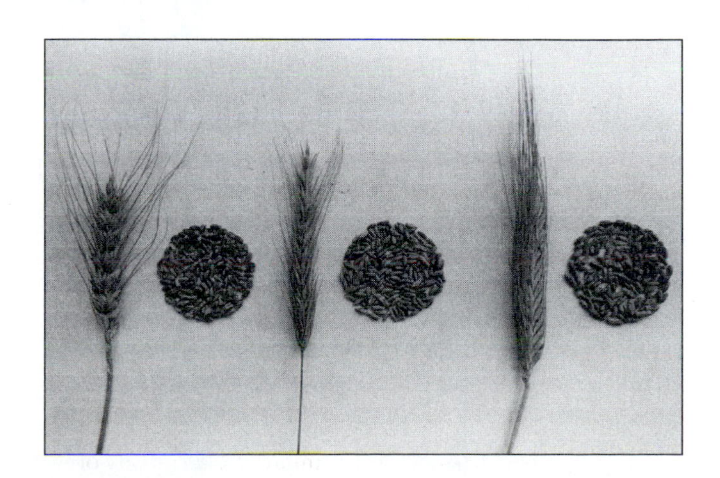

FIGURE 5.14

Wheat and rye are crossbred to produce triticale. This manufactured hybrid combines the high yields of wheat with the hardiness of rye. The grains of triticale are larger than those of either of its parents. (Left) wheat; (center) rye; (right) triticale.

the first truly human-made cereal (Fig. 5.14). Both hexaploid triticale, resulting from the doubling of the chromosome complement of hybrids between tetraploid wheat and diploid rye, and octaploids, resulting from the doubling of the chromosome complements of hybrids between hexaploid wheat and diploid rye, are now being grown.

One might ask why such a cross was attempted or why triticale has become a commercial crop. Like most crosses that have been found to be useful to humans, triticale combines desirable attributes of its parents and exceeds both of them in some characters. Triticale produces higher yields than does either parent in marginal grain-growing regions such as those with acid soils.

Oats

Like rye, oats are generally considered a secondary crop that originally grew as a weed in barley, wheat, or even rye fields. In Ethiopia today oats are simply gathered and resown with wheat. In most other areas they are now cultivated alone. Of all the major Eurasian grains, oats appear to have been domesticated last, perhaps as recently as 3000 years ago. Domestication also probably occurred first in Europe rather than in the Near East. Early cultivars were diploid or tetraploid, but by the first century A.D. the hexaploid, *Avena sativa,* had been established as the major cultigen (Fig. 5.15).

Until recent times oats were a major cereal in temperate areas. Part of this prominence was due to the use of oats as feed, particularly for horses. Horses were introduced into Europe as draft animals in about 200 B.C. but initially did poorly because the native grasses did not provide good forage, and in wet southern regions horses suffered from foot problems. After the eighth century, several developments led to the adoption of horses as the primary draft animals (replacing oxen) in Europe. First, horseshoes, which provided protection against foot injuries and diseases, began to be employed. Second, a type of harness that did not choke horses when they were hitched to a plow or wagon was introduced from China. Peasants preferred horses to oxen because horses were much faster. However, horses needed a source of good feed, and oats served this purpose (Fig. 5.16). The Romans took up the cultivation of oats for their animals but referred to the Germans as "oat-eating barbarians," reflecting the Roman opinion that the grain was unsuitable for human consumption. Oats are high in protein and fat, and the stems provide a straw that makes good roughage. Oat cultivation thus became an integral part of the European agricultural system.

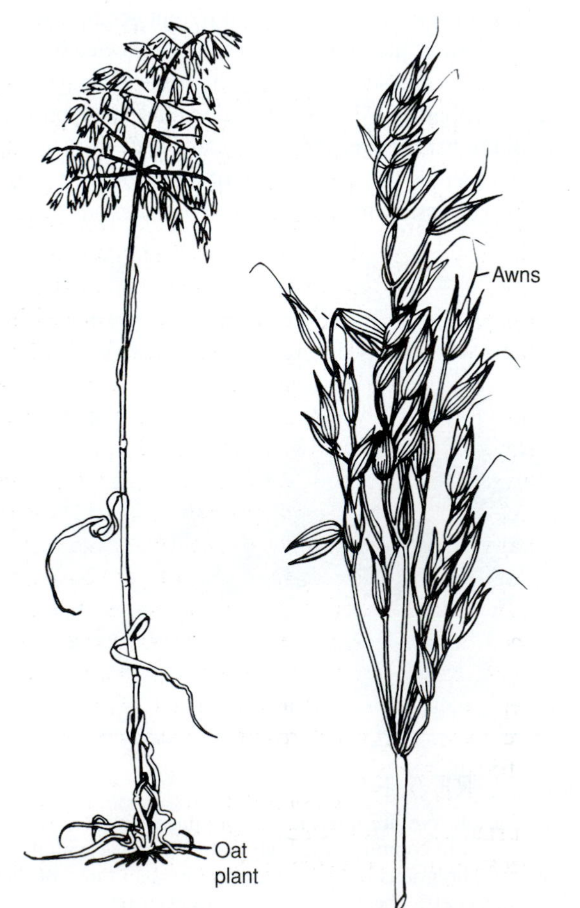

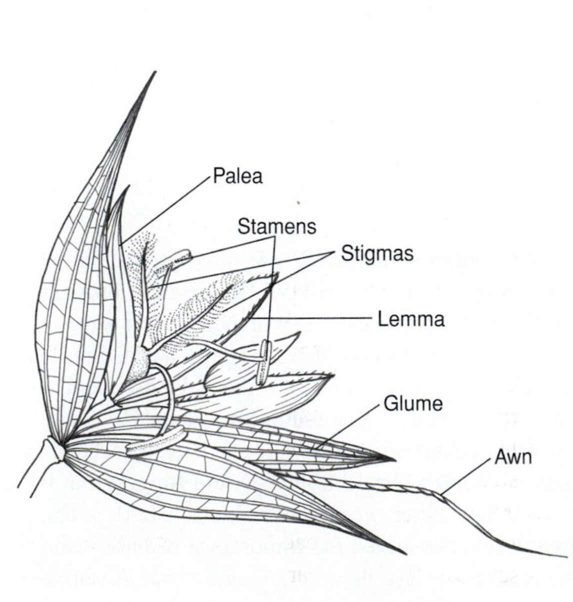

FIGURE 5.15
An oat plant, inflorescence, and flower.

FIGURE 5.16
A common sight in former times was a horse with a feed bag filled with oats.

In the United States as elsewhere, oats, like rye, were extremely important until the 1920s, when mechanized equipment replaced farm draft animals after the 1920s. Vestiges of the former widespread cultivation and use of oats are evident in the American love of oatmeal and the popular song of the 1940s:

Mares eat oats, and
Does eat oats, and
Little lambs eat ivy;
A kid'll eat ivy too,
Wouldn't you?

Rice

Rice (*Oryza sativa*) (see cover) can properly be called the world's most important crop. Wheat outranks it in terms of the number of tons produced per year (Table 5.2), but more people eat rice. It has been estimated that 1.7 billion people depend on rice, with over 588 million metric tons harvested each year. Rice has been an integral part of eastern cultures for thousands of years. Rice is considered sacred and a symbol of fertility in the Orient, a feeling expressed by the throwing of rice at weddings. Many Asians do not feel they have completed a meal unless rice is among the items eaten.

Oryza sativa has been a major crop for at least 8000 years in China, but recent phytolith data from the Yangtze River region suggest it was originally domesticated between 9000 and 12,000 years ago.

Although rice must have already been an important food item in Asia several thousand years ago, it was unknown in ancient Egypt and is not mentioned in the Bible. Alexander the Great's conquests led to the first mention of the grain in Europe—about 320 B.C. Over the next 300 years rice cultivation spread across the Middle East and into Egypt. By the fifteenth century Spain and Italy were both growing rice. Shortly thereafter the Portuguese introduced rice into Brazil and western Africa. In the sixteenth century the English started importing rice from the island of Madagascar off the east coast of Africa. One hundred years later (1695) a British ship sailing home loaded with rice was blown off course and landed in Charlestown, South Carolina. This unplanned visit led to the first plantings of rice in the United States, but commercially profitable operations began only much later (1912) in California. Australia, now a large rice producer, began commercial cultivation in 1925. In the United States today rice constitutes a major crop in some areas of California, Arkansas, Texas, Louisiana, and Mississippi.

There are currently several major kinds of rice and numerous methods of rice cultivation. Rice is often divided into two or three races, each of which produces a slightly different cooked product. One of the two principal races, commonly called the *indica* type, is labeled in the United States as long-grain rice. When cooked, the grains are relatively dry and separate easily from one another. The second major race, the *sativa* or *japonica* type, is called short-grain rice in the United States. When it is steamed or boiled, it becomes quite soft and often slightly gluey. The third type, *javanica* (not recognized by some authorities), is grown only on a very limited scale in the equatorial part of Indochina.

Grains of rice have properties that make them especially desirable as food. The endosperm is almost completely composed of starch granules surrounded by cellulosic cell walls. When white rice is cooked, the granules gelatinize and the walls rupture. The result is a grain that retains its shape but has a soft, chewy texture. In addition, rice is able to absorb flavors into the starch through the ruptured cell walls. It can therefore be cooked in variously flavored broths, lending itself to a variety of dishes.

Rice needs large quantities of water to grow well, but it does not need to grow in standing water. Rice grown without standing water is known as **upland rice,** but upland rice will produce a harvestable crop only in areas where there is 1.5 to 2 m (5 to 6.5 ft) of rain per year. Brazil is now the world's largest producer of upland rice. Huge acreages of humid tropical rain forest or savanna in Brazil have been cleared and planted with rice. Unfortunately, there are usually not enough nutrients in the soil to support rice cultivation for more than a year or two. After the soil's potential has been depleted in areas such as the cerrado of central Brazil, fields are converted to pastures that are often seeded with tropical African grasses.

Most rice is wet, or **paddy,** rice, meaning that it is grown in standing water. Rice is able to grow partially submerged because its specialized stem and root anatomy maintain an

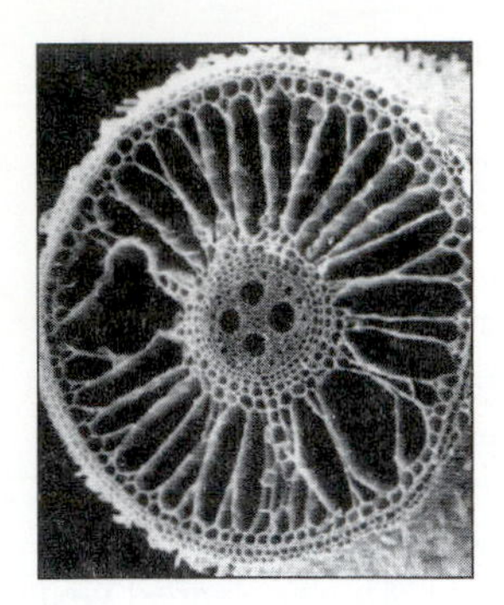

FIGURE 5.17
A cross section of the root of a rice plant shows the specialized anatomy that allows rice to grow in anaerobic (oxygen-free) conditions. Like all cells, those of submerged rice stems and roots need oxygen to carry out respiration. At depths of more than a few centimeters under water there is little oxygen dissolved in the water. The cortex of rice roots has large air spaces formed by the dissolution of cells, which allows oxygen to diffuse downward from aerial parts of stems (compare Fig. 8.9). In addition, the epidermis of rice roots does not allow the diffusion of oxygen outward from the root into the water.
(Courtesy of M. C. Drew.)

oxygen supply for the underwater parts of the plant. People tend to call flooded fields rice paddies (Fig. 5.18), but the word *paddy* is also applied to unhusked rice grains. If temperatures are favorable, wet rice cultivation can permit almost year-round farming. The standing water rots the plant material remaining after harvest and harbors blue-green algae that provide green fertilizer and add nitrogen. Flooding the fields also prevents the growth of terrestrial weeds. In southeastern Asia rice cultivation often involves extensive hand labor. The seed is sown in nursery beds, and the seedlings are individually transplanted into the fields for maturation (Fig. 5.19). In the United States seeding is often done by planes that broadcast the grains directly across the fields. Special harvesters have also been designed that mechanically harvest the crop when it is ripe.

A third method of rice cultivation makes use of natural floodwater. In floodplain areas, seeds are sown on the plain at times of low water. As the water rises, the rice grows apace with it. In some cases the plants can reach 4 to 5 m (13 to 16 ft) in height before they finish growing. Harvesting in such areas is often done by boat.

With the advent of modern milling procedures, the world switched from brown rice to polished rice. The loss of the bran and germ was a subtle dietary change that had profound consequences. In the nineteenth century a number of men in the Japanese Navy experienced a loss of muscle tone in their arms and legs as a result of nerve inflammation. The disease, which was later diagnosed as **beriberi,** soon broke out among

FIGURE 5.18
Rice paddies in Indonesia.
(Photo by G. Fox.)

FIGURE 5.19
More than 90 percent of the rice cultivation in Asia is done by women. Here they are shown transplanting rice in India.
(Photo by G. Bizzari, FAO.)

thousands of people in Asia and Africa who subsisted on a rice diet. A Dutch doctor, Christian Eijkmann, working in Java was among those who concluded that the disease results from a dietary deficiency that is aggravated by eating foods high in carbohydrates. It was later shown that a lack of vitamin B_1 is the cause of the illness. The brans of grains are naturally high in vitamin B_1, and brown rice is particularly rich, containing 0.40 mg per 100 g of grain. By contrast, polished rice contains only 0.04 mg per 100 g of grain. Several ways have been devised to get around the vitamin deficiency. In Japan, fish oils compensate to some degree. Parboiling unmilled rice causes the outer layers of the endosperm to absorb some of the vitamin before polishing. Rice can also be enriched by spraying it with synthetically produced vitamin B_1 after polishing. Puffed rice (Fig. 5.20) was an American invention that uses enriched, partially milled rice.

For many years, rice production in the wet tropics was unsuccessful because of lodging or because the soils in such regions are poor and require large amounts of nitrogenous fertilizers. The development by the International Rice Institute in the Philippines of dwarf varieties such as IR8, which do not lodge, expanded cultivation in tropical wet areas. Newer varieties, such as IR64, developed by the institute combine the qualities of IR8 with increased pest resistance and a wide tolerance of poor soil conditions.

The genus *Oryza* has other species, several of which occur in Africa. One of them, *O. glaberrima,* was independently domesticated in western Africa, where it is still grown, although its cultivation is being replaced by that of *O. sativa.*

Wild Rice

Wild rice is not a member of the genus *Oryza* but a New World species, *Zizania palustris.* The species, unlike its rare relative, Texas wild rice (*Z. texana*) (Fig. 5.21), was used as a source of grain by Native Americans, although it was never really domesticated by them. The inflorescences of wild rice shatter, and the grain was traditionally collected by beating mature inflorescences while they were held over canoes. Native peoples roasted the grain, poured it into deerskin-lined pits, and trampled it to remove the husks. Today, wild stands of rice are harvested green and cured by rotating the seeds under water and then drying them in roasters.

FIGURE 5.20
"The cereal that's shot from guns!" Quaker Oats promoted its new puffed rice cereal at the 1904 World's Fair by loading several cannon barrels from the Spanish-American War with rice. After being rotated in gas-fired ovens, the sealed barrels were unplugged. With the sudden release of pressure, the grains exploded and shot out of the cannon with a loud bang. In modern cereal factories, automatically loaded, self-firing multiple-barrel devices are used to puff grains.

Wild rice cultivation in the United States began in 1959 after the diking and flooding of areas of poor farmland in Wisconsin and Minnesota. Plant breeders were finally successful in selecting a nonshattering strain that facilitated harvesting. Nonshattering varieties raised yields from 100 to 700 pounds per acre because they reduced the losses incurred during harvesting of the older shattering types. Combines equipped with oversized wheels now harvest the grain. Over 17,000 acres, primarily in California, are currently under cultivation, but the demand for this gourmet treat still exceeds the supply, and prices remain very high.

Sorghum

Sorghum (*Sorghum bicolor*) is a grain native to tropical Africa that ranks fourth after rice, corn, and wheat in terms of importance for human nutrition. It is the major food grain in many parts of India and in much of Africa, where it is used primarily to make unleavened bread. Sorghum appears to have been domesticated about 5000–7000 years ago in Ethiopia. By 2000 years ago its cultivation had reached the Nile Valley. Sorghum has been subjected to intensive divergent selection,

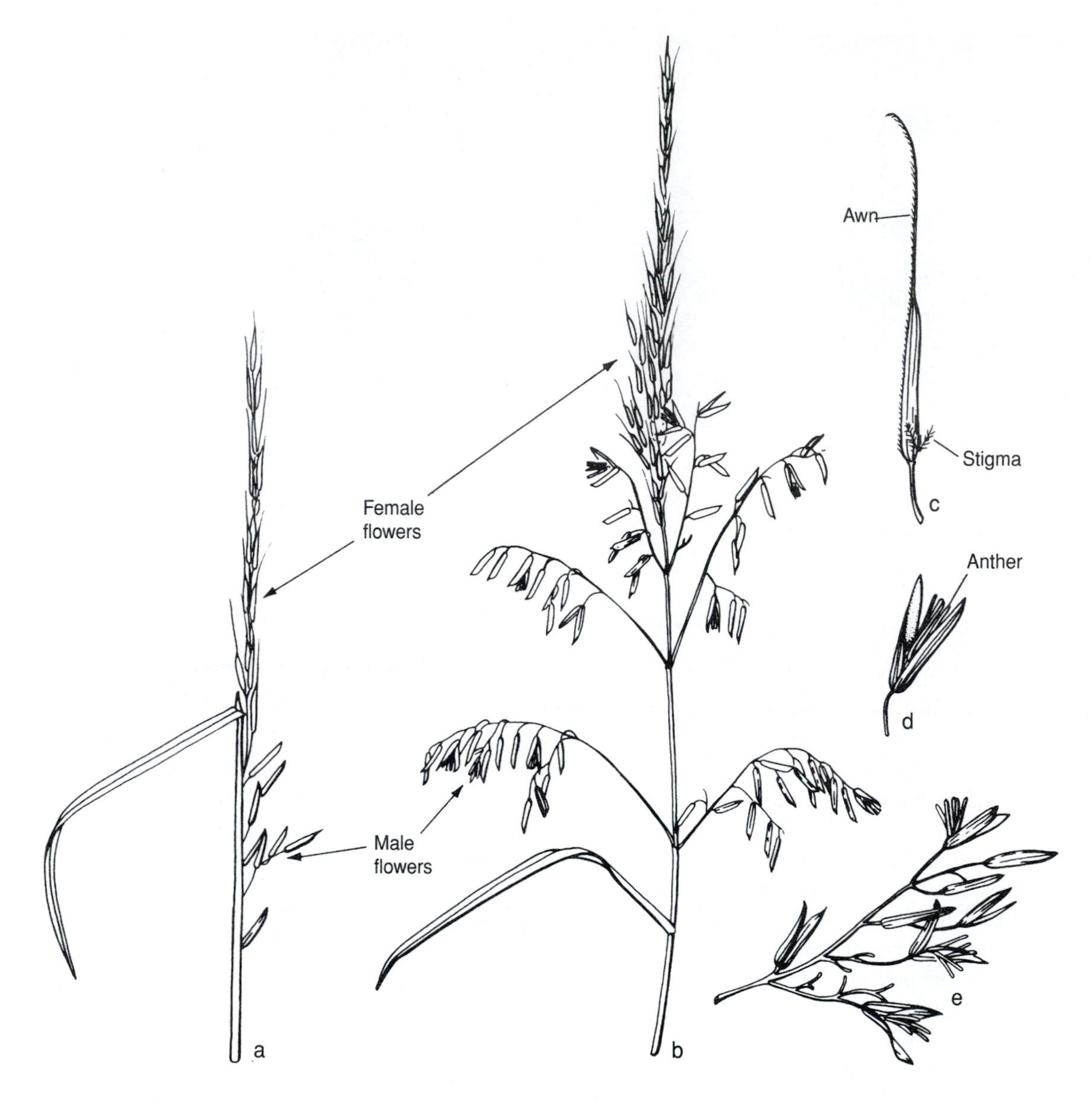

FIGURE 5.21

Wild rice and its relative Texas wild rice (shown here) are among the few monoecious grain crops (a, b). It bears female flowers on the top of the flowering stalk and male flowers below. The female flowers emerge before the male flowers, as shown in the (a) young stalk versus the (b) mature stalk. Close-ups of (c) the female flower and (d) the male flower and (e) of male flowers at the time of pollen release. Texas wild rice, used in former times by Native Americans like its more familiar northern relative, is on the endangered species list because its populations have been so reduced by destruction of its habitat by humans.

leading to numerous distinctive types (Fig. 5.22) over the last few thousand years. There has been much discussion but little agreement about the best method for grouping these types. Cultivated sorghums are now grouped into five main races. Numerous intermediates between these races occur because of crossing practices. Nevertheless, cultivated sorghums are usually grouped by farmers into four main types based primarily on the use of the sorghum: (1) grain sorghums, (2) sweet sorghum or sorgo (used for animal feed), (3) Sudan grass (a different but related species), and (4) broomcorn or broom millet. The hybrid sorgos grown in the southern part of the United States are crosses between members of two of the races (Kafir and Durra).

Traditionally, sorghum has been a grain of hot regions with too little rainfall (500 to 1500 mm, 12.7 in. to 38.1 in., per year) for other grains. There has recently been an increased interest in sorghum because of its drought tolerance, potential use on land that is marginal for other grains, and ability to yield alcohol via fermentation. The waxy leaves of *S. bicolor* roll up to retard evaporation and thus aid in coping with water stress. In India and Africa dried grains are often ground and used to make flatbreads. There are even pop sorghums similar to popcorn. The grain is also used to make a beer. If sorghum is cut while in the milk stage (the same stage at which people pick sweet corn), it can be used for silage (see Forage Grasses later in this chapter). In the

FIGURE 5.22
A sorghum field in Texas.

FIGURE 5.23
Gambian villagers holding up millet and corncobs.
(Photo by Liba Taylor, Corbis.)

United States almost all the sorghum grown is used for livestock feed, although in some localities special syrup varieties are cultivated. Syrup, or sorghum "molasses," is made by harvesting the sorghum before flowering and crushing the stalks. The expressed juice is not purified, so the final thick, sweet syrup resulting from the evaporation of most of the water retains a characteristic flavor. Broomcorn is a type of sorghum with an open inflorescence. The tillers are cut and stripped of grain, and the stiff inflorescence "branches" are used for making whisk brooms.

FIGURE 5.24
Common broomcorn millet or hog's millet (*Panicum miliaceum*) as it appeared in Fuch's *De Historia Stirpivum,* 1542.

Millets

The name *millet* does not refer to a single cereal but to several edible grasses (Figs. 5.23 and 5.24). Numerous forage grasses and even sorghum are often called millet. A list of some of the most widely grown millets is presented in Table 5.4. Two of the species in this list, *Eleusine coracana* and *Pennisetum glaucum,* are more important than are the other millets as human food. The first, known most commonly as finger millet, was domesticated in northeastern tropical Africa and is now a staple in the eastern and central parts of the continent. *Eleusine* is able to withstand exceedingly long periods of storage, up to 10 years or more. Weevils avoid the grain, and the grain itself does not deteriorate. Finger millet is higher in protein and the sulfur-containing amino acids than are rice, maize, and pearl millet. In Africa the grain is generally ground into flour and made into large, flat breads

TABLE 5.4 Common Millets of the World*

SCIENTIFIC NAME	COMMON NAME	USES
Echinochloa crus-galli	Japanese barnyard millet	Food
E. colona	Shama millet	Food
Eleusine coracana	Finger millet, ragi, African millet	Food, primarily Africa and southern Asia
Eragrostis tef	Teff	Food
Panicum		
P. miliaceum	Common millet, proso, broomcorn millet, hog millet, Hershey millet	Food, feed
P. ramosum	Brown top millet	Feed
P. virgatum	Switch grass	Pasture grass
P. obtusum	Vine mesquite	Pasture grass
P. texanum	Texas millet	Pasture grass
P. maximum	Guinea grass	Pasture grass
P. purpurascens	Pará grass	Pasture grass
Pennisetum glaucum	Pearl millet, cattail millet, bajra, bajri	Food
Paspalum scrobiculatum	Koda millet	Forage grass
Setaria italica	Foxtail millet	Food

*Some sorghums are also called millets.

FIGURE 5.25
Ethiopian woman cooking *injera,* a pancake-like bread made from millet flour.
(Photo by Arba Minch, Corbis.)

(Fig. 5.25) or eaten as porridge. Once ground, however, the flour has to be consumed quickly because the germ is not removed and the flour rapidly turns rancid. Finger millet can also be malted and used to make beer.

The second species that constitutes an important human food *Pennisetum glaucum,* is known as pearl, bullrush, spiked, or cattail millet (Fig. 5.23). Pearl millet was originally domesticated in sub-Saharan Africa, where it has been cultivated for 4000 years. This grain is more drought-resistant than any of the other cereals and produces economically profitable yields on exhausted and nutrient-poor soil. Consequently, pearl millet has become the food on which millions of Africans living on the edges of the Sahara now depend. Like finger millet, pearl millet stores well, but it is susceptible to weevil infestation and bird predation. It can also be ground into flour, eaten like rice, or malted.

Corn

In contrast to the Old World, the New World has provided only one major domesticated cereal, corn or maize (*Zea mays*) (Fig. 5.26). Among the important modern cereal grains, corn is the most efficient in converting water and carbon dioxide into foodstuffs. Moreover, it can be cultivated in both tropical and temperate areas. Maize formed the basis of all the major New World civilizations (the Maya, Aztec, and Inca), although *Amaranthus* species (Amaranthaceae) were very important in some regions in pre-Columbian times.

As children, we learned that Native Americans (Fig. 5.27) showed the Pilgrims how to grow corn and beans, using a fish as a fertilizer. The planting of beans and corn together (usually without the fish) was a common practice throughout North and South America by the time Europeans reached the New World. Presumably, the practice developed without any real knowledge of the mutual benefits that the system provides. An Indian legend says that the association came about when man-corn was looking for a wife. Squash asked to be considered, but she was rejected because she had the habit of wandering lasciviously over the ground. Bean, by contrast, clasped the corn stalk so dearly that corn was assured of her fidelity and consequently chose her for his bride. Whatever the origin of the biculture, the association proved highly successful in terms of both agronomy and nutrition. All the legumes consumed as dry seeds (pulses) have associations with nitrogen-fixing bacteria (Chapter 6), which provide nitrogenous compounds that the corn can use. The amino acid profiles of grains and legumes also complement one another and provide all the building blocks needed for humans to synthesize proteins.

There has been great interest in the origin and subsequent selection of corn because its history has been so difficult to determine. Modern corn (Figs. 5.26 and 5.28) is morphologically unlike any wild grass. Plants have one or several large stalks or culms, each terminated by a tassel of male flowers. The female flowers are tightly packed together, forming the cob inside sheathing leaves called husks. A long single style emerging from each female flower protrudes from the enveloping husks. The remnants of these styles are the "silks" that are removed from fresh corn when

FIGURE 5.26

Cultivated corn. (a) A corn plant. (b) Tassels of male flowers. (c) An individual male flower. (d) A young ear of female flowers at the silk stage. (e) A magnification of the top of an individual silk stigma to which pollen grains adhere.

it is shucked for cooking. The ear of domesticated corn is larger than that of any potential relative and has characteristics different from those of all potential relatives.

Because of its economic importance and unique, perplexing morphology, there has been great interest in the origin and subsequent evolutionary history of corn (Box 5.2). Although this has been a subject of intense controversy over the past century, many scientists have come to the conclusion that corn was initially domesticated in southern Mexico, with its cultivation spreading north and southeast from there. Charred cobs now dated to be 4700 years old have been found in southern Mexico. The tiny archaeological cobs, which were excavated from dry pre-Columbian cave deposits near Tehuacan, east of Mexico City, are similar in appearance to some kinds of popcorn that have hard, pointed grains. These popcorns are still cultivated in a few remote areas of Mexico (Fig. 5.29d). These land races also have relatively large, soft glumes with wide and deep **cupules** (cup-shaped segments of the axis) around the bases of their kernels. Recent work in Panama with phytoliths that

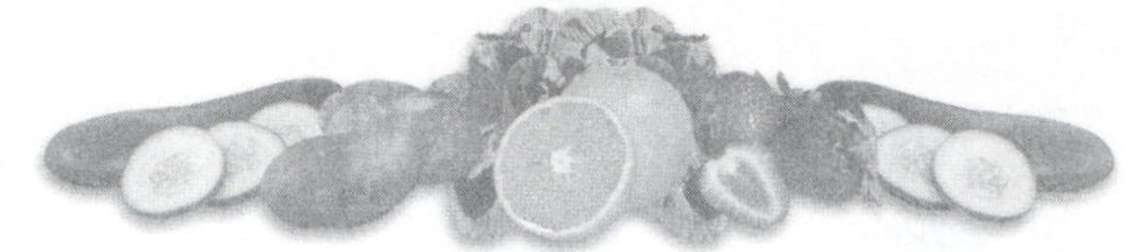

Box 5.2

Corn: An Evolutionary Enigma

Of all the cultivated crops, corn has remained the most difficult to understand in terms of its origins. There is no wild plant, or any evidence of one, that resembles even primitive corn with its giant husked ears. When one looks beyond the ears, however, corn and teosinte are difficult to distinguish.

It has been recognized for over 70 years that cultivated corn ears differ from teosinte in five main characters. First, the cobs of corn do not shatter, whereas those of teosinte fall apart even when mature. Second, each teosinte grain is nestled into a hard, deep floral axis segment (the cupule) and completely covered by a hard glume, forming a bony cupulate fruit case. The grains of corn are naked, borne outside the collapsed cupule, and have soft glumes. Third, each teosinte cupule contains only one fertile spikelet (Fig. 5.30) that produces a single grain. The cupules of corn each produce two fertile spikelets, each with a fertile grain. Fourth, the cupules of teosinte are arranged in only 2 ranks, whereas those of corn are 4- to 10-ranked. Fifth, teosinte has long primary branches, each of which terminates in a male tassel. Numerous tiny ears occur laterally along the branches. Corn has extremely short primary branches, each of which ends in a single terminal ear. The sole surviving male tassel occurs at the apex of the primary culm.

The first attempts to assess the genetic nature of the cob differences in a natural setting were done by George Beadle in 1972. Beadle crossed a primitive Chapalote corn with a Chalco race of annual teosinte and scored the morphology of 50,000 offspring from that cross. He found that the segregation of traits in the progeny suggested that the five major differences between corn and teosinte are controlled by five simple independent Mendelian genes.

John Doebley repeated this experiment, using molecular markers to locate the genes responsible for the distinguishing traits on particular chromosomes. He found that each of the major differentiating traits is controlled by a primary gene located on one of 5 of corn's 10 chromosomes. However, he also found that each of the traits is affected by other modifier, or pleiotropic, genes located on different chromosomes. For example, the gene primarily responsible for the change from hard to soft glumes is located on chromosome 4, but genetic material that affects this trait is also located on chromosomes 1, 2, 3, and 5. Similarly, the gene primarily responsible for the difference between single or paired spikelets per cupule is on chromosome 1, but genes on chromosomes 2, 3, and 4 also play a role in the expression of this character. Another important result of Doebley's analysis was the discovery that the primary and pleiotropic genes that govern the major morphological traits are clustered together in blocks on five chromosomes, leading to linked units that tend to be inherited together. This linkage leads to a pattern of inheritance that Beadle attributed to five simple genes.

While complicating the situation somewhat, Doebley's results confirm the idea that few traits are responsible for the "conversion" of a teosinte ear into an ancient form of corn and add strength to the "orthodox theory" that humans selected corn from teosinte and natural mutants were gradually selected to create what eventually became modern corn.

This exciting research explains much of the change from teosinte to corn but does not address the question of the positional change of the female inflorescences between teosinte and maize. Teosinte branches are always terminated by a male tassel, whereas all lateral corn branches end in a cob. To explain this change, Hugh Iltis suggested a sexual shift in the lateral branches during the domestication of teosinte. This shift would have occurred in two steps in the evolutionary sequence and would have allowed the dramatic differences now seen in the ears of the two plants. The first step was the suppression of all the lateral ears on the tertiary branches, the shortening of these branches, and the conversion of a male tassel at the end of a secondary branch into a female ear. This was followed by condensation of secondary branches on all primary branches except the main culm and the feminization of a male spike on the end of these primary branches. Condensation of the lateral primary branches and dominance of the apical ear led to suppression of the cobs of the secondary branches. The remnants of the cobs on the secondary branches are now visible as nubbins at the bases of the leaves forming the husks around the giant cob of modern corn.

The argument that corn arose so suddenly from teosinte that the differences between the two could not have been caused by artificial selection was formulated when archaeological evidence suggested no traces of corn before the Tehuacan cobs originally dated to be 7000 years old. A redating of this material has now shown that it is no older than 4700 years. In addition, recent paleobotanical work in the lowlands of Panama by Smithsonian scientists has suggested that corn was present in the lowlands of Mesoamerica as early as 7700 years ago. These findings, based on pollen and features of leaves, do not provide evidence about the nature of the cobs present on the plants, but they do indicate the use of corn over a much longer time span than was previously thought. It is therefore possible that teosinte was originally collected and used in the lowlands of the Río Balsas, Mexico (where it is native), and over a thousand years or so acquired the primary features of the modern corncob under the influence of human selection. By the time the cultivation of corn spread to Tehuacan on the central plateau region of Mexico, corn plants were essentially small versions of modern corn. All that remained was to increase cob size and change kernel characters to suit particular needs (Fig. 5.28).

FIGURE 5.27

Sand paintings, an art form developed by the Native Americans of the Southwest, often portrayed the four plants sacred to them. Clockwise from the upper left: corn, squash, beans, and tobacco.

appear to have come from cultivated corn suggests that corn was cultivated there 7700 years ago. By the time of Columbus's arrival in 1492, over 300 local land races were grown from Canada to Argentina and Chile, many of which were quite similar to the large-eared cultivars people plant today. Several of these land races survived until recently under the care of indigenous peoples, but the numbers are dwindling.

Corn's gigantic husked ears seem to defy explanation and have initiated much speculation (Figs. 5.26 and 5.28) about their origin. Scientists have tried to locate a potential wild ancestor, and there has been a heated debate about the origins of modern corn (*Zea mays* subsp. *mays*) and its relationship to two other somewhat similar plants, *Tripsacum* and teosinte. *Tripsacum* is a New World genus of about 12 perennial species, and teosinte consists of 4 species of giant annual and perennial grasses of Mexico and Guatemala that are all now placed together with corn in the genus *Zea.*

Over 50 years ago a group of researchers led by Paul Mangelsdorf and Edgar Anderson produced a theory of corn (the "tripartite theory"). This theory postulated that cultivated maize originated from a wild form of corn that once was, and perhaps still is, indigenous to the lowlands of South America. Teosinte, the closest relative of maize, is a recent product of the natural hybridization of *Z. mays* and *Tripsacum* that occurred after cultivated maize had been introduced by humans into Central America, and new types of maize originating directly or indirectly from such crosses and backcrosses with *Tripsacum* led to the majority of Central American and North American varieties of domesticated corn.

Extensive fieldwork to test this hypothesis was not done until the 1960s, when Mangelsdorf's student Garrison Wilkes showed that teosinte must be a relatively ancient complex of distinct localized races or species that are unlikely to have arisen as a consequence of recent hybridization. In view of these data and the finding that *Tripsacum* ($2n = 34$ or 72) will not cross with corn ($2n = 20$), Mangelsdorf eventually abandoned this theory and suggested that a perennial species of teosinte (described in 1979 by Hugh Iltis and colleagues) hybridized with corn to produce annual teosinte.

Today it is clear that modern corn is a direct result of human selection from an annual species of Mexican teosinte (Fig. 5.30) and that a naturally occurring many-rowed, round-eared wild corn never existed. When the basic taxonomy of the annual teosintes and corn became better understood, corn biologists High Iltis and John Doebley determined that both grasses should be considered as subspecies of *Z. mays.* On the basis of cytological, biochemical, and molecular work, one of the annual teosintes, *Z. mays* subsp. *parviglumis* (Rio Balsas or Guerrero teosinte), is now considered to be the wild ancestor of cultivated corn.

Acceptance of a teosinte as the ancestor has not settled the problem of how the minute, few-seeded ears of teosinte (Fig. 5.30) changed into the large, multiseeded ears of corn

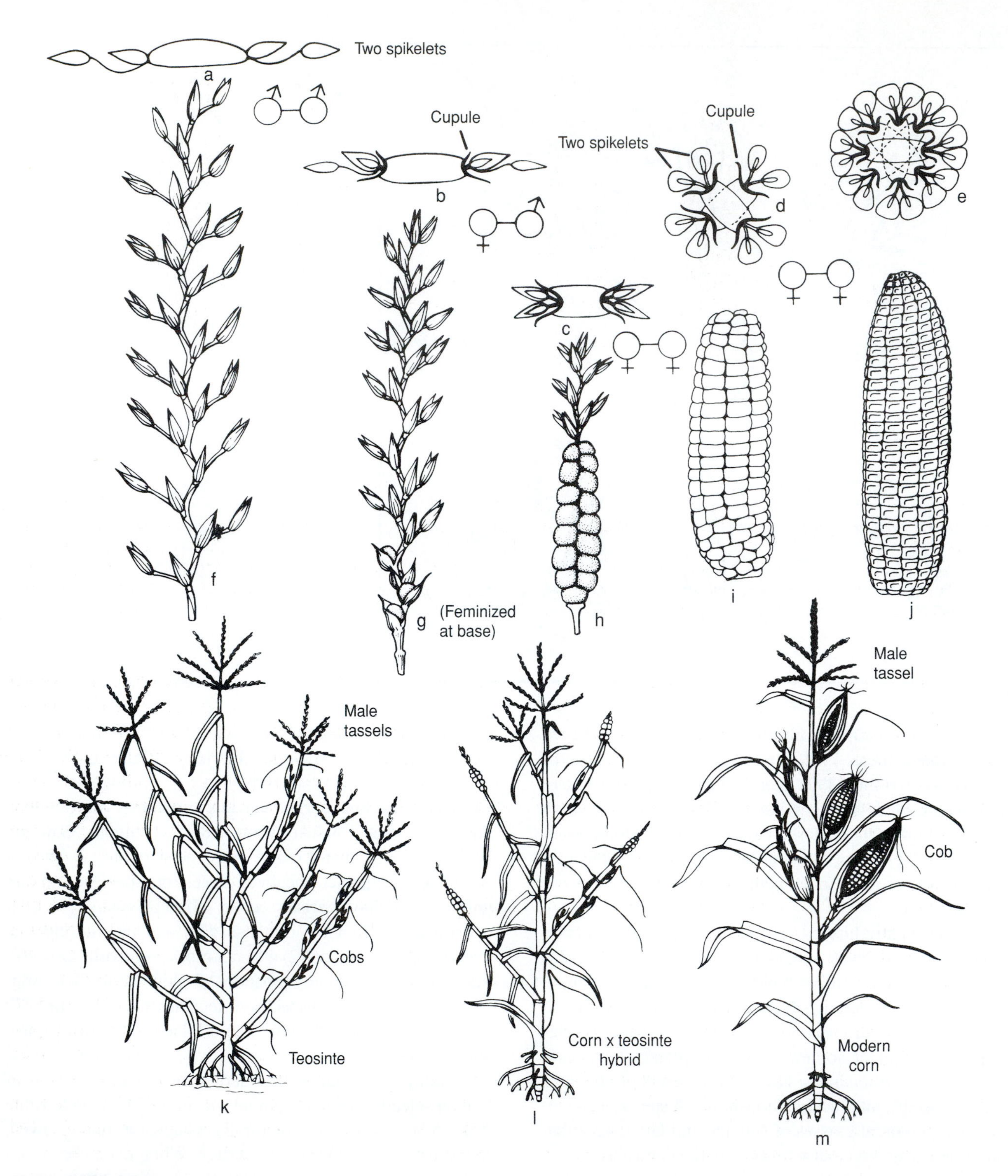

FIGURE 5.28

As does the female cob of corn (d, e), the male tassels of teosinte (a, f) and maize have two fertile spikelets with soft glumes at each node on the axis. The cupule below female flowers (b, d, e) is formed by the main axis at the point where the spikelets diverge from it. The cob of modern corn (e, i, j) is formed by the twisting of a female inflorescence with opposing cupules of paired fertile spikelets (d). Unlike corn, the female inflorescences of teosinte have a single fertile spikelet inside hard cupules, each sealed by a bony glume (see Fig. 5.30). Occasionally, male tassels of corn produce kernels at the base (b, g). Similarly, cobs can be found that bear a portion of a male tassel at the tip (c, h). Hybrids between corn and teosinte (l) sometimes produce these abnormal cobs. The natural occurrence of sexually mixed inflorescences, combined with the fact that the corncob has paired spikelets and soft glumes, led in part to the "catastrophic sexual transmutation theory," which hypothesizes that the cob of modern corn arose by conversion of the central spike of a male tassel on a lateral branch of teosinte (k, m) into a female inflorescence. Progressive shortening of these lateral branches and the suppression of all other inflorescences on the lateral branches were hypothesized to have led to modern corn plants with axillary lateral cobs and a male tassel only at the top of the main stem (m).

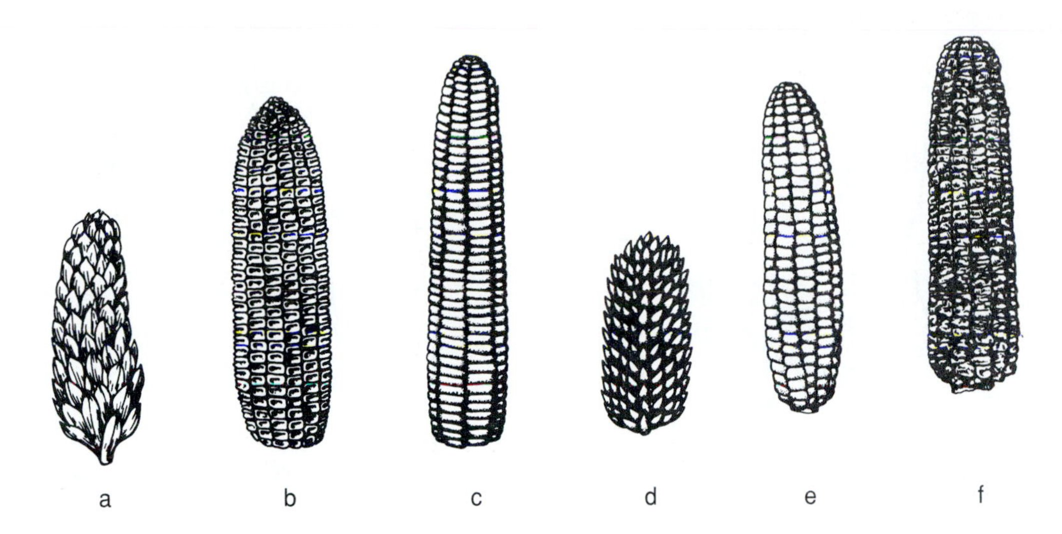

FIGURE 5.29
The six main corn varieties are (a) pod corn, with entire, husklike glumes surrounding each kernel; (b) dent corn, with a small dent in the back of each dried kernel that is caused by the differential drying of the soft and hard endosperm in the grain; (c) flint corn, with long kernels composed primarily of hard endosperm, widely grown in former times in northern regions of North America; (d) popcorn, which contains mostly dense endosperm inside a tough coat (as the kernels heat and the starch expands, the pericarp holds firm until the core of dense starch suddenly swells with a burst that ruptures the wall); (e) flour corn, which contains only soft endosperm and formerly was widely grown in South America; and (f) sweet corn, the variety eaten as a vegetable in the United States because some of the sugars in the endosperm are not converted to starch until some time after picking. The differences in the corns are often caused by differences in the starch in the grains.

(Fig. 5.28). The "orthodox teosinte theory," proposed in the 1970s by George Beadle, Walter Galinat, and others, assumed that an ear of corn is directly homologous to the tiny ear of teosinte and was derived from it as a result of mutations and artificial selection over a long period. More recently Hugh Iltis suggested that the female ear of corn is not only homologous to, but is derived from, the central branch of a male tassel terminating a primary lateral branch. He proposed that this change occurred through a feminization of the male spike. This change, he reasoned, would have resulted in a rapid, catastrophic reallocation of nutrients within a branch and the production of a single, large dominant ear of corn at the ends of lateral branches. Iltis asserted that this hypothesis, which he called the "catastrophic sexual transmutation theory," was more consistent with the sudden appearance of corn in the archaeological record than was the "orthodox theory" described above (Box 5.2). His theory was also consistent with the occasional appearance of a female ear in the center of a tassel of modern corn and the common occurrence of a tassel spike on top of an abnormal ear of corn.

Molecular studies by Iltis's student John Doebley confirmed and extended earlier genetic studies by George Beadle that showed that most of the morphological differences between corn and teosinte are controlled by a few genes. These genetic studies suggest that modification of a teosinte cob into a corncob not only is possible but could perhaps occur relatively rapidly. In light of recent morphological studies of teosinte by J. Camara and colleagues and genetic and molecular studies by John Doebley, Iltis has accepted the orthodox theory but has provided a mechanism by which the female inflorescence is translocated to the terminal position on the lateral branches (Box 5.2).

Fortunately, the history of corn over the last 4000 years is comparatively easy to follow. Under the influence of continued selection, corn acquired several other features that reflected its increased dependency on humans and allowed its spread across most of North and South America. Teosinte, like most wild grasses, has seeds that show variable dormancy. Only a portion of the seeds produced during a given year germinate the next year, but farmers using seeds to sow a crop want 100 percent germination. Since cobs of only the plants that germinated the year after they were produced are collected as seed sources for the next year, selection for absence of dormancy would have proceeded rapidly after domestication began. Another feature of cultivated corn that became more pronounced under domestication is the enclosure of the head, or cob, in increasingly tenacious layers of leaves ("shucks"). The fruits of modern corncobs not only lack the shattering properties of their progenitors but also are irrevocably locked inside the persistent husks of the ear, guaranteeing their complete harvesting.

As corn spread into North and South America, several other changes occurred. First, in Mexico there appear to have been two lines of selection leading to kernels that protruded from the cupule by (1) a lengthening of the individual kernel stalks or (2) an elongation of the kernels themselves. The first line of selection led to the flint corns that spread into the southwestern United States, and the second to the

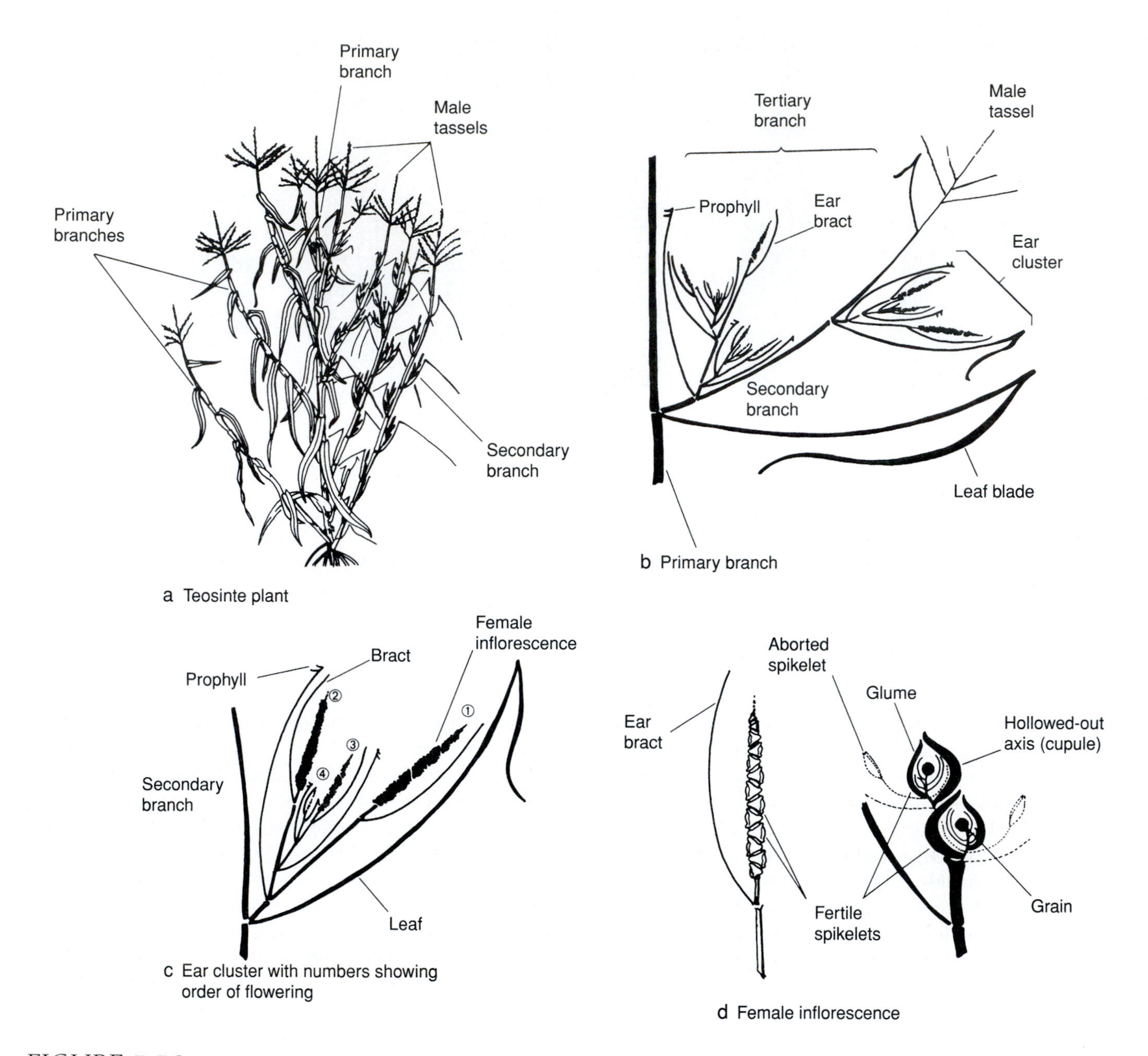

FIGURE 5.30

Teosinte differs from modern cultivated corn in a number of ways. (a) Vegetatively, adult plants look like primitive races of corn. However, corn lacks the males tassels on secondary branches and does not have numerous females inflorescences (ears) borne laterally in clusters along secondary branches. (b) Tertiary branches occur along the primary branches. Each of these, like the primary branch and the secondary branches, is terminated by a male tassel. Each tertiary branch is borne in the axil of a prophyll, a leaflike bract. (c) Clusters of ears occur along inflorescences on the secondary branches. Each cluster is borne in the axis of a leaf. Each individual inflorescence has a bract at the base. The individual inflorescences in a cluster are matured sequentially, as shown by the differing sizes of the cobs. (d) An individual ear consists of 5–12 cupulate fruit cases stacked on top of one another. The cupule fruit case is formed by the hollowed-out axis of the stem. Each cupule is sealed with a hardened glume. While the cob seems to be a simple stack of single spikelets, each with a fertile floret, it is actually much more complicated. Like corn, each cupule has two spikelets, but one always aborts. Each fertile spikelet has only one fertile flower that produces a single grain. The palea, lemma, and glumes of the fertile spikelet are shown as lines within the hard shell.

(Illustrations courtesy of Hugh Iltis and modified from Camara and Gambino.)

gourd-seed, shoe-peg, and dent corns that spread into eastern North America. Before the northern spread of either type, however, another mutation had to occur. Teosinte, and presumably early corn, has a photoperiod requirement that precludes flowering and fruiting in temperate regions that have long days and short nights in the summer growing season. A loss of sensitivity to day length was necessary before corn could be grown as an annual in temperate regions.

Although there are now many thousands of corn cultivars, many of them hybrids, there are six main historical types of corn: pod, dent, flint, pop, flour, and sweet (Fig. 5.29). For many years flints and dents were the mainstays of the native populations of North America and subsequently of the European settlers. In 1812 a cross was made between these two varieties that led to a hybrid that was to become the ancestor of the modern Corn Belt dent corn. Subsequent hybridizations and selection in the 1930s led to sweet eating corn and high-yield field corn.

Corn was taken back to Europe on the first voyage of Columbus but did not immediately gain much attention. Even today, corn is less popular in Europe than in other parts of the world, partly because most of Europe lies too far north for corn to grow well. The European prejudice against corn was so strong that the Irish are said to have refused to eat corn when dying of hunger after the decimation of the potato crop in the 1840s (Chapter 7). Europeans insisted on grinding dried corn into cornmeal, which has limited uses compared with wheat flour. Although this was the same method of preparation practiced by the early European settlers in the New World (and used to make the classic corn bread, johnnycakes, and so on), Native Americans had more interesting and nutritionally superior ways to use the grain. The Aztecs ate maize porridge or they boiled corn with a little charcoal to form an alkaline solution that separated the grain coats from the endosperm. The endosperm was then washed and mashed, and the resultant dough was formed around pieces of meat or a mixture of beans, peppers, and meat. The mass was wrapped with leaves and steamed to produce tamales. Corn that has been treated in this manner, dried, and ground into flour is sold in the United States as masa harina or "instant tortilla mix." Ground, dried corn was used to thicken the Aztecs' chocolate drink (Chapter 13).

Corn does not have the same nutritional value as other whole grains. It is deficient in the amino acids tryptophan and lysine and is relatively low in total protein. Since corn has no gluten, it can produce only flat breads such as tortillas or crumbly cakes. The nutritional deficiencies manifest themselves wherever corn has become the main, or the sole, part of the diet. **Pellagra,** a disease characterized by dermatitis, diarrhea, and dementia, was originally thought to be caused by a niacin (a member of the vitamin B complex) deficiency. When pellagra became a common problem of people in parts of Europe, northern Africa, and the Americas who were subsisting on practically pure corn diets, scientists were puzzled. It was known that niacin deficiencies caused pellagra, but corn does not have particularly low levels of niacin. It was subsequently discovered that tryptophan deficiencies are also a factor in causing pellagra. The diet of many rural Americans of molasses, corn bread, and salt pork made pellagra a serious problem in the South until 1914, when yeast (which is naturally high in tryptophan) was introduced as a food supplement. Concerted efforts to improve dietary practices in this region led to a 74 percent reduction in pellagra in some southeastern states between 1928 and 1938.

FIGURE 5.31

One of the most unusual uses for corn is as an art medium, demonstrated by the Corn Palace in Mitchell, South Dakota. Since 1892 the residents of Mitchell have decorated this building with 2000 to 3000 bushels of corn along with other grasses. Through the use of variously colored ears of corn ,which are sawed in half lengthwise and nailed onto tar paper designs, a different motif is created each year. The last week in September is set aside for the Corn Palace Festival.

(Photo courtesy of Mitchell Chamber of Commerce.)

Usually between 80 and 90 percent of the corn produced each year is used to feed animals, primarily hogs. Animal food can also be obtained from corn plants chopped and fermented into silage. Corn is also used for a wide variety of other purposes (Fig. 5.31). Cornstarch can be hydrolyzed to make corn syrup, which contains glucose and fructose, and is used in baby formulas and intravenous solutions. Purified ground corn endosperm yields cornstarch that is used to thicken cooking liquids and even to powder baby bottoms. Maize beer, or chicha, is produced by fermenting hydrolyzed cornstarch. Traditionally, Central and South American Indians chewed corn kernels before fermenting to introduce salivary amylase to break down the starch. Corn is an important adjunct in modern brewing operations (Chapter 14), a major ingredient in the production of bourbon, and it is used to produce industrial alcohol.

The cytology and genetics of corn are today probably better understood than are those of any other plant species. Breeders now produce superior hybrids, often by making crosses using inbred, or highly homozygous, lines (see Fig. 1.31). Because corn is monoecious, hybridization is relatively easy. Plants to be hybridized are planted in alternate rows. The rows that serve as the female parents (those with ears to be used for seed) are detasseled by workers riding through the fields. Self-pollination is thus prevented, and all the seeds produced are of hybrid origin. In the late 1960s it was discovered that a male-sterile gene could be bred into corn lines. With the use of male-sterile plants, detasseling became unnecessary. At first the use of male-sterile lines saved money, but it soon became dramatically clear that the cytoplasmic sterility gene carried with it a susceptibility to a new race of the southern corn blight fungus, *Helminthosporium maydis* (Dematiaceae). Once this fungus started infesting fields, it spread rapidly. In 1970 warm, moist weather encouraged fungal growth, leading to devastating crop losses amounting to over $4 billion in the United States. Luckily, the next season was dry, averting a second major outbreak. New immune hybrids without this particular male-sterility gene were redistributed during the next few years. This episode forced plant breeders to implement programs for the conservation of genetic reserves of this and other major crop species (Chapter 19).

Forage Grasses

It would be impossible to discuss all the forage grasses used throughout the world. In many cases statistics are lacking because forage grasses are usually consumed on the farm and thus do not enter into trade. Nevertheless, it is estimated that in the United States forage crops (which include forage legumes as well as forage grasses) cover five times the acreage of all cereal grain crops combined.

Land that is not well suited for high-intensity agriculture or that is "resting" between periods of cultivation is often allowed to become grassland. In addition to renewing the humus in the soil, grasses help prevent erosion and increase soil drainage.

Many of the major forage grasses of the United States were introduced from Europe or Africa (Fig. 5.32), although native species are also used. Usually forage grasses are perennials, but some are annuals. Table 5.5 lists many of the major forage grasses of the United States.

To maximize the food value of forage land and to improve the soil, farmers normally sow grasses with legumes. This co-cultivation provides a natural source of nitrogen for the soil and provides grazers with a mixture of high-roughage grasses and nitrogen-rich legumes. Forage can be grazed directly, made into hay, or used for silage.

Hay is produced by dehydrating green forage to a moisture content of 15 percent water or less. Hay quality is determined by the plants that are incorporated into the hay, the amount of leaf material relative to stem material, the time when the forage was harvested, and the amount of weathering and handling to which it has been subjected. The form in which the hay is fed to animals (loose, pelletized, etc.) also affects its suitability for different animals.

Grasses have a high cellulose content, and for most animals the cellulose has to be digested via bacterial fermentation before the hay can be utilized. In addition to reducing cellulose to simpler compounds that can be digested by vertebrates, the intestinal bacteria produce amino acids and vitamins. Ruminants such as cows have large gastrointestinal tracts that contain abundant bacteria. The rumen, the primary area where bacterial fermentation occurs, is at the beginning of the digestive tract. In nonruminants, or animals with simple stomachs, fermentation takes place near the end of the digestive tract. Whereas ruminants can digest hay with a high percentage of grass relative to legumes, nonruminants cannot take full advantage of bacterial fermentation and their feed must contain proportionately higher quantities of proteinaceous foods. Since over 95 percent of the hay produced in this country is for ruminants, most hay has a high proportion of grasses relative to legumes.

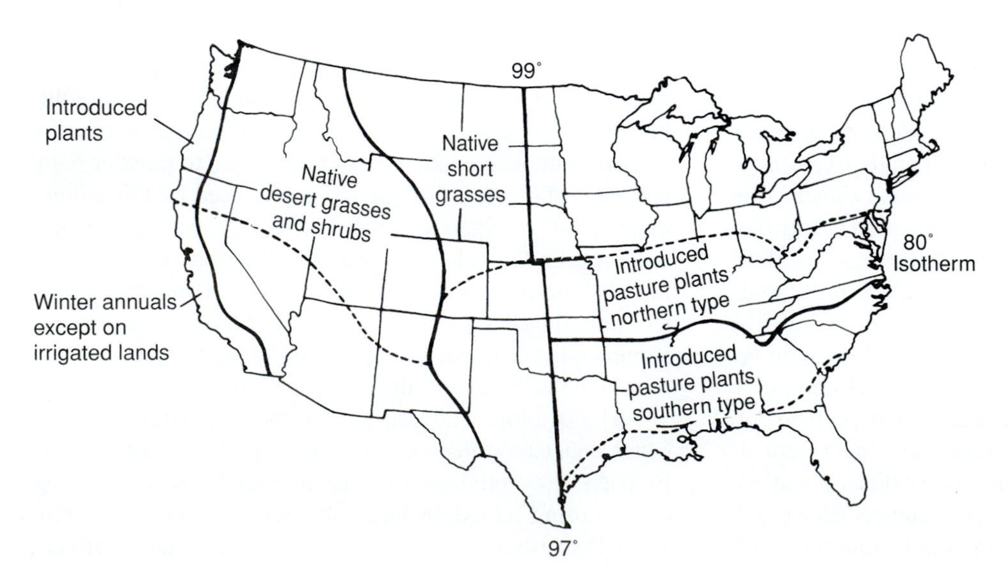

FIGURE 5.32
Comparative distributions of introduced and native forage grasses important in the United States today. Many of the introduced grasses are from Africa or Europe.

TABLE 5.5 Important Forage Grasses of the United States

SCIENTIFIC NAME	COMMON NAME
	Perennials
Agropyron cristatum	Crested wheat grass
A. intermedium	Intermediate wheat grass
Andropogon gerardi	Big bluestem
Bouteloua curtipendula	Sideoats grama
B. gracilis	Blue grama
Bromus inermis	Smooth brome grass
Cynodon dactylon	Bermuda grass
Dactylis glomerata	Orchard grass
Digitaria decumbens	Pangola grass
Elymus junceus	Russian wild rye
Eragrostis curvula	Weeping love grass
Festuca arundinacea	Tall fescue
F. pratensis	
F. rubra	
Lolium perenne	Perennial rye grass
Paspalum dilatatum	Dallis grass
P. notatum	Bahia grass
Pennisetum americanum	Pearl millet
Phalaris arundinacea	Reed canary grass
Phleum pratense	Timothy grass
Poa pratensis	Kentucky bluegrass
Schizachyrium scoparium	Little bluestem
	Annuals
Lolium multiflorum	Italian rye grass
Sorghum sudanensis	Sudan grass

Source: Modified from H. J. Hodgson. 1976. Forage crops. *Scientific American* 234(2):61–68.

Silage is the product of controlled anaerobic fermentation of undried forage or the stalks of corn, sorghum, or cane. The process of making silage is called **ensiling,** and the place where silage is made is known as a **silo.** Silage can be classified into high moisture (over 70 percent water), wilted (between 60 and 70 percent moisture), or low moisture (40 to 60 percent water), but the basic process involved in the production of all three types is the same. The cut plant material is loaded into a silo. For awhile after the plants are cut, aerobic bacteria present on the plant surfaces multiply, using carbohydrates exuded by the cut plants. Both heat and carbon dioxide are produced by the respiration of these bacteria. Once these microorganisms have exhausted the oxygen supply, a type of bacteria that can grow without oxygen begins to multiply. These bacteria produce lactic acid and some additional organic acids. The acid levels in the silo eventually reach 8 to 9 percent. This concentration of acid eventually "pickles" the plant matter and keeps it from rotting. If the acid content is high enough and air is excluded, silage can be kept for several years.

6

Legumes

The legume family is second only to the grass family in terms of its importance to humans. Almost every major civilization since the development of agriculture has had a legume as well as a grain as part of its support system. Barley and lentils, rice and soybeans, and corn and beans are classic grain-legume combinations that not only taste good but make good nutritional sense as well.

Legumes are members of the bean family (Fabaceae or Leguminosae). The family is large and diverse, containing over 16,000 species that botanists usually group into three subfamilies on the basis of their floral characteristics (Fig. 6.1). Despite differences in their flowers, almost all members of the Fabaceae share the same kind of fruit formed by a single carpel that splits along the two opposite margins to release its seeds. The fruit botanically called a **legume,** is more commonly known as a pea (Fig. 6.2) or bean pod, although some members of the family have fruits so modified that they resemble a common bean only slightly.

Of the three subfamilies, the Faboideae is the most important as a source of food crops. Members of the other subfamilies are occasionally used for food, but more commonly they are employed to improve soils, as forage, for their gums (Chapter 10), as sources of dyestuffs (Chapter 15), or as fuel (Chapter 19). Here we deal primarily with **pulses** (dried legume seeds used for human food) and forage crops, but we also include tamarind and carob, two foods obtained from the Caesalpinioideae.

An important characteristic of most species of the Faboideae is their propensity to form root associations with various bacteria. This interaction has been hailed as a classic example of a symbiotic relationship. **Symbiosis** involves two organisms that live together to the mutual benefit of both. In the legume association, *Rhizobium* and related bacteria infect the roots of different species, causing the production of swollen areas called **nodules** (Fig. 6.3). The bacteria live within the nodules and absorb nutrients from the host plant. In return, they incorporate, or **fix,** atmospheric nitrogen into a form usable to the plant host. Although nitrogen gas is the most abundant substance in the earth's atmosphere, flowering plants cannot use elemental nitrogen. Instead, they must absorb it through their roots in the form of ammonia, nitrite, or nitrate ions or perhaps as organic compounds from decayed organisms (Fig. 6.4). It is the supply of these nitrogenous ions that is often lacking in the soil. Hence, in cultivated fields, one often has to add fertilizers high in

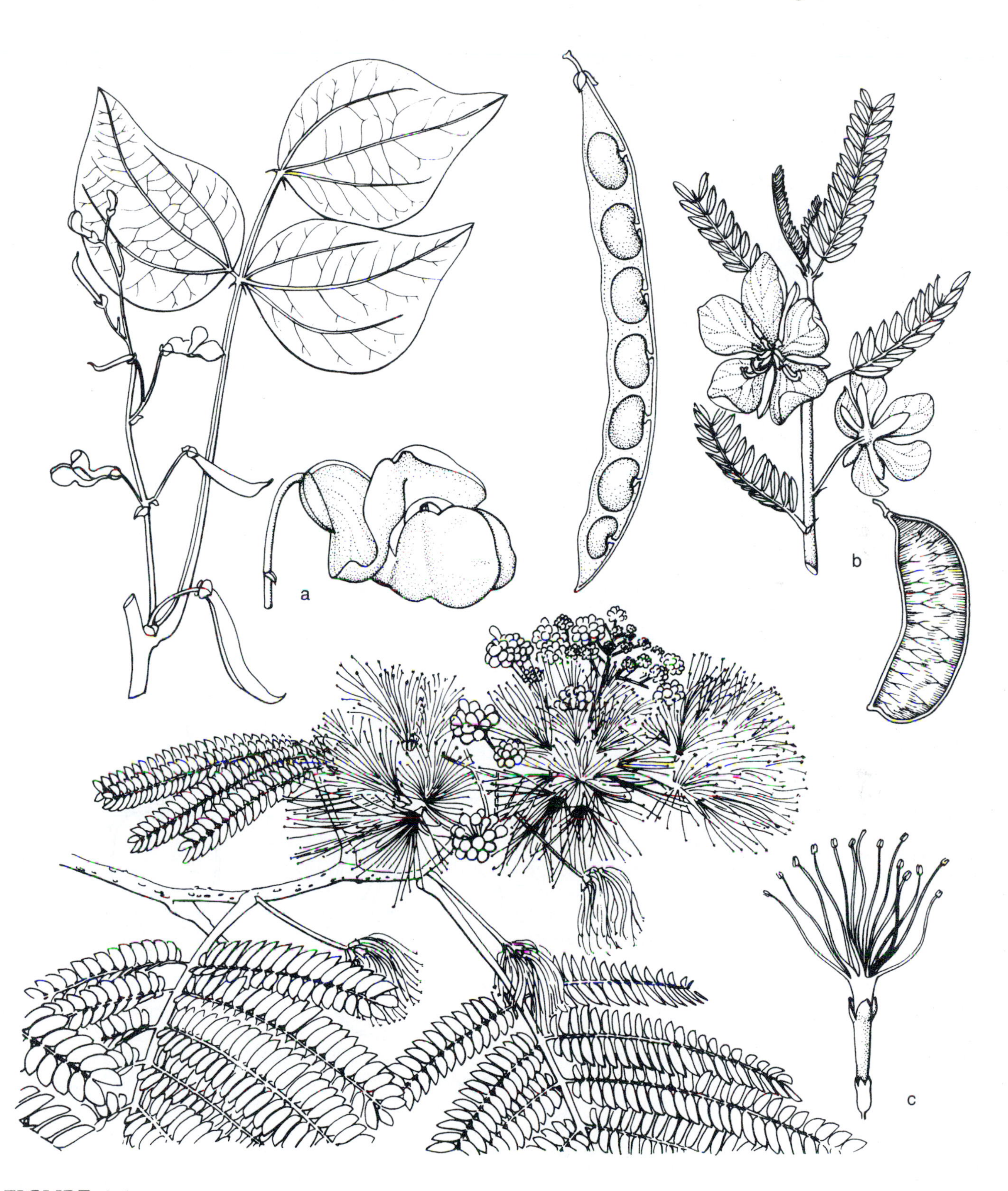

FIGURE 6.1

The subfamilies of the Fabaceae differ in floral characters. (a) The common bean, like most members of the Faboideae, has bilaterally symmetrical flowers with the upper petal, or standard, exterior to the wing petals. The keel, formed by the joining of the two lower petals, houses the style and the variously joined stamens. (b) The Caesalpinioideae, illustrated by *Senna,* has irregular flowers in which the upper petal is inside the others and the stamens are usually free. (c) *Albizzia,* similar to other members of the Mimosoideae, has tiny regular flowers borne in clusters. The stamens are often the showy part of the flowers.

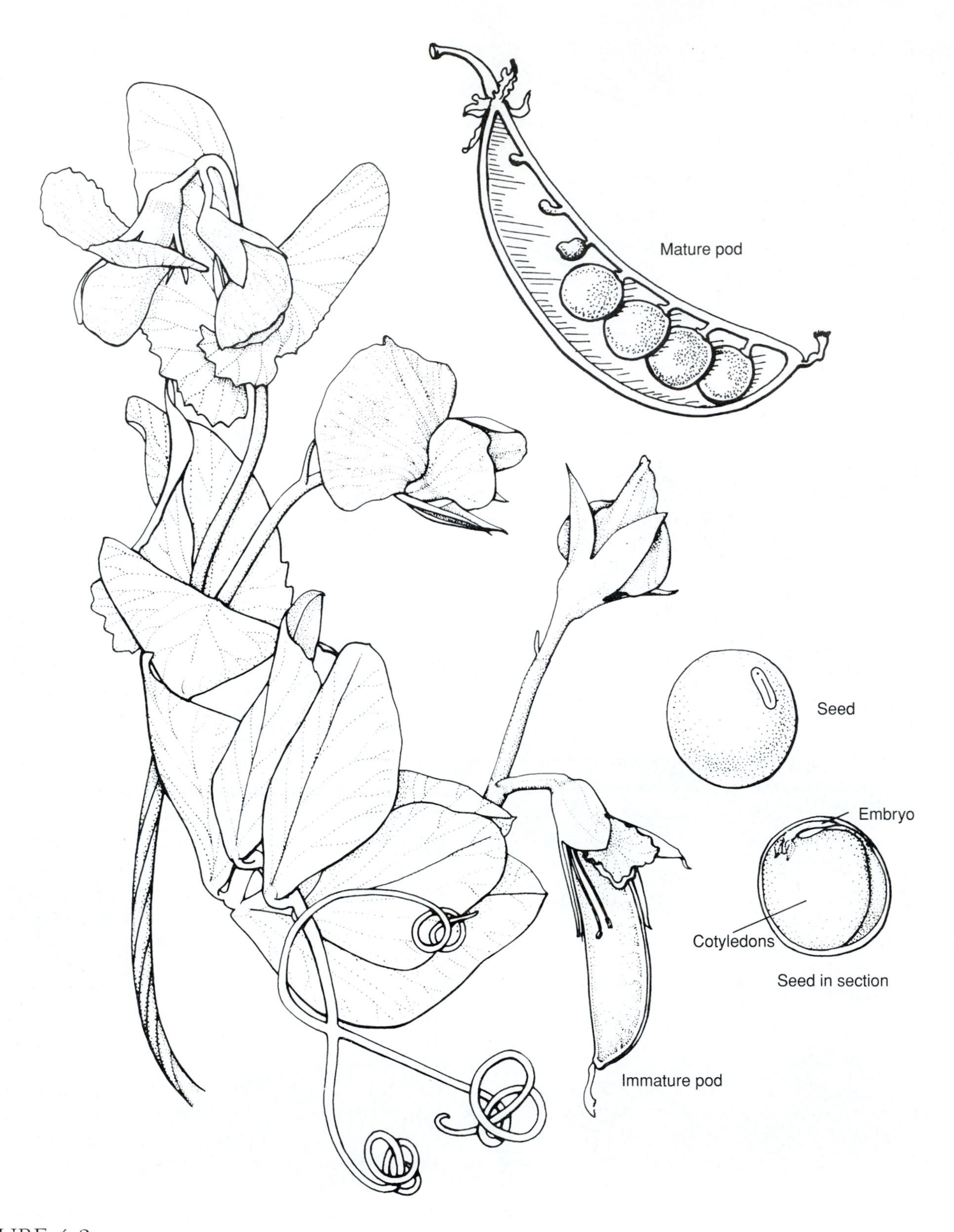

FIGURE 6.2
Snow peas are simply a variety of the common pea that are picked while they are immature. Sweet green peas are also picked before the pod is completely mature but after the seeds have expanded to their final size.

FIGURE 6.3

A close-up of a peanut plant shows the root nodules where nitrogen is fixed. While it was originally thought that all of the legume-associated bacteria were in the genus *Rhizobium,* molecular analyses have now shown that at least three and maybe four genera *Rhizobium, Brachyrhizobium, Azorhizobium* and possibly *Sinorhizobium,* are involved. Interestingly, these genera do not even appear to be closely related to one another. For any particular association, however, there is often striking specificity between the species of bacteria and its host.

(Photo by P. Cenini, FAO.)

nitrogenous compounds (Fig. 6.5). Plants with nodulating bacteria have overcome the problem of obtaining usable nitrogen because the bacteria convert atmospheric nitrogen into ammonia (Box 6.1). The cost incurred by a plant in supplying its nodulating bacteria with nutrients appears to be more than offset by the benefits gained from having a guaranteed source of nitrogen ions. For this reason, farmers sometimes inoculate legume seeds with an appropriate bacterium before sowing them into a new field.

It has been estimated that over 93 percent of the Faboideae are associated with bacteria that fix nitrogen. The percentage appears to be smaller in the other major groups of legumes. The ability of legumes to form associations with bacteria that fix atmospheric nitrogen into a usable form leads to two additional features that are of importance to humans. First, since nitrogen is not limiting to them, these legumes seem to be more lavish in its incorporation into tissues than are other angiosperms. For example, proteins all contain nitrogen, and legume seeds and foliage are especially protein-rich compared with those of other plant species. A comparison of Tables 5.3 and 6.1, shows the varying protein levels of important grains and legumes. It should be noted that, although legumes are good sources of protein, they tend to be low in some of the amino acids necessary for humans, such as isoleucine, lysine, and the few amino acids that contain sulfur. By eating other kinds of plants containing the amino acids that are deficient in legumes, one can obtain a complete spectrum of the amino acids necessary for human nutrition (Box 6.2). The eating of grains and legumes together provides a combination of nutrients and calories that is so beneficial that different pairs have supported the major cultures of the world.

The second feature of legumes that is of particular value to humans is the fact that the legume-bacteria association usually produces an excess of usable nitrogen in the soil. This means that growing legumes can provide a food crop and simultaneously fertilize the soil. Because of this property, farmers historically rotated (alternated) legume crops with other crops that would normally require the application of nitrogenous fertilizers. After nonlegume crops have been grown for 1 or 2 years, the supply of nitrogen ions in the soil is depleted and legumes are replanted.

Like grasses, legumes have undergone a number of changes associated with domestication. Some are parallel to the changes that accompanied the domestication of grains. Like most cultivated grains, the major cultivated legumes are annuals and normally produce seeds by self-pollination and self-fertilization. Consequently, the energy put into seed production by individual plants is high. Selection has been directed at characters that increase yield and maximize efficient harvesting. These characters include reduction of seed scattering, an increase in the size of seeds, and more synchronous fruiting within a plant. In addition, there is a decided difference in dormancy of seeds between the wild ancestors of cultivated pulses and the domesticated varieties. Seeds of wild populations have variable dormancy that leads to germination within a year of only a portion of the seeds of a given plant. In domesticated legumes, virtually all seeds germinate as soon as they are planted.

One issue concerning dormancy that has prompted considerable discussion is whether suppression of dormancy in lentils and perhaps some other southwest Asian pulses preceded or followed domestication. The case has been made that it would be unprofitable to harvest and plant legumes that had the normal wild type of delayed dormancy. Yields of legumes are comparatively low, and if only a small fraction of the seeds collected would be able to germinate when planted, the probability of obtaining a useful crop if a random sample

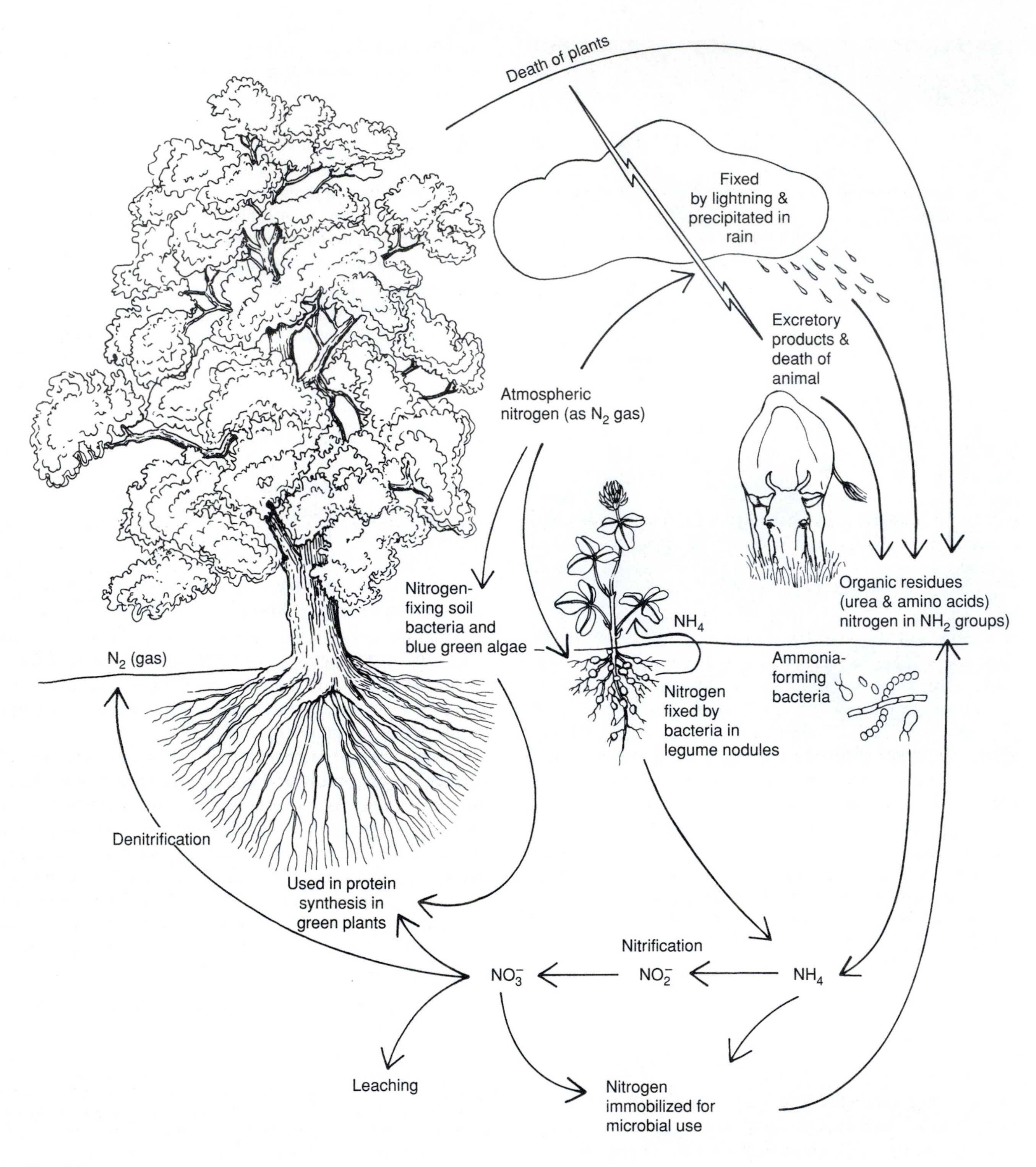

FIGURE 6.4
The nitrogen cycle, showing the role of legumes and their associated nitrogen-fixing bacteria.

of seeds were sown would have been small. One group of researchers thus argues that early farmers would not have been interested in cultivating such legumes. These scientists contend that legumes would have been an acceptable crop only after suppression of dormancy occurred in the wild. These sorts of arguments fall short even when low yields are not a factor, as in the case of grains. Moreover, the procuring of a crop has to be weighed against nutritional value and importance in balancing the diet. Conventional wisdom therefore favors the hypothesis that many cultivated legumes had

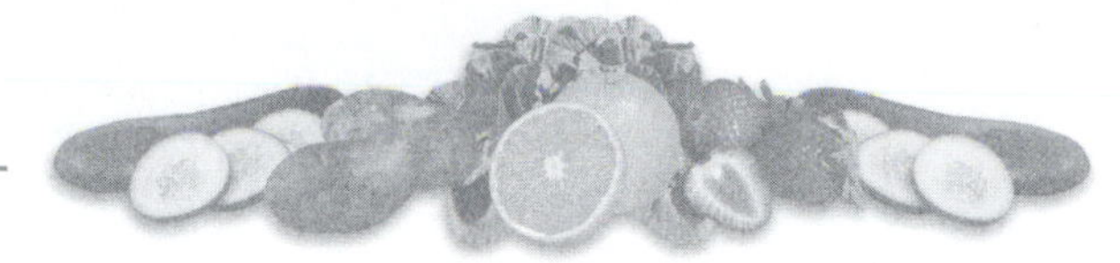

Box 6.1

Can We Turn Nonlegumes into Nitrogen Fixers?

The morphological changes that occur in legume roots during the process of nodule formation have been known for several decades. With the advent of molecular engineering techniques, researchers began to investigate the possibility of turning nonlegume crop plants into nitrogen fixers. All efforts to date have failed because so many different genes are involved. A look at the step-by-step set of interactions involved in nitrogen fixation illustrates the magnitude of the task faced by genetic engineers.

The association between the legume host and its bacterial symbiont begins with the infection of root hairs by bacterial cells that are already in the soil. Nitrogen-fixing bacteria can live as free cells in the soil, but in that state they do not fix nitrogen and must obtain their food saprophytically from decaying organic matter. After a legume seed germinates, the roots of the seedling secrete compounds that stimulate multiplication of free-living bacterial cells, increasing the probability of contact. As a root grows and produces root hairs, it comes into contact with bacterial cells that become embedded in mucilaginous compounds that it secretes. The mucilage contains chemicals, probably lectins, that are recognized by the "correct" bacterium species for that legume. If recognition occurs, the bacteria in turn release compounds that cause a root hair to curl. The tip of the cell invaginates, forming a channel by which the bacteria enter the root.

Once the bacteria, carried inside the channel as they grow inward, reach the root cortex, they invade individual cells by pinching off the channel membrane, which leaves them trapped within cell vacuoles. At no point are the bacteria in direct contact with the cytoplasm of their hosts. Once inside the cells, the bacteria release compounds that cause their host cells to divide and rapidly produce a nodule. The bacteria then change into a form no longer capable of saprophytic living, and they then take up the business of nitrogen fixing, using sugars provided by the host as metabolic fuel for the production of adenosine triphosphate (ATP), a compound that stores chemical energy. The bacteria use the stored energy to break apart the two atoms of nitrogen tightly held together in nitrogen gas (N_2).

Two major enzymes or enzyme complexes coded for by two sets of genes are involved in the cleavage of nitrogen gas and the subsequent addition of hydrogen to form ammonia (NH_4). The first is nitrogenase, a complex of two enzymes. One of these enzymes splits the bonds of nitrogen gas, and the other adds the hydrogen ions to form ammonia. The second enzyme complex produces leghemoglobin, which consists of an iron-containing, or heme, portion supplied by the bacterium and a protein part produced in the host's cytoplasm. Leghemoglobin shields the bacteria from direct contact with oxygen yet provides oxygen for cellular respiration. Nitrogen fixation is a process that must occur at very low oxygen levels.

The ammonia produced by nitrogen fixation is released into the plant cell as ammonium ions, which enter into one of the various biochemical pathways that ultimately form either amino acids or purines. Either type of these nitrogen-containing compound can be transported through the xylem to various parts of the plant for protein synthesis or as a source of nitrogen for compounds such as alkaloids.

From this description, one can see that the problem of introducing such an extensive repertoire of genes into plants that lack them is formidable. Scientists must transfer the genes involved in nodulation (including recognition genes), those that code for the polysaccharides and membranes, nodulation (including recognition genes), those that code for the polysaccharides and membrances, and those that are involved in producing hemoglobin protein. For the moment, therefore, there seems to be little hope of converting grains and other nonlegume crops into nitrogen-fixing species.

FIGURE 6.5
The numbers printed on bags of fertilizer refer to the relative proportions of nitrogen, phosphorus, and potassium. Thus, 10-20-10 has 10 percent available nitrogen, 20 percent phosphorus, and 10 percent potassium by weight. The other information on this label is required by law.

variable dormancy when they began to be grown as crops and that the loss of dormancy occurred quickly under artificial selection.

Breeding programs have been unable to increase yields of legumes to match those of grains such as corn or sorghum. Between 1930 and 1990, yields of the grains increased 11-fold while those of beans averaged only a 3-fold increase. These differences have been attributed to the difference in the fruiting strategies of the two kinds of

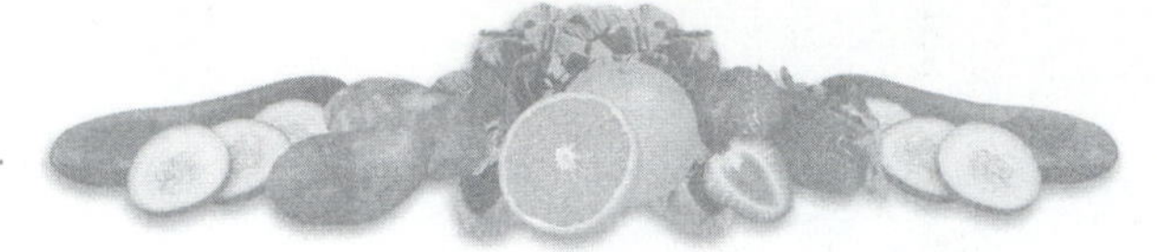

Box 6.2

Nutritional Aspects of Legumes

It has long been known that legumes are good for you and that bean and grain combinations are even better. Thanks to their association with nitrogen-fixing bacteria, legumes naturally excel at producing nitrogen-containing compounds such as amino acids and alkaloids. The family can thus provide both nutritious foods and deadly toxins.

Most dried legumes are high in protein and contain a full complement of the eight essential amino acids. These are the amino acids that human adults are unable to manufacture and that must be obtained from an external source such as meat or protein-rich plant foods. The two amino acids in which legumes are comparatively deficient, methionine and cysteine, are those in which grains are highest.

Legumes contribute other things to our diet as well. They are high in carbohydrates and, in some cases, fat. The ratio of protein to carbohydrate to oil is usually about 22:70:8, although there is variation between different legumes. Dried legumes also contain comparatively high levels of fiber, and it has therefore been suggested that their consumption is associated with reduced levels of intestinal and colon irritability and cancer.

In recent years, there has been increasing interest in the relationship between legume consumption and low levels of cholesterol, particularly among people who eat soybeans. Originally, it was thought that the association stemmed from the high levels of fiber, but interest shifted from fiber to proteins because soy-textured protein (which is fiber-free) had the same effect. Later work suggested that a group of pigments known as isoflavones, particularly genistein and daidzein, and not fiber or protein were the beneficial agents. They appear to act by preventing the oxidation of "bad" cholesterol (LDL) and thus the formation of plaque. Experiments showed, however, that the isoflavones alone, without the legume proteins, were ineffective in lowering LDL or cholesterol in general.

The association of soybean consumption and the relief of menopause symptoms in Asian women led to the discovery of other bioactive effects of legume isoflavones. These pigments appear to have a weak estrogenic effect and are considered a source of plant estrogens, or phytogestrogens, that are potentially useful in relieving the symptoms of menopause and, like synthetic estrogens, in lowering incidences of cancer.

Along with their reputation for being nutritious, legumes are also notorious for their toxic factors. These include protease inhibitors, antivitamins, cyanogens, saponins, and various factors to which people are allergic. The first two can be rendered harmless with cooking. Concentrations of the other have, for the most part, been reduced through breeding. Many legumes, especially the common bean, are associated with flatulence (see Box 6.3). Peanuts can cause a severe allergenic reaction in some people—so severe, in fact, that airlines have been asked to have "peanut-free" areas for allergic passengers. Recently, a vaccine has been developed that shows promise in eliminating the allergenic response.

TABLE 6.1 Nutritional Composition of Food Legumes, Based on a 100-g Edible Portion

					CARBOHYDRATE	
LEGUME	WATER, %	CALORIES	PROTEIN, g	FAT, g	TOTAL, g	FIBER, g
Beans						
Pinto, raw	8.3	349	22.9	1.2	63.7	4.3
White, raw	10.9	340	22.3	1.6	61.3	4.3
Broad beans, dry	11.9	338	25.1	1.7	58.2	6.7
Carob						
Flour (pulp only)	11.2	180	4.5	1.4	81.0	7.7
Flour and germ	9.8	326	47.3	2.8	34.0	3.8
Chick-peas	10.7	360	20.5	4.8	61.0	5.0
Cowpeas, dry or raw	10.5	343	22.8	1.5	61.7	4.4
Lentils, raw	11.1	340	24.7	1.1	60.1	3.9
Lima beans, raw	10.3	345	20.4	1.6	64.0	4.3
Peanuts, raw, with skins	5.6	564	26.0	47.5	18.6	2.4
Peas	11.7	340	24.1	1.3	60.3	4.9
Pigeon peas	10.8	342	20.4	1.4	63.7	7.0
Soybeans, raw	10.0	403	34.1	17.7	33.5	4.9
Tamarind, raw	31.0	239	2.8	0.6	62.0	5.1

Source: Composition of Foods. Agricultural Handbook No. 8. 1975 printing, Washington, D. C., USDA.
Note: In all cases, mature seeds are considered.

TABLE 6.2 Legumes Discussed in Chapter 6

COMMON NAME	SCIENTIFIC NAME	NATIVE REGION
Bean		
Broad	*Vicia faba*	Near East
Common	*Phaseolus vulgaris*	Mexico, Colombia (?), Peru
Lima	*P. lunatus*	Mexico, Peru
Carob	*Ceratonia siliqua* (Caesalpinioideae)	Mediterranean
Chick-pea	*Cicer arietinum*	Near East
Cowpea	*Vigna unguiculata*	Africa
Lentil	*Lens culinaris*	Near East
Pea, green	*Pisum sativum*	Near East
Peanut	*Arachis hypogaea*	Central eastern South America
Pigeon pea	*Cajanus cajan*	India
Soybean	*Glycine max*	China
Tamarind	*Tamarindus indica* (Caesalpinioideae)	Tropical Africa

Note: Unless otherwise noted, species are in the subfamily Faboideae. Forages discussed in this chapter are listed in Table 6.4.

crops and to the fact that legume seeds with their high nitrogen content are more expensive to produce than grains.

We could have included many of the legume species used by humans, but we have chosen first to discuss 10 of the most important pulses and 2 caesalpinoid fruit crops (Table 6.2). We then look at several of the major legume forages grown today. In our discussion of the important pulses, we begin with native Old World species and finish with American contributions to the world's supply of these protein-rich plant foods. Table 6.3 shows the production figures for seven of the 10 pulses we discuss that are of importance in international trade.

Pulses

Lentils

Lentils are among the most ancient plants known to have been cultivated by humans. Carbonized lentils have been found in Neolithic villages in the Middle East and dated as being between 8000 and 9000 years old. Analysis of the fossilized seeds seems to indicate that a period of domestication had occurred years before the time when the seeds were deposited. Part of the evidence for the much earlier domestication is that the seeds found in the fossil excavations are larger than the seeds of wild lentils that still grow in the area. After initial cultivation of the crop in the Middle East, lentil use began to spread around the Mediterranean. By 4200 B.C., lentils began to appear in Europe. Lentils are the first pulse to be mentioned in the Bible. According to the story in Genesis (Genesis 25), Esau, the firstborn twin of Rebecca and Isaac, sold his birthright to his brother Jacob for a meal of red lentils.

Both the common name *lentil* and the scientific generic name *Lens* refer to the flattened shape of the seed, which is similar to that of the human eye lens (Fig. 6.6). One to three seeds are produced in each pod, a smaller number than produced by most other pulses. However, as shown in Table 6.1, lentils rank fifth among the major legumes in protein content and are among the most digestible of the commonly eaten pulses. Over the many thousands of years during which they have been cultivated, lentils have been selected by humans for a range of colors and shapes. Today lentils are particularly important in India, where black, green, yellow, and red lentils form the basis of most **dals** (or dhals), purees derived from several pulses but usually from lentils or pigeon peas. Dals are an extremely important component of almost every Indian meal.

Because of their relatively high drought resistance, lentils are grown in semiarid regions scattered throughout the world. In the United States they are grown primarily in the dry portions of the Northwest, where they serve as a rotational crop with wheat.

Peas

Peas, along with lentils, barley, and wheat, form the oldest complex of cultivated foods yet discovered. Fossil seeds unmistakably identified as peas (*Pisum sativum*) have been recorded from excavations in the Near East and Europe and have been dated as being between 8000 and 9500 years old. These seeds are similar to wild peas, and thus it is impossible to tell if they were grown as a crop or gathered. Fossil seeds showing smooth seed coats characteristic of domesticated peas appear by 5850 to 5600 B.C. in Near Eastern archaeological sites. The original domestication of peas is now considered to have occurred somewhere in the area that now includes Turkey and Syria.

The nutritive value of pulses must have been known (at least subconsciously) since early biblical times. In the Book of Daniel (Daniel 1), Daniel was taken into the

TABLE 6.3 1999 Annual World Production of Pulses

COMMON NAME	TOP 5 COUNTRIES	1000 METRIC TONS	TOP 5 CONTINENTS	1000 METRIC TONS	WORLD 1000 METRIC TONS
Beans, common	India	4,550	Asia	9,312	19,363
	Brazil	2,928	South America	3,659	
	Mexico	1,514	North and Central America	3,604	
	China	1,112			
	United States	1,440	Africa	2,152	
			Europe	275	
Beans, broad	China	1,500	Asia	1,600	3,037
	Ethiopia	339	Africa	907	
	Egypt	307	Europe	278	
	Australia	115	South America	94	
	Morocco	100	North and Central America	43	
Chick-peas	India	6,500	Asia	8,304	9,108
	Pakistan	699	Africa	323	
	Turkey	600	North and Central America	211	
	Iran	248			
	Mexico	211	Oceania	180	
	Ethiopia	137	Europe	78	
Lentils	India	900	Asia	2,090	3,009
	Canada	610	North and Central America	727	
	Turkey	586			
	Bangladesh	165	Africa	75	
	Nepal	132	Europe	29	
			South America	25	
Peanuts (groundnuts)	China	12,084	Asia	21,849	32,814
	India	7,300	Africa	8,111	
	Nigeria	2,783	North and Central America	1,941	
	United States	1,736			
	Indonesia	988	South America	857	
			Oceania	41	
Peas	France	2,577	Europe	6,308	11,335
	Canada	2,262	North and Central America	2,536	
	China	900			
	Russian Federation	846	Asia	1,669	
	Ukraine	680	Oceania	441	
			Africa	290	
Soybeans	United States	59,780	North and Central America	61,881	154,902
	Brazil	30,821			
	Argentina	18,000	South America	53,059	
	China	13,704	Asia	22,598	
	India	6,100	Europe	2,608	
			Africa	890	

Source: FAOSTAT Database Gateway. Online at http://apps.fao.org/ (under "All databases" then Crops, Primary).

court of Nebuchadnezzar with several other young Jewish boys so that the Israelites could learn the ways of the ruling Babylonians. The young Daniel and a few of his fellow Jews refused to eat the meat served to them because it had not been prepared according to Hebrew specifications and the eating of it thus would have been a violation of their dietary laws. Daniel asked that they be allowed to eat pulses (probably peas) rather than meat. The story continues that the Babylonians granted the request despite their fears that the boys would show ill effects from the "poor" diet and hence bring punishment from the king. However, when they were finally brought for an audience at the royal court, Daniel and the other youths who had refused to eat the king's meat had fatter and more beautiful faces than those who had forsaken their religion.

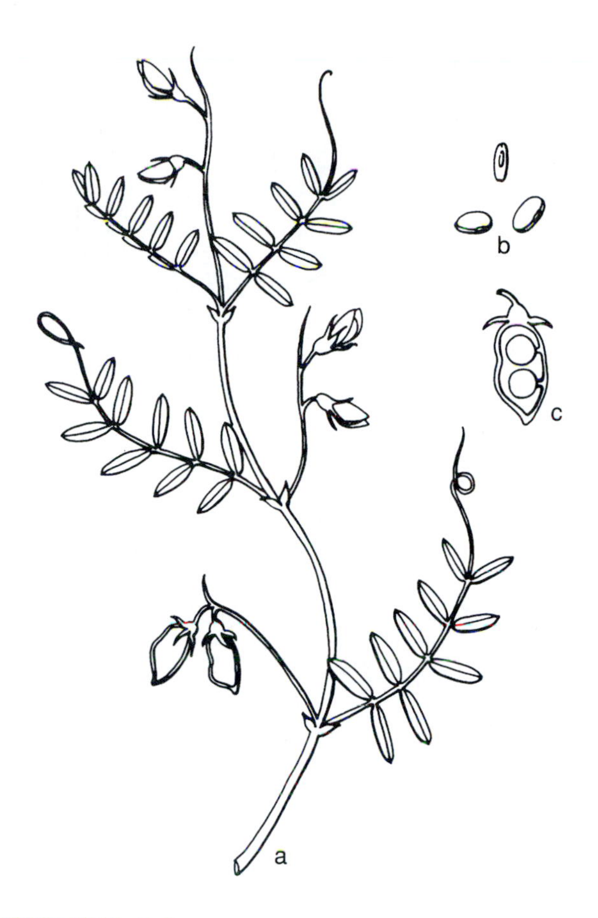

FIGURE 6.6
Lentils. (a) A branch, (b) seeds, and (c) a seed pod in section. Unlike those of its wild relatives, the pods of lentils do not split when they are ripe.

During the Middle Ages in Europe, dried peas were the mainstay of the peasant population. The familiar nursery rhyme

> Pease porridge hot
> Pease porridge cold
> Pease porridge in the pot
> Nine days old

refers to a thick broth or steaming pudding made from dried peas that, if we are to believe the text of the rhyme, could be served from the same pot for over a week.

Peas were brought to the New World by Columbus on his second voyage in 1493 and planted in the West Indies. English settlers brought the pea to New England in the beginning of the seventeenth century. Of the two introductions, the latter was by far the more successful because of the inability of peas to grow well under conditions of high night temperatures.

Peas were not eaten as a fresh green vegetable until the seventeenth century, when a Dutch horticulturist bred and offered in trade the first "garden" varieties. Green peas were enthusiastically accepted by members of the court of King Louis XIV even though at that early time fresh peas would probably have been much inferior to modern fresh peas. Peas are still one of the most important pulses on a worldwide basis, ranking fourth behind soybeans, common beans, and peanuts (Table 6.3). Breeding in recent times has sought an increase in pods per node and synchronous maturation to facilitate mechanical harvesting.

Chinese snow peas and the new sugar snap peas are pea varieties eaten when the pods are still immature and tender (Fig. 6.2). Peas that are eaten in their entirety must have tender shells. Edible pea pods lack the fibrous layer of the pericarp that is characteristic of dry pulses. Tender forms similar to modern snow peas had probably been selected for by the early part of the seventeenth century, but the fat, sweet sugar snap peas enjoyed today were developed and placed on the market only as recently as 1979.

Broad Beans

Today broad, or faba beans, *Vicia faba* (Fig. 6.7), are associated with the Mediterranean region believed to be the original home of the species. Nevertheless, a wild form sharing a common ancestor with the faba bean has not been identified.

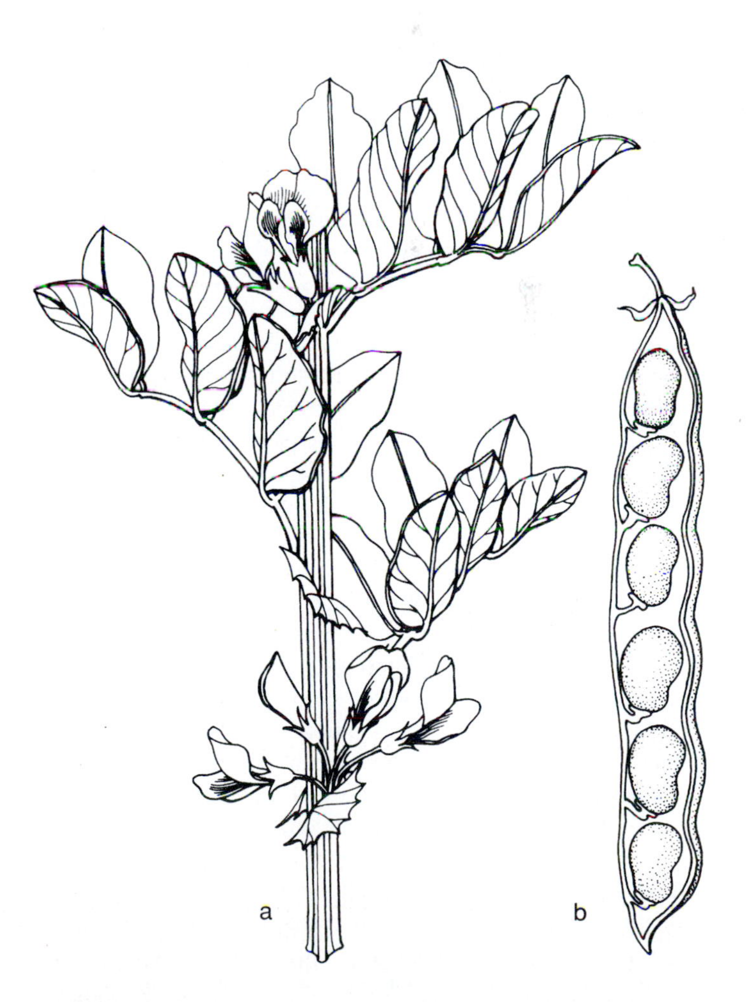

FIGURE 6.7
Foliage, flowers, and pod of a broad bean. (a) A shoot, (b) longitudinal section of a pod.

Fossil broad beans date to 8800 years before present, and cultivation was widespread in the eastern Mediterranean region in prehistoric times. Writings attest to the cultivation of broad beans by the Egyptians, Greeks, and Romans. From southern Europe, broad beans spread to Asia and later to the New World. The second governor of Colombia brought the beans with him when he arrived in South America in 1543 from Spain, but as is indicated in Table 6.3, broad beans are primarily a crop of cool regions. China is now the world's largest producer (Table 6.3), and in North America, Canada is a larger producer than the United States. In the United States production has declined to such an extent in recent years that most of the canned and dried beans offered in supermarkets are imported.

The seeds of broad beans are said to produce favism, a disease involving hemolytic anemia. The illness, which results from the breakdown of red blood cells, is most commonly found in people of Mediterranean origin. Although exposure to the beans was long thought to cause the disease, it is now known that the disease is due to a genetic disorder that results in a lack of an enzyme (glucose-6-phosphate dehydrogenase). When oxidative agents such as faba bean alkaloids are ingested by individuals with this disorder, the anemia is aggravated. Oddly, before the advent of modern medicine, the disease was often advantageous to the individuals who had it because it provided resistance to malaria in much the same way sickle-cell anemia confers malaria resistance to blacks in Africa.

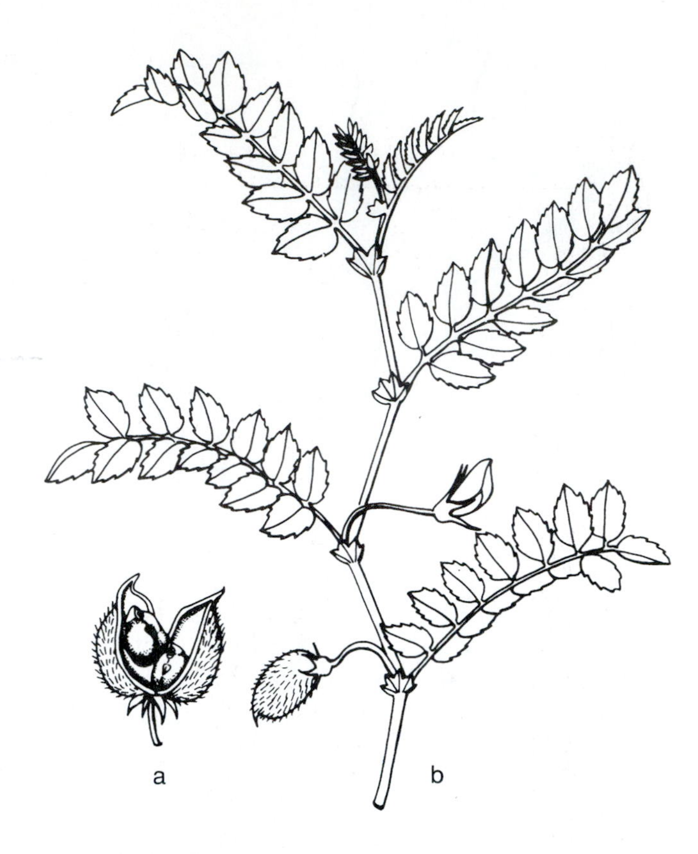

FIGURE 6.8
Chick-peas have pods that contain two or three seeds each. (a) pod, (b) plant.

Chick-Peas

Although chick-peas, *Cicer arietinum* (Fig. 6.8), are part of the ancient complex of Near Eastern domesticates that includes barley, wheat, lentils, and peas, they do not grow well in cool climates and their subsequent spread was to the east and west rather than to the northern parts of Europe. East of the Mediterranean, chick-peas replace peas and broad beans and serve, along with lentils, as a basic component of the diet of millions of people. The first certain records of cultivated chick-peas are from Iran and are dated to be 8500 years old. The species is assumed to be native to northeastern Africa. By 2000 B.C., chick-peas had been introduced to India, which now produces two-thirds of the world's crop (Table 6.3). Cultivation also spread around the Mediterranean, and today chick-peas form a part of the cuisines of Italy, Spain, Morocco, and Algeria. In India there is a preference for a smaller, darker variety of chick-pea over the globular beige European variety. Brought to Mexico by the Spanish, who call them garbanzos, they now constitute, along with common beans, a Mexican nonmeat source of protein. The quality of chick-pea protein, both in terms of amino acid content and digestibility, is among the highest of the pulses. The peas can be used to produce a liquid infant formula that is equal to that produced from soybeans and less likely to cause diarrhea.

Soybeans

Among the legumes, soybeans (*Glycine max*) must be crowned the king (Fig. 6.9). In China, soybeans are referred to as "poor man's meat" or "cow-without-bones," indicating their importance in that populous country. Although all the legumes we are discussing can be considered as partial substitutes for animal protein, the soybean has more protein (Table 6.1) and greater versatility than any of the others. Soybeans also have less carbohydrate per unit weight than any other pulse except peanuts. The amino acid content of soybeans, especially that of the sulfur-containing amino acids, is particulary impressive. These factors combine to make soybeans the most important legume crop in the world (Table 6.3). In the Western Hemisphere, soybeans are used primarily for oil extraction and animal food. Only 5 percent of the huge U.S. crop is consumed directly as food. In the east, the use of the beans is quite a different story.

Soybeans are thought to be native to northeastern China, where domestication is hypothesized to have occurred over 3000 years ago. Brought from the Orient to the Netherlands in 1737, soybeans did not become popular outside China and Japan until after 1890. Now soybeans are considered by many experts to be one of the principal plants that will help feed the world's future population. Included in the repertoire of possible foodstuffs from soybeans are

FIGURE 6.9
A new giant variety of soybeans.
(Photo by Scott Bauer, courtesy of Agricultural Research Service.)

FIGURE 6.10
The array of soybean products on the U.S. market just keep growing! Besides their obvious inclusion in infant formulas and tofu products, soybeans provide the most widely used edible oil in the United States, soy proteins are integrated into everything from breakfast bars to potato chips, and soy lecithin is incorporated into most of our chocolate products. Nonfood uses of soybeans include inks, plastics, and stain-removing cleaners.
(Photo by Scott Bauer, courtesy of Agricultural Research Service.)

curds, cheese, drinks, sauces (soy), "greens" (sprouts) for salad, and oil. The dried or fresh seeds of the soybean are bitter and contain trypsin, a compound that inhibits digestion. Roasting or cooking helps alleviate this problem, but soybeans are still comparatively little used as cooked beans. They are more often processed, especially in the Orient, to provide a variety of nutritive products. Of these, miso, tofu, tempeh, soy milk, and soy sauce (shoyu) are the most important (Fig. 6.10). Miso is a paste of cooked soybeans fermented by *Aspergillus,* yeasts, *rhizopus,* and bacteria. Red, black, and white miso are made by using different color varieties of soybeans and amounts of salt.

In making tofu, dried soybeans are soaked, rinsed, and then crushed in water. The slurry is heated to near boiling, and the liquid is then decanted. This liquid, known as soy milk, can be used as a drink and forms the basis of nondairy infant formulas used in the United States, but soy infant formulas need to be supplemented in order to equal milk in nutritional quality. The solid portion left after decanting the soy milk is called okara, a spongy mass that can be eaten like cottage cheese. The soy milk is used to make tofu. The liquid is boiled with magnesium or calcium salts that initiate the production of curds (coagulated protein). Once curd formation has started, the mixture is allowed to stand until almost all the protein has coagulated. In a manner similar to that of fresh cheese making, the curds are scooped from the now thin liquid and drained as much as possible. After sufficient draining, the curds settle into firm, smooth-textured tofu cakes. Tofu is extremely nutritious, very digestible, and quite bland, qualities that allow it to be used in a variety of dishes. If the curds are fermented, they produce tempeh.

Soy sauce is another product made from the beans. In the process, okara or crushed soybeans, usually mixed with wheat, are formed into cakes. The cakes are innoculated with *Aspergillus* to start fermentation. The starter mix is then placed in salt brine with *Lactobacillus* to complete fermentation. Traditional soy sauce needs from 1 to 3 years to reach a proper color and flavor. The filtered, now flavored salt water is soy sauce. Before you recoil from the thought of pouring juice from moldy soybean paste onto your food, remember that most cheeses are made by allowing fungi to act on milk curds (in Roquefort, blue Danish, and Stilton cheeses, the blue veins of the fungi are still visible). In fact, the fermented sediment left after straining off the soy sauce is eaten in China like jam. Nontraditional American soy sauce is often made simply by hydrolyzing vegetable protein, coloring it with caramel coloring, and flavoring it with salt water.

One of the conspicuous nutrients that most beans lack is vitamin C. The Chinese, so dependent on soybeans, have managed to overcome this shortcoming by sprouting the beans. Soy sprouts contain abundant vitamin C and are eaten in salads. It is little wonder that soybeans have been held in such high esteem in the Orient for so long and are now coming into their own in the west as well.

Soybeans were introduced into the United States in 1765 and into the Corn Belt region in 1851. However, their rise to prominence as a U.S. crop came only after 1920 when there was a change in use from growing the plants as a hay crop to the use of the fruits as a major animal feed. Before the twentieth century soybeans were not used as livestock feed because of their trypsin content. Once it was discovered that heating destroys the inhibitor, soybeans became a major component of animal feeds and worldwide demand for the crop soared. Since 1924, a meteoric rise in soybean production has made the United States the world's largest soybean supplier and has led to the legume's being called a "Cinderella crop."

One of the first Americans to become interested in soybeans was Henry Ford, who thought that the bean had as much potential as a raw material for manufactured goods as it had as a source of protein. The Ford Motor Company built three factories in the 1930s where soybean oil was extracted and used in the production of paints and plastics. Ford ate soybeans at every meal and had a suit made from "soy fabric" that he wore to a convention. His company also sponsored a 16-course soybean dinner at the 1934 "Century of Progress" World's Fair in Chicago.

Despite Ford's efforts, little of the huge American soybean crop is consumed directly. Soybeans have a bad flavor when cooked simply as "beans" because they contain a foul-tasting enzyme, lipoxidage. The beans must be soaked and blanched before cooking if they are to be eaten directly. Over half of the U.S. production is exported, and most of the rest is fed to animals or processed into oil, plastic, paints, and adhesives. This pattern may change as a result of the process of spinning soy proteins into texturized vegetable protein (TVP). Soy protein processed in this way can be flavored to imitate various meats, used as a meat substitute, or added as a filler to canned or processed meats. Many dry dog foods produced in the United States contain processed soybeans as well as grain by-products.

Pigeon Peas

Although the names pigeon pea and red gram (*Cajanus cajan*) may not be familiar ones, you have probably eaten these legumes without realizing it. In the United States pigeon peas are often used as the basis of split pea soup, although true peas are also used. In other parts of the world, however, they constitute a common and important part of the diet. India appears to be the place of origin of these small yellow pulses when the modern geographic distribution of related wild species is considered. The oldest fossil remains of domesticated pigeon peas are only 2000 years old. This legume is now of major importance in India, which produces 95 percent of the world's crop. Pigeon peas can grow on poor soil, making them a good crop for farmers using marginal agricultural land. Unlike most other pulses, pigeon pea plants are perennial shrubs, not annual herbs. In India, the dried peas are used like lentils in the preparation of dals. Since pigeon peas have remained a crop grown on a small scale by poor farmers, few large-scale efforts have been made to produce new varieties and better yields. Feeding trials with animals have shown that the protein quality of pigeon peas is lower than that of chick-peas or soybeans, but these other pulses are too expensive for many people.

Black-Eyed Peas (Cowpeas)

Africa is usually considered to be the original area of domestication of black-eyed peas or cowpeas, *Vigna unguiculata* (Fig. 6.11). It is possible that these peas formed, along with sorghum, a cereal-based agricultural system that developed in eastern Africa by 3000 B.C. There is still an extraordinary amount of diversity in the peas across Africa. Their cultivation seems to have spread both westward within Africa and eastward to Asia about 2000 years ago. Most of

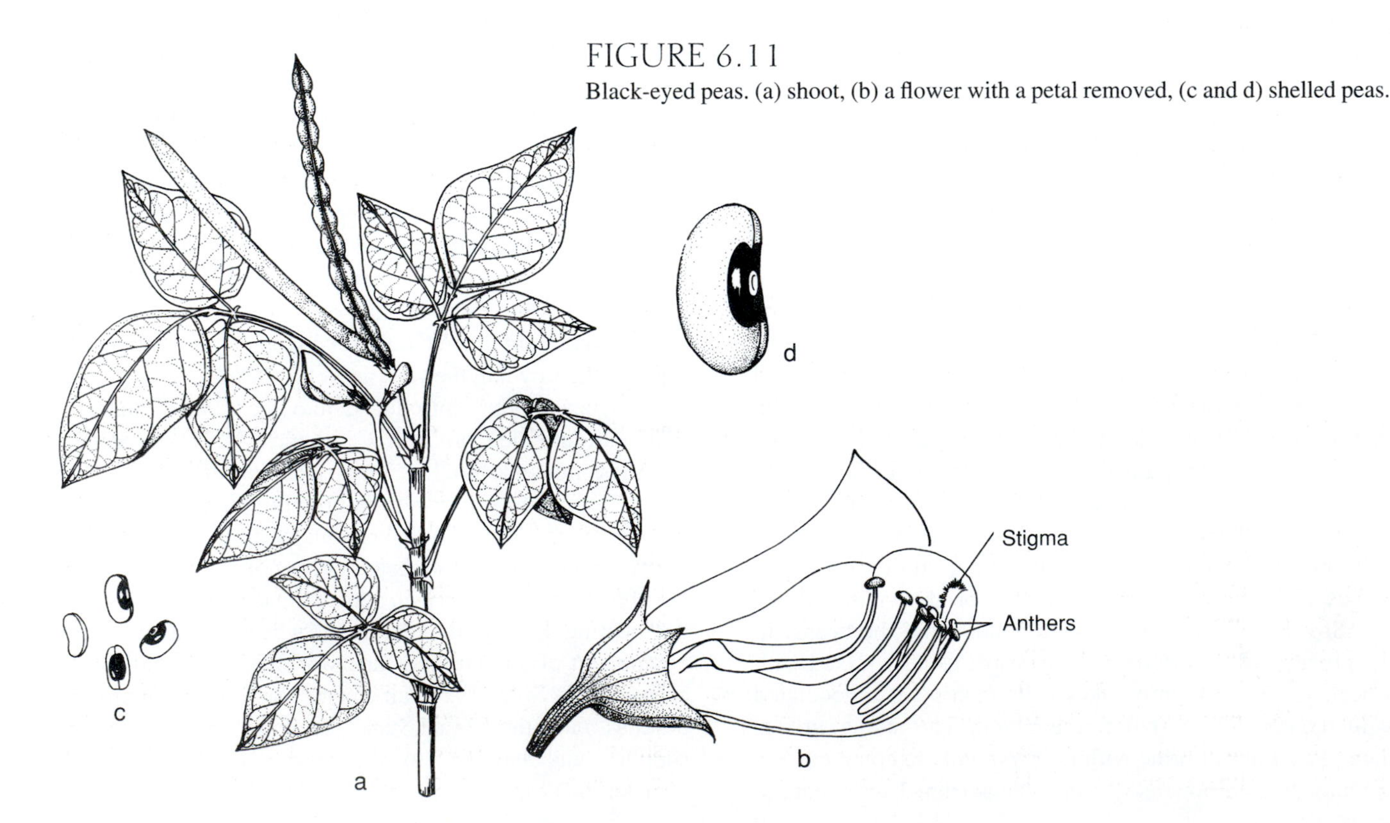

FIGURE 6.11
Black-eyed peas. (a) shoot, (b) a flower with a petal removed, (c and d) shelled peas.

the crop, however, is still grown in Africa, where the seeds, sprouts, and leaves are eaten. Black-eyed peas were introduced to the United States with slaves brought from western Africa. Today they are part of the regional cooking of the American south. In parts of the southern United States hoppin' John, a mixture of rice, black-eyed peas (or cowpeas as they are often called), and salt pork, is a customary dish on New Year's Day. Tradition says the stew must be eaten before noon on January 1 to ensure good luck during the new year.

Lima Beans

The lima bean, *Phaseolus lunatus,* now has a natural wild distributional range from Mexico to Peru and is considered to have been independently domesticated in the central Andes and in Mexico. Archaeological sites containing fossilized beans in layers with material dated to be 5600 years old have been found on the northern coast of Peru. In Mexico the oldest limas found in archaeological deposits are only 571 years old. The common name of the bean, *lima,* comes from the place from which the beans were originally shipped to Europe: Lima, Peru. As in the case of peas, lima beans have historically been used in dried form. In many tropical areas shelled and dried limas are still the primary product for which the crop is grown. In the United States dried limas are used in soups, but most lima beans are consumed as shelled, immature fresh (including frozen and canned) seeds.

Some cultivars of lima beans contain cyanogenic compounds that release cyanide when the beans are chewed or ground. The amount of these compounds allowable in beans sold for foods is set by the U.S. Food and Drug Administration (FDA) in the United States. Consequently, the amounts present in commercial forms are negligible, and cooking destroys the enzymes that liberate the cyanide.

FIGURE 6.12
Varieties of *Phaseolus vulgaris* eaten as shelled beans include (left to right from the top): Bertolli, Yellow eye, Cannellini, Appaloosa, Navy, Black, Pinto, Flageolet, Anasazi, Northern, Red.

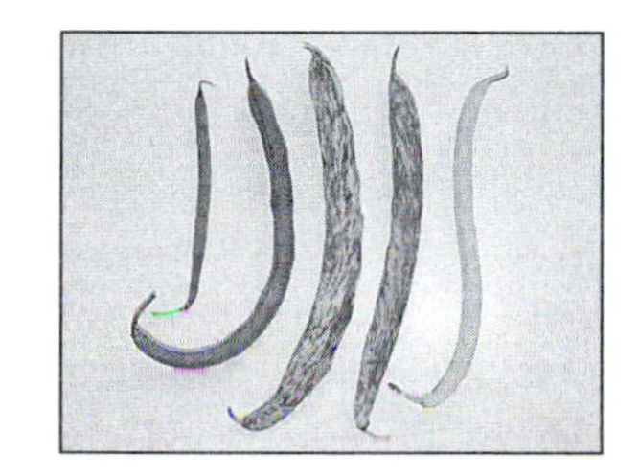

FIGURE 6.13
The variety of common beans eaten whole includes (left to right) French haricot verts, snap beans, dragon tongue beans, cranberry beans, and yellow wax beans.

Common Bean

In the United States, one can find a wide variety of beans (Fig. 6.12) such as kidney beans, navy beans, shell beans, pinto beans, pea beans, Mexican beans, black beans, runner beans, green beans (Fig. 6.13), string beans, wax beans, and snap beans. Despite the apparent differences in these commonly encountered beans, all belong to the same species, *Phaseolus vulgaris.* The common bean appears to have been independently domesticated in at least two regions: Mexico and Central America and the south-central Andean region of South America. The independent domestications of legumes have led to two separate gene pools of cultivars that are morphologically distinctive and cross only with difficulty. Early cultivated fossil material from Mexico is about 2500 years old. Material from the coastal valleys of Peru has been dated to 4400 years ago. By the time of European arrival in the New World, common beans were an important dietary item for native peoples throughout North, Central, and South America. Today *P. vulgaris* is the most widely cultivated species of legume in the world. Soybeans outrank common beans in terms of tons of legumes produced each year (Table 6.3), but common beans are an important crop in more countries than soybeans. Common beans do, however, lead to flatulence more readily than other pulses (Box 6.3).

Because plants of the common bean are naturally vines, they were historically often grown with another crop, such as corn, that provided the vines with a support. As was emphasized early in this chapter, this mixed cropping was also advantageous for the corn, since beans are associated with nitrogen-fixing bacteria. Native Americans and European settlers, without realizing why, must have found that growing the two crops together increased the production of corn. The Native American dish succotash reflected the method of mixed agriculture, as it was made from dried corn and dried common beans. Our modern idea of succotash as being a mixture of corn and lima beans comes from the name on packages of frozen vegetables.

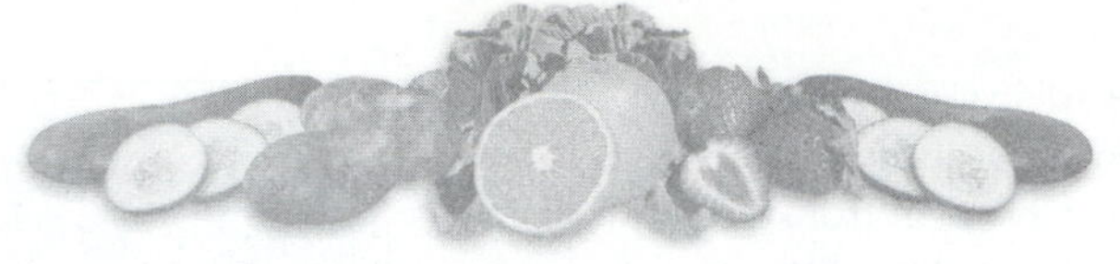

Box 6.3

Preventing Legumes From Tooting Their Own Horn

It is no secret that some people have a problem with beans, even the most innocent-looking forms. The tasty soup or dip goes down easily, but after a few hours it can turn the large intestine into a gas-generating machine that in its most docile form makes one feel bloated and in a more aggressive mode leads to explosive and embarrassing eruptions. For years the potential social problems were outweighed by the nutrition and tastiness of beans. However, with other sources of protein at hand and an emphasis on politeness, the pros and cons of bean consumption are now often weighed by Americans who subsequently opt to forgo the potentially offending pulses.

Embarrassment was eclipsed by medical problems once people began to fly, first at high elevations and then in spaceships. Intestinal gas, like all gas, expands under reduced pressure, and the air pressure in airplanes and spacecraft, even those with pressurized cabins, is lower than that on the surface of the earth. Severe pain and distension can result when there are large amounts of gas in the intestines of people who fly. Consequently, the U.S. military began to investigate the causes and potential solutions of the bean gas problem. By the 1960s it had been determined that the sources of the problem were low molecular weight sugars, or oligosaccharides, primarily raffinose and stachyose. Beans and some other foods, such as broccoli and its relatives, contain enough of these compounds to cause problems. Unlike simple sugars, oligosaccharides are not broken down by human digestive enzymes, so they arrive at the end of the intestinal system more or less intact.

Although inert to us, the oligosaccharides constitute a usable food supply for the bacteria in the large intestine, which metabolize them readily with a corresponding release of carbon dioxide (CO_2). It is this gaseous product of the bacterial metabolism, mixed with odoriferous compounds such as hydrogen sulfide, that produces foul flatulence.

Recently, an enzyme, alpha-galactosidase, has been found that is able to break down these oligosaccharides. The enzyme was isolated from *Aspergillus,* a fungus that has other uses in human food processing, such as the production of sake (Chapter 14). Only a few drops will sever the long oligosaccharide chains in a serving of beans into shorter sugar chains that humans can digest. An enzyme solution recommended for people plagued by problems with broccoli and beans is marketed under the trade name Beano and is touted as "a scientific and social breakthrough." Like most enzymatic proteins, alpha-galactosidase denatures if temperatures are too high. Consequently, the enzyme cannot be added before or during cooking. It should be mixed in when the food reaches a temperature of less than 54°C (130°F).

At this point the question is legitimately raised whether the use of the enzyme increases the caloric content of beans and other such flatuliferous foods. In theory, the answer is yes. The fundamental reason for adding the enzymes is to cleave the oligosaccharides into pieces that are digested before they reach the end of the digestive tract. If the normal situation is for a certain percentage of the sugars to be indigestible, the obvious answer is that there must be an increase in calories. However, the amount of oligosaccharides in these foods is so small—although the effect is large—that dumping them into the digestible pool is negligible in terms of calories.

Peanuts

Peanuts are legumes that are more often thought of as nuts or as an oil seed crop (see Chapter 9) than as a pulse, but in many parts of the world the seeds are cooked and eaten much like any other grain legume. The species from which we obtain peanuts, *Arachis hypogaea,* is native to central South America, perhaps eastern Peru. Domestication probably occurred first in southern Bolivia and northwestern Argentina. By the time Columbus reached the New World, peanuts were cultivated throughout the warm regions of South America. The Portuguese took peanuts to Africa, where their cultivation was quickly adopted. Peanuts are now an important dietary item in west African countries. Peanuts were also taken to Southeast Asia via the Philippines by the Spanish.

Peanuts are called by different names in various parts of the world. The British name, groundnut or ground pea, refers to the way in which peanuts bear their fruits (Fig. 6.14). Like other legumes of the subfamily Faboideae, peanuts bear pea type flowers above the ground. Self-pollination occurs within the flowers. After fertilization, instead of producing a crop of legumes on the stems, the pedicels, or flower stalks, curve downward. The cells under the ovary begin to divide, forcing the ovary into the ground. As the peg develops, a cap of cells forms next to the withered style. This cap protects the ovary as it is pushed into the soil in much the same way the root cap protects a root. After the expanding ovary is pushed a few centimeters into the soil, downward growth ceases. The ovary turns sideways and matures underground. Many people unfamiliar with the taxonomy or flowering process of peanuts think the fruits are roots or tubers (Fig. 6.14).

Peanuts are also called goobers, a name that was brought with the peanut back to the New World by African slaves. The widespread production of peanuts in the

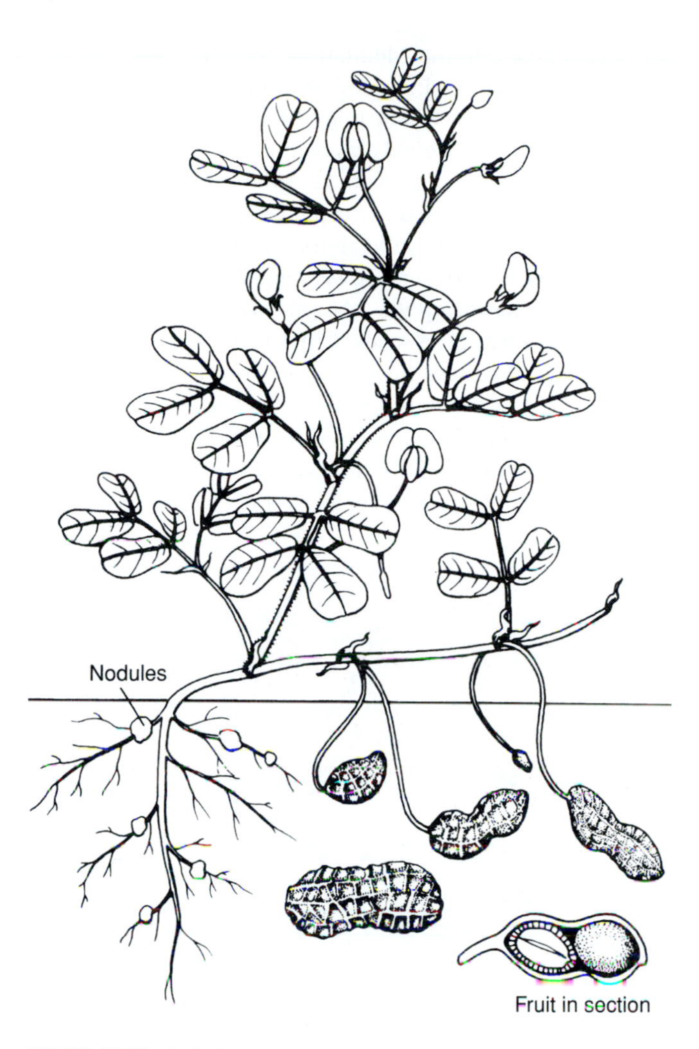

FIGURE 6.14
After fertilization, the flower pedicels of the peanut curve downward, and the developing fruit is forced into the ground by the proliferation and elongation of cells under the ovary. The legume subsequently develops underground.

FIGURE 6.15
George Washington Carver is often credited with establishing the peanut industry in the south after the Civil War had destroyed the cotton empire.
(Courtesy of USDA.)

American Southeast can be attributed not to its being native in the New World but to the popularity of the crop in black Africa and its secondary introduction to North America by slaves on southern plantations. With the decline of the cotton empire, peanuts have partially replaced cotton as a major crop in that part of the United States (Fig. 6.15). The popularity of peanuts now extends to all parts of the American population.

Most of the peanut consumption in the United States today occurs in the form of peanut butter, although a substantial part of the crop is used for hog feed. South American Indians ground peanuts and produced a product similar to modern peanut butter, but "development" of the creamy paste now used is often attributed to a physician thought by some to be John Harvey Kellogg (see Box 5.1). The doctor, a health food advocate, was searching for an easily digestible, highly nutritious food that required little effort (chewing) to eat. Peanut butter gained notice at the 1904 St. Louis World's Fair, but it became popular only after 1922, when J. J. Rosefield developed a smooth paste that did not separate into an oil and solid layer. Rosefield licensed his process to Swift and Co., which started selling Peter Pan® in 1928. Rosefield began marketing his own peanut butter, Skippy®, in 1933.

Peanuts have been the subject of much agricultural research because they are a good protein source, grow on poor soils, and thrive in tropical regions. However, they are susceptible to infestation by a fungus that produces aflatoxins as by-products of metabolism. Aflatoxins are chemicals that are deadly poisonous to humans. Deaths from eating contaminated peanuts have been greatly reduced by instituting proper methods of handling and storage that can virtually eliminate the danger of infestation.

Tamarind and Carob

Tamarinds, *Tamarindus indica* (Fig. 6.16), have long been used in Africa, where wild plants grow in the tropical dry savannas, and in southern Asia. The ripe long brown pods borne on tall, spreading trees are used primarily for their tart, rather sticky mesocarp, but the seeds can also be roasted or boiled and eaten after removal of the seed coat. Tamarinds are rarely grown on plantations. The plants are usually grown from seed, and the pods are harvested from semiwild trees. In the United States and Mexico tamarinds are used primarily as flavoring in sauces. In India the fruit pulp is an integral part of chutneys and sauces.

Carob is another caesalpinioid legume that produces pods used primarily for their pulp. The species (*Ceratonia siliqua*) is native to the Mediterranean region. Its common name, St. John's bread, comes from the fact that it constituted the "locust" on which John the Baptist fed. In ancient times carob seeds were used as weights for small quantities of precious substances such as gold because they are extremely uniform in size. Our modern unit the carat, used for gold and jewels, is a reflection of this former use.

Traditionally, carob pods were gathered from wild trees and the sweet mesocarp was chewed from the endocarp surrounding the seeds. The seeds themselves have been used to make a coffeelike beverage. Today carob trees are propagated by seed or by grafting. Since the species is primarily dioecious, grafting ensures a large proportion of female trees. Production is still highest in the Mediterranean region, with the tiny island of Cyprus being the world's largest producer.

FIGURE 6.16
(a) Tamarind pods hanging from a branch. (b) An individual flower showing the characters of the flowers of the subfamily Caesalpinioideae (compare with Fig. 6.1). (c) A longitudinal section of a flower shows the simple ovary that develops into the legume.
(After Baillon.)

The two most important uses of carob in the United States are as a chocolate substitute and as a source of locust gum, a polymer extracted from the endosperm of the seeds (Chapter 10).

Forages

Just as legumes complement grains for human consumption, legume forages supplement forage grasses and feed grains for domesticated herbivores. In addition to affording the best mixture of foods for grazing animals, legume-grass herbage associations are beneficial for other reasons. Research has shown that legumes alone do not build soil as well as herbaceous mixtures do. One of the best ways to improve soil is to plant a grass-legume mixture and use the area prudently for pasturage. Simply harvesting the hay from such a field removes much of the potentially available nitrogen, and the plowing under of the cover without cutting precludes any direct profit. Grazing animals on the field allows conversion of part of the vegetable biomass into a cash crop (meat), permits the roots of the grass and legumes to build up the soil, and provides the addition of considerable nitrogen in the form of animal urea. Where conditions permit, such a system is superior to letting the fields go fallow or using a simple crop rotation system.

The growing of grasses with legumes also circumvents problems presented by pure fields of forage legumes. Ruminants do poorly on a diet consisting principally of legumes because the fermentation of large amounts of fresh leguminous material in the rumen causes bloat. Forage legumes also generally do not grow well in pure stands because they can be overrun with weedy grasses and herbs. The growing of legumes with desirable grasses alleviates the problem of weed invasion. If the crop is to be cut for hay, growing a suitable mixture of grasses and legumes permits direct baling of high-quality hay.

Several hundred legumes are used for forage on a worldwide basis, but a few predominate, especially in the United States. Unlike grasses, native legumes are used less for forage in the United States than are introduced domesticated species. Table 6.4 lists some temperate forage legumes that are important in the United States. Of these, alfalfa and the clovers are by far the most important. We therefore discuss these two kinds of forage in some detail and then mention some of the less important American legume forages.

Alfalfa (*Medicago sativa*), or lucerne, as it is called in most of the Old World, is king of the forage legumes. Alfalfa was probably the only forage crop cultivated in prehistoric times. It is now extensively grown on every continent except Antarctica, with an area of cultivation that exceeds 33 million hectares. This acreage is a little larger than the amount of land on which the common bean is planted (approximately 24 million hectares, acres, in 1981). Although cultivated alfalfa is tetraploid, primitive diploids occur in Iran and other parts of the Near East believed to be the area of original domestication. It is likely that the spread of alfalfa cultivation followed the adoption of horse husbandry, extending to the north, east, and west from the Near East after 300 B.C. Introduction to the United States occurred in about 1850, although alfalfa had been cultivated earlier in Mexico and South America. An initial stumbling block to the use of alfalfa in north temperate areas was its lack of winter hardiness. After

TABLE 6.4 Major Legume Forage Plants of North Temperate Regions

COMMON NAME	SCIENTIFIC NAME	CYCLE	NATIVE REGION
Alfalfa, leucerne	*Medicago sativa*	Perennial	Near East
Bird's-foot trefoil	*Lotus corniculatus*	Perennial	Eurasia
Clover	*Trifolium*		
Alsike	*T. hybridum*	Perennial	Eurasia, Asia Minor
Arrowleaf	*T. vesiculosum*	Annual	Eurasia
Crimson	*T. incarnatum*	Annual	Southeastern Europe
Egyptian (berseem)	*T. alexandrinum*	Annual	Near East
Kura	*T. ambiguum*	Perennial	Near East
Persian	*T. resupinatum*	Perennial	Near East
Red	*T. pratense*	Perennial	Eurasia
Strawberry	*T. fragiferum*	Perennial	Mediterranean
Subterranean	*T. subterraneum*	Annual	Mediterranean
White	*T. repens*	Perennial	Near East
Lespedeza	*Lespedeza*		
Korean	*L. stipulacea*	Annual	East Asia
Sericea	*L. cuneata*	Perennial	East Asia
Striate	*L. striata*	Annual	East Asia
Sweet clover	*Melilotus officinalis*	Biennial	Eurasia
	M. alba	Annual	Eurasia

1900 breeding efforts overcame this problem, and the acreage devoted to alfalfa in the northern United States and Eurasia escalated rapidly.

Alfalfa is a perennial that is grown from seed. It can be used as a pasture crop or harvested for hay. In warm regions with an adequate moisture supply, up to six to nine cuttings can be made each year. In dried form, alfalfa makes an excellent fodder, but pure alfalfa hay is comparatively expensive. Consequently, alfalfa is usually grown with forage grasses so that the baled harvest will contain a proper mixture of the two kinds of forage.

True clovers, species of the genus *Trifolium,* are the next most important group of forage legumes. In the United States red clover (*Trifolium pratense*) and white clover (*T. repens*) are the most widely grown species (Fig. 6.17). Both are perennials that are native to the Old World (Table 6.4). The two clovers were independently introduced to North America by European colonists. Red clover is the most extensively grown of all the clovers on a worldwide basis, but white clover is the most important in many temperate regions. Like alfalfa, clovers are often planted with grasses and can be used as pasturage or cut for hay. The other three major clovers grown in the United States are alsike clover (*T. hybridum*), arrowleaf clover (*T. vesiculosum*), and crimson clover (*T. incarnatum*).

Sweet clovers, primarily *Melilotus officinalis* and *M. alba,* constitute the other important "clovers" grown in the United States. One reason for the popularity of sweet clovers is their superior ability to improve soils. Both species produce long nodulated roots that aerate the soil well during growth. Once senescence begins, the roots rapidly decay, releasing precious nutrients at considerable depths in the soil.

The lespedezas (*Lespedeza* spp.) are another major group of pasture and hay legumes grown in the United States. All are native to eastern Asia and were successfully introduced into this country only after 1919. Despite this rather late start, their cultivation rose rapidly. Annual lespedezas make excellent pasture forage, and plants can be harvested for hay at the end of the summer.

Bird's-foot trefoil, *Lotus corniculatus,* is a forage legume that is extremely adaptable to a wide range of climate and soil conditions and is very persistent. It has proved to be especially useful in the northeastern United States, where heavy soils predominate.

FIGURE 6.17

Clover. (a) An entire flower. (b) A sectioned flower. (c) A shoot showing the compound leaves typical of legumes and the flowers. The clover inflorescence, often mistaken for a "flower," is made of a cluster of the flowers shown in a and b.

(After Baillon.)

Foods from Leaves, Stems, and Roots

One enterprising analogy considers plants as animals turned inside out, with the leaves representing lungs and the roots, with all their ramifications, representing the intestinal system. This comparison, which was made by Baron Justus von Liebig in 1829 in his publication *Natural Laws of Vegetation,* reflected the views of some of the early investigators of plant physiology. Quaint notions such as interpreting roots as the digestive organs of a plant may seem ridiculous today, but they are not much more erroneous than the common modern impression that any portion of a plant growing underground is a root and that all the aboveground green or woody portions are stems or leaves. In this chapter we discuss stems, leaves, and roots in more detail than was done in Chapter 1 and relate their diverse natural structures and functions to the uses people make of them as food crops.

The Structure and Function of Stems and Leaves

In Chapter 1 we mentioned that stems and leaves are basic organs of flowering plants although both are parts of the shoot system. Functionally, it is useful to recognize a distinction between stems and the leaves they bear: stems generally provide support for the aboveground plant parts and house the conduction system, whereas leaves are the usual site of photosynthesis. When one looks at the anatomy and cellular structure of stems and leaves, the distinctions between the two become even more clear.

Technically, a stem consists of nodes where leaves diverge and the intervening internodal regions. In some monocotyledons, such as bananas, the structure called the stem appears to be a collection of leaf bases (Fig. 7.1). In the uppermost parts of a banana stem (Fig. 7.2) the same type of arrangement is evident, although it is obscured lower in the stem. In these plants, as in all monocotyledons, the vascular system consists of separate clusters or bundles of xylem and phloem cells scattered throughout the stem (Fig. 7.3). The scattered appearance is a reflection of the fact that the stem is composed of layers of vascular traces from the leaf bases. In contrast, most herbaceous dicotyledons have vascular bundles that are arranged in a ring near the periphery of the stem (Fig. 7.3). In woody

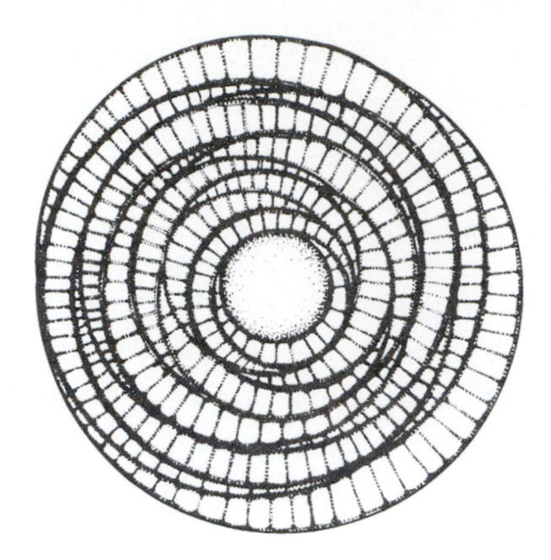

FIGURE 7.1
A cross section of a banana tree shows the overlapping leaf bases surrounding the flowering stalk; together, they form a false stem.

FIGURE 7.2
The stem of this banana plant is actually formed by leaf bases.

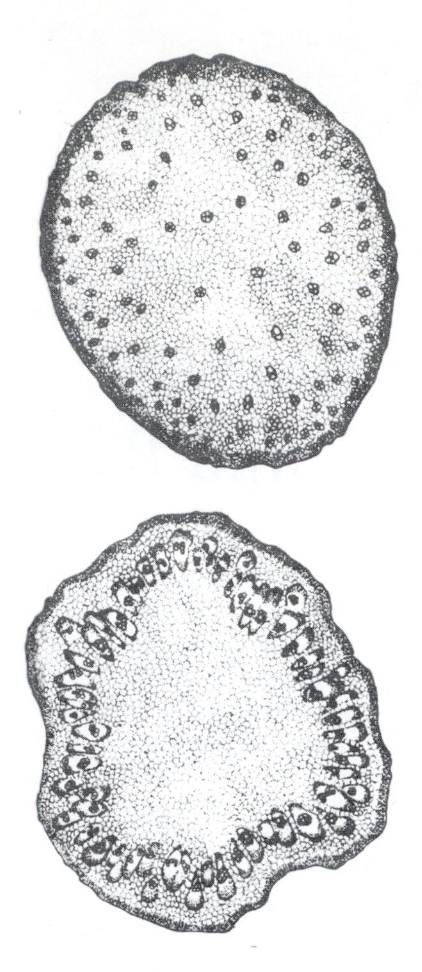

FIGURE 7.3
Vascular bundles are scattered in herbaceous monocotyledonous stems such as corn (top), whereas they are arranged around the perimeter of the stems of herbaceous dicotyledons such as the sunflower (bottom).

perennial dicotyledons, the cylinder of conductive tissue opens for a short distance as lateral bundles diverge into branches or leaves (Fig. 7.4).

Meristems

In both stems and leaves, growth takes place at a number of rather precise places. Unlike animals, in which almost all living cells can divide, plants have only certain cells that retain the ability to divide and produce new tissue under natural conditions. A mature plant cell becomes encased in a rigid cell wall, preventing it from undergoing further division. Regions composed of plant cells that remain immature and are able to divide continuously are called **meristems** (from the Greek word meaning "divisible").

Meristems can be classified in several different ways. One method is to group them into primary and secondary meristems. **Primary meristems,** such as the ground, or apical, meristem, are regions of cell division throughout adult

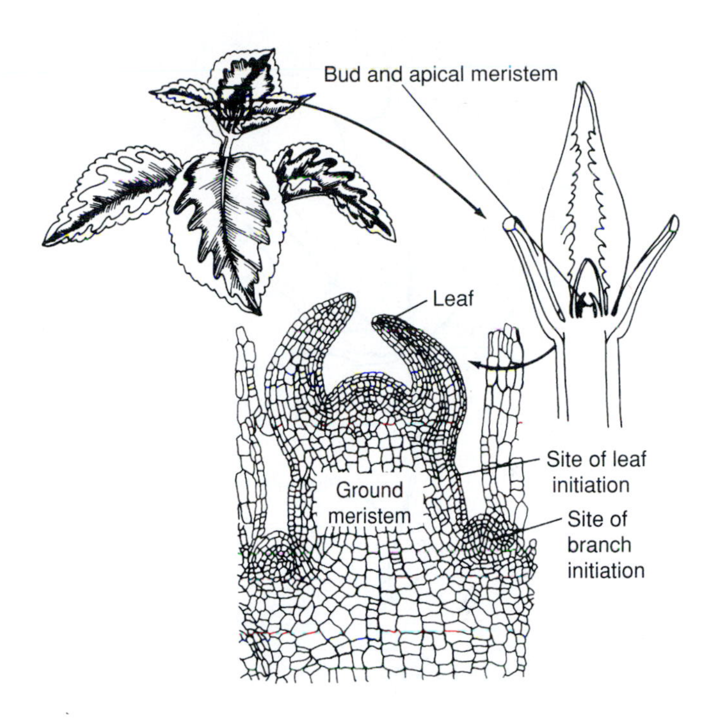

FIGURE 7.4
The apical (ground) meristem of a *Coleus* seen under different degrees of magnification has a precise arrangement of cells.

life (Fig. 7.4). **Secondary meristems** are derived from cells produced by primary meristems and generally develop in areas of comparatively mature tissue such as bark (Chapter 16). Another way to classify meristems is as terminal or lateral meristems, depending on where they occur in the plant. **Terminal,** or **apical meristems** occur at the ends of stems, branches, and roots. They mainly produce growth in length. **Lateral meristems** are bands or patches of tissue located parallel to the sides of the plant parts in which they occur. During cell division, lateral meristems cause growth in width or circumference (Fig. 7.5). We will discuss lateral meristems in more detail when we treat fibers (Chapter 15) and wood (Chapter 16). Here we are concerned with shoot terminal meristems.

Terminal meristems are found in buds at the tips of stems or roots and in buds located in the axils of branches (secondary stems). Knowledge of the location of axillary buds allows one to determine where a leaf "begins" since there is always a bud between the stem and the base of a leaf but not at the juncture of a leaflet and the axis of a compound (several-parted) leaf (Fig. 7.6). When a leaf is attached to a branch, the axillary buds are usually dormant, but once the leaf has been shed, the buds can become active and produce a branch with leaves, each of which will have in turn an axillary bud. Axillary buds can also produce flowers or flowering shoots. In contrast to leafy branches, flowers or flowering shoots usually appear while the leaf is still attached. Axillary buds of stems are generally found above crescent-shaped scars that are the remnants of aborted leaves. In fact, the presence of such buds on organs growing in the soil (for example, in Fig. 7.31) indicates that the organ is a stem rather than a root. In natural situations the sprouting of new shoots from the buds of underground stems constitutes a method of vegetative reproduction.

At this point you might ask how leaves and flower parts expand if the only areas of cell division are the meristems mentioned previously. These structures expand as they reach maturity by means of cell elongation and cell division in localized regions of a leaf or flower. Cells in these patches lose the capacity to divide once the leaf or flower part has fully expanded.

The Structure and Function of Roots

The fallacy of defining roots simply as the underground parts of a plant is evident when one considers that rhizomes, tubers, corms, and bulbs, all modified stems (see Fig. 1.11b, d, f), and even some fruits, such as peanuts (see Fig. 6.14), can develop underground. Moreover, not all roots are limited to growing in the soil. Some can row underwater or even in the air. Commercially important edible roots are often lumped together with tubers (potatoes) and rhizomes (Jerusalem artichokes) into "root crops" because all these structures can be similarly shaped, fleshy, and full of stored nutrients. However, roots lack the telltale leaf scars and axillary buds of underground stems (Table 7.1). Although this superficial difference in the two kinds of underground organs constitutes the easiest way to distinguish roots from stems, the two structures actually differ in many other aspects of their anatomy and physiology. Since roots do not carry out photosynthesis, they lack **stomates,** openings that allow gas exchange. Because they must absorb water and dissolved nutrients, they have root hairs, each of which is an extension of a single cell but is capable of probing between soil particles as it takes up aqueous solutions (Fig. 7.7).

Like stems, roots have apical meristems, but the pattern of growth is different from that of shoots. When the root grows downward, pushing its way through abrasive soil, cells from its **root cap** are torn off. The root cap provides protection for the meristematic region where primary xylem, phloem, and cambium and new cap cells are produced (Fig. 7.8). Instead of occurring as discrete bundles or rings inside the phloem (as in stems), the primary xylem in roots forms radial strands. In dicotyledons, the strands merge in the center of the root and the xylem forms a core with radiating arms. Clusters of phloem cells occur in the spaces between the xylem arms (Fig 7.8). In the roots of woody plants, the cambium can eventually form a cylinder

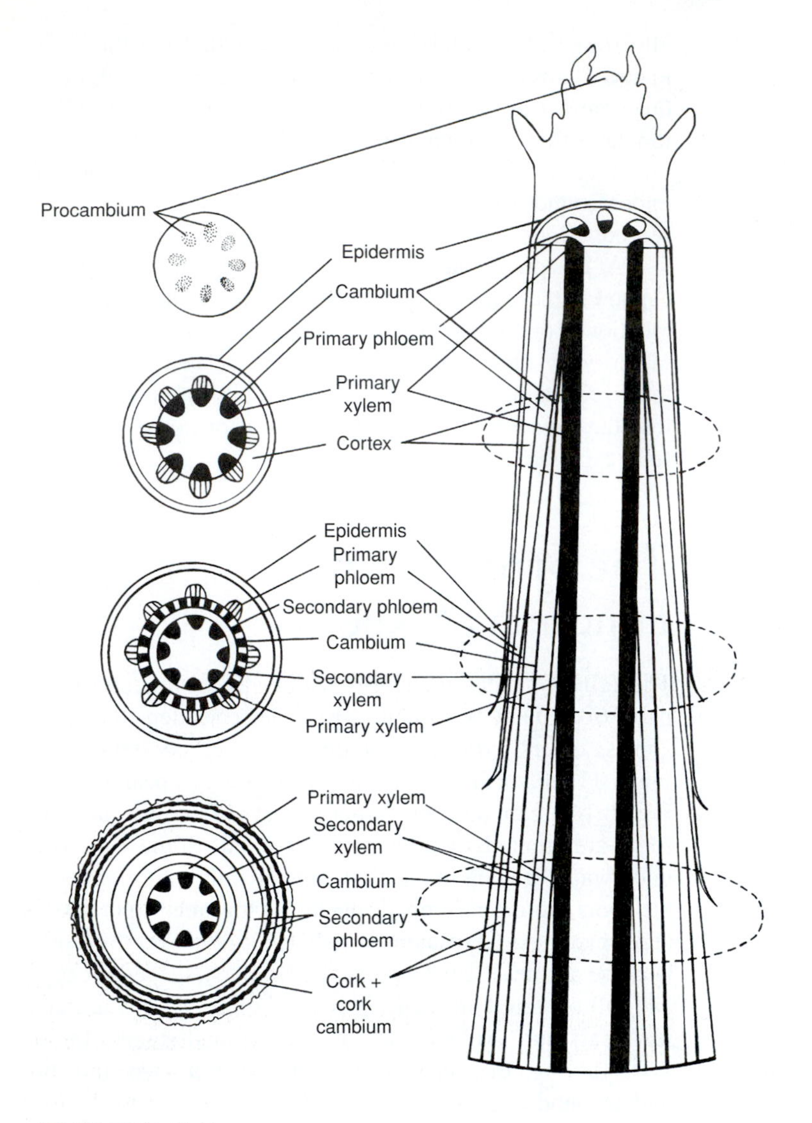

FIGURE 7.5

A diagrammatic representation of a plant's vascular system. In the top section, newly formed cells are differentiating to form a procambium. Farther down the stem these cells form the primary xylem and phloem separated by a cambial layer which is responsible for the subsequent production of xylem and phloem cells. This cambial activity results in lateral growth in woody plants.

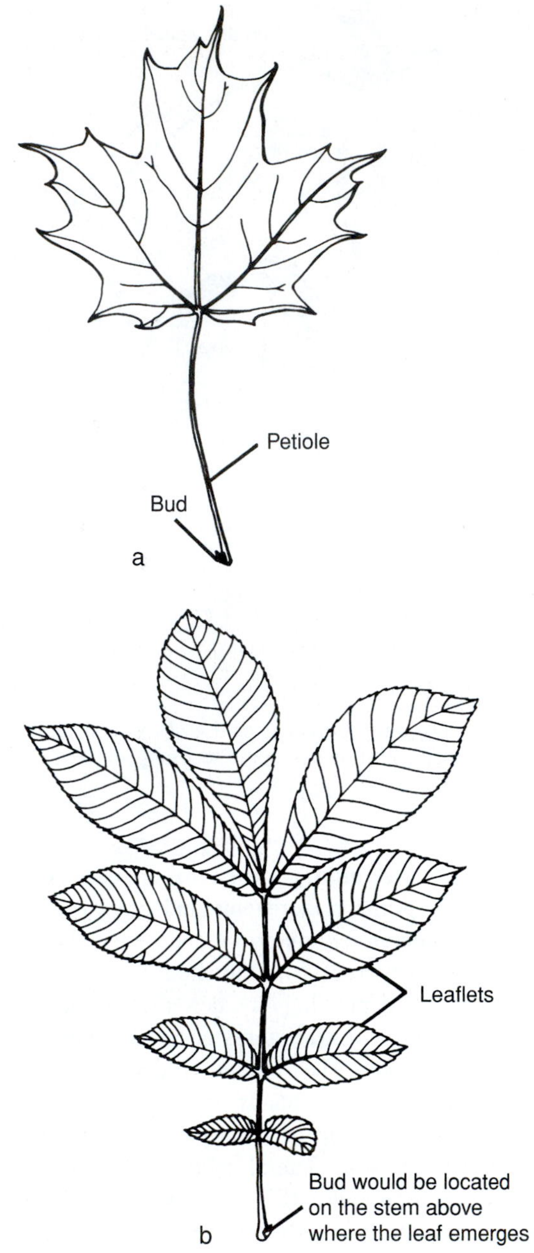

FIGURE 7.6

Both the simple leaf of a sugar maple (a) and the compound leaf of a hickory (b) have axillary buds only where the leaf emerges from the stem. It is easy to tell that the leaflets of a hickory leaf are parts of a compound leaf because they do not have buds at their bases.

by growing laterally around the arms of the xylem. In some cases, primarily among the monocotyledons, the xylem strands do not extend to the center of the root. In such situations, there is some pith inside the xylem.

Encircling the conductive tissues is a layer of cells called the **pericycle** (Figs. 7.7 and 7.8), which is important in the formation of new roots. Since roots have no axillary buds, the formation of lateral roots must proceed by a process different from that of branch formation. Lateral roots arise from specific patches of cells within the pericycle that retain the ability to produce new cells. As they divide, these cells form protuberances that push through the cortex and epidermis (Fig. 7.9) as they elongate. The initiation of lateral organs from an area near the middle of a root thus contrasts with stem development in areas where lateral branches arise from the peripheries of shoots. Once they are sufficiently mature, lateral roots can themselves give rise to other lateral roots, eventually forming large, ramified systems.

TABLE 7.1 Major Morphological and Anatomical Differences between Roots and Stems

CHARACTER	STEM	ROOT
Nodes and internodes	Nodes, regions where bundles of vascular tissue diverge from the stem into a leaf or branch, and intervening internodes present	Nodes and internodes lacking
Cuticle	Present over all aerial surfaces	Lacking
Location of vascular tissue	In herbaceous plants, in bundles around the stem periphery (dicotyledons) or scattered in the stem (monocotyledons); in woody plants, a cylinder with the xylem surrounded by the cambium and phloem	In a more or less solid mass in the center of the root with the xylem usually forming projecting arms alternating with the phloem
Stomates	Present on all aerial tissues	Lacking
Apical meristems	Not covered by a cap of cells	Covered by a root cap
Lateral structures	Leaves are initiated peripherally by the apical meristem, and branches from meristems in the axils of leaves, both at nodes	Generally originate deep within the root from the pericycle
Root hairs	Lacking	Formed from epidermal cells on the zone just behind the region of cell division
Endothermal cells	Rarely distinct from other cells of the cortex	In the absorbing region of the root, these innermost cells of the cortex have suberinized bands forming a Casparian strip

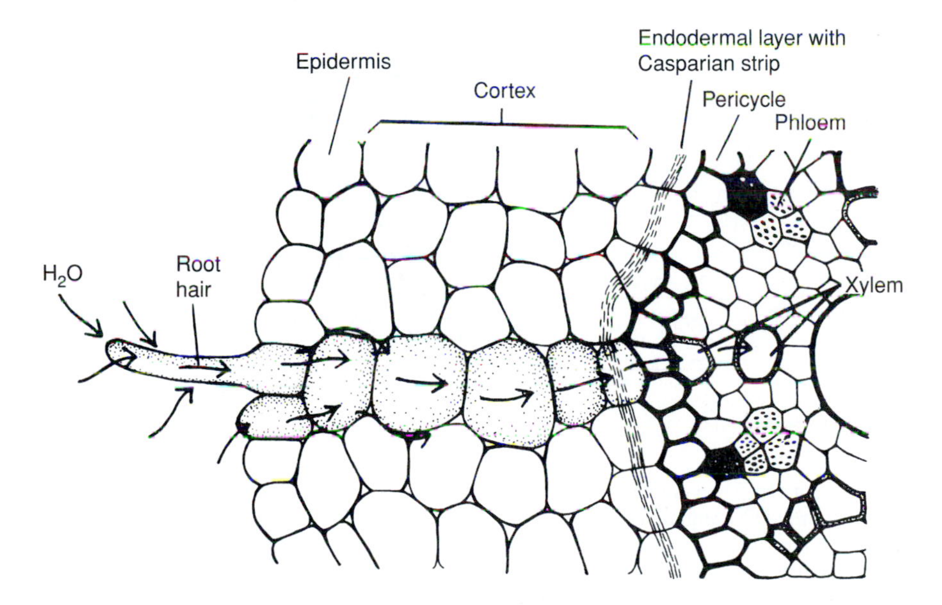

FIGURE 7.7

Each root hair is formed by the extension of a single epidermal cell. In cross section one can follow the path a water molecule might take from the root hairs into the xylem. There are numerous intercellular spaces between the cortical cells, allowing water movement and oxygen exchange, but water can also move across the cortex by passing through one cell and into the next. In either case, it must cross the endodermal cells before it enters the vascular cylinder. Water is blocked from passing between the endodermal cells because their radial cell walls and the spaces between them are impregnated with suberin. Water is forced to cross the endodermal cells, passing through a cell membrane when it enters and leaves the cell. Passage across cell membranes allows selective uptake of ions carried in the water.

Surrounding the pericycle is the **endodermis** (Fig. 7.7), a ring of cells characterized by having radial and lateral walls impregnated with **suberin,** a complex waxy molecule that repels water. Surrounding the endodermis is the **cortex,** a relatively thick layer of unspecialized cells. These cells can serve as storage sites by accumulating sugar or starch in their vacuoles. Commercially important edible roots have particularly large accumulations of starch in the cortex. Surrounding the entire root is a layer of cells forming the **epidermis.** Just behind the root tip, epidermal cells can produce the **root hairs** where water absorption takes place. New absorbing regions with root hairs are produced continually as the root grows downward and the hairs of older portions wither. Figure 7.8 gives an indication of the arrangement of progressively older tissues behind the root apical meristem.

The delicate appearance of growing root tips contrasts with the rapidity with which they grow and their ability to

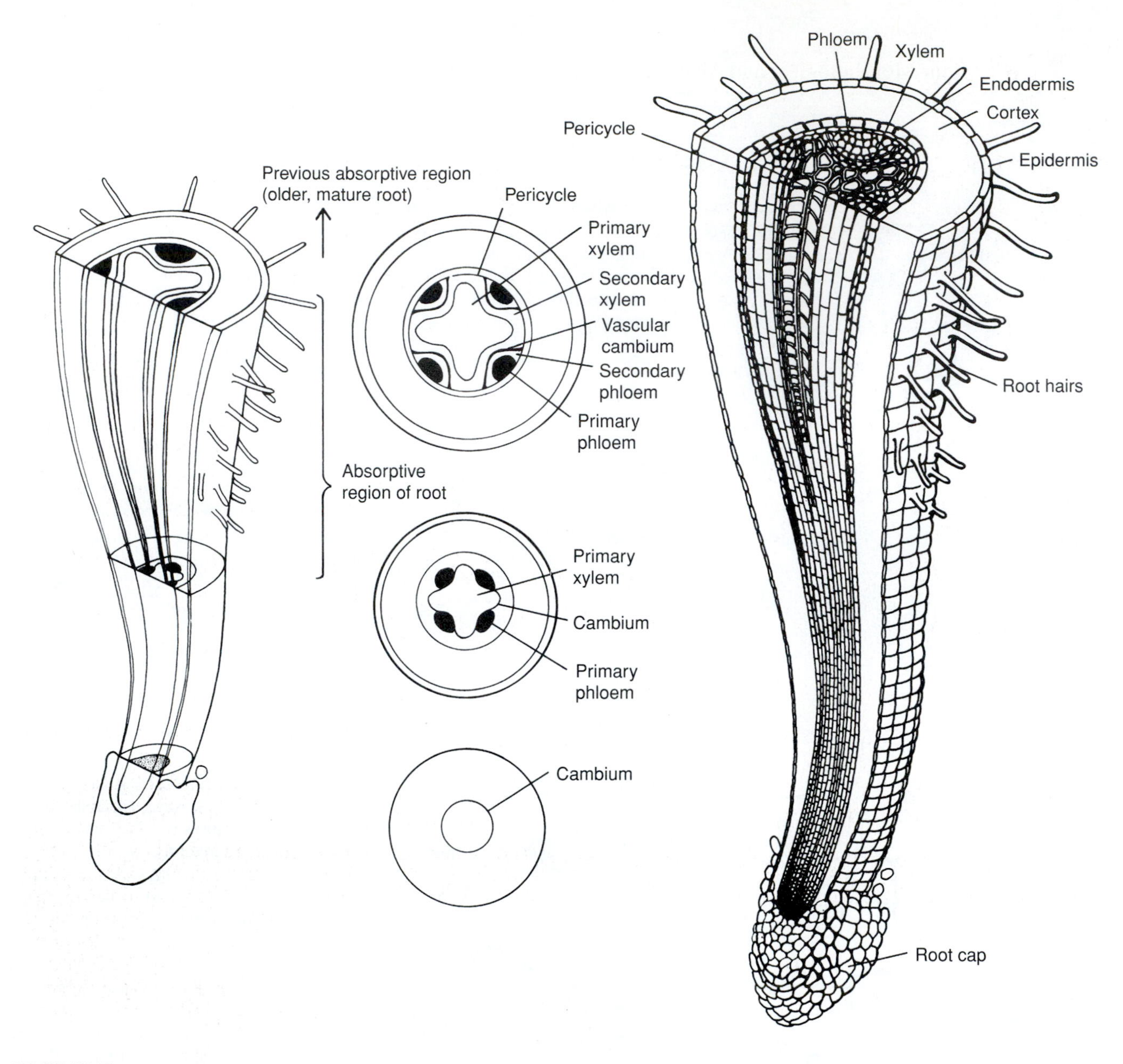

FIGURE 7.8

The apical meristem of the root lies protected beneath the root cap. The youngest, least-differentiated cells produced by the apical meristem lie just above it. The oldest root cells are farthest from the root tip. Differentiated cells lie some distance above it. The root of a dicotyledon typically has a central mass of xylem with projecting arms between which lie clusters of phloem cells. The cambial cells lie between the two kinds of conductive tissue. In old roots the vascular cambium can grow around the xylem arms and eventually form a ring of meristematic tissue.

push through rock and concrete. It has been estimated that at the end of only 4 months of growth the root system of an annual rye plant has a total length of 11,000 km (6830 miles) and a surface area of 630 m^2 (6780 ft^2). Since only the tips of primary and lateral roots carry out absorption, continuous root growth is essential. The size of the root system does not expand indefinitely, however, because not all roots are maintained from year to year or season to season. In temperate regions there is extensive root elongation and initiation of lateral roots throughout the spring and summer. During the winter many roots wither and die.

How Water and Nutrients Move into Plants

If nutrients were drawn with water into the plant solely by capillary action or by suction caused by evaporation from the leaf and stem surfaces, chemicals would move through the plant body in the same concentrations in which they occurred in the soil water. Exhaustive studies have shown that most compounds and ions are selectively moved into the xylem for transport. Consequently, the mechanism of nutrient

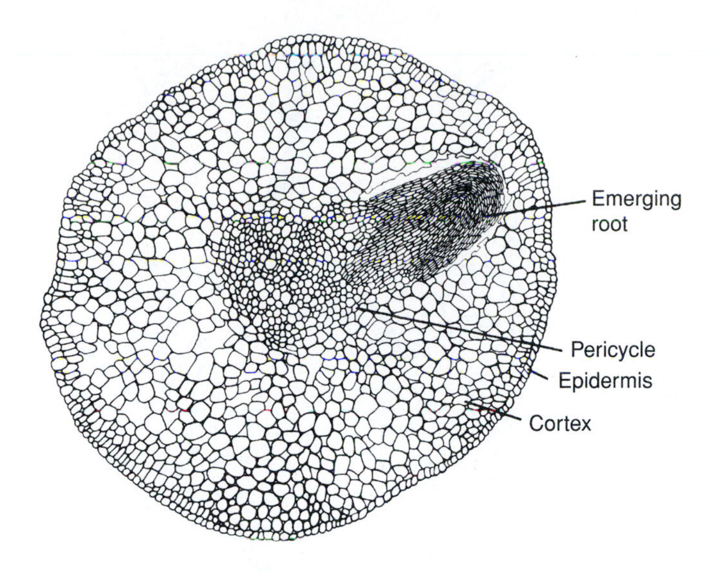

FIGURE 7.9
Lateral roots are initiated by cells of the pericycle near the middle of the root.

absorption is not strictly passive. Water absorbed by the root hairs can flow toward the center of the root by passing from one cell to the next through tiny cytoplasmic bridges. It can also move inward between cells until it reaches the endodermis (Fig. 7.7). The spaces between the endodermal cells are sealed by suberin bands. These bands, which are called **Casparian strips,** prevent water from moving between cells and force it to cross the endodermal cell membranes. Cell membranes contain proteins that can act as carriers that actively transport certain ions against a concentration gradient. This active transport provides a mechanism for the selection of mineral nutrients that enter the xylem.

Shoots, Roots, and Plant Habit

Almost all plants have leaves, stems, and roots, but their sizes, shapes, and arrangements differ among species. Often these differences are correlated with the life history or habitat of the species. For example, although all roots serve the same basic function, those of many temperate biennial species serve as storage organs as well. This function is linked with the 2-year life cycle of these species.

In some temperate biennial species storage in roots is so pronounced that their taproots have become dramatically fleshy (Fig. 7.10). Humans have taken advantage of the biennial habit and have learned to harvest the roots at the end of the first year of growth, after storage is complete (Fig. 7.11). Sometimes these biennial species are sources of both a root crop and a leafy crop, but it is difficult to obtain both from the same plant. For leafy crops, farmers want mature but tender leaves or immature shoots, so plants are harvested when they

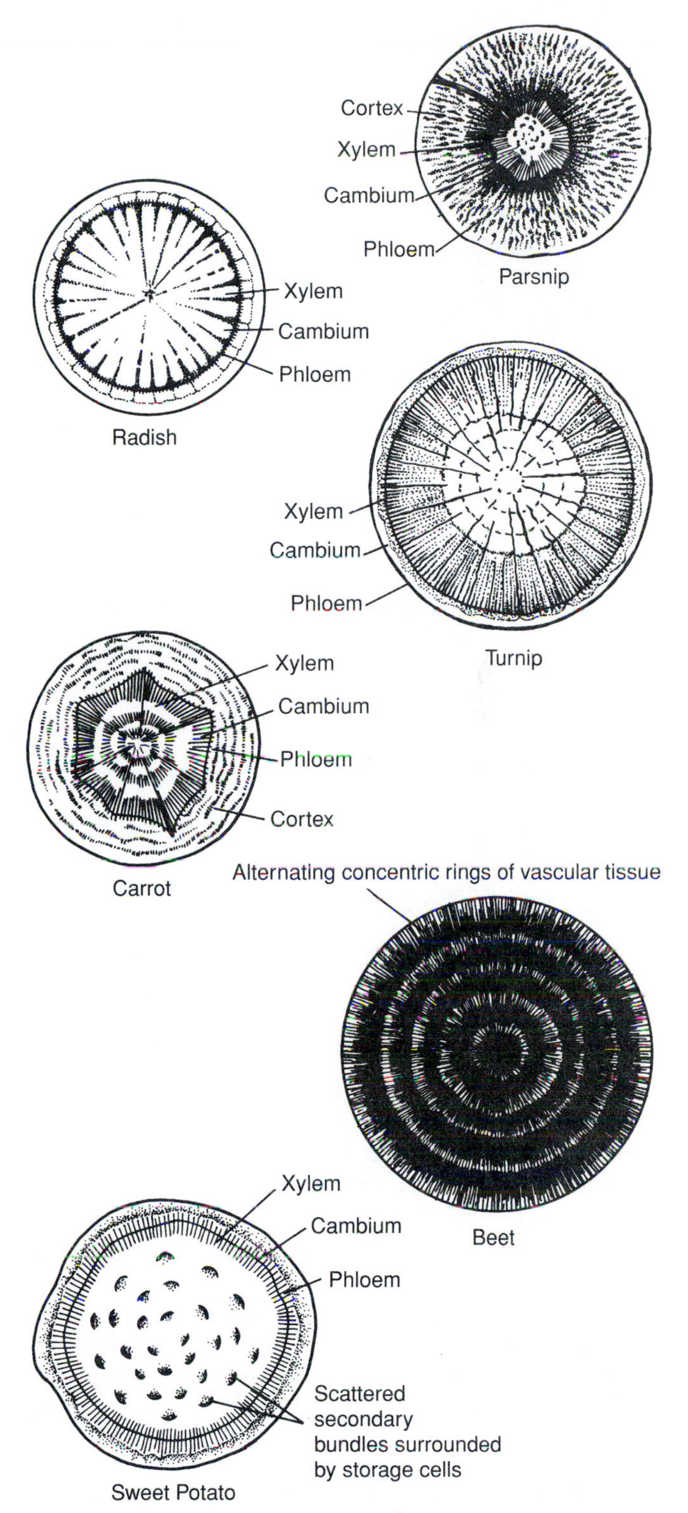

FIGURE 7.10
Cross sections of storage stems show that in different species, different areas are the sites of starch accumulation. In the radish and turnip, parenchyma tissue in the xylem proliferates and stores starch. In the parsnip and the carrot, starch is stored in the cortex. In beets, there are successive bands of vascular tissue each with storage cells. The sweet potato, the only perennial shown, stores starch in parenchyma in the center of the root in which there are numerous secondary xylem bundles.

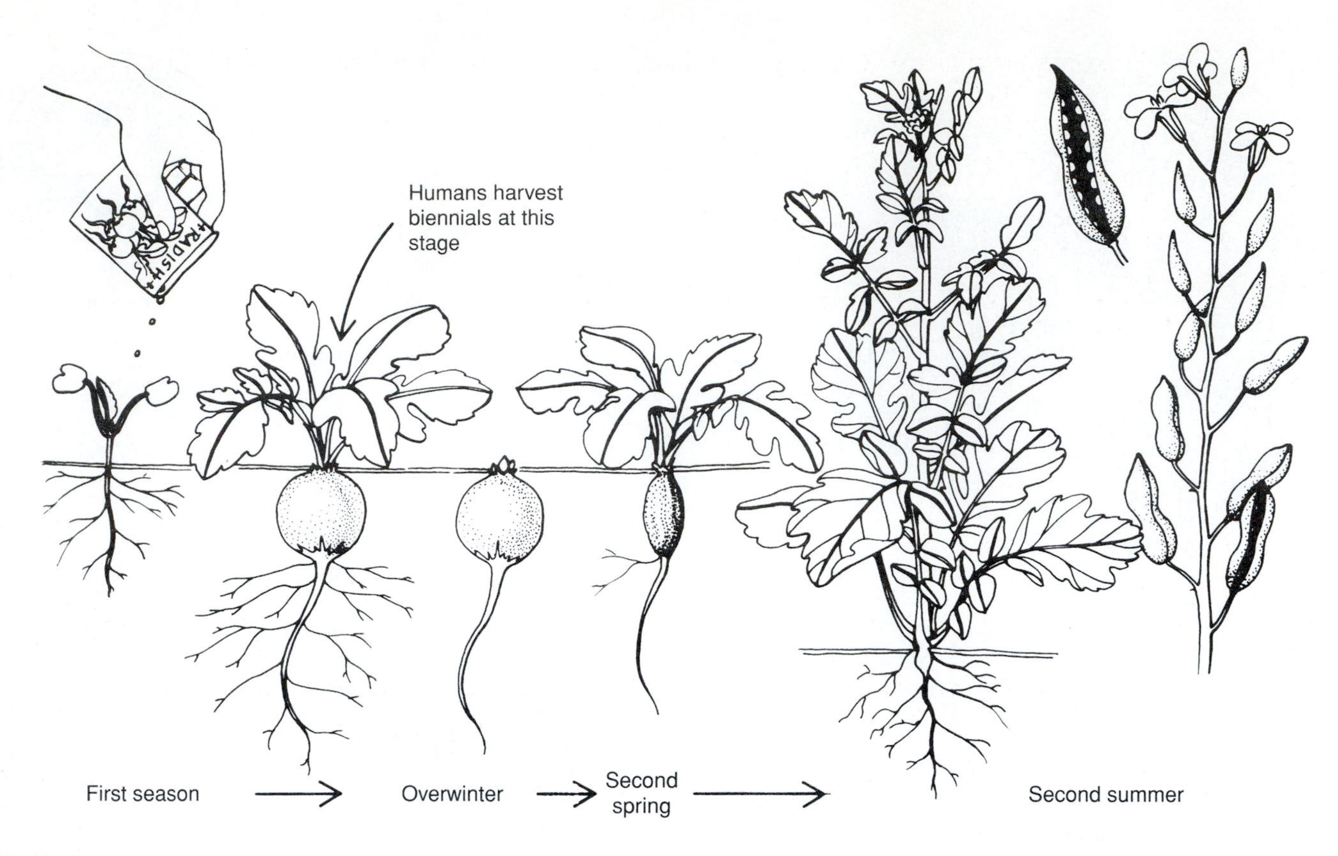

FIGURE 7.11
Humans take advantage of the biennial habit of many species by harvesting the roots after storage has been completed at the end of the first year's growth. During the second year the storage material is used by the plant for growth, flowering, and fruiting.

are relatively young. For root crops, farmers wait as long as possible before harvesting so that the greatest amount of nutrients, often including those once held in the leaves, will be translocated to the roots. Storage usually occurs in **parenchyma** cells, which are unspecialized cells characterized by having thin walls. The roots swell as parenchyma tissue in the xylem proliferates (radish), or additional layers of parenchyma tissue arise in successive rings of conductive tissue (beets) or by production of parenchyma in the cortex (carrot) (Fig. 7.10). The swollen roots are dug at the end of the first year, sometimes after the leaves have withered. There has usually been divergent selection, leading to one set of cultivars producing the leafy crops and another set producing the root crops in biennial species that are cultivated for both a leaf crop and a root crop (e.g., beets and chard).

In many regions woody plants also store food in their roots. Plants that lose their leaves during times of freezing temperatures or extreme aridity need food reserves to produce a new set of leaves when conditions become favorable. Toward the end of the growing season these plants transport photosynthetic products from the leaves to the stems and roots and store them in parenchyma cells in the secondary xylem. In temperate woody species these nutrients, stored as starch, are broken down to sugar at the end of the winter to provide energy for the production of new leaves that will subsequently manufacture the glucose needed for growth and maintenance. In arid areas, particularly the dry tropics, herbaceous perennials store appreciable amounts of nutrients in modified roots. The major sources of starch in many tropical areas (e.g., manioc) are the roots of such species.

Annual species have roots that are less developed than those of either biennials or perennials. Since growth and reproduction occur in less than a year, there is no appreciable storage of nutrients in the roots. Food crops that come from annuals are therefore restricted to stems, leaves, fruits, and seeds.

Composition and Production of Edible Stems, Leaves, and Roots

In our discussion of edible products from stems, leaves, and roots, we first deal with crops from biennial species. Many of these species produce both a leafy crop and a root crop. Leafy vegetables have few calories and comparatively little protein (Table 7.2). They are consumed

TABLE 7.2 Composition of Leafy, Root, Tuber, Rhizome, and Bulbous Crops, Based on a 100-g Edible Portion

						Vitamin	
CROP	WATER, %	CALORIES	PROTEIN, g	FAT, g	CARBOHYDRATE, g	A*	VITAMIN C, mg
Leafy Crops							
Brussel sprouts	85	45	5	0.4	8	550	102
Cabbage	92	24	1	0.2	5	130	47
Cauliflower	91	27	3	0.2	5	60	78
Celery	94	17	1	0.1	4	240	9
Chard	91	25	2	0.3	5	6,500	32
Collards	87	40	4	0.7	7	6,500	92
Endive	93	20	2	0.1	4	3,300	10
Kale	88	38	4	0.1	6	8,900	125
Kohlrabi	90	29	2	0.1	7	20	66
Lettuce	96	13	1	0.1	3	330	6
Spinach	91	26	3	0.3	4	8,100	51
Biennial Root Crops							
Beets	87	43	2	0.1	10	20	10
Carrots	88	42	1	0.2	10	11,000	8
Parsnips	79	76	2	0.5	18	30	16
Radish	95	17	1	0.1	4	10	26
Rutabaga	87	46	1	0.1	11	580	43
Turnips	91	30	1	0.2	7	Trace	36
Perennial Stem Crops							
Artichokes	86	ca. 20	3	0.2	11	160	12
Asparagus	92	26	3	0.2	5	900	33
Bulbs							
Leeks	85	52	2	0.3	11	40	17
Onions	88	45	1	0.2	11	Trace	25
Starchy Rhizomes, Tubers, Roots							
Manioc (tapioca)	13	352	1	0.2	86	0	0
Potato, sweet	71	114	2	0.4	26	8,800	21
Potato, Irish	80	76	2	0.1	17	Trace	20
Taro	73	98	2	0.2	24	20	4
Yam	74	101	2	0.2	23	Trace	9

Source: Data from *Composition of Foods,* Agricultural handbook No. 8. 1975. Washington, D.C., USDA.

*Values for vitamin A are in international units.

primarily as salads or "additions" to meals in developed countries but are often important as pot herbs in developing countries. In China they provide up to half the dietary protein. In cultures that suffer from food shortages they are luxuries used primarily to flavor dishes or stretch a meal. The fleshy roots of biennials have higher caloric values than do leafy food crops, but they are low in carbohydrates compared with the storage roots, rhizomes, and tubers of many perennial species (Table 7.2). After discussing biennials, we describe a few important leafy vegetable crops obtained from annual and perennial species, many of which are important for human health (Box 7.1). We then turn to the starch crops obtained from the rhizomes, tubers, and roots of tropical perennials. These crops constitute some of the world's most important food plants because of their high caloric value and impressive yields. Last, we discuss the species from which people obtain sweeteners, primarily sugar. Table 7.3 gives the worldwide production of most of the crops discussed in this chapter. Table 7.4 lists the plants discussed in this chapter.

Box 7.1

You Really Should Eat Your Vegetables

"Eat your vegetables" is a phrase familiar to many children whose protests have usually been met with the monotonous response that "they are good for you." We all know that vegetables contain essential vitamins, but now evidence is mounting that they provide benefits never dreamed of before. Leading the pack of so-called wonder foods are the vegetables that contain carotenoids such as beta-carotene. Among these are broccoli, carrots, pumpkin, squash, and dark-green leafy vegetables such as spinach and kale. Cooked vegetables yield greater quantities of carotenoids than raw vegetables because the cooking makes the cells containing them easier to rupture when chewing. Beta-carotene, it turns out, not only protects against blindness, but it also appears to help ward off heart attacks and strokes by reducing the amount of bad cholesterol (low-density lipoprotein, or LDL) in the blood. Other studies suggest that carotenoids can even help avert cancer by preventing the oxidation of scavenging cells.

Broccoli and its relatives have received additional attention as anticancer agents with the discovery that the ring-shaped sulfurous compounds they contain may inactivate estrogen. A reduction in estrogen has been correlated with a decrease in some cancers, particularly breast cancer. In addition, it now appears that sulforaphane, another glucosinolate, stimulates the production of enzymes that help guard against cancer.

Onions and other alliums have also been extensively studied during the last 10 years because of their sulfur-containing (glucosinolate) compounds. Garlic has been revered as a medicinal herb in Asia for centuries, but its benefits have been slighted in the Western world. Recent investigations have suggested that some of its sulfur-containing compounds (chiefly allicin, diallyl di- and trisulfide, and ajoene) prevent cancer by intercepting cancer-causing mutagens that the body ingests or manufactures. Other compounds contained in garlic (perhaps prostaglandins) lower blood pressure and thus help protect against hypertension. Garlic extract also seems to retard blood clotting, slow tumor development, and reduce cholesterol levels in the blood. Extracts also inhibit bacterial growth and have been shown to thwart proliferation of fungi responsible for human "yeast" infections such as thrush. Even celery, that seemingly innocuous dieters' delight, now seems to have medicinal properties. Researchers from Chicago have provided substantiation for an ancient Oriental use of celery in treating hypertension. They found that the stalks contain a compound that relaxes smooth muscles of blood vessels, thereby helping to maintain low blood pressure.

However, before you start consuming vast quantities of vegetables or reaching for synthetic supplements, some caution is necessary. Beta-carotene is an oil-soluble "vitamin" that can be toxic in large quantities, and a high consumption can lead to yellow-orange skin. In combination with vitamins A or D it can increase the probability of developing lung cancer in heavy smokers. Sulfur-containing compounds can likewise be mutagenic. The flip side of celery's benefits is that it contains sodium, which can exacerbate high blood pressure if consumed in large quantities. Undoubtedly future studies will find ways in which many other plant secondary compounds play a role in promoting human health.

TABLE 7.3 1999 Annual World Production of Major Leaf, Stem, and Root Crops

COMMON NAME	TOP 5 COUNTRIES	1000 METRIC TONS	TOP CONTINENTS	1000 METRIC TONS	WORLD, 1000 METRIC TONS
Artichoke	Italy	508	Europe	901	1,296
	Spain	285	Africa	153	
	Argentina	85	South America	112	
	France	84	North and Central America	73	
	Egypt	57			
			Asia	41	
Cabbage	China	18,504	Asia	31,840	49,205
	India	4,200	Europe	6,113	
	Korea, Republic	2,992	North and Central America	2,611	
	Russia	2,850	Africa	902	
	Japan	2,700	South America	571	

TABLE 7.3 *Continued*

COMMON NAME	TOP 5 COUNTRIES	1000 METRIC TONS	TOP CONTINENTS	1000 METRIC TONS	WORLD, 1000 METRIC TONS
Carrots	China	4,478	Asia	6,817	18,169
	United States	2,201	North and	2,779	
	Poland	925	Central America		
	Japan	720	Europe	1,301	
	United Kingdom	618	South America	917	
			Africa	907	
Cassava (manioc)	Nigeria	33,060	Africa	91,286	165,986
	Brazil	20,171	Asia	45,767	
	Congo, Dem. Rep.	17,100	South America	27,744	
	Thailand	16,930	North and	1,006	
	Indonesia	14,728	Central America		
			Oceania	183	
Cauliflower	India	5,000	Europe	2,339	13,810
	China	4,621	Asia	10,408	
	Italy	527	North and	427	
	Spain	360	Central America		
	Italy	402	Africa	197	
			Oceania	111	
Onion, dry	China	11,290	Asia	25,511	41,528
	India	4,428	Europe	6,851	
	United States	2,994	North and	3,370	
	Turkey	2,300	Central America		
	Japan	1,240	South America	2,795	
			Africa	2,780	
Potato, Irish	China	43,477	Europe	143,350	288,523
	Russian Federation	30,300	Asia	92,058	
	Poland	26,000	North and	28,132	
	India	22,100	Central America		
	United States	21,840	South America	14,008	
			Africa	9,072	
Potato, sweet	China	100,207	Asia	107,764	119,919
	Uganda	2,520	Africa	9,052	
	Indonesia	1,927	South America	1,426	
	Nigeria	1,560	North and	1,233	
	Vietnam	1,517	Central America		
			Oceania	569	
Sugar cane	Brazil	338,486	Asia	569,052	1,276,911
	India	282,249	South America	420,838	
	Mexico	46,000	North and		
	Cuba	35,000	Central America	162,184	
	United States	34,106	Africa	82,970	
			Oceania	41,722	
Sugar beets	France	32,181	Europe	175,128	259,833
	United States	30,294	Asia	45,363	
	Germany	25,668	North and		
	Turkey	20,000	Central America	31,294	
	Ukraine	16,500	Africa	4,945	
			South America	3,102	
Yams	Nigeria	25,077	Africa	34,934	36,545
	Ivory Coast	2,923	South America	492	
	Ghana	2,703	North and		
	Benin	1,770	Central America	474	
	Togo	696	Oceania	271	
			Asia	234	

Source: FAOSTAT Database Gateway. Online at http://apps.fao.org/ (under "All databases" then Crops, Primary).

TABLE 7.4 Plants Discussed in Chapter 7

COMMON NAME	SCIENTIFIC NAME	FAMILY	NATIVE REGION
Arrowroot	*Maranta arundinacea*	Marantaceae	New World tropics
Artichoke	*Cynara scolymus*	Asteraceae	Mediterranean
Asparagus	*Asparagus officinalis*	Liliaceae	Mediterranean
Beet, sugar beet	*Beta vulgaris*	Chenopodiaceae	Mediterranean
Bok-choi	*Brassica chinensis*	Brassicaceae	Asia
Broccoli	*Brassica oleracea* var. *botrytis*	Brassicaceae	Europe
Brussels sprouts	*Brassica oleracea* var. *gemmifera*	Brassicaceae	Europe
Cabbage	*Brassica oleracea* var. *capitata*	Brassicaceae	Mediterranean
Canna	*Canna edulis*	Cannaceae	West Indies, South America
Carrot	*Daucus carota*	Apiaceae	Mediterranean
Cauliflower	*Brassica oleracea* var. *cauliflora*	Brassicaceae	Europe
Celery	*Apium graveolens*	Apiaceae	Mediterranean
Chicory	*Cichorium intybus*	Asteraceae	Mediterranean
Chinese cabbage	*Brassica campestris* subsp. *pekinensis*	Brassicaceae	Asia
Chive	*Allium schoenoprasum*	Alliaceae	North temperate
Cocoyam	*Colocasia esculenta*	Araceae	Asia
Collard	*Brassica oleracea* var. *acephala*	Brassicaceae	Mediterranean
Dasheen	*Colocasia esculenta*	Araceae	Asia
Date palm	*Phoenix dactylifera*	Arecaceae	Mediterranean
Eddoe	*Colocasia esculenta*	Araceae	Asia
Endive or escarole	*Cichorium endiva*	Asteraceae	Mediterranean
Garlic	*Allium sativum*	Alliaceae	Asia
Jicama	*Pachyrhizus erosus*	Fabaceae	Mexico, Central America
Kale	*Brassica oleracea* var. *acephala*	Brassicaceae	Mediterranean
Kohlrabi	*B. oleracea* var. *gongyloides*	Brassicaceae	Mediterranean
Leek	*Allium ampeloprasum*	Alliaceae	Near East
Lettuce	*Lactuca sativa*	Asteraceae	Eurasia
Malanga	*Xanthosoma sagittifolium*	Araceae	West Indies
Manioc	*Manihot esculenta*	Euphorbiaceae	South America
Maple, sugar	*Acer saccharum*	Aceraceae	North America
Onion	*Allium*	Alliaceae	Asia
Bulb	*A. cepa*		Asia
Branching	*A. fistulosum*		Asia
Parsnip	*Pastinaca sativa*	Apiaceae	Mediterranean
Potato			
Irish	*Solanum tuberosum*	Solanaceae	South America
Sweet	*Ipomoea batatas*	Convolvulaceae	South America
Radish	*Raphanus sativus*	Brassicaceae	Western Asia
Rutabaga	*Brassica napus*	Brassicaceae	Europe
Shallot	*Allium cepa*	Alliaceae	Asia
Sorghum	*Sorghum bicolor*	Poaceae	Africa
Spinach	*Spinacia oleracea*	Chenopodiaceae	Asia
Swiss chard	*Beta vulgaris* var. *cicla*	Chenopodiaceae	Mediterranean
Sugar cane	*Saccharum officinarum*	Poaceae	New Guinea
Taatsai	*Brassica campestris* subsp. *narinosa*	Brassicaceae	Asia
Tannia	*Xanthosoma sagittifolium*	Araceae	West Indies
Taro	*Colocasia escialenta*	Araceae	Asia
Turnip	*Brassica campestris*	Brassicaceae	Eurasia
Yam	*Diascorea*		
cush-cush	*D. trifida*		South America
water	*D. alata*	Dioscoreaceae	Asia
white	*D. rotundata*		Africa
yellow	*D. cayenensis*		Asia
	D. esculenta		Asia
Yautia	*Xanthosoma sagittifolium*	Araceae	Tropical Americas

Biennial and Annual Stem, Leaf, and Root Crops

Cabbage and Its Relatives

The array of vegetables produced by modifications of the shoots and roots of members of the Brassicaceae, or mustard family (Fig. 7.12), exceeds that of any other plant group. The food crops of the Brassicaceae, including mustard (Chapter 8) and canola oil (Chapter 9), share a pungent flavor imparted by a class of compounds known as the mustard oil glycosides, or glucosinolates. Although not toxic to humans in the quantities in which people consume them, mustard oils have been shown to be toxic to insects.

In the United States the most commonly consumed vegetables in the mustard family belong to the genus *Brassica.* These vegetables are grouped together as the cole crops, a name taken from the Latin *caulis,* meaning "stem." Interestingly, almost all the leafy vegetables in this genus belong to one species, *Brassica oleracea,* native to the north Atlantic coast of Europe. Each of these vegetables—cabbage, collard greens, Brussels sprouts, kale, kohlrabi, broccoli, and cauliflower—is produced by a different modification of the leaves or stems, and each has its own history of domestication. Table 7.4 gives the scientific names for the various varieties. All of these leafy cultivars are now grown as annuals, but the wild ancestor was undoubtedly biennial or perennial as is their close relative, the turnip.

FIGURE 7.12
The flowers of wild mustard (left) and cabbage (right) are typical of the flowers of all the species in the Brassicaceae. The four petals form a cross, and the six stamens are arranged in two groups, with four of them long and two short.

Fossil evidence for the domestication of *B. oleracea* is lacking, but historical documentation of this Mediterranean native suggests that it was cultivated by the Greeks as early as 650 B.C. These writings describe a type of leafy vegetable with a head that might have been cabbage and another that sounds from the description rather like kohlrabi. Selection within the species was advanced at an early date, and the development of the various modern morphological forms shown in Figure 7.13 was under way. Presumably, the plants first taken into cultivation were rather loose-leafed, rank herbs. Forage kales (the European black cabbage) are probably the modern cultivated forms that most closely resemble the wild ancestral types (Fig. 7.13). These early sea kales were adapted to coastal habitats, where they were frequently exposed to salt spray. The waxy layer that protected these plants from salt damage is still evident in modern cole crops and contributes to their resistance to drought and cold.

The common cabbage includes red and green cabbages, with a tight head and smooth leaves (Fig. 7.14). Headed cabbages are formed when the lateral meristems of the primary shoot do not elongate and the "inner" leaves do not expand. The terminal bud (Fig. 7.15) remains buried deep inside the closely appressed leaves. Several kinds of headed cabbages were developed in Germany, where both red and white (= green) color forms were grown 600 years ago. Savoy cabbage, considered by many researchers to be a distinct variety, has loose crumpled leaves and is less commonly eaten in the United States than the more compact type.

Cabbages have always been an important item in the diets of European peasants. They could be harvested late in the fall and stored for considerable periods. Moreover, since about 200 B.C. the process of packing shredded cabbage leaves and salt into earthenware crocks and allowing the mixture to ferment into sauerkraut has been used as a method of preservation. Sauerkraut and its Asian cousin kimchee historically provided sources of vitamin C when fresh fruits and vegetables were unavailable. However, the cold tolerance that has made cole crops so popular in the north temperate zone has also imposed limits on their cultivation in warm regions. In areas where summers are hot, these hardy vegetables should be planted early in the spring or fall because they mature best under cool temperatures. In addition, if cabbages are not picked at the proper time, they can **bolt,** or produce a flowering shoot. The expression "gone to seed" comes from the tattered appearance of a bolted leafy crop. In the case of cabbage, the leaves become flaccid and bitter after bolting, and the plants assume a ragged, weedy appearance. Other major *Brassica* leafy crops include bok-choi (*B. chinensis*) and Chinese cabbage (*B. campestris*).

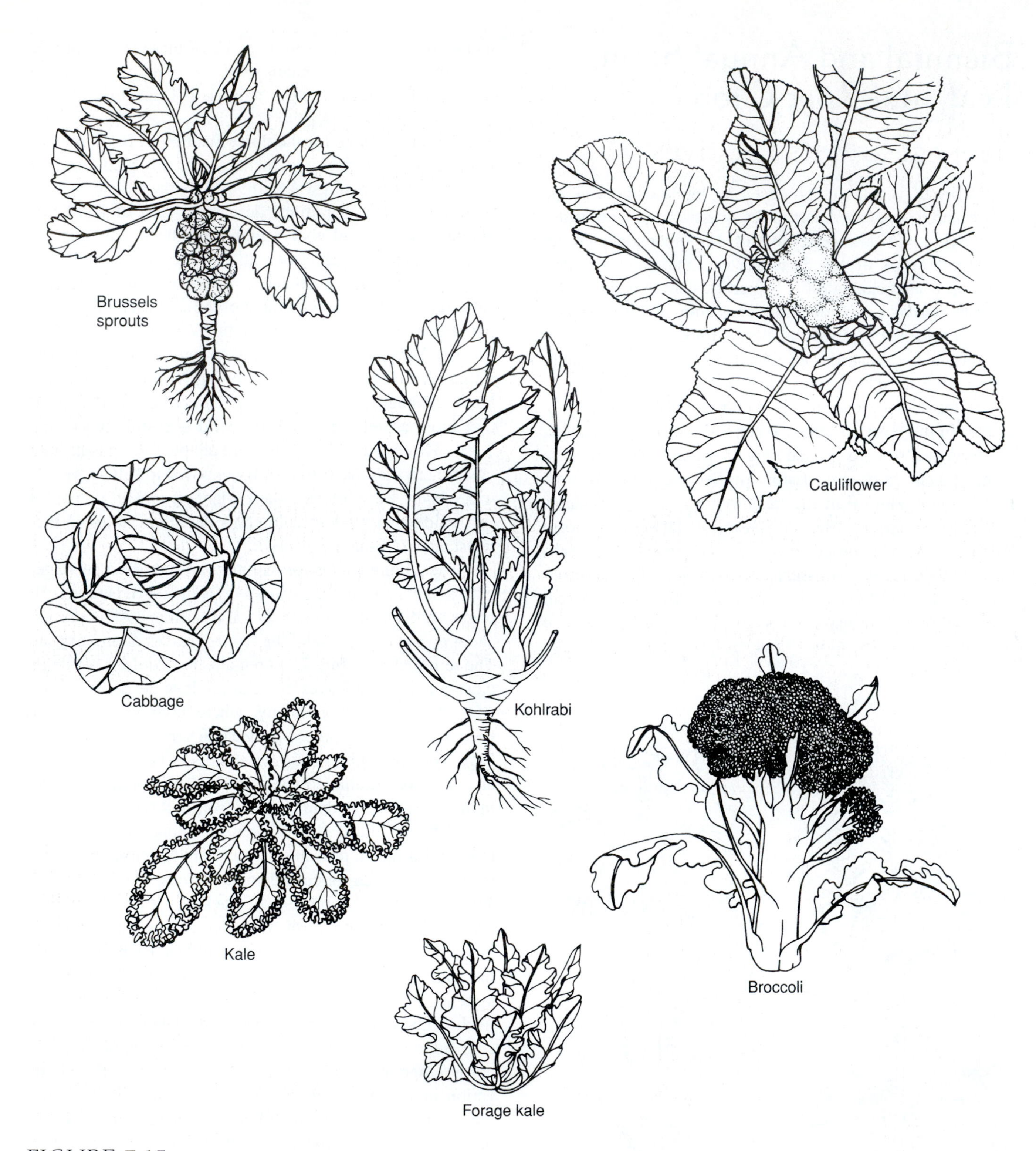

FIGURE 7.13
The proliferation of the various edible brassicas. The ancestral type of the modern crops is presumed to have been similar to the forage kales shown at the bottom. Broccoli and cauliflower are considered to be the most recently selected forms.

FIGURE 7.14
A cabbage field in El Salvador, showing a new system where crops are planted in rows flanked by irrigation ditches.
(Photo by L. Dematteis, FAO.)

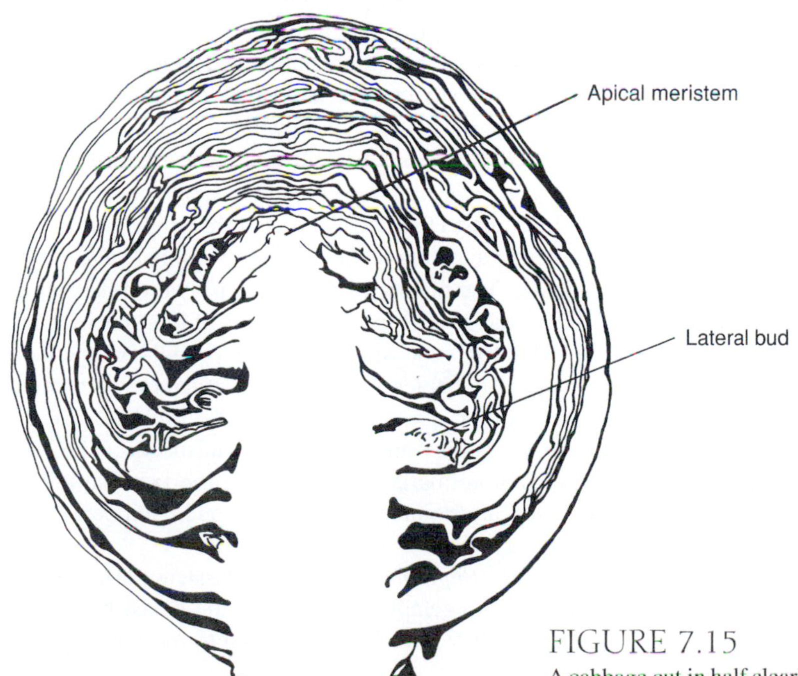

FIGURE 7.15
A cabbage cut in half clearly shows that the head is formed by suppression of the terminal bud. Lateral buds remain undeveloped in the leaf axils.

Brussels sprouts look like miniature cabbages but are formed when axillary buds form lateral "heads." Unequivocal evidence of this form dates only to 1818. Because lateral shoots are smaller and more tender than the primary stem, Brussels sprouts are eaten in their entirety. The "core" (main axis) of a cabbage is usually removed because it is woody and fibrous.

Kohlrabi (Fig. 7.13) is a bizarre form of the cabbage plant in which the swollen stem base is harvested. A sketchy early description in Pliny's works seems to apply to kohlrabi, but a definitive drawing does not appear until the sixteenth century. Nevertheless, Pliny's account of a cabbage with an expanded edible portion just above the roots would seem to correspond to few other vegetables.

Both cauliflower and broccoli are relatively recent selections from cabbages or kales. Descriptions of a sprouting broccoli date to the middle of the sixteenth century; cauliflower appears to have been developed somewhat later. Despite the fact that both vegetables are derived from inflorescences of the cabbage plant, they are considered different varieties. Broccoli consists of a mass of fertile flower buds (Fig. 7.13) that turn into an inflorescence of yellow blossoms if they are not harvested soon enough. Cauliflower, by contrast, is formed by a proliferation of the stem tips, forming a mass of tissue called "curds." The curds of early-headed cauliflowers are produced by a massive expansion of the shoot apices. Late-flowering varieties produce curds that also contain undeveloped floral buds. Since much of the cauliflower produced in the United States is of the early, or "snowball," type, we never see cauliflower flowers. Even in late-flowering forms the flowers are too young to develop when the heads are picked. Seeds are produced only in late-flowering varieties that are allowed to remain on the plants. Originally, the pure white heads of cauliflower were produced by tying the leaves borne below the flowering head over the cluster of flowers to prevent chlorophyll synthesis, which produces a green color. Modern varieties lack the ability to produce chlorophyll in the apical regions and do not turn green even when exposed to sunlight. There are varieties that produce purple curds or that retain the ability to turn green (sometimes labeled as broccoflower in supermarkets).

The two major cole root crops are turnips and rutabagas. These two vegetables are often confused even at the produce counter in a grocery store because they can look somewhat similar once they have been harvested. Nevertheless, they are produced by different species. The part of turnips and rutabagas we eat is the **hypocotyl,** the portion of the plant where the stem and root meet. Since most of the mass consists of root tissue, the vegetables are generally considered root crops rather than stem crops.

Turnips are the roots of *Brassica campestris,* thought to have grown originally wild in Europe and Asia (Fig. 7.16). There are references to a plant that appears to have been an oilseed type of turnip in Indian writings almost 4000 years old. European cultivation of turnips as a source of seed oil seems to have begun in the thirteenth century. Different subspecies of this species yield Chinese cabbage (pe-tsai) and taatsai. The English name *turnip* comes from the same etymological base as the verb "to turn" because turnips are so smooth and perfectly formed that they appear to have been turned on a lathe. The flesh of turnips is usually white. The roots themselves are flat on top and generally are tinged with purple. Yellow-fleshed varieties are grown but are not commonly found in stores.

Turnips have for some reason always been held in low esteem. In Roman times they were spoken of in derogatory terms and were a favorite item to throw at miscreants. The Aryans disdained them because they were eaten by Indian races. Young German women in some areas would present suitors they wanted to reject with a plate of boiled turnips. The majority of Europeans, however, consumed great quantities of turnips throughout the Middle Ages and still include them as a part of many winter meals. Yet this dependency has not changed the opinions of most people about this vegetable. Even in the United States children are led to believe that nothing could be worse than having to eat turnips. Most people believe that they have a strong flavor, but they have perhaps never had a good turnip or have mistakenly confused the vegetable they ate with its relative the rutabaga.

FIGURE 7.16

These freshly dug turnips clearly show the flat top of the hypocotyl, which differs from the pointed top of a rutabaga. In speaking of turnips, Thomas Elyot said, "Boiled it nourysheth moch, augmenteth the sede of man and provoketh carnal lust."

Rutabagas, or Swedish turnips, come from *Brassica napus* and are more nutritious but have a more pronounced flavor than their smaller cousins (Fig. 7.17). Genetic and molecular studies have clearly shown that rutabagas ($n = 19$) resulted from polyploidy after hybridization between cabbage (*Brassica oleracea,* $n = 9$) and turnip (*B. campestris,* $n = 10$). Chloroplast DNA work has also shown that the turnip was the maternal parent in the original cross that gave rise to the rutabaga. The history of rutabaga cultivation has paralleled that of turnips except that cultivars of this species have been selected for use as an oilseed crop, producing rapeseed, or canola, oil (Chapter 9). In Europe rutabagas are often used as food for domesticated animals. In U.S. markets rutabagas are yellow and are larger than turnips. The tops are cut off before marketing, obscuring the fact that unlike turnips, they are pointed at the upper end. Shippers often coat rutabagas with wax to prevent drying.

Radishes, *Raphanus sativus* (Brassicaceae), are an important root vegetable on a worldwide basis, but in the United States they have been relegated to the position of a garnish carved to resemble roses and serving as a decoration on relish trays and platters. Other cultures hold the

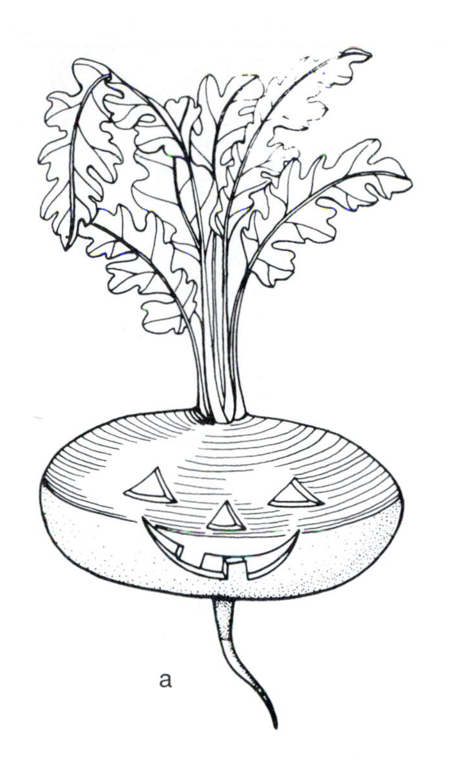

FIGURE 7.17

The first jack-o-lantern was carved out of a turnip. The Irish, who started our Halloween tradition, explain it this way. One night Satan came to a local pub to claim the soul of Jack, a drunken miserly fellow. Jack suggested that they "have one for the road," but when the Devil turned himself into a sixpence to pay the barkeeper, Jack grabbed the coin and put it in his wallet. After some bickering, Jack agreed to set the Devil free with the promise that he be left alone for another year. Each year thereafter Jack managed to outwit the Devil. Finally, Jack grew old and died of natural causes. Heaven refused to admit him and the Devil was not about to offer his adversary a resting place, so Jack's soul was stuck between heaven and hell. Satan gave Jack a burning coal to light his way, and Jack fashioned a lantern from a turnip he was eating to hold it. The Irish have used (a) turnips, (b) rutabagas, and potatoes to make lanterns for All Saints' or All Hallow's Day. Americans introduced the practice of using pumpkins for jack-o-lanterns.

peppery roots in much higher esteem. Inscriptions from Egyptian tombs dated to be almost 4000 years old show that radishes were important in that ancient civilization. Egyptians not only ate the roots but also valued radish seed oil. In China selected cultivars are still grown as an oilseed crop. In Japan and many other east Asian countries radishes are among the most important vegetables. In the Orient, however, the roots are not the small red spheres that are consumed in the United States but rather are elongated white or giant black-skinned forms known as daikons. Although the ancestor of the domesticated radish is not known with certainty, the large straight-rooted types are probably similar to the wild types from which modern variants have been selected. The globular form was not developed until the eighteenth century and was initially white. The bright solid red color resulted from a later mutation that was perpetuated by people.

Lettuce and Its Allies

Lettuce, chicory, and endive belong to the Asteraceae, one of the largest angiosperm families, containing over 25,000 species. The members of this family are easily recognized by the characteristic heads of many small flowers (see Fig. 1.37). Yet despite its large number of species, the family has not contributed many plants to the human diet, although there are a few notable exceptions. The most important of these is lettuce (*Lactuca sativa*). In ancient times it was thought that the milky juice exuded from a cut lettuce possessed medicinal virtues similar to those of opium latex. Although such ideas were quickly dispelled, Linnaeus chose the generic name *Lactuca,* derived from the Latin *lac,* for "milk," because of the milklike sap. Some forms of lettuce may have been cultivated as early as 4500 B.C., and lettuce is depicted on Egyptian tombs. The Romans later enjoyed various greens tossed with olive oil and vinegar as preludes to their renowned feasts. There now seems to be little doubt that modern lettuce is derived from *L. serriola,* a species of the Mediterranean region.

Throughout history lettuce has been subjected to divergent selection. By the fourteenth century, salads had become the rage in England, and housewives had over 50 different types of greens, many of them varieties of lettuce, at their disposal. A virtuous homemaker arranged her assortment of buttery, crisp, sweet, and bitter greens with snippings of herbs, violets, and other flowers to give them aesthetic

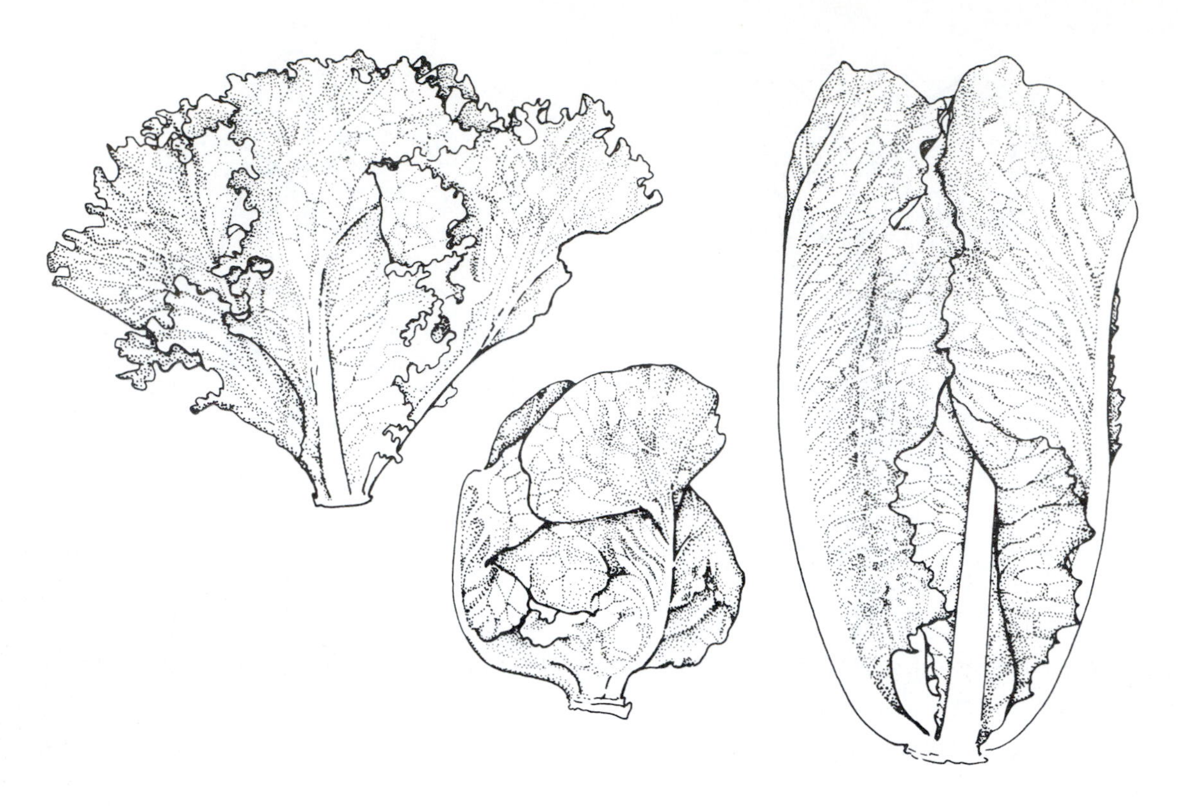

FIGURE 7.18
The lettuce types encountered today can be categorized into loose-leaf, heading, and cos types. Loose-leaf types such as the leaf lettuce shown on the left include salad-bowl, red-tipped, and oak-leaf lettuces. Boston lettuce (center) is a headed type, as is iceberg lettuce, which has become the most common form sold in the United States because its tight, round heads ship well. The primary cos type, with stiff, rather elongated leaves, is Romaine lettuce (right).

appeal as well as an interplay of tastes. The salad greens in our modern grocery stores seem banal in comparison (Fig. 7.18). However, there has been increased consumption of a wider variety of salad greens and of red-pigmented forms of loose-leaf and Romaine lettuce.

Like cabbage, lettuce heads are formed by suppression of the lateral buds, and they will also bolt if not picked soon enough. Flowering lettuce is a bitter-tasting, rank-looking plant about 1 m tall (3.275 ft) with small yellow heads resembling those of dandelions.

Endives and chicory have graced salad bowls as long as lettuce, but both belong to the related genus *Cichorium,* not to *Lactuca.* In the United States the salad delicacies known as Belgian endives come from the perennial species *Cichorium intybus.* To produce these small, tender heads, seeds are sown in the spring and the plants are allowed to grow freely until the first frost, after which they are dug and topped. The thick rootstocks are then planted in sand and forced to initiate new shoots in the absence of light, yielding small, mild-flavored torpedo-shaped heads. If the plants are allowed to leaf naturally, they produce large, rather bitter leaves that are also used as salad greens. *Cichorium intybus* plants that have escaped and become wild in the United States are known as chicory. In Europe and in parts of the American south the roots are roasted and ground. The powder is mixed with coffee or used as a coffee substitute (see Figs. 13.10 and 13.11). Adding to the confusion is the fact that in American supermarkets the salad green known as chicory or escarole comes from another species, *C. endiva.* The heads of this annual species can also be blanched by tying the large outer leaves around the inner "heart" of the head. Gourmet cooks have recently been exposed to radicchio, a strong-flavored, red-hued salad "green" that turns out to be nothing more than an Italian form of *C. endiva.*

Celery and Its Relatives

The carrot family, or Apiaceae, is known primarily for the herbs it contains (Chapter 8), but celery (*Apium graveolens*) and a few spices are also used as commonly grown vegetables. When one examines the stalks (Fig. 7.19), it becomes apparent that the edible parts of the plant are the petioles and swollen leaf bases. On the inside, at the bottom of each stalk, is an axillary bud, indicating the juncture of a leaf and a stem. What are commonly called celery leaves in cookbooks are therefore simply the partially expanded leaf blades. Celery occurs wild today in temperate Eurasia. It was cultivated by Greeks and Romans, but our modern forms have been developed within the last 300 years. Celeriac (Fig. 7.20), the swollen hypocotyl of a celery plant,

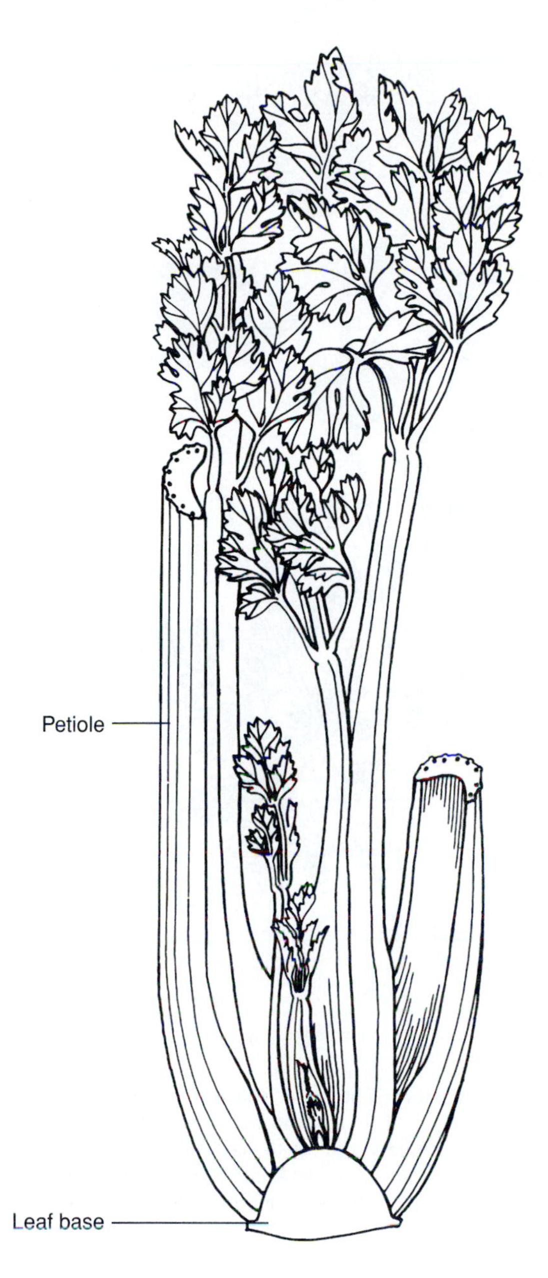

FIGURE 7.19
The unexpanded leaflets of celery are often discarded. The portion people usually eat is the expanded petiole.

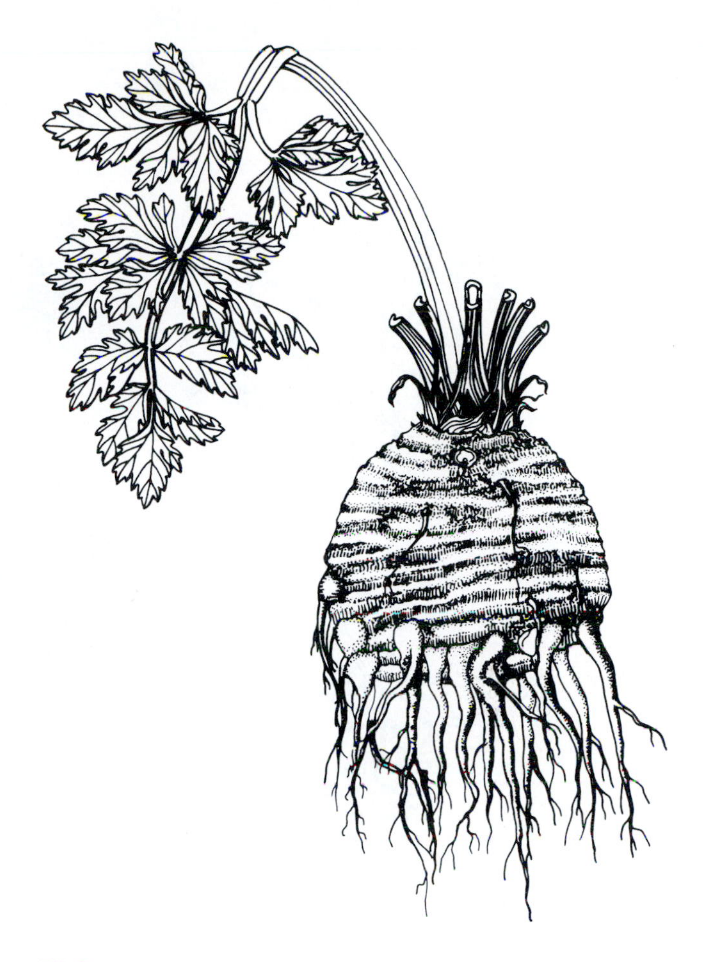

FIGURE 7.20
Celeriac, also called celery root, celery knob, and turnip-rooted celery, is highly prized in Europe for its swollen root.

is rather popular in Europe, where it is eaten raw in salads or cooked as a side dish.

In addition to culinary herbs and celery, the Apiaceae yield two root vegetables of some importance: carrots (Table 7.3) and parsnips. Of the two, carrots (*Daucus carota*) are the more popular in this country. Like most biennials, carrots store reserves in a main taproot during the first summer, but in the cortex, not in expanded regions of vascular tissue (Fig. 7.10). One seldom sees flowers on harvested carrots, but if one waited for a second year, one would see a display of lacy white flowers showing that carrots are nothing more than the common weed Queen Anne's lace (Fig. 7.21). Originally, carrot roots, like those of the wild Queen Anne's lace, must have been rather woody and branched. Human selection led to the conical, unbranched versions that we now consume. A study of the cultivars grown in Asia Minor and information from historical writings have revealed that the original domesticates were purple, although a yellow mutant devoid of anthocyanin pigment commonly occurred. Since the purple pigment was water-soluble, the use of carrots in soups and stews often produced a purple-brown, rather unappetizing dish. It was soon seen that the yellow forms tasted the same as the purple and circumvented the bad color problem. Consequently, they eventually became the most commonly grown type in western Europe. Later selection intensified the color. Carrots are now eaten not only cooked but as a raw vegetable.

The pigment responsible for the orange color of carrots is beta-carotene. Chemically, this pigment consists of two vitamin A molecules joined end to end. During digestion, carotene is immediately broken down to vitamin A by the mucus lining of the intestine and by the liver; hence, the vitamin A content rather than the beta-carotene content is

FIGURE 7.21

Carrot plants that have escaped from cultivation are commonly known as Queen Anne's lace. The plant is said to have been named after Queen Anne of England (1665–1714), the daughter of James II and the sister of Mary Queen of Scots, who was apparently fat and plain but excelled at needlework. Anne is said to have challenged her ladies-in-waiting to an embroidery contest to see who could produce lace as beautiful as the cluster of flowers of the wild carrot. No one was able to outdo the queen's exquisite work. Anne thus won the contest, and the plant has since borne her name. (a) A whole flower. (b) A flower in longitudinal section. (c) Bunches of small flowers make up the delicate, flat-topped inflorescences characteristic of the Apiaceae (d).

(After Baillon.)

given in nutrition tables (Table 7.2). Vitamin A in turn is almost identical to the chemical retinal present in the rod cells of the retina of the human eye. Rod cells are important in vision because they are sensitive to light at low intensities. Thus, among its other functions, vitamin A supplies the precursor for the light-absorbing pigment of the rod cells. Consequently, there is some truth to the old wives' tale that eating carrots can help a person see in the dark. However, eating carrots merely helps ensure that an individual receives enough vitamin A. Beyond a threshold, additional vitamin A cannot improve vision.

In appearance, parsnips (*Pastinaca sativa*) resemble carrots except that they are pale yellow instead of orange. Although they are sweeter than carrots, they are not as popular in the United States. They seem to fall into the category of "old-fashioned" vegetables. Like carrots, they are native to the eastern region of the Mediterranean and were eaten throughout the Middle Ages in Europe. The Pilgrims brought them to the New World, where they became one of the few European crops enthusiastically adopted by Native North Americans.

Beets

Beets, *Beta vulgaris* (Chenopodiaceae) (Fig. 7.22) might seem to be an odd vegetable to number among the most important crops in the world (see Fig. 19.7) but this apparent anomaly is due to the fact that most people think of beets only as the small, round, deep red roots that turn up as a side

FIGURE 7.22
Swiss chard (left) and beets (right) come from different cultivars of the same species. Chard is a leafy vegetable, whereas beets are a root crop. Different varieties are grown for each of the crops. Beet greens can, however, be eaten as a green vegetable.
(Courtesy of G. Fox.)

FIGURE 7.23
This statue of Popeye stands in Crystal City, Texas, reputed to be the spinach capital of the world.
(J. L. Neff.)

dish, pickled at salad bars, or as a primary ingredient in borscht soup. The species also provides three other crop plants: the mangel-wurzel, Swiss chard, and the sugar beet. The first is a cultivar with a yellowish white root that is used as cattle feed. Chard (Fig. 7.22) is a leafy vegetable. The most economically important of the products obtained from the species, the sugar beet is discussed in detail later in this chapter along with sugar cane. The three "root crops" differ in the position of the hypocotyl relative to the ground and the amount of root versus transition tissue in the swollen structures. Fodder beets (mangel-wurzels) protrude from the ground for about a fourth of their length and have a substantial component of root tissue. Sugar beets consist primarily of root tissue and grow almost completely beneath the soil surface. The red vegetable beet, in contrast, consists primarily of transition tissue and is mostly borne above ground.

It is believed that chard (*Beta vulgaris* subsp. *cicla*) or a chardlike green has been eaten by humans since prehistoric times in the Mediterranean region where beets are native. Early Greeks mentioned varieties of chard in their writings. The roots of Swiss chard plants are stringy and more branched than those of the subspecies that yields edible beets and consequently are not generally eaten.

Spinach, an Annual Leafy Crop

Spinach (*Spinacia oleracea*), like beets, is a member of the goosefoot family (Chenopodiaceae). Spinach is now often substituted for lettuce in salads, but it would not have been included among the salad greens of the Romans because the species is native to western Asia and appears to have been domesticated after the fall of the Roman empire. Today spinach is popular around the Mediterranean. The cartoon character Popeye the Sailorman (Fig. 7.23) was created in 1929 by E. C. Segar and has served ever since to cajole children into eating spinach because of the large amounts of iron it contains. Spinach is the species from which folic acid, a vitamin necessary for the synthesis of nucleic acids, was originally isolated. Since nucleic acids are manufactured primarily in cells that rapidly multiply, such as those that produce red blood cells, folic acid is used to treat some anemias and is now recommended for pregnant women to help prevent neural tube defects that lead to spina bifida. Spinach is second only to liver in the amount of folic acid it contains. It may have been the folic acid, therefore, as well as the iron that helped give Popeye his energy.

Perennial Vegetables: Artichokes, Asparagus, and Jicama

The Perennial Habit

Farmers follow a different regime when planting a perennial crop than when cultivating annual or biennial plants. Great care is taken in the preparation of the soil for perennial crops because once the plants start to grow, they remain in place for many years. Unlike the harvesting of annual or biennial species, the harvesting of crops from perennial vegetables is delayed until 2 or more years after planting to allow the plants to become well established and store reserves in the root system. However, as the years progress, the farmer reaps increasingly larger harvests until the plants begin to senesce. Perennial temperate vegetable crops are generally replanted every 15 to 25 years.

Artichokes

The Arabic word *ardischauki,* meaning "earth thorn," is the origin of the name given to one of the most distinctive vegetables. Artichokes (*Cynara scolymus*) are members of the Asteraceae, but unlike lettuce and the other leafy family members, they are cultivated for their immature flowering heads (Fig. 7.24). When one eats an artichoke, one eats the carbohydrates stored in the tender portions of the fleshy bracts surrounding the unopened flowers (Table 7.2). The undeveloped flowers and the chaff intermingled with them form the inedible "choke" of the artichoke. The savored artichoke "heart" is the swollen part of the stem (receptacle) on which the "choke" is borne. If allowed to mature, an artichoke becomes a beautiful large blue thistle (Fig. 7.24). Artichokes are generally planted using suckers, or basal branches, to ensure uniformity and to hasten plant establishment.

Asparagus

Asparagus, *Asparagus officinalis* (Liliaceae), is a dioecious perennial (Fig. 7.25) that is native to the temperate scrub communities of southern Europe, western Asia, and northern Africa. These habitats are subject to periodic burning, and many of the plants that occur natively within them can sprout rapidly from a robust underground system of rhizomes. Humans have made use of this resprouting ability by growing plants in habitats where the plants die back in the winter and send up new shoots in the spring. The vegetable consists of young, unexpanded shoots that are cut where they emerge from the ground. Although shoot production occurs for only a limited period in the spring and only a certain proportion of the new shoots can be harvested without

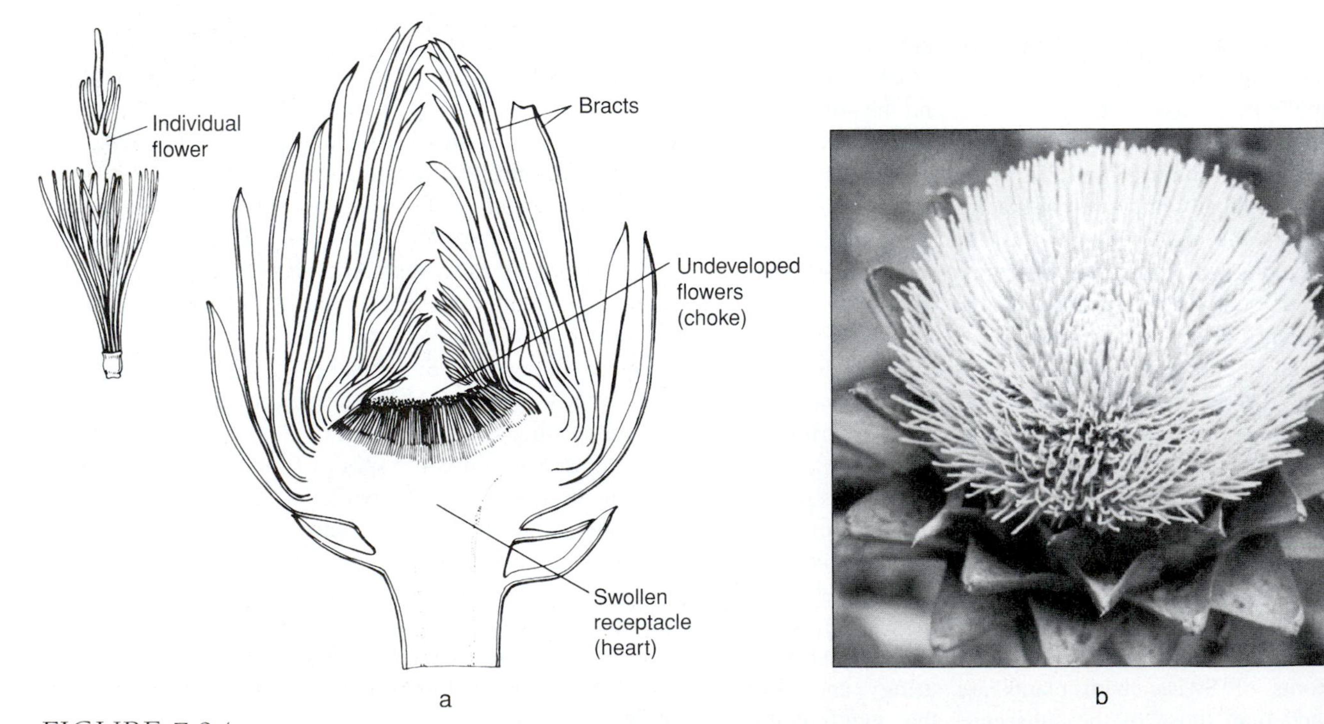

FIGURE 7.24
(a) An artichoke cut in half shows that the vegetable is really an immature flowering head. (b) If allowed to bloom, the artichoke will produce a beautiful purple or blue thistle head.

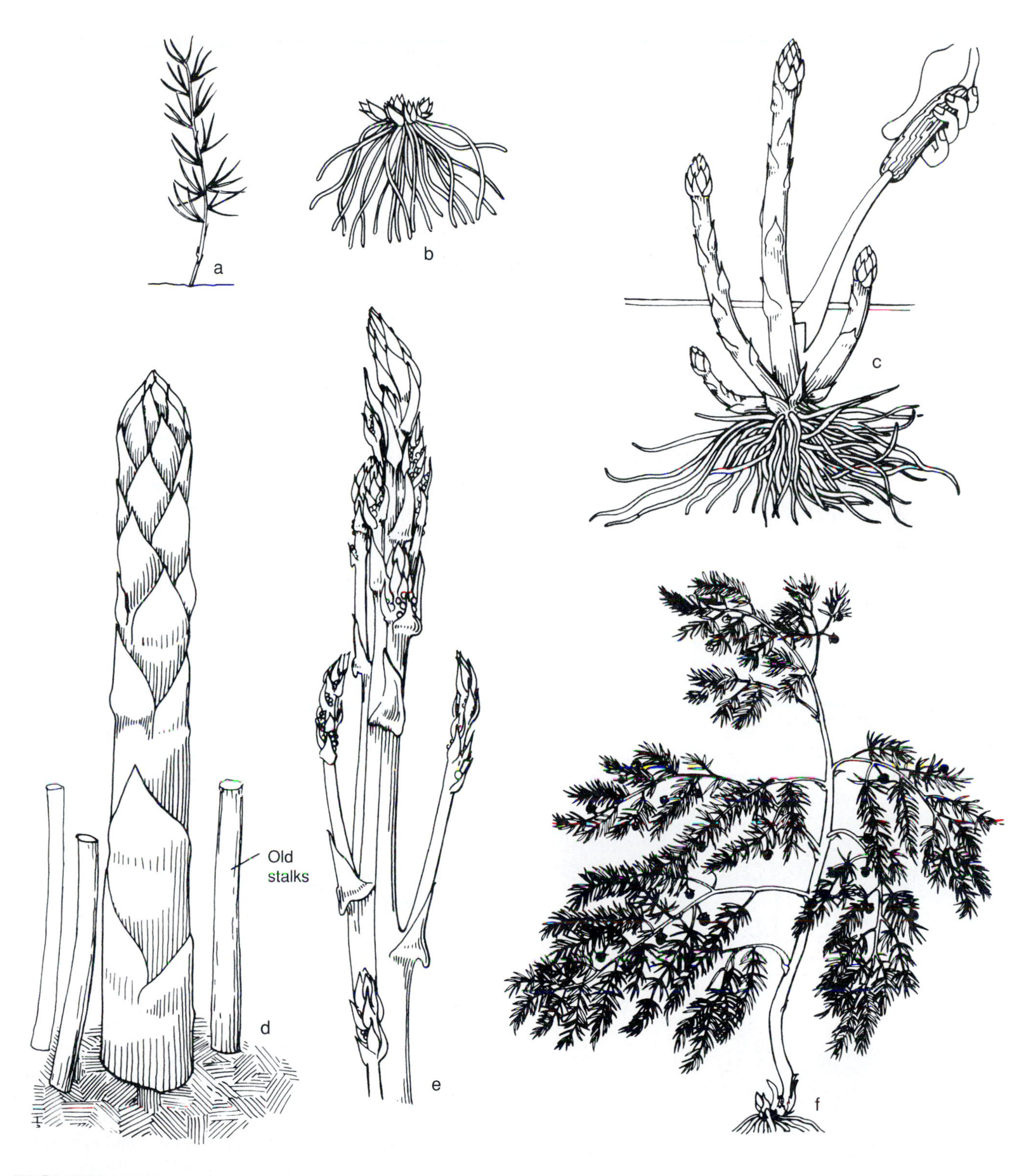

FIGURE 7.25
Asparagus is a perennial herb that produces a crop for many years. Because of the long life of individual plants, planting and harvesting procedures differ from those for annual crops. (a) Seedlings are established and transferred to permanent beds. (b) Alternatively, rootstocks (crowns) are purchased and set into permanent beds. (c) After a plant is 3 years old, some of the shoots produced in the spring can be harvested using an asparagus knife. (d through f) Some shoots are left to develop into fernlike branches (cladophylls) that will produce photosynthetic products, some of which will be transported to the rootstock for storage.

damaging the plants, asparagus is now available almost all year because it is grown in temperate regions of the Northern and Southern hemispheres. Shoots that are allowed to mature expand into ramified branch systems that resemble feathery leaves. In the horticultural trade, some varieties of asparagus plants are called "ferns" because of their green, highly dissected vegetative systems. The true leaves of an asparagus are reduced to scales at the base of each cladophyll. Asparagus is usually propagated using pieces of rhizomes called crowns because seed production is generally poor.

Jicamas (*Pachyrhizus erosus,* Fabaceae) (Fig. 7.26) are an increasingly common item in U.S. supermarkets. The vines that bear the large, turnip-shaped roots, 5–13 cm (3–8 in.) in diameter, are also known as yam beans. However, the beans are poisonous when mature, although young pods can be eaten. In Mexico, where jicama is native, street vendors sell peeled slices with sugar or salt. In the United States, the roots are also eaten raw, but they are usually cut into strips that provide a refreshing crunch in salads or as appetizers.

FIGURE 7.26
Jicama is a legume that produces a vine with pealike flowers and pods that are poisonous when mature.

Vegetables from Bulbs

Onions, Leeks, Garlic, Shallots, and Chives

The onion family's reputation for ornamental plants is rivaled only by its contribution to the culinary arts, principally the contribution made by members of the genus *Allium* (Fig. 7.27), which includes the onion and its relatives. Most of these species were cultivated, or at least eaten, before recorded history. Like most lilies, these pungent herbs are bulbous plants. People have made use of the stored nutrients in the bulbs and in all cases (except chives) consume the bulb and sometimes the basal portions of the flattened leaf blades. Onions and shallots (both now considered to belong to *A. cepa*) and garlic (*A. sativum*) probably originated in central Asia, and the leek (*A. ampeloprasum*) in the Near East. All (Fig. 7.27) were cultivated in Egypt, however, by 3200 B.C. Onions are grown for their single, large bulb, and garlic for its cluster of bulbs, each of which is called a clove. Leeks produce a bulb that is only slightly differentiated from the basal portions of the leaf blades. Consequently, a substantial portion of the leaf blades of leeks and the swollen basal portions are eaten. Elephant garlic, now commonly seen in stores and sold as a mild form of garlic, is a variety of leek that forms a large bulb with many cloves. Chives (*A. schoenoprasum*) are eaten solely for their leaves and appear to have been domesticated in the sixteenth century. Bunching onions (*A. fistulosum*), which are eaten for their leaves rather than their bulbs, are the most popular onions in the Orient.

Almost all members of the *Allium* genus have been reputed to exhibit medicinal properties. At one time or another various species have been prescribed for bad eyesight (Hippocrates), voice improvement (Nero), and baldness (Gerarde) or as a cure for the common cold, acne, arthritis, hypertension, or poor digestion. General Ulysses S. Grant was so convinced that a supply of onions was needed to protect his troops from dysentery that he declared, "I will not move my army without onions." Onions contain more sugar than the other alliums, and their sugar content leads to their being eaten raw and fried. There is still controversy about the real effects of onions and garlic, with many health food enthusiasts advocating their use and others denying any therapeutic benefit (Box 7.1).

The pungent quality common to all the alliums is linked with the compounds that make people cry when cutting onions. These are volatile sulfur compounds (including methyl di- and trisulfides and *n*-propyl di- and trisulfides) that are released when the cells of an onion are ruptured. When these compounds dissolve in the fluid covering the eyes, they form sulfuric acid. The tears brought on by the burning sensation only provide more water into which the compounds can dissolve. Since the chemicals are water-soluble, their effects can be diminished by submerging an onion in water while it is being sliced or by chilling or freezing the onion before slicing.

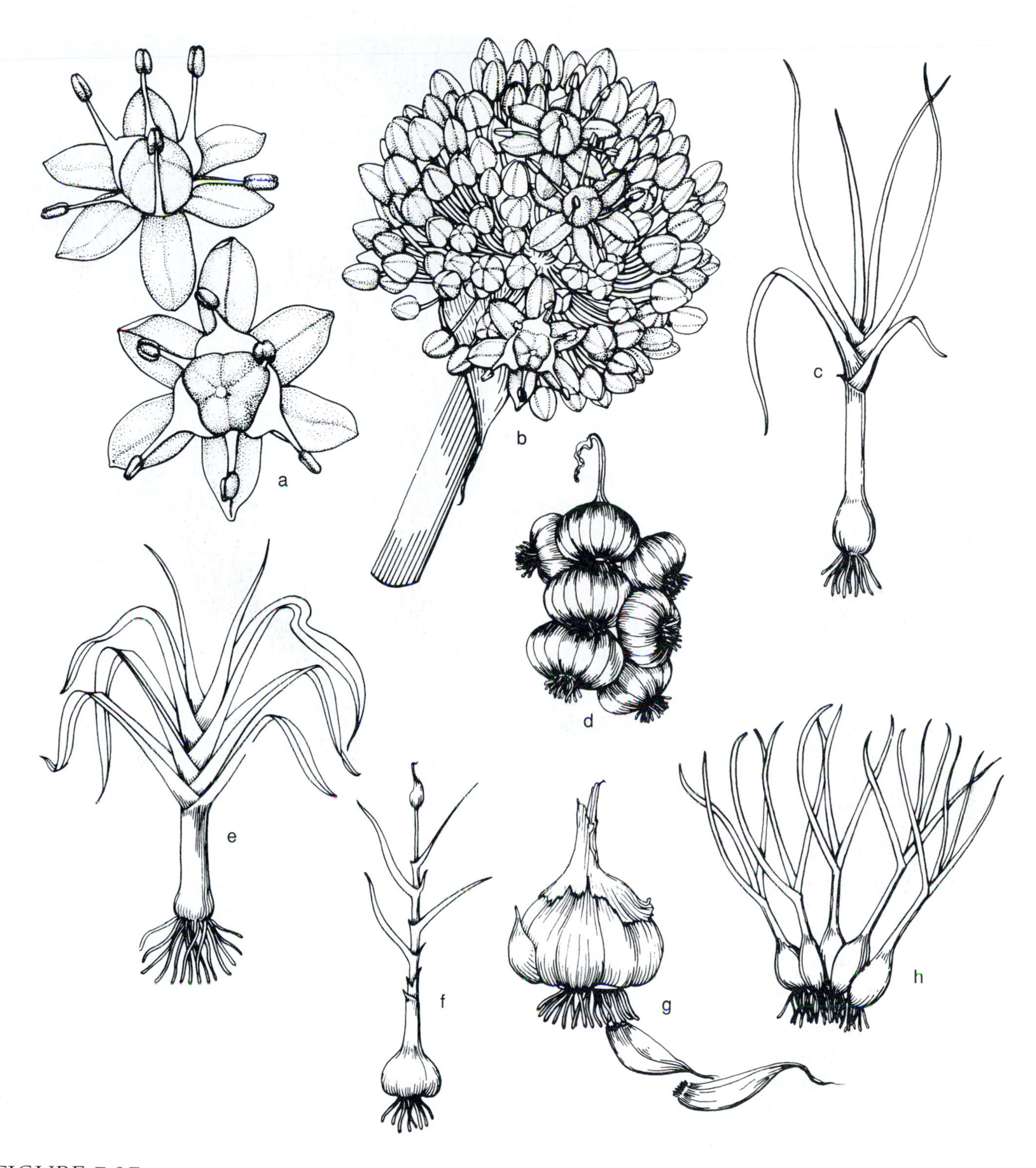

FIGURE 7.27

The lily family provides people with the onion and its relatives. (a) Onion flowers, (b) onion inflorescence, (c) young onion, (d) dried onion bulbs, (e) leek, (f) garlic, (g) garlic bulbs, (h) shallots.

Starches from Tubers, Rhizomes, and Roots

Functions of Tubers and Rhizomes

As was pointed out in Chapter 1, one important modification of stems has been for underground growth. Underground stems serve as overwintering organs or are used for asexual reproduction or serve both purposes. Particularly useful for humans are those underground stems that store nutrients. A seedling draws the nutrients it needs for initial growth from either the cotyledons or the endosperm, as described in Chapters 3 through 6. If a plant reproduces vegetatively from underground stems, the newly emerging shoot can derive nutritive material directly from the "parent" plant's vascular system until it has established roots of its own or can pull reserve nutrients from an underground storage organ such as a fleshy rhizome or tuber (see Fig. 1.11).

People have made abundant use of storage rhizomes and tubers since prehistoric times. Undoubtedly, wild tubers, along with fleshy fruits, were among the principal foods gathered by preagricultural people. They are still an important item in the diets of the hunter-gatherer tribes that exist today. Humans have also taken advantage of the characteristics of rhizomes and tubers by using them as propagules. People can use segments of storage stems or a few "seed" rhizomes or tubers to produce plants that provide an abundant supply of food plus enough material for further propagation.

FIGURE 7.28
A Bolivian and his daughter stomp on potatoes that are thawing after freezing overnight. This primitive method of drying produces chuño, a former staple on the Andean Antiplano.
(Photo courtesy of the United Nations.)

Irish Potatoes

Although the vast majority of the world is fed by wheat, corn, and rice (Chapter 5), one nongrain food item, the tuberous potato, ranks fourth as a major food staple. When one considers the modern importance of potatoes, *Solanum tuberosum* (Solanaceae), it is surprising that they have been known outside the Americas for only a few hundred years.

Spanish explorers found potatoes being grown by Indians all along the Andes from Colombia to Chile. There is archaeological evidence that wild potatoes were eaten 13,000 years ago in Chile and have been cultivated for at least the last 7000 years. In high-elevation areas of the central Andes Native Americans preserved potatoes by a primitive method of drying (Fig. 7.28). Temperatures in these areas can fluctuate each 24 hours from below freezing to almost 21°C (85°F) during the day. In addition, the climate is dry for most of the year. These combined climatic factors allow Native Americans to spread harvested potatoes on the ground and leave them overnight to freeze. The next day they stomp on them to squeeze out as much water as possible. After a few days this process yields a light, dry mass of cellulose and starch known as *chuño.* This method is not as rapid as modern freeze-drying procedures and ruptures the plant cells. Consequently, chuño often tastes moldy, but it keeps for months once it is completely dry.

Although chuño never became a favorite of Europeans, the Spaniards quickly saw that potato tubers had potential as a food source. They introduced them into Spain and used them for food on return voyages to Europe. The rest of Europe learned about potatoes from their independent introduction to England. Some authors give Sir Walter Raleigh credit for bringing them to Britain; others, Sir Francis Drake. One version even states that Drake initially carried potatoes to England from Colombia. The story then suggests that he subsequently grew them in his garden and gave some to Sir Walter, who in turn presented them to Queen Elizabeth I. Another folktale says that Queen Elizabeth I accused the good knight of trying to poison her when her cooks served a salad of potato leaves rather than a dish of cooked tubers. Potato leaves, like the skins of the tubers if they are allowed to turn green, contain solanine, a poisonous glycoalkaloid.

J. G. Hawkes, the leading authority on potato evolution, contends that neither of these knighted gentlemen should be credited with importing the potato into England (Box 7.2). All that can be said with certainty is that potatoes arrived in Spain about 1570 and in the British Isles before 1590. Whatever

Box 7.2

One Potato? Two Potatoes? Three Potatoes? Four?

The botanical origin of potatoes, like their introduction into England, has been the subject of much speculation. Almost all the cultivated potatoes belong to eight species native to western South America. Four of these species are diploid ($n = 12$), and four are polyploid. The common potato, *Solanum tuberosum,* is a tetraploid ($n = 24$) consisting of two subspecies, *S. tuberosum* subsp. *tuberosum.* and *S. tuberosum* subsp. *andigena.* The potatoes taken to Europe in the sixteenth century all belonged to the andigena subspecies widely cultivated in the Andean highlands (Bolivia and Peru south to northern Argentina). The origin of this tetraploid was debated for years, but recent molecular work suggests that it resulted from the hybridization of a diploid Andean species, *S. stenotomum,* with another, undetermined diploid species (possibly *S. sparsipilum*), followed by polyploidy. Bouts of hybridization with other species led to the introduction of new genes into the subspecies at various central Andean localities. This subspecies was virtually the only potato grown in northern Europe before 1840. The origin of the subspecies native to Chile is even more controversial. Several botanists maintain that it was derived solely from the andigena subspecies. Others contend that it is the product of a hybridization between an unreduced egg from a still unknown diploid female parent and *S. tuberosum* subsp. *andigena.* This suggestion is based on the fact that the chloroplast DNA of the tuberosum subspecies was originally thought to be completely distinct from that of the andigena subspecies. However, recent work shows that there is considerable variation in the chloroplast DNA of both subspecies and that both contain the range of variation found in the species as a whole. After the virtual destruction of the European potato crop by the blight of the 1840s, tuberosum varieties that are more resistant to fungi than the andigena type virtually replaced the andigena type. Today the potatoes most commonly planted throughout the world are derived from Chilean tuberosum ancestors, with andigena cultivation primarily restricted to Andean South America and Mexico.

route it took to northern Europe, the potato was destined after 150 years of experimentation to become a great success in England and soon an even greater success in Ireland.

Ireland's enthusiastic reception of the potato in the middle of the eighteenth century was related not only to its ability to grow well in that country's cool, moist climate but also to the fact that a small amount of land could produce enough calories to meet the needs of an entire family. In addition, Ireland was in the middle of a rebellious conflict with England, and potatoes could be buried to hide them from British tax collectors or from the enemy troops that periodically swept into Irish towns. By the 1840s not only was Ireland fond of potatoes; its population was completely dependent on them. The entire country had essentially adopted a potato monoculture. A week's food for an average family of five typically consisted of about 40 lb of potatoes, several pounds of meat (probably mutton), a few loaves of bread, a few quarts of beer, and bits of butter, sugar, and tea. Between 1843 and 1844 disaster struck. Potato blight caused by a fungus (*Phytophthora infestans*) reached Europe, and within 5 years virtually all the Irish (and British) potato crops were destroyed. It is estimated that during those 5 years at least 1 million people (some say twice that number) died of starvation and over another million emigrated, many to the United States. After this lesson, farmers began to grow a different subspecies of potato and to pay attention to selecting for disease resistance and to the dangers of dependency on a single crop. Luckily, at about the same time a fungicide (Bordeaux mixture) was developed that helped control the blight. Since 1985, virulent, fungicide-resistant strains of the fungus have spread across the United States, spurring increased research on the pathogens (Fig. 7.29).

Today potatoes are grown throughout the temperate parts of the world and in upland tropical areas (Table 7.3). Planting is exclusively by seed tubers (small, deformed tubers) or by pieces of tubers that contain dormant buds (Fig. 7.30). Although the white potato is still a mainstay in the United States, new varieties such a blue potatoes and the yellow-fleshed Yukon Gold and Yellow Finn are gaining in popularity.

Potatoes have traditionally been grown from pieces of tubers because the plants are almost always male-sterile. Using tubers, however, creates problems because they carry pathogens, particularly nematodes, fungi, and viroids. Gardeners can help reduce infestations in their crops by buying only "certified disease-free" seed tubers, which are produced under carefully controlled conditions. Recently, a new variety of potato called Pioneer has been produced that can be grown from seed. This variety has caused some excitement because using seeds not only will reduce pest problems but also will allow new breeding programs to be undertaken.

Yams

The second most important tuber in terms of world production is the yam (Fig. 7.31). Although the sweet potato, *Ipomoea batatas* (Convolvulaceae), is often called a yam in the United States, true yams belong to the genus *Dioscorea*

FIGURE 7.29
Plant geneticists and pathologists worked together to develop two new potato varieties resistant to strains of potato blight.
(Photo by Scott Bauer courtesy of Agricultural and Research Service, USDA.)

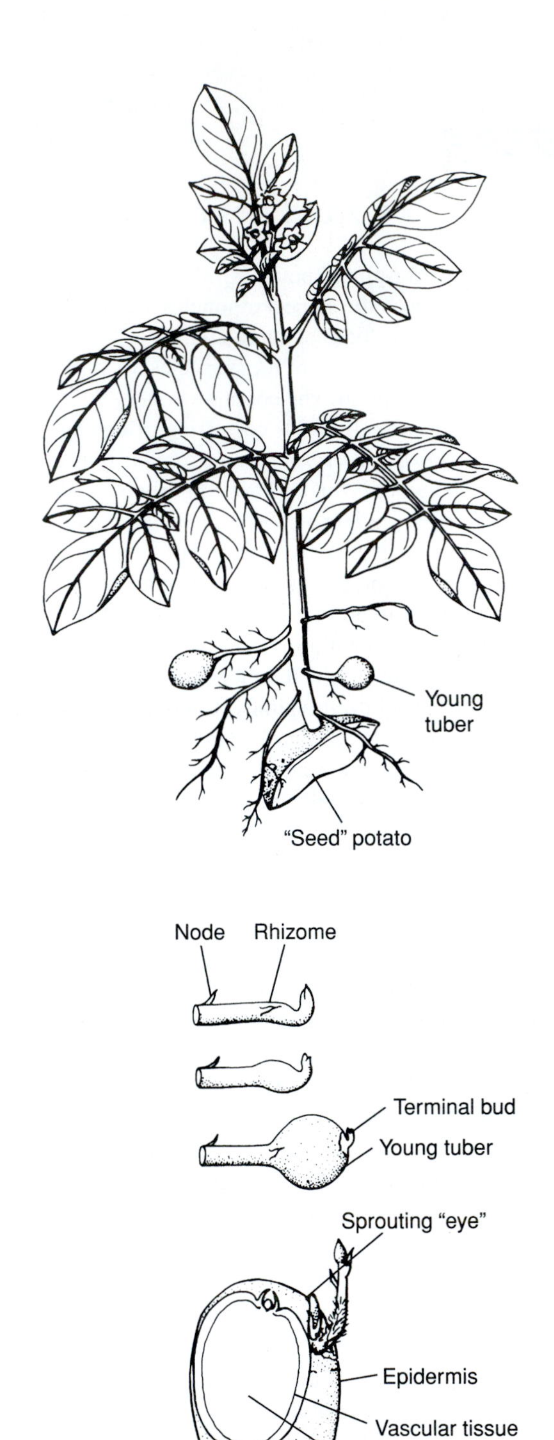

FIGURE 7.30
Tuber formation in potatoes.

(Dioscoreaceae). There is some debate about whether a yam is derived from stem or root tissue; most authorities consider the underground storage organs to be tubers and thus modified stems. However, it is possible that some of the species have storage organs that are derived from roots rather than from stems. Different species of yam are native to Africa, Asia, and South America, and domestication appears to have occurred independently on each continent. The most important species traditionally cultivated in the three native areas are *D. rotundata* (white yam) and *D. cayenensis* (yellow yam) in Africa, *D. alata* (water yam) and *D. esculenta* in Asia, and *D. trifida* (cush-cush yam) in the New World. Today, however, *D. rotundata* is grown throughout the world in the largest quantities.

In many parts of Africa (Fig. 7.32) and Asia yams are important in all aspects of the culture. In both New Guinea and Melanesia yams are central to certain ceremonies, and ceremonial yams are planted in special gardens. The size of these yams, which can reach up to 50 kg (122 lb), is thought to reflect the grower's status in the community. The ceremonial yams are used for gifts in ritualized exchanges. Yams are held in such high esteem that only men are allowed to cultivate both eating and ceremonial yams. A yam festival is held at harvesttime, during which tubers are dressed or covered by masks (Fig. 7.33).

FIGURE 7.31
Many tubers and rhizomes, such as the (a) yam, (b) arrowroot, (c) taro or dasheen, and (d) yautia, are food staples in tropical areas.

Propagation of yams occurs by asexual means, using pieces of tubers with buds. Many of the cultivated species are poisonous or semipoisonous because of the presence of oxalic acid (or oxalates) in cells just below the skin of the tubers. Peeling and boiling eliminate most of the potential problems caused by these crystals. Yams are higher in protein than some of the other starchy tuber crops (Table 7.2), but unfortunately, their cultivation has been replaced in many areas of the tropics by that of other tuberous and root crops, such as cassava (manioc), that are easier to grow but are often of lower quality in terms of protein.

FIGURE 7.32
Selling yams in Nigeria.
(Photo by B. L. Turner.)

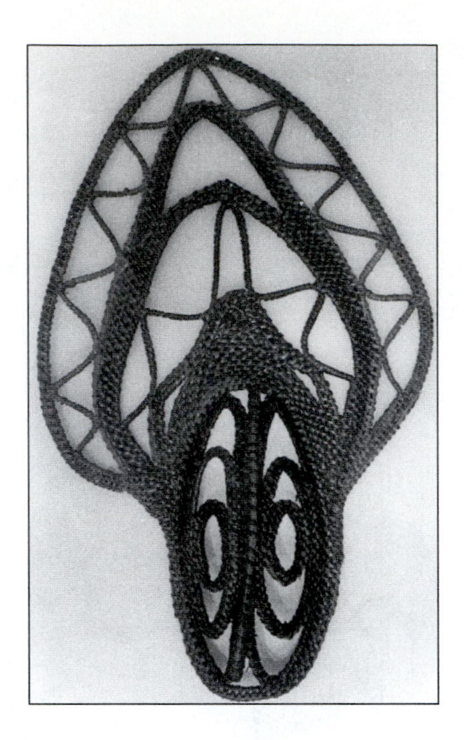

FIGURE 7.33
This Tumbuan mask from the Maprik district in northeastern New Guinea was made for a yam to wear during the harvest ceremonies. Like most yam masks, the figure depicted is animistic and has two prominent eyes separated by a nose or beak. Woven masks are also worn by young male dancers during yam festivals.
(Mask a gift of Molly Whalen.)

Other tubers grown on a more limited scale include the taro, eddoe, dasheen, and "old" cocoyam (all *Colocasia esculenta,* Araceae); malangas, yautias or tannias (*Xanthosoma* spp., Araceae); arrowroot (*Maranta* spp., Marantaceae); and canna (*Canna edulis,* Cannaceae) (Fig. 7.31). Only the first two are important crop plants. Dasheens are staple foods in tropical Pacific areas and are widely cultivated in western Africa. Anthropologists have been able to link the spread of dasheen to Hawaii from its native home in Asia with the migration of the Polynesians from Asia eastward across the Pacific. Tannias, in contrast, are native to the New World, where they have been cultivated since ancient times. Arrowroot has a limited area of cultivation restricted to its presumed area of origin in the West Indies and northern South America. The tubers are generally not eaten but are used to make a fine, easily digested powdered starch. Adulterants, consisting of starch from other tuberous crops, are often sold under the name "arrowroot," but they are inferior in quality.

Cannas are also native to South America, but their propagation has now spread throughout the tropics. The showy large-leafed plants are also grown as ornamental canna "lilies." The tuber is somewhat purplish and, when dried and powdered, produces a starch known as purple arrowroot or Queensland arrowroot.

Starchy Root Crops

Manioc

Most of the root crops of temperate regions are vegetables that are used as side dishes during a meal, but the major root crops of tropical regions often constitute the entire meal. The most important tropical root crop has many names, including manioc, tapioca, and cassava, *Manihot esculenta* (Euphorbiaceae) (Fig. 7.34). Although practically unknown in temperate regions, manioc is the staple food of over 500 million people. It contributes about 37 percent of the calories consumed in Africa and 11 percent of the caloric intake in Latin America. As is true of all the world's staple foods, manioc is predominantly starch. The storage roots are produced by monoecious perennial plants native to tropical areas of the New World. The species was formerly known only as a cultigen but has recently been found in a truly wild state in Brazil. Cassava has been grown for so many thousands of years by small tribes of Native Americans that there are now hundreds of local varieties.

The variability of the species and its widespread cultivation have created problems in interpreting the evolutionary history of manioc and determining the area of initial cultivation. Recent molecular work indicates domestication along the southern border of the Amazon River in Brazil. Griddles for baking manioc bread dated to the year 1992 B.C. have been uncovered in South America. However, the wild species that appear to be most closely related to manioc grow in southern Mexico.

Manioc seems like an ideal crop for tropical regions. It grows well in both arid and wet climates, produces storage roots in poor soil, is relatively resistant to insect and fungal pests, requires a minimum of agricultural labor, and has a high yield per unit area. Planting consists simply of clearing the ground and inserting into it 30 cm long (11.79 in.) pieces of the stem that have several axillary buds. Each plant

FIGURE 7.34

The manioc, or cassava, plant. According to a Tupi legend, there was once a mother with no food who had to watch her starving child die. Sadly, she buried the child under the floor of her hut. That night a wood spirit, or "mani," came and transformed the child's body into the roots of a plant that grew up to feed future generations of Indians. The plant was called mani "oca" (root) after the root that the wood spirit brought.

produces several storage roots that can be harvested in about 18 months or left in the ground until needed.

Manioc does have drawbacks, however. Despite their high carbohydrate content (about 86 percent by weight; Table 7.2), the roots contain very little protein (1 percent) and essentially no vitamins. Manioc roots often contain poisonous compounds that must be removed before the roots can be eaten by humans. These compounds are two cyanogenic glycosides—linamarin and lotaustralin—that hydrolyze in the presence of the enzyme linamarase to form, among other compounds, hydrogen cyanide. Any bruising, cutting, or grating of the roots brings the enzyme and the glycosides into contact and leads to cyanide production. Roots that have little or none of the glycosides are known as sweet maniocs; those with appreciable amounts of the glycosides are called bitter maniocs. If present, the glycosides are located primarily in the outer cortical cells of sweet forms, whereas the toxic precursors are dispersed throughout the roots of bitter varieties. Unfortunately, it is impossible to tell without prior knowledge of the kind of manioc that was planted whether it is of the bitter or sweet type. Environmental factors may also affect the quantity of cyanogenic compounds.

This lack of certainty about the toxicity of the roots may account for the fact that the two forms are often processed similarly, particularly in South America. In Brazil and Venezuela, both important countries for manioc production and consumption, the roots are often processed by a mechanized version of an ancient hand process. The roots are first shredded and then pressed to ensure that all the cells are ruptured and the enzymes and glycosides have been released. The soggy mass is drained, and the resultant wet fibrous residue is allowed to sit overnight. Chemical studies have shown that right after shredding and pounding there are still large amounts of toxic compounds in the mass. After the mass has stood for a day or more, the concentrations are reduced since some of the volatile hydrogen cyanide gas has had time to escape. The wet, day-old pulp can be used directly to make a kind of flatbread by spreading it on a large circular griddle about a meter in diameter (Fig. 7.35). The circular manioc pancakes that are produced do not contain cyanide because the cooking drives off the last of the gas. These flat manioc breads, which taste something like shredded wheat, can be stored for long periods (Fig. 7.36). The drained manioc pulp can also be dried and powdered. It has been suggested that the main reason for the elaborate grating and pressing is not to detoxify the material (since boiling would be equally effective) but to produce a flour that can be kept for a long period. If the flour is toasted, it is known as farofa. Farofa often constitutes the only food eaten during an entire day by many natives of the Brazilian tropical lowlands. To those unaccustomed to it, the powder tastes like a cross between plain cornmeal and flour, yet large numbers of people eat it by the handful.

Where factory processing is not available, various other primitive methods are still employed. The roots are often ground by hand, and the wet pulp is poured into basketry tubes 2 or more meters (6.5 ft) tall. The baskets, which are hung from the top and closed at the bottom, are constructed in such a way that they act like Chinese finger puzzles. The weight of the wet manioc causes the tubes to constrict and squeeze out the moisture, which runs down the baskets into vessels below.

Although the juice squeezed from the grated manioc mass is usually discarded, it contains a modest amount of starch that can be recovered by letting it settle out of the solution and decanting the liquid, which can be fermented into a beer. The starch cake left behind can be dried and eaten raw, baked into a breadlike product, or used commercially like cornstarch or potato starch. Purified manioc starch is a better sizing material (Chapter 10) than those two starches, but it is not produced on a large enough scale to challenge them commercially. If the starch is heated on a hot metal plate or tossed in a hot metal drum, it becomes gelatinized into pellets known as tapioca. Because tapioca is such a good thickening agent, it is a common ingredient in fruit pies and puddings.

FIGURE 7.35
A Venezuelan woman preparing cassava bread.
(Photo by J. L. Neff.)

FIGURE 7.36
The large, flat circles of cassava bread store well but are a little tough and dry to eat by themselves.
(Courtesy of J. L. Neff.)

In Africa, where primarily sweet types were introduced and are the most extensively grown, the roots are often simply peeled, boiled, or dried. During the boiling process, the few cyanogenic glycosides are broken down and cyanide gas is driven off. The roots are then sold as dry, heavy starchy lumps. These detoxified lumps can be ground into manioc flour to be used in breadlike products or can be reboiled to yield a rather gelatinous, flavorless starchy vegetable reminiscent of boiled potatoes. This cursory preparation can cause a disease known as *konzo* (tired legs).

Sweet Potatoes

The idea that humans crossed the Pacific from the New World to the Old World long before Columbus discovered America has intrigued people for years. Thor Heyerdahl built his balsa raft, the *Kon-Tiki,* fashioned after the reed rafts of the Oru Indians living on Lake Titicaca in Bolivia (Fig. 7.37), and sailed westward from the coast of Peru in 1947 to prove that such a journey was possible. His success heightened the controversy about whether prehistoric exchanges had occurred between South America and the islands of Oceania. Among the evidence mustered to support the hypothesis that they had occurred is the cultivation of sweet potatoes, *Ipomoea batatas* (Convolvulaceae), by native peoples of Malaysia and Polynesia as well as by those of South America.

There is now little doubt that the species is native to South America. Fossilized sweet potatoes from the Andes have been dated as being between 8000 and 10,000 years old. The unsolved question thus becomes when and how the sweet potato reached the Old World. There are no fossil or archaeological remains from the Pacific Islands, but the sweet potato forms part of an agricultural complex in

FIGURE 7.37
The reed rafts of the Oru Indians on Lake Titicaca, Bolivia, provided a model for the *Kon-Tiki.* Thor Heyerdahl sailed from Peru in a craft similar to this to prove that there could have been pre-European exchanges between South America and Oceania.

Polynesia thought to date from A.D. 1200. Ancient pits believed to have been used to store sweet potatoes have also been found in New Zealand, and the roots figure prominently in the mythology of primitive New Zealand peoples in much the same way that true yams (Dioscoreaceae) do in Melanesia and New Guinea. Was the sweet potato really carried across the Pacific by humans before Europeans began to traffic the oceans? The answer currently seems to be no. This conclusion will undoubtedly stand unless fossil evidence of sweet potatoes establishes their presence in Oceania before the seventeenth century. Even if such evidence is found, it will not explain how the roots crossed the ocean.

Today the sweet potato (Fig. 7.38) is cultivated throughout the world. In temperate areas it is grown as an annual, and in the tropics it persists as a perennial. The plants themselves are trailing vines with morning glory flowers. Although sweet potatoes resemble tubers and were confused in the eighteenth century with Irish potatoes, they are true roots, as evidenced by the fact that they lack leaf scars. A whole sweet potato has to be planted to get a new plant. Plantings are therefore made using pieces of the aerial stems. The process of swollen root formation in sweet potatoes has been studied in some detail. Regular roots become tough and hard as they mature because of lignin deposition in the cell walls. Lignification appears to be prevented in the storage roots by hormones that also initiate the swelling process. During swelling, the storage roots produce additional cambium layers in the xylem that make parenchyma cells in which starch is stored (Fig. 7.10).

Unlike manioc, sweet potatoes contain a modest amount of protein (about 2 percent by weight), some sugar, and almost as much beta-carotene (tabulated as vitamin A) as carrots (Table 7.2). They are grown less extensively than manioc in the tropics because they are much less resistant to postharvest rot and insect attack. Improvements resulting from breeding programs have been slow because of the difficulty of producing flowers under temperate summer conditions.

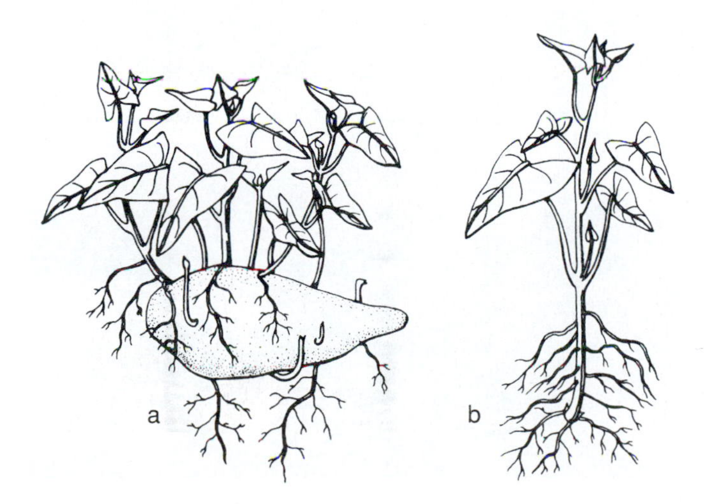

FIGURE 7.38
(a) Sweet potatoes sprout from a storage root, but most shoots sprout from the "top" end. (b) A seedling.

In the United States sweet potatoes are eaten like Irish potatoes or are "candied" (cooked with sugar). In the southern and western portions of the country the roots are often called yams, and this terminology causes some confusion with true yams. Over half the annual crop of sweet potatoes is used for starch, wine, and alcohol in other parts of the world, such as Japan. A relatively large percentage is also used for animal food. In view of the fact that the species is native to the Americas and to tropical regions, it is interesting that China is now the world's leading producer (Table 7.3). Sweet potatoes are often grown in China as a second crop after a summer crop of rice.

Sweets from Stems and Roots

When people eat candy or pour sugar into coffee, they seldom think about the fact that sugar comes from the stem or root of a plant. Historically, most sweeteners used by humans (other than honey) have been syrups or crystallized sugar obtained by boiling down the sap or juice expressed from the stems or fruits of various plants.

What people commonly call sugar is sucrose, a **disaccharide** (two sugars) produced by the chemical binding of glucose and fructose (see Fig. 1.5). Most sugar is obtained from cane, but in some areas sugar beets are used for its production. The final product, refined (white) sugar, is identical whether cane or sugar beets are used. Consequently, the processing of both is discussed below. Sucrose is the primary sugar carried in the phloem; it is not commonly stored in fruits, roots, or tubers.

Sugar Cane

Sugar cane is a perennial grass, *Saccharum officinarum* (Poaceae), believed to have been selected from wild species of *Saccharum* by primitive peoples of Oceania, probably in New Guinea. Several other species of *Saccharum* and even species of related genera believed to have been involved in the production of *S. officinarum* are also grown for their sugar content. In fact, the common sugar cane is known only as a cultivated species, and it has been argued that the species consists of a variable array of clones that are considered to belong to a single species only because they all produce thick-stemmed canes and have low fiber and a high sucrose content. The process of producing unrefined sugar from various canes has been known in India since 2992 B.C. Alexander the Great brought back to Europe stories of a honey obtained from reeds, but since the Far East was the only area of the Old World where cane grew well, very little arrived in Europe. The Ottoman Turks produced refined (white) sugar in the fourteenth century, but honey was the primary sweetener in Europe until after 1500. Because sugar was initially so scarce in Europe, it was used primarily to make medicines more palatable (a spoonful of sugar still makes medicine go down more easily). Half a kilogram (about a pound) of sugar cost a third of an ounce of gold (equivalent to US$110) in A.D. 1500, but in 1999 it cost less than US$0.40. The discovery of the New World provided Europeans with land where they could grow tropical and subtropical crops such as sugar cane. Columbus brought sugar cane to the New World on his second voyage (1493) and successfully established its cultivation in the West Indies (Fig. 7.39). Sugar rapidly became an important American export, and its production was intimately linked with the history of the New World (Box 7.3).

The famous American sugar triangle of the late seventeenth and early eighteenth centuries involved New England, Africa, and the West Indies. Raw sugar or molasses produced in the West Indies was shipped to Connecticut, where some was used to make rum. The rum was sent to Africa to buy slaves, who were brought to the West Indies to provide labor for the cane fields. The British triangular trade differed only in that England was the northern point of the triangle and trinkets and cloth were sent to Africa to buy slaves. West Indian landowners paid for slaves with cane products that went to New England or England. In 1764 the British Parliament imposed a duty on sugar with the passage of the sugar taxes. As a result, people in Connecticut started to smuggle sugar into the colonies. When British customs vessels began to patrol the waters, one, the *Gaspee,* was burned and sunk. This overt act of aggression resulting from the Sugar Act predated the Boston Tea Party (1773) as the initial violent action that finally led to the American Revolution.

One of the reasons for the successful establishment of cane once suitable areas had been found is the method by which it is grown. Virtually all sugar cane is propagated vegetatively by means of setts. A **sett** is a part of the stem that includes lateral buds and a circle of cells that give rise to a number of adventitious roots. The initial adventitious sett roots absorb nutrients until new "shoot" roots are formed. Once it is established, the cane puts out tillers that give rise to new stalks. This method of growth allows the harvesting of a crop each year. Yields per plant decrease, and a new field is planted after several years.

FIGURE 7.39
Sugar cane harvest in the West Indies over 100 years ago.
(From Histoire et Legendes des Plantes, J. Rambosson, 1868.)

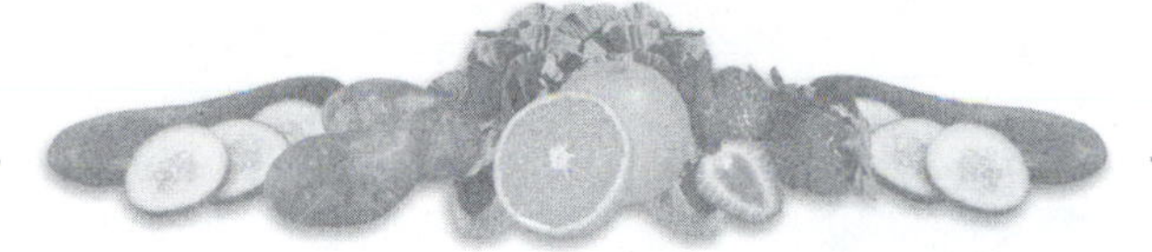

Box 7.3

Sugar, Slavery, and Sweet Teeth

Sugar cultivation had a profound impact on history, particularly that of the New World. It is hard to believe that fortunes were amassed, empires built, and uncountable human suffering caused by a substance that we associate with pleasure and love. Columbus brought sugar cane to the New World and successfully established its cultivation in the West Indies. Europeans were quick to see the economic potential of plantations in their warm western colonies. However, sugar growing required vast amounts of manual labor, work that Native Americans were either not willing or, owing to extremely high mortality from introduced diseases, were unable to do.

In the United States slavery is commonly associated with the southern cotton plantation system, but that was a relatively late development. The American slave trade began because of sugar, with Europe being the initial instigator and perpetrator. Slaves were brought from Africa, primarily western Africa, and taken to the West Indies, northern South America, and Brazil, where they were used for hot, backbreaking, agricultural labor. It is estimated that between 10 million and 15 million slaves were brought from Africa between A.D. 1517, when the Spanish officially began importing them to Hispaniola, and 1888, when the last vestiges of legal slavery were finally abolished in Brazil. Until the late 1700s, when the rise of the cotton plantations opened a huge new market for slaves, the vast majority of slaves were imported to work in sugar-growing regions. Portugal, France, Holland, and England all had colonies engaged in sugar production, and all were involved in the slave trade. Once slavery was abolished, sugar-producing countries encouraged the immigration of workers from China and India, but the need for vast armies of manual laborers dwindled with the advent of modern technologies.

Sugars normally constitute a relatively small part of "natural" human diets. The modern increase in sugar consumption, resulting from the convergence of our genetic predisposition for sweets with greatly increased availability due to sugar cultivation, has fostered a variety of societal problems. On average, each American now directly eats 20 kg (43 lb) of sugar, corn syrup, and other sweeteners each year. Indirect consumption has soared even higher, with 30 kg (67 lb) of sugar, 39 kg (85 lb) of corn sweeteners, and 0.5 kg (1 lb) of honey and other syrups added to the food supply for every American each year. Obesity, kidney stones, osteoporosis, heart disease, and dental caries have all been linked to high sugar intakes. Sugar-induced diabetes is also on the rise. Modern excessive consumption and resultant health problems are perhaps retribution for the shameful history of sugar cane cultivation in the New World, but the tragic irony is that the most affected groups in the New World are African Americans and Native Americans.

Sugar Beets

Beet greens and beet roots, *Beta vulgaris* (Chenopodiaceae) (Fig. 7.40), were eaten by the Romans, but in 1747 Andreas Margraff pointed out that sugar beets also contained sucrose and were a potential source of sugar. The first sugar beet extraction factory began operation in 1802.

Napoleon realized the value of a domestic source of sugar and ordered research to be conducted on the plant. Artificial selection increased the sugar content so dramatically that other temperate countries began to take notice. This initial work and subsequent selection raised the sugar content from a lowly 6 percent in the eighteenth century to over 20 percent today. In the 1890s, sugar beet cultivation expanded rapidly throughout Europe and the United States. Today, sugar beets provide most of the sugar for many European countries. There has been less pressure to expand sugar beet production in the United States because of the inexpensive supply of cane sugar from Hawaii, Puerto Rico, and the Gulf Coast states. The predominant use of cane sugar in the United States is therefore a result of economic factors, not any inherent superiority of cane over beet sugar.

Raw Sugar Production and Refining

Crystallized sugar is made by processing the juice of cut sugar cane (Fig. 7.41) or harvested sugar beets. Cane is ready to harvest 22 to 24 months after the setts have been planted. The canes can simply be cut or the fields can be burned to remove the leaves and evaporate much of the water in the stems, reducing the time needed later to concentrate the expressed juice. Sugar beets are harvested in the fall of the year in which they are planted. When they are harvested, they are often "topped" (the top of the root and the rosettes of leaves are severed from the lower thin root). The tops are used as animal feed. Once canes reach the mills, they are fed through rollers that crush them and express the juice. Sugar beets are washed, shredded, and crushed. From this point on, the process of sugar production from the sweet juice of both crops is essentially identical.

Figure 7.41 diagrams the steps in producing raw sugar from cane (or sugar beets), showing boiling, clarification with calcium hydroxide (lime), filtration, evaporation, and

FIGURE 7.40
A white sugar beet growing at Saginaw Research Farm
(Photo courtesy P. Horny.)

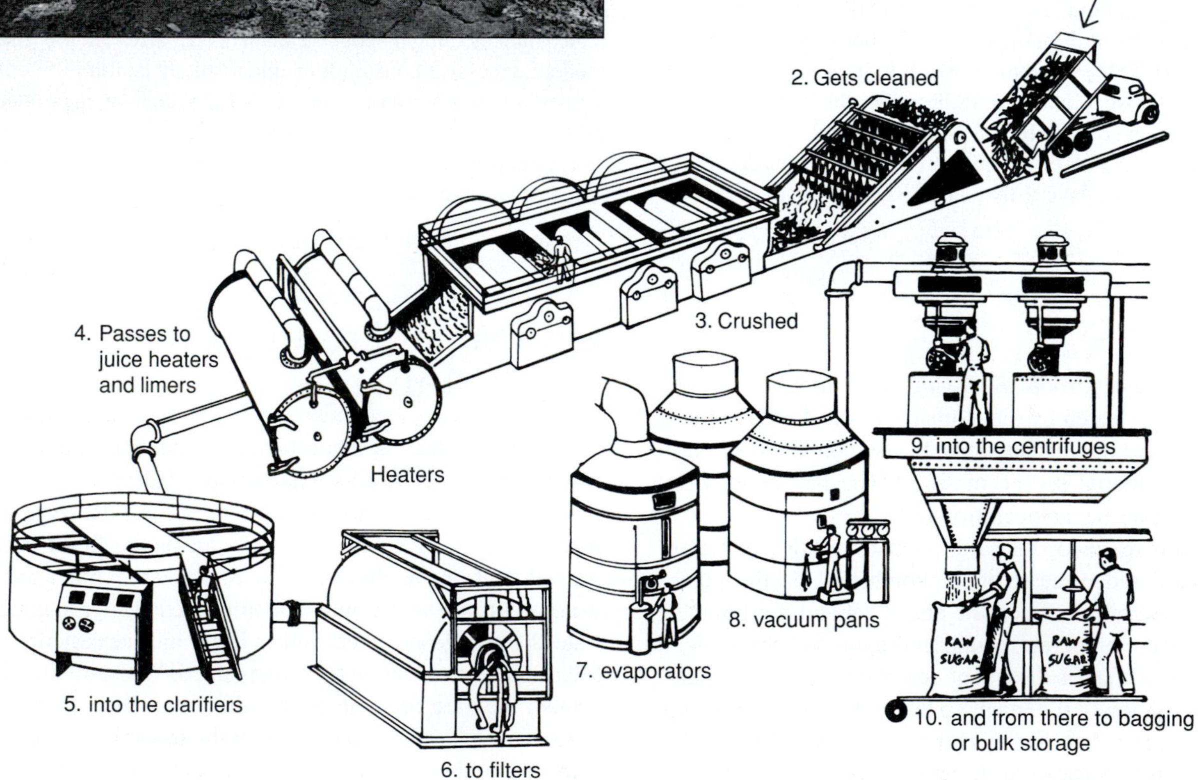

FIGURE 7.41
The major steps in sugar cane processing. After the cane is brought to the mill, it is washed and then crushed under rollers. The expressed juice is boiled to kill any microorganisms, and lime (calcium hydroxide) is added to reduce acidity and "clarify" the juice. The clarified juice is siphoned from the sludge collected on the bottom of the tank, and filtered to remove any remaining particles. The juice is heated to evaporate as much water as possible, and the thick liquid is then poured into pans and subjected to a vacuum, causing crystallization. When crystallization has reached a maximum, the sticky mass is centrifuged to separate the crystals of raw brown sugar from the uncrystallized liquid molasses. Raw sugar is shipped to factories for refining.
(Redrawn by permission from Sugar, 1949, Hawaiian Sugar Plantation Association.)

finally separation of crystals by centrifugation. The brown liquid separated from the crystals is molasses. Most sugar-producing areas ship raw sugar or use part of it to make rum. Consequently, the modern rum-producing countries are also major sugar cane producers (Table 7.2). Raw cane sugar is compact, brown and not free-flowing. The refining of raw sugar into white table sugar is accomplished in refineries in the sugar-importing countries.

The refining of raw sugar involves several steps. First, the crystals are redissolved into a heavy syrup and centrifuged to remove some of the remaining unwanted chemicals. Next, a filtering agent such as diatomaceous earth is added to the syrup to clarify it as it passes through the liquid. Another filtration, usually through charred bone, produces enough clarity to allow final crystallization and production of the various forms of white sugar. Granulated sugar is made by crystallization of the syrupy mass in large pans, followed by spinning of the rough crystals in hot, revolving drums. Powdered sugar is made by finely grinding the crystals. Because it prevents clumping, cornstarch is added to powdered sugar to produce confectioner's sugar. Sugar cubes are formed by slightly moistening the sugar and pressing it into forms to dry.

Despite the claims that raw sugar is "better" for one's health than refined sugar, the major ingredients used in the refining process—diatomaceous earth and bone char—are innocuous substances. Most of the unhealthy effects of sugar are caused by the ingestion of too much sucrose at the expense of other necessary foods. Raw sugar still contains many of the chemicals present in the cane stalks, none of which have ever been shown to have beneficial effects. Sugar sold as "brown sugar" is often refined sugar to which molasses has been readded.

FIGURE 7.42
In the spring, as the ground thaws, the sap rising in maple trees is tapped and evaporated down into maple syrup. Trees are no longer slashed to allow the sap to accumulate. Instead, a shallow hole is bored into the trunk, and a tube or spigot is inserted.
(Photo courtesy of USDA.)

Other Plant Sources of Sweeteners

The making of sorghum "molasses" is similar to the making of sugar from cane except that the sugar is not crystallized from the syrupy juice. Sorghum (*Sorghum bicolor*) is a grasslike cane, but it is grown primarily for its grain (Chapter 5).

In the Old World, primitive people of the Mediterranean region collected the sap from *Phoenix dactylifera* (Arecaceae), a palm, and condensed it by boiling off part of the water to produce a sweet syrup. Similarly, early American settlers used the sap of native plants for syrup production. Importation of sugar was costly, and honey bees were not present in North America until hives were established by Europeans. Learning from the Native Americans, the settlers began to slash the bark of various trees, particularly the sugar maple, *Acer saccharum* (Aceraceae) (Fig. 7.42), and collect the sap in the spring as it exuded from the cut. The sap was then boiled down until it reached the desired consistency. The amount of sap from temperate trees needed to produce syrup is enormous compared with the amount of juice from crushed cane used to make molasses because the sugar content of the tree sap is much lower than that of cane juice (maple sap has 8 percent sugar; cane juice, about 22 percent). It takes approximately 40 gallons of sugar maple sap to produce a gallon of syrup.

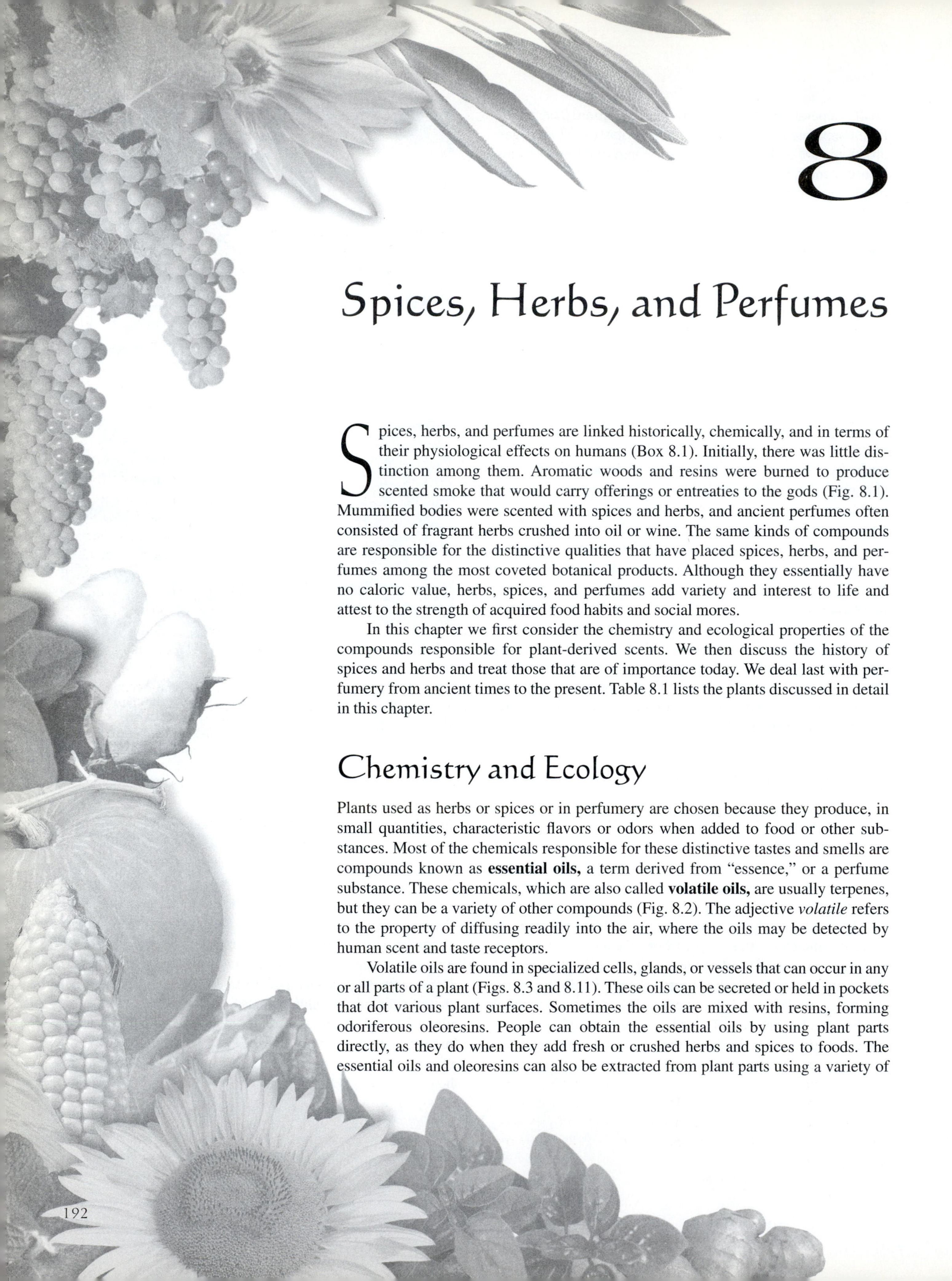

8

Spices, Herbs, and Perfumes

Spices, herbs, and perfumes are linked historically, chemically, and in terms of their physiological effects on humans (Box 8.1). Initially, there was little distinction among them. Aromatic woods and resins were burned to produce scented smoke that would carry offerings or entreaties to the gods (Fig. 8.1). Mummified bodies were scented with spices and herbs, and ancient perfumes often consisted of fragrant herbs crushed into oil or wine. The same kinds of compounds are responsible for the distinctive qualities that have placed spices, herbs, and perfumes among the most coveted botanical products. Although they essentially have no caloric value, herbs, spices, and perfumes add variety and interest to life and attest to the strength of acquired food habits and social mores.

In this chapter we first consider the chemistry and ecological properties of the compounds responsible for plant-derived scents. We then discuss the history of spices and herbs and treat those that are of importance today. We deal last with perfumery from ancient times to the present. Table 8.1 lists the plants discussed in detail in this chapter.

Chemistry and Ecology

Plants used as herbs or spices or in perfumery are chosen because they produce, in small quantities, characteristic flavors or odors when added to food or other substances. Most of the chemicals responsible for these distinctive tastes and smells are compounds known as **essential oils,** a term derived from "essence," or a perfume substance. These chemicals, which are also called **volatile oils,** are usually terpenes, but they can be a variety of other compounds (Fig. 8.2). The adjective *volatile* refers to the property of diffusing readily into the air, where the oils may be detected by human scent and taste receptors.

Volatile oils are found in specialized cells, glands, or vessels that can occur in any or all parts of a plant (Figs. 8.3 and 8.11). These oils can be secreted or held in pockets that dot various plant surfaces. Sometimes the oils are mixed with resins, forming odoriferous oleoresins. People can obtain the essential oils by using plant parts directly, as they do when they add fresh or crushed herbs and spices to foods. The essential oils and oleoresins can also be extracted from plant parts using a variety of

Box 8.1

Making Sense of Tastes and Smells

Inhaling a wine's bouquet, the aroma of fresh bread, or the perfume of a freshly peeled fruit reveals that the delight we take in food is derived not only from taste but also from smell. Our sensing of tastes and of smells are separate physiological events, but the resultant impulses they produce combine in the brain to produce the flavors we commonly associate with different foods. Surprisingly, smell is 80 percent of what we call taste. For example, if we had no sense of smell, all gumdrops would taste sweet, but we could not distinguish among pineapple-, lemon-, and cherry-flavored gumdrops.

Taste is divisible into four main components: sweet, sour, salty, and bitter. These qualities are perceived in the taste buds, which are located primarily across the surface of the tongue but also on various other parts of the mouth. Each bud is formed by about a hundred long cells that cluster together, with their tips protruding from openings that occur in the layer of smaller cells that coat the mouth. As food swishes around the tongue and teeth, compounds in the food are washed into and out of the pores. Buds on different parts of the tongue tend to be receptive to one of the four basic types of flavor. Each of the four types produces a different chemophysiological reaction in specific taste cells. Salt contains sodium, which flows into taste cells through channels in the cells' membranes. Sour substances cause the membrane channels to close, preventing potassium flow out of the cells. Chemically, therefore, salty sensations are registered by an increase of sodium in the cell and sour tastes by a buildup of potassium. Sweet molecules seem to be picked up by receptors on the protruding surface of the taste cells. Only molecules of certain configurations, including natural ones such as sucrose and fructose and artificial substances such as saccharin, bind to the receptors, producing a signal that the food is sweet. Bitter substances are also detected by specific receptors. In addition to these classic taste receptors, we can perhaps now add another kind, pain receptors. It is now known that hot peppers and other similar spices activate the same sensory neurons in the mouth as liquids and foods that are noxiously hot. The ultimate effect of all of these processes is the production of electrical currents in the taste cells that are transmitted to nerve endings that surround the basal parts of the cells. The induced nerve impulses are carried to various areas of the brain, where they are translated into our perceptions of sweet, sour, bitter, salty, and hot.

Smells are captured by a different group of specialized cells located in the top of the nasal canal. Only molecules that have a molecular weight of less than 300–400, are rather neutral in pH, and have a high vapor pressure produce a reaction in these cells. Consequently, we can perceive as odors only molecules small enough to diffuse into the air and then into the mucous layer covering the smell receptor cells. These cells are elongate like those of the taste buds, but they have numerous hairs on their tips that float in the mucus that covers them. These hairs possess receptor sites to which fragrant molecules adhere for a long enough time to cause a reaction in the cell. The hairs of different receptor cells have sites for different molecules. Once a smell molecule has attached to a receptor site, it produces a change in the cell membrane that causes a movement of charged ions, possibly calcium, into, or out of, the cell. This ion movement establishes an electrical current that stimulates nerves attached to the smell cells. The nerve cells carry the impulses to the main olfactory bulb just above the nasal cavity, where different signals converge on specific areas and allow discrimination of the odor signal.

When you eat any food or combination of foods, the ultimate flavor sensation is produced by the brain's integration of the combined perceptions of the many different impulses received from the taste buds and the smell receptor cells. The actual recognition and total impact of the food emerge from these impressions plus those received visually and tactilely.

methods. These isolated oils are used for flavorings and in perfumes. Many of the essential oils used in perfumes are produced by plants to attract animals that will serve as pollinators or fruit dispersers (Chapters 1 and 3). Other perfume oils, such as thymol, are secreted by glands on plant leaves and stems (Fig. 8.3). The role of the oils found in leaves, seeds, and even roots is less obvious than the function of such compounds in flowers and fruits and has therefore been the subject of some discussion. Likewise, there is considerable controversy about why humans use spices and why certain kinds and quantities of spices are used in different parts of the world (Box 8.2).

One possible explanation for the presence of volatile oils in stems, leaves, and roots is that they are waste products that are sequestered to prevent their interfering with normal plant processes. However, although some volatile compounds are produced as parts of metabolic processes, others are not. Instead, they seem to be synthesized by specialized pathways. This fact suggests that there has been selection for their production, perhaps as deterrents to the growth of competing plants. This effect, which is called **allelopathy,** requires that essential oils in the leaves be washed onto the soil around the plant. The chemicals in the soil might subsequently prevent the germination of seeds or cause the death

FIGURE 8.1

The king offers the falcon-headed god Selchmet burning incense and a libation that he pours into a bowl of lotus flowers. In return for these offerings, he hopes to receive bravery and strength.

(Redrawn from an Egyptian wall sculpture at Abydos, ca. 2300 B.C.)

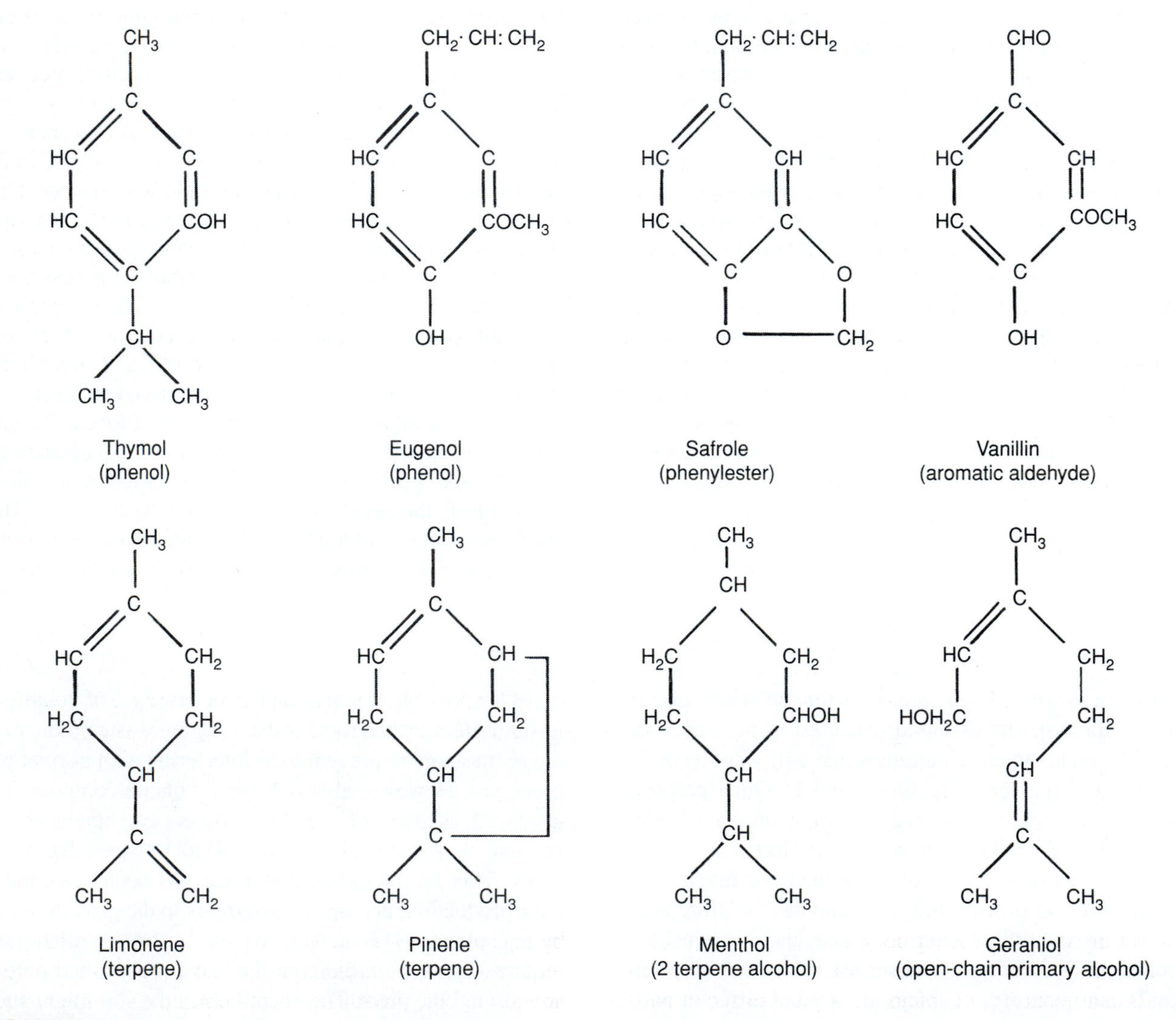

FIGURE 8.2

Structures of some common volatile oils responsible for the flavors and aromas of herbs, spices, and perfumes. Note that several classes of compounds, aldehydes, alcohols, phenols, and terpenes are used.

TABLE 8.1 Plants Discussed in Chapter 8

COMMON NAME	SCIENTIFIC NAME	FAMILY	NATIVE REGION
Allspice	*Pimenta dioica*	Myrtaceae	West Indies, Central America
Anise	*Pimpinella anisum*	Apiaceae	Mediterranean
Annatto	*Bixa orellana*	Bixaceae	Tropical Americas
Basil	*Ocimum basilicum*	Lamiaceae	Asia
Bay	*Laurus nobilis*	Lauraceae	Mediterranean
Caper	*Capparis spinosa*	Capparaceae	Mediterranean
Capsicum peppers	*Capsicum annuum*	Solanaceae	Mexico and Central America
	C. fructescens		Central and South America
	C. baccatum		South America
	C. pubescens		South America
	C. chinense		South America
Caraway	*Carum carvi*	Apiaceae	Mediterranean
Cardamom	*Elettaria cardamomum*	Zingiberaceae	Sri Lanka
Cassia	*Cinnamomum cassia*	Lauraceae	Sri Lanka
Cayenne	*Capsicum annuum*	Solanaceae	Central America, South America
	C. fructescens		Central America
Cedar	*Juniperus* sp.	Cupressaceae	Northern Hemisphere
Celery seed	*Apium graveolens*	Apiaceae	Mediterranean
Chervil	*Anthriscus cerefolium*	Lamiaceae	Russia
Cilantro	*Coriandrum sativum*	Apiaceae	Mediterranean
Cinnamon	*Cinnamomum zeylanicum*	Lauraceae	India, Sri Lanka
Cloves	*Syzygium aromaticum*	Myrtaceae	Moluccas
Coriander	*Coriandrum sativum*	Apiaceae	Mediterranean
Cumin	*Cuminum cyminum*	Apiaceae	Near East
Dill	*Anethum graveolens*	Apiaceae	Mediterranean
Epazote	*Chenopodium ambrosioides*	Chenopodiaceae	Mexico, Central America
Fennel	*Foeniculum vulgare*	Apiaceae	Mediterranean
Galanga	*Languas galanga*	Zingiberaceae	Asia
Ginger	*Zingiber officinale*	Zingiberaceae	Tropical Asia
Horseradish	*Armoracia rusticana*	Brassicaceae	Eurasia
Kaffir lime	*Citrus hystrix*	Rutaceae	Asia
Lemon grass	*Cymbopogon citratus*	Poaceae	Asia
Mace	*Myristica fragrans*	Myristicaceae	Moluccas
Marjoram	*Origanum majorana*	Lamiaceae	Mediterranean
Mustard	*Brassica nigra*	Brassicaceae	Mediterranean
	B. alba (hirta)		Mediterranean
Nutmeg	*Myristica fragrans*	Myristicaceae	Moluccas
Oregano	*Origanum vulgare*	Lamiaceae	Mediterranean
Parsley	*Petroselinum crispum*	Apiaceae	Eurasia
Pepper, black	*Piper nigrum*	Piperaceae	India, Sri Lanka
Pink	*Schinus molle*	Anacardiaceae	South America
Peppermint	*Mentha piperita*	Lamiaceae	Mediterranean
Rosemary	*Rosmarinus officinalis*	Lamiaceae	Mediterranean
Saffron	*Crocus sativus*	Iridaceae	Mediterranean
Sage	*Salvia officinalis*	Lamiaceae	Mediterranean
Spearmint	*Mentha spicata*	Lamiaceae	Mediterranean
Tarragon	*Artemisia dracunculus*	Asteraceae	Southern Russia
Thyme	*Thymus*	Lamiaceae	Mediterranean
Common	*T. vulgaris*		Mediterranean
Lemon	*T. citriodorus*		Mediterranean
Turmeric	*Curcuma domestica*	Zingiberaceae	Southern Asia
Vanilla	*Vanilla fragrans*	Orchidaceae	Central and South America

of encroaching plants, reducing competition for water, nutrients, and sunlight. Recent studies do not, however, support the idea that allelopathy is a major ecological factor in most plant populations. Instead, there is growing evidence that the chemicals responsible for the pungent flavors and odors of spice and herb plants deter predation by insects or infestation by pathogens. Fortunately, there is no evidence that spices in the quantities in which people consume them have an adverse effect on humans. In small amounts, some spices or herbs may even be beneficial because they can aid digestion or calm smooth muscles. In large doses, a few spices (e.g., mace and nutmeg) can have harmful effects.

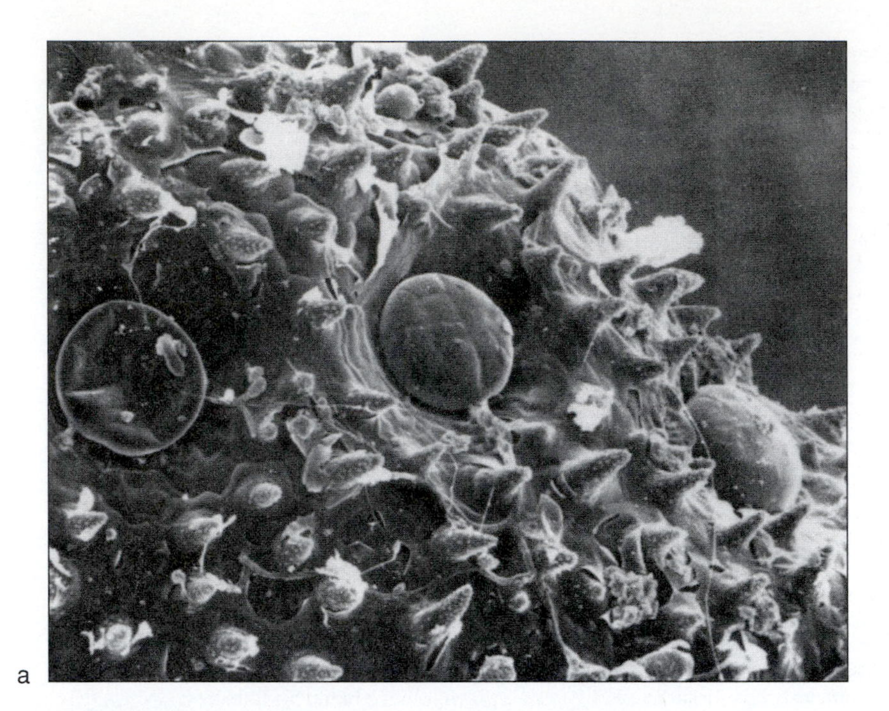

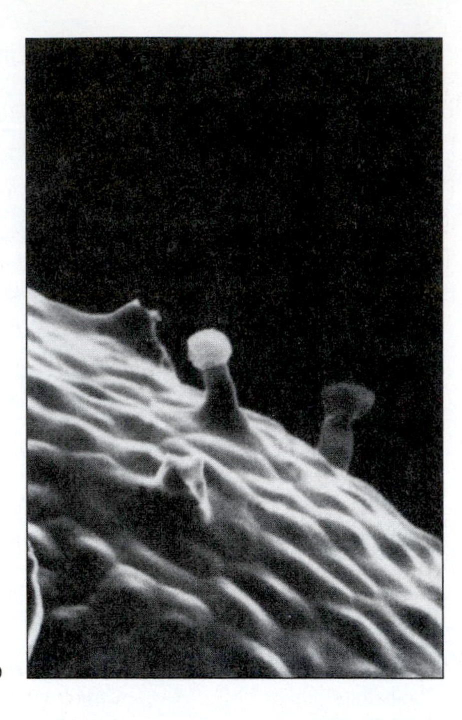

FIGURE 8.3
Specialized hairs or glands on the leaves or stems can secrete volatile compounds. (a) A magnification of a thyme leaf showing the sunken secretory glands. (b) Trichomes on the leaves of rosemary have glandular cells at their tips which secrete pungent oils. When the leaves of these species are crushed (or chewed), the glandular cells are ruptured, liberating the oils.

Herbs and Spices

To the epicure, spices and herbs bring to mind delightful smells and evoke thoughts of delicious foods. Spices are associated with feistiness, piquancy, and pungency, whereas herbs somehow make one think of pastoral scenes, old-fashioned settings, home remedies, and subtle flavors. However, it is difficult to draw any real distinction between spices and herbs. As is the case with fruits and vegetables, there is a discrepancy between the botanical and common definitions of the terms. To a botanist, the word *herb* refers to a nonwoody plant. It can be an annual or a perennial, as long as the stems do not form any appreciable wood. Consequently, there is a tendency to associate herbs with leafy plant parts. The word *spice,* in contrast, has no botanical definition. Instead, it is loosely applied to an assortment of dried barks, roots, seeds, fruits, and flower parts used for their scents and flavors. Some people define herbs and spices in terms of their places of origin: herbs from temperate regions and spices from tropical areas. In practice, this way of grouping plants used for flavorings and condiments matches well with a grouping based on the parts of the plants used. "Herbs" tend to come from temperate regions, whereas most "spices" come from tropical regions. The main exceptions to the notion that herbs come from leaves are found in an important group of temperate herbs, including dill, coriander, caraway, and cumin, that yield aromatic fruits that are used to flavor foods. Similarly, saffron, usually called a spice, comes from a flower part and is native to the Mediterranean. Spices come from a variety of plant parts but comparatively rarely come from leaves. When used medicinally, both are referred to as herbs.

Herbs and Spices in History

People take for granted the large repertoire of spices and herbs commonly found in supermarkets and used in most American kitchens. Today the average American consumes almost 1 kg (2.2 lb) of herbs and spices annually, yet only a few hundred years ago many common spices would have been rare or exceedingly expensive. Several of them would have been unknown in Europe. Although spices and herbs have lost their former importance as items of commerce, the value placed upon them was once high enough to promote exploration and set into motion a struggle for power between European nations.

We do not know when humans first began to use spices and herbs as flavoring agents, but there are records of the use of garlic and onions as far back as 4500 years ago. Before the advent of refrigeration, humans used spices and herbs to help preserve food, cover up the flavors of spoiling food, and make otherwise monotonous meals more interesting. Spices were also used at a very early time for religious ceremonies, for embalming, and to produce fragrant smoke during ritualized cremation of the dead. The practice of embalming goes back 5200 years in Egypt (Fig. 8.4). Herodotus provided a

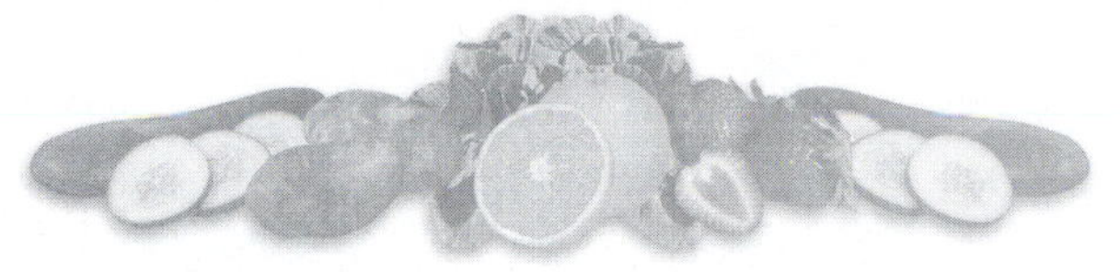

Box 8.2

The Spice of Life

Why do humans add spices and herbs to their foods and quickly adopt and add new combinations to their foods and drinks? This proclivity is especially puzzling since many spices, such as cilantro, which contains the same compounds as stinkbugs, and epazote, which resembles skunk odor, are not particularly pleasant by themselves. Many different, but not mutually exclusive, explanations have been offered for humans' initial use of spices and herbs. The major explanations are that spices: mask the odor of rotting food, neutralize unpleasant odors, act as antibacterial and antifungal agents, provide nutritional components, yield a physiological "high," and simply make otherwise dull food more sensually interesting.

Various authors have argued that it would be unwise to use spices simply to mask rotting food since such food is unhealthy. This is a valid point, but perhaps spices could extend the acceptability of food that is not as fresh as it once was. There seems to be little doubt that some spices can neutralize odors such as fishy smells, and they are used commonly for raw seafood dishes. Some spices and herbs do supply vitamins such as vitamins A, C, and B_1, and some such as garlic and onions contain compounds that lower blood pressure. Likewise, many spices such as red pepper and other similar "hot" spices create a physiological flush that many people learn to relish.

The idea that spices were originally used to make food safer has received considerable publicity. Studies have shown that almost all spices and herbs have some antibacterial effects, with cloves, mace, and garlic showing the strongest effects. An analysis in 1998 suggested that more spices were used in the tropics than in cool temperate regions because diseases are more prevalent in tropical areas. However, the study did not control for the fact that species diversity is higher in the tropics and the higher absolute number of spices might simply reflect the fact that more species are available for use. Likewise, the study did not concentrate on traditional methods of food preparation. Another study found that 2.5 percent of a hamburger's weight had to be ground cloves in order for 90 percent of the *Escherichia coli* added to it to be killed. A hamburger this spiced up would be inedible, and normal levels of spices in foods would be a fraction of this amount. Nevertheless, before modern sanitary methods of food preparation, any reduction in disease would have been advantageous enough to promote the use of spices.

Certainly, the incorporation of spices and herbs into a regional cuisine would initially depend on the potential supply of native species. The synthesis of flavors produced by adding various spices or herbs to particular foods subsequently determined which were adopted and for what purposes. Onions, for example, are not suitable for addition to coffee or sweet liquors, whereas cinnamon is not compatible with beef or shellfish dishes, so such combinations have not become popular.

Once spices began to be traded around the world, spices with similar qualities could be substituted for others in regional cuisines. For example, capsicum peppers have supplanted ginger or wasabi in many Oriental dishes. Coriander and cumin are substituted for epazote in Mexican cooking, and rosemary for pine resins.

Certainly today, people use spices and herbs to create colorful, distinctive, and novel foods. The appreciation of many of the spices and herbs commonly consumed today involve culturally conditioned tastes, but, with increasing globalization, people around the world are able to share many flavors that were once restricted to particular regions or cultures.

FIGURE 8.4

Anubis, a jackal-headed son of Osiris who brought the dead to judgment, preparing a mummy. Cassia, cinnamon, and a variety of other herbs and myrrh were used in preparing bodies for interment.

(Redrawn from an Egyptian wall fresco.)

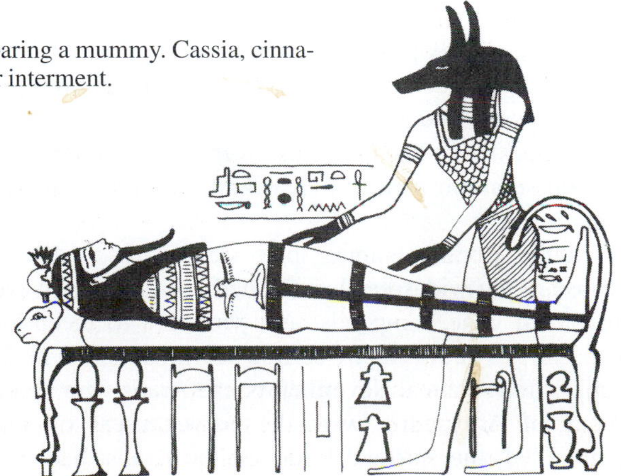

description of the mummification process and the use of myrrh, cassia, cinnamon, cumin, anise, and marjoram, among other plants, as materials for stuffing body cavities. Many of the spices and herbs used in ancient embalming are not native to Egypt or even to the Mediterranean region. Demand for them in Egypt eventually led to an important set of trade routes to Southeast Asia and China that crisscrossed the Middle East, Arabia, and India by 1400 B.C. (Fig. 8.5).

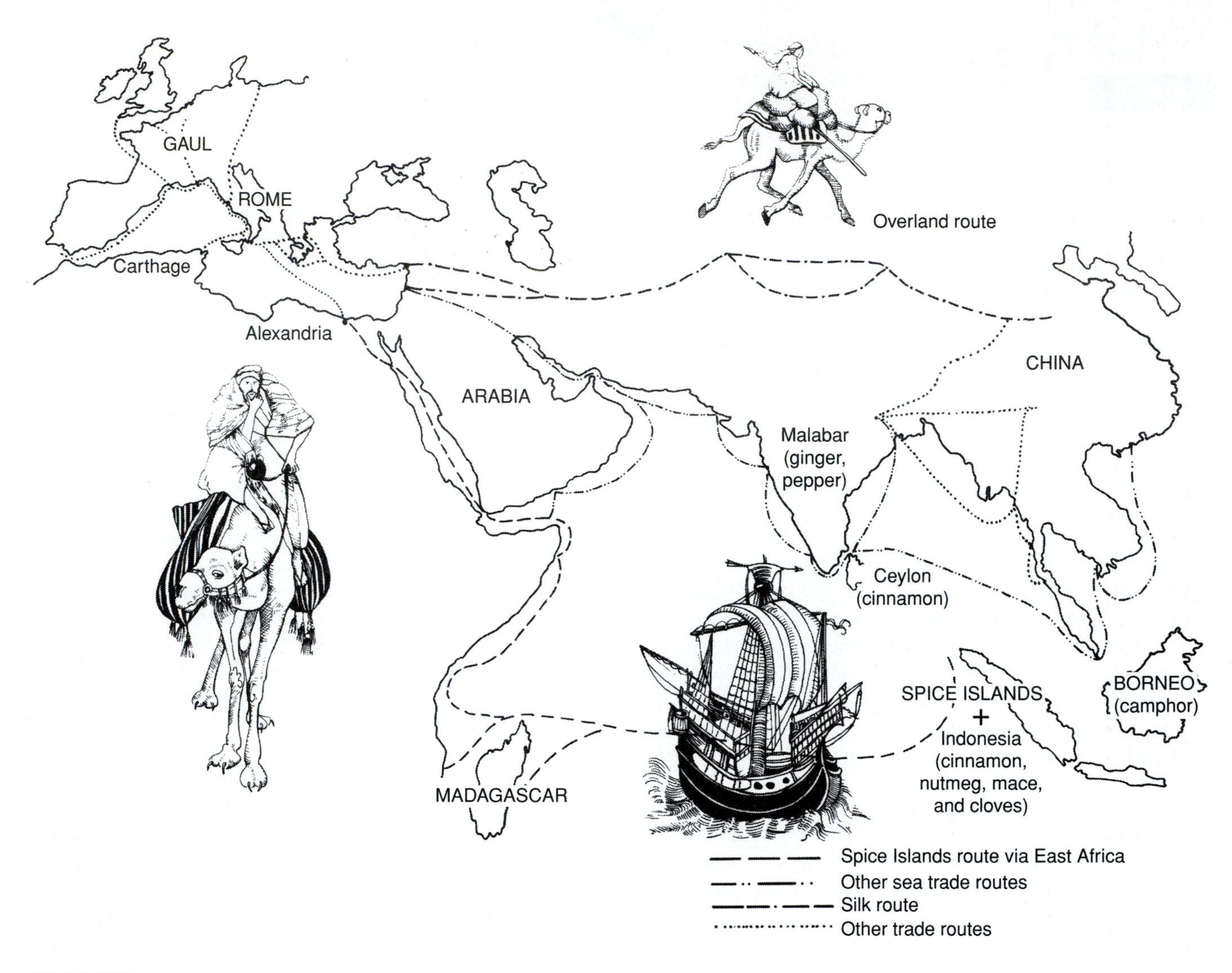

FIGURE 8.5

A map of the ancient trade routes that provided the Mediterranean region with silk and spices from about 4000 B.C. until after 1498, when Vasco da Gama reached India by sailing around the African Cape of Good Hope.

The Greeks expanded the spice trade routes, establishing Alexandria (Cairo) as the trade center for the Mediterranean. By the times of Hippocrates (approximately 400 B.C.) and Theophrastus (approximately 300 B.C.) herbs and spices were commonly used in medicine as well as for other purposes, and many were accurately described botanically (Fig. 8.6). Theophrastus even noted a correlation between semiarid and arid areas and the predominance of aromatic herbs.

In the Roman empire spice trading reached a peak because of the widespread use of herbs and spices in everything from wine, lamp oils, and perfumes to incense and food. The fall of the empire carried down with it the network of spice trade routes. Nevertheless, the Arabs who assumed control of Alexandria were themselves users of spices. Mohammed, the spiritual leader of the Arabs, had been a spice trader as a young man.

Europe's supply of spices was sharply reduced during the Dark Ages (A.D. 641 to 1096). The few herbs available were generally cultivated in monasteries. Many spices reappeared in Europe only after the beginning of the Crusades (after A.D. 1096). The crusaders fighting in Palestine and Syria were exposed to new and exotic flavors and acquired a taste for them. Upon their return to Europe, they helped regenerate a demand for imported spices. Venice and Genoa gradually became centers of trade, linking Europe to the Near East and Far East. The wealth brought to those Italian cities by their trading activities helped foster the cultural rebirth known as the Renaissance. By the end of the Crusades (approximately 1300) Europe was a changed continent. Spice merchantry had become a legitimate profession, and spicers were given a charter by England's King Henry IV in 1429 to sell spices *en gros* (from which our word *grocery* is derived), or wholesale.

PLANTÆ CVIQVE SVAS VIRES DEVS INDIDIT, ATQVE
PRÆSENTEM ESSE ILLVM, QVAELIBET HERBA DOCET.
CAROLI CLVSI ATREBATIS,
IMPP. CÆSS. AVGG.
MAXIMILIANI II.
RVDOLPHI II.
Aulæ quondam familiaris,
RARIORVM PLANTARVM
HISTORIA.
Quæ accesserint, proxima pagina docebit.
ANTVERPIÆ
Ex officina Plantiniana
Apud Ioannem Moretum.
CIↃ. IↃCI.
VIRTVTE ET GENIO

FIGURE 8.6
Frontispiece from Clusius's herbal. Clusius (1526–1609) was a Dutch botanist famous for growing plants from exotic places. Among the novel plants he grew were potatoes from the Andes and beans from Mexico. While he was a professor in Leiden, he cultivated tulips from Asia, effectively initiating the Dutch tulip industry.

The search for new trade routes and sources of spices escalated after the Crusades. In 1453 Constantinople (Istanbul) was captured by the Turks, and once again Europe was severed from eastern spice sources. Europe needed a new way to reach India and the Orient, and this time, it was technologically ready to venture into alien territory. The Portuguese were among the first explorers. In 1498 a Portuguese ship captained by Vasco da Gama managed to reach India by sailing around the southern tip of Africa. Both the navigation of a direct sea route to India and the eventual reopening of the overland routes to the Orient reestablished the flow of Asian spices into Europe by 1560. Columbus, of course, had set out earlier on his Spanish-sponsored voyage to find a route to the East Indies. Instead, in 1492 he inadvertently set the stage for the influx of a previously unknown set of spices from the New World (see Fig. 8.26).

Because of their early explorations, the Portuguese dominated the spice trade for some time, but the Dutch eventually managed to take over the Spice Islands (islands in the Malay Archipelago now known as the Moluccas) between 1605 and 1621. Holland then virtually controlled the spice trade for over 200 years, until overproduction of some spices and the development of the British Navy brought an end to the Dutch monopoly in about 1796.

The New World introduced few spices into commerce, but even so, the United States had a small place in the history of spice trading. Salem, Massachusetts (usually associated with whaling), served for many years as a center for black pepper trading between Sumatra and England. Still, spices never played as important a role in American history as they had in that of Europe, in part because by the time America was settled by Europeans, travel was much improved from that of the Middle Ages. In addition, plantations of Old World spices were eventually established in the West Indies, reducing the necessity for long, expensive trips to procure them. Increased availability diminished the value of most spices, and the importance of spice trading gradually faded. Since 1900 there has been an increase in the use of artificial flavorings, and this has caused a further reduction in the trading of spices. Nevertheless, spices and herbs give the foods of various countries their distinctive characteristics, and the use of both is increasing in the United States.

It is interesting to reflect on the flavors people now associate with different cuisines and the native homes of the spices and herbs that produce those flavors. For example, we associate the hotness of red peppers with Indian and Szechwan Chinese cooking, but red peppers could not have been used in Asia before 1492. Likewise, cumin and coriander are integral parts of Mexican cooking, but both were brought to the New World by Europeans. In our discussion of these spices and herbs, we treat first the classical herbs of Europe, then the spices of Asia, and finally the spices and herbs native to the New World.

Herbs and Spices of the Mediterranean Region

In terms of the definition of herbs as flavorful plant leaves, the mint family (Lamiaceae or Labiatae) (Fig. 8.7) stands out. Almost all the members of this family have fragrant herbage, and many are dominant species of the type of scrub vegetation that occurs around the Mediterranean Sea and the coast of California. Almost all the green herbs that readily come to mind—rosemary, thyme, marjoram, oregano, basil, sage, and mint—belong to this family and are native to the northern and eastern edges of the Mediterranean. These herbs have been used for millennia, and during the course of history many have acquired symbolic meaning (see Table 17.5). In cooking, the leaves of any of these plants can be used fresh, but most commonly they are dried and sold in whole, crushed, or powdered form.

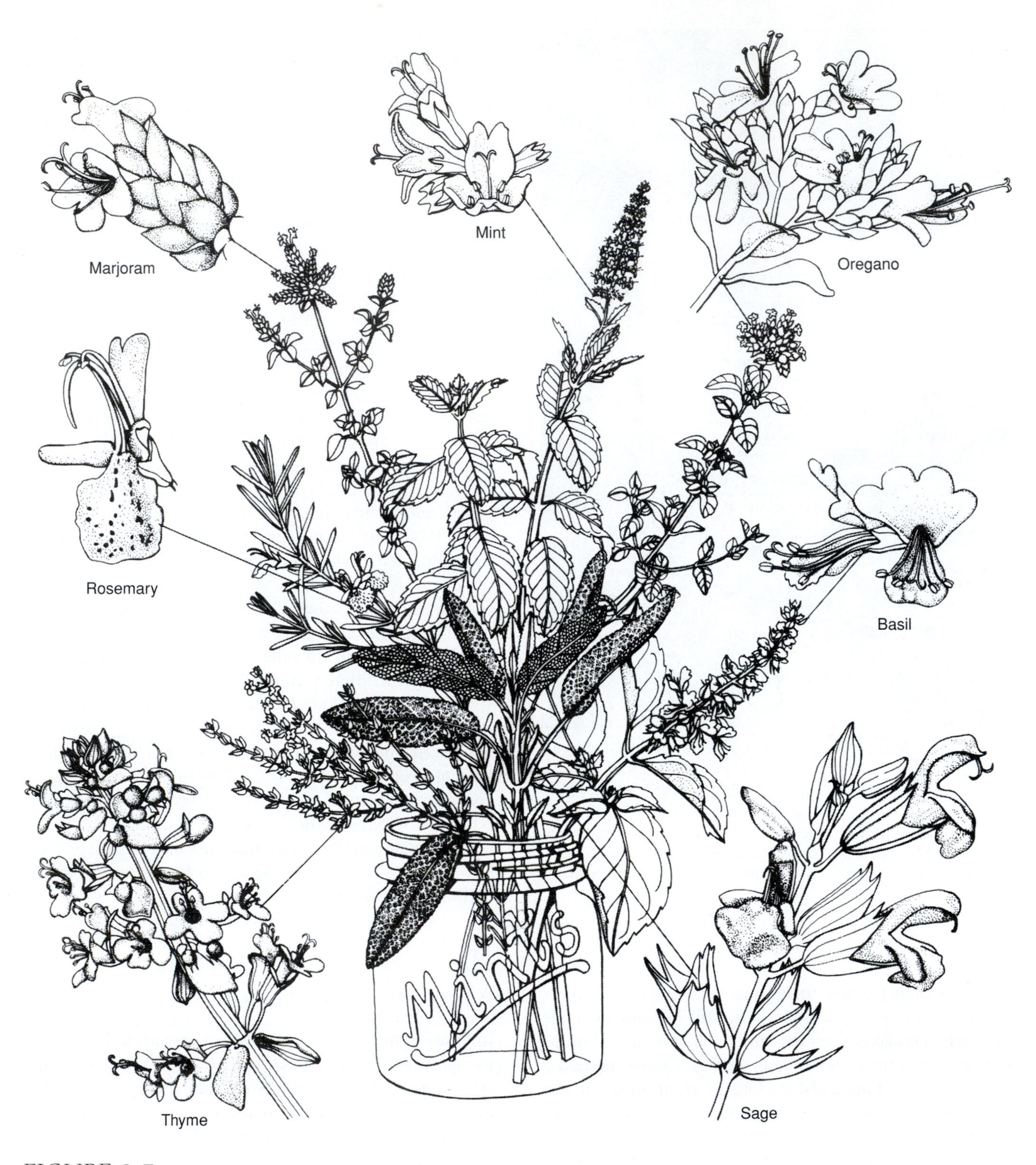

FIGURE 8.7
Herbs of the Lamiaceae (mint family) are characterized by flowers with bilabiate (two-lipped) corollas, square stems, and opposite leaves.

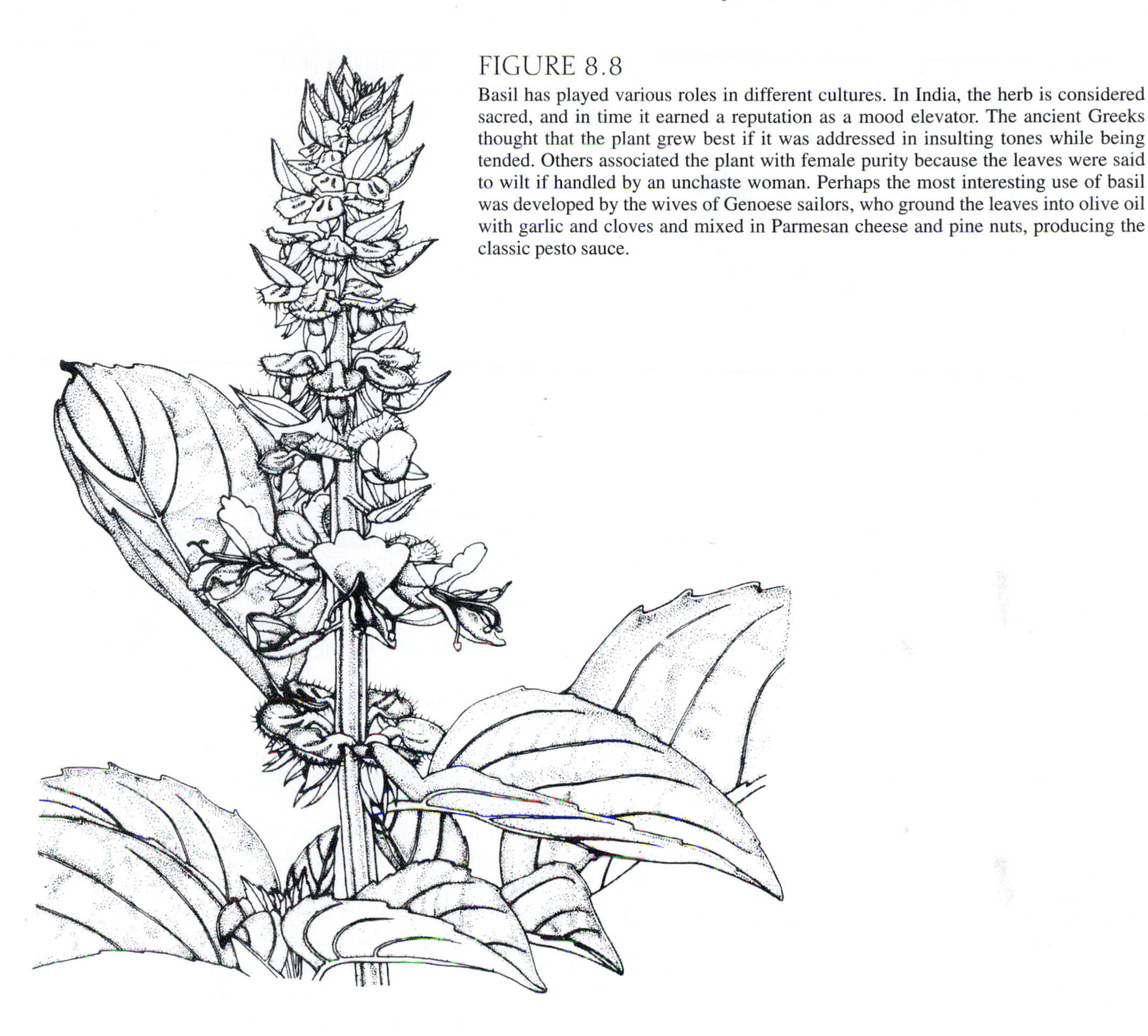

FIGURE 8.8

Basil has played various roles in different cultures. In India, the herb is considered sacred, and in time it earned a reputation as a mood elevator. The ancient Greeks thought that the plant grew best if it was addressed in insulting tones while being tended. Others associated the plant with female purity because the leaves were said to wilt if handled by an unchaste woman. Perhaps the most interesting use of basil was developed by the wives of Genoese sailors, who ground the leaves into olive oil with garlic and cloves and mixed in Parmesan cheese and pine nuts, producing the classic pesto sauce.

Herbs of the Mint Family Basil, *Ocimum basilicum* (Figs. 8.7 and 8.8), is probably native to Africa or Asia, but it was well known in the Mediterranean region by the early Greeks, and its spread around the world was achieved by Europeans. Today it is an herb often associated with Italian cooking. Part of basil's use in Italian food reflects its particular affinity for tomatoes, a component of many pasta sauces. Before tomatoes were introduced to the Old World, this herb was used in a variety of dishes and beverages. Basil is an important ingredient of Chartreuse, a green or yellow sweet cordial manufactured by Carthusian monks in France (see Fig. 14.27).

Mint (Fig. 8.7) is a name that can be applied to a number of species of the genus *Mentha.* In ancient times spearmint, *M. spicata,* was probably the most commonly used, but *M. piperita,* peppermint, is the now most frequently cultivated. The genus is named for Menthes, a young woman of classical mythology who apparently provoked Persephone's jealousy and was consequently turned into a mint plant. Today mint is extremely important in home cooking and as a commercial flavoring for mint jelly, candies, medicines, toothpastes, and mouthwashes. Mint is cultivated in the United States in California, Washington, Oregon, and Indiana. Like other herbs of the Lamiaceae, the mints grown in the United States are native to the Mediterranean, where they are also commercially produced. Although various mints are widely used today in herbal teas, there is no substantiation of claims that they have antiseptic or antispasmodic properties.

Although they are distinctly different in flavor, marjoram and oregano are closely related to one another (Fig. 8.7). Both are perennials belonging to the genus *Origanum* (Lamiaceae). Oregano plants tend to be robust and grow to about 1 m (3.275 ft) tall. Marjoram (*Origanum majorana*) plants are much more delicate in appearance and are about 30 cm (11.79 in.) tall. The modern fondness for oregano

FIGURE 8.9

A rosemary sprig. A poem by Sir Thomas Moore (1779–1852) shows how herbs were used to express sentiments in nineteenth-century England and Ireland:

> As for rosemary, I lette it runne all over my garden
> walls,
> not onlie because my bees love it,
> but because it is the herb sacred to remembrance and to
> friendship, whence a sprig
> of it hath dumb language.

(*O. vulgare*) in the United States can be attributed to a rise in the popularity of Italian food—particularly pizza—after World War II. Soldiers fighting in Italy learned to like the flavor of oregano, and as a result it is now one of the most common herbs in the United States. Marjoram has a milder flavor than oregano and never experienced the rapid rise to popularity achieved by its relative.

Rosemary (*Rosmarinus officinalis*) comes from a shrubby plant with needle-like leaves (Figs. 8.7 and 8.9). Like many herbs in this family, it has long been used to brew a tea that is thought without substantiation to have calming properties. The oil responsible for its flavor in cooking can also be extracted and is sometimes used as a constituent of perfumes and hair conditioners. When rosemary plants are grown for commercial leaf production, the plants are continually divided and pruned to maintain a low, dense mat that frequently sends up new harvestable shoots.

Among all the members of the Lamiaceae used as herbs, sage, *Salvia officinalis* (Fig. 8.7), was the one most commonly used in medicine from classical Greek times through the Middle Ages. Nevertheless, modern testing has not found it to have medicinal value. The leaves were the most popular dried cooking herb in the United States until the spectacular rise of oregano after 1945. Sage is considered an indispensable part of the "stuffing" that usually accompanies a Thanksgiving turkey and other poultry dishes. Since sage oils mix particularly well with fats, sage is commonly used in processed meats and sausages.

Plants of thyme are generally smaller than those of rosemary. Although there are several hundred species of the genus *Thymus, T. vulgaris* (Lamiaceae) (Fig. 8.7) is the one most commonly grown as an herb. The major part of the world's production is still centered in the Mediterranean, but both the common thyme and lemon thyme, *T. citriodorus,* are also grown in California. Thymol, a crystallized phenol constituent of thyme oil, has some antiseptic and fungicidal properties and is used in mouthwashes and cough drops.

Herbs of the Carrot Family The carrot family, or Apiaceae, is second in importance to the Lamiaceae in terms of the number of herbs it contains. Members of this family are easily recognizable because of their flat-topped clusters of flowers (see Fig. 7.21 and Fig. 8.10). The older family name, Umbelliferae, is derived from the word *umbel,* the name of this type of inflorescence. Some members of the family are used for their leaves or roots (Chapter 7), others for their fruits, and a few for both parts of the plant. Chervil (*Anthriscus cerefolium*) and parsley (*Petroselinum crispum*) are used exclusively for their herbage. Parsley, the often neglected garnish laid aside on the plate, is the most commonly used culinary herb in the United States today. Unfortunately, it is more often used as a decoration than as a flavoring agent. Rather than ignoring it, people would do well to eat it because it is rich in vitamins A and D and supposedly dispels the odors of garlic and onions if chewed after a meal containing them. Breath fresheners with parsley as the active ingredient are now marketed. Enormous quantities of dehydrated parsley leaves (estimated at over 0.5 million pounds a year, which would require more than 18 million pounds of fresh parsley) are produced annually in the United States. Chervil, a much less common herb, is one of the few members of this group of herbs that is not native to the Mediterranean. The species occurs naturally in Russia and western Asia.

The other two members of the Apiaceae used for their leaves are dill (*Anethum graveolens*) and cilantro (*Coriandrum sativum*) (Fig. 8.10). Dill "weed" is indispensable as a flavoring in pickled items. Cilantro leaves, which come from the same plant as coriander, have had a recent surge in popularity because they are a component of many Mexican

FIGURE 8.10

Coriander and cilantro come from the fruit and foliage of the same member of the Apiaceae. (a) Flowering and fruiting plant, (b) close-up of a flower, (c) longitudinal section of a flower, (d) fruit. Dill has the same kind of two-parted fruit (usually called a seed in cookbooks) as other herbs in the family. (e) Dill flower and fruit, (f) inflorescence, (g) infructescence. We do not often see the fruits (i) of parsley since it is grown primarily for its foliage (h).

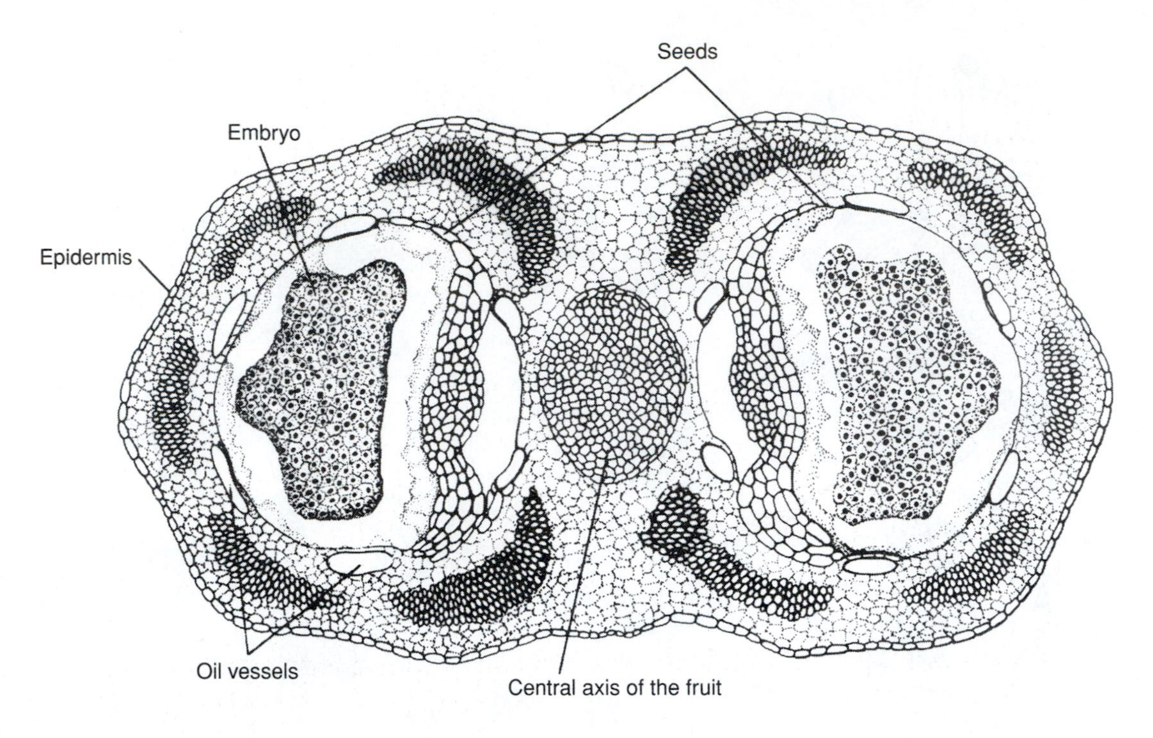

FIGURE 8.11
Oil vessels that contain the volatile compounds found in many members of the Apiaceae appear as a ring of empty spaces in this fruit cross section.

dishes, such as *pico de gallo* and salsas. Since 1995, salsas have surpassed catsup as America's favorite condiment.

Features common to almost all the herbs of the carrot family include a musky flavor and aroma. Although rather hard to define, these features are characteristic of both leafy herbs and those used in the form of dried fruits. Included in this last group are anise (*Pimpinella anisum*), caraway (*Carum carvi*), celery (*Apium graveolens*), cumin (*Cuminum cyminum*), coriander, dill, and fennel (*Foeniculum vulgare*). Although the part of the plant used in all of these cases is generally called a seed, it is in fact half a fruit (see Fig. 8.10), composed of both the pericarp and the seed. The oils that give flavor to the fruits are usually located in vessels in the fruit wall (Fig. 8.11). These oils can be extracted and used as flavorings for medicines and candies.

Herbs of the Mustard Family Because of two genera in the mustard family (Brassicaceae), Europeans were not deprived of piquant flavors during the Middle Ages despite their lack of access to exotic spices. One of these genera is mustard itself (*Brassica*), and the other is horseradish. The same types of sulfur-containing compounds that give cabbage and turnips their sharp flavors produce the biting tang of these two herbs. The traditional "hot dog" mustard Americans use as a condiment is a mixture of the dried seeds of two *Brassica* species, *B. nigra* and *B. alba.* The first, known as black mustard, is native to Europe and Africa. Black mustard has flavored food since Greek and Roman times, and a few references to it are found in the Bible. The second, white mustard, is a European species that probably came into use at a later date. In Europe today usage of the two species differs in various countries. The French and Germans prefer mustard made from black mustard seeds, whereas the British prefer the milder yellow or white species. The difference in potency of the two species is due in part to the fact that each produces a different oil when mixed with water. Mustard gas used in chemical warfare does not come from mustard plants; it belongs to a group of disagreeable-smelling compounds called thioethers that contain sulfur, an element also found in natural mustard oils. Mustard gas is made synthetically by a number of procedures and contains chlorine as well as sulfur.

Horseradish, *Armoracia rusticana* (Fig. 8.12), was enjoyed by early Scandinavians and Germans as a condiment for meats and fish, but it was not until the seventeenth century that the rest of Europe began to appreciate the pungent flavor of the grated root of this plant. Although the roots are yellowish white, the English refer to the plant as "red cole," perhaps alluding to the hot sensation one gets when ingesting even small pieces. The root itself is only about 2.5 cm (1 in.) in diameter but may reach a length of 1 m (3.3 ft). The smell and flavor of horseradish come from the glycoside sinigrin, which decomposes in the presence of water to form a mustard oil.

Other Herbs from Temperate Europe and Asia

Tarragon, *Artemisia dracunculus* (Asteraceae), a native of southern Russia and neighboring areas, is a newcomer as a flavoring agent compared with many of the herbs of the

FIGURE 8.12

Horseradish plant. Horseradish belongs to the mustard family and contains the same sulfur-containing compounds that give cabbages and turnips their sharp taste.

(After Fuchs.)

FIGURE 8.13

A flowering branch of bay from Clusius's herbal published during the fifteenth century. *Mirtus laurifolia* was a pre-Linnaean name for the species that is now considered to belong to the genus *Laurus* in the family Lauraceae.

Lamiaceae and Apiaceae. It is mentioned in the literature of the Middle Ages but was apparently unknown to the Greeks and Romans. Now, however, it is one of the most popular herbs in Europe. In the United States it is perhaps best known as the flavoring in tarragon vinegar.

Bay leaves are often considered to be a spice rather than an herb, even though they are obviously leaves. One reason for the frequent inclusion of bay leaves with the spices may be that they come from a tree, *Laurus nobilis,* which belongs to the same family (Lauraceae) as cinnamon. The classical species of bay used since antiquity is native to the Mediterranean region. Other species that occur in Central America are sometimes substituted. Laurel (or bay) sprigs were used in Rome to crown the winners of athletic events (Fig. 8.13). Our word *baccalaureate* comes from the Latin *bacca* for "berry" and *lauri,* meaning "of the laurel." Today a baccalaureate signifies the completion of one's studies, just as in Roman times being crowned with branches of laurel celebrated the successful completion of a sporting event.

Another noteworthy flavoring agent from the Mediterranean region is saffron (Fig. 8.14), the most costly of all herbs and spices. The reason for its high price lies in the fact that saffron comes from the stigmas of a crocus, *Crocus sativus* (Iridaceae). No other part of the plant is used. An acre planted solely with saffron will yield only 3.5 to 5.5 kg (8 to 12 lb) of dried spice per year; 462,000 stigma branches from 154,000 flowers are needed for each kilogram (about 210,000 stigmas per pound) of dried spice. Nevertheless, there is always a demand for saffron because of its taste and ability to produce an intense yellow color. The robes of ancient Irish kings were dyed with saffron, and classic Spanish and Indian dishes such as paella are flavored and colored with saffron. Often a much less expensive, southern Asian spice, turmeric (*Curcuma domestica,* Zingiberaceae), is substituted for saffron, but with a corresponding difference in flavor.

FIGURE 8.14
Because only the red stigmas of *Crocus sativus* are used to produce saffron, saffron is the most expensive spice.

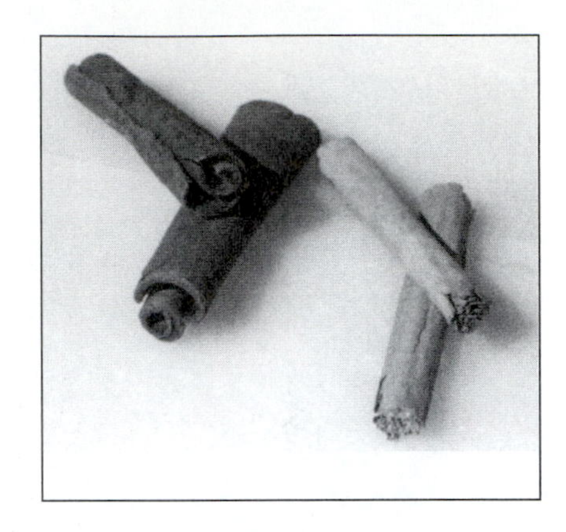

FIGURE 8.15
Cassia quills (left) are thicker than those of true cinnamon (right) because they are made from the entire bark. Fine cinnamon quills are made from only the inner layers of bark.

Spices of the Asian Tropics

The exotic spices that fostered European exploration of the world came primarily from India and Southeast Asia. Two of the most important from a historical point of view are cinnamon and cassia (Fig. 8.15). Although the name *cassia* may seem unfamiliar, most of what you have eaten that was labeled cinnamon was probably cassia. True cinnamon, *Cinnamomum zeylanicum* (Lauraceae) (Fig. 8.16), is considered to have a more delicate and fine flavor than cassia, which comes from several other species of *Cinnamomum* (primarily *C. cassia* but also *C. burmannii* and *C. loureirii*). Under the guidelines of the Federal Trade Commission, all these spices can be sold under the name cinnamon. Both true cinnamon and cassia are among the few spices obtained from bark. The quills, or curls, of cinnamon that are sold whole or ground are produced by stripping the bark from the trunks, branches, and shoots of trees. Initially the primary trunk is cut, but in later harvestings shoots that have sprouted from the stump are used (Fig. 8.17). A vertical slice is made in the bark between two incisions that encircle or partially encircle the branch or trunk, and the outer bark is stripped away. A second incision is made, and the inner bark is then pulled free. The strips of inner bark are cut and rolled by hand, often with irregular bits and pieces of bark inside the rolls. Cassia and some cinnamon are produced by stripping the entire bark from the trunk in the initial operation. The bark is then rolled in matting and allowed to ferment chemically for a day. The coarser outer portions of the bark are scraped away, leaving the inner bark with parts of the outer bark attached. Cassia is therefore a much coarser product than most cinnamon (Fig. 8.15). Sri Lanka and several countries in Southeast Asia are the major suppliers of both cinnamon and cassia.

Other spices that come from tropical trees of the Old World include cloves, nutmeg, and mace. Cloves are rather unusual in that they come from unopened flower buds (Fig. 8.18). The only other "spice" that comes from flower buds is capers, *Capparis spinosa* (Capparaceae), a relatively minor crop from the Mediterranean.

Cloves, *Syzygium aromaticum* (Myrtaceae), are native to the Spice Islands (the Moluccas) and nearby areas in Indonesia. They were one of the major spices imported into Alexandria for distribution throughout the Greek empire. The flower buds are harvested when they are just turning pink (Fig. 8.19). If a flower has opened before the clove is picked, it produces an "empty" clove when dry. An empty clove, which is considered inferior in commercial trading, has lost the corolla, stamens, and style. Round, fat unopened cloves have superior flavor and are particularly attractive compared with empty cloves. The collected buds are dried, cleaned, graded, and packed for export. About half the world's production of cloves is used in the manufacture of Indonesian kretek cigarettes. Clove oil commercially extracted from the buds, stems, and leaves is used as a source of eugenol and in the perfume and soap industries.

Nutmeg and mace are two differently flavored spices that come from separate parts of the fruit of *Myristica*

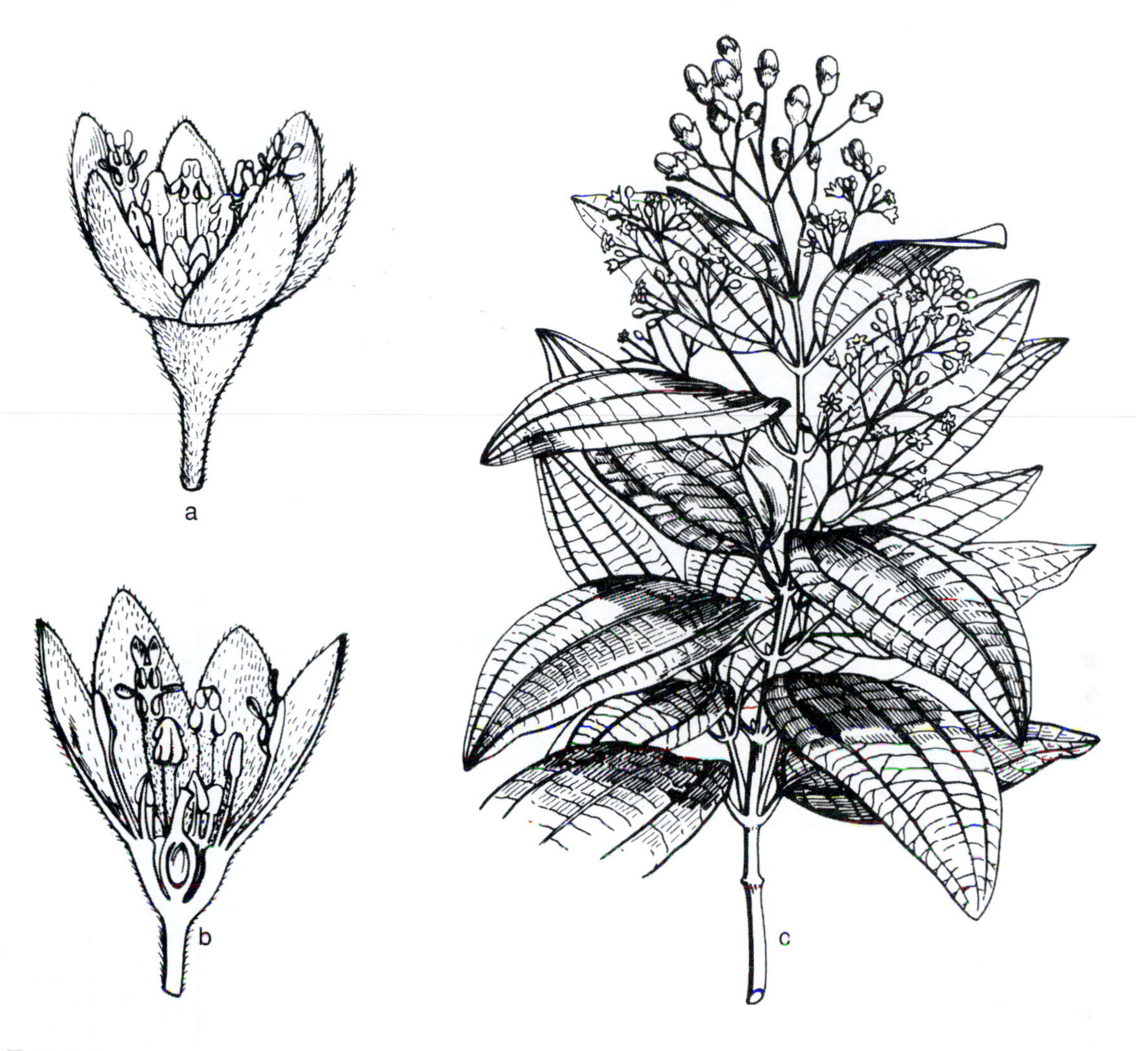

FIGURE 8.16
True cinnamon comes from the ground bark of *Cinnamomum zeylanicum.* Like other members of the laurel family, cinnamon has simple leaves, and its flowers have anthers with pores covered by flaps from which pollen is shed. (a) Flower, (b) flower longitudinal section, (c) branch.
(After Baillon.)

FIGURE 8.17
Peeling cinnamon bark in Sri Lanka ca. 1924.
(Corbis photo from the Hulton Deutsch collection.)

fragrans (Myristicaceae) (Fig. 8.20). The fruit can be considered a dehiscent berry that splits open at maturity, exposing a seed surrounded by a red, slightly fleshy network, the aril. Once peeled off and dried, the aril becomes mace (Fig. 8.21). The seed inside is also dried. Unlike true nuts, nutmeg is composed primarily of a mass of endosperm, not cotyledonous tissue. Nutmeg trees are dioecious and produce flowers only after they are about 6 years old. They only sure way to determine the sex of a tree grown from seed is to wait until it has flowered. In the past, once the sex of the trees had been determined, male trees were decapitated, leaving a ratio of male to female trees of about 1:10, which is sufficient to ensure pollination. However, this method led to irregular fields with large open spaces. Most plants today are raised from cuttings that allow the perpetuation of superior genetic lines and ensure that the plantation manager knows the sex of the trees before they are planted in orchards. Nutmeg, which is toxic in large quantities, produces hallucinogenic effects in intermediate doses (Chapter 11). The

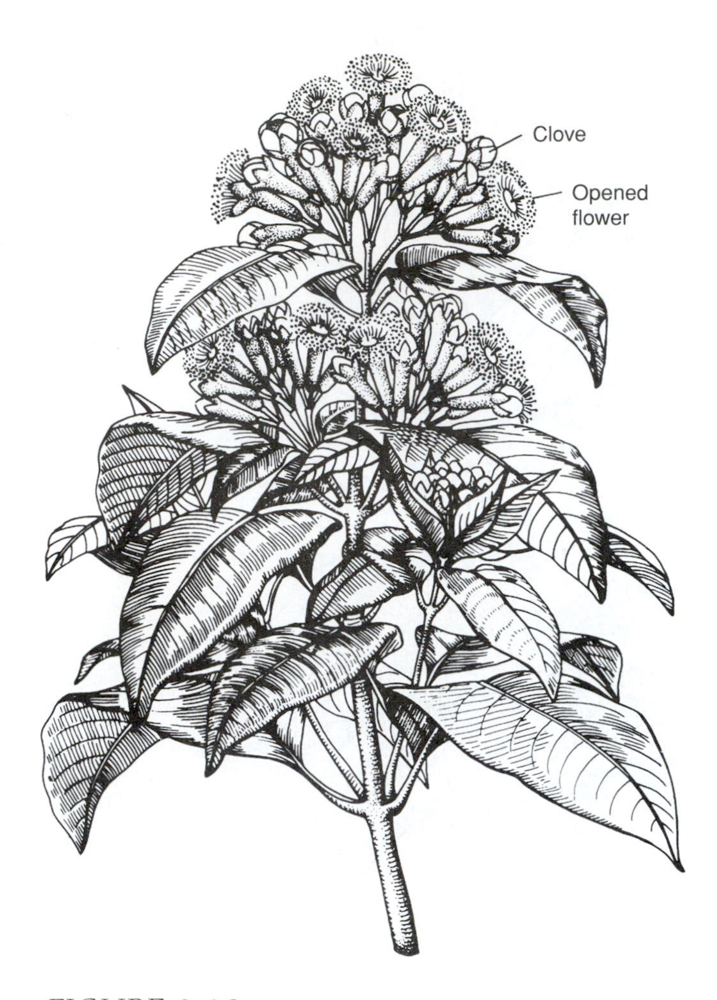

FIGURE 8.18
Cloves are the unopened flower buds of *Syzygium aromaticum*. Cloves have been used for over 2000 years in India to fasten packets of crushed betel nuts, which are chewed for their stimulating effect.
(After Baillon.)

FIGURE 8.19
A Zanzibari woman picking cloves.
(Corbis photo by B. Brecilj.)

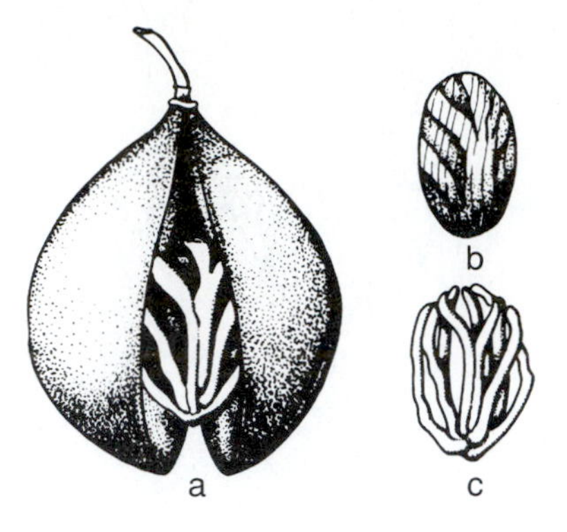

FIGURE 8.20
(a) Nutmeg and mace come from different parts of the same fruit. (b) Nutmeg is the dried endosperm of the seed. (c) Mace is produced from the dried red, fleshy net (aril) around the seed.

FIGURE 8.21
Women in Grenada separating mace from nutmegs. Nutmegs were first planted in the New World in 1802, when the British established trees on Saint Vincent in the West Indies. Grenada, today one of the world's largest producers of nutmeg and mace, started plantings in 1843.
(Corbis photo by D. Conger.)

spice has been used (erroneously) to induce abortions, but it does appear to have some effectiveness as an aid to digestion. The mace used in modern riot control is a synthetic compound unrelated to the spice obtained from *M. fragrans.*

Two other spices of great importance in ancient cultures are cardamom and ginger, both members of the ginger family, or Zingiberaceae. Both are native to southern Asia but were introduced to Europe by Arabs over 2000 years ago. Cardamom (*Elettaria cardamomum*) comes from the dried seeds of an exotic-looking plant (Fig. 8.22). Oils extracted from cardamom seeds were formerly used for a variety of purposes, including medicines.

Ginger is obtained from the fresh or dried rhizome (see Fig. 1.11d) of *Zingiber officinale.* It is usually propagated vegetatively by planting pieces of the rhizomes because the plants rarely produce viable seeds (Fig. 8.23). Ginger yields

FIGURE 8.22
Harvesting cardamom in India.
(United Nations photo by John Isaac.)

FIGURE 8.23
A ginger plant showing the rhizome that is grated to produce a fiery spice used in many cuisines. Before the discovery of the New World, ginger and black pepper provided the hotness characteristic of Indian and other Southeast Asian foods.

an oleoresin that is used to flavor soft drinks and ginger beer. Turmeric (*Curcuma domestica*) is an Asian member of the ginger family that produces rhizomes that are dried and ground and then incorporated into curry powder or used as a source of a yellow dye.

A less well known but increasingly common member of the ginger family is galanga, or galangal (*Languas galanga*), an essential ingredient in many Southeast Asian dishes. The spicy rhizomes of galanga look like those of ginger, but the flavor is more resinous.

Another herb that is finding its way into more and more Western meals is kaffir lime, or magrut (*Citrus hystrix,* Rutaceae), a member of the same genus as the common lime. The leaves, or occasionally the rinds, are used as a flavoring agent. Tearing or shredding the leaves before adding them to soups or curries releases the citrus aroma and flavor.

The grass family also yields a spice common in foods of southeastern Asia and growing in popularity in hybrid cuisines. Lemon grass (*Cymbopogon citratus,* Poaceae) comes from the same genus as citronella, a perfume oil commonly used to repel mosquitoes and to scent cleaning solutions. The portion of the lemon grass that is used is the slightly swollen base of the tiller, which is more tender than the leaves. The flavor is reminiscent of a combination of ginger and lemon.

In terms of quantities traded, black pepper is one of the most important spices in the world today. Like cinnamon, nutmeg, cloves, cardamom, and ginger, pepper was an important item in early east-west trade. Black pepper comes from *Piper nigrum,* a viny member of the Piperaceae (Fig. 8.24) native to India and Sri Lanka. Various forms of pepper, including traditional black pepper, white pepper, green pepper, and red or rose pepper, can be obtained from the drupes. Black pepper is produced by stripping whole, unripe, and still green or yellow but mature drupes from the clusters. The fruits are then heaped into piles in full sun to stimulate a chemical enzymatic fermentation that begins the darkening process and helps develop the flavor. The fermented drupes are then spread out and dried in the sun. Sometimes the fruits are dipped into boiling water for about 10 minutes before drying. The entire dried fruit is then ground or left whole as peppercorns. White pepper is obtained from ripe drupes that have become somewhat fleshy. The fruits are placed in running water for 1 to 2 weeks. This treatment causes the fleshy portion to rot away from the harder endocarp. After most of the pericarp has rotted away, crushing removes any remaining exocarp and mesocarp. The fruit,

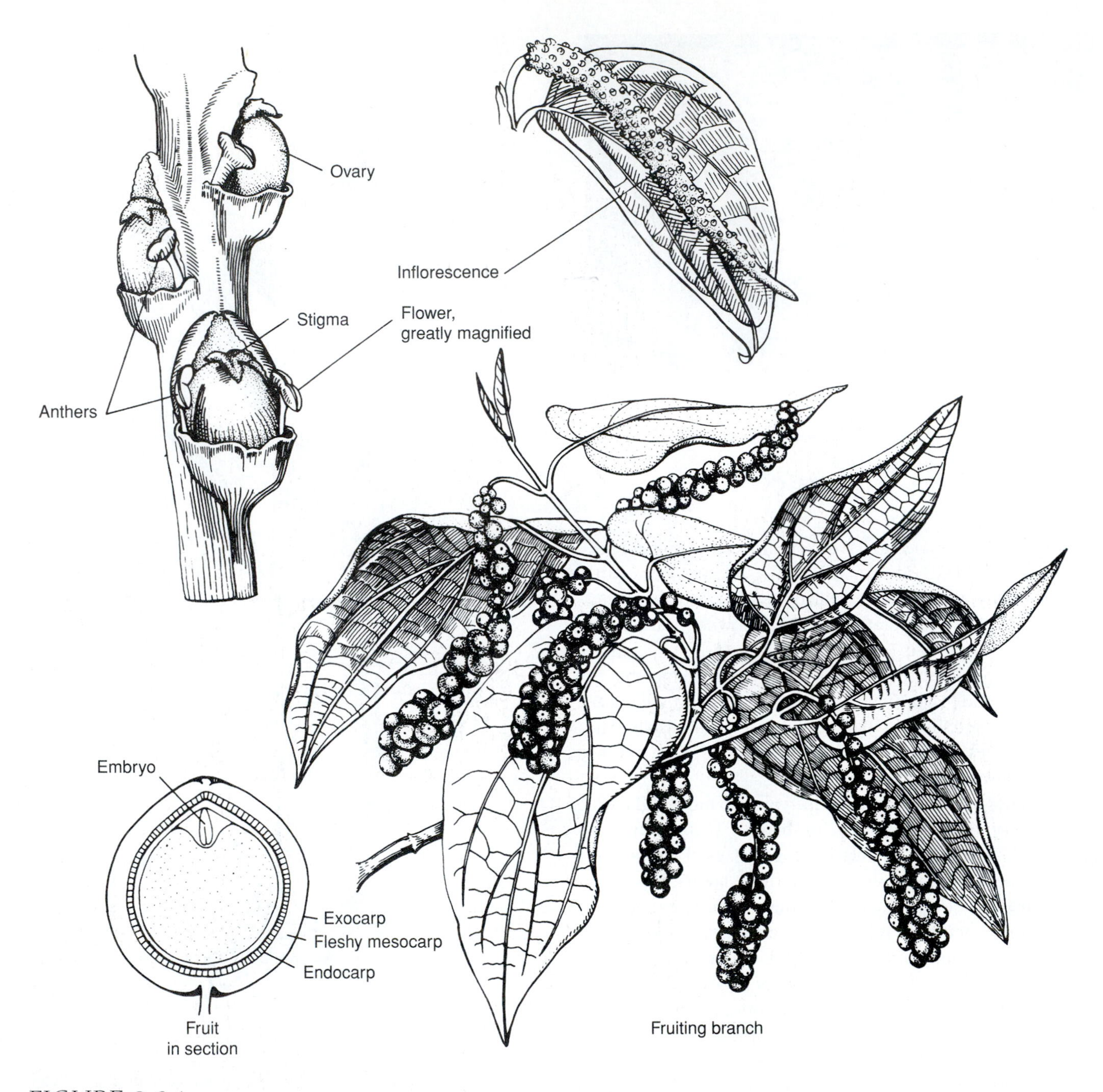

FIGURE 8.24
A fruiting pepper vine, inflorescence, flower, and peppercorn in longitudinal section.
(After Baillon, in part.)

with the white endocarp exposed, is then dried. Green peppercorns are simply pickled in brine or vinegar in the green stage or are freeze-dried. True red or rose pepper is prepared in the same way except that the fruits are allowed to ripen until they turn red. Much of the so-called pink pepper available in the United States is obtained from another species, *Schinus molle,* a member of the cashew family, Anacardiaceae. Most black pepper is still produced in southeastern Asia, primarily India and Indonesia, but Brazil is also a major world producer. The United States is the world's largest consumer of black pepper.

Spices from the New World

The New World has contributed only three significant spices to our repertoire: allspice, capsicum peppers (red peppers), and vanilla. The last two have assumed prominent places in the world spice trade. Two other New World herbs, epazote and annatto, are used on a more limited scale.

Allspice, *Pimenta dioica* (Fig. 8.25), comes from the same family (Myrtaceae) as cloves but is native to Central America and the West Indies. The name *allspice* refers to the flavor of the dried green berries, which is rather like a combination of

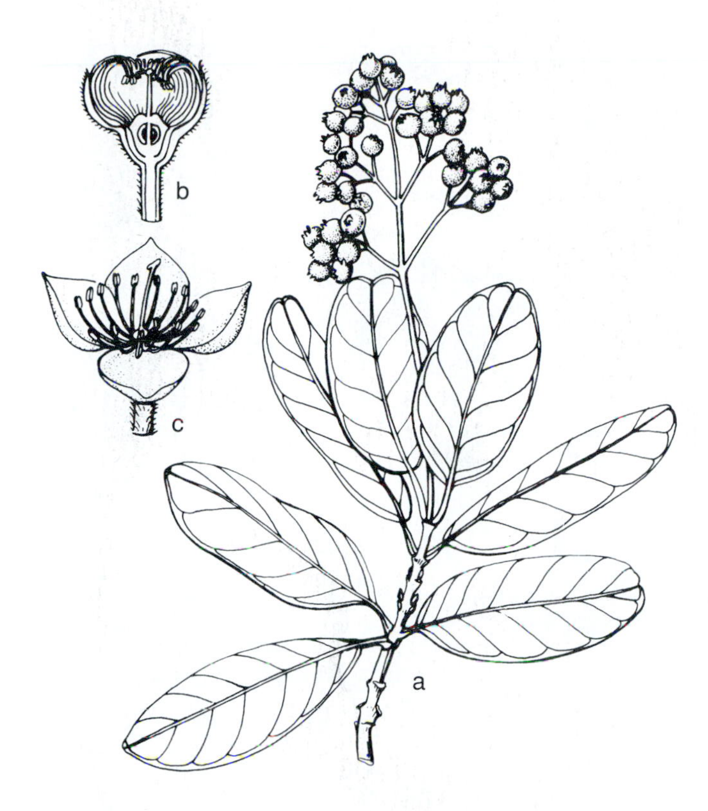

FIGURE 8.25
The clovelike flowers with a whorl of stamens show that allspice belongs with cloves in the Myrtaceae. The spice itself comes from the dried false berries.

FIGURE 8.26
Allspice was one of the New World treasures that Columbus brought back to his sponsor, Queen Isabella of Spain.

(This photogravure is from *Great Men and Famous Women* by V. Brozik, 1894, edited by C. Howe.)

cinnamon, cloves, and nutmeg. Allspice was among the few treasures Columbus was able to present to the court of his sponsors (Fig. 8.26). Among all the American spices, it has the somewhat dubious distinction of being the only one that is still exclusively grown in the New World. Almost all the modern production occurs in Jamaica.

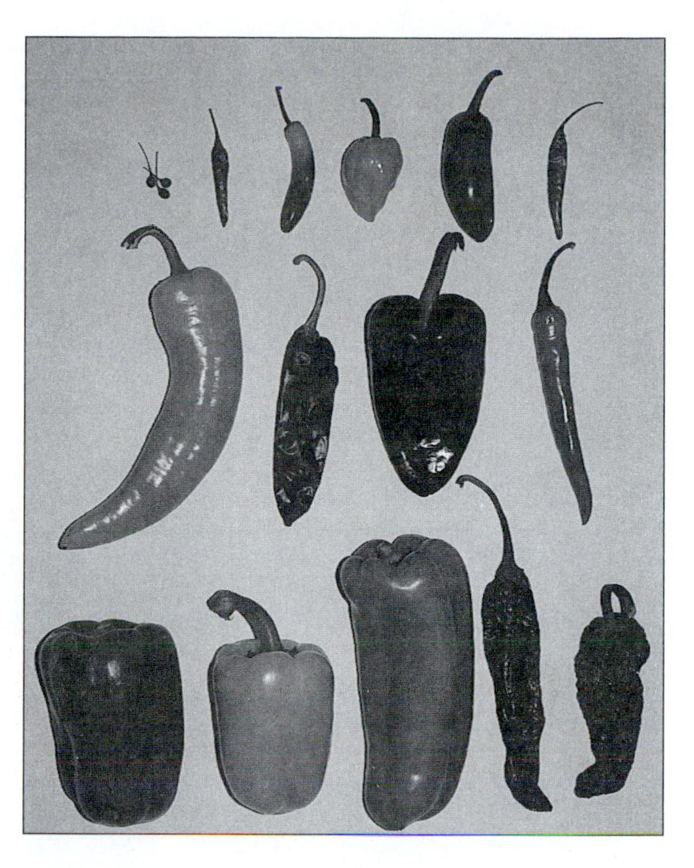

FIGURE 8.27
Among the fruits of capsicum peppers are (left to right from the top): Bird, Japone, Serrano, Habanero, Jalapeño, Thai hot, Anaheim, Guajillo, Hungarian Wax, Finger Sweet Green, Orange and Red Bell, Pasilla and Ancho.

In contrast to allspice, capsicum peppers, spices of *Capsicum* (Solanaceae), are now grown in many areas of the world. One species, *Capsicum annuum* (Fig. 8.27), produces both sweet edible bell peppers and the majority of the other common peppers of various shapes and degrees of hotness. Paprika, the flavor base of the paprikashes of the eastern European countries, is obtained from dried, powdered peppers of this species, as is the chiltepín, a tiny, very hot pepper occasionally found in U.S. supermarkets. A single gene mutation separates the sweet varieties from the fiery types. Other pungent kinds of peppers are often obtained from *C. frutescens,* the pepper used for Tabasco sauce. *Capsicum annuum* and *C. frutescens* are native to Central and South America, but it is probable that both arose in South America and spread to Central America, where they were originally domesticated (Fig. 8.28).

There is still disagreement about exactly how many species are involved in the vast array of domesticated capsicum peppers, but three in addition to *C. annuum* and *C. fructescens* are generally recognized today. *Capsicum baccatum* var. *pendulum* and *C. pubescens* from South America have been used for thousands of years in Peru and other parts of the Andes. *Capsicum chinense* is native to northern South America and the West Indies and is the fiery

FIGURE 8.28
This spice vendor in Mexico offers an array of peppers and garlic.

FIGURE 8.29
The flower of the orchid that will produce a vanilla pod after pollination.

pepper traditionally used in Caribbean pepper pot stews. The common names applied to capsicum peppers and their products are even more confusing. For example, the name *cayenne pepper* refers simply to dried, pungent red peppers, regardless of the species involved. Red pepper (as a label on a can of spice) can be anything from cayenne to paprika in terms of hotness. Paprika itself can vary in color from practically yellow to deep brick red and in degree of pungency. The coarsely ground red pepper flakes seen in shakers at fast-food restaurants are often a mixture of species of *Capsicum.* Finally, the chili powder that is sold in cans or jars is a mixture of oregano, cumin, garlic powder, and dried red peppers.

The compound in some red peppers that causes the burning sensation is capsaicin, which is concentrated in the fruit placenta on which the seeds are borne. Removal of the

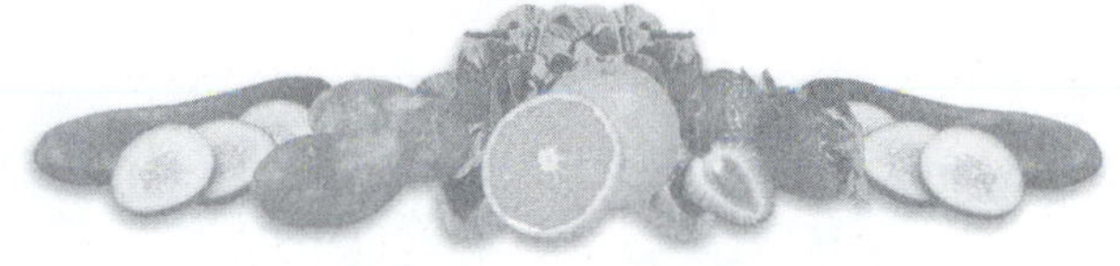

Box 8.3

Hot Stuff

Anyone who has inadvertently eaten a jalapeno pepper can attest to the searing pain, watering eyes, and blinding sensation that occur moments after the pepper hits the mouth. Why would anyone deliberately inflict such pain on himself? Apparently, humans like a little excitement in their dining; we are the only mammals who have overcome an aversion to these sensations. Evolutionary biologists think that the fiery fruits evolved to deter mammalian predators. On the other hand, birds that serve as seed dispersers readily consume the fruits.

The masochistic craving for fiery foods is universal. Before the discovery of America, horseradish satisfied this need in Europe, wasabi in Japan, black pepper in India, and ginger in Asia. Once capsicum peppers were discovered in the New World, they spread worldwide. In addition to adding zest, these spices have some slight antibacterial properties that help to preserve food. Capsicum peppers have now partially replaced other hot spices or have been combined with them in "traditional" cuisines. The reasons for the popularity of peppers and new discoveries about their properties and uses are as fascinating as our affinity for them.

Part of the transcultural success of hot peppers may lie in the fruits' versatility and ease with which they can be grown. Varieties of capsicum peppers can vary from about 25 mm (1 in.) to 20 cm (8 in.) long. They range in color from green to orange, red, yellow, and purple and vary in potency from fruits so strong that their juice can cause blistering of exposed skin to mild forms so sweet and juicy that they can be munched like apples. The ripe red or orange forms are also high in vitamin C, a relatively rare contribution from a spice.

The property for which capsicum peppers are most prized, however, is their fire. The principle compound responsible for the pungency, capsaicin, is an amide with part of its structure similar to vanilla. It takes 454 kg (1000 lb) of hot peppers to yield 137 mg (1 oz) of the compound, but capsaicin is so potent that one drop diluted in 100,000 drops of water still produces a burning sensation. Hotness is measured in Scoville units, with green peppers registering a 0, jalapenos 2500–5000, and habaneros a stunning 300,000 units. It is perhaps not surprising that black pepper and ginger also have a vanilloid structure as part of their molecular configuration. All of these substances are now believed to stimulate pain receptors in the mouth—the same receptors that perceive substances that are dangerously hot in temperature.

Capsaicin has recently been studied for reasons other than culinary innovation. It appears to have beneficial effects on the heart, and it has the ability to suppress pain. Taken internally, capsaicin relaxes arteries, thereby reducing blood pressure. It also seems to decrease the amount of harmful triglycerides in the blood.

In the last few years, the U.S. Food and Drug Administration has approved several drugs that contain capsaicin. The drugs are topical creams that are applied locally to block pain generated by arthritis, shingles, diabetes, neuralgia, or surgery. Neurologists believe that capsaicin works by interacting with particular nerves of the peripheral nervous system that initiate pain messages to the brain. These nerves release a substance when something happens that affects their equilibrium. This compound, known as substance P, is received by nearby neurons that send a message to the brain that there is pain at the initiation spot. Capsaicin depletes the initial nerve cells of their substance P and prevents them from accumulating more. Nevertheless, although native Mexicans have traditionally used chilies to treat toothaches, it is dangerous to experiment with rubbing cut jalapenos on sore regions of the skin. Prepared drugs have carefully measured doses of purified capsaicin, often at concentrations as low as 0.025 percent. Direct applications of cut chilies can cause redness, burning pain, and temporary blindness.

---

seeds and placenta can render an otherwise scorching red pepper somewhat less fiery. Since their introduction to Europe, Asia, and Africa, capsicum peppers have become integral parts of the cooking of Italy, Spain, Hungary, India, China, Indonesia, and many parts of Africa. Peppers thus rank with corn and tobacco as one of the most important crops to have originated in the Americas (Box 8.3).

Vanilla can be considered one of the strangest spices. It comes from one of the largest families of flowering plants, the Orchidaceae (Fig. 8.29), but, except for horticultural flowers, vanilla is the family's only crop. Vanilla plants are perennial vines native to Central and South America. The Aztecs used vanilla (*Vanilla fragrans*) as a flavoring for chocolate, and the Spanish carried it to Europe, where it was first used in the same way. The recognition of vanilla as a delicious flavoring in its own right came later. The part of the plant used to produce the flavoring is the fruit, usually called a vanilla bean. However, strictly speaking, the fruit is a false capsule (the ovary is inferior) despite its similarity in appearance to a legume pod (Fig. 8.30). Inside the fruit are millions of small seeds embedded in a slightly pulpy mass. The tiny brown or black specks in "real" vanilla ice cream are vanilla seeds. Their presence is supposed to indicate that the manufacturer used real vanilla fruits to flavor the ice cream.

Like the flowers of many orchids, the flowers of vanilla are intricately shaped and adapted for pollination by specialized insects (Fig. 8.29). Consequently, for commercial

FIGURE 8.30
Woman in Madagascar picking mature vanilla beans.
(Photo courtesy of the Vanilla Information Bureau.)

FIGURE 8.31
Spreading vanilla beans out to heat in the sun, followed by wrapping the warm fruits in blankets, is one of the steps in the lengthy curing process.
(Photo courtesy of the Vanilla Growers Association.)

production of vanilla, particularly in areas where it is not native, the flowers are hand-pollinated by workers who carefully transfer pollen to about 1500 flowers a day. The fruit takes 9 months to mature (Fig. 8.30). The mature pods then undergo a lengthy process that brings out the flavor. The process involves several steps that stop the development of the fruit, allow an enzymatic fermentation to occur, and finally dry and condition the fruit, which ultimately becomes dark brown to black, flexible, and highly aromatic. The first step that stops any further maturation of the fruit involves placing the pods in the sun for up to 5 hours. When they have reached their hottest temperature, they are tightly wrapped in blankets and placed in dark airtight boxes to sweat. After a day, they are checked to determine if they have started to turn brown. If the first step has been successful, the pods are exposed to daily sunnings (Fig. 8.31) and sweatings for another 25 to 36 days. Those that are still green after the first day are oven-wilted and then subjected to the daily heat-cool treatment. The flexible and almost black fruits are then dried indoors for a month and finally are placed in boxes to condition for up to 3 additional months. Other, shorter methods of treatment have been devised, but none produces the complex flavor and aroma of the traditional processing.

In view of the laborious process of producing aromatic vanilla, it is little wonder that vanilla ranks second to saffron as the most expensive spice. Vanilla extract is made by chopping the fermented beans and continually percolating an ethanol-water mixture over them to dissolve out the pure vanillin and other flavoring agents. Any product labeled pure vanilla extract must come from vanilla beans. Vanillin can be made synthetically from a number of products, including wood pulp. Artificial vanilla is extensively used as a flavoring, but for most people synthetic vanilla lacks the subtle qualities of real vanilla.

In the last few years an American biotechnology firm has developed a method of obtaining vanilla by culturing vanilla plant cells. This procedure could reduce the cost of natural vanilla from $1200 to $25 a pound. However, this alternative source would disrupt the economy of Madagascar, which relies on vanilla exporting for much of its income.

A native American herb with properties much like coriander is epazote (*Chenopodium ambrosioides,* Chenopodiaceae). Its pungent flavor is reflected in its common name, derived from a Nahuatl word that means "an animal with a skunklike odor." The leaves are most commonly used to flavor beans.

Although annatto or achiote (*Bixa orellana,* Bixaceae) is more commonly used as a dye in the United States than as a flavoring, it is an important component of many seasoning pastes in Mexico and northern Latin America. The flavor is mild and faintly resembles paprika. The spice comes from the bright red arils surrounding the seeds borne in fuzzy pods on small trees native to tropical regions of the New World (Fig. 15.31).

Perfumes

Among all the products obtained from plants, perfumes have played the most provocative role in human affairs, yet it is impossible to tell when people first began to use them. Prehistoric humans must have tried many ways to capture the pleasing scents of flowers and herbs before true perfumery techniques were developed. By 3000 B.C., however, the Egyptians had become skilled perfumers. They taught the art of perfumery to the Hebrews, and the frequent references to perfumes and incenses in the Torah reflect the importance that perfumes subsequently achieved in Judaic culture. In eastern cultures as well perfumes were used in temples and private homes. The Chinese used scents in ceremonies and even developed an incense clock that indicated the hour as it burned. The Japanese modified the Chinese clock by making sticks with different odors. As they burned, these "joss sticks" provided scents appropriate for the different parts of the day.

The wealthy Greeks employed perfumes in many aspects of their lives, but it was in the Roman empire that perfumery reached its pinnacle. The word *perfume* itself comes from the Latin *per,* meaning "through," and *fumus,* meaning "air" or "smoke," referring to the fact that fragrances diffuse through the air. Incense was commonly used in temples and homes. Perfumes were added to baths (Fig. 8.32), to oils for anointing the body, and even to wine. The value of scents at this time is reflected in the story of the birth of Christ. Two of the three gifts presented to Jesus were used as incense (Chapter 10).

After the fall of Rome, perfumery remained important in the Near East and Asia, but Europe seemed to have forsaken the pleasures of sweet smells. As in the case of spices, it was the returning crusaders who reintroduced perfumes to Europeans. Shortly thereafter, the popularity of perfumes soared. The French led in the recognition of the importance of perfumery as a trade by granting a charter to perfumers as early as 1190. Today, rising labor costs have shifted the cultivation of plants used for scent extraction to Asia and Africa. Moreover, there has been a steady increase in the use of synthetic compounds from 30 percent synthetic fragrances 30 years ago to over 75 percent today. Increased technology has made it possible to reconstitute natural odors with more fidelity than ever before.

How Perfumes Are Made

Originally, all perfumes were derived from natural sources: flowers, fruits, leaves, roots, wood, resins, and occasionally animal secretions. Unless a substance was to be used directly, as in the case of an oleoresin that was to be burned, the trick of the perfumer was to extract fragrant substances from plants or animals and present them in a usable form. The most successful extraction procedures are those that destroy or alter the natural fragrances as little as possible. Since the compounds responsible for fragrances are oils, they are not water-soluble. Consequently, extraction methods must use substances in which volatile oils dissolve. In the manufacture of a perfume, substances known as **fixatives** can also be added to help retard rapid dissipation of the volatile compounds.

FIGURE 8.32

Lavender and the generic name *Lavandula* come from the Latin *lavare,* meaning "to wash," because in Roman times the herb was used to add fragrance to wash water. Lavender is among the plants cultivated commercially in the perfume region in the south of France. (Redrawn from Matthiolus, 1565.)

The basic ingredients that are used in the perfume trade are known as **odorants.** Odorants fall into five groups: concretes, absolutes, distilled and fractional distilled essential oils, resinoids, and tinctures, including those of animal-derived substances. These classes of odorants are based on the way in which the oils are extracted and the substances from which the oils are obtained. **Concretes,** the purest of the natural odorants, are obtained by immersing fragrant plant products in a hydrocarbon solvent that penetrates the tissues and dissolves out the oils and other lipid compounds. **Resinoids** are obtained in the same manner, but instead of flowers, leaves, bark, or seed tissues, solid or semisolid plant secretions such as resin are dissolved in organic solvents. The solvent is then evaporated from the oils under reduced pressure. When a concrete is extracted to a more concentrated state (because waxes and glycerides are left behind)

FIGURE 8.33
One way to capture plant fragrances is to macerate scented parts in hot fat or oil. The mixture is usually allowed to stand in copper pots for a period of time before the oil or fat is extracted with alcohol.

FIGURE 8.34
Each day this woman removes faded jasmine flowers from a layer of fat and replaces them with fresh flowers. This process, called enfleurage, is used to capture particularly delicate scents. In this procedure, flower petals are pressed onto a coating of pure lard (usually pig lard). Each day the petals are changed until the fat is saturated with absorbed oils. The essential oils are then extracted from the fat with alcohol. Once the oils have been removed, the fat is practically odorless and is often later used for soap manufacture. Enfleurage is a time-consuming and expensive operation because it requires extensive hard labor. Consequently, it is rarely used for commercial perfumery today. However, it produces extremely true fragrances and is able to capture scents that elude other techniques.

with alcohol and the alcohol is then evaporated, an **absolute** is produced. Most perfumes consist of mixtures of absolutes.

There are two other, more traditional, ways of making concretes and absolutes. One is to **macerate,** or chop, fragrant substances in hot fat or oil and subsequently extract the fat or oil with alcohol (Fig. 8.33). The second is a cold fat process called **enfleurage** (Fig. 8.34). Oils from citrus peels used to be extracted by maceration, but today hydraulic presses are used to press the oils from pockets that dot the fruits.

Distillation and **fractional distillation** are the most common methods employed today for the extraction of natural fragrances (Fig. 8.35). Both involve exposing plant parts to steam, often superheated steam. The volatile oils are carried off in the steam. Because volatile oils are insoluble in water, they float on the surface of the water produced by the cooled steam. The layer of oil is easily skimmed from the top of the water column. Fractional distillation takes advantage of the fact that different volatile oils vary in their solubilities in steam. The most soluble are carried away first, and the less soluble later. When different fractions of the steam are condensed, the fragrant compounds can be separated from one another and collected individually. Distillation has an advantage over other extraction methods because it is inexpensive and rapid. However, heat is detrimental to many fragrances.

Tinctures have been used since ancient times and are still employed to extract medicinal compounds as well as fragrant oils. They are produced by extracting the oils from a macerated substance in 95 percent ethanol (the ancient Egyptians used wine). Tinctures may seem very similar to absolutes, but absolutes are obtained by the extraction of concretes, whereas tinctures are obtained by direct extraction of plant or animal material with alcohol.

In the production of a synthetic odorant, the chemical structure of a natural fragrance is usually determined and then procedures are devised for its synthesis. It is often cheaper to synthesize a compound than to extract it from a natural source. In addition, synthetic fragrances are pure,

FIGURE 8.35
Distilling fragrances.
(Photo courtesy of Parfumerie Fragonard, Grasse, France.)

whereas natural extracts are mixtures of volatile oils. If some of the compounds in an absolute are undesirable, the mixture must be fractionally distilled to separate them from the desirable components. In some cases, however, it is so inexpensive to use natural sources that there has been little attempt to produce a synthetic compound. This is the case with cedar (*Juniperus*) oil. In other cases, attempts by chemists to reproduce a natural fragrance such as true jasmine have failed, with the result that natural sources must be used.

As in other aspects of plant use, perfumers are experimenting with modern methods to produce cheaper and novel substances. Standard selection and modern biotechnological techniques are being employed to improve scent quantity and quality in plants. Microbes that biotransform compounds into alternative molecular configurations have been used to increase the production of some components of rose oil and jasmine. As was described earlier for vanilla, biotechnology holds promise for the production of some compounds from cultured plant cells.

From Oils to Perfumes

Early in the profession of perfumery it was discovered that the blending of scents could provide new, sometimes superior fragrances. In fact, all natural "perfumes" are mixtures of odoriferous compounds. In some cases, one scent dominates, but all are needed to provide a complexity of odor. In other cases, there is a main fragrance but the other compounds present are of major importance in producing the particular fragrance. Finally, in some instances all the odoriferous compounds are of equal importance in determining the final sensory effect of the mixture. Perfumers try to mimic complex natural fragrances or "invent" new ones by mixing absolutes and tinctures with fixatives to retard diffusion into the air and balance the different scents. The creation of a perfume is thus an art that requires the careful blending of different fixatives and odorants. Today perfume houses employ not only chemists but also at least one master perfumer whose job can be compared to that of a music composer. This analogy is carried through the fragrance lexicon, where different scents, called **notes,** are used to create masterpieces. As in music, the same notes are recognized all over the world.

Once a perfume has been produced by a skillful blending of diverse fragrances, it is classified by its dominant series, or note, if one is particularly evident. In choosing a perfume, a gift giver should consider the personality of the potential wearer in terms of the dominant note. Also, there is truth in the adage that a person should "try on" a perfume. The odorants in a perfume are chemicals that blend and react differently with different human body chemistries.

9

Vegetable Oils and Waxes

The expression "cholesterol-free" seems to leap from every package on grocery store shelves. This statement implies that cholesterol has been removed from certain products, making them healthier than they were before. However, vegetable oils contain virtually no cholesterol, so the statement merely means that these products contain plant-derived oils and fats rather than butter or lard. Furthermore, the use of vegetable oils is not a recent innovation but one that dates to at least the beginnings of agriculture. The importance of vegetable oils in ancient cultures is highlighted by the naming of ancient Greece's capital city. According to legend, the capital was to be named for the deity who gave the Greeks the most valuable tribute. As her offering, Athena flung her spear into southern Greece, and an olive tree sprang up on the spot where it touched the ground. This tree, with its oil-laden fruits, was declared the most valuable of the gods' gifts, and the capital was named Athens. The Greeks held the olive in high esteem because it provided food, their most important cooking and lubricating oil, and a popular cleansing agent (Fig. 9.1).

Most vegetable oils are expressed (pressed) from seeds. Within seeds, oils can be stored in the endosperm (castor bean, coconut), the cotyledons (peanut, soybean, cotton), or the **scutellum,** the single modified cotyledon of grains such as corn. In the majority of these cases the stored oils are the primary or sole food material for the germinating embryo. Since oil contains more calories than does starch (9 versus 3.7 kcal/g, respectively), oils are excellent sources of energy for emerging seedlings. Seeds with large amounts of oil usually do not contain large amounts of starch.

In a few cases, such as olives and palm fruits, oils are pressed from fruit pulps. It is not easy to understand why fruit pulps contain large amounts of oil. The pulp is not used by the embryo for nutrition; rather, it is used by the animals that disperse the fruits. Although these animals benefit from the rich food sources, the plant must expend considerable energy in their production. People have capitalized on several plants that produce oil-rich fruit pulps and have incorporated them into the human repertoire of edible plants since early times.

Saturation is a chemical property of oils that affects human nutrition, the ways in which different oils can be most profitably used, and to some extent the way in which an oil is extracted, refined, and processed. Although correlations have been established between the consumption of saturated fats and oils and an increased incidence of heart disease (Box 9.1), the prevalent belief that all vegetable oils are

FIGURE 9.1

The cleaning properties of vegetable oils, palm and olive in particular, were known to the ancient Greeks, as this 1915 advertisement for Palmolive soap emphasizes.

unsaturated and therefore healthier than animal fats is incorrect. Vegetable oils such as palm and coconut oil are more saturated than are butter and lard. Before we explore the ancient and modern uses of oils (Fig. 9.2) and the ways they are obtained and processed, we will describe the chemistry of vegetable oils. We also briefly note the major waxes obtained from plants because they share some of the chemical properties of vegetable oils. Table 9.1 lists the plants discussed in this chapter.

The Composition of Seed Oils

Saturated fats, unsaturated fats, monounsaturated fats, *trans*-fatty acids, essential fatty acids, 3-omega fatty acids—the terminology of fats and oils seems to go on endlessly. The interest in oils and fats and their various forms stems from our fondness for edible oils and the relationship between fat and health concerns (Box 9.1). Although the word *oil* can refer to a number of substances that are insoluble in water, only one group of substances—acylglycerides (glycerides)—is expressed in large quantities from plants. Acylglycerides are called **fatty, neutral, fixed,** or **nonvolatile** oils because they feel "greasy," are comparatively nonreactive chemically, and do not diffuse readily into the air. These oils are constructed from two kinds of compounds (Fig. 9.3): **glycerol,** a three-carbon alcohol, and fatty acids indicated by the prefix **acyl.** The glycerol "backbone" is the same for all acylglycerides. Differences between oils result from the various combinations of fatty acids attached to the three available sites along the backbone.

A **fatty acid** is a hydrocarbon chain with a carboxyl (COOH) group at one end that causes it to interact chemically as an acid. When such an acid combines with a hydroxyl (OH) group of the glycerol molecule, water is released. If only one acid is attached to the backbone, the reaction produces a monoacylglyceride. If two acids are attached, a diacylglyceride is formed. In a triacylglyceride, fatty acids are attached at all the hydroxyl positions on the glycerol molecule. Fatty acids can also be cleaved from a glyceride by **hydrolysis,** or the breaking apart with the addition of water. Alkalis, or "bases," are often used to promote this separation. The most commonly used alkalis—sodium hydroxide and potassium hydroxide—produce glycerol and sodium or potassium salts of the fatty acids. These metal salts of fatty acids are commonly called **soaps** (Fig. 9.5). As their name implies, they are the primary ingredients in common soap, one of the major products derived from seed oils.

Since fatty acids are long-chain organic compounds, there are several ways in which the chains can be constructed. The structure of fatty acids determines their chemical properties and ultimately those of a vegetable oil. The primary ways in which the fatty acids differ from one another involve the number of carbon atoms (the length of the chain), the position of the carbon atoms relative to one another (straight chain or variously branched), and the number of double bonds between carbon atoms (the degree of saturation). In general, the longer the hydrocarbon chain of a fatty acid, the higher the melting point. Triacylglycerides with predominantly long-chain fatty acids have higher melting points than do those composed of shorter-chain acids with the same saturation levels.

In commercial vegetable oils most of the fatty acids have straight chains. In vegetable oil jargon, fatty acids with only single bonds are considered **saturated.** When adjacent carbon atoms share two pairs of electrons and produce a double bond, the bond between the carbons is called **unsaturated** because each carbon atom is capable of accepting another atom with which it can share a pair of electrons. A given fatty acid can have no, one, or several double bonds. In the first case, it is called a **saturated fatty acid.** If it has one double bond, it is called **monounsaturated.** If it has more than one double bond,

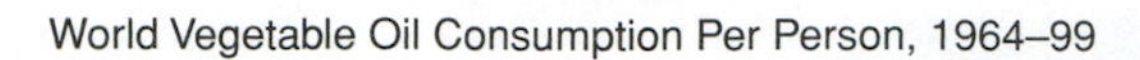

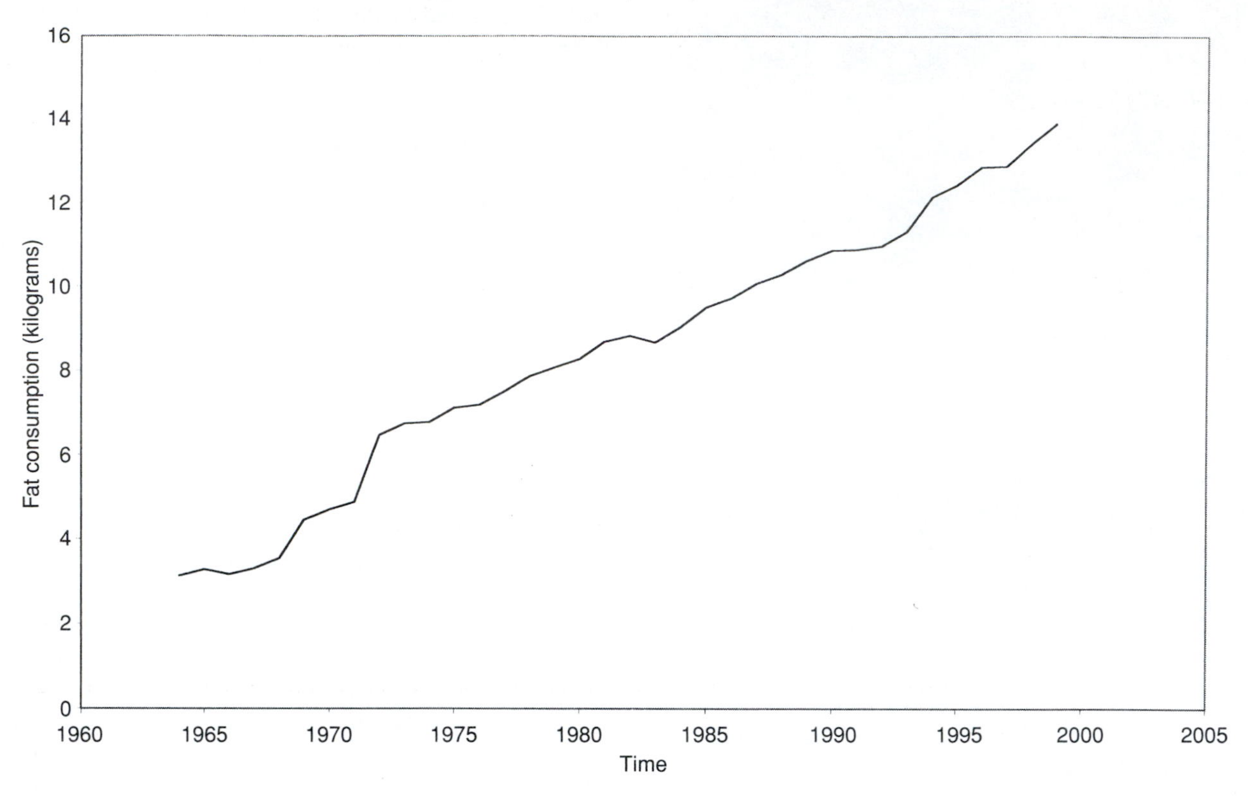

FIGURE 9.2
Increased consumption of fats has been linked to obesity and heart disease.
(From Worldwatch Institute, *State of the World 2000,* p. 66.)

TABLE 9.1 Plants Discussed in Chapter 9

COMMON NAME	SCIENTIFIC NAME	FAMILY	NATIVE REGION
Bayberry	*Myrica pensylvanica*	Myricaceae	North America
Candelilla	*Euphorbia antisyphilitica*	Euphorbiaceae	North America
Carnauba	*Copernicia cerifera*	Arecaceae	Brazil
Canola, rapeseed	*Brassica napus*	Brassicaceae	Mediterranean
Castor	*Ricinus communis*	Euphorbiaceae	Africa
Coconut	*Cocos nucifera*	Arecaceae	Cosmopolitan
Corn	*Zea mays*	Poaceae	Mexico
Cotton, upland	*Gossypium hirsutum*	Malvaceae	New World
Jojoba	*Simmondsia chinensis*	Simmondsiaceae	North America
Linseed	*Linum usitatissimum*	Linaceae	Southwestern Asia
Olive	*Olea europea*	Oleaceae	Europe
Palm, oil	*Elaeis guineensis*	Arecaceae	Africa
Peanut	*Arachis hypogaea*	Fabaceae	South America
Safflower	*Carthamus tinctorius*	Asteraceae	Mediterranean
Sesame	*Sesamum indicum*	Pedaliaceae	Africa
Soybean	*Glycine max*	Fabaceae	Asia
Sunflower	*Helianthus annuus*	Asteraceae	North America
Tung	*Aleurites fordii*	Euphorbiaceae	China

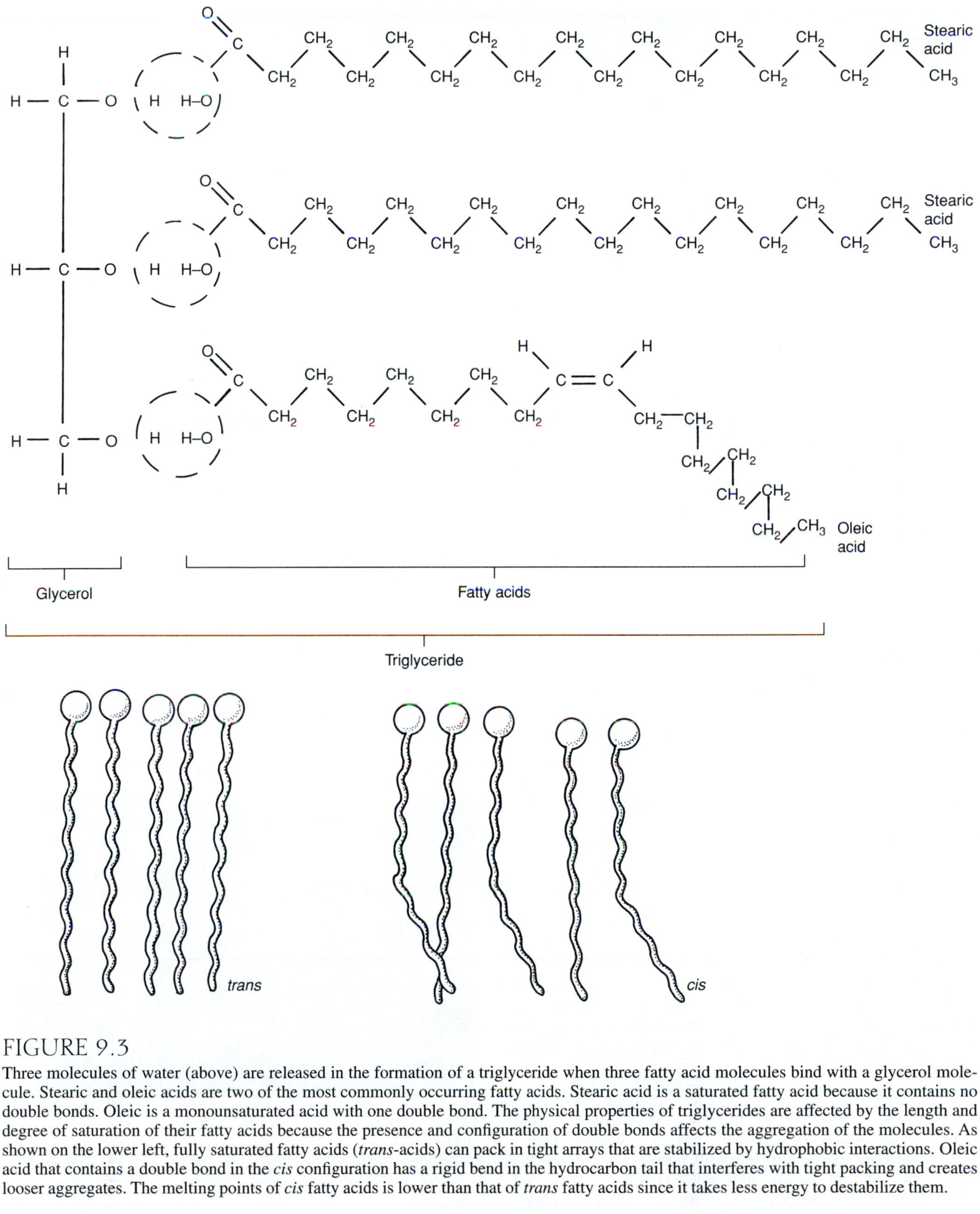

FIGURE 9.3

Three molecules of water (above) are released in the formation of a triglyceride when three fatty acid molecules bind with a glycerol molecule. Stearic and oleic acids are two of the most commonly occurring fatty acids. Stearic acid is a saturated fatty acid because it contains no double bonds. Oleic is a monounsaturated acid with one double bond. The physical properties of triglycerides are affected by the length and degree of saturation of their fatty acids because the presence and configuration of double bonds affects the aggregation of the molecules. As shown on the lower left, fully saturated fatty acids (*trans*-acids) can pack in tight arrays that are stabilized by hydrophobic interactions. Oleic acid that contains a double bond in the *cis* configuration has a rigid bend in the hydrocarbon tail that interferes with tight packing and creates looser aggregates. The melting points of *cis* fatty acids is lower than that of *trans* fatty acids since it takes less energy to destabilize them.

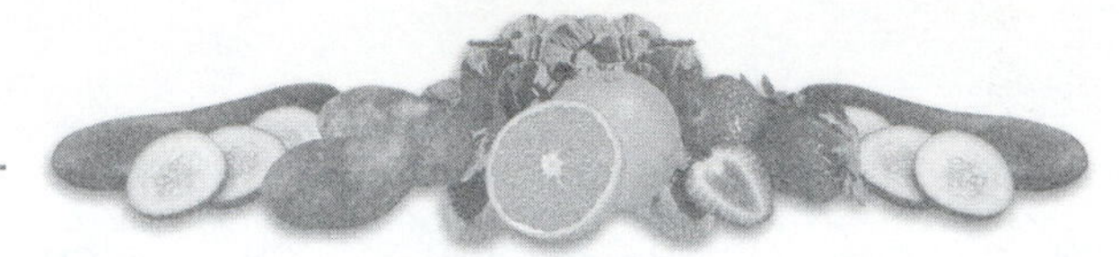

Box 9.1

Fats, Cholesterol, and Health

It seems a cruel twist of fate that an appetite for fats and oils, which so benefited our early ancestors, for whom food was often scarce, should create major health problems today. Although it is natural for us to crave fatty foods, overconsumption in our age of abundance is correlated with heart disease, the predominant cause of mortality in the Unites States today. Heart disease alone was responsible for 43 percent of all deaths in the United States in 1990. Researchers have been aware for 40 years of the correlation between high levels of cholesterol in the blood and the probability of heart disease. The reason behind this correlation and the role of different oils and fats in the accumulation of cholesterol have taken decades to decipher.

Cholesterol is a complex lipid similar in structure to many human hormones (see Fig. 11.5). It is required by the human body for use in cell membranes, the production of vitamin D, and the synthesis of most hormones. Part of our supply is manufactured by the body itself; the rest comes from eating foods that contain it. No plants are known to contain appreciable amounts of cholesterol, so all external sources must come from animal products such as meat, milk, cheese, and other dairy products. Normally, cholesterol does not create a problem, but if it builds up within the walls of the arteries, it can lead to narrowing of the channels, a condition known as arteriosclerosis. These constricted areas can reduce blood flow to the heart or can provide places where blood clots can be trapped, resulting in a subsequent heart attack. In the 1970s it was discovered that cholesterol is carried to sites in the body by proteins attached to lipid molecules called **low-density lipoproteins** (LDLs). Shortly afterward, it was found that another such protein complex, called **high-density lipoproteins** (HDLs), removes cholesterol from parts of the body and delivers it to the liver for breakdown. HDLs consequently gained a reputation as "good" lipoproteins, and LDLs as "bad" lipoproteins.

The process of forming fatty accumulations within an artery wall is now known in detail. The syndrome often begins with an injury or strain at a point on the wall. Such a trauma causes the cells underlying the internal wall to divide. White blood cells known as macrophages accumulate in the area to ingest particles of damaged cells. LDLs carrying cholesterol gather in the area to provide materials for repairing the wall. Unfortunately, while waiting to be used, cholesterol molecules can oxidize, and the macrophages engulf the oxidized molecules. After gobbling several such oxidized molecules, macrophages die and become deposited as fatty foam cells in the area of the wall being repaired. A substantial layer of these fatty deposits mixed with fibrous muscle cells can build up creating **plaque.** It is accumulations of plaque that eventually narrow the diameter of the artery to a point where blood flow to the heart is insufficient to maintain its operation.

Even without injury, plaque can form in arteries if cholesterol levels in the blood are sufficiently high. Excessive amounts of cholesterol increase the probability that some of the molecules will oxidize and be ingested by roving macrophages. Similar to the case of artery repair, macrophages poisoned by the ingestion of too many oxidized molecules can lead to fatty deposits in the arterial wall.

Any substance that lowers LDLs in the blood or increases HDLs tends to reduce the probability of plaque formation. Lowering the intake of animal fats can reduce the total cholesterol level, but it does not alter the ratios of the lipoproteins. Various kinds of fatty acids associated with vegetable oils are now known to alter this balance. Saturated fatty oils that are found in tropical oils or hydrogenated oils in margarine typically increase the amount of LDL in the blood. However, stearic acid, a saturated acid found in high quantities in beef and chocolate, does not show this effect. Linolenic acid has also been shown to increase the amount of LDL and lower the concentration of HDL in the blood.

Current research has indicated that the best kinds of fatty acid for a good balance of the lipoproteins are the monounsaturated acids (Fig. 9.4). These oils also appear to withdraw cholesterol from circulation in the blood by increasing the number of cholesterol receptor sites on cells and promoting uptake by the liver. Olive oil and canola oil are among those highest in monounsaturates, leading to their recent rise in popularity among health-conscious consumers.

FIGURE 9.4
Agricultural researchers pollinating sunflowers in a breeding program to develop plants that produce oils that are high in monounsaturates.
(Photo by R. Hanson, courtesy of Agricultural Research Service, USDA.)

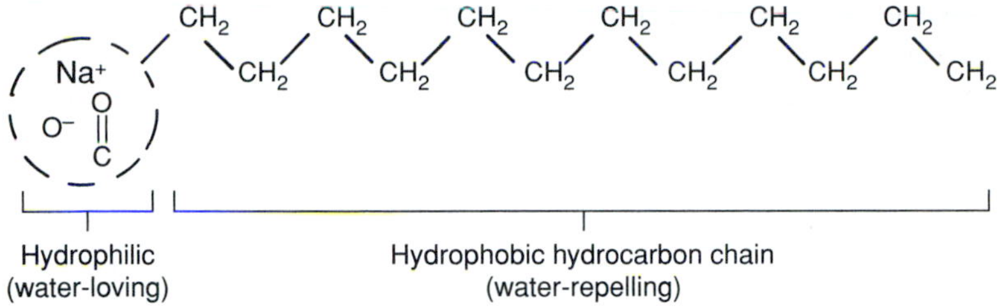

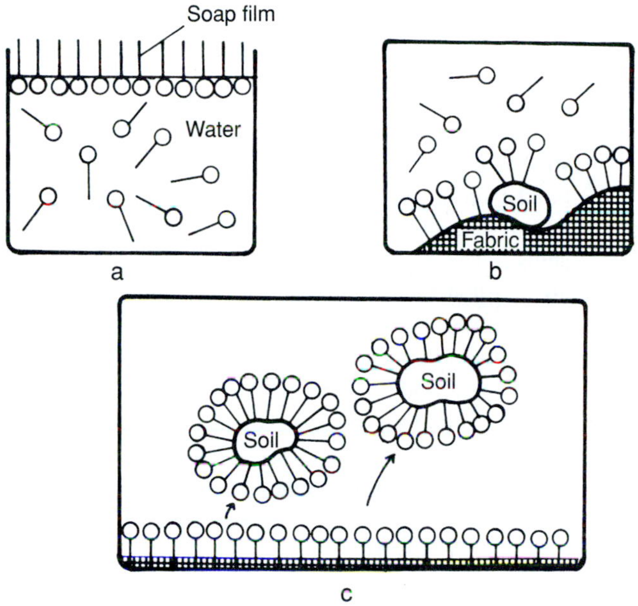

FIGURE 9.5

Soaps are sodium or potassium salts of fatty acids. The carboxyl (COO^-) end (circular head) of the fatty acid ion carries a positive charge and acts hydrophilically, while the long hydrocarbon chain (straight tail) is hydrophobic. (a) This duality of the soap molecule causes soap to form a film on water but also allows soaps (b) to attract oily dirt particles with their hydrocarbon tails and then (c) disperse them into the water so that they can be rinsed away.

it is usually referred to as **polyunsaturated.** However, since all vegetable oils are mixtures of triacylglycerides and some free (not attached to an alcohol) fatty acids, the terms *saturated, monounsaturated,* and *polyunsaturated* refer to the relative proportions of saturated and unsaturated fatty acids that make up an oil. Studies over the last 20 years have shown that there is a correlation between the consumption of saturated fats and heart disease. Polyunsaturated fats, which originally were thought to lack detrimental effects, have been shown to be linked with the production of free radicals, which produce carcinogens by oxidizing compounds. Currently, oils comparatively high in monounsaturated fatty acids appear to be the least injurious kind to consume.

The fatty acids most commonly found in vegetable oils are palmitic, stearic, and oleic acids (see Table 9.2). Palmitic acid has a chain length of 16 carbons. Stearic and oleic acids have 18 carbon atoms but differ because stearic acid has no double bonds and oleic acid has one. Linolenic acid is a fatty acid that has 18 carbons but three unsaturated positions. Linolenic acid leads to off flavors in oils and has been implicated in arteriosclerosis (see Box 9.1)

TABLE 9.2 Examples of Naturally Occurring Fatty Acids

CARBON SKELETON: DOUBLE BONDS	STRUCTURE	COMMON NAME*	MELTING POINT (°C)
12:0	$CH_3\ (CH_2)_{10}COOH$	Lauric acid (from the Latin *Laurus,* the bay tree)	44.2
14:0	$CH_3\ (CH_2)_{12}COOH$	Myristic acid (from the Latin *Myristica,* nutmeg)	53.9
16:0	$CH_3\ (CH_2)_{14}COOH$	Palmitic acid (from the Greek *palma,* palm tree)	63.1
18:0	$CH_3\ (CH_2)_{16}COOH$	Stearic acid (from the Greek *stear,* hard fat)	69.6
18:1	$CH_3\ (CH_2)_7CH{=}CH(CH_2)_7COOH$	Oleic acid (from the Greek *oleum,* oil)	13.4
18:2	$CH_3\ (CH_2)_4CH{=}CHCH_2CH{=}CH(CH_2)_7COOH$	α-Linoleic acid (from the Greek *linon,* flax)	–5
18:3	$CH_3CH_2CH{=}CHCH_2CH{=}CHCH_2CH{=}CH(CH_2)_7COOH$	Linolenic acid (from linoleic)	–11

*As indicated by the common names, many of these oils were originally characterized structurally from oils of economically important plants.

The double bonds of an unsaturated fatty acid can be broken by the addition of reactive atoms such as those of hydrogen, iodine, and chlorine. The degree of unsaturation is measured by allowing iodine to combine with a standard amount of a particular oil. Iodine ions are incorporated into the oil's fatty acid chains at positions where there were double bonds. After all chemical reaction has ceased, the amount of iodine that has been absorbed can be calculated, giving a value called the **iodine value.** The iodine values of edible oils range from about 7 to over 200 (Table 9.3). Oils with values below 70 are usually referred to as fats because they are solid at room temperature.

Another way of grouping vegetable oils that reflects the degree of saturation is into drying, semidrying, and nondrying oils. The term *drying oil* refers to the fact that oils with many double bonds tend to link together, forming polymers. If these oils are spread in a thin film over a substance, they dry into impervious coatings. Nondrying and most semidrying oils do not form such coatings. A rule of thumb is that drying oils have iodine values higher than 150; semidrying oils, between 100 and 150; nondrying oils, between 70 and 100; and fats, below 70 (see Table 9.3)

In addition to the degree of unsaturation, or number of double bonds between the carbons of the fatty acids of a triacylglyceride, other properties of fats and oils affect the healthfulness of particular oils. A simple, saturated fatty acid with each carbon linked to its adjacent carbon has a slightly zig-zagged structure but is basically straight from top to bottom (Fig. 9.3). When a double bond occurs between adjacent carbons, the chain normally turns away from the double bond, forming an almost U-shaped bend. Fatty acids with such a bend are said to be in the *cis*-configuration. If a glyceride has saturated fatty acids attached to it, the fatty acids can form regular arrays, or crystallize. This crystallization can cause an oil to be solid or semisolid at room temperature. If some or all of the fatty acids on a glyceride are in the *cis*-configuration, they stick out at various angles that prevent crystallization. These oils tend to be liquid at room temperature. When a liquid oil is hydrogenated, hydrogen gas is pumped into the oil in the presence of a nickel catalyst. As a result, several of the double bonds are eliminated, causing a "straightening" of the chain. This process also tends to twist the carbons that are still flanking double bonds so that portions of the chain on either side of the double bond face in opposite directions. This pattern is called the *trans*-configuration. Research has shown that *trans*-fatty acids lower HDL cholesterol and raise LDL cholesterol levels and that their consumption is correlated with incidence of heart disease. *Trans*-fatty acids also cause alterations in compounds that normally function to break down chemical carcinogens and drugs in the body.

Essential fatty acids are those that humans cannot make out of other ingested fatty oils or simpler molecules. There are two of these: linoleic acid, with two double bonds, and

TABLE 9.3 Saturation Indexes (Iodine Value) and Oil and Protein Content of Vegetable Oil Seeds and Fruits

CATEGORY/VEGETABLE OIL	IODINE VALUE	OIL, %	PROTEIN, %
Polyunsaturated (drying)			
Linseed	165–204	35–45	33
Tung	160–175	16–18	Not used*
Unsaturated (semidrying)			
Safflower	140–150	25–37	16–22
Soybean (shelled)	103–152	12–24	30–40
Sunflower	113–143	22–36	36–38 (meal)
Corn	103–133	2–5	10
Sesame	103–118	44–54	40
Cottonseed	90–117	19	21
Canola	94–105	2–49	20
Moderately saturated (nondrying)			
Peanut	84–100	45–55	25–28
Olive	78–88	75	—†
Castor	81–89	50	Not used*
Vegetable fats			
Palm (fruit)	46–60	50	—†
Palm (kernel)	14–22	44–53	17
Coconut	7–10	65–68	21

Sources: Iodine values from E. A. Weiss. 1971. *Castor, Sesame, and Safflower.* New York, Barnes & Noble; and D. Swem. 1979. *Bailey's Industrial Oil and Fat Products,* vol. 1, 4th ed. New York, Wiley.

Note: Percentage composition figures based on dry weights.

*Indicates that the meal is toxic and therefore the protein content is not important.

†Indicates fruit pulps that are low in protein and not used for feed.

linolenic acid, with three (Table 9.2). In addition to serving as precursors for other fatty acids, these fatty acids are needed for oxygen transfer, hemoglobin production, transportation of substances across membranes, and controlling the damage that saturated fats can cause.

Cholesterol is an oil that consists of a complex series of rings. Although needed for normal body function, excessive amounts of cholesterol can cause health problems (Box 9.1). We obtain cholesterol in two ways. Some is manufactured within our bodies, and some is acquired by eating meat, eggs, and dairy products. Vegetable oils lack cholesterol.

The omega fatty acids are unsaturated fatty oils of comparatively high carbon chain lengths (e.g., C_{20} and C_{22}). These oils are particularly abundant in fish oils, but some can be obtained from the seeds of plants such as evening primroses, borages, and black currants. These fatty acids reduce triaclyglyceride levels in the blood, reduce blood pressure, and may have anticarcinogen properties.

The presence of double bonds in many fatty acids allows another kind of chemical reaction to occur. A fatty acid can absorb oxygen at positions where carbons are linked by double bonds. The resultant oxidation of the fatty acid leads to cleavage of the acid into smaller units such as aldehydes and other unpleasant volatile compounds. People perceive this change as obnoxious and describe the oils as rancid. Rancidity is prevented by adding antioxidants such as BHT, BHA, and polysorbate 80 to oils or processed oil-containing foods such as crackers, cookies, cereals, and margarine. These antioxidants are called **preservatives.** Natural antioxidants also occur in association with fats and oils. Some of them, such as vitamin K, are necessary for human nutrition.

Oils contribute in other ways to the human diet. Vitamins A, D, and E (alpha-tocopherol) are oil-soluble and are absorbed by the body only when they are dissolved in fats and oils. Oils also aid in the accumulation of the B vitamins, provide linoleic acid (see Fig. 9.3), and are a major source of calories in the United States. In addition, fats and oils contribute to the flavor and texture of many baked and fried foods (Box 9.2).

Nonfood Uses of Vegetable Oils

Although most vegetable oils end up in foods, large quantities are used by industry for the manufacture of nonedible products. In some cases they serve as lubricants, but most often they are used to create an array of products from paints to plastics.

It was mentioned earlier how "soaps" are made from triacylglycerides by the addition of alkali. The old-fashioned method of soap making involved boiling animal fat (lard) and ashes together. When combined with water, the ashes formed an alkali that helped cleave the fatty acids from the triacylglycerides of the lard. Soap has cleaning properties because the COO^- groups of fatty acids have an affinity for water, while the remaining parts of the molecules are attracted to grease or "dirt" (Fig. 9.5). Today vegetable oils are used more extensively than lard in soap making.

Oils, particularly highly unsaturated drying or semidrying oils, are important ingredients in the manufacture of paints and varnishes. Linseed oil and tung oil polymerize (form long molecular strings) easily and can be applied to surfaces without processing. Color can be added to the oil to form a decorative coating. The Flemish were the first to combine pigments with vegetable oils and produce what are now called oil paints. In the fifteenth century they perfected a technique of glazing paintings with translucent colors over an undercoating to create an illusion of depth and texture never before achieved in painting. Oils are now used in paints as the binder in which pigments are dispersed. In the manufacture of paints, oils are first boiled with compounds containing heavy metals, such as magnesium, cobalt, and formerly, lead. These compounds help the oils absorb oxygen after they are applied, resulting in hard films. Varnishes are made by mixing boiled oils with resins or gums (Chapter 10). Enamels are made by mixing pigments with varnishes. Paints differ from varnishes and enamels because they do not contain gums or resins; they have traditionally been prepared from mixtures of boiled oils, pigments, and thinning agents such as turpentine. Modern latex paints are made from alkyd resins, which are polymer resins manufactured from fatty acids cleaved from vegetable oils. Because of their ease of application and cleanup (they are water-soluble), latex paints now account for about 60 percent of commercial paint production.

The original linoleum, as implied by the name (from *linum* for "flax" and *oleum* for "oil"), was another product for which seed oils once were used. Oxidized linseed oil was mixed with ground cork or wood and various resins and pigments, pressed onto a backing (usually felt or burlap), and heated for a few days. True linoleum has given way to plastic floor coverings in the United States, but even these products are often made from polymers derived from seed oils.

Vegetable oils have become increasingly important as starting points for the synthesis of organic compounds formerly obtained from petroleum. Fatty acids separated from a glycerol backbone can be converted chemically into alcohols and other compounds used in the manufacture of detergents, resins, and special types of industrial oils. Molecular fragments from vegetable oils also become incorporated into adhesives, fabric softeners, plastics, and synthetic fibers. To a limited extent, vegetable oils have a place in medicine and in the cosmetics industry, although only castor oil is employed medicinally today. Finally, some seed oils, such as sunflower oil, have been proposed as potential substitutes for diesel oil.

Box 9.2

Fats and Fat Substitutes in Processed Food

Cooks around the world know that the only way to produce succulent, crispy foods such as French fries or fried chicken is to immerse them in hot bubbling oil. However, the role of fats and oils in other foods is less obvious. Both play a subtle role in our perception of the tastes of numerous foods, and they are necessary for the texture we expect in many baked goods. Fatty oils in foods often carry the essential oils that are the basis for the smell and consequently "taste" of most foods. In addition, fats and oils cause food to appear juicy because of their own lubricating properties and because they stimulate saliva production. Lean cuts of meat are perceived as dry and tough not only because they lack fat but also because they fail to generate a moist environment in our mouth. Our love of the effects of fats and oils in foods is reflected by the fact that their average per capita consumption in the United States is about 31 kg (69 lb) per year.

Fats can be heated to high temperatures and transfer this heat quickly to foods. Consequently, items that are submerged or sauteed in very hot oil become deliciously crispy and brown on the outside. The temperature to which a particular oil can be heated is limited by its smoke point, or the temperature at which the oil degrades and begins to burn or smoke. Mono- and diacylglycerides have lower smoke temperatures than triacylglycerides. Likewise, longer chain length is positively correlated with higher smoke temperature. Because they smoke at comparatively high temperatures, soybean and sunflower oils are now preferred to cottonseed oil for high-temperature deep fat frying.

Fats and oils also play a key role in producing the texture of baked foods. In these products, oils, often initially beaten with sugar, surround air bubbles in the batter. As the item is baked, the fat melts from the air bubble and disperses evenly throughout, creating a richness, while the batter solidifies into a tender spongy mass. In pastry, fats and oils coat flour proteins and prevent the formation of elastic gluten strands, resulting in a tender, flaky texture that "melts" in the mouth. In the vernacular, we call foods made with high fat content "short," from the Old English meaning "easily crumbled." The term *shortening bread* or *shortbread* is used for a type of friable cookie with a high level of fat. Although the name was applied long before the chemistry of flour was well known, it is extremely suitable because we now know that the effect of the fat is to keep the gluten chains in the flour from forming the interlocking network they produce in bread dough. In other words, the protein molecules remain short.

Before 1900, all pastries, cookies, and cakes were made with butter or animal lard. With the advent of hydrogenation, vegetable fats began to be substituted. These fats are cheap to produce and do not become rancid as readily as animal-derived products. The use of vegetable fats accelerated when it was discovered that high consumption of animal fats was correlated with high levels of cholesterol in the blood. Hydrogenated vegetable oils seemed to circumvent this problem, and the sales of margarine and solid vegetable shortenings soared (see Fig. 9.2). Unfortunately, hydrogenation has created problems of its own that we now know result from the structural differences between natural saturated oils and artificially hydrogenated oils.

Fats and oils are absorbed through the lining of the small intestine. In order for these water-insoluble molecules to be absorbed, they must first be emulsified into tiny particles so that lipases can break them down into smaller units such as monoacylglycerides, diacylglycerides, fatty acids, and glycerol. After they pass into the lining of the intestine, they are reconverted to triacylglycerides, which move through the lymph system or the blood to various tissues. When they reach their destination, they are cleaved into fatty acids and glycerol. The fatty acids can be metabolized as a source of energy or reattached to glycerol molecules for storage.

The search for fat substitutes that satisfy our craving for fatty foods without padding our waistline or clogging our arteries has recently met with some success. Food technologists have developed a variety of compounds that behave like fats in various foods but are chemically different and generally contain many fewer calories. Olestra ™, made by Procter & Gamble is a fat substitute consisting of eight fatty acids attached to a sugar molecule. The sugar–fatty acid polyesters are so large that they cannot be broken down by lipases and consequently they are never absorbed by the intestinal wall. As a result, some individuals experience cramping and diarrhea after eating foods containing this fat substitute. Olestra™ does not break down when heated, so it can be used for frying, and it produces the same mouth feel as regular fats when used in bakery products.

Simplesse™, developed by Monsanto, is made from milk or egg whites, or a combination of the two, that are reduced to microparticles that produce the same silky feel in the mouth as fats and oils. Because it is protein-based, Simplesse™ cannot be used for frying or in most baked products. However, it is a good substitute for hydrogenated oils in margarines, cheeses, and ice cream substitutes. Several other products such as Maltrin™, Paselli SA2™, and N-oil™ are manufactured from potato, corn, or manioc starch. All provide a creamy, fatlike texture in low-fat frozen desserts, salad dressings, and snacks. Like Simplesse™, these products have fewer calories than true fats and oils.

A more recent innovation is the addition of sterol esters obtained from soybeans or pine needles to margarine or butter substitutes such as Promise™ or Take Control™. The sterol esters prevent the absorption of cholesterol by the small intestine and thus promote a better ratio of cholesterol to high-density lipoproteins (HDLs). With continued documentation of the harmful effects of fats in the diet, other fat substitutes will undoubtedly be developed.

Methods of Extracting Vegetable Oils

An examination of a sesame seed or a castor bean give little indication that over half its weight (see Table 9.3) is due to oil held in special cellular organelles. Humans have devised various methods of removing oils from seeds and fruits. One of the earliest methods of extraction involved grinding with stones to crush the tissues and release the oils (Fig. 9.6). The

FIGURE 9.6
Tunisian workers use a traditional stone press to extract oil from the pulp of olive fruits.
(Photo by F. Botts, WPF/FAO.)

crushed mass was often suspended in a cloth bag that allowed the oil to drip into a container. Sometimes heat was applied to facilitate the flow of oil from the crushed debris. Both a mortar and pestle arrangement, often with a deep inverted cone and a revolving pestle, and a press with a wedge that was manually pounded have been used for oil expression since early times. Various types of manual or animal-driven presses are still employed in many countries. Almost all the oil extracted by these methods is consumed locally.

The first mechanized presses, produced in the seventeenth century, were steam-driven and operated on the same principle as the manual press. In 1795 the hydraulic press was developed in England, and in 1900 the first screw presses were used to extract seed oil. Before the widespread adoption of the screw press, pressing was done either at air temperature or by applying heat. In the first case, the procedure was called **cold pressing,** and the second, **hot pressing.** Although cold pressing extracted less oil than did hot pressing, the oil obtained was purer because high temperatures promoted the extraction of other compounds along with the glycerides. Today these distinctions mean little because the process of extracting with a screw press generates heat that can reach temperatures between 65° and 72°C (149° and 162°F).

The screw press was adopted in industrialized countries because it made large-scale oil production economically feasible. A screw press, or **expeller** (Fig. 9.7), allows the continuous feeding of seeds into one end of a constantly turning screw. The oil released during the crushing flows out of the press through slots along the cylinder enclosing the screw. The residue, or cake, is expelled through the end of the press, which is fitted with a conical choke. The size of the opening

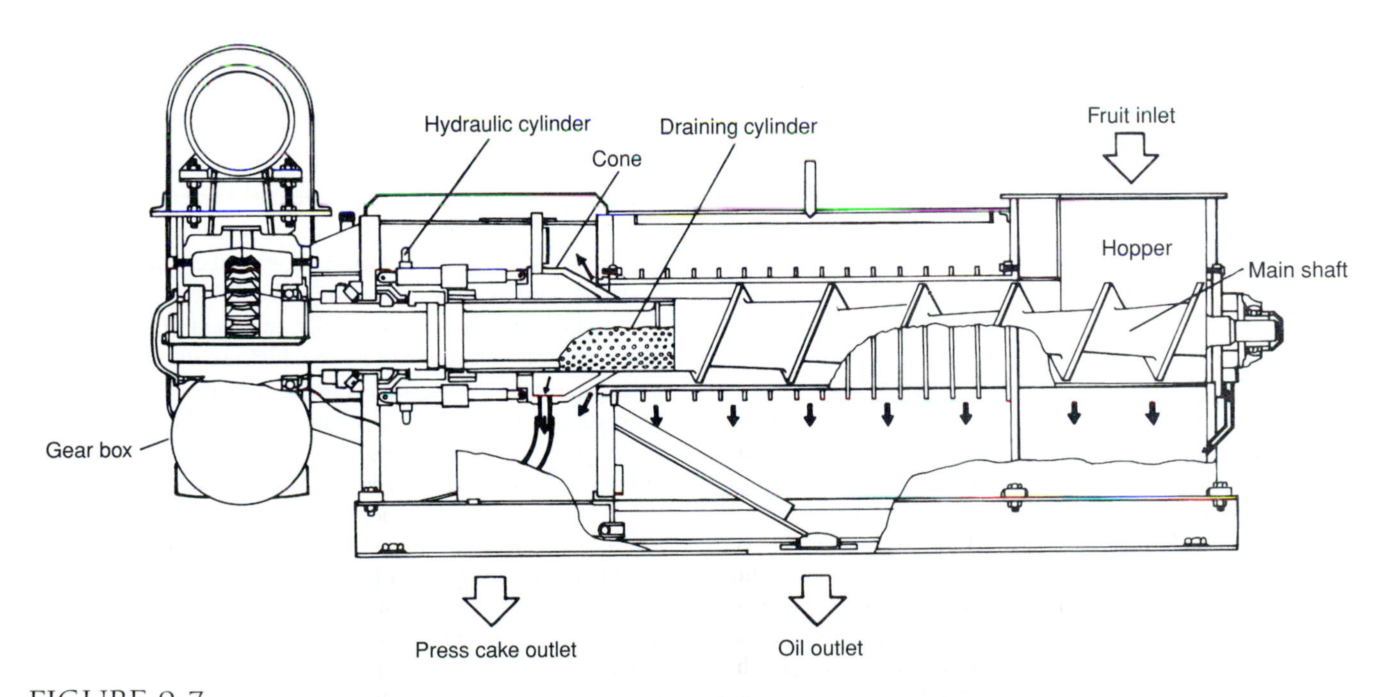

FIGURE 9.7
A screw press, or expeller, used to express oil from palm fruits. Similar designs are used for seeds.
(After a diagram supplied by the French Oil Mill Machinery Co.)

through which the residue is ultimately forced is controlled by varying the extent to which the choke is fitted into the screw cylinder. Because the residual cake is often high in protein (see Table 9.3), it is frequently used as animal food. Today oilseed meal constitutes an important commercial by-product.

Because expeller pressing leaves a residue that contains 2 to 4 percent oil, a further refinement using organic solvents such as hexane was developed. Solvent extraction, which leaves a scant 0.5 to 1 percent of the oil in the cake, has now been adopted in most large operations. During solvent extraction the husked seeds or fruits are crushed, sometimes heated, and then agitated in an organic solvent. The solution containing the oils is separated from the residue, and then the solvent is driven from the mixture by distillation. Hexane boils off at temperatures between 63° and 65°C (145° and 149°F), leaving the crude oil behind. The solvent, recovered by chilling, is reused.

Figure 9.8 shows the steps followed in most modern commercial extraction operations. Procedures that are unique to individual seed or fruit crops are discussed below under the treatments of those crops. Usually, seeds are cleaned and dried before they are sold to oil-refining companies or just after they arrive at the mill (Fig. 9.8). The seeds or fruits are then recleaned and dehusked. *Husk* in this case is a general term referring to the fruit wall, the seed coat, or both. In the case of sesame seed or cottonseed, the seed coat is removed. Sunflower and safflower "seeds" are fruits with a "shell" composed of the ovary wall. The seeds inside the fruit have a thin membranous coat. Both the shell and the seed coat are removed in the decortication of these oil crops.

The naked seeds separated from all covering material are usually pretreated by breaking and flaking them with a series of rollers. The flakes are cooked or subjected to high temperatures to reduce their moisture content. This process, known as **conditioning,** also facilitates oil release and helps coagulate proteins so that they are not extracted with the oil. The mass of prepared, flaked, and conditioned seeds may be simply pressed, sent directly to a solvent extractor, or subjected to an initial pressing followed by solvent extraction. The wet meal left after pressing or solvent extraction is often sold for animal feed. Meal from solvent extraction is heated with steam to evaporate the toxic solvent before sale.

Oil Purifying and Processing

After extraction, oils are usually refined and then subjected to additional procedures. **Refining** (neutralizing) is a process that produces a neutral oil by using alkali to remove unbound (free) fatty acids. In addition to being refined, the oil can be degummed, bleached, deodorized, or winterized or any combination of those processes. **Degumming** is effected by mixing the oil with water at 32° to 49°C (90° to 120°F) and centrifuging the mixture. The process removes any mucilaginous material that might have been extracted with the oil. Lecithin, a phospholipid sold as a cholesterol-reducing aid in health food stores, is recovered from the water fraction produced during the degumming of soybean and other seed oils. Triacylglycerides are usually odorless and colorless. Both the taste and color of vegetable oils are caused by other compounds that are extracted with the oil. These compounds include volatile oils and pigments such as gossypol and the carotenoids. **Bleaching,** or the removal of pigments, is carried out by adding agents, such as diatomaceous (Fuller's) earth, to which color particles adhere. The oil is filtered, leaving behind the solid agents and coloring matter. Many oils are also **deodorized** by being heated with steam under a vacuum. **Winterizing** is a process used to circumvent the clouding of oil under low-temperature conditions by chilling the oil to about 7°C (45°F) and filtering out all the particles that crystallize.

An oil can undergo a number of additional processing steps, depending on its eventual use. One of the most common is **hydrogenation,** during which hydrogen gas is bubbled under pressure through an oil in the presence of a catalyst, usually nickel. The hydrogen binds to carbon atoms previously linked by double bonds and causes the unsaturated fatty acid components of the oil molecules to become completely or partially saturated. Hydrogenation thus lowers the iodine value of an oil and raises its melting point. Most hydrogenated oils are therefore solid at room temperature. The discovery of hydrogenation in the early 1900s led to an enormous industry that now produces vegetable fats, margarines, and cheese substitutes. Despite earlier claims that margarine and nondairy cheeses are much healthier than butter and lard, there is now concern about the effect of hydrogenated oils on human health (Box 9.2).

Purifying and processing tend to produce oils and fats that have little individual character. Although gourmets relish the flavor of olive oil, sesame, and peanut oils and the roles they play in the characteristic flavors of ethnic cuisines, oils such as cottonseed oil, soybean oil, and palm oil are purposely treated to render them virtually tasteless and identical. The U.S. Food and Drug Administration allows the substitution of several such oils in food manufacturing (Fig. 9.9). This interchangeability permits manufacturers to purchase whichever oil is least expensive at a given time.

Using this background of vegetable oil chemistry, extraction, and processing, we can turn to the different plant species from which oils are commercially extracted. Table 9.3 lists seeds and fruit oils in the order in which we discuss them and gives their iodine values and classification according to drying properties. This table also shows the relative amounts of oil and protein per dry weight of the different vegetable oil crops discussed. Table 9.4 gives recent worldwide production figures.

Polyunsaturated Oils

Linseed Oil

Linseed is probably the oldest domesticated oilseed crop and forms part of the ancient complex of wheat, barley, peas, lentils, and chick-peas that formed the early domesticates

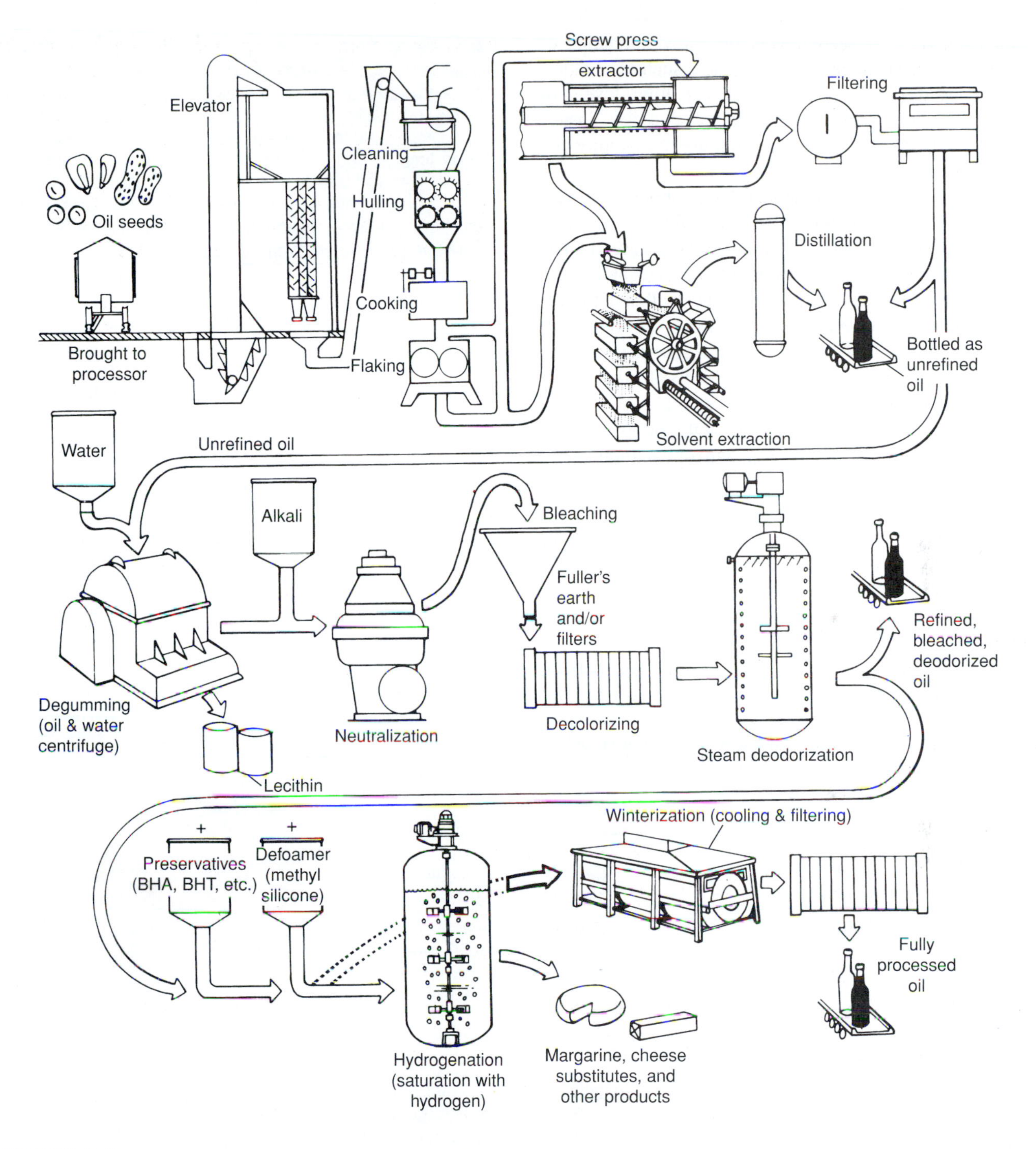

FIGURE 9.8

A diagrammatic representation of the extraction and processing of vegetable oils. The top right portion shows that seeds can be expeller-pressed only, solvent-extracted only, or prepressed before solvent extraction. The alkali treatment (center) removes free fatty acids and produces a neutral oil. Processing with steam (center right) removes odors from the oil. After the addition of preservatives and defoaming agents, the oil can be hydrogenated to produce a solid vegetable fat or winterized to prevent clouding.

in the Near East (Fig. 9.10). Fossil linseeds clearly show signs of human selection by 6000 B.C. and, combined with evidence from other crops, indicate that well-developed agriculture had been established by that time. The same species, *Linum usitatissimum* (Linaceae), that produces seeds for linseed oil is the source of flax (see Chapter 15), but oil extraction predated fiber use by several thousand years.

Because of its high level of unsaturation, linseed oil is a drying oil, and it has been applied to surfaces to produce water-repellent glazes since at least classical Egyptian times, when coffins were coated with a mixture of linseed oil and resin. Today linseed oil is incorporated into paints or used alone to protect natural wood surfaces such as shingles, decking, and fences. The comparatively high levels of linolenic acid and unsaturation (Table 9.2) that give linseed oil good drying properties also lead rapidly to rancidity and a pungent, unpleasant taste.

Linseed plants are erect annual herbs about 1 m (39.3 in.) tall (see Fig. 15.17) with blue "flax" flowers that give way to capsules each containing about 10 elliptic seeds. The cultivars used for oil extraction are shorter and more branched than are those used for fiber. The decline in the use of linseed oil as an edible oil has resulted from its unpleasant taste, the rapidity with which the flavor degenerates because of oxidation, and the increased availability of other, more palatable oils. In the United States linseed oil is used almost exclusively for inedible products, although bottles of cooking linseed oil can be found in health or "natural" food stores.

Ingredients: Corn, wheat, and oat flour; sugar; partially hydrogenated vegetable oil (one or more of: coconut, cottonseed, and soybean); salt; calcium carbonate; corn syrup; sodium ascorbate and ascorbic acid (vitamin C); red #40; yellow #6; niacinamide; zinc oxide; reduced iron; natural orange, lemon, cherry, raspberry, blueberry, lime, and other natural flavors; turmeric color; annatto color; blue #1; red #40 lake; blue #1 lake; blue #2 lake; blue #2; yellow #5 lake; pyridoxine hydrochloride (vitamin B_6); riboflavin (vitamin B_2); vitamin A palmitate; thiamin hydrochloride (vitamin B_1); BHT (preservative); folic acid; vitamin B_{12} and vitamin D.

FIGURE 9.9

The U.S. Food and Drug Administration considers coconut, cottonseed, soybean, and palm oils interchangeable in packaged foods because they are indistinguishable once refined. This ruling has led to the common statement on food packages that the product may contain any or all of the three oils.

TABLE 9.4 1999 World Production of Vegetable Oil Crops

COMMON NAME	TOP COUNTRIES	1000 METRIC TONS	TOP CONTINENTS	1000 METRIC TONS	WORLD, 1000 METRIC TONS
Canola (as rapeseed)	China	9,700	Asia	16,065	42,615
	Canada	8,798	Europe	14,750	
	India	5,800	North and Central America	8,798	
	France	4,483			
	Germany	4,211	Oceania	2,103	
			Africa	185	
Castor (bean)	India	900	Asia	1,124	1,213
	China	200	South America	48	
	Brazil	25	Africa	36	
	Paraguay	18	North and Central America	3	
	Thailand	6			
			Europe	1	
Coconuts	Philippines	11,000	Asia	39,822	46,554
	Indonesia	10,000	Oceania	1,931	
	India	13,000	North and Central America		
	Sri Lanka	1,850		1,906	
	Thailand	1,400	Africa	1,825	
	Vietnam	1,133	South America	1,068	
Cottonseed (as seed cotton)	China	12,000	Asia	33,916	54,543
	United States	9,517	North and Central America	10,208	
	India	7,720			
	Pakistan	4,485	South America	2,687	
	Uzbekistan	3,680	Africa	4,419	
			Oceania	1,728	
Linseed	Canada	1,049	North and Central America	1,249	2,838
	China	420			
	India	320	Asia	840	
	United Kingdom	295	Europe	572	
	United States	200	South America	101	
			Africa	69	

TABLE 9.4 *Continued*

COMMON NAME	TOP COUNTRIES	1000 METRIC TONS	TOP CONTINENTS	1000 METRIC TONS	WORLD, 1000 METRIC TONS
Oil palm fruit	Malaysia	42,600	Asia	76,398	98,408
	Indonesia	29,500	Africa	14,182	
	Nigeria	8,050	South America	5,171	
	Colombia	2,700	Oceania	1,130	
	Ivory Coast	1,100	North and Central America	1,525	
Olive oil	Spain	602	Europe	1,533	2,260
	Italy	554	Asia	358	
	Greece	333	Africa	357	
	Tunisia	218	South America	10	
	Syria	129	North and Central America	2	
Peanuts (in shell)	India	7,300	Asia	21,873	33,039
	China	12,068	Africa	8,294	
	Nigeria	2,783	North and Central America	1,963	
	United States	1,756	South America	854	
	Indonesia	990	Oceania	41	
Safflower seed	India	430	Asia	487	901
	United States	183	North and Central America	355	
	Mexico	171	Africa	36	
	Ethiopia	36	South America	10	
	Australia	8	Oceania	8	
Sesame seed	India	650	Asia	1,699	2,444
	China	550	Africa	619	
	Sudan	220	North and Central America	85	
	Myanmar	210	South America	40	
	Uganda	93			
Soybeans	United States	71,928	North and Central America	74,888	153,857
	Brazil	30,904	South America	53,107	
	Argentina	18,000	Asia	22,572	
	China	13,701	Europe	1,999	
	India	6,100	Africa	902	
Sunflower seed	Argentina	6,500	Europe	12,899	28,018
	Russian Federation	3,500	Asia	4,545	
	France	1,930	South America	6,929	
	Ukraine	2,750	North and Central America	2,092	
	China	1,550	Africa	1,342	
	United States	1,969			
Tung oil (as nuts)	China	450	Asia	450	527
	Paraguay	42	South America	72	
	Argentina	30	Africa	4	
	Malawi	1			

Source: FAOSTAT Database Gateway. Online at http://apps.fao.org/ (under "All databases" then Crops, Primary).

Tung Oil

Most people have never heard of tung oil since it is never sold in grocery stores, never added to commercial food products, and rarely sold under its own name. Tung oil is not eaten because it is poisonous, but its high level of unsaturation makes it an excellent oil for many other purposes. The oil is extracted from the seeds of several species of the Old World genus *Aleurites* (Euphorbiaceae), a family with many species that contain poisonous alkaloids. Species of the genus are tall [15 m (49 ft)] deciduous trees that bear fruits each containing three oil-rich seeds. Most of the oil produced comes from the seeds of *A. fordii,* a native of China which has been employed for hundreds of years as a source of oil for paints, waterproof

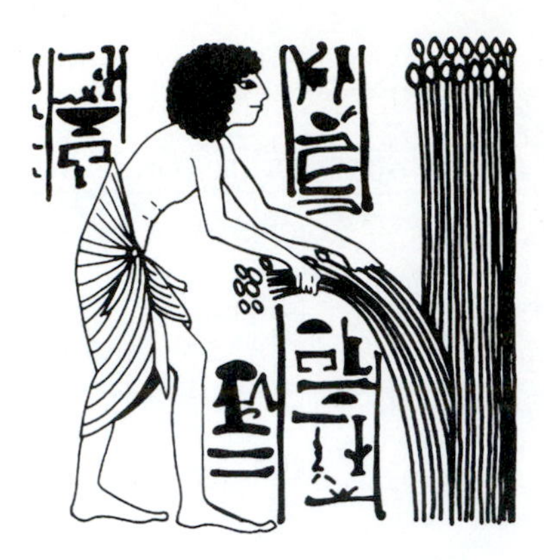

FIGURE 9.10
Linseed oil was used extensively in ancient Egypt. A copy of a scene on an Egyptian tomb shows the harvesting of flax, from which both fiber and linseed oil were obtained.

FIGURE 9,11
This photograph of a safflower head shows its similarity to its relative the thistle.

coverings, and caulking. Tung oil is an ingredient in "India" ink, an important item in China's traditional calligraphy. The so-called teak oil sold for fine furniture is usually refined tung oil, which is perfect for conditioning wood and providing a thin protective covering.

In the early part of the twentieth century tung oil was a relatively important oil in the United States, but all the oil used was imported. When the U.S. supply of tung oil was cut off during World War I, attempts were made to start tung plantations in the southern portions of the country, primarily in the Gulf Coast region. By 1950 production reached a level of almost 50,000 tons of oil per year. Subsequent crop losses caused by frost and a resumption of importation of the oil led to an abandonment of virtually all the plantations. As a result, no significant quantities of tung oil are now produced in this country. Most tung oil today comes from China.

Unsaturated Oils

Safflower Oil

Safflower oil received considerable attention in the United States during the 1970s because it is the most unsaturated of the commonly used edible oils. After an initial flurry of interest, however, its use declined because of evidence indicating that extreme unsaturation is not desirable in edible oils. It seems likely that safflower, *Carthamus tinctorius* (Asteraceae), was originally domesticated in the eastern Mediterranean region, but this origin is difficult to prove because the species is known only in cultivation. Safflowers were originally grown for their flowers, which produce the deep orange pigment carthamin. Ancient writings suggest that this pigment was used for dyeing 2000 years ago, but Hebrew writing mentioned it only in the second century A.D. Natives of India seem to have been the first to have extracted oil from the seeds, but they initially employed it only for medicinal purposes.

Safflower plants are spiny annuals (Fig. 9.11) with a colorful thistlelike head and large achenes [6 to 7 mm (0.25 in.) long]. Until recently the major use of safflower oil in developed countries was as a commercial oil for paint, varnish, and alkyd resin production. Safflower oil has the highest linoleic acid content of any known seed oil and is now exploited as a source of that acid in nutrient supplements.

Today safflower oil is found in cooking oils and manufactured products with an oil base. Since the oil tolerates cold well, it is especially well suited for use in the manufacture of salad dressings and margarines that are kept at low temperatures to preserve their flavor. Because of its high level of unsaturation the oil is not recommended for heavy, long-term use because it may reduce HDL levels and alter the composition of cell membranes. The cake remaining after pressing is used for feed. If the shells are removed before pressing, the cake can contain up to 52 percent protein.

Soybean Oil

Since the origin of soybeans, *Glycine max* (Fabaceae) (see Fig. 6.9), and their history and cultivation were discussed in Chapter 6, we will concentrate here on soybeans as an oilseed crop. Although soybeans are not widely used for oil in the Orient, Europeans were pressing them for oil and meal in the 1700s. Before 1940 the principal North American use of soybeans was for forage, but by 1947 the acreage devoted to it had greatly increased and 85 percent of the crop was processed for oil. By 1978 this figure had risen to 98 percent. Fifty percent of the vegetable oil produced in the world comes from soybeans. The United States accounts for half of this global production.

Before processing, the pods are shelled and the beans are cleaned and dried. Like peanuts, soybeans can serve as hosts to *Aspergillus flavus,* a fungus that excretes deadly toxins (aflatoxins). With proper handling, fungal infestations can be eliminated almost completely. In addition, the seeds contain enzymes that interfere with protein digestion. Consequently, soybeans are always heat-treated before being consumed or extracted for oil. After recleaning, hulling, and heat-treating, the oil-filled cotyledons are flaked and, in the United States, extracted with organic solvents. The cake is exceedingly high in protein and can be used as animal feed or further processed and used as an additive in human food. Soybean oil is always highly processed because it has a high percentage of linoleic (53 percent) and linolenic (7 percent) fatty acids that in combination cause the oil to foam when used for frying. Manufacturers have overcome this problem by partially hydrogenating the oil to reduce the linolenic content. The "fishy" smell caused when linolenic acid breaks down is also circumvented by this procedure. During the degumming of soybean oil, appreciable amounts of lecithin are recovered. This by-product is an important agent in wetting solutions and is used to help suspend particulate matter in colloidal solutions. The cake is exceedingly high in protein and can be used as animal feed or further processed and used as an additive in human food. The majority of the products that go under the label "vegetable oil" consist of processed soybean oil. Most soybean oil is used for salad or cooking oils and in artificially fluffy or creamy products such as whipped toppings, mixes for icing, and instant cheesecakes.

Sunflower Oil

Among all the plants used for seed oils, only the sunflower can claim North America as its original home. Wild, weedy populations of the species, *Helianthus annuus* (Asteraceae), still occur throughout Canada, the United States, and Mexico. Wild plants are branched and bear numerous relatively small heads. Domesticated sunflowers usually have one large head atop a single flowering stem (Fig. 9.12). Fossil achenes with features of domesticated varieties dated to be about 2800 years old have been found in eastern North America. Native Americans ate the seeds and crushed them for oil.

The first firm dates for sunflower cultivation in Europe fall in the middle of the sixteenth century, when specimens were grown as horticultural novelties. In the last part of the nineteenth century Russians began to cultivate sunflowers. Throughout the twentieth century Russia remained a major producer of sunflower seeds (Table 9.4). Although fossil evidence indicates that Americans had already selected for large-headed, large-seeded sunflower plants before the arrival of Europeans in the New World, the Russians continued to select for giant inflorescences and now have varieties that can produce up to 1000 seeds on each head. Selection has also produced dwarf oil-yielding varieties that are easily harvested by machine.

The achene coat of sunflowers can vary in color, but two types predominate: black, and gray with brown or black stripes (Fig. 9.13). These two color variants are maintained in cultivation and generally indicate whether the seeds are to be used for candy or snacks (striped shells) or for oil (black shells). Achenes for eating or candies are larger than those used for oil and have a lower oil content (about 30 versus 40 percent oil on a dry-weight basis). Most of the crop is used as animal feed in the United States.

Sunflower seeds yield an oil that is subsequently moderately refined. The presence of waxes in the oil that cloud the oil under cold conditions necessitates winterizing. The oil has been used without processing in a few areas of the northern United States and has been mixed with diesel fuel (75:25 oil:diesel ratio) in farm machinery. This mixture performs as well as pure diesel fuel, and studies indicate that if free fatty acids are removed from the oil, the mixture can outperform diesel fuel. Refined sunflower oil is used as a salad, cooking, and commercial oil in prepared foods. In Russia, it is an important source of oil for paints, varnish, and synthetic resins.

Corn Oil

Like soybeans, corn, *Zea mays* (Poaceae), is a plant for which humans have found innumerable uses, including the production of oil. However, oil production constitutes a minor use of the grain (see Chapter 5 for the history of corn use). Corn oil production is basically a by-product of the corn-milling industry that began late in the nineteenth century. Corn can be milled wet or dry. In the wet process, the cleaned corn is soaked or steeped in water and then lightly macerated to separate the germ (embryo) from the endosperm (Fig. 9.14). The pressed wet grains are washed into flotation tanks where the germs, which are lighter, float to the top and are removed. The solution containing valuable soluble proteins is evaporated down and sold for use as a culture medium for yeasts and other microorganisms. In the dry milling process, the grains are first tempered by adding just enough moisture to dry corn to make it pliant. The germ and seed coat of the grains are then removed by friction. The oil-rich germs recovered from the milling operations are dried, preexpelled, and extracted with solvents. Starch recovered from the endosperm of either milling operation is used in brewing (see Chapter 14) and in the textile and paper industries (see Chapters 10 and 16).

Since corn is used for both oil and starch production and starch is the principal product, there has been little selection for high oil content in corn destined for commercial use. For feed, however, a premium is place on high oil content since oil contains many more calories than does starch. As a result selection for better feed corns, oil contents as high as 10 percent of the grain (dry weight) have been produced. In general, values ranging from 2 to 5 percent of the dry weight of the kernels.

Most of the refined corn oil that is marketed is used for salad dressings and margarine. The oil contains natural antioxidants that give it good stability, but it is not a good frying oil because it smokes when heated to a high temperature.

FIGURE 9.12

(a) In the course of domestication, the sunflower, naturally a much-branched annual with numerous small heads, became a plant with unbranched stems, each topped by a single large head (b, c). (d) Each head is composed of numerous flowers (florets) that mature sequentially from the outside of the head to the inside. (e) An individual disk floret first sheds its pollen and then exposes its stigma; if successfully fertilized, a floret will produce an achene.

FIGURE 9.13
Sunflower seeds used for oil extraction are usually solid black (left), whereas those used for confectionery or eating purposes are striped (right).

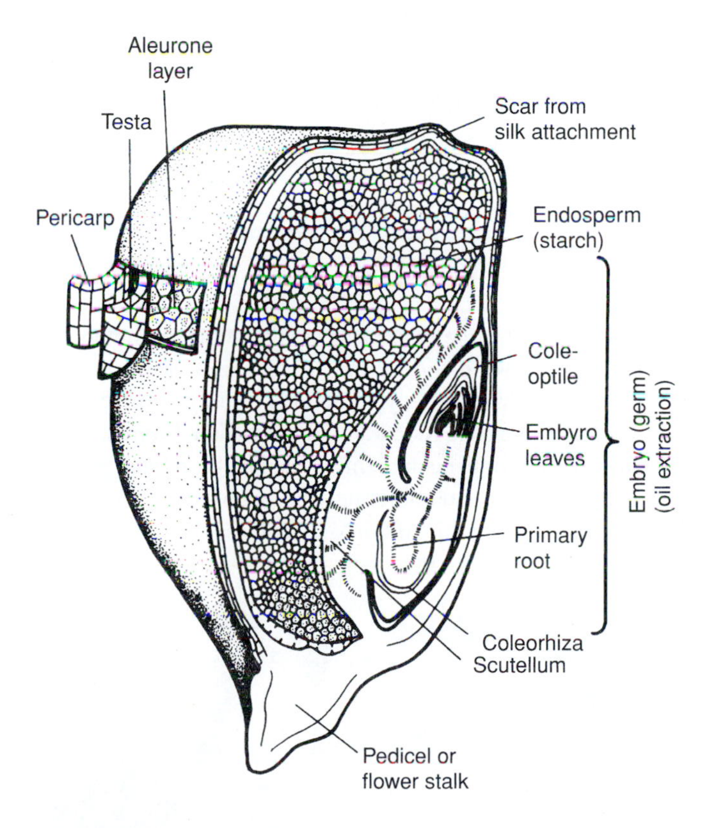

FIGURE 9.14
Corn oil is extracted from the scutellum of the embryo, or germ, which constitutes a small portion of the kernels compared with the endosperm.

Sesame Oil

Although less important than the other edible oils included in this chapter, sesame has historical importance and is still the major seed oil source in some Asian countries. There has been some suggestion that the use of sesame, *Sesamum indicum* (Pedaliaceae), as a source of oil predated the use of linseed oil, but since the earliest archaeological finds of seeds from Palestine and Syria date from only 3000 B.C., there is little firm evidence for the hypothesis. Indications of its early use come from Sumerian writings from about 2350 B.C. Dates from western Pakistan are about the same. Although the early records are all from the eastern Mediterranean area, botanists have proposed that the species, now known only as a domesticated crop, was initially taken from the wild in Ethiopia. Most sesame oil is consumed at or near areas of production in Africa, the Middle East, India, and China.

Sesame oil is not mentioned in the Bible, but it appears to have been important in non-Hebrew cultures 2000 to 4000 years ago. Loans were negotiated in both silver and sesame seeds. The expression "open sesame" made famous by the story of Ali-Baba in the *Arabian Nights* may be based on the sesame seed. Some authors have suggested that the expression was adopted by the writer of the stories because the capsules must be tapped precisely to open them without crushing the valuable seeds.

Sesame plants are herbaceous annuals with attractive white, pink, or purple flowers (Fig. 9.15) that have led to their use as ornamentals. The fruit is a flat capsule that contains a variable number of white oval seeds. The seeds can be pressed with or without the seed coats removed. Most sesame oil extraction is accomplished by relatively unsophisticated cold pressing methods that leave the characteristic nutty flavor. The yellow oil pressed from the seeds is simply filtered before local consumption. In large-scale operations, a combination of pressing and solvent extraction is employed, and the oil is generally refined. Sesame is considered a semidrying oil because of its high iodine value, but it is one of the oils most resistant to oxidation because it contains the powerful natural antioxidants sesamolin and sesamium. The cake remaining after pressing or extraction makes an excellent livestock food and in times of famine is eaten by people as well. The seeds can be eaten toasted or crushed and sweetened to make the Turkish candy known as halva.

Cottonseed Oil

It is doubtful whether cottonseed oil would have ever become a major product if cotton, *Gossypium* spp. (Malvaceae) (Chapter 15), were not so important a fiber crop. The ginning of cotton produces tons of seeds that have only recently been commercially exploited on a large scale.

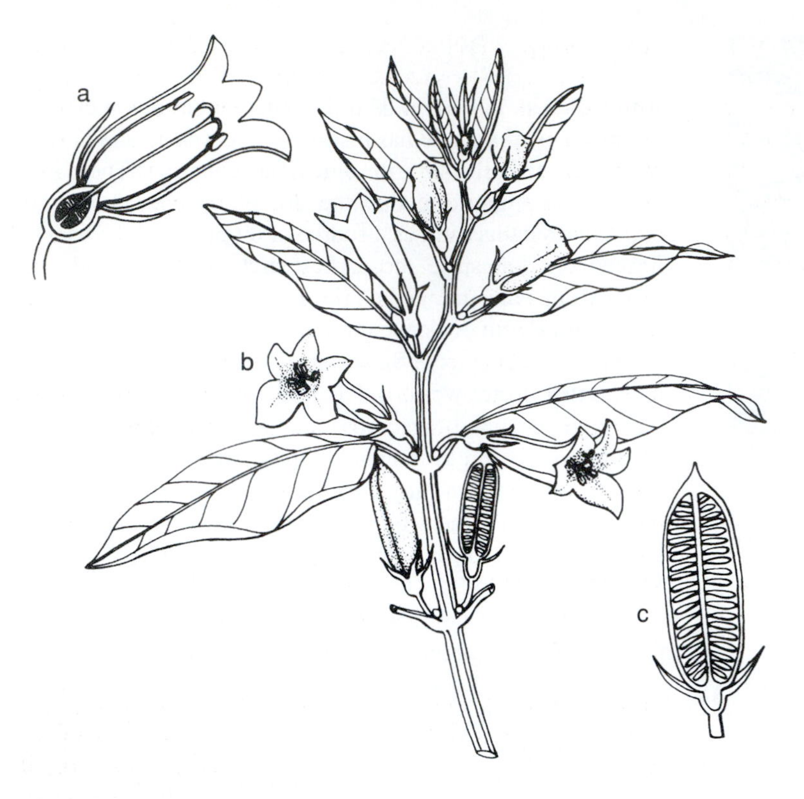

FIGURE 9.15
Flowers of the sesame plant, usually grown for seed production, are sometimes cultivated in this country as ornamentals. (a) A longitudinal section of a flower, (b) a sesame branch, (c) a fruit in longitudinal section.

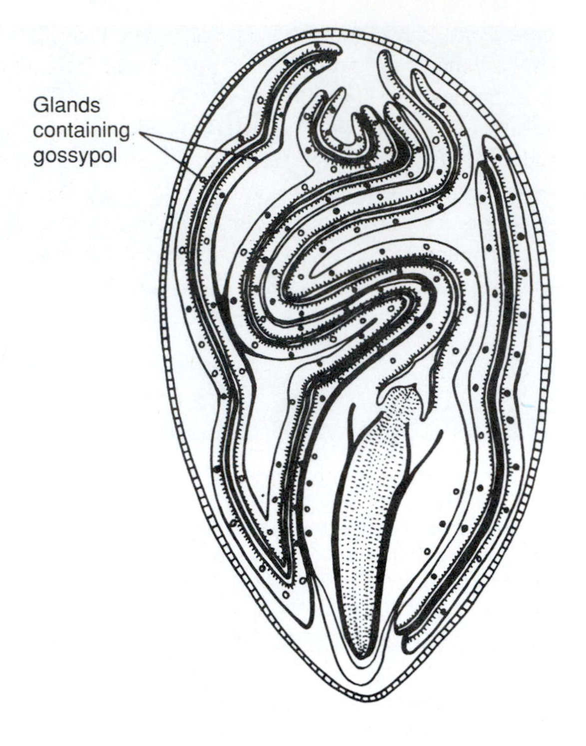

FIGURE 9.16
A longitudinal section of a cottonseed reveals the gossypol glands dispersed throughout the folded embryo.

The ancient Hindus extracted oil from Asian cottonseeds by first pounding them and then boiling the pounded meal. The oil was subsequently skimmed from the surface of the water and used in medicine and as lamp oil. However, until almost 1900 cottonseed oil was considered essentially inedible because it contained, among other things, gossypol, a bitter pigment produced in glands located throughout the cotyledons (Fig. 9.16). Consequently, most cottonseed was historically used for fertilizer.

In 1899 David Wesson began using caustic soda and diatomaceous earth to purify seed oils. By 1900 the young chemist has perfected the process by using steam in combination with a decolorizing powder such as diatomaceous earth. In the same year the first edible cottonseed oil was marketed under the name Wesson Oil. The procedures developed for this oil essentially started the modern vegetable oil industry that today includes cottonseed, soybean, canola, and palm oils.

Cottonseed oil experimentation led to another landmark discovery. In 1911 Procter & Gamble adopted and improved on a British technique for hydrogenation and introduced the first American vegetable shortening—Crisco—made from cottonseed oil.

Cottonseeds present a unique problem in the oil extraction process because bits of fiber called **linters** adhere to the seed coats (Fig. 9.17). In modern factories the linters are removed by machines with fine rasping teeth. The removed linters are used in papermaking and for the production of semisynthetic fibers such as rayon and acyl-celluloses (see Chapter 15). The delinted hulls are used for cattle food or mulch. Modern refining includes filtering, neutralizing, washing, bleaching, winterizing, deodorizing, and refiltering. This processing changes the dark-colored, semitoxic oil into a colorless, tasteless, and

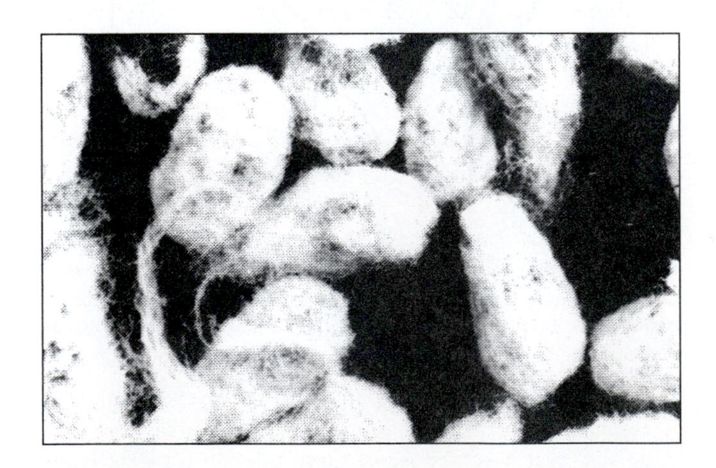

FIGURE 9.17
After the long fibers have been removed, cottonseeds are still covered with fine hairs that must be removed before processing.

stable product. The refined oil can be used in liquid form or hydrogenated to produce various types of shortenings and margarines. The cake remaining after pressing is often used as cattle food. Although cottonseed residues seem to have no adverse effect on cattle, they can cause death in rabbits, rats, guinea pigs, swine, and presumably, humans. The toxic principal, gossypol, has been tested in China as a potential birth control agent, but the adverse effects of the chemical have outweighed its potential medicinal value.

Canola Oil

Canola, colza, or rapeseed oil has shown a spectacular rise in use since 1985. It is today the third most important edible oil in the United States after soybean oil and cottonseed oil. The oil is extracted from the seeds of *Brassica napus,* rape, or *B. campestris,* turnip (both Brassicaceae) (see Chapter 7). Historically, the oil was employed both for cooking and as a lubricant, but its use for edible purposes was limited because it contains sulfur compounds, an unusual long-chain fatty oil called erucic acid that imparts an acrid taste, and a high percentage of linolenic acid. Even after deodorizing, the oil often reacquired an unpleasant taste because of the two undesirable fatty acids. Erucic acid is harmful to humans because it causes heart muscle lesions. Extensive breeding programs in Canada led to the development of a variety that differs from traditional rapeseed oil in its reduced erucic content. In view of the various negative connotations of the word *rape* and to draw attention to the Canadian origin of the oil, the term *canola* for *Ca*nadian *oil* was coined. The problem of the sulfur compounds was circumvented when it was discovered that preheating can destroy the enzymes responsible for the breakdown of the compounds into noxious smaller products. The procedures for extracting the oil are similar to those used for sesame, safflower, and sunflower oils. The cake can be used as livestock food if the seeds are heat-treated before extraction.

Canola has a low level of saturated acids, a level of monounsaturated fatty acids second only to olive oil, and an intermediate level of polyunsaturated fatty acids. It also has a good ratio (2:1) between linoleic and linolenic acids, which are precursors of omega-6 and omega-3 fatty acids. Canola oil consumption has been shown to lower plasma levels of cholesterol.

The seeds of canola, similar to regular mustard seeds, are small and borne in capsules clustered at the tips of branches. The fruits in these clusters tend to mature very quickly once ripening begins at the base of the inflorescence. If the fruits on an inflorescence become completely ripe, they rapidly shed their seeds, which are then lost to the farmer. In some growing areas, farmers can accurately gauge the ripening and simply harvest the fruits with a combine like that used for grain harvesting. In Canada, determining the stage of ripening is more difficult. Harvesting too soon results in green seed and harvesting too late causes losses from prematurely shed seeds. To avoid these problems, farmers often cut the crop in the field when about 10 percent of an inflorescence is mature and allow it to remain there until the seeds have lost their immature green color and have absorbed enough moisture for eventual ripening. After this initial harvesting, a combine is adjusted to gather up the stalks, separate the fruits from them, and thresh the seeds from the capsules.

As low a level as possible of erucic acid is desirable in edible oils, but high levels of the acid are desirable for industrial purposes. Consequently, there has been parallel breeding for high–erucic acid varieties. In addition to being used for lubrication, the oil is used to cool steel plates after they have been forged, and indication of its ability to withstand high temperatures.

Moderately Saturated Oils

Peanut Oil

In Chapter 6 we discussed the origin, dispersal, and rise of the peanut, *Arachis hypogaea* (Fabaceae), to its modern position as a major world crop. Peanuts were not used for oil until about 135 years ago, when a French firm began to crush peanuts imported from Africa in a factory in Marseilles. After this humble beginning, oil extraction spread to other areas. The United States today is unique in that only one-fourth of the crop goes into oil production, with most of the rest being turned into peanut butter.

Before pressing, shelled peanuts must be carefully dried. The thin seed coats (skins) are removed, and the oil is expressed by crushing and solvent extraction. Although the oil will solidify if kept in a refrigerator, it is considered a premium cooking oil because it does not smoke when heated to a high temperature and does not retain odors. The cake makes a valuable livestock feed, as would be predicted by the high protein content of the seeds (see Table 9.3).

Olive Oil

As was mentioned at the beginning of this chapter, olives, *Olea europea* (Oleaceae), have sustained cultures nutritionally and spiritually since ancient times (Fig. 9.18). However, despite the Greek legend, the first records of olive cultivation are from Crete, where there is archaeological evidence of olive culture dating from 3500 B.C. Egyptians (approximately 3100 B.C.) used olive oil as a cleanser, as did the Greeks (Fig. 9.1), to anoint bodies, and as a lamp oil, medicine, foodstuff, and religious accoutrement. It has even been suggested that the Egyptians slid the enormous blocks used in the construction of the pyramids over one another by placing olives or olive oil between them.

Olive oil is also one of the few oils obtained from a fruit pulp rather than from seeds. The seeds yield little or no additional oil. Because only the pulp is used, the procedures for

FIGURE 9.18
This drawing, copied from a classical Greek vase, shows how olives were picked in ancient Greece. It takes about 1300 to 2000 olives to produce 1 liter (approximately 1 quart) of olive oil.

extraction differ from those of most oils. The fruits are macerated, the seeds are removed, and the pulp is pressed. The first press, a cold press, yields "virgin" olive oil. Subsequent pressings, usually with heat, yield oils of lower grades. Olive oil is one of the few oils that is not generally processed to tastelessness and thus remains a delight for gourmets. Its high level of monounsaturation has led to its recent increased use. It also has a long shelf life, during which it retains its subtle flavor.

Castor Oil

Although it is rarely used medicinally today, castor oil, *Ricinus communis* (Euphorbiaceae), was forced down the throats of many children in the past by their well-intentioned parents. The laxative action for which the oil was administered is caused by ricinoleic acid, the predominant fatty acid in the oil. The seeds, the hulls, and even the unrefined oil taken in large quantities are toxic to humans. Three compounds are responsible for the poisonous effects: ricinine, a mildly toxic alkaloid; ricin, a highly toxic protein that leads to hemorrhaging; and a protein-polysaccharide mixture called CB-A, which causes violent allergic reactions in people sensitive to it. Some varieties have been developed that lack the poisonous compounds.

There is no agreement among botanists about the original area of distribution of the species. By the time it was domesticated, it appears to have already spread across Africa and southern Asia. Investigations of the taxonomy of the genus suggest that the species arose in tropical Africa, but the oldest records of its use come from Egypt, where seeds in tombs have been dated to almost 6000 years ago. Ancient uses of the oil seem to have been as a lamp oil, in medicines, and for religious ceremonies. India is now the world's largest producer, processing more than four times as much as China (Table 9.4).

Castor plants (Fig. 9.19) are monoecious perennials that can reach 13 m (42.6 ft) height in tropical areas with the male flowers at the bottom and the females at the top or the inflorescences (Fig. 9.19). The fruits are three-lobed, usually spiny capsules that split open at maturity to release the three seeds (Fig. 9.20), which are dried, pressed, and extracted. If the oil is to be used for medicinal purposes, the seeds are pressed at temperatures lower than 50°C (122°F) so that ricin will not be removed with the oil. The oil is rarely refined as such but is often bleached. Because of its high level of saturation, the oil can be stored for long periods with no change.

Today castor oil primarily goes into paints, varnishes, synthetic polymers and resins, and cosmetics. Hydrogenated castor oil is used as a lubricant for airplanes and rocket engines. The cake left after extraction contains enough toxic compounds to render it unsuitable as a livestock food. If detoxified by heat, the meal can be fed to domesticated animals that chew their cuds, such as cows; otherwise, it is used for fertilizer.

Vegetable Fats

Palm and Palm Kernel Oil

Oil palms, *Elaeis guineensis* (Arecaceae), differ from all the other important oil crops except olives in that distinct oils are obtained from the fruit pulp and the seed. The fruits of the palm (Fig. 9.21) are 2.5 cm (1 in.) long with an oil-rich mesocarp (pulp) surrounding a hard endocarp. In commerce, the seed and fruit pulp are extracted separately. Palm oil consists of 51 percent saturated fats. Palm kernel (seed) oil is similar to coconut oil in composition (see Table 9.3), and they are often substituted for each other.

A native of Africa, the dioecious oil palm is also now an integral part of the culture of coastal Brazil, where other species of the genus are native. In many areas where they are grown, the palms are only semicultivated. In both South America and Africa fruits are picked when ripe and sold in local markets (Fig. 9.22). Customers press out their own oil by fermenting the fruits for 2 to 4 days, boiling them, and pounding the mass. The oil is recovered by stirring the pulp in water and skimming off the oil as it floats to the top. The recovered oil, which has a characteristic orange-red color caused by the presence of carotenes, is heated to drive off any remaining traces of water. The celebrated cooking of Bahia, Brazil, liberally uses unprocessed palm oil, which goes under the name of *dende* oil. In commercial operations the pulp is steam-sterilized to prevent enzymatic breakdown of the flesh before extraction. The oil is high in saturated fatty acids, primarily palmitic acid.

In commercial operations palm fruits are processed as soon as possible after picking. The fruit pulp is mashed off

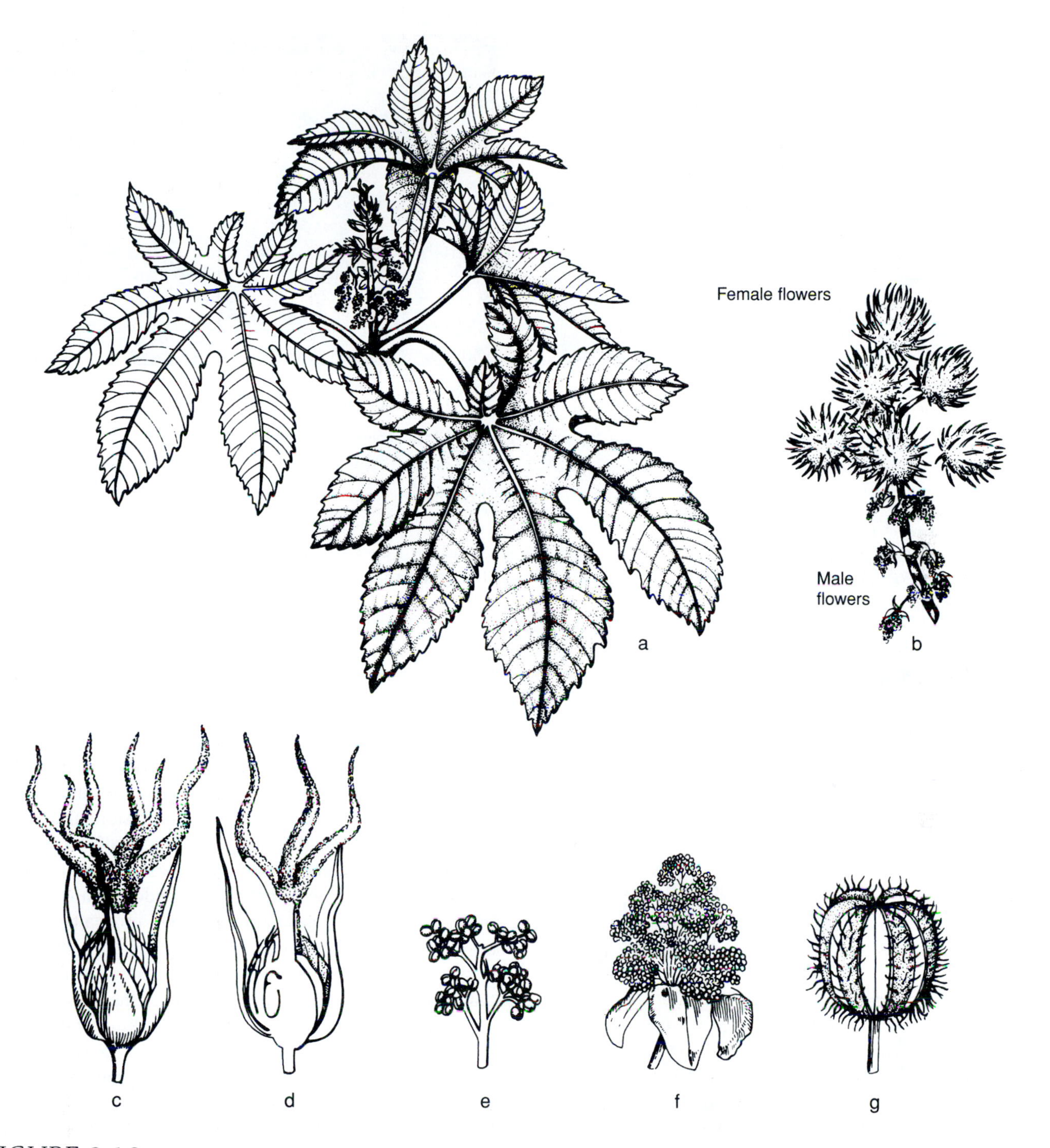

FIGURE 9.19

(a) Because of its attractive shiny foliage, the castor bean plant is sometimes grown as an ornamental. The female flowers (c, and d in section) of the monoecious plant occur in clusters at the top of the inflorescences (b), with the male flowers (f, and e in detail) in clusters below them. (g) The mature fruits are spiny and three-parted with a single seed in each locule.

(a and b from nature; c–g after Baillon.)

the seeds and heated into a homogeneous mass. The oil is extracted rapidly once the fruits have been picked because the fruits contain natural lipases that begin to break down the oil after they are removed from the trees. Therefore, once picked, the fruits are sterilized with steam and pressed with hydraulic presses, and the oil is filtered and stored for possible future processing. The endocarps containing the seeds are dried, graded, and cracked. The kernel, or seed, is then dried and sacked for large-scale extraction operations.

Palm oils have been used primarily for soap and candles, but their importance in margarine and solid shortenings rose rapidly during the 1960s and 1970s because baked

FIGURE 9.20
Castor beans are attractively mottled but contain several toxic compounds, among them ricin, a lethal protein.

FIGURE 9.22
Palm oil fruits being sold in an African market.
(Photo courtesy of B. L. Turner.)

products made with them have a long shelf life. Concern about the health risks of consuming oils with a high level of saturation has led to a recent decline in their use in foods.

Coconut Oil

Coconuts seem to have limitless uses, among them the production of an oil or, perhaps more correctly, a fat, since the triacylglyceride is solid at room temperatures (below 24°C, or 75°F). There is little documentation of the early uses of coconut oil, and we lack fossil or even historical evidence with which to assess the origin and spread of the coconut, *Cocos nucifera* (Arecaceae) (Fig. 9.23). However, since scrapers used for processing coconut endosperm were common tools in Polynesia when the Europeans arrived, coconuts presumably were used there as oil sources. Oil constitutes such a high percentage of the coconut endosperm that it can be extracted simply by pressing the dried meat.

Until the nineteenth century almost all coconut oil was consumed locally. In 1841 a patent was issued for a process that used coconut oil in soap manufacture. Since that time coconut oil production has risen dramatically as new applications have been found. Soon after the discovery of the process of hydrogenation, coconut oil found its way into margarine. The oil is naturally so saturated that little additional hydrogenation is needed. The flavor of coconut oil also mixes well with a variety of flavors, both sweet and salty; as a result, it is used in many processed foods. Macadamia nuts are roasted in coconut oil to give them an additional subtle flavor. Coconut oil is little used in paint and varnish manufacture but is a common constituent

FIGURE 9.21
Palm oil is extracted from the fleshy mesocarps of the palm fruits. A different oil—palm kernel oils—is extracted from the seeds of the same fruits.

FIGURE 9.23
Picking coconuts in Ghana.
(Photo by P. Cenini, courtesy of FAO.)

of detergents and resin products. In the Asian tropics, where most coconuts are grown, the cake is used as cattle and poultry food.

Coconut oil was at one time prominent in cosmetics and nondairy "dairy" products, but its use, like that of other "tropical oils" in food products, has declined. Coconut oils are often said to enhance the effectiveness of shampoos, hand lotions, and suntan creams. It is commonly used as a liquid carrier for vitamins, hormones, and antibiotics and to coat particles in delayed-action pills.

Waxes

Like vegetable oils, waxes are made from alcohol and fatty acids, forming an ester linkage. However, "seed oils" are combinations of the trialcohol, glycerol, and fatty acids, whereas waxes are made by joining simple alcohols and fatty acids. All flowering plants manufacture waxes that cover their aerial surfaces to prevent drying or injury, yet few plants have pronounced accumulations or readily obtainable amounts of wax. Species that produce substantial quantities have been exploited as sources of wax, but only four are used to any extent in the United States today.

FIGURE 9.24
The belief that jojoba oil can duplicate the lubricating qualities of sperm whale oil seemed to offer a way to save the mammal from extinction, but yields are too low for jojoba to be used for industrial purposes on a large scale.

The first of these is jojoba (pronounced *ho-ho-ba*), *Simmondsia chinensis* (Simmondsiaceae). The chemical nature of jojoba "oil" emphasizes the similarity between oils and waxes. Jojoba received much publicity as an important future oilseed crop in the 1970s and 1980s. The excitement about jojoba stems from several of its properties. First, the shrubby plants grow naturally in the Sonoran Desert of North America. Consequently, it is a crop that can be grown on land unsuitable for other purposes. Second, jojoba oil is chemically different from any other seed oil because it lacks glycerol. It is consequently a liquid wax rather than an oil. Since the chemistry of the oil is like that of sperm whale oil, there were hopes that this renewable oil would replace the almost depleted supply of sperm whale oil (Fig. 9.24) as a lubricant for fine machinery. Jojoba is tolerant enough of high temperatures to be an industrial substitute for sperm whale oil, but yields are low and its use in lubrication of machinery has been limited. Jojoba oil is particularly effective in penetrating the outer layers of human skin, and this property has led to its use in cosmetics.

FIGURE 9.26
Candelilla plants uprooted from the desert are carried on the backs of burros to a camp where they will be boiled for wax extraction. (Photo courtesy of J. L. Neff.)

FIGURE 9.25
Carnauba wax is harvested from the leaves of these palms. (Photo courtesy J. L. Neff.)

Carnauba wax, a solid "traditional" wax, is obtained from the leaf surfaces of *Copernicia cerifera* (Arecaceae), a native of northeastern Brazil. The wax is obtained by collecting new or immature leaves of wild palms (Fig. 9.25) and allowing them to dry. Once dry, the leaves are beaten to dislodge the wax, which is shipped in particulate form and processed by importing countries. Carnauba wax is used primarily in car waxes and shoe polishes. Although extraction methods are laborious and crude, carnauba wax is still collected because it is harder than beeswax and many synthetic waxes.

Another wax often substituted for carnauba wax is candelilla wax, obtained from the succulent stems of *Euphorbia antisyphilitica* (Euphorbiaceae). The stems are gathered (Fig. 9.26) and boiled in water. The wax that melts and floats to the surface is scooped off and allowed to harden. Candelilla plants are native to the Chihuahuan Desert along the United States–Mexico border. Plants collected for wax extraction are all wild. Entire plants are uprooted by bands of nomadic candelilla workers, who leave behind a trail of devastation. In the United States collection of the plants is forbidden because the species is considered endangered.

The fourth natural plant wax that is still available in the United States is bayberry. Early settlers in the Northeast collected the fruits of bayberry, *Myrica pensylvanica* (Myricaceae), and boiled the berries to melt off the wax. The wax, scooped from the surfaces of the boiling cauldrons, cooled and was later remelted for use in candles. Bayberry is still used for novelty candles because of the pleasant fragrance when they burn.

10

Hydrogels, Elastic Latexes, and Resins

The various plant products discussed in this chapter (Table 10.1) do not come from the same plant organs, are not produced in similar ways, and do not contain the same kinds of chemicals. The one quality they have in common is that they are all sticky substances exuded or extracted from plants. Because of this feature, they are often confused with one another. For example, many people chew gum, but chewing gum is not a gum; rather, it is a latex. Likewise, pine pitch or pine gum is a resin. By discussing true gums with latexes and resins, we highlight the differences between them. All these plant products have been used for thousands of years, but natural resins and latexes are now used less frequently than in the past because similar or superior synthetic products have become available. The use of hydrogels, in contrast, has exploded in the last 50 years, and the demand has only been partially filled by synthetically produced compounds.

Hydrogels

Hydrogels, or **hydrocolloids,** defined as water-modifying substances, include all the products collected, extracted, or synthesized to alter the behavior of water. A familiar illustration of how a hydrogel works is provided by the steps one takes when making gravy. The cook starts with a paste made from flour and fat. She or he then adds a thin water-based stock to the hot paste while stirring continuously. After a few minutes, the liquid becomes thick and the stock turns into "gravy." The thickening occurred because the starch molecules of the flour associated with the water molecules of the stock. Once they became associated with starch molecules, the water molecules could no longer move freely. Their sluggishness was manifested as a thickening of the solution (Fig. 10.1). Starch, pectin, and their animal-derived counterpart, gelatin, are the most common household hydrogels. Commercial operations use a much wider array of water-modifying agents. Plant-derived hydrogels fall into three main classes: gums, pectins, and starches, all of which are important in modern industry.

TABLE 10.1 Plants and Plant Products Discussed in Chapter 10

COMMON NAME	SCIENTIFIC NAME	FAMILY	NATIVE REGION
African rubber	*Funtumia elastica*	Apocynaceae	Africa
Bambong rubber	*Ficus elastica*	Moraceae	India
Castilla	*Castilla elastica*	Moraceae	Central America
Ceara rubber	*Manihot glaziovii*	Euphorbiaceae	Brazil
Chicle	*Manikara zapota*	Sapotaceae	Mexico, Central America
Copal	*Copaifera* spp.	Fabaceae	Tropical America
	Hymenaea spp.	Fabaceae	Tropical America
	Agathis spp.	Araucariaceae	Southeast Asia, New Zealand, Australia
Dammar	*Shorea* spp.	Dipterocarpaceae	Asia
	Bursera spp.	Burseraceae	Tropical America
Frankincense	*Boswellia sacra*	Burseraceae	Africa
Guar gum	*Cyamopsis tetragonolobus*	Fabaceae	Africa
Guayule	*Parthenium argentatum*	Asteraceae	North America
Gum arabic	*Acacia senegal*	Fabaceae	Africa
Gum ghatti	*Anogeissus latifolia*	Combretaceae	India, Sri Lanka
Gum karaya	*Sterculia urens*	Sterculiaceae	India
Gum tragacanth	*Astragalus* spp.	Fabaceae	Near East, Asia Minor
Gutta-percha	*Palaquium gutta*	Sapotaceae	American tropics
Hevea or para rubber	*Hevea brasiliensis*	Euphorbiaceae	Brazil
	H. benthamiana		
	H. nitida		
Indian rubber	*Ficus elastica*	Moraceae	India
Kauri resin	*Agathis australis*	Araucariaceae	Southeast Asia
Lacquer	*Rhus verniciflua*	Anacardiaceae	China
Landolphia	*Landolphia* spp.	Apocynaceae	Africa
Larch gum	*Larix occidentalis*	Pinaceae	North America
Locust gum	*Ceratonia siliqua*	Fabaceae	Mediterranean
Mastic	*Pistacia lentiscus*	Anacardiaceae	Mediterranean
Myrrh	*Commiphora abyssinica*	Burseraceae	Africa
Turpentine	*Pinus elliotii*	Pinaceae	North America
	P. palustris		North America
	P. pinaster		Europe
	P. sylvestris		Europe

Gums

Nature and Uses of Gums Plant gums are solids consisting of a mixture of polysaccharides composed of sugars other than glucose (primarily arabinose, galactose, mannose, and xylose) (Fig. 10.2) that are either water-soluble or capable of absorbing water. These polymers can be linear, branched, or cross-linked. They are insoluble in oil and in organic solvents. Mixing these polysaccharides with water leads to the formation of a gel. Economically important plant gums can be exudate gums or extractive gums. Exudate gums are not always present as gums in plants that produce them, and there are no special cells, tissues, or organs for their production. They are produced only at specific times and in response to injury or wounding and appear to come from the breakdown of the injured cells. In nature, these substances seal wounds and help prevent invasions by fungi and bacteria. In addition to exudate gums, many modern commercially important gums are extracted from plant tissues, primarily seeds and wood.

Gums are only partially digested by humans and, with a few exceptions, have no adverse effects when eaten. From the point of view of human metabolism, gums can thus be considered inert substances. For this reason, they are particularly well suited for use in foods, diet products, and medicines. The other major uses of gums are in the paper, textile, cosmetic, and petroleum industries.

In the food industry, gums are used to texturize products, provide "body," improve the feel of foods in the mouth, stabilize emulsions, retain moisture, thicken liquids, and suspend particles. Most dairy products use gums to disperse fat and protein molecules evenly in a water base. In frozen products gums help prevent the formation of ice crystals (Fig. 10.3), and in sauces and syrups they produce a thick, rich consistency. Gums also increase the shelf lives of many products because they prevent solid particles from settling out of suspensions and help bakery products retain moisture. The powdered "crystals" used to make instant drinks are often sprayed with a thin film of gum to prevent water absorption before use and to help disperse the particles once they are mixed with

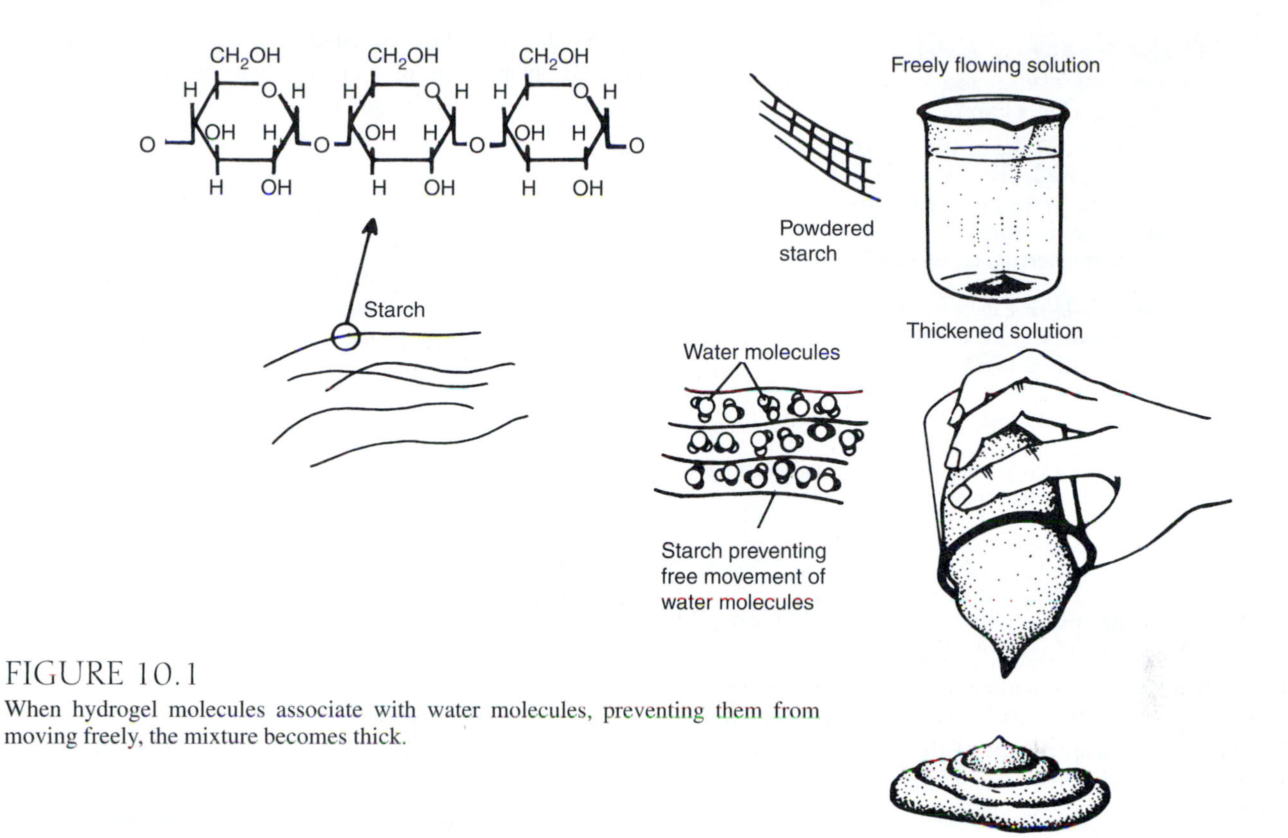

FIGURE 10.1
When hydrogel molecules associate with water molecules, preventing them from moving freely, the mixture becomes thick.

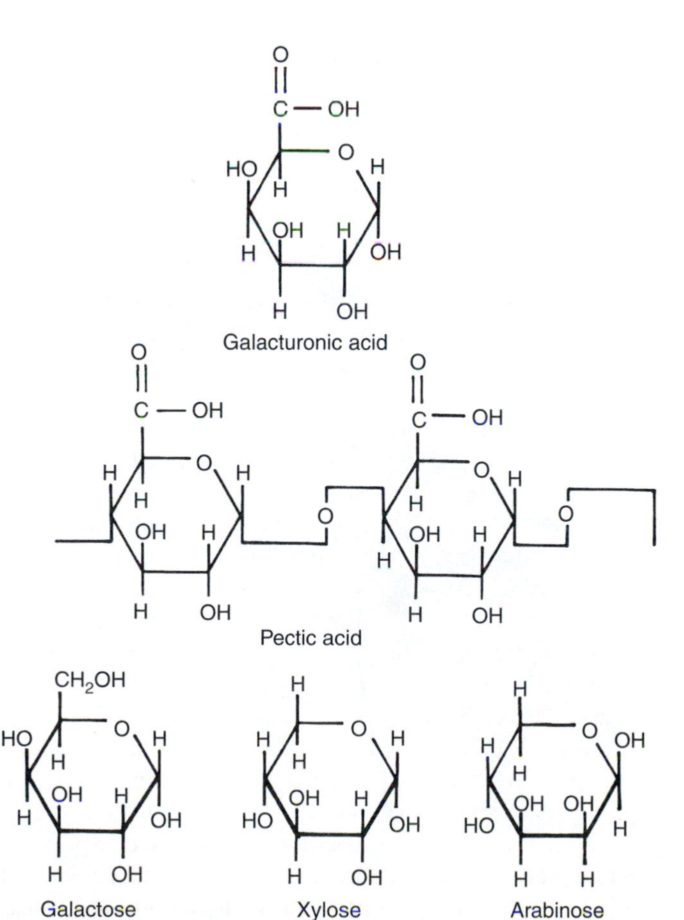

water or milk. The minute amounts of gum used also thicken the drink slightly, making it seem smoother and richer. This smoothness improves "mouth feel" and has led to the widespread use of gums in ice cream, whipped toppings, commercial frostings, and cream fillings. Finally, gums are often added to sandwich spreads and luncheon meats to bind together the processed protein particles.

In medicine, gums hold tablets together and keep the particulate matter in oral antibiotic solutions such as penicillin dispersed in liquid. Some gums can act as laxatives and have been used alone or in combination with other products. Gums also make toothpaste a "paste" (or a gel) and allow the fats and oils in hand and body lotions to maintain their smooth, creamy consistency. The pigments in crayons are bound to the wax with gums.

The paper and textile industries are the primary consumers of many gums because of their use of sizing agents. **Sizings,** or sizes, are substances spread on cloth or paper to fill in the pores and irregularities and to make the material

FIGURE 10.2
True gums are salts of polysaccharides composed of sugars such as galactose, xylose, and arabinose. In some cases, the CH_2OH group of these sugars has been oxidized to COOH as in galacturonic acid or pectic acid. Pectins are chemically related to gums and are polymers of pectic acid.

Coffee Ice Cream is rich, creamy ice cream made with natural coffee beans.

Ingredients: Milk, cream, sugar, skim milk, high fructose corn syrup, corn syrup, natural coffee, cellulose gum, vegetable gums (guar, carrageenan, carob bean).

Made with loving care by Blue Bell Creameries, L.P., Brenham, Texas 77833. ©1993

FIGURE 10.3
The gums in this "coffee" ice cream prevent the formation of ice crystals and make the ice cream feel smooth in the mouth.

stiff or smooth. In the textile industry (Chapter 15) sizing stiffens and strengthens threads during weaving and occasionally during garment manufacture. Gums also provide the stiffening agents for felt hats. After weaving or stitching, the sizing is usually washed out of the fabric and flushed into a stream or river flowing near the mill. Gums have properties that make them particularly good sizing agents, and only small quantities are needed to produce adequate sizing. As was previously mentioned, they are not readily metabolized by many organisms. Consequently, they have relatively minor effects when they are washed into waterways.

In the paper industry a dilute solution of gums is spread on paper products to produce both a smooth texture and a surface that highlights a crisp printed image. Gums have traditionally been mixed with glycerin and water to form the "glue" on the flaps of envelopes and on stamps. The dried gum mixture forms a mucilage when licked.

The petroleum industry is a gum consumer because gums have characteristics that facilitate two complex operations involved in modern drilling procedures. The soil particles through which a drill moves are potentially very abrasive but act like lubricants if gums are added to the water-soil mixture surrounding the bit. As in the case of food products, gums keep the particles well dispersed in the water and give the mixture a smooth consistency. The other major application involves secondary recovery operations. Gums are added to the water or brine that is pumped into the ground to maintain enough pressure to keep gas and oil rising to the surface. If pure water were pumped in, it would quickly penetrate through the rock and flood the well. Gums thicken the water and slow its movement. Although plant gums were originally used, almost all gums used in drilling operations today are synthetically produced.

In 1993, the estimated value of the world market for hydrocolloids used as food additives was US $10 billion. Hydrogels obtained from flowering plants accounted for 12 percent of those used. The remainder were obtained from algae (Chapter 18) or consisted of starch, pectin, or animal-derived gelatin. In this chapter we discuss only gums obtained from flowering plants.

Sources of Vegetable Gums

Exudate Gums The classic and still most widely employed exudate gum is gum arabic. This gum, exuded from wounded trees of *Acacia senegal* (Fabaceae) (Fig. 10.4) has been used since at least ancient Egyptian times. The small spindly trees of this acacia are native to northeastern Africa, primarily Sudan, but because the gum was sent to Arabian ports for shipment to Europe, it acquired the English name *gum arabic.* Ninety percent of the world supply is still obtained from wild trees that are slashed or punctured to induce a wound reaction. Natives then collect the dried globs or beads of gum that exude from the wounds (Fig. 10.5). Most of the collecting takes place between October and June

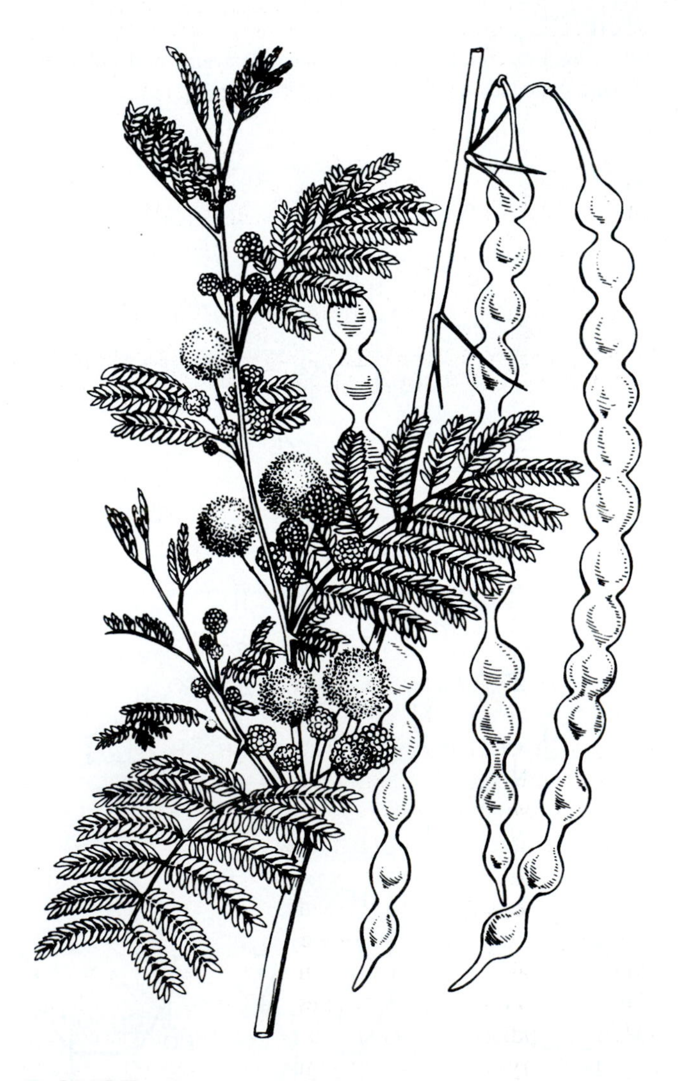

FIGURE 10.4
A branch of *Acacia senegal,* the small tree that provides gum arabic. (After Baillon.)

a

b

FIGURE 10.5
The tapping and collecting of gum arabic. The worker on the left is slashing branches causing the formation of gum. The woman on the right is picking globules of accumulated gum from previously wounded branches.
(Photos courtesy of the Gum Arabic Co., Ltd., Khartoum.)

during the dry season. Particles of dry gum are carried to a central locality, graded, and dried further before being shipped to processing plants around the world (Fig. 10.6). The final processing consists of regrading the particles and cleaning and blending them, if necessary, before they are milled or dissolved and spray-dried into powder.

Gum arabic is almost ubiquitous in people's daily lives. The "lace curtain" left on the sides of a freshly drained beer glass is created by the gums added to the beer to stabilize the foam. When one licks a postage stamp, the water-soluble glue is probably gum arabic. Gum arabic is the gum most often used to coat flavor particles, and small quantities are added to candies and confections with a high sugar content to prevent crystallization. This gum also emulsifies fats in many foods, hand lotions, and liquid soaps. The Egyptians mixed gum arabic into paint to suspend the pigments. The gum is still a component of fine watercolors. Gum arabic is also the most important gum used in the manufacture of ink, but it is less used in paper (except tissue paper) and fabric manufacture than are other gums.

Gum tragacanth is another silent partner in people's lives. Tragacanth is one of the substances first recorded as an emulsifier. The most important characters of gum tragacanth are its ability to avoid degradation in acid solutions (in contrast to most other gums) and its ability to emulsify oil without a surfactant (a wetting agent). It is also one of the few gums to which some people have an allergic reaction.

Gum tragacanth comes from any of several species of *Astragalus* (Fabaceae), but primarily *A. gummifer* (Fig. 10.7), which is native to the Near East and Asia Minor. The gum is obtained by tapping shrubby bushes of wild *Astragalus* species by carefully making incisions in the top part of the root or the bases of the larger branches. The name *tragacanth* comes from the appearance of the exuded gum,

FIGURE 10.6
Cleaning and grading gum in Port Sudam on the Red Sea.
(United Nations photo issued by the FAO.)

which tends to form ribbons similar in appearance to a goat horn (*tragos* meaning "goat" and *akantha* meaning "horn" in Greek). The gum is collected as partially dried flakes or ribbons, which are brought to a trade center and sold. During processing, the irregular pieces are cleaned and ground as needed. Gum tragacanth is used primarily in food preparations such as mayonnaise, sandwich spreads, and pre-prepared milk shakes. It is also used in toothpaste and hand lotions and medicinally as a binder for tablets or as a suspending agent for oral penicillin.

Gum karaya, or sterculia gum, is unique among the gums because it can form a strongly adhesive gel when mixed with only small quantities of water. Although it is the least soluble commercial gum, its adhesive properties and resistance to bacterial and enzymatic breakdown have led to its use as a dental adhesive and a binder for the fibers in bologna and other lunch meats. The gum is collected from trees of *Sterculia urens* (Sterculiaceae), a widespread species native to the rocky hills and plateaus of India. The trees are tapped or blazed during the dry season, and the exuded gum is collected before the monsoons arrive. Some gum karaya is obtained from trees grown on commercial plantations that are tapped year-round. The collected drops of gum are sorted, cleaned, blended, and granulated or powdered. Granulated gum karaya is sometimes used as a bulk laxative. The powdered, dried gum is also used throughout the food industry, primarily in salad dressings, ice creams, cheese spreads, and whipped toppings. Hair-setting gels owe their texture to gum karaya, not to gelatin. The gum often replaces gum arabic in the tissue paper industry and was formerly used in drilling operations for thickening mud or plugging wells.

The last of the important natural gums is gum ghatti, named because it was originally brought to ports in India across mountain passes, or *ghats*. The plants that yield this gum are trees of *Anogeissus latifolia* (Combretaceae), which are native to the dry deciduous forests of India and Sri Lanka. As in other gum collection operations, the trees are wounded and the exudate is collected, sun-dried, sorted by hand, and shipped. Gum ghatti maintains a constant place in the spectrum of available gums because it has properties intermediate between those of gum arabic and those of gum karaya, but it is a better oil emulsifier than either one because it has high viscosity. Consequently, it is used as an emulsifier in liquid and paste waxes and for fat-soluble vitamins. Like many other gums, it has also been used in drilling operations.

Extractive Gums Exudate gums come from wounded woody tissue; similar substances can be obtained from the endosperm of seeds of some legume species or can be

FIGURE 10.7
A branch and flower of *Astragalus gummifer,* the major source of gum tragacanth.
(After Baillon.)

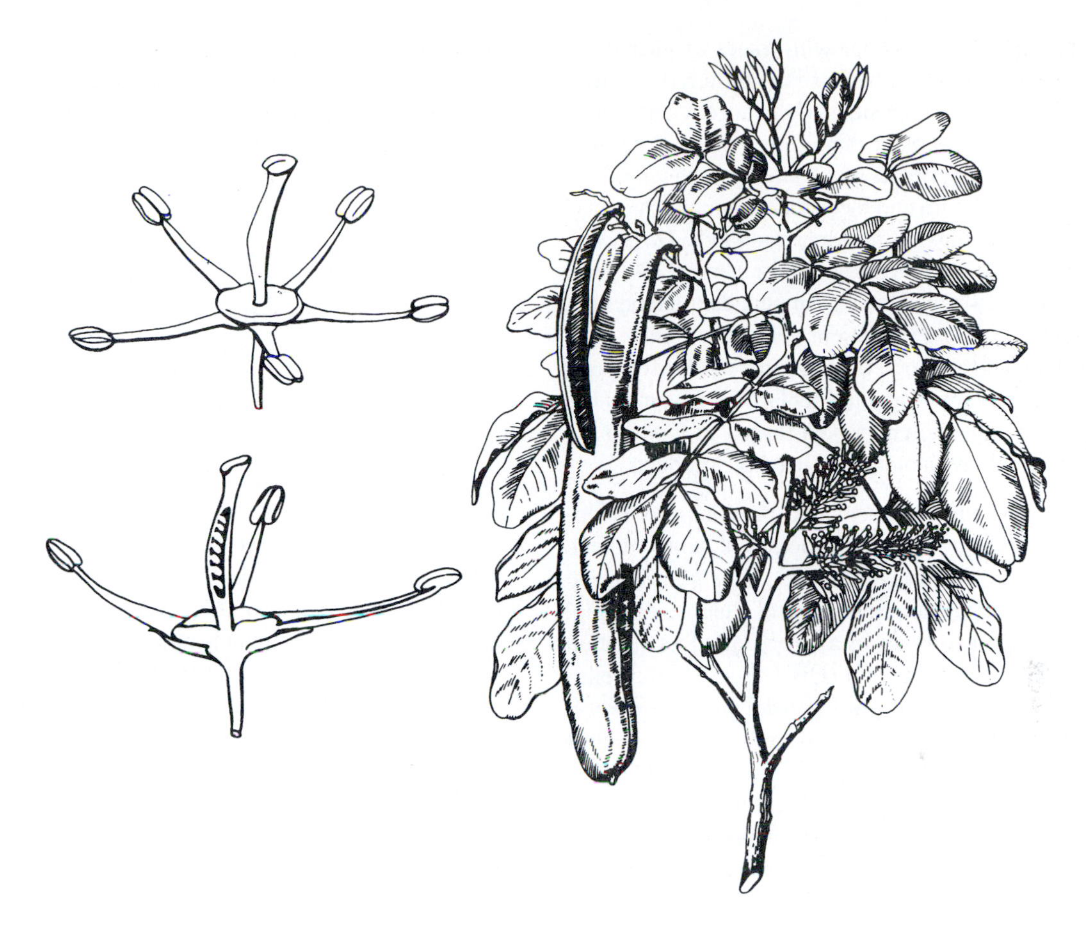

FIGURE 10.8
The seeds of the carob tree are the source of locust bean gum, and the powdered, sweet mesocarp of the fruit is used as a chocolate substitute. Shown here are a branch (right) and individual flowers, whole (top left) and in section (bottom left).
(After Baillon.)

extracted from the wood of others. The two most important seed gums are locust gum and guar gum. Locust bean gum, or carob gum, was used by the Egyptians as an adhesive for mummy bindings, but the tree from which the gum-producing seeds are obtained, *Ceratonia siliqua* (Fabaceae) (Fig. 10.8), is best known in the Western Hemisphere as a chocolate substitute (Chapter 6). The importance of extractive gums has risen recently in the United States partially because their availability and price are more reliable than those of exudate gums.

Locust bean gum is obtained by shaking the pods from the 3- to 7-m- (6- to 23-ft-) tall trees. The seeds are removed from the pods, and the seed coats and the hard, yellow-green embryo are abraded from the translucent endosperm. The endosperm, which constitutes about one-third of the seed, is dried and ground into powder. Locust bean gum is used almost exclusively in the food industry, primarily in ice creams (see Fig. 10.3), salad dressings, and pie fillings.

Guar gum is extracted from the seeds of *Cyamopsis tetragonolubus* (Fabaceae), a species known only as a cultivated plant. Guar gum was apparently domesticated in India, although all the wild species of the genus are native to Africa. Guar is an herbaceous perennial that can be used as cattle food. It was originally introduced into the United States in 1903 as a cover crop for grazing. During World War II, when locust gum was difficult to obtain, alternative sources of gum were sought. Domestic production of guar escalated, and by 1953 guar gum was replacing locust gum in many operations, particularly in the paper and textile industries.

Guar plants are 1 to 2 m (3 to 6 ft) tall with vertical stalks bearing clusters of pods. The way in which the pods are borne led to the common name cluster bean in Texas and Oklahoma, where almost all the guar in the United States is grown. Guar is the only commercially exploited source of gum that lends itself to mechanized agriculture. Its pods can be harvested with a grain harvester, a definite advantage over the laborious hand wounding and collection of true gums. Once collected, the seeds are separated from the pods and the seed coats are removed. The endosperm is then dried and ground to a fine powder.

Guar gum is used principally in the paper industry to improve paper quality, durability, and folding strength. It also serves as a waterproofing agent in explosives. A recently publicized use of guar gum involves its addition to water pumped through fire hoses. Water with traces of guar flows faster than does regular water because the gum reduces friction between the water and the sides of the hose. Guar gum is still less extensively employed than gum exudates, particularly in the food industry, but its use is steadily growing.

In contrast to guar gum and locust bean gum, larch gum is extracted from wood chips of the American larch, *Larix occidentalis* (Pinaceae). It has been known for almost 100 years that the wood of this species houses an abundant supply of a gumlike substance, but only in recent years has it been commercially exploited as a source of gum. In its behavior, larch gum is similar to gum arabic and has replaced it in some lithographic operations. Although approved for use in foods, it has not been used to an appreciable extent in edible products.

Semisynthetic Gums A recent and promising development in the gum industry is the production of semisynthetic cellulose gum (carboxymethylcellulose). This gum is made by taking purified cellulose, a polymer of glucose (see Fig. 10.9), and allowing it to react in a strong alkaline solution with a compound that adds acetyl (CH_3CH_2—) to the cellulose backbone. The resultant gum is used in more diverse ways than any other water-soluble polymer. The most prominent use for carboxymethylcellulose is in detergents, where it provides a whitening and brightening effect. In a washing machine, part of the dirt that has been flushed from clothes can be redeposited on them during the spin-dry cycle. Cellulose gum binds with the dirt, helps prevent redeposition, and thus reduces graying of the fabric.

Cellulose gum is also important in the paper industry and is the major replacement for starch in the textile industry. Small quantities are indispensable in latex paints because they give the paint the proper viscosity to flow evenly from brushes and rollers. Modern processed foods often contain cellulose gum rather than or in combination with one of the other gums to texturize, stabilize, thicken, and improve their sensory qualities (see Fig. 10.3).

Pectins

Pectin is the name given to a special group of plant polysaccharides that form gels under particular conditions. Pectins are composed of unbranched chains of 200–1000 saccharide molecules (see Fig. 10.2) or their methyl esters. The number of methoxyl groups is important in determining the properties, and thus the uses, of different pectins. Pectins are found between cells and as components of the primary cell wall. The pectin found in cell walls is associated with calcium and forms molecules of extremely high molecular weight that are highly insoluble. During fruit senescence or in the presence of acid or alkali, the large molecules break down into smaller, water-soluble units. Pectin also often occurs in a layer of the epidermal cells between the cuticle and the inner portion of the cell wall. Some species have a particularly pronounced layer in the epidermal cells of their fruits. The peels of such fruits therefore constitute a convenient source of pectin. People cannot digest pectins, but the bacteria that live in human intestines can. Although these bacteria catabolize pectins and use the products as carbohydrate sources, their human hosts are still unable to receive any nutritional value from them.

Over 75 percent of the world's pectin goes into the manufacture of jams and jellies. Some fruits contain enough pectin to "gel" simply by being cooked in water, but others do not. Commercially prepared pectins are added to these fruits to produce a thick, sticky, or gelatinous consistency. For the same reason, a small amount of pectin is used in antidiarrheal medicines such as Kaopectate.

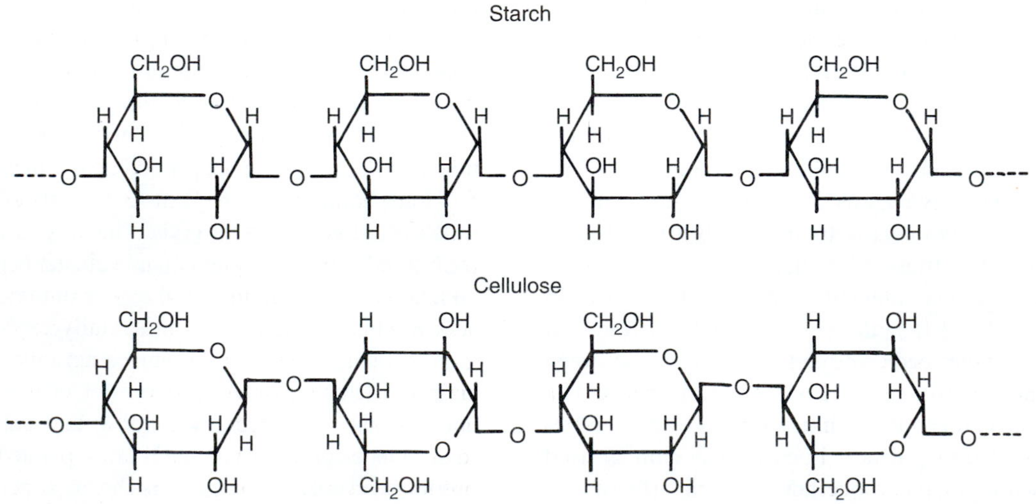

FIGURE 10.9

Cellulose is composed of glucose molecules joined by beta bonds. The exposed OH groups on either side of the cellulose chain facilitate cross-links between cellulose chains. Starch is also composed of glucose molecules, but the molecules are linked with alpha bonds. The exposed OH groups all lie on the same side of the starch polymer, causing the molecules to coil with the OH groups to the inside.

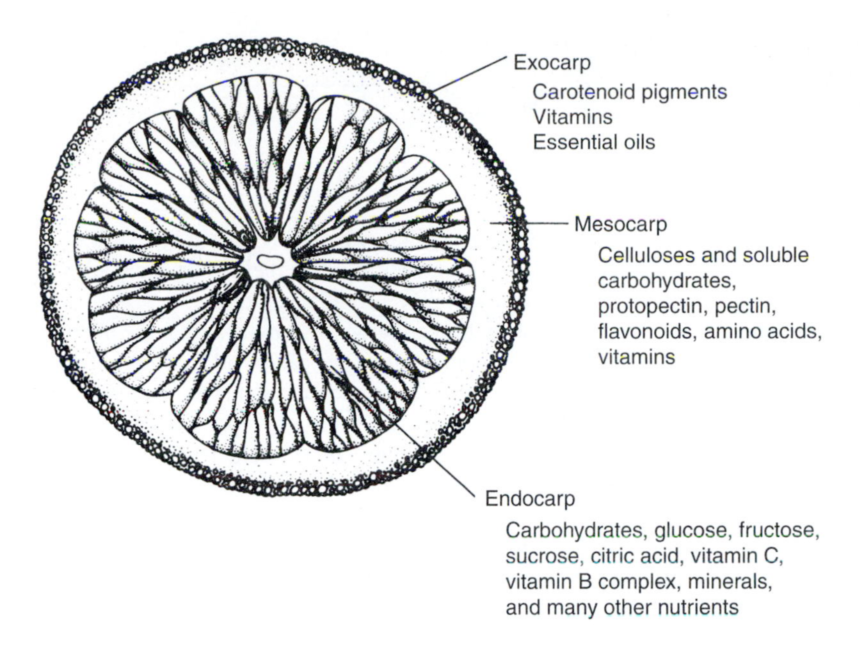

FIGURE 10.10
A cross section of a lemon shows the region of the peel richest in pectin. Because their skins are much thicker than those of other citrus fruits, lemons and grapefruits are most often used as pectin sources.

Since there are many potential sources of pectins, economics has determined which have become commercially important. The most widely used sources are apple pomace (the residue left after pressing apples) and citrus peels (Fig. 10.10). Apple residues contain 10 to 15 percent pectin on a dry-weight basis, and lemon, orange, lime, and grapefruit peels contain about 20 to 30 percent. Because these substances are by-products of the apple and citrus fruit juice industries, pectin extraction constitutes a profitable use of what would otherwise be waste.

Pectins are extracted from the pomace and peels by heating them in water between 60° and 95°C (140° and 203°F) at carefully controlled acidity levels (about pH 2.5). The high acidity breaks down the insoluble pectins and allows them to dissolve in the warm water. The pectin is separated from the water by centrifugation or filtration or both.

Starches

Starch (Box 10.1) constitutes the last category of commercially important plant-derived hydrogels. In contrast to gums and pectins, starches are linear or branched polymers of glucose (Fig. 10.11). Starch is easily broken down by enzymes known as amylases because the molecules of glucose in the starch polymer are linked in such a way that the enzyme can readily hydrolyze the bonds. Glucose molecules in starch are linked by alpha bonds; in cellulose they are linked by beta bonds (Fig. 10.9). Beta bonds, unlike alpha bonds, are exceedingly difficult to break.

Starch is found in almost all plant parts but is commercially extracted only from seeds, roots, and a few stems that store large quantities of it.

We have already discussed most of the major crop plants exploited for their large amounts of starch (Chapters 5 and 7). The most common sources of commercial starches are corn, wheat, sorghum (Chapter 5), arrowroot, cassava (Chapter 7), and sago, a gymnosperm that has large quantities of starch in its pith. Although our earlier emphasis was on the growing of these plants as carbohydrate food sources, people also employ many of them as hydrogels or substrates that are converted into sugar for fermentation (Chapter 14). For hydrogel use, starches are always extracted from the seeds, tubers, or roots in which they were stored. For conversion into sugar, starches are usually not purified.

Starch exists in plants in two forms: linear polymers called **amyloses** and branched forms known as **amylopectins** (Fig. 10.11). Amyloses usually contain about 150 glucose units, whereas amylopectins can contain up to 600,000 glucose units. Amylopectin, which is so branched that the molecules form flattened disks, generally constitutes about 70 percent of a starch granule, but high-amylose starches have been selected for in specific crops such as some varieties of corn. Within plant cells, the polymers are packed into insoluble granules that have such characteristic shapes and sizes that a particular starch source such as wheat, corn, rice, or taro can be identified simply by looking at the starch granules (Fig. 10.12).

Starch granules are typically extracted from plant material by flushing ruptured cells with cool water. The water is then evaporated, leaving a solid residue that is easily dried and powdered. It is in this form that people buy cornstarch, potato starch, and arrowroot.

Most of the starch produced in the world is consumed as food or food additives, but one-third is used industrially for nonfood purposes (Fig. 10.13). About 60 percent of this industrial starch is used for paper and cardboard. Starch has traditionally been employed in papermaking because it is inexpensive and readily available. However, it has a few serious drawbacks: it must be used in large quantities, and it is highly digestible by many organisms. Disposal of starch sizing wastes into waterways has upset the ecology of many aquatic systems. Starch does not kill organisms and is not toxic to humans. Instead, it leads to **eutrophication,** or nutrient enrichment, of the water. This unnatural abundance of nutrients causes population increases of algae. When the algae die, they provide abundant food for bacteria. As populations of bacteria balloon, they cause a depletion of the oxygen in the streams and ultimately kill most aquatic plants and animals.

Box 10.1

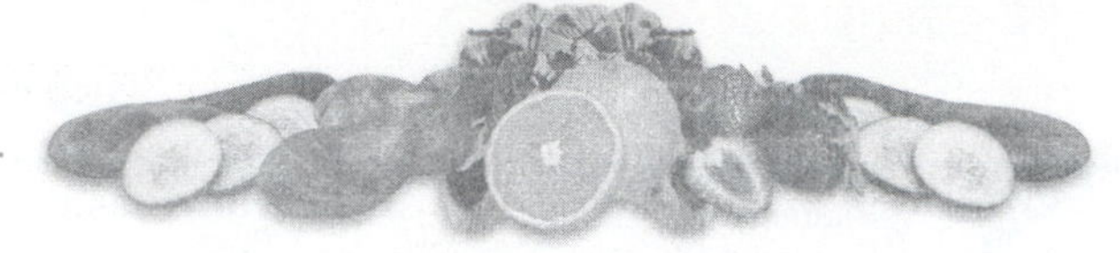

The Wonders of Starch

Like cellulose, starch is one of the most abundant compounds in nature. It occurs in all plant parts but can accumulate in large quantities in seeds, roots, and stems. The properties of starch that make it so useful in foods are its hydrogel qualities, its digestibility, and its ease of transformation into sugars. The hydrocolloidal properties of starch lie in the fact that it can form a gel with water. Starch granules suspended in cool water can swell and absorb up to 30 percent of their dry weight in moisture. Much of the starch in a granule is in crystalline form that is retained even with this degree of water absorption. If the temperature of the solution is raised to between 55° and 88°C (131° and 176°F) the granules swell to an irreversible degree of gelatinization, forming a starch-water complex. Instantaneous gelatinization, consisting of lumps of converted starch, is avoided by first dispersing starch granules in a cool liquid and then heating and stirring it. Alternatively, the granules can be mixed with fats (a procedure that also helps prevent clumping of the granules), and then liquid is later added. As a starch-thickened solution ages, or if it is frozen and thawed, it can separate into a watery and a thick, rubbery layer because of the precipitation of the linear amylose molecules that align with one another in increasingly larger complexes, release their hold on their water molecules, and finally separate from the colloidal solution.

Starch is easily broken down by amylases that are abundant in the human mouth. Amylases can be added to starch to break the molecules into sugar polymers of various lengths. Complete hydrolysis leads to glucose syrups. The treatment of starch with 3 percent hydrochloric acid leads to partial hydrolysis and the production of a polymer with a lower molecular weight that behaves differently from natural starch. These smaller polymers more readily gelatinize and produce more rigid gels that are preferable for some candies such as jelly beans.

Starch molecules can also be cross-linked by treating them with phosphoric acid. These larger molecules can withstand very high temperatures and thus can be boiled without gelatinizing. They are also less affected by acid than natural starch and thus are suitable for pastries that contain acidic fruit juices. If starch is treated with hypochlorite, it can oxidize; oxidation improves its binding properties to various substrates such as paper.

Starch (amylose)

Starch (amylopectin)

Coiled amylose

FIGURE 10.11
Starch molecules can be either linear or branched. Amylose (right) contains more than 1000 glucose molecules in unbranched chains that form coiled molecules. Amylopectin (left) is composed of 48 to 60 glucose molecules in branched chains that do not coil.

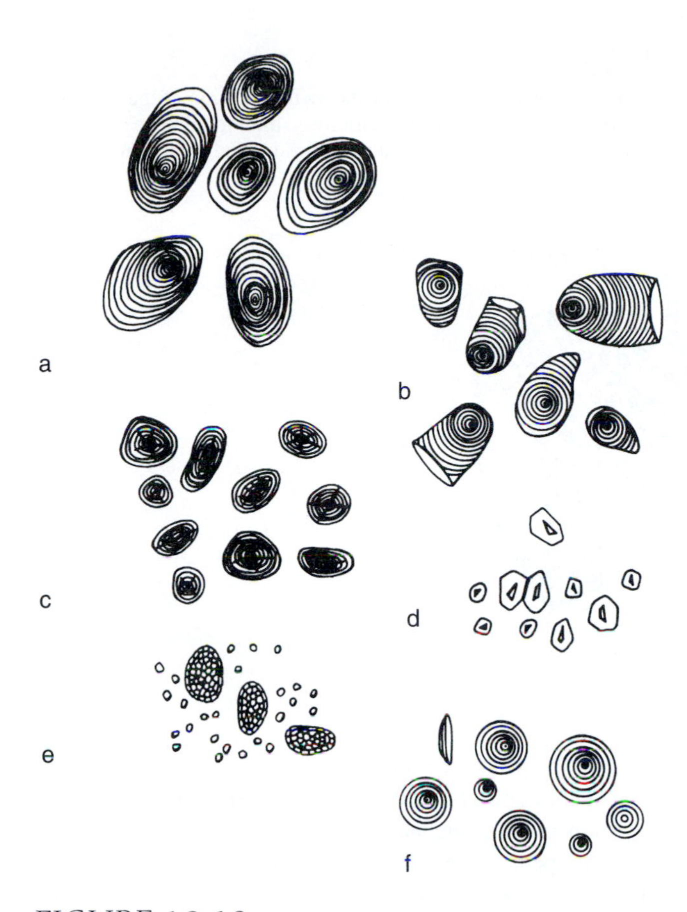

FIGURE 10.12
Starch granules of potato (a), sago (b), bean (c), corn (d), rice (e), and wheat (f).

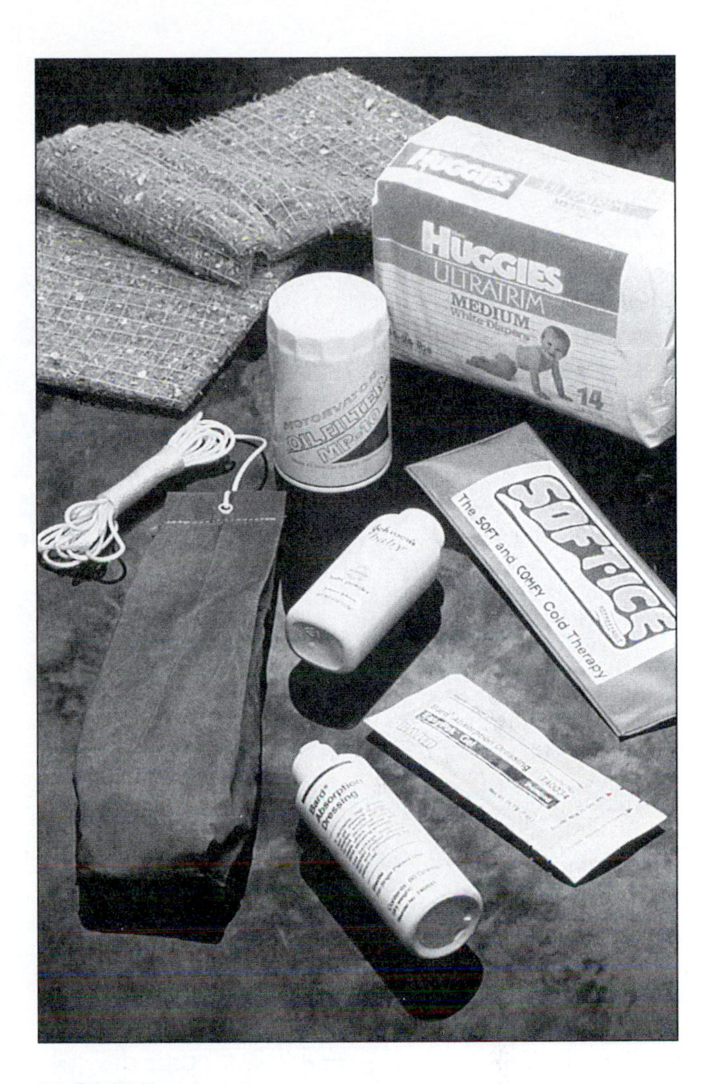

FIGURE 10.13
Starch has many household and industrial uses. One product made by linking cornstarch to a synthetic compound is called SuperSlurper because it can absorb several hundred times its own weight in water. Such products are now commonly used in diapers, wound dressings, and fuel filters and as conducting agents in batteries.

(Photo by S. Bauer, courtesy of Agricultural Research Service, USDA.)

Rubber-Producing Latex

Whereas gums come predominantly from Old World species, elastic latexes tend to come from the New World. Apparently, rubber was used in preindustrial times only in Central and South America. Strictly speaking, a **latex** is any mixture of organic compounds produced in laticifers. **Laticifers** are single cells (nonarticulated laticifers) or strings of cells (articulated laticifers) that form tubes, canals, or networks in various plant organs (Fig. 10.14). The walls between the cells of articulated laticifers can remain intact to varying degrees. In mature laticifers, the cellular organelles have usually disintegrated. Laticifers are not known in gymnosperms but occur sporadically throughout the angiosperms, primarily among dicotyledons.

Latex is an emulsion of a variety of compounds. The particular composition varies among species, but since long-chain hydrocarbons predominate in elastic latexes, they are all insoluble in water. Alkaloids, resins, phenolics, terpenes, proteins, and sugars can be mixed with these hydrocarbons. Not all latexes have elastic properties. Some, such as opium poppy latex (see Chapter 11 and 12) and papaya latex (Chapter 4), are essentially inelastic. The "latex" in latex paints is not latex at all. Latex paint consists of synthetic plastic particles dispersed in water with a binding agent. When the water evaporates, the particles fuse, producing a latexlike coating.

The functions of latexes in plants are not known. They may serve as antiherbivore agents, but many plant anatomists maintain that they are composed of by-products of primary chemical reactions in a plant that are secreted into the laticifers to keep them from interfering with normal cell functions. A recent suggestion is that laticifers serve as reservoirs that can supply surrounding tissues with compounds for further synthesis.

In this discussion we deal only with latexes that exhibit elastic properties. In common language, these latexes are lumped together as "rubber," a name given to hevea latex by Sir Joseph Priestley in 1770, when he discovered that it could be used for rubbing errors from a page. Today the term

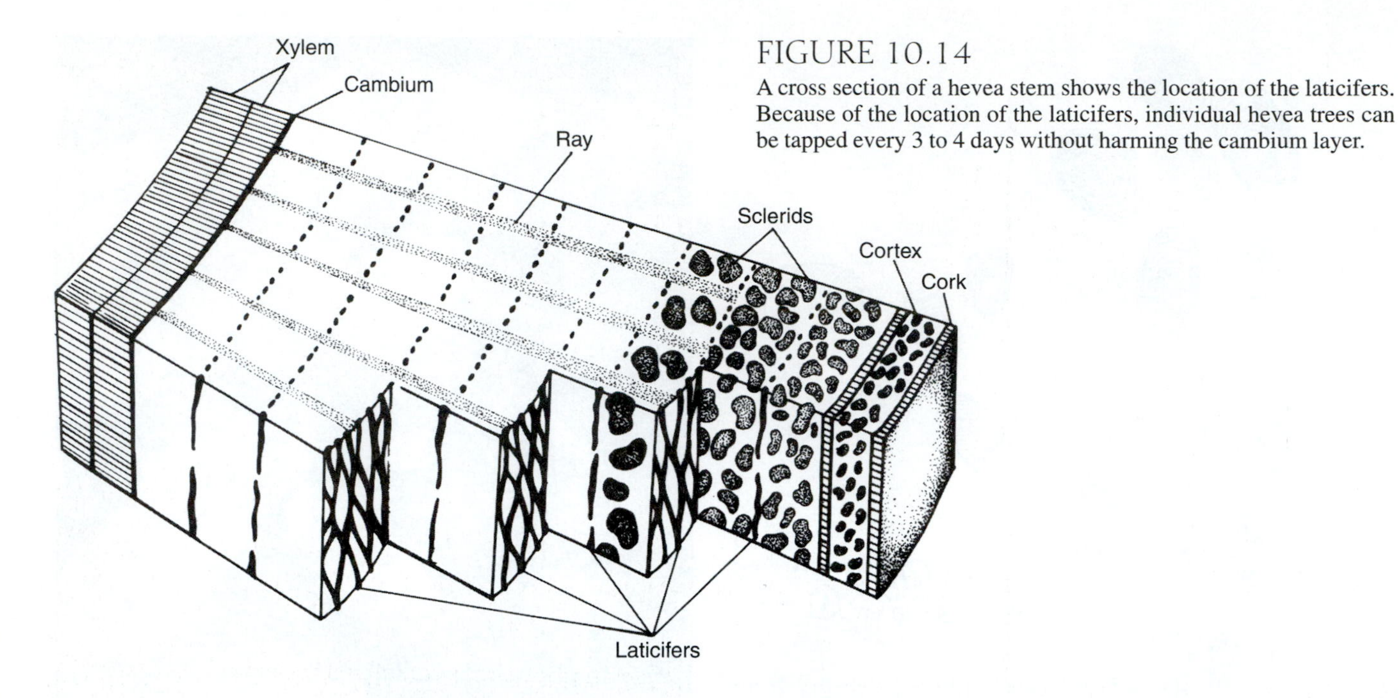

FIGURE 10.14
A cross section of a hevea stem shows the location of the laticifers. Because of the location of the laticifers, individual hevea trees can be tapped every 3 to 4 days without harming the cambium layer.

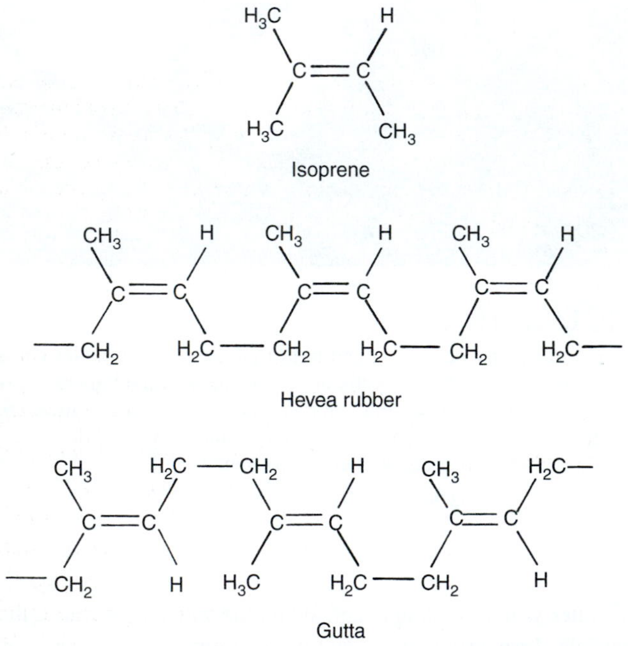

H_3C H
C=C
H_3C CH_3
Isoprene

CH_3 H CH_3 H CH_3 H
C=C C=C C=C
— CH_2 H_2C— CH_2 H_2C— CH_2 H_2C—
Hevea rubber

CH_3 H_2C — CH_2 H CH_3 H_2C—
C=C C=C C=C
— CH_2 H H_3C H_2C— CH_2 H
Gutta

FIGURE 10.15
The five-carbon isoprene molecule is the basic building block of natural rubber. Both hevea and guayule rubbers are composed of up to 6000 of these units. Less elastic natural rubbers such as gutta-percha contain fewer branched units. Most synthetic rubber is manufactured by polymerizing either butadiene or styrene molecules obtained from coal, petroleum, or alcohol.

rubber is restricted to terpenoid polymers (based on isoprene or similar units) with elastic properties (Fig. 10.15).

Rubber-containing latexes were independently discovered by native peoples in many parts of the world and were employed in various ways. In 1510 a chaplain from Queen Isabella and King Ferdinand's court wrote that the Aztecs played a vigorous game using balls made from the juice of an herb (Fig. 10.16). The balls were made from the latex of *Castilla elastica,* a member of the fig family (Moraceae). Because the latex from which the balls were made flowed from "veins" in the trees, the Aztecs considered it to represent blood. They consequently used the rubber not only for game balls but also for symbolic figurines. Objects of rubber were burned during religious ceremonies. Similarly, early explorers in the Amazonian rain forest reported that the Indians had the curious custom of dipping their feet in "sap" collected from a tree and then holding them in the smoke of a fire. The tree in this case was *Hevea brasiliensis* (Euphorbiaceae), the species from which almost all natural rubber is now obtained. By holding their feet in the smoke, the natives coagulated the rubber on the soles of their feet, producing a pair of perfectly fitted "proto" tennis shoes. Coagulated latex was something of a curiosity in Europe, but the Spanish in South America were quick to see that it had possibilities. They began to dip their hats and cloaks in latex and smoke them to render them waterproof.

In 1823 Charles Macintosh discovered that hevea rubber is soluble in hexane. This discovery led to new uses because coagulated latex could be shipped, redissolved, and then applied to specific substances under controlled industrial conditions. When the solvent was evaporated, a thin coating of rubber was left. The famous mackintosh, a sort of rainproof "slicker" named after Macintosh, was made by applying dissolved rubber onto a piece of fabric and covering it with another piece of fabric. Nevertheless, coatings made in

FIGURE 10.16
A Mayan ballplayer was carved on the walls of the ball court at Chichen Itza in the Yucatán state in Mexico. Similar games were later played by the Aztecs, who considered the rubber balls used in the game so valuable that they were used to pay taxes to Aztec chiefs.

FIGURE 10.17
Charles Goodyear had been experimenting with rubber for several years when, in 1839, he accidentally came upon the most important development in rubber history. Goodyear spilled a mixture of rubber and sulfur on a hot stove and noted that when cooled, the rubber lost its stickiness and retained its elasticity. The process was named vulcanization after Vulcan, the Roman god of fire.
(Photo courtesy of the Goodyear Tire Company.)

this way had numerous problems. In very cold weather they cracked, and in hot weather they became sticky. These problems were overcome when Charles Goodyear discovered **vulcanization** in 1839 (Fig. 10.17). Vulcanization is the addition of sulfur with lead oxide to rubber, which leads to cross-linking of the molecules of the isoprene chains. This relatively simple change makes the latex impervious to weather and improves its elasticity. It has recently been discovered that vulcanization can be accomplished without sulfur by irradiating the latex with cobalt-60.

Until the 1880s all hevea rubber was extracted from wild trees. The latex was collected by natives who were each assigned individual areas within the forest. The hevea trees within this area were tapped by diagonally slashing the bark (Fig. 10.18). Cups placed at the lower ends of the slashes were periodically collected, and the day's load of latex was brought to a smoking house. In the smoking house it was filtered, mixed with water, and poured over a paddle placed above a smoky fire. When latex was continuously poured over the paddle, a ball of coagulated latex was eventually produced. The rubber was, and still is sometimes, shipped in this form. Today dilute acetic acid or formic acid rather than smoke is often used to coagulate the rubber, which is formed into sheets instead of balls. At a later time sheets or balls are shredded and dissolved in an organic solvent. Complete purification of the latex is rare.

In 1876 seeds from a highly productive population of hevea known to produce large quantities of latex were taken to Kew Gardens in London and from there to Southeast Asia, where they were used to establish plantations. Attempts to grow hevea in plantations in the Amazon basin had failed, largely because of the South American leaf blight that invariably destroyed plantations once they were established. The seeds taken to Southeast Asia were free of the fungus, and by the turn of the century rubber production in the Old World began to rival that in South America. By World War II 90 percent of the world's natural rubber was collected from Asian plantations. Today rubber is primarily vegetatively propagated by grafting buds of superior trees onto stocks grown from seed in nurseries.

During World War II fighting operations effectively cut off the supply of rubber to the United States, threatening disaster (Fig. 10.19). The United States began serious experimentation with other sources of rubber and simultaneously discovered a process for producing a synthetic substitute. The first synthetic developed was styrene butadiene rubber. This synthetic rubber still predominates, but several others have been developed, including polybutadiene and polyisoprene, which is chemically similar to the major

FIGURE 10.18
A tapped rubber tree. In the method shown here, sloping horizontal cuts are made into the bark without injuring the cambium. These cuts drain the flowing latex to a central vertical cut with a spigot, on which a cup is hung, at the bottom. Once an area has been utilized for several years, the tapper moves around the tree as indicated by the scars of old cuts near the cup.
(Photo courtesy C. Todzia.)

FIGURE 10.19
On October 24, 1927, Harvey Firestone (left) received this message: "Suppose war was declared; embargo shuts off rubber—in one year our rubber is exhausted—but still guayule is obtainable. This being the case, can inner tubes and shoes be made of guayule so that a fair mileage would be possible? Yours very truly, Thomas Edison" (right). The two men are shown here examining the results of their research, a tire made from guayule grown in Edison's Florida gardens. Guayule has since been shown to be chemically identical to hevea rubber.
(Correspondence and photo courtesy of Firestone Tire and Rubber Company.)

component of natural rubber (see Fig. 10.15). By the end of World War II synthetic rubber accounted for 75 percent of the market.

All the synthetic rubbers are compounds made from the polymerization of dienes. The procedures used for polymerization differ among the types of synthetic rubber, but all the units, or monomers, used for commercial synthesis come from petroleum. In view of increases in petroleum costs and the impending depletion of petroleum sources, the synthetic rubber industry will have to find new sources of basic compounds to use for synthesis. Suggested sources include shale oil, old tires, and plant-derived carbohydrates. As in all business operations, the ultimate consideration will be the profit margin.

After 1970, the demand for natural rubber grew as the production of radial tires increased. Over two-thirds of all rubber, either synthetic or natural, goes into the manufacture of tires. Radial tires provide better handling and run cooler and more safely in foul weather than conventional bias-ply tires, but they require more resilience than is provided by synthetic rubber. Consequently, manufacturers incorporate about 30 percent natural rubber into these tires, so there is a relatively stable market for the natural latexes. The need for natural rubber has led to searches for alternatives to hevea rubber and to a potential additional source, guayule, *Parthenium argentatum* (Asteraceae) (Fig. 10.20).

Guayule is a shrub native to the Chihuahuan Desert of southwestern Texas and adjacent Mexico. The species was not botanically described until 1859 and was not used as a source of latex by native peoples, presumably because the latex occurs in individual, thin-walled single cells dispersed throughout the cortical and ray tissues of the stems and roots of the small, brittle bushes (Fig. 10.21). The stems also contain resin ducts, but the resin has never been considered a primary recovery product.

Guayule was among the plants studied during World War II as a source of rubber. Many wild plants were harvested and used as auxiliary rubber sources. There were high hopes for guayule because the rubber is virtually identical to hevea rubber and because the plants from which it can be extracted grow naturally in semiarid regions. Furthermore, shrubs can produce up to 20 percent of their dry weight in rubber. With breeding, this amount probably

FIGURE 10.20
Flowering heads of guayule.
(Photo by J. Dykinga, courtesy of Agricultural Research Service, USDA.)

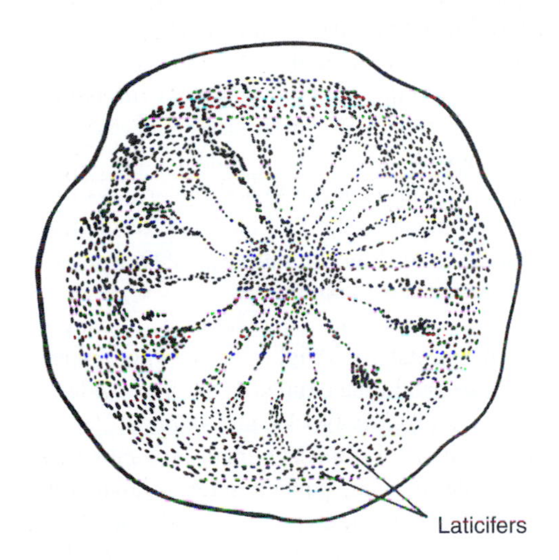

FIGURE 10.21
A cross section of a guayule stem showing the distribution of laticifers. Because the latex is produced in cells throughout the cortex of the roots and stems, entire plants are harvested for rubber extraction.

could be increased substantially. The most important drawback of guayule is that plants cannot be profitably harvested until they are about 7 years old. Mexico already has a functional guayule rubber production plant.

The shrubs are collected either by cutting them at the base or, more often, by uprooting them. They are dipped in hot water to coagulate the rubber and help in the removal of leaves. The woody tissues are then pulped in water, and the resins and latex are skimmed from the bagasse. The resins are separated from the latex, and the latex is purified.

Several other kinds of rubber were extracted from New World species before 1900. Among them were other species of *Hevea,* Ceara rubber from *Manihot glaziovii* (Euphorbiaceae),

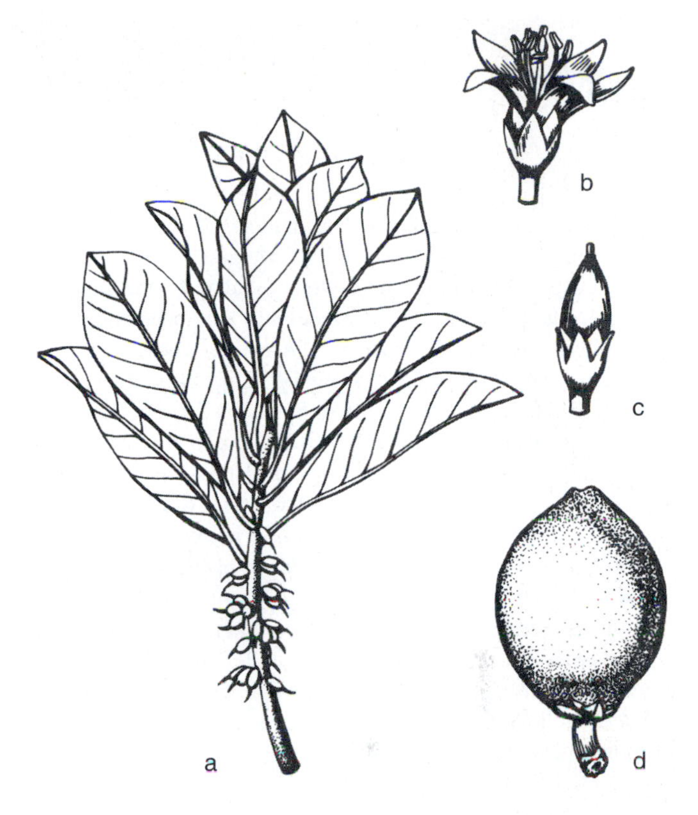

FIGURE 10.22
(a) A branch, (b) flower, and (c, d) fruit of *Palaquium gutta,* the source of gutta-percha rubber. This rubber disintegrates in the air but is a good insulator and is resistant to seawater. Gutta-percha was used in the past for the rubber strands wound into a ball in the center of golf balls and to coat undersea cables.

and Panama or caucho rubber from *Castilla elastica* (Moraceae). Gutta-percha, *Palaquium gutta* (Sapotaceae) (Fig. 10.22), was extracted from trees in both the Old and New Worlds. Indian fig or bambong rubber, *Ficus elastica* (Moraceae); African rubber, *Funtumia elastica* (Apocynaceae); and landolphia, *Landolphia gummifera* (Apocynaceae) were all extracted from species native to the Old World. The primary source of rubber, however, has always been *H. brasiliensis* because of the high percentage of rubber in its latex. An unusual elastic latex put to an entirely different use is chicle, harvested from *Manilkara zapota* (Sapotaceae) (Fig. 10.23). Chicle has historically provided the base for chewing gum (Box 10.2).

Resins

Natural resins have played a varied role in the history of many cultures. Early paints, incense for religious services, and caulking for ships all came from resins. Now almost all resins are synthetically produced, but the demand for resins provided the impetus for the development of synthetic replacements.

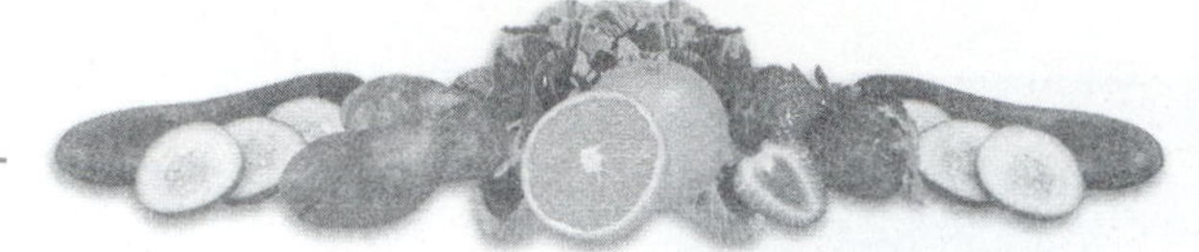

Box 10.2

Santa Anna and the Origin of Chewing Gum

Chiclets are one of the few reminders of the relationship between chewing gum and one of Mexico's most notorious politicians. Chicle, the milky-white exudate of the sapodilla tree, *Manikara zapota* (Sapotaceae), has long been chewed by the Mayans of Mexico and Central America, where the species is native. However, modern chewing gum can be traced back to an ill-fated visit to the United States by Antonio López de Santa Anna. Santa Anna (1794–1876) was an eccentric political leader from Veracruz who served 11 times as president of Mexico. Stories about his life abound, and few agree in detail. In the United States Santa Anna is best known as the general who led the Mexican troops in the battle of the Alamo (1836), during which Davy Crockett and Jim Bowie were killed along with the other defenders of the fort. After Santa Anna's last presidency he spent some time in exile in the West Indies, where he was visited by William Seward, the U.S. Secretary of State. Believing that he had secured Seward's trust, Santa Anna allowed himself to be duped by a group of scoundrels who convinced him to go to New York to gain support from important Americans that would allow him to wrest control of Mexico from the French. He sailed for New York in 1866. Unfortunately, his new friends were unscrupulous and stole his money while pretending to carry out negotiations.

Santa Anna, like many Mexicans, had chewed chicle when he was young, but he also recognized that it was similar to other elastic substances obtained by wounding trees. Several attempts had been made to produce a usable rubber from the latex, but all had been unsuccessful. Santa Anna brought a little chicle with him, believing that he might find an inventor who could make rubber from the latex and thus finance his return to power. During his visit he employed a secretary, Rudolph Napegy, who became acquainted with Thomas Adams, an innovative if not very successful inventor. Napegy suggested to Santa Anna that Adams was the man who could produce rubber from chicle.

Adams was willing to experiment with the latex, and so Santa Anna arranged to have 2 tons of chicle shipped to New York for experimentation. In March 1867 Santa Anna left for Mexico, never realizing that during his few months in New York he had inadvertently set the stage for the discovery of a product quite different from industrial rubber. Over the subsequent few months Adams tried unsuccessfully to vulcanize the substance. He also attempted to combine it in various ways with true rubber, but these attempts did not yield a usable elastic product. Adams appeared to be stuck with hundreds of pounds of useless chicle. According to his son's account, Adams entered a drugstore at this time to make a purchase and overheard a child asking for chewing gum. Gum chewing was fairly common, but most gum was made of sweetened paraffin, which was not very chewy. Remembering a comment by Santa Anna about chewing chicle, Adams went home and with his sons heated the latex to make it soft, added sugar, and formed it into small balls that they offered to druggists for sale. The new treat was an instant success. After all, this was the heyday of chewing tobacco for adult men. Shortly afterward Adams began to add flavorings to his confection and soon patented the first machine for the manufacture of chewing gum. Later additions included corn syrup, flavorings, and hard sugar coatings.

Chicle was originally gathered by tree-climbing Mexicans called chicleros (Fig. 10.23), who tapped wild sapodilla trees to obtain the exuded latex. The crude exudate was collected, molded into blocks, and shipped. Before heating and mixing, it was cleaned. A small amount of chewing gum is still made by mixing chicle latex in giant vats with sugar and flavoring, although synthetic vinyl resins as well as sugar substitutes are now used to make almost all regular chewing gum and specialty gums such as bubble gum.

In plants, resins are actively synthesized and secreted into specialized canals or ducts. The ducts are either simple intercellular spaces or canals formed by the disintegration of a series of adjacent cells. Like latexes, resins seem to have an antiherbivore function in the plant. They are known to deter herbivory by many insects, and some have been shown to have antibacterial properties. Resin canals and production occur in the xylem, phloem, and bark of many gymnosperms and in some dicotyledons. Chemically, natural resins are a rather heterogeneous group of compounds, but all are polymerized terpenes that generally are mixed with volatile oils. Resins are also insoluble in water. The largest number of natural resins used by people come from species native to the Old World, but numerous New World species have historically been significant resin sources as well.

One of the most ancient uses of resin was as incense. It is not known when the practice of burning substances that produce a sweet-smelling smoke began, but the earliest writings attest to their common use in religious services. Unlike flowers or even most aromatic herbs, incense does not simply smell like burned leaves when ignited. Instead, it gradually releases volatile oils that diffuse outward in the smoke. Two classic resins (or gum-resins since they have gums mixed with them) used in incense are frankincense, *Boswellia sacra* (Burseraceae), and myrrh, *Commiphora abyssinica* (Burseraceae). The value placed on these resins in early Egyptian, Greek, and Roman times is underscored by the biblical story of the wise men who presented these two resins, along with gold, to the infant Jesus. Both frankincense and myrrh are natives of Ethiopia and were brought

FIGURE 10.23

A chiclero tapping a chicle tree.
(Courtesy of Wm. Wrigley, Jr. Co.)

FIGURE 10.24

Queen Hatshepsut sponsored what scholars consider to be the world's first plant-collecting expedition in around 1400 B.C. She sent her expedition to the land of Punt, a spur on the eastern coast of Africa that projects into the Indian Ocean. Besides ivory, gold, ebony, and cinnamon, the party returned with monkeys and baboons for the queen's zoo and 31 frankincense trees that were planted at the temple of Karnak on the banks of the Upper Nile in Egypt. Carvings, such as the one reproduced here, on the walls of one temple show how frankincense shrubs were loaded onto a ship and transported in wicker baskets.

to the Mediterranean region from the Ethiopian coast (Fig. 10.24). The resins were collected, much like gums, by scoring the shrubs that produced them and collecting the exudate. Neither is an important commodity today, and their use is restricted to some church services.

Resins from these and other sources were also used in embalming (Fig. 10.25). Records from early Egypt indicate that myrrh and frankincense were used in the body cavity or for anointing the head of the deceased. Once coffins began to be used in the third Egyptian dynasty (2770–2670 B.C.), resins, probably obtained from pines and junipers, were used to varnish them.

Other resins that have been extensively used include mastic and lacquer. In countries bordering the Mediterranean, mastic, *Pistacia lentiscus* (Anacardiaceae), has been employed for centuries as a sealing material, a masticant to sweeten the breath, and, more recently, an adhesive for dental caps. Lacquer, *Rhus verniciflua* (Anacardiaceae), is collected by tapping the trunks of trees that are native to China and Japan. The liquid resin is filtered and kept in a dark, tightly closed container.

The use of lacquer was developed into an art form in China before the Christian era, but the Japanese perfected its use during the Ming Dynasty (A.D. 1368–1644). The production of a good piece of lacquered work took months. The wooden surfaces were smoothed, coated with a thin layer of lacquer, dried, rubbed with charcoal powder, and polished. Up to 300 layers of lacquer were applied and smoothed in

FIGURE 10.25
The process of embalming as depicted in Ridpath's *Cyclopedia of Universal History* (1890).

this way. Designs of gold leaf or rice paper were placed on the surface and coated with numerous additional layers of lacquer. The products have exquisite designs embedded in glasslike surfaces.

Two other groups of resinous substances of limited economic but major aesthetic importance in people's lives are copals and dammars used today primarily in artists' paints. Copal usually refers to recent or fossilized resins, most of which come from species of *Copaifera* or the related genus *Hymenaea* (Fabaceae). Fresh material is obtained by slashing the resin-bearing trees and collecting the resin once masses of it have accumulated over the cut. Copal is also collected from gymnosperm species of the genus *Agathis* (Araucariaceae). Despite the disparity of the sources, the resins of all the species are soluble in nonpolar solvents and, when dry, are lustrous and transparent. Dammars likewise turn shiny and transparent when dry. They are obtained in the same way as copals but come from members of the Dipterocarpaceae, primarily species of *Shorea*.

Because of their importance in caulking and rendering wood and ropes resistant to seawater, pine resin products acquired the name **naval stores** (Fig. 10.26). The principal products of the naval stores industry are pitch, turpentine, and rosin. The pines most commonly used as sources of the resins used in this industry are *Pinus pinaster* and *P. sylvestris* in Europe and *P. palustris* and *P. elliottii* (all Pinaceae) in the southeastern United States. Pine pitch was well known in Egypt, Greece, and Italy centuries before the birth of Christ as an effective sealant and waterproofing substance. It was smeared on the inside of Grecian clay wine urns to prevent

FIGURE 10.26
Pine pitch, obtained by tapping several species of pines, is used to produce rosin and turpentine. This worker is extracting resin from a pine tree at an Agroforestry cooperative in Hondouras.
(Photo by G. Bizzari, FAO.)

FIGURE 10.27
Amber often has insects or other items trapped inside. A crane fly trapped in Baltic amber.
(Photo from the collection of the Field Museum of Natural History, Chicago.)

leakage and imparted a resinous or piney taste, for which the Greeks developed a fondness. Pine flavoring is now added to reproduce this taste in Greek retsina wine.

In the production of turpentine and rosin, trees are tapped and the resin is collected and allowed to stand. Rosin precipitates from the crude resin, and the liquid is distilled to produce turpentine. Rosin is a brittle, friable substance when dry but becomes sticky when heated. The rosin bag of baseball players contains powdered rosin that becomes sticky when the pitcher's hand warms it. The slight stickiness helps the ballplayer grip the ball and thus improves the accuracy of the pitch. The bows of stringed instruments are drawn across blocks of rosin to increase friction between the bow and the strings; this treatment intensifies the tone of the music produced. Rosin also finds its way into printer's ink, paper coatings, varnishes, and sealants.

Turpentine is important as a solvent and cleaning agent for oil-based paints. It is used today mostly as a source of organic compounds for further synthesis. Everyday items such as deodorants, shaving lotions, and some medicines contain turpentine or chemicals derived from it. Limonene is a commonly used lemon flavoring made from turpentine precursors.

Natural resins played a role in the development of linoleum in 1860. During his experiments with linseed, the English scientist Frederick Walton discovered that a mixture of linseed oil, resin, and cork particles could be rolled onto fabric to produce a solid, easy-to-clean, and durable surface. The resin initially preferred was kauri resin obtained from a New Zealand gymnosperm, *Agathis australis* (Araucariaceae). Modern linoleum-like floor coverings primarily contain synthetic resins.

Finally, we mention the only jewel of plant origin, amber. Amber was gathered by peoples of both the Old World and the New World and was used for ornamentation or, in a few cases, for arrow points. It is lightweight, brilliant, and easily carved or drilled and can range in color from yellow to dark brown. Some ambers are opaque; others are transparent (Fig. 10.27). Amber is now known to be fossilized terpenoid resin ranging in age from 1.5 to 300 million years old. It was originally thought that amber came only from gymnosperm resins, but the work of Jean Langenheim showed that other trees, especially species of legumes that now provide copal, are important sources of amber in the New World.

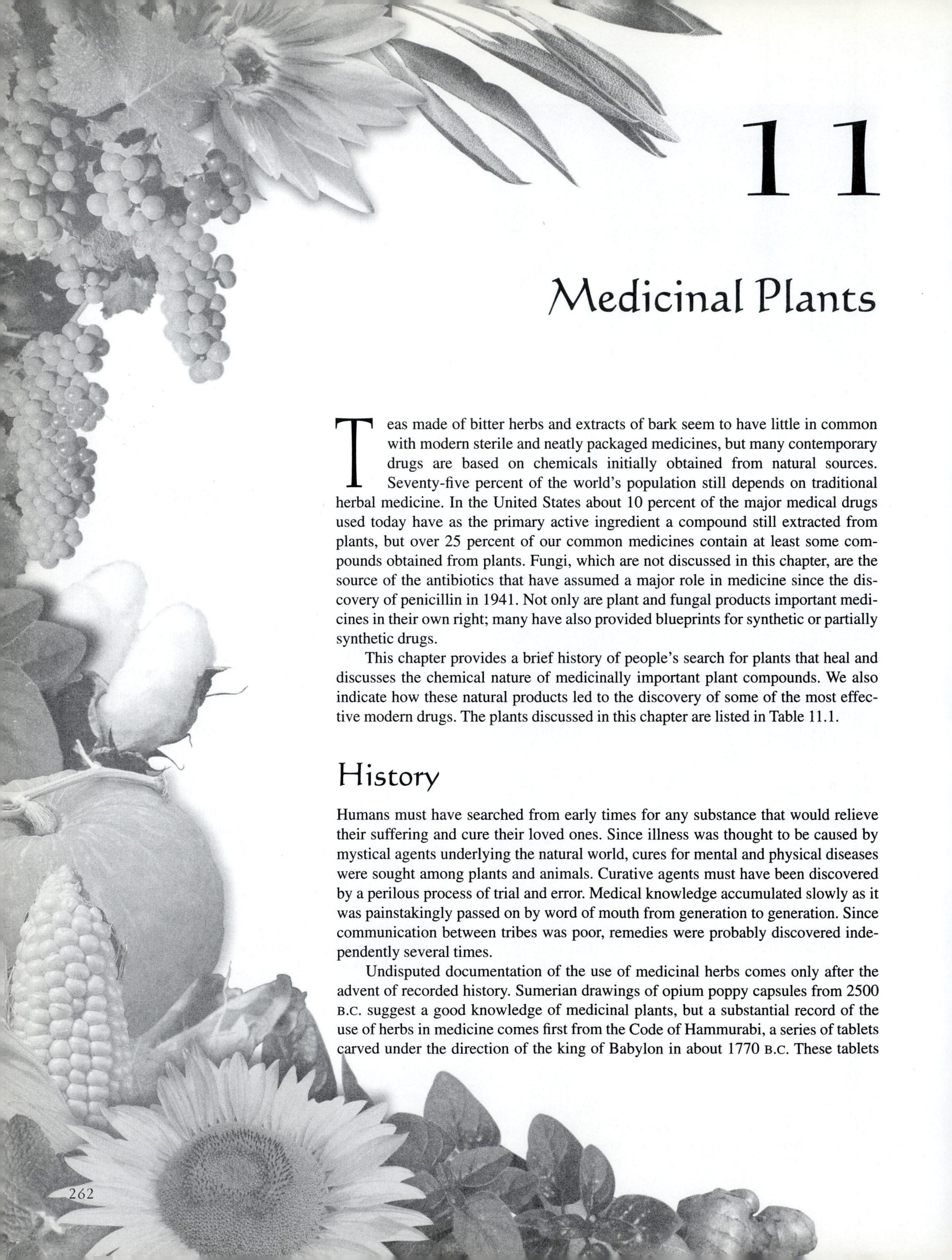

11

Medicinal Plants

Teas made of bitter herbs and extracts of bark seem to have little in common with modern sterile and neatly packaged medicines, but many contemporary drugs are based on chemicals initially obtained from natural sources. Seventy-five percent of the world's population still depends on traditional herbal medicine. In the United States about 10 percent of the major medical drugs used today have as the primary active ingredient a compound still extracted from plants, but over 25 percent of our common medicines contain at least some compounds obtained from plants. Fungi, which are not discussed in this chapter, are the source of the antibiotics that have assumed a major role in medicine since the discovery of penicillin in 1941. Not only are plant and fungal products important medicines in their own right; many have also provided blueprints for synthetic or partially synthetic drugs.

This chapter provides a brief history of people's search for plants that heal and discusses the chemical nature of medicinally important plant compounds. We also indicate how these natural products led to the discovery of some of the most effective modern drugs. The plants discussed in this chapter are listed in Table 11.1.

History

Humans must have searched from early times for any substance that would relieve their suffering and cure their loved ones. Since illness was thought to be caused by mystical agents underlying the natural world, cures for mental and physical diseases were sought among plants and animals. Curative agents must have been discovered by a perilous process of trial and error. Medical knowledge accumulated slowly as it was painstakingly passed on by word of mouth from generation to generation. Since communication between tribes was poor, remedies were probably discovered independently several times.

Undisputed documentation of the use of medicinal herbs comes only after the advent of recorded history. Sumerian drawings of opium poppy capsules from 2500 B.C. suggest a good knowledge of medicinal plants, but a substantial record of the use of herbs in medicine comes first from the Code of Hammurabi, a series of tablets carved under the direction of the king of Babylon in about 1770 B.C. These tablets

TABLE 11.1 Plants and Plant Products Discussed in Chapter 11

COMMON NAME	SCIENTIFIC NAME	FAMILY	MAJOR ACTIVE PRINCIPLES	NATIVE REGION
Agave	*Agave sisalana*	Agavaceae	Hecogenin	Mexico
Aloe	*Aloe barbadensis*	Liliaceae	Mucilaginous gel, barbaloin	Africa
Belladonna	*Atropa belladonna*	Solanaceae	Hyoscyamine	Europe
Chaulmoogra	*Hydnocarpus* spp.	Flacourtiaceae	Chaulmoogric acid, hydnocarpic acid	Southern Asia
Cinchona	*Cinchona officinalis* *C. succiruba* *C. ledgeriana*	Rubiaceae	Quinine, quinidine	South America
Coca	*Erythroxylum coca* *E. truxillense*	Erythroxylaceae	Cocaine	South America
Crocus, autumn	*Colchicum autumnale*	Liliaceae	Colchicine	Europe, Northern Africa
Duboisia	*Duboisia myoporoides* *D. leichtardtii*	Solanaceae	Hyoscyamine, hyoscine	Australia
Foxglove	*Digitalis purpurea* *D. lanatus*	Scrophulariaceae	Digoxin, lanatoside C	Europe
Hellebore	*Veratrum viride*	Liliaceae	Steroidal alkaloids	North America
Henbane	*Hyoscyamus niger* *H. muticus*	Solanaceae	Hyoscyamine, hyoscine	Eurasia
Ipecac	*Cephaelis ipecacuanha*	Rubiaceae	Emetine	South America
Ma-huang	*Ephedra sinica*	Ephedraceae	Ephedrine	China
Mayapple	*Podophyllm peltatum*	Berberidaceae	Podophyllin	North America
Opium poppy	*Papaver somniferum*	Papaveraceae	Codeine, morphine, noscopine, papaverine	Eurasia
Papaya	*Carica papaya*	Caricaceae	Papain	Tropical Americas
Periwinkle	*Catharanthus roseus*	Apocynaceae	Vincristine, vinblastine	Madagascar, Europe
Plantain	*Plantago ovata* *P. psyllium*	Plantaginaceae	Mucilage	Europe, Asia, Africa
Snakeroot	*Rauwolfia serpentina,* *R. tetraphylla*	Apocynaceae	Reserpine	Tropical Asia, North and South America
Willow, white	*Salix alba*	Salicaceae	Salicin	Europe
Yam	*Dioscorea floribunda* *D. composita*	Dioscoreaceae	Diosgenin, hecogenin	Central America
Yew, Pacific	*Taxus brevifolia*	Taxaceae	Taxol	North America

mention plants, such as henbane, licorice, and mint, that are still used in medicines. Later, the Egyptians recorded their knowledge of illnesses and cures on temple walls and in the Ebers papyrus (1550 B.C.), which contains over 700 medicinal formulas. Many of these recipes contain plant substances from species now known to have therapeutic value. These include *Cannabis,* aloe, castor, mandrake (Fig. 11.1), and numerous gum- and resin-producing shrubs.

The golden age of Greece was a time of great advancement in medicinal and biological knowledge. Several prominent figures of this era stand out for their contribution to **pharmacology,** the study of drugs (Fig. 11.2). Hippocrates (approximately 460–377 B.C.) earned his reputation as the

FIGURE 11.1

An Egyptian queen holding a mandrake fruit in her right hand and a water lily in her left. Mandrake preceded ether as an anesthetic by 2000 years. Because atropine and scopolamine are present throughout the plant, mandrake extracts can kill pain and produce a dreamlike sleep.

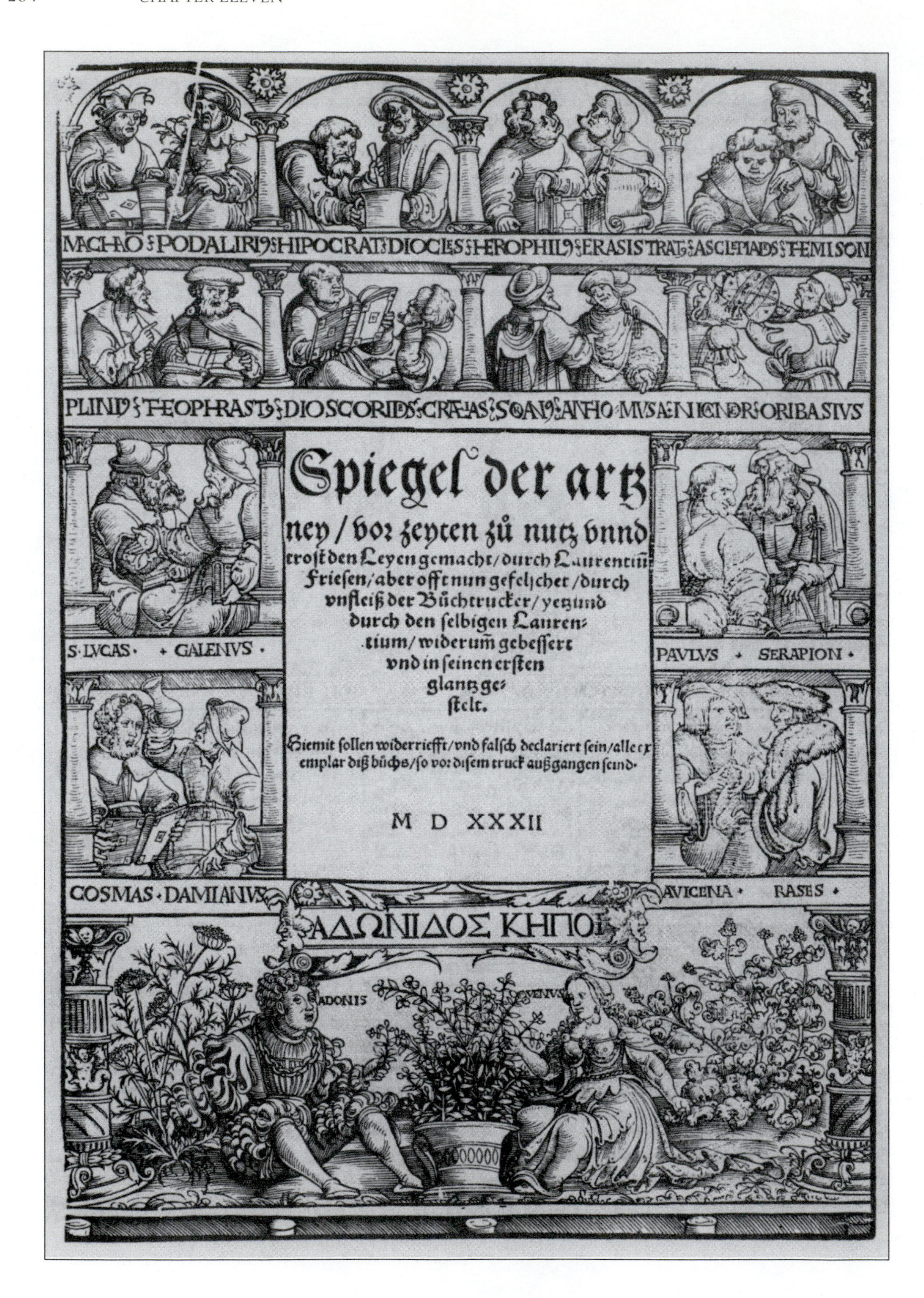

Spiegel der artz
ney / vor zeyten zů nutz vnnd
trost den Leyen gemacht / durch Laurentiū
Friesen / aber offt nun gefelschet / durch
vnfleiß der Bůchtrucker / yetzund
durch den selbigen Lauren-
tium / widerum gebessert
vnd in seinen ersten
glantz ge-
stelt.

Hiemit sollen widerriefft / vnd falsch declariert sein / alle ex
emplar diß bůchs / so vor disem truck außgangen seind.

M D XXXII

FIGURE 11.2

When Hans Weiditz designed his frontspiece for a medical handbook (*Spiegel der Artzney,* Strassbourg, 1532), he included 24 portraits of the most significant contributors to medical knowledge up to that time. Although the drawings do not necessarily provide true likenesses of these famous men, they provide and outline of medical history. For this reason and because their names are abbreviated in the drawings, we provide an annotated translation, beginning at the top left:

Machaon and Podalarius, mentioned by Homer in the *Iliad* as "healers both and skillful."

Hippocrates (460–377 B.C.), the father of medicine, for whom the Hippocratic oath, the traditional code of physicians, was named.

Diocles, considered Greece's second best physician.

Herophilus (300 B.C.) introduced dissection as part of the study of anatomy.

Erasistratus (310–250 B.C.), the founder of the science of physiology.

Asclepiades (ca. 124 B.C.), a Greek physician who moved to Rome, where he espoused that medical treatment be given *cito, tuto, et jucunde* (promptly, safely, and pleasantly).

Themison (123–43 B.C.), a pupil of Asclepiades, who founded the "Methodist" school of medicine in Rome.

Pliny (A.D. 43–79), author of *Natural History,* a well-known compendium of ancient scientific knowledge.

Theophrastus (372–287 B.C.), considered Aristotle's finest pupil and the founder of botany with his *Enquiry into Plants.*

Dioscorides (ca. A.D. 40–90) compiled known medical knowledge in *De Materia medica.*

Cratuas (120–63 B.C.) introduced the concept of polyvalent drugs and serums.

Soranus (A.D. 98–138) produced the text *Diseases of Women,* which laid the foundation for obstetrics and gynecology.

Nicander (185–135 B.C.) wrote a book about poisons called *Theriaca.*

St. Luke, the biblical apostle who was also a physician.

Galen (A.D. 138–201) established the science of experimental physiology.

Paul of Aegina (A.D. 625–690), author of a comprehensive work on surgery.

St. Cosmas and St. Damian practiced medicine and were killed in the Christian persecutions of A.D. 303. As a result, they became the patron saints of physicians and pharmacists, respectively.

Avicenna (A.D. 980–1037), an acclaimed Arabian physician who produced a classic medical text that was used in medical schools into the seventeenth century.

Rhazes, a Persian physician (A.D. 865–925) who studied smallpox and is credited with introducing animal gut as suture material.

Last, but not least, Adonis and Venus represent regeneration and fertility. As they sit clutching potted herbs beside them, we are reminded of the close ties between botany, medicine, and mythology until more scientific ways of curing disease were adopted.

(Reproduced from *Medicine and the Artist* [*Ars Medica*] by permission of the Philadelphia Museum of Art, Carl Zigrosser, Dover Publications, Inc., 1970.)

father of medicine by being the first chronicler to discuss illnesses and their treatment in a rational way. Unlike most of his predecessors, Hippocrates did not believe that sickness is caused by evil spirits, but rather by bodily problems. He therefore prescribed sound nutrition, purgatives, and, in certain cases, botanical drugs. The number of effective medicinal plants he discussed came to between 300 and 400 species. The philosopher Aristotle (384–322 B.C.) also compiled a list of plants of medicinal value a few years later, and his best pupil, Theophrastus (372–287 B.C.), started the science of botany with detailed descriptions of the species growing in the botanical gardens in Athens. Theophrastus also provided the first thorough account of opium and its effects. The most significant Greek contribution, however, was made by Dioscorides, whose legacy was a five-volume work entitled *De Materia medica.* This encyclopedic work described the preparation of about 1000 simple drugs. Although poorly organized and often inaccurate, it became the prototype for future pharmacopeias (books listing medicines and their preparation) and was accepted without question by Europeans until the fifteenth century.

Historians often say that the writings of Dioscorides and his contemporaries set back medicinal science 1500 years because Europeans slavishly followed their detailed but often incorrect ideas about diseases and cures. It is probably more accurate to ascribe the stagnation of medicine during the Middle Ages to the dormancy of intellectualism in Europe rather than to the stifling effects of the works of the early Greeks.

During the Middle Ages the studies of botany and medicine became more closely linked than ever before. Because of the lack of a central power in Europe, society became more and more provincial. Virtually all reading and writing were carried out in monasteries, where monks laboriously copied and compiled manuscripts. Following a Greek precedent, monks produced herbals (see Fig. 8.6), or manuals, for the identification and preparation of plants of purported medicinal value. Until the invention of the printing press in 1439, herbals were available to only a handful of people. Once the printing press allowed the wide circulation of ideas and recipes for medicines, remedies could be explained, compared, and discarded if they were found to be ineffective. The general availability of printed herbals with information about the properties, collection, and preparation of useful plants (Fig. 11.3) was thus the prelude to the discovery (or rediscovery) of some truly effective plant-derived medicines, many of which are still in use today.

In the fourteenth century the Renaissance arrived, leading Europe into a new era of intellectualism, scientific inquiry, and artistry. Studies of the human body were renewed, and surgical procedures were duly improved. The widespread extent of the search for knowledge during the next three centuries is exemplified by the drawings made by Leonardo da Vinci (1452–1519). Da Vinci believed that the

Der Apotecker.

FIGURE 11.3
This old lithograph entitled *The Apothecary* shows a pharmacy at the end of the sixteenth century. At that time pharmacists often sold spices as well as medicines. Shown in the back on a shelf are cones of raw sugar. The sugar was used to make medicines more palatable. From Hans Sach's "Beschreibung aller Stande," Frankfurt, 1574.

(Reproduced from *Medicine and the Artist [Ars Medica]* by permission of the Philadelphia Museum of Art, Carl Zigrosser, Dover Publications, Inc., 1970.)

task of the artist, like that of the scientist, is to observe one's subjects as carefully as possible and transmit those visual perceptions faithfully.

A contemporary of da Vinci's, Paracelsus (1493–1541) demonstrated his break with tradition by publicly burning the works of Theophrastus, Dioscorides, and Galen while proclaiming that he had superior insights into the prevention and cure of illness. Plants, he explained, were placed on the earth by God for human use. Consequently, God had provided signs embodied in the plants to indicate their potential uses (Fig. 11.4). For example, if a plant has red sap, that is a sign that the plant is intended for the treatment of blood disorders. The brainlike convolutions of a walnut are an indication that walnuts are effective for treating brain ailments. This idea, called the **doctrine of signatures,** seems absurd now but received great acclaim when it was proposed. Luckily, it was soon displaced by less subjective and more secular methods of determining a plant's medicinal efficacy. Paracelsus did, however, contribute to the advancement of medicine by writing an enlightened textbook on surgery and introducing the use of chemicals such as mercury for the treatment of syphilis.

In the seventeenth and eighteenth centuries science and philosophy advanced to the stage of hypothesis testing. The experimental approach to medicine led to an improved understanding of physiology and provided a framework for the careful testing of medicines that was manifested in the work of Edward Jenner who discovered the process of vaccination (1796), and that of Dr. William Withering (Box 11.1), who experimented with foxglove extracts as remedies for heart problems (1775). During the nineteenth century significant progress was made in surgical procedures. Anesthesia was introduced, and Joseph Lister promoted the use of chemicals to prevent infection.

The first half of the twentieth century was a period of tremendous advancement in medicine as causes of diseases were uncovered and new "miracle" drugs were produced. Initially, many modern medicines were isolated, purified products from traditional plant-derived extracts. Such medicines included morphine, quinine, and ephedrine. Medical chemists then began the task of determining the structures of these compounds and their subsequent synthesis. In addition, once the structure of a naturally occurring active principle was determined, it could serve as a "lead" for the synthesis of chemically related compounds that were potentially better medicines than the original product. A synthetic counterpart might retain the active site on the molecule but have different side groups that facilitate entry into the system or negate some of the adverse effects of the original chemical.

Natural products have been less studied for their potential medicinal value in the United States than elsewhere over the last 50 years for several reasons. First, there has been disdain for what many workers perceive as old-fashioned or folkloric medicine. Many pharmacologists believe that over the years humans have ferreted out and tested most of the plants that might have medicinal value. Second, pharmaceutical companies have been reluctant to take on the massive screening and testing programs involved in successfully producing a new useful plant drug, especially in the face of low returns on investment. In addition, many medical researchers have argued that advances in the biochemistry of diseases make it easier to design synthetic drugs that address specific biochemical problems without relying on plant-derived compounds or precursors or blueprints. However, the probability of obtaining an effective compound from natural sources is still much greater than that of designing a completely new one.

In light of the fact that plant biodiversity is disappearing at an alarming rate, there is a new sense of urgency behind the search for plants that cure. New and more virulent strains of malaria, continued losses from the AIDS epidemic, and

FIGURE 11.4

The doctrine of signatures holds that a plant's appearance contains clues to its use. Thus, the heart-shaped flowers of the bleeding heart (b) indicate the plant's usefulness in treating heart problems. (a) Dutchman's breeches was thought to be useful in treating syphilis. (c) The leaves of hepatica were used to treat liver ailments, and (d) the convoluted surface of walnuts gave a sign that the nuts were useful for head and brain problems.

the increasing proportion of deaths due to cancer make the loss of potential cures all the more significant. In addition, although only 1 compound of every 10,000 extracted during the random screening of plants may make it to market, improved rapid testing methods make this kind of search more feasible than it was in the past. All these factors have led the National Cancer Institute to expand its testing program for natural products. This program now has teams of botanists collecting plants in the major rain forest areas of the world: in Central America, the Amazon basin, Southeast Asia, and Africa.

The Western emphasis on diagnosis of explicit diseases and treatment with one or a few specific compounds is different from the practice of medicine in many other parts of the world, where illness is treated holistically. An ailment is viewed as a manifestation of an imbalance, either within a person's body or between an individual and some external element. In many such systems of treatment, whole plants or infusions or decoctions of plants are used to help restore the body's balance. Since 1990, the use of alternative medicines and herbal remedies has soared in the United States and Europe (Box 11.2). Some of the most common types of alternative medicines that incorporate herbal products are listed in Table 11.2.

The Chemistry of Plant-Derived Medicines

People are only beginning to realize the complex nature of plant secondary compounds and their roles in plant physiology. Moreover, it now appears that other animals exploit plants for medicinal purposes (Box 11.3). Fatty acids and essential oils, gums and resins, and alkaloids and steroids are all plant products that have found their way into modern

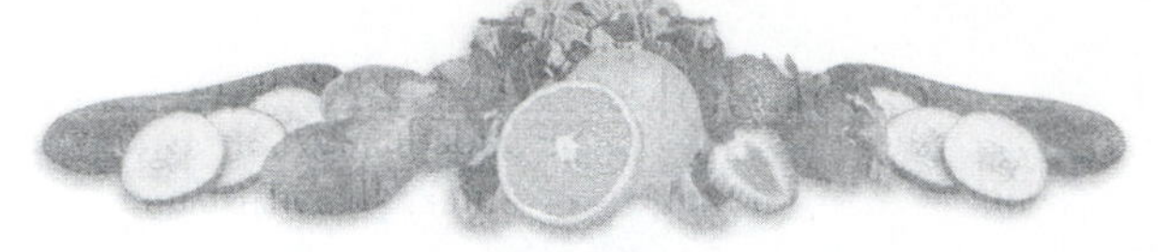

Box 11.1

From Folk Cure to Wonder Drug: The Story of Digitalis

The story of William Withering's discovery of the effectiveness of *Digitalis* for the treatment of dropsy is considered by many as the beginning of modern pharmacology. Withering (1741–1799), the son of a pharmacist, set up a medical practice in Birmingham, England, after completing his studies. The young physician became curious after a patient who was suffering from a supposedly incurable case of dropsy recovered after receiving a local folk cure. A curative tea made of plant parts included the leaves of the purple foxglove, *Digitalis purpurea.* Since foxglove had a reputation for being so poisonous that even its nectar would kill a child, Withering's first job was to ascertain what constituted a nonlethal dose of the dried, powdered leaves steeped in measured amounts of water.

Withering's treatments and documentation of amounts administered to patients and their effects were carefully recorded in his book *An Account of the Foxglove and Some of Its Medical Uses: With Practical Remarks on Dropsy and Other Diseases.* Its publication in 1785 eventually led to the widespread use of *Digitalis* even though the connection between dropsy and heart malfunction was not known until much later.

Although the glycosides responsible for the action of *Digitalis* were identified in 1928, the mechanism by which they act was not discovered for another half century. Congestive heart failure is caused by a reduction in the force of the contractions of the heart as a result of a reduced blood supply, injury to the heart muscle, or genetic abnormalities. One result of this reduction of the force with which blood is pumped from the heart is stagnation, or congestion, of blood in the veins. This congestion causes fluid to accumulate in the extremities and often in the lungs. The fluid retained in the body causes the bloating and shortness of breath that led to the old name of "dropsy."

It is now known that the muscles of the heart, like those of other parts of the body, are triggered by electrical impulses produced by changing balances of positive ions across membranes. During the resting stage, the concentration of positive ions is higher outside the cells than inside because three sodium ions (Na^+) are actively pumped out of the cells for every two potassium ions (K^+) pumped in. A stimulus from nerve cells or other heart muscle cells can cause channels in the cell membranes to open, allowing sodium ions back into the cells. This change negates the electrical potential across the cell membrane, producing an electrical impulse that also depolarizes the membrane of the sarcoplasmic reticulum, a membrane-bound compartment unique to muscle cells. Normally, calcium ions are actively pumped from the cytoplasm into the sarcoplasmic reticulum, but depolarization causes their release back into the cytoplasm. Once in the cytoplasm, the calcium ions activate heart muscle proteins, causing the heart to contract. Once the sodium-potassium pumps repolarize the cell membrane, the calcium pumps on the sarcoplasmic reticulum can remove calcium ions from the cytoplasm. The muscle proteins again become inactive, and the heart relaxes, allowing the heart's chambers to fill with blood before the next pumping cycle.

The action of cardiac glycosides inhibits the pumping of sodium ions out of the cell. It takes longer after the heart contracts for the cell membranes to repolarize and for calcium ions to be removed from the cytoplasm. This slowing down results in longer and more complete contractions. More blood is pumped from the heart, and near normal circulation is restored. *Digitalis* compounds are the most commonly used medications for weak heart contractions, but they are difficult to tolerate, and doses must be carefully determined for the individual.

drugs. Oils and gums are used as purgatives (substances that promote rapid elimination of material from the intestinal tract) and as carriers or emulsifiers in many drug preparations. Volatile oils and resins are often used to help penetrate tissues and as antiseptics. Because we have discussed oils, gums, and resins in Chapters 8, 9, and 10, we concentrate here on alkaloids and steroids, the two major classes of plant-derived compounds used in human medicine today. Compounds in both classes can occur in forms with one or more sugar molecules attached. Such forms, called **glycosides** (Fig. 11.5), are often the medicinally active form of a compound.

Steroids are complex chemical compounds that all have the same fundamental structure of four carbon rings (Fig. 11.5a), called the backbone. The addition of diverse chemical groups at different places on the backbone (the positions are numbered as shown in Fig. 11.5a) leads to the production of a variety of steroidal compounds. For example, the addition of sugar molecules produces steroidal glycosides.

Both steroids and steroidal glycosides occur in many unrelated groups of angiosperms. No direct physiological functions for steroids have been found in plants, so these compounds are classified as secondary products. However, steroids often have a pronounced effect on animals, particularly vertebrates. Consequently, many biologists believe that the purpose of these compounds is to deter herbivory. Support for this view comes from studies of the monarch butterfly (Fig. 11.6), which as a caterpillar feeds on milkweeds. Milkweeds, which are members of the genus *Asclepias* (Asclepiadaceae), are toxic to

Box 11.2

Alternatives to Conventional Western Medicine

It is estimated that as of 1998, 47 percent of the people in the United States had tried, or regularly used, some form of alternative medicine. Often, the alternative involved little more than drinking an herbal tea or a dietary supplement pill such as those listed in Table 11.2, but in many cases an entirely different approach to medicine was sought. Plants are an integral part of many of these alternative therapies. We provide here a brief synopsis of several of the more common healing therapies in which plants play an integral role.

Herbalism relies on plants for the treatment of illness, but how it is practiced depends upon whether it is in the form of Western, Chinese, or Ayurvedic herbalism. In the West, herbalism is similar to conventional medicine in that remedies are directed at specific problems or diseases, but whole plant extracts rather than single, purified active ingredients are used for treatment. This form of alternative medicine has become extremely popular in the United States with sales of herbal supplements reaching $4 billion in 1997.

In Chinese herbal medicine formulas are chosen on the basis of patterns of illness or imbalance, not just symptoms. For example, if the imbalance has a "cold" pattern such as cold hands and feet, dull abdominal pain and loose stools with undigested food, then the formula will contain plants that are warming or hot in nature (e.g., cinnamon or ginger). Conversely, in cases exhibiting hot flushes, fever, hot, odorous stools, constipation, or heartburn, the pattern is deemed "hot," and herbs that dispel heat, such as rhubarb, coptis rhizome, or scutellaria root are prescribed. The Chinese pharmacopoeia contains almost 6000 herbs usually formulated in mixtures of up to 20 herbs.

Ayurvedic medicine is a 5000-year-old tradition from India that classifies individuals according to three hereditarily determined body types, or doshas: vata, pitta, and kapha. Ayurvedic physicians determine the dosha (often a combination of body types) of a patient through observations, interviews, and pulse diagnosis. The goal of treatment is to maintain the optimum balance of health of the mind and body of a particular dosha to heighten disease resistance. This balance is achieved through diet, yoga, massage, aroma and music therapy, and combinations of 20 different herbal remedies.

Homeopathy is one of the most widely practiced alternative therapies in the Western world. Its basic principle, that "like cures like" dates as far back as Hippocrates (fifth century B.C.), but the modern incarnation began in the late eighteenth century when Samuel Hahnemann promoted this approach and gave it its name from the Greek *homoios* (same) and *pathos* (suffering). The idea behind this therapy is that a substance that causes certain symptoms when ingested by a healthy person can be used to treat or cure diseases that produce similar symptoms. For example, belladonna, when ingested by a healthy person, causes flushed face, red ears, fever, nausea, dry mouth, and dilated pupils. Since scarlet fever produces a similar host of symptoms a potentized dose of belladonna would be administered to cause the body's vital force to resolve the symptoms and overcome the disease.

Hahnemann employed the term *potentization* to refer to the dilution of the therapeutic substance. To the homeopath, "less is more," meaning that the more a substance is diluted, the more effective it becomes. Specific remedies are made from plants about 65 percent of the time, but animal (25 percent) and mineral (10 percent) extracts are also used. The material to be used is initially extracted in an alcohol-water solution called the "mother tincture." Further sequential dilutions are made by placing one part in 99 parts of alcohol until the desired potency is achieved. In some cases, dilution is carried out to such an extent that no detectable trace of the initial compounds can be found in the final solution even using sophisticated chemical procedures. However, for the homeopath the solution has been "potentized" and can serve as a curative agent. Some homeopaths prescribe remedies based on the "constitutional type" (personality profile) of a patient; others focus on specific illnesses.

Naturopathy is a word coined in 1895 by Dr. John Scheel for another therapy that strives to maintain the body's natural balance, or homeostasis, by stimulating the body's natural self-healing powers. Treatments consist of a variety of movement therapies such as yoga, massage, and hydrotherapy combined with a diet rich in fruits and vegetables and prescribed herbal remedies.

Two other alternative therapies that rely on plants are aromatherapy and Bach flower therapy. In the first case, it is believed that essential oil molecules absorbed through the skin or by inhalation affect the hypothalamus, the part of the brain that controls mood. These same compounds can also be used to treat a variety of physical disorders. Almost all of the oils used in aromatherapy are obtained from plants, and many are also employed in the perfume and flavoring industries.

Bach flower therapy is similar to aromatherapy in that plant-derived compounds are used to alter mood. Dr. Edward Bach believed that physical illness is a manifestation of emotional problems and developed this treatment in the 1930s. It is based on liquid extracts of wildflowers. Bach based his work on the "life force" of the individual flower, looking at such things as where and how it grew (harsh conditions, heat or cold) to help determine the "signature" of each flower and its potential for treating patterns of disharmony in individuals.

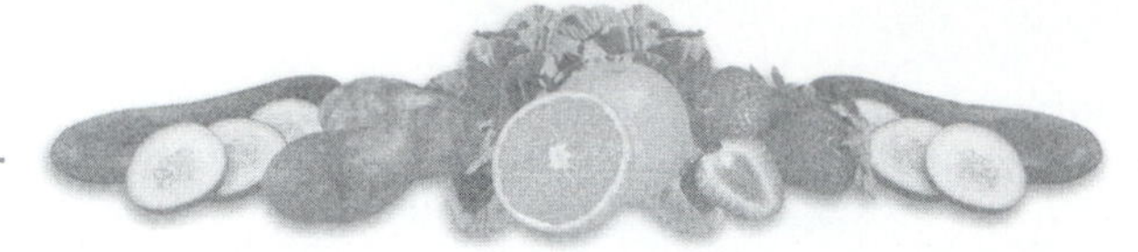

Box 11.3

From Plant Medicine to Animal Pharmacology

Botanists have had a crude understanding of plant defensive systems for a long time. The waxy cuticle that covers the aerial parts of a plant provides a barrier against water loss and the invasion of tissues by viruses, bacteria, and fungi. If a tissue is punctured, gums, latexes, or resins can flow outward, inhibiting further encroachment. Some of these exudates have antibacterial or antifungal properties. Plants can also "seal off" wounded areas so that invading microorganisms are deprived of food and cannot use the vascular system to invade other parts of the plant. In addition to these long-understood mechanisms, plant pathologists have now found that plants produce enzymes that digest some kinds of invading organisms, particularly fungi. Plants can also produce specific antibiotics against pathogens such as the bacterium *Erwinia amylovora,* which attacks pears. The recent finding that this bacterium is a relative of the microorganism that causes bubonic plague and diarrhea in humans has spurred interest in plant defensive compounds and has provided support for the contention that many vertebrate animals seek out and use specific plants for medicinal purposes. It now appears that people may have something to learn from animals' expertise.

Legends suggest that some people have long suspected that animals use plants for medicinal purposes. A Navajo story, for example, tells of how the bear taught the Indians to use *Ligusticum* (Apiaceae) for a number of illnesses. Recent work with caged Kodiac bears has shown that they do chew the roots of this plant, spit out the masticated mass, and rub the mixture on their feet and fur. Analyses of the compounds present in *Ligusticum* have revealed a variety of coumarins, substances used by humans to treat external skin conditions, heart disease, and stroke. These chemicals may also have insecticidal properties.

Numerous other studies have recently documented the nonfood use of plants by a variety of other animals. In Tanzania workers found that chimpanzees seek out leaves of *Aspilia mossambicensis* (Asteraceae) when ill. Extraction of the leaves yielded thiarubrine-A, a compound that kills bacteria, parasitic worms, and fungi. In another study listless Tanzanian chimps were observed chewing on the leaves and shoots of *Vernonia amygdalina* (Asteraceae). Although therapeutic compounds have not been isolated from this species, local African people use it for intestinal problems. Some birds also appear to exploit plants medicinally. The hoatzin is a Venezuelan bird that subsists almost entirely on a diet of leaves, which it can digest because its crop houses bacteria similar to those in the stomachs of ruminant animals. Unlike most birds, the hoatzin is practically free of internal parasites. Biologists suspect that among the leaves ingested are some that contain antifungal and antiworm compounds.

Perhaps the most unusual cases of plant use by animals involve plants that appear to be eaten to regulate aspects of reproduction. Observations of mantled howler monkeys in Central America revealed that these monkeys eat certain plants before or after mating but not at other times of the year. Other work had documented that births in particular troops were not random but were biased toward male offspring or, more rarely, toward females. Biologists have speculated that the plants ingested contain estrogen-like compounds or chemicals that alter the pH of the vaginal secretions, which can alter sex ratios in humans. Other anecdotal evidence involves an African elephant that was observed to change her pattern of food foraging one day and walk over 20 km (12.4 mi) farther than normal to feed on the leaves of a particular plant in the borage family (Boraginaceae). Four days later the elephant delivered a healthy baby. Coincidentally, a tea made from the leaves of the same species is used by residents of Kenya to induce labor. In Brazil miriqui monkeys feed on leguminous plants that have been shown to contain stigmasterol, a compound used by pharmaceutical companies as a precursor to human steroids. It has been speculated that eating these leaves may regulate a monkey's fertility or fight intestinal parasites.

As more observations are made and coordinated with chemical analyses, they will undoubtedly greatly expand people's knowledge of animal pharmacopoeias. In some cases animals may provide clues to previously uninvestigated drug plants.

---

humans because they contain abundant steroidal glycosides. Monarch larvae are not poisoned by the glycosides when they eat milkweed leaves; instead, they store the compounds in parts of their body. When the caterpillars metamorphose into butterflies, the stored glycosides end up in the wings, causing the butterflies to be toxic to vertebrate predators such as birds. One swallow of such a butterfly causes a bird to become ill or to eject the piece of the insect. Birds quickly learn to avoid monarchs. Hence, the butterflies have made a secondary use of a plant product that might have been produced by the plant to ward off predation.

The second major group of medicinally important plant products are **alkaloids** (Fig. 11.7), a diverse group of multicyclic chemicals lumped together because they contain nitrogen and usually exhibit an alkaline reaction. Alkaloids were formerly considered secondary products, but unlike steroids, they have recently been shown occasionally to enter into the primary metabolism of plants.

Many alkaloids are extremely poisonous to humans, and many, as we shall see in Chapter 12, have been used as poisons. Locoweed is a familiar example of a plant toxic to range animals. Its toxicity is due to its alkaloid constituents.

TABLE 11.2 Top 10 Dietary Supplements Used in the United States in 1999

COMMON NAME*	SCIENTIFIC NAME	FAMILY	DISORDER	NATIVE REGION
Bilberry	*Vaccinium myrtillus*	Ericacee	Diarrhea, microcirculatory problems	Eurasia
Cranberry	*Vaccinium* spp.	Ericaceae	Urinary tract infections	North America
Echinacea	*Echinacea purpurea*	Asteraceae	Colds and respiratory tract infections	North America
Evening primrose	*Oenothera biennis*	Onagraceae	Eczema, arthritis, premenstrual syndrome	Eastern North America
Ginkgo	*Ginkgo biloba*	Ginkgoaceae	Memory loss, depression	North America, China
Ginseng	*Panax ginseng*	Araliacea	Loss of energy, stress	U.S.A.
Goldenseal	*Hydratis canadensis*	Ranunculaceae	Colds and respiratory tract infections	North America
Kava rhizome	*Piper methysticum*	Piperaceae	Anxiety, stress, sleep disorders	South Pacific
Milk thistle fruit	*Silybum marianum*	Asteraceae	Liver damage	Europe
St. John's wort	*Hypericum perforatum*	Hypericaceae	Depression	Europe
Saw palmetto berry	*Serenoa repens*	Arecaceae	Urinary problems associated with prostrate enlargement	North America
Valerian root	*Valeriana officinalis*	Valerianaceae	Sleep disorders	Eurasia

*These herbs are listed in alphabetical order, not in terms of sales, and their common uses do not imply that they are effective agents for the problems listed. Garlic, often ranked among the top 10 in sales has been excluded because its use as a culinary herb may inflate its importance.

Source: M. Blumenthal. 1997. *Popular Herbs in the U.S. Market.* Austin, Tx., American Botanical Council. Report. Herbalgram 47:64–65.

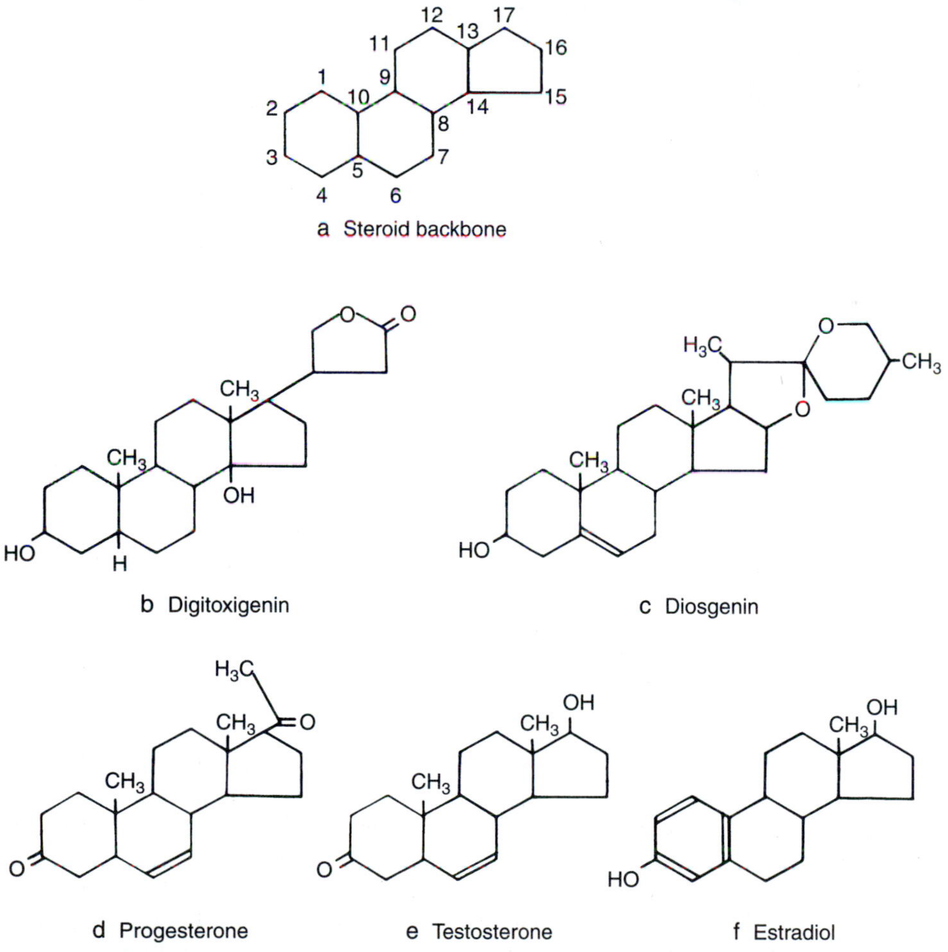

a Steroid backbone

b Digitoxigenin

c Diosgenin

d Progesterone

e Testosterone

f Estradiol

FIGURE 11.5

With the addition of particular side chains or extra rings to the steroid backbone, various cardiac glycosides and steroid hormones are produced. (b) Cardiac glycosides like digitoxigenin from *Digitalis* have a unique ring attached to the seventeenth carbon of the steroid backbone (a). (c) Steroidal saponins like diosgenin from *Dioscorea* are common in the Liliaceae, Agavaceae, and Dioscoreaceae. They differ from other steroids because they have a particular combination of additional rings attached to the steroid skeleton. This complex makes them especially useful as precursors of human steroid hormones (d,e,f).

FIGURE 11.6
Glycosides ingested by the caterpillar of the monarch butterfly while feeding on milkweed are stored in its body and, after metamorphosis, appear in the adult's body. These sequestered chemicals act as a defense against predation.

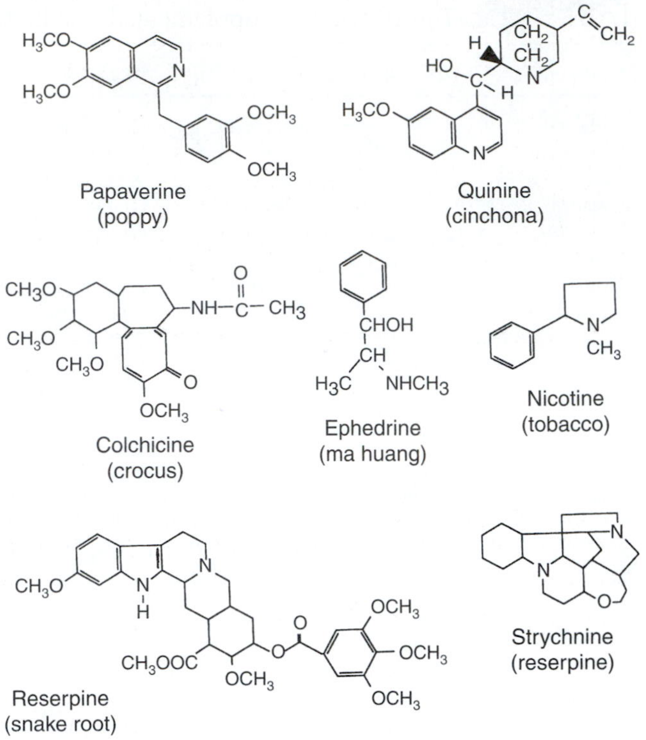

Papaverine (poppy)

Quinine (cinchona)

Colchicine (crocus)

Ephedrine (ma huang)

Nicotine (tobacco)

Reserpine (snake root)

Strychnine (reserpine)

FIGURE 11.7
All the compounds shown here are classified as alkaloids, although the nitrogen in both colchicine and ephedrine is not contained in one of the rings. The other compounds belong to a variety of kinds of alkaloids, depending on the type of ring in which the nitrogen atom is located and its position within the ring.

Other plants containing poisonous alkaloids include deadly nightshade and hemlock. Some natural insecticides, such as nicotine, are alkaloids. Many alkaloids that are toxic to humans in large or moderate amounts can alleviate physiological problems when ingested in small quantities. Consequently, one of the most important steps in the development of medicines from alkaloids (and steroids) is the determination of standardized safe doses. Without proper administration of the medicines, patients in former times were as likely to die from the cure as from the disease. Figure 11.7 gives examples of a variety of alkaloids that are used medicinally or in poisons.

Plants Formerly of Importance in Medical Treatment

Among the worst diseases that have historically afflicted humans are leprosy and malaria. Leprosy was one of the most dreaded diseases of ancient times because of the terrible disfiguration it caused and the slow and painful way in which it led to death. Malaria has been called the world's greatest killer because it has caused more deaths throughout recorded history than any other disease or all wars combined. The medicines first used with any success in treating these two diseases came from plants.

The horror with which people react to leprosy is still evident in the expression "to treat someone like a leper." For centuries leprosy was considered an incurable disease and lepers were shunned or confined to colonies designated for them. It is now known that leprosy is caused by a bacterium related to the microorganism that causes tuberculosis and is apparently transmitted by contact between susceptible individuals. Although the Vedas mentioned over 2000 years ago an oil called chaulmoogra that helped in curing leprosy, Europeans did not pay serious attention to it until the middle of the nineteenth century. Once chaulmoogra reached Europe, its effectiveness was evident but the source of the seed oil was not. Europeans and Americans received fruits without any substantial knowledge of the plants from which they had come. Joseph Rock, a naturalized American with a reputation for being able to

FIGURE 11.8
Joseph Rock had taught Arabic in Austria and Chinese in Hawaii before being chosen to lead the USDA's search for the source of chaulmoogra oil. Knowing only that the fruits appeared in native markets in China, Burma, and India and that a description indicated that the plant was a member of the tropical family Flacourtiaceae, Rock journeyed thousands of miles through the wilds of Asia. Following linguistic and botanical clues, he finally tracked down the woolly fruits on 25- to 35-m-tall trees in Burma. He is shown here in 1925 in a Hawaiian plantation of chaulmoogra (*Hydnocarpus*) established with the seeds he collected.
(Photo courtesy of the USDA.)

do the impossible, was sent to Asia in the 1920s to find the species that produced the fruits (Fig. 11.8). After a journey of thousands of miles in all types of conveyances, Rock found the source of chaulmoogra in trees of *Hydnocarpus* species (Flacoutiaceae) native to India and the surrounding countries. Rock sent viable seeds to Hawaii, where the cultivation of *Hydnocarpus* began. From shortly after the time of Rock's discoveries until the production of sulfa drugs in 1946, chaulmoogra oil provided the only effective treatment for leprosy.

The drastic effects of malaria as a cause of death have been documented since ancient times. Examinations of skulls from Bronze Age Greece have shown that malaria was a major cause of mortality. During World War I more people died from malaria than from enemy fire. The Centers for Disease Control estimated that in 1999, 310–500 million people worldwide were infected with malaria. Over 1000 cases are reported in the United States each year. About 1 million people, primarily children, died of the disease in 1999. Before the cause of malaria was known, people believed that it was transmitted through the air. The name *malaria* comes from the Italian *mal'aria,* "bad air," reflecting this idea of aerial transmission. Since it is now known that the disease is transmitted by mosquitoes, the early belief that malaria moves through the air does not seem so farfetched.

Malaria is caused by a single-celled organism belonging to the genus *Plasmodium.* When bitten by a mosquito that has previously had a blood meal from a malaria victim, an individual becomes inoculated with the microorganism. The disease is characterized by fever, chills, anemia, and spleen enlargement. Usually a person who has contracted malaria has spells that come and go. Attacks occur when large numbers of blood cells simultaneously rupture, releasing a form of the organism that had been multiplying within the cells.

For countless centuries there was no way to control the effects of malaria. Finally, in the middle of the seventeenth century, Jesuits in South America discovered a native remedy for the disease consisting of an infusion made by steeping pieces of the back of cinchona, *Cinchona officinalis* and other species (Rubiaceae) (Fig. 11.9), in water. The Indian name for the tree from which the bark was obtained was *quina* or *quina quina* (hence *quinine*). The Jesuits proclaimed that they had found a cure for malaria, but that religious order was so hated and feared in Europe that large segments of the population, believing promotion of the drug to be a conspiracy to kill Protestants, would not try it. Oliver Cromwell died of malaria rather than take the "Jesuits' powder." Cinchona was not universally accepted as an efficacious treatment for malaria until 1681, and it was not until 1820 that quinine, its active ingredient, was isolated from crude extracts.

By the middle of the nineteenth century it was recognized that individual trees of cinchona differ in the quality of the alkaloids they contain. Knowledge of particularly potent trees was carefully guarded by the Dutch, who were finally able to acquire seeds of a high-yielding related species, *Cinchona ledgeriana,* taken from a plant near Lake Titicaca, Bolivia. The Indian who divulged the location of the tree died in prison as a result of his trading of secrets. After overcoming many agronomic problems, the Dutch finally established productive plantations from those seeds in Java. Their efforts quickly gave them a monopoly of the world supply of quinine. With the onset of World War II, Europe and the United States were cut off from supplies of the drug, causing the United States to send expeditions to Bolivia to discover new sources. Although one of the missions was successful, its accomplishment was overshadowed by the synthesis of quinine in 1944. Soon afterward, chemically similar alkaloids such as atabrine, chloroquinine, and primaquine were synthesized, reducing the need for natural quinine.

FIGURE 11.9
A flowering and fruiting branch of *Cinchona officinalis,* the bark of which yields quinine, a remedy for malaria.

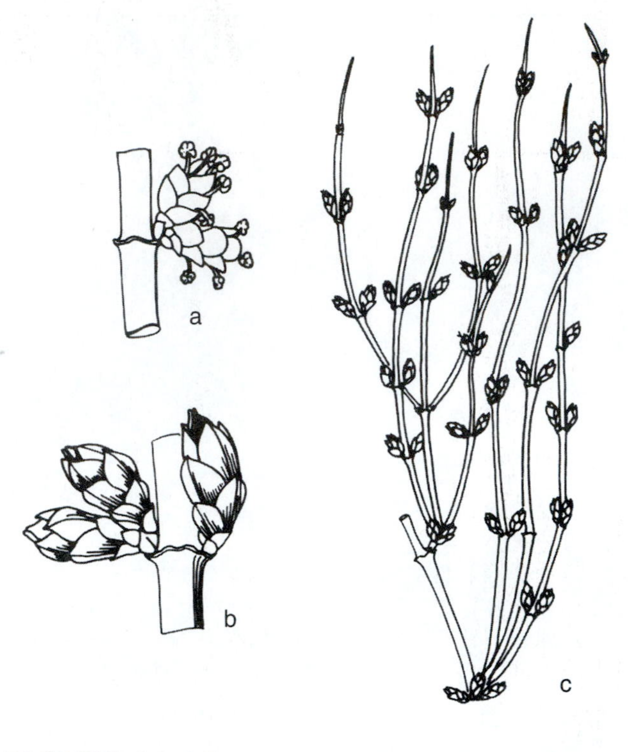

FIGURE 11.10
Ephedra, from which ephedrine is obtained, is a gymnosperm, as indicated by the presence of cones rather than flowers. (a) Staminate cones, (b) ovulate cones, (c) plant. *Ephedra,* or ma-huang, is the active ingredient of herbal asthma medicines that have recently been implicated in deaths when improperly used.

For many years virtually all the antimalaria drugs have been synthetically produced. Nevertheless, strains of *Plasmodium,* particularly in the Far East and Africa, have recently become resistant to many of the synthetic quinine analogues. As a result, there is renewed interest in natural quinine, which seems to have retained much of its effectiveness against the new strains. One legacy of the former use of natural quinine to prevent malaria is the evening gin and tonic, a drink originated by the British, whose tours of duty included tropical countries where the threat of malaria caused them to take prophylactic doses of quinine, the bitter component of tonic water.

Recently, a potent antimalarial compound has been isolated from *Artemisia annua* (Asteraceae). This species, a wormwood related to the plant from which absinthe is made, has been used in China for various medicinal purposes since 168 B.C. The compound, artemisinin, is not an alkaloid and does not even contain nitrogen. Unfortunately, in the last few years its effectiveness against malaria has waned.

Although their impact was less dramatic, three other drugs initially derived from plants and now produced synthetically are worth mentioning. One is ephedrine, originally obtained by soaking the dried stems of ma-huang, *Ephedra sinica* (Ephedraceae, a gymnosperm) (Fig. 11.10). Infusions of the plant stems have been prescribed in China for centuries as a stimulant and for the treatment of high blood pressure, asthma, and hay fever. It was only after 1920 that Westerners accepted *Ephedra* extract as a decongestant and isolated the alkaloid, ephedrine, responsible for its action. However, when it was first prescribed, ephedrine caused the death of many users because it led to cardiac arrest, especially in children. The structure of ephedrine was later shown to be similar to that of adrenaline, and its method of action was discovered to be similar. New synthetic drugs based on the ephedrine-adrenaline ring structure are now marketed as effective and much safer bronchodilators.

The second plant of note is the white willow, *Salix alba* (Salicaceae). Since the time of Dioscorides solutions made by soaking willow leaves (decoctions) were often placed on areas of the body that throbbed or ached. The active ingredient in willow that alleviated pain was isolated in 1827 and called **salicin.** Salicin and its derivative salicylic acid are extremely irritating to the stomach, but an acetylated form, acetylsalicylic acid, originally produced in Germany in 1898, can be ingested easily and provides relief for all types of pains. This compound, named **aspirin** by its original producers, is now the most widely used medicine in the world, but its physiological action on the body is only now being unraveled.

People seldom think of cocaine as a medicine, but it has historically been used as a calmative and a local anesthetic.

FIGURE 11.11
A branch and flower of a cocoa plant, which yields the alkaloid cocaine.

Coca, *Erythroxylum coca* (Erythroxylaceae) (Fig. 11.11), is a native of the South American Andes, where Indians have chewed the leaves mixed with lime (calcium oxide) for thousands of years. The alkaloids extracted during the chewing process reduce feelings of hunger, pain, and fatigue.

The potential use of cocaine in medicine was first discovered in 1884 when an assistant to Sigmund Freud placed a solution of cocaine, which was isolated from coca in 1858, on his tongue and found that it produced a numbing sensation. A later series of experiments showed that a cocaine solution could be used as a local anesthetic in eye surgery, dentistry, and other operations where only a part of the body needs to be desensitized. Cocaine has never been synthesized, but similar compounds have. The most important of these similar alkaloids are lidocaine and procaine, better known by their trade names, Novocain. Synthetic alkaloids similar in action to cocaine have virtually replaced it in medical use in the United States.

Plants Important in Medicine Today

Although natural quinine, cocaine, and chaulmoogra oil are rarely used today to treat disease, many plants are still of great importance as sources of medicinal compounds (Fig. 11.12). In our discussion of plants important in modern medicine, we treat first those yielding steroids and then those from which alkaloids are obtained.

Steroids

As was point out earlier, steroids have pronounced effects on humans. One of the primary reasons for these effects is the fact that many animal hormones have a steroidal skeleton like those shown in Figure 11.5. A **hormone** can loosely be defined as a substance produced in one part of the body that affects the functioning of tissues or organs in other parts. In humans most steroidal hormones are produced by the pituitary gland, the adrenal cortex, and the sex organs. Among the sex hormones are estrogen, androgen, progesterone, and testosterone (see Fig. 11.5). Cortisone is another steroidal hormone produced by the adrenal cortex. It is possible to synthesize these hormones from bile acids of cattle or to extract them from excised adrenal cortexes, but such methods are time-consuming and expensive. The most cumbersome part of the chemical synthesis of these hormones is the complex steroid backbone. Consequently, if a steroid skeleton can be obtained from a relatively inexpensive, accessible source, the final synthesis of a human hormone by means of the addition of peripheral components to the molecule can be accomplished relatively easily.

During the years between 1936 and 1940 it was discovered that certain members of the yam genus *Dioscorea* (Dioscoreaceae) (Fig. 11.13; see also Chapter 7) contain particular kinds of steroids called **saponins** (see Fig. 11.5). The name *saponin* refers to the fact that these compounds make a soaplike foam when shaken with water. The medicinally important saponins have an additional ring added to the steroid backbone (Fig. 11.5), but otherwise they are similar to human sex hormones. A number of plant species, such as *Agave sisalana* (Agavaceae) (see also Chapter 15), several species of *Dioscorea,* and a few species in the Liliaceae and Leguminosae contain such compounds in the form of sapogenins, which are saponins to which one or more sugars are attached. Two Central American species of *Dioscorea, D. floribunda* and *D. composita,* have been found to be the most productive sources. Both species yield diosgenin, a good starting point for chemical synthesis. This sapogenin is extracted from the tubers, which are washed and chopped or chopped and dried for later extraction. Human hormones are made using microorganisms that cleave side groups from the plant compounds and add others, producing synthetic hormones at a relatively inexpensive price.

The majority of the hormones synthesized from diosgenin are used in birth control pills, for the production of hormones that regular the menstrual cycle, or as a component of fertility drugs. Cortisone and hydrocortisone are two other important hormones that are synthesized from diosgenin. They are used for the treatment of severe allergenic

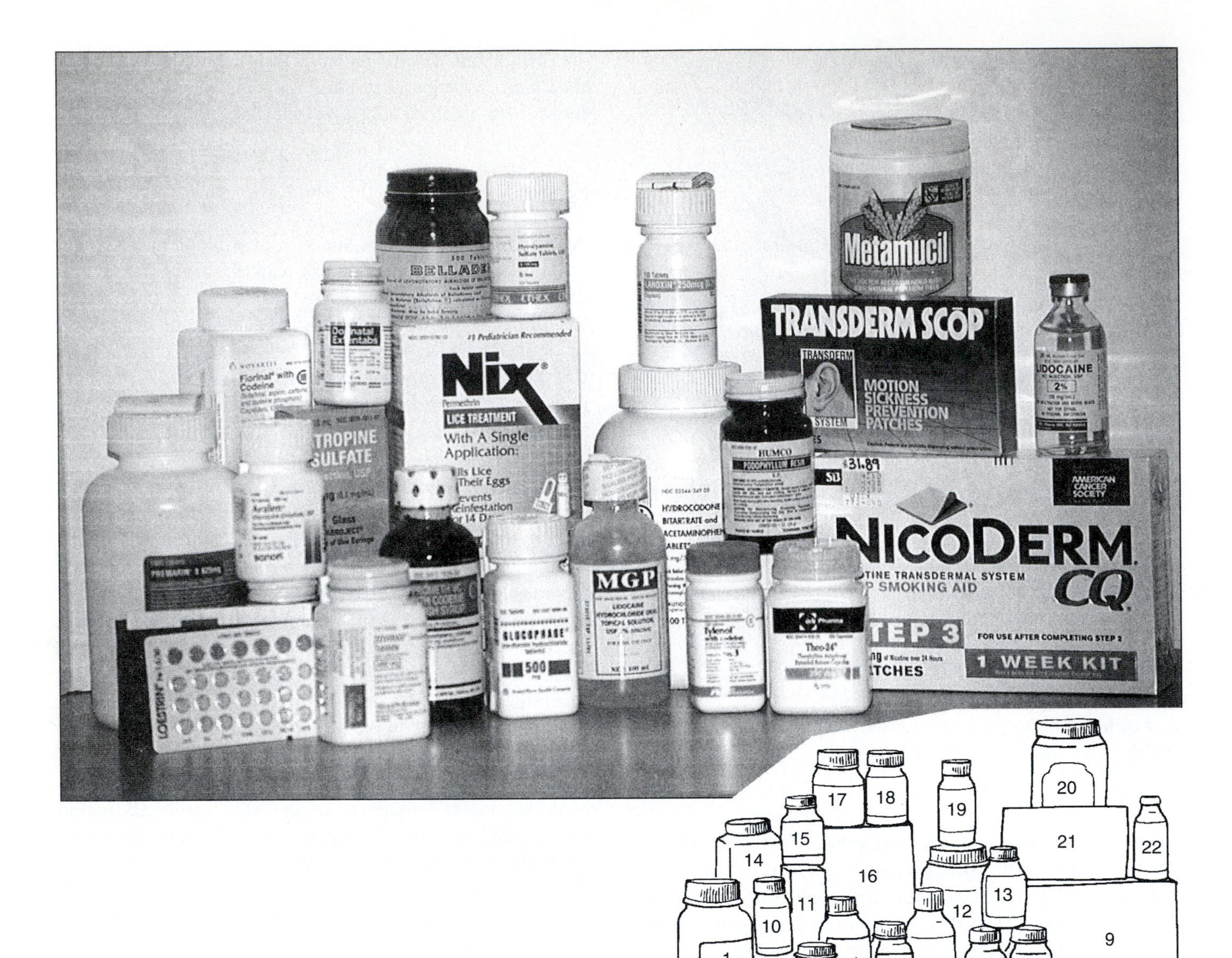

FIGURE 11.12

A sample of prescription and common medicines with plant compounds as active ingredients or which contain compounds for which natural products served as chemical models.

1. Premarin is made from natural sources of steroid precursors and is used to treat symptoms of menopause, hypoestrogenism and to prevent osteoporosis.
2. Loestrin is a brand of birth control pills that contain progesterone-estrogen manufactured using a sterol backbone derived from plants.
3. Zovirax is an antiviral synthetic purine nucleoside prescribed for the treatment of herpes.
4. Promethazine hydrocloride is used in the management of allergenic conditions and for motion sickness, nausea, and as a sedative.
5. Glucophage is used to manage diabetes.
6. Lidocaine hydrochloride, a local anesthetic, is based on cocaine.
7. Tylenol with codeine has the codeine added to enhance the pain killing action of Tylenol.
8. Theo 24 contains theophylline and is a smooth muscle relaxant and bronchodilator used in the treatment of chronic asthma and other lung diseases.
9. Nicoderm, available as a nicotine-containing patch, helps people break the smoking habit.
10. Aralen, a synthetic based on quinine, is effective in suppressing and treating acute attacks of malaria.
11. Atropine sulfate acts as a neuromuscular relaxant.
12. Hydrocodone bitartrate is a narcotic analgesic similar in structure to codeine.
13. Podophyllum resin has long been used as a purgative. The resin is also painted on soft venereal warts.
14. Fiorinal has an added kick with the addition of codeine.
15. Donnatal used to treat irritable bowel syndrome, spastic colon, and acute enterocolitis contains the tropane alkaloids scopolamine, hyoscyamine, and atropine.
16. Nix based on naturally occurring pyrethrins is used to treat for head lice.
17. Belladonna, a historical medicine from an extract of *Atropa* which contains atropine. Used today to alleviate stomach and bladder cramps, in the treatment of urinary infections and to dilate pupils during eye exams.
18. Hyoscyamine sulfate helps to heal peptic ulcers and control visceral spasms and other intestinal disorders.
19. Lanoxin, a trade name for digoxin, treats mild to moderate heart failure.
20. Metamucil contains psyllium extracted from plantain seed husks and is used as a laxative.
21. Transderm scop contains scopolamine and is sold as patches placed on the skin to prevent nausea associated with motion sickness and recovery from surgery.
22. Lidocaine is a topical anesthetic based on novocaine that provides temporary relief of pain from cuts, burns and abrasions.

(Photo taken with the help of Department of Pharmacy, University of Texas.)

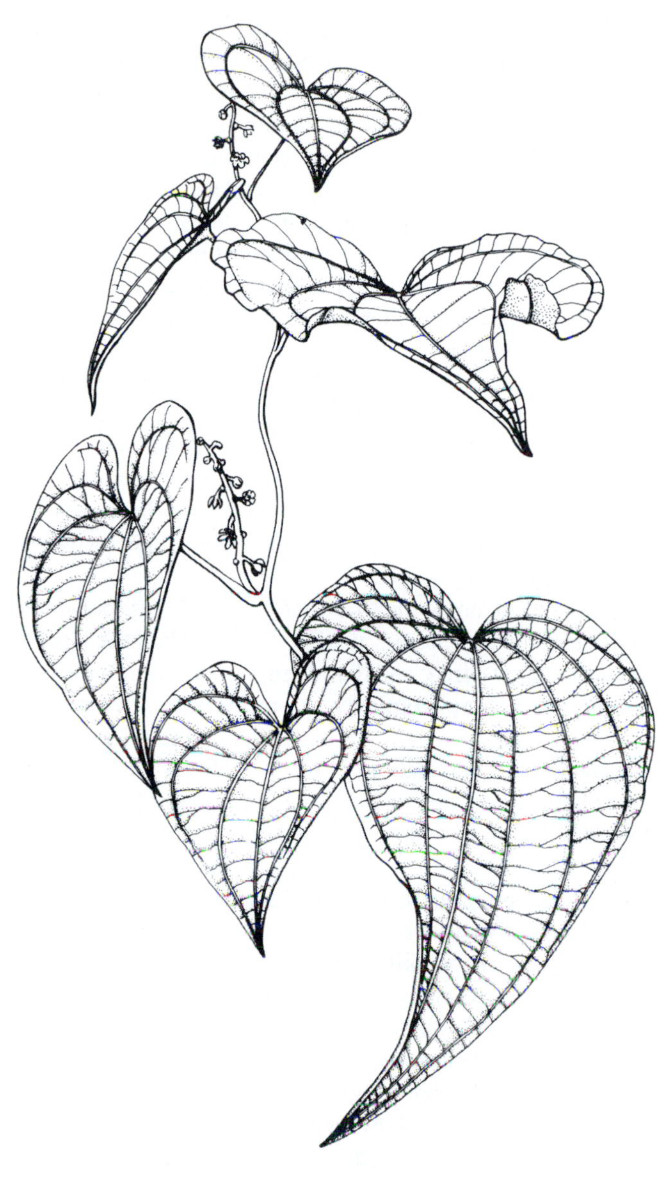

FIGURE 11.13

The tubers of yam vines (shown here) have been a source of steroid precursors used to produce active compounds in oral contraceptives and to treat hormone imbalances and heart disease.

FIGURE 11.14

The leaves of the garden ornamental foxglove yield digitoxin, a steroidal glycoside effective in stabilizing the heart's action. The scientific name *Digitalis* was given to the plant by a sixteenth-century German botanist who knew the plant by its common name *Fingerhut,* or thimble. He therefore based the formal name for the plant on the Latin word *digitus,* for "finger."

reactions, arthritis, and Addison's disease, which is caused by malfunction of the adrenal glands.

Heart disease is now the leading cause of death in the United States, with 1999 figures showing that 2500 Americans die of heart disease each day. Although the incidence is dropping, two of every five Americans still die of heart-related causes. The apparent absence of heart disease as a source of mortality in former times was probably due to the fact that most individuals died of other causes before reaching the age at which heart disease becomes a problem. It is also possible that heart disease was not recognized. For example, dropsy was a common disease 100 years ago but is unheard of today. Dropsy is an ailment characterized by the retention of fluid in the tissues. It is now known that the problem is caused by poor circulation due to congestive heart failure. It is little wonder that scientific investigation into the cure for dropsy led to the detection of one of the most important modern treatments for certain heart problems. William Withering, a prominent British physician, is credited with discovering that the leaves of the purple foxglove, *Digitalis purpurea* (Scrophulariaceae) (Fig. 11.14), a common garden ornamental, were effective in treating this form of heart disease (see Box 11.1).

In 1928, almost 150 years after Withering's work, digitoxin and digitalin, the chemicals responsible for the effectiveness of *Digitalis purpurea,* were isolated and characterized. Two years later, digoxin was isolated from *D. lanatus.* These compounds foam in water like saponins but have different

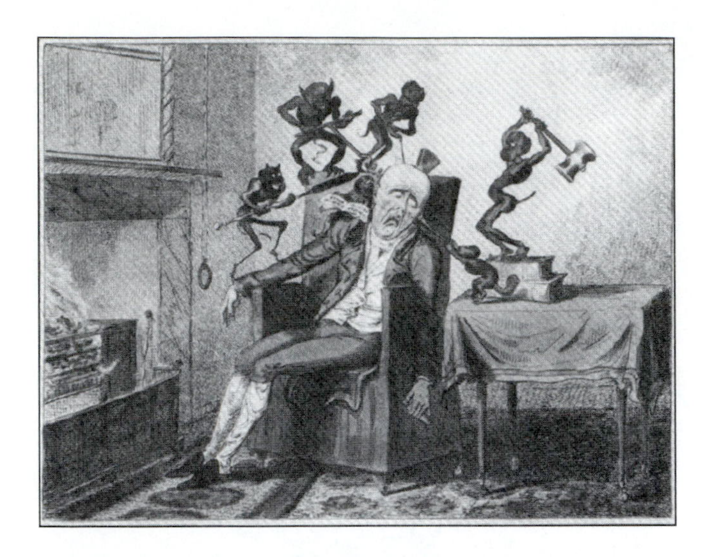

FIGURE 11.15

The Headache by George Cruikshank, published in London in 1819.

(Reproduced from "Medicine and the Artist" by permission of the Philadelphia Museum of Art.)

rings attached to the steroid backbone (see Fig. 11.5). They are named cardiac glycosides because they have a strong effect on cardiac (heart) muscle. Cardiac glycosides used medicinally can also improve circulation in general, relieve fluid retention, and help kidney function.

Alkaloids

Pain is a seemingly horrible curse that humans have to endure (Fig. 11.15). The feeling of pain may be caused by physical or psychological factors or a combination of the two. Despite the horrible sensation it causes, pain is a necessary warning of potentially harmful disruptions of people's bodies and minds, but once warned, people seek alleviation from the agony. Often the same drugs are effective against physical pain and psychological anguish because they numb the senses and produce a feeling of disassociation with the real, conscious world. Drugs that relieve pain without causing unconsciousness are called **analgesics.** One of the oldest and still predominant source of analgesics is the opium poppy, *Papaver somniferum* (Papaveraceae) (Fig. 11.16; see also Chapter 12). Fossilized poppy capsules have been found in prehistoric settlements around the Mediterranean, and representations of opium poppies can be found in Egyptian, Greek, and Roman art. As happens today, early use was probably largely for psychological purposes (see Chapter 12), but the use of opium as a calming agent and to help quell dysentery dates from at lest 2600 years ago.

The opium poppy is native to the region that includes eastern Europe and western Asia. Most modern production is in Nepal, India, Laos, and Cambodia. Because only about 4 percent of the opium harvested is used for legitimate medical purposes, cultivation of the crop is prohibited in many countries, including the United States. The depiction of the poppy capsule in ancient cultures (see Fig. 2.10) shows that the same method of obtaining opium latex from the fruits has been employed for many centuries. The fertilized, swollen capsules are slashed by a worker backing through the fields to avoid brushing against capsules that are already cut. The incision across the capsule must be deep enough to cut the latex vessels running through the outer fruit wall but shallow enough to prevent severe injury. With care, a capsule can be tapped 3 to 10 different times. After the incision is made, the capsule is left for a day while the latex oozes from the cut and dries slightly in the air. The next day the tacky mess is scraped from the capsule, stored, and eventually dried (see Fig. 12.8). In some modern operations capsules remaining after poppy seeds have been harvested are extracted with organic solvents and opiates are recovered from the solvents.

In ancient times the dried latex was simply powdered and dissolved in a liquid, usually wine. Today, the latex is also dried, powdered, and dissolved, but the solution is then subjected to chromatography, a process that separates the chemical components of a mixture. For medicinal purposes, the important chemicals present in the latex are alkaloids. Over 26 different alkaloids have been isolated from opium, but only 3 of them—morphine, codeine, and papaverine (see Figs. 11.7 and 12.10)—are used extensively. Morphine, isolated from opium in 1803 by the German pharmacist F. W. Serturner, is the most abundant of the three in the latex and the most potent painkiller. It is also the most dangerous because it is exceedingly habit-forming. Administration is usually by means of a hypodermic needle or a suppository in carefully regulated amounts. Even when it is administered medicinally, the patient can become addicted. Nevertheless, one should appreciate how important morphine was before the advent of modern anesthetics, when all surgery was performed while the patient was awake (or drunk) and strapped to a table. Morphine provided practically the only relief from the pain of operations as well as from wounds incurred in wars. It was also used, mixed with scopolamine, during the 1930s and 1940s to produce "twilight sleep," a common anesthetic used during labor and delivery. For the last 60 years morphine has tended to be administered only in cases of extreme pain such as that endured by terminal cancer patients because of fears of addiction. Recently, however, it has been shown that such fears are unfounded and that patients recover much more quickly if surgical or trauma pain is alleviated than they do if they are forced to suffer great pain. For pain management, administration of morphine into the sheath around the spinal column (epidurally) has proved to be especially effective.

In contrast to morphine, codeine is a milder and usually non-habit-forming painkiller. It is used in small amounts in some over-the-counter pain relief medicines and many prescription drugs. The alkaloid can be isolated from opium or synthesized from morphine.

FIGURE 11.16
The opium poppy is the source of morphine, papaverine, and codeine.

The last of the three major alkaloids obtained from opium is papaverine, used primarily in drugs for the treatment of internal spasms, particularly those of the intestinal tract. Paregoric, a camphorated tincture (solution) of opium used historically for diarrhea and cramps, is effective in large part because of the papaverine it contains. Although paregoric is now difficult to obtain because of strict laws governing the dispensing of opiates, it forms a component of several commercial products used to treat simple diarrhea.

Another group of alkaloids used for controlling a variety of muscle spasms and in psychiatry is obtained from members of the potato family (Solanaceae). Like the chemicals isolated from opium, these compounds are considered analgesics because they relieve pain by soothing muscle spasms. The solanaceous alkaloids (see Table 11.1) most commonly used are hyoscyamine (and *l*-hyoscyamine), atropine, and scopolamine (or hyosine). Technically, these are known as **tropane alkaloids** because they were first identified from *Atropa.* All contain a similar complex nitrogen-containing ring structure (Fig. 11.17). Although they occur in several species, the tropane alkaloids are most commonly extracted from *Atropa belladonna* (Figs. 11.18 and 11.19), which is native to central and southern Europe; *Hyoscyamus niger* and *H. muticus* from Eurasia; and *Duboisia myoporoides* and *D. leichtardtii,* native to Australia and the neighboring islands (see Chapter 12 for a different view of these plants).

The first of these, commonly called belladonna or deadly nightshade, has been used medicinally since the time of Dioscorides. Its scientific name comes from the name of the Greek god Atropos, one of the three fates responsible for cutting the thread of life. Needless to say, the plant is very poisonous. The common name *belladonna* comes from the medieval use of eye drops prepared from dried leaves of the plant. Women would use the drops to dilate their pupils. The wide-eyed, totally absorbed look produced was considered charming (hence, *bella,* meaning "beautiful," and *donna,* "lady"). Species of *Hyoscyamus,* or henbane (Fig. 11.19), also figured prominently in ancient European medicine and

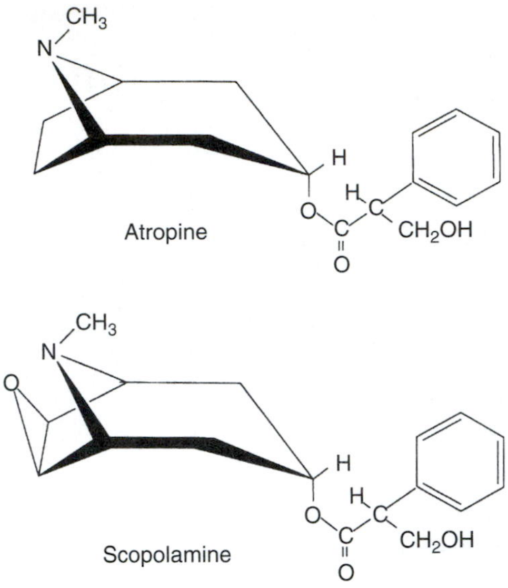

Atropine

Scopolamine

FIGURE 11.17

Tropane alkaloids, found predominantly in members of the Solanaceae, are dicyclic nitrogen compounds formed by combinations of ornithine and molecules of acetic acid. Atropine and scopolamine are extracted from belladonna and henbane.

FIGURE 11.18

Stamps commemorating bioactive plants. From top left, a Czechoslovakian stamp showing a poppy, a Polish stamp of *Datura* and another with belladonna, and a pyrethrum flower and plantation from the Republic of Rwanda (lower right).

sorcery, although they apparently were never used to enhance beauty. Species of *Duboisia* have been a source of alkaloids in Western medicine only since about 1877, but the Australian bush people recognized the potency of these plants long before then.

Although all parts of these solanaceous species contain alkaloids, usually only the leaves are dried and placed in solvents. The extracted alkaloids are then separated and purified. The alkaloids soothe the smooth muscle system and thus control painful cramping. Atropine also affects the nerves that increase the heart rate. Consequently, tropane alkaloids are used in many medicines prescribed for cardiac problems. Scopolamine and atropine are also used to dilate pupils during eye examinations. Atropine is used mixed with methylene blue, phenyl salicylate, and benzoic acid to combat infections and pain in the urinary tract.

The belladonna alkaloids are prescribed for stomach and bladder cramps and to prevent nausea and vomiting caused by motion sickness. They are prescribed for victims of Parkinson's disease to decrease stiffness and tremors and are often given to patients before surgery as a relaxant and to reduce salivation. Finally, these alkaloids, especially atropine, are helpful in cases of nerve gas (organophosphate insecticides) or mushroom poisoning.

Not all alkaloids used to treat heart disease are obtained from members of the potato family. Another important source of different compounds is a lily, hellebore, *Veratrum viride* (Liliaceae), a species indigenous to North America. The most important chemicals obtained from these plants are veratramine and jervine. These alkaloids have hypotensive properties and are used to treat high blood pressure. American hellebore was used by North American Indians and adopted by settlers from Europe. Undoubtedly, the adoption of the plant for medicinal purposes was facilitated by the fact that a related species, *V. album,* had been used in Europe for centuries to treat cholera and other diseases. Alkaloids are extracted from the bulbous roots of both species after they have been cleaned and dried.

Cinchona, the genus known primarily as a source of quinine, produces about 30 other alkaloids, among them quinidine, a compound useful in treating heart disease. Quinidine inhibits abnormal rapid contractions of the upper right chamber of the heart and corrects improper heart rhythms.

A common illness in the United States today that is often associated with or results from heart disease is high blood pressure (hypertension), which is caused by difficulty in pumping blood through the arteries. A plant that was first

FIGURE 11.19
Henbane (left) and belladonna (right) are the sources of alkaloids historically used for both medicine and murder. Hamlet's father was poisoned by a solution of henbane poured in his ear, and Juliet drank a potion of deadly nightshade to feign death.

FIGURE 11.20
The snakelike root of *Rauwolfia* contains an alkaloid once used in the treatment of hypertension and now used to treat schizophrenia.

mentioned in the Vedas of India several centuries before Christ was used at one time to treat hypertension. These scientific poems noted that snakeroot, *Rauwolfia serpentina* (Apocynaceae) (Fig. 11.20), was useful for treating snake bites. Hindis applied the name *chandra* ("moon") to this plant since it was also used to treat "moon disease," or lunacy. Indian sages have chewed on snakeroots for centuries because of its calming effects, and Mahatma Ghandi was extremely fond of snakeroot tea. Despite the abundant folk wisdom about *Rauwolfia,* it was not until 1949 that Indian scientists realized its potential for regulating hypertension. *Rauwolfia* works by relaxing the heart muscle and thus lowering blood pressure. However, a common side effect is depression and the production of tremors similar to those experienced by patients with Parkinson's disease. Consequently, *Rauwolfia* is not longer used to treat hypertension, but it has found a seemingly unrelated use in the treatment of mental illness.

The most important chemical obtained from snakeroot is reserpine (Fig. 11.7), which was isolated in Switzerland in 1952. Although the alkaloid occurs throughout the plant, it is most concentrated in, and thus primarily extracted from, the root. The dramatic effects of this drug completely altered practices in mental institutions once it was released for general use. Before reserpine was administered, schizophrenics were treated with electrical shock or injections of insulin, both of which cause violent reactions (Fig. 11.21). The injection of relatively large amounts of reserpine produces a

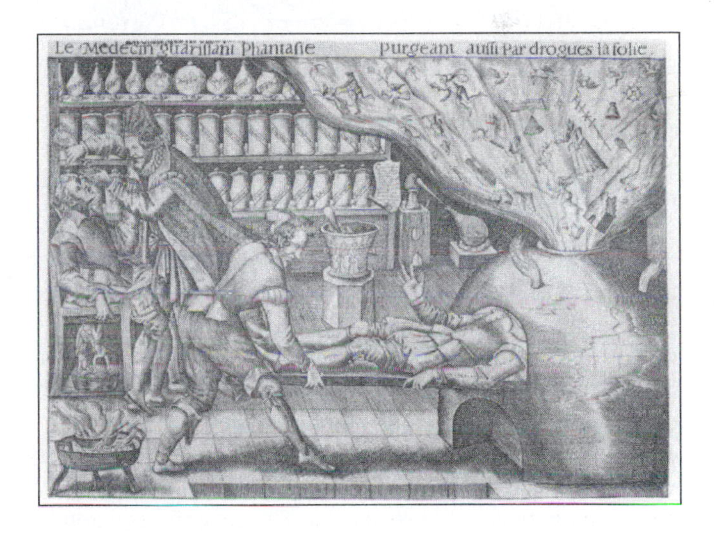

FIGURE 11.21
Before reserpine was used, treatments for mental illness were barbaric and only moderately successful. This anonymous illustration from seventeenth-century France shows a physician curing "fantasy" by using drastic measures to purge the demons causing the madness. The drawing also shows another physician administering wisdom (*sagesse*) from a bottle. On the shelf reside bottles labeled reason (*raison*), truth (*verité*), humility, etc., implying that personality qualities could be dispensed in liquid form. Although this drawing was meant as a parody, drugs have since been shown to have profound effects on mental illnesses.

(Reproduced from "Medicine and the Artist" by permission of the Philadelphia Museum of Art.)

pronounced calming effect with relatively little trauma. A related New World species, *R. tetraphylla,* is also used as a source of alkaloids, but most of the world's supply still comes from India, Pakistan, and Java.

Cancer is one of the most dreaded diseases of our time. Drug companies continually search for anticancer agents, and people afflicted with cancer and their relatives grasp at

FIGURE 11.22
The alkaloids obtained from the periwinkle (shown in flower) have been found effective in the treatment of some leukemias.

any cure. The search for effective anticancer agents has not been without success, however. In a few cases substances that are able to arrest cancer turned out to be plant alkaloids. To date, almost all effective plant-derived anticarcinogenic drugs act by interfering with mitotic cell division. Since cancer cells multiply at the expense of normal cells, compounds that inhibit cell division affect cancer cells proportionally more than they affect normal cells. The general cessation of cell division is also reflected in the lowering of blood cell counts and the loss of hair experienced by individuals undergoing chemotherapy.

The common periwinkle, *Catharanthus roseus* (Apocynaceae) (Fig. 11.22), has been used in its native range in Europe for hundreds of years as a folk treatment for diabetes. Because of its history, the plant was tested for effectiveness against various illnesses. In 1957 it was used to treat leukemia and was found to be effective in curing some forms of the disease, especially those that commonly afflict children. The active chemicals, identified as vinblastine and vincristine, were marketed shortly after 1957 under several trade names. In addition to inhibiting cell division, the periwinkle alkaloids slow cell (primarily tumor cell) growth. Rates of successful cures using drugs manufactured from these alkaloids ranged from 50 to 70 percent in some lymphomas. Remission occurred in 99 percent of the cases of lymphocytic leukemia in which vinblastine derivatives were used. Patients subsequently exhibited a 50 percent survival rate after 3 years. This treatment has also been effective for Hodgkin's disease. Continued experimentation has indicated that mixtures of the alkaloids with other chemicals may produce even better results.

A second plant found to contain antitumor alkaloids is the mayapple, *Podophyllum peltatum* (Berberidaceae). Plants of this species are herbaceous perennials that flower early in the spring in the deciduous forests of Canada and the eastern United States. The active compounds podophyllin and alpha- and beta-peltatin are particularly abundant in the rhizomes. Extracts of the roots were used by Native Americans as purgatives and for skin disorders and tumorous growths. Today mayapple alkaloids are used as the basis of VM-26 (teniposide), a drug used to treat testicular tumors and, with other agents, breast and lung cancer.

Another plant used with some success in the treatment of cancer is the autumn crocus, *Colchicum autumnale* (Liliaceae) (Fig. 11.23). The alkaloid colchicine extracted from the corms of this species interferes with mitotic cell division. Its primary use, however, is for the reduction of inflammation and pain due to gout.

Looking for cures of intractable diseases, whether they are derived from plants, animals, or synthesized, can sometimes seem like looking for a needle in a haystack. Folklore and molecular biology were both instrumental in leading to a new drug that has been a major success in treating breast and uterine cancer. The drug is commonly known as taxol, a compound most abundant in *Taxus brevifolia* (Taxaceae), the Pacific yew. In the Pacific Northwest and throughout the world, yews are considered the "tree of death." The generic name *Taxus* comes from the Greek *toxin,* meaning poison. Taxol is a complex multicyclical alcohol that can not be synthesized at an economically reasonable cost but can be isolated from the bark of the yew. In 1971 taxol's structure was published, and in the mid-1970s the U.S. National Cancer Institute decided to test it in preclinical trials. In 1979 it was shown to inhibit mitotic spindles, and it became a popular compound for biologists working on cell division. Since many chemotherapeutic agents work by inhibiting mitosis and thus effectively stopping the rampant cell divisions that are characteristic of cancer, taxol became a candidate for anticancer research. It was started through a series of clinical trials in 1983, and by 1985 it had been shown to be effective in treating ovarian and breast cancers that did not respond to other drugs. In 1992, taxol was approved as a drug for the treatment of ovarian cancer by the U.S. Food and Drug Administration.

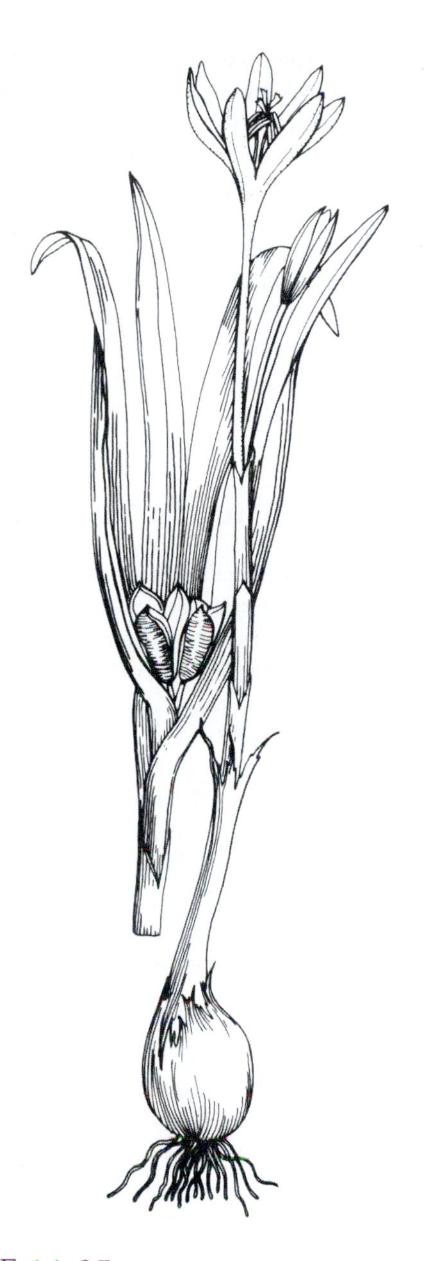

FIGURE 11.23
The autumn crocus yields colchicine, which is used to treat gout and cell malignancies.
(After Baillon.)

Despite the fact that taxol is extracted from plants, supposedly renewable resources, there were soon major problems of supply because the demand could not be met by accessible yew trees. The western yew grows naturally in old-growth forests that had been increasingly logged. The yews were not the trees the loggers sought; in fact they were often cut down and destroyed because they were considered worthless. Moreover, for taxol harvesting, trees must be large (ca. 100 years old) before their bark can be profitably exploited. When the demand escalated for yew bark, conservationists rose up to protect the disturbance of the few remaining old-growth forests. Fierce arguments ensued that resulted, in part, for the search for alternative sources. Plant taxonomists were able to provide assistance by pointing out close relatives of the yew. Now, taxol is extracted from the needles of related species of *Taxus,* and a precursor is extracted from another species and chemically converted into taxol. In 2000, taxol, known generically as paclitaxel, became the number one–selling cancer drug, with annual worldwide sales of about $3 billion.

Other Plants of Medicinal Use

Many other plants are employed in the production of the enormous array of medicines available today. Several mucilaginous compounds of plant origin (see Chapter 10) are used in soothing ointments and as carriers for other medicines. Two species of plantain, *Plantago ovata* and *P. psyllium* (Plantaginaceae), are sources of psyllium, a colloid mucilage used for intestinal problems. These species are annual rosette herbs native to northern Africa, the Canary Islands, and southern Europe east to Pakistan and India. The seed husks of these species, which can contain up to 30 percent psyllium, are removed, ground, sieved, and screened. The resultant powder is administered and absorbs water in the intestinal tract, providing a smooth bulky mass that is unaffected by bacteria. As the mass moves through the intestine, it provides relief from irritations caused by constipation or chronic diarrhea.

Species of *Aloe* (Liliaceae), primarily *A. barbadensis,* have long been used for their soothing gels. In the Mediterranean region *A. barbadensis* has been prescribed since ancient Greek times for a long list of disorders but primarily for skin problems such as burns, ringworm, and venereal sores. The soothing agent derived from aloe is the gelatinous pulp scraped from under the epidermis or the juice of the perennial succulent. The fresh gel is applied topically to inflamed areas. In addition to the gelatinous pulp, aloes produce a latex that contains anthraquinone glycosides. A mixture of these glycosides isolated from the water-insoluble parts of the latex is known as "aloin" (barbaloin). The latex has been used as a powerful purgative because it irritates the intestine. Today aloe gel is incorporated into shampoos, conditioners, and hand lotions and mixed with lanolin into salves that provide relief from burns caused by fire, overexposure to sunlight, or radiation. Aloin is often used in combination with the belladonna alkaloids in medications for intestinal ailments.

Another plant-derived drug that can be found in many homes is ipecac syrup. Ipecac is obtained from *Cephaelis ipecacuanha* (Fig. 11.24), a member of the coffee family (Rubiaceae). It contains several alkaloids that cause vomiting. Although this may appear to be more of an illness

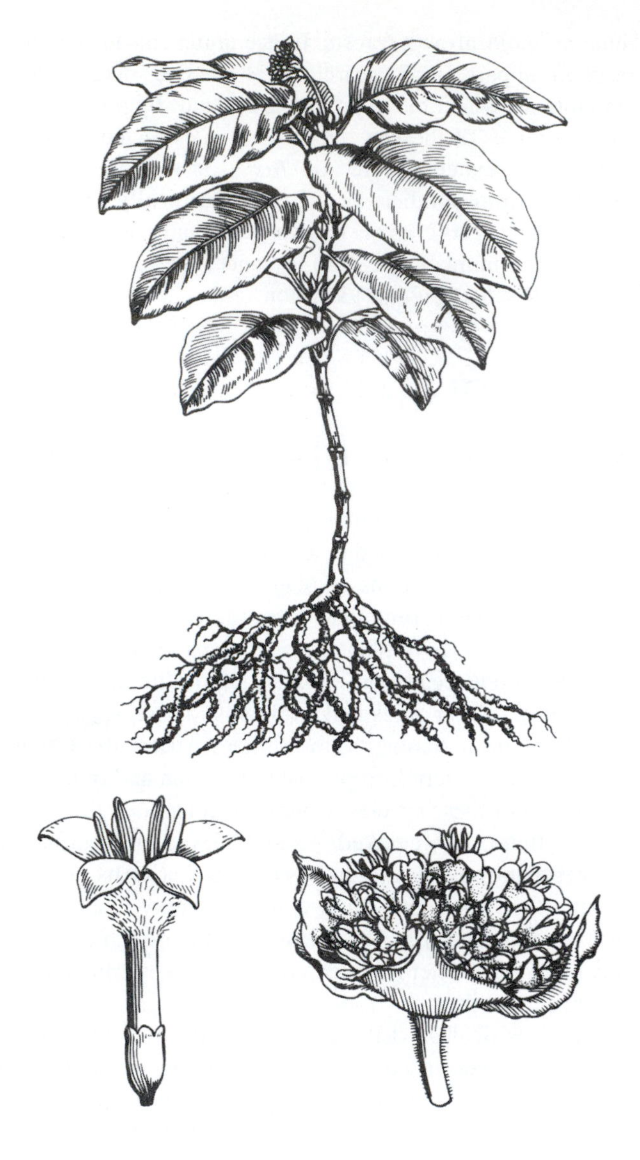

FIGURE 11.24
Ipecac, an emetic, is extracted from the roots of *Cephaelis,* a small shrub native to South America.
(After Baillon.)

inducer than a cure, ipecac is a standard item in poison control kits. When poison ingestion occurs, vomiting can cause expulsion of the substance before it is assimilated. Although ipecac is effective only for poisonings that do not involve acids or alkalis, it can often safely and quickly avert a tragic accident.

A final plant-derived drug that has recently received considerable attention is **chymopapain,** an enzyme obtained from papaya, *Carica papaya* (Caricaceae) (see Fig. 4.24). This enzyme is a sister to the more familiar papain (see Chapter 4) used for tenderizing meats. In medicine, chymopapain is used because it exhibits specificity in its dissolution of proteins. Formerly, surgery was the recommended treatment for severe back pain caused by the slippage of a disk in the spinal column. The pain caused by pressure from the displaced cartilaginous disk required drastic action, and the excision of the misplaced piece of cartilage seemed to be the only medical option. Now doctors have found that injection of chymopapain into the soft central area of a deformed disk dissolves a large part of the disk and relieves the pressure on adjacent nerves. Care must be taken not to dissolve the whole disk, but when properly done, the procedure is much simpler and less costly than major surgery. Since drawbacks include the sensitivity of some individuals to the enzyme, anyone contemplating this procedure needs to be thoroughly screened for potential allergic reactions.

Dietary Supplements

Between 1970 and 1997 the use of herbs for the treatment of illness, fatigue, or depression has escalated dramatically. Sales during the same period rose from $25 million to $4 billion. As interest and profits escalated, so did issues of commercialization and regulation.

Heightened interest in therapeutic herbs is probably the result of a number of factors. Prescription medicines are increasingly expensive as development, testing, and approval costs soar. Herbal remedies have an aura of being more natural and less invasive. Moreover, people can purchase and use them without seeking the advice of a physician. Some people turn to them for their prophylactic effects and others as a last resort when conventional treatments have failed.

How and when to regulate the production and sale of these products has engendered heated debates. Producers of herbal supplements claim that they are safe, are extensively used in many parts of the world and that federal regulation would lead to undue complications and increased cost and would deprive many people of their only source of treatment. Advocates of regulation point out that there is no standardization of products and processes that supposedly contain the same herb; products differ widely in content and sometimes contain potentially antagonistic ingredients. In 1994, the Dietary Supplement and Health and Education Act was passed, which allowed the sale of herbal products if no claims were made for medicinal efficacy. These products are therefore labeled as dietary supplements rather than as herbal medicines, and they are exempt from regulation by the U.S. Food and Drug Administration. Although most dietary supplements have caused few problems and evidence suggests that several of those listed in Table 11.2 are effective for minor problems, this policy has created a situation in which testing has been put in the hands of the public.

Herbiogenic deaths and scores of serious complications underscore the hazards of consuming unregulated bioactive compounds. Several deaths stemming from improper use of ma-huang (*Ephedra*) as a diet aid highlight the dangers inherent in making physiologically potent agents available over the counter. Another concern of physicians is the fact that many patients take herbal supplements without informing their doctors. Interactions between prescribed medication and self-prescribed herbal treatments are possible. For example, St. John's wort (*Hypericum perforatum*, Hypericaceae), taken for depression, has been shown to reduce the efficacy of birth control pills, opening the possibility of unexpected pregnancies. Since St. John's wort affects the same cytochrome P450 pathway as some prescription drugs that treat heart disease, seizures, transplant rejections, cancer, and HIV complications, doctors are being forewarned about potential interactions.

Lawmakers and providers of herbal supplements who are searching for a model of how to integrate conventional medicine with phytotherapy may want to look to Germany, where, since 1978, the Commission E has regulated bioactive compounds and protected the public while allowing access to supplements that prove to be beneficial and nontoxic.

It is possible that more useful botanical drugs will be found as different plants are tested. In China herbal medicine has been practiced for thousands of years, and today Chinese scientists are actively investigating substances in plants that have traditionally been used in folk cures. Several of those being investigated show promise and are being clinically tested. In the United States private and government agencies also carry out large-scale screening programs in an attempt to locate medicinal plants. This search probably will continue despite the destruction of natural vegetation and the extinction of species caused by population increases and rampant exploitation of natural resources.

12

Psychoactive Drugs and Poisons from Plants

In Chapter 11 we discussed drugs that are used medicinally to relieve pain or to control disease. Many of the same drugs can also alter an individual's perceptions by producing feelings of tranquility, invigoration, or otherworldliness. Sometimes people want to escape from, or "alter," reality and thus seek out these drugs, but in excessive quantities, most are toxic. This chapter focuses on plant substances that are used because of their psychoactive properties and on the overlapping group of poisons obtained from plants. Table 12.1 lists the plants discussed in this chapter.

There are many ways to group psychoactive drugs. They can be classified in terms of the chemical nature of the compounds involved, the effects they produce, or the source from which they are obtained. None of these classification schemes is perfect. The psychological effects of the compounds sometimes differ between chemically similar substances, and a classification based on chemical structure is difficult for a nonchemist to remember. Attempts to place psychoactive drugs into categories of stimulants or depressants, hallucinogens or narcotics sometimes lead to confusion because many drugs act in a combination of ways. For example, many so-called depressants, such as the narcotic analgesics (opiates) and solanaceous alkaloids, can also act as hallucinogens. Stimulants such as nicotine and cocaine can cause visions when taken in sufficient quantities. The use of the word *narcotic* itself is fraught with problems because some people use it to refer strictly to drugs obtained from the opium poppy (as we do), while others employ it for any habit-forming drug or even for any drug used illicitly. Table 12.2 provides a way of grouping psychoactive drugs by their primary effects. Since this is a botany text, we will follow another system by discussing the major psychoactive drugs—marijuana, heroin, cocaine, and tobacco—used in the United States today and then briefly reviewing several others of lesser importance. The two most widely used psychoactive compounds—caffeine and alcohol—are treated separately in Chapters 13 and 14. They merit individual attention because of their widespread economic and cultural importance and because each is obtained from a variety of plant sources.

TABLE 12.1 Plants and Plant Products Discussed in Chapter 12

COMMON NAME	SCIENTIFIC NAME	FAMILY	MAJOR ACTIVE PRINCIPLES	NATIVE REGION
Ayahuasca	*Banisteriopsis caapi*	Malpighiaceae	Harmine, harmoline	South America
Barbasco	*Derris elliptica*	Fabaceae	Rotenone	South America
Belladonna	*Atropa belladonna*	Solanaceae	Hyoscyamine	Europe
Caapi	*Banisteriopsis caapi*	Malpighiaceae	Harmine, harmoline	South America
Coca	*Erythroxylum coca*	Erythtoxylaceae	Cocaine	South America
Curare	*Strychnos* spp.	Loganiaceae	Strychnine	South America, Southeast Asia, and Indonesia
	Chondodendron tomentosum	Menispermaceae		South America
Datura	*Datura* spp.	Solanaceae	Hyoscyamine, hyocine	Europe and North and South America
Hemlock, poison	*Conium maculatum*	Apiaceae	Kavalactones, especially methysticin	Eurasia
Henbane	*Hyoscyamus niger*	Solanaceae	Hyoscyamine, hyoscine	Eurasia
Kat	*Catha edulis*	Celastraceae	Cathinone	Ethiopia
Kava	*Piper methysticum*	Piperaceae	Methysticin	Southeast Asia and Oceania
Mandrake	*Mandragora officinarum*	Solanaceae	D-norpseudoephedrine	Europe
Marijuana	*Cannabis sativa*	Cannabaceae	Tetrahydrocannabinol	Central Asia
Mescal bean	*Sophora secundiflora*	Fabaceae	Cytiscine	North America
Neem	*Azadirachta indica*	Meliaceae	Azadirachtin	Asia
Ololiuqui	*Rivea corymbosa*	Convolvulaceae	D-lysergic acid	Mexico
	Ipomoea tricolor	Convolvulaceae		Mexico
Opium poppy	*Papaver somniferum*	Papaveraceae	Morphine, codeine, papaverine	Eurasia
Peyote	*Lophophora williamsii*	Cactaceae	Mescaline	Southwestern United States and Mexico
Pyrethrum	*Chrysanthemum cinerariifolium*	Asteraceae	Pyrethrin	Asia
Snakeroot	*Rauwolfia serpentina*	Apocynaceae	Reserpine	Tropical Asia, North and South America
Tobacco	*Nicotiana tabacum* *N. rustica*	Solanaceae	Nicotine	Tropical Americas
Tuba	*Lonchocarpus nicou*	Fabaceae	Rotenone	South America

Sources: W. H. Lewis and M. P. F. Lewis. 1977. *Medical Botany.* New York, Wiley; W. C. Evans. 1996. *Trease and Evans' Pharmacognosy.* Philadelphia, W. B. Saunders; M. Blumenthal. 2000. *Herbal Medicine.* Newton, Mass. Integrative Medicine Communications.

TABLE 12.2 Psychological Classification of the Major Plant-Derived Psychoactive Drugs

KIND OF PSYCHOLOGICAL EFFECT	PLANT-DERIVED COMPOUNDS	PHYSIOLOGICAL ACTION
Behavior stimulants	Cocaine	Blocks reuptake of norepinephrine
	Caffeine	Activates intracellular metabolism
	Nicotine	Stimulates acetylcholine receptors
Convulsant	Strychnine	Blocks inhibitory synapses
Narcotic analgesics (opiates)	Opium, morphine, codeine	Mimic endogenous neurotransmitters that relieve pain
Psychedelics	Tetrahydrocannabinol	Mimic an endogenous neurotransmitter
	Mescaline, myristicine, elemicin	Mimic norepinephrine
	Atropine, scopolamine, ololiuqui, harmine	Block acetylcholine action; mimic serotonin
Antipsychotic agents	Reserpine	Depletes norepinephrine

The Chemistry and Pharmacology of Psychoactive Drugs

Almost all chemicals that have psychoactive properties contain nitrogen, and most belong to one of the classes of alkaloids. The most notable exception is the active compound of marijuana, delta-*trans*-tetrahydrocannabinol, or THC (see Fig.12.3). Some other alcohols (including ethanol) and terpenes have psychoactive effects and have been designated as hallucinogenic agents.

Before they can act, psychoactive drugs must be absorbed into the bloodstream and transported to sites where they can exert their effects. Like medicinal drugs, psychoactive drugs can be taken orally, injected into the body, or absorbed through membranes such as those lining the nose, rectum, vagina, and lungs.

Once the active compounds enter the bloodstream, they are transported to all parts of the body. A drug may exert its influence in only one place, but it is present at least initially throughout the body because the blood and the products it carries circulate continuously. As blood passes through the liver, some drug substances are removed, chemically degraded, and later excreted. Other drugs are degraded in the places where they act, and the breakdown products are picked up by the blood and transported to the kidneys for excretion. It is this process of circulation, ending in degradation and excretion, that accounts for the initial "rush" followed by an eventual wearing off of a drug's effect.

Some psychoactive drugs act only on cells of the central nervous system (the brain and spinal column), whereas others act by affecting both the central nervous system and the

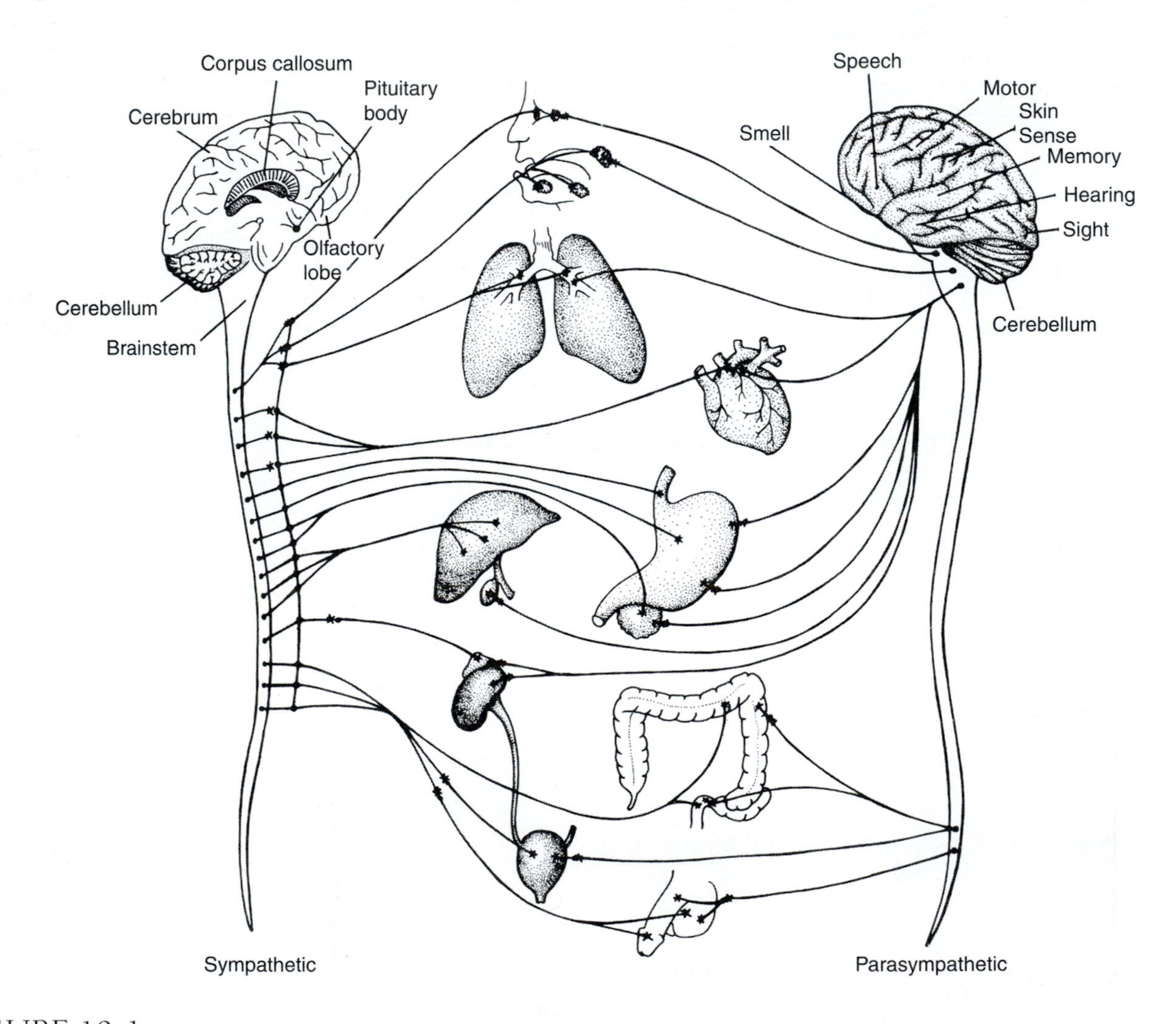

FIGURE 12.1

The central nervous system (CNS) consists of the brain and the spinal column. Neurons from the CNS extend out to form the peripheral nervous system which controls both sensory and motor functions. Neurons of the sensory system carry signals into the CNS, whereas the motor functions are controlled by impulses sent from the CNS to affected organs. Within the motor system are the sympathetic and parasympathetic divisions that control the cardiac muscle, glands, and smooth muscles such as those found in blood vessels, lungs, the digestive system, and the reproductive system. The sympathetic and parasympathetic divisions differ in the place of origin of the nerve fibers, the distances of their synapses from the organs affected (Fig. 12.2), and, to some extent, the neurotransmitters released at the organ.

peripheral nervous system (Fig. 12.1). Once these drugs reach the nervous system, they act by altering the natural interactions between **neurons,** or nerve cells. These cells (Fig. 12.2) release chemical substances called **neurotransmitters** in response to stimulation. Once released, these transmitters flow across the space, or **synapse,** between a transmitting neuron and a receiving neuron. Specific sites on the receptor recognize the transmitted compound and bind briefly with it. Once the compound has been accepted, it triggers a response in the receptor neuron. Among the receptor neurons affected by psychoactive drugs are those responsible for the perception of pain and emotion as well as the control of smooth muscles and many organs and the interpretation of audio and visual stimuli. Psychoactive drugs for the most part alter or mimic the behavior of five kinds of natural neurotransmitters: acetylcholine, norepinephrine, serotonin, dopamine, and the neuropeptides. Drugs that mimic the actions of peripheral nervous system neurotransmitters are called agonists. Drugs that inhibit the actions of these neurotransmitters are called antagonists. Table 12.3 summarizes the major neurotransmitters of the central and peripheral nervous systems, their actions, and the effects of drugs that mimic or inhibit their actions. In contrast to these drugs, caffeine (Chapter 13) and alcohol (Chapter 14) do not affect the action of particular neurotransmitters.

Five types of natural neurotransmitters are found in different parts of the nervous system and produce different physiological and psychological effects. Acetylcholine is released by the neurons of the brain and the peripheral nervous system. It stimulates contraction of skeletal muscle and smooth muscle in the digestive tract, but slows the heart rate. It is synthesized within the transmitting neuron and released into the synaptic region (region between neurons) when the neuron is triggered. After the neighboring neuron has reacted, the acetylcholine is broken down by an enzyme so that it can no longer produce an effect. Drugs can affect this sequence by blocking the transmission of acetylcholine (atropine and scopolamine), preventing its breakdown, or mimicking its action (nicotine). Drugs that block its action prevent rapid or jerky muscle reactions and thus produce a relaxed sensation. Compounds that prevent acetylcholine breakdown or mimic its effects act as stimulants because they cause the receiving neurons to fire at an increased or continuous rate.

Norepinephrine is synthesized in transmitting neurons and is released when those cells are stimulated. However, after it exerts its effect, norepinephrine is reabsorbed, not broken down. Apparently, the same molecules are reused over and over. Some drugs (reserpine from *Rauwolfia*) deplete norepinephrine. Others prevent its reabsorption (cocaine) or mimic its action (mescaline).

Serotonin stimulates the cells regulating body temperature, sensory perception, and sleep. Some psychoactive drugs, such as the LSD-type compounds, appear to alter the functioning of the neurons that transmit serotonin, thus producing illusions or strange images.

Dopamine is found in the brain and appears to influence areas that control pleasure responses. Lower than normal

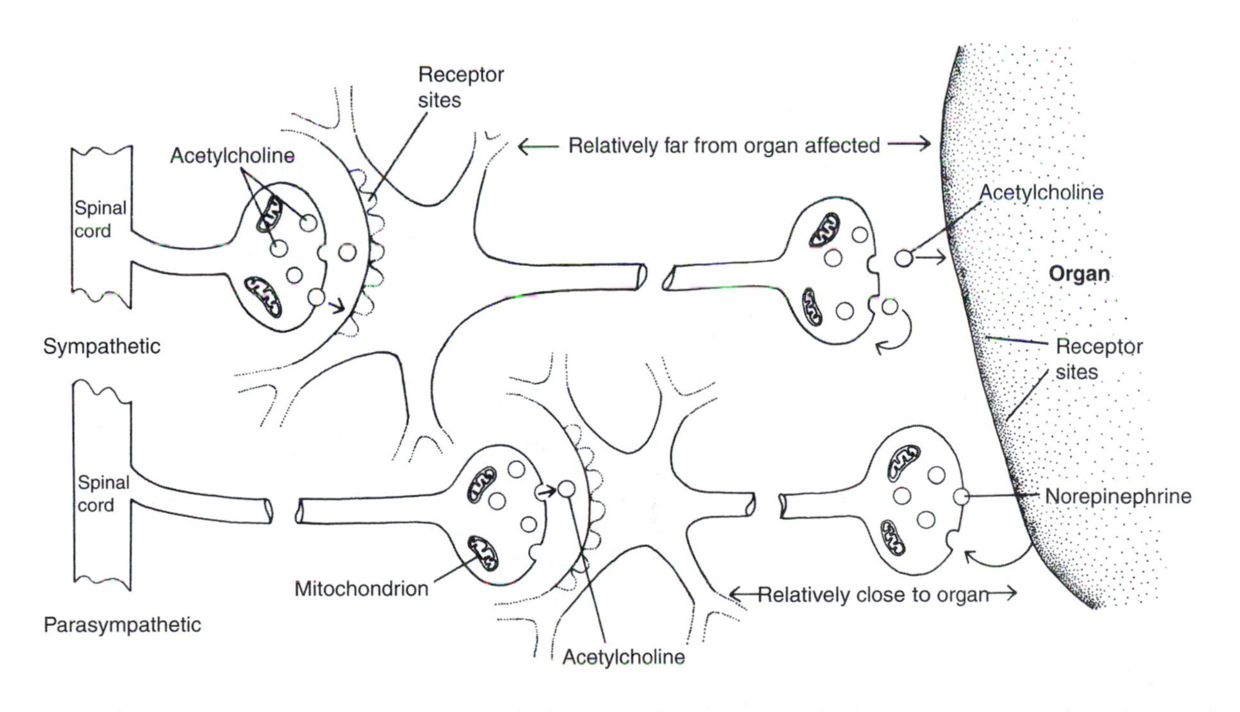

FIGURE 12.2

Neurons of the sympathetic and parasympathetic divisions of the motor system release neurotransmitters when stimulated. On the left are the preganglionic synapses which release acetylcholine in both divisions. On the right are the postganglionic synapses which release acetylcholine in the sympathetic division and norepinephrine in the parasympathetic division which, unlike acetylcholine, is resorbed, not broken down.

(After a National Institutes of Health Clinical Center photo.)

TABLE 12.3 Major Neurotransmitters and Drugs that Mimic or Inhibit Their Actions

NERVOUS SYSTEM NEUROTRANSMITTER	ACTION	KINDS OF DRUGS OR EXAMPLE	EFFECT
Central Nervous System			
Amines			
Dopamine	Influences areas that control pleasure centers. Low concentrations are associated with Parkinson's disease, high concentrations with schizophrenia	Cocaine prevents normal uptake	Prolongs feelings of well-being
		Reserpine depletes dopamine	Reduces effects of schizophrenia
Serotonin	Regulates sensory perception, sleep, and body temperature	Harmine and LSD mimic action	Produce psychedelic visions
Anandamine	Binds to specific brain receptors	THC	Produces feelings of well-being, reduces nausea
Neuropeptides	Bind to specific brain receptors	Opiates: morphine, codeine, heroin	Produce feelings of well-being, block pain
Peripheral Nervous System			
Sympathetic			
Norepinepherine	Accelerates heartbeat, stimulates secretion of adrenaline, relaxes bronchial passages and bladder, inhibits stomach activity	Agonists (mimics): cocaine, epinephrine, mescaline	Stimulate heart muscle, increase blood pressure, relax bronchial muscles and decrease nasal congestion, reduce superficial bleeding. In large doses, mimics such as mescaline can cause sensory hallucinations
		Antagonists (inhibitors): reserpine	Reduces blood pressure, helps heart irregularities. In large doses, causes depression, sedation
Parasympathetic			
Acetylcholine	Slows heartbeat, constricts bronchial passages, contracts bladder, stimulates stomach activity	Agonists (mimics): muscarine, nicotine	Bronchoconstrictors, increase peristaltic movement and relax sphincter. In excess, cause hallucinations and convulsions
		Antagonists (inhibitors): atropine, turbocurarine	Decrease muscle actions and secretions of the intestinal tract, relax bronchial constriction, decrease muscle activity, block central nervous system receptors, constrict anal sphincter. In excess, produce visions

concentrations of this substance are found in patients with Parkinson's disease, and elevated levels are associated with schizophrenia, some cases of hyperactivity, and Giles de la Tourette's syndrome (a heredity disease that causes multiple tic disorder).

The most recently discovered neurotransmitters are peptides, polymers of amino acids. About 50 neuropeptides are now known, including enkephalin, beta-endorphin, and oxytocin. These chemicals are produced in minute quantities and are received by very specific receptors. Some of them appear to act as the body's own painkillers. It is believed that opiates affect the same receptor sites that endogenous painkillers affect and thus produce a dull, relaxed sensation. The evidence for this hypothesis is that opiates are very specific in their effects, effective in low concentrations, and blocked by antagonistic agents. Opiates influence receptor cells in the brainstem (see Fig. 12.1), the thalamus, and several places in the spinal cord. Tetrahydrocannabinol has recently been shown to be another mimic of a naturally occurring substance in the body that binds to specific receptors in the brain. In this case, the natural molecule, or **ligand,** that binds to the receptor is not a peptide but an amide (Fig. 12.3).

History of Drug Use

People tend to think that drug use is a new phenomenon, but the use of mind-altering drugs is ancient. What perhaps distinguishes this culture from previous ones is the modern

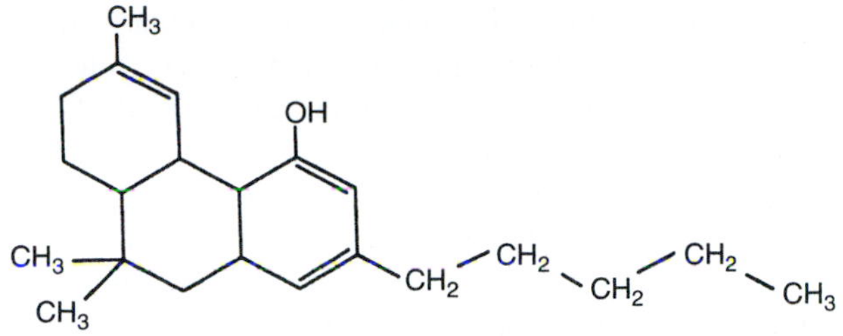

Delta-*trans*-Tetrahydrocannabinol

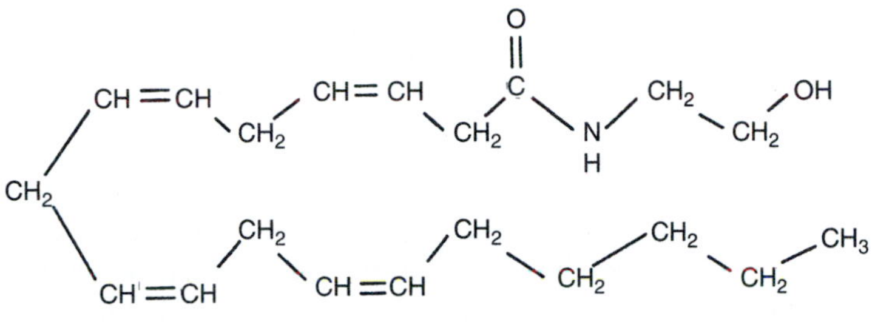

Anandamide

FIGURE 12.3
THC, or delta-trans-tetrahydrocannabinol (top), a complex alcohol responsible for marijuana's intoxicating effects, is now known to bind to specific receptors in the brain. One compound that would ordinarily bind to the receptors appears to be an anandamide (bottom).

dissociation of drugs from formal cultural or religious customs. It is easy to understand why humans would appreciate a substance that alleviates pain or reduces hunger and fatigue. In such cases, the psychoactive properties of drugs merge into medicinal usage. Nevertheless, many drugs have always been taken in excessive quantities when there was no physical need for them. These large doses often lead to audio and visual hallucinations. The original appeal of these kinds of psychological effects was quite different from the modern enjoyment of the sensations as recreational experiences.

Primitive peoples without rational explanations for the phenomena they observed around them ascribed events to the actions of supernatural agents. Because people could not see or speak with such powers under natural conditions, they sought agents that would transport them to another world where communication with gods, demons, and even the dead was possible. In most cultures only specific individuals were allowed to ingest psychedelic substances. Such people, usually men, were variously called healers, medicine men, witches, or shamans. *Shaman,* originally a Siberian word, has been adopted by anthropologists for this group of diviners because it lacks the connotations attached to most of the other designations.

The initiation of an individual into the role of shaman was often painful and dangerous. Initiation rites sometimes involved starvation, self-mutilation, and the consumption of psychoactive drugs. The combined effect produced a physical condition that enhanced the illusion of visiting the supernatural realm and acquiring the insight necessary to serve as an intermediary between the earthly and spiritual worlds. The shaman repeated the ingestion of drugs when he needed the advice of ancestors or divine guidance to solve a problem.

Anthropologists and botanists who have studied psychoactive drug use have found that New World peoples employed more species of plants for their psychoactive properties than did their counterparts in the Old World (40 versus 6 species). Assuming that the distribution of psychoactive plants is the same in the two hemispheres, some authors have suggested that the explanation for the use of fewer plants in the Old World lies in the cultural superiority of Eurasian civilizations. This explanation assumes that shamanism is a primitive trait and that the cultural environments of the ancient European and Asian civilizations provided a milieu that replaced the need for spiritual intermediaries. Such authors forget, however, the localized distributions of many of the plants used by New World peoples and the early and overwhelming dominance of two psychoactive drug plants—marijuana and the opium poppy—in the Old World. In the New World before European contact, the only drug that was used as widely as those two plants were used in Eurasia and Africa was nicotine.

Marijuana

According to a Neolithic Chinese legend, the gods gave humans one plant to fulfill all their needs. The plant was *Cannabis sativa* (Cannabaceae), *ma,* marijuana, or hemp. This assertion of *Cannabis's* usefulness is not so far-fetched. Its durable fibers can be turned into ropes, fishnets, and clothing (see Chapter 15); its seeds are highly nutritious, and the oil expressed from them can be used for lamps or in paints and varnishes. Ten-thousand-year-old potsherds imprinted with twisted hempen fibers constitute part of the evidence that *Cannabis* was among the oldest cultivated plants. Today the species is best known for the psychoactive chemicals it produces. With approximately 5.1 million weekly users estimated in 1993, marijuana rivals alcohol, caffeine, and nicotine as the most widely used nonmedical drug in the United States today. It is estimated that 69 million Americans, a third of the U.S. population, have tried marijuana at some time in their lives. How it came to occupy this position is a fascinating story involving cultures spanning almost every continent.

Cannabis, native to central Asia, has seeds that can be dispersed by water, wind, birds, and mammals. Still, it is humans who have spread the species worldwide, often without knowledge of its psychoactive properties. The Chinese were the first to use *Cannabis,* and they had such high respect for the plant that they referred to their country as the

FIGURE 12.4
Because the Vedas, sacred Indian texts, credited the god Siva with giving humans *Cannabis,* he became known as the Lord of Bhang. As the plant acquired sacred meaning, the devout started calling it *indracanna,* literally, "food of the gods."

"land of mulberry and hemp." The ingenious Chinese found many uses for *ma.* It was hemp fiber that T'sai Lun used to make the first true paper in A.D. 105. The ancient Chinese *Book of Rites* instructed mourners to wear hempen clothes to show respect for the dead. The Chinese understood the biology of the species and apparently knew about hemp's psychoactive properties, but they were interested primarily in its medicinal virtues. The legendary Chinese emperor Shen Hung is said to have observed that the female plants contained a greater proportion of the creative (yin) principles than did the male (yang) and recommended cultivation of the former and their use in correcting spiritual imbalances.

The ancient Vedas, sacred Hindu writings, describe Siva, the Lord of Bhang, bringing *Cannabis* down from the Himalayas for the enjoyment of the Indian people (Fig. 12.4). India was the country in which the hallucinogenic properties of *Cannabis* were first exploited. Whereas the Chinese grew their plants tightly spaced to discourage branching and therefore improve fiber production, the Bengalis (people from "Bhangland") developed a cultivation strategy that maximized production of the psychoactive compounds. The Indians also realized that marijuana is dioecious (Fig. 12.5) and that female plants are more potent than males. The resins exuded on the upper leaves and bracts of the female inflorescences are particularly rich in psychoactive substances. Since the resins protect vulnerable plant parts from drying, they are produced in greatest abundance when the plants are exposed to heat and sun. By growing the plants widely spaced and removing the male plants (to prevent fertilization and prolong female flowering), growers can enhance resin production. The stimulation of resin production by dry conditions explains why plants grown in semiarid climates are more potent than those from cool temperate regions. Likewise, **sinsemilla** (unfertilized, or without seeds) marijuana is one of the strongest forms.

Indians classify *Cannabis* products into ganja, consisting of the potent female flowers and upper leaves, and hashish (charas), which is relatively pure resin. The most common way of ingesting *Cannabis* in India is in the form of bhang, a milk-based beverage made with ground *Cannabis* leaves, sugar, and an assortment of spices. Bhang is widely drunk and is commonly offered as a gesture of hospitality.

There are many colorful stories about traditional hashish collection. In Nepal naked men are said to have run through fields of flowering female plants and then to have scraped off the clinging resin globules. In Persia harvested plants were beaten on rugs that were subsequently washed to dislodge the resins.

Marijuana may have also played a role in Western civilizations, but it is generally believed that the species did not spread farther west than Turkey in ancient times. The history of the Arabs' use of hashish is not documented, but legends about the plant reached Europe as early as the first part of the thirteenth century. One of them, the story of the Old Man of the Mountain, was to be recounted for the next five centuries and was resurrected in modified form by opponents of the drug in the 1920s and 1930s (Box 12.1).

In the fifteenth and sixteenth centuries Arab traders introduced marijuana to Africa, where its use quickly spread. The drug was commonly given to calm women during childbirth and was fed to babies when they were weaned. The kafirs called the plant dagga, a name later applied to a culture that made widespread use of *Cannabis.* Africans originally mixed the dried plant parts into beverages or chewed them. Smoking became popular only after the Dutch colonized the continent (Fig. 12.6).

By the time of the Crusades, when European contact with the Arabs was reestablished, marijuana use was common throughout Africa and Asia. *The Tales of the Arabian Nights* (written sometime between A.D. 1000 and 1700) was widely read in Europe and provided a romantic introduction to hashish for many people. "The Tale of the Hashish Eater" in particular described the effects of the drug on a degenerate who used it to delude himself that he was wealthy and handsome. When Napoleon's army was stationed in Egypt in 1798, recruits experimented with the plant and returned to France with stories and samples. French students suddenly felt that a visit to northern Africa was a necessary part of their education. They returned home with glowing reports of marijuana use.

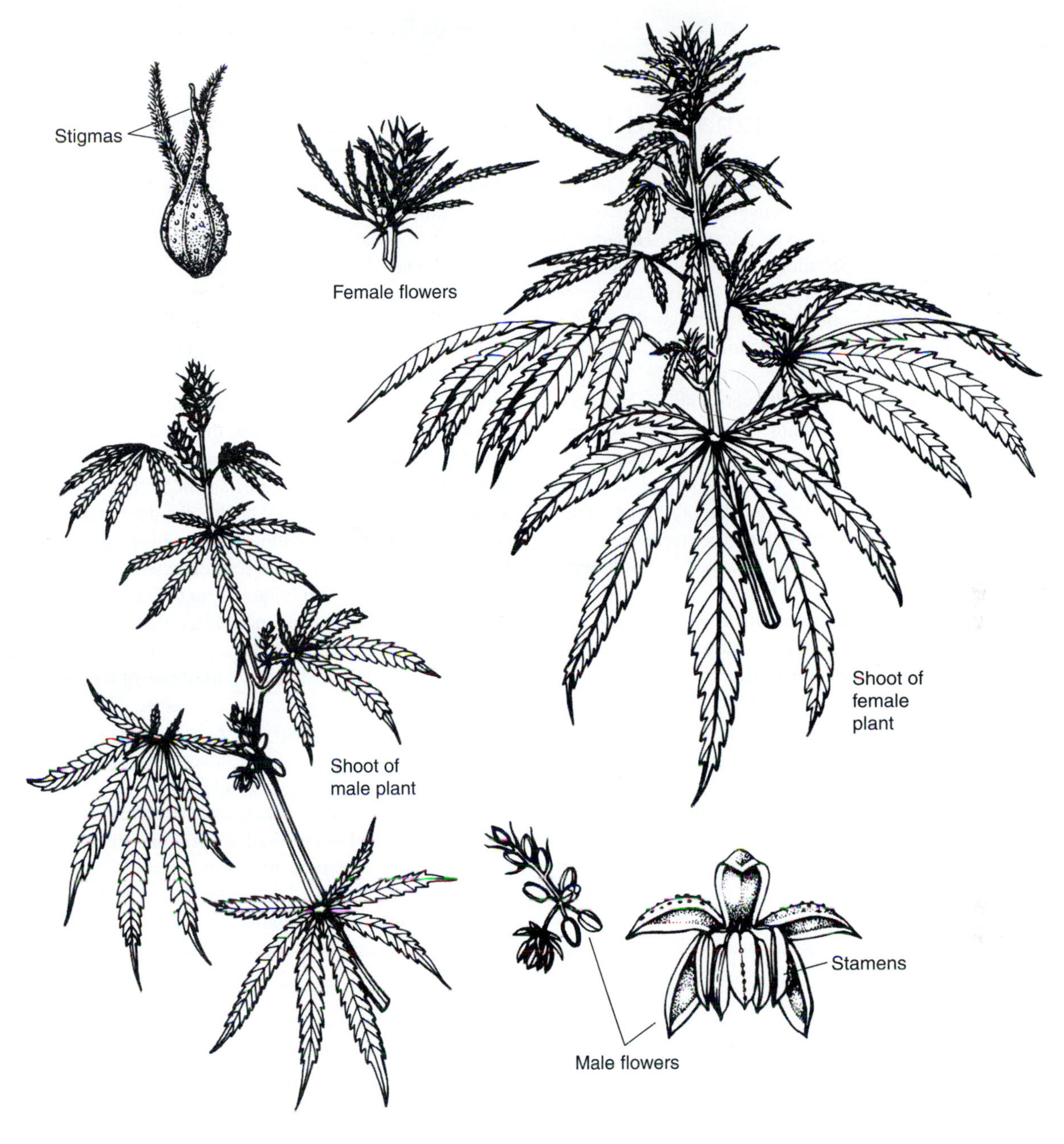

FIGURE 12.5
Cannabis sativa. Marijuana is primarily a dioecious species with both male and female plants.

Marijuana's introduction into the Americas was sporadic. The Spaniards were apparently the first to bring *Cannabis* to the New World when they started hemp cultivation in Chile in 1545. The British, hoping to produce a lucrative fiber crop, brought hemp to Jamaica around 1800. Although their commercial attempts failed, *Cannabis* escaped cultivation and spread around the island. When African slaves were brought to Jamaica in the middle of the century to harvest sugar cane, they found their familiar drug already established.

Before 1800 the British had sent formal orders for the colonists in North America to produce hemp. Even Thomas Jefferson and George Washington realized the military need for rope-making fibers and took up hemp cultivation. An appreciation of the other uses of *Cannabis* did not develop in the United States until the 1800s. The introduction of hemp's psychedelic effects followed the pattern previously established in France and England. First there was tremendous curiosity, primarily by intellectuals and artists, then a rash of drug-inspired art, and finally a general spread of marijuana use. In fact, the 1876 Centennial Exposition had a Turkish bazaar that featured hashish smoking as a special attraction (Fig. 12.7). Despite the sensationalized stories that soon began to appear about decadent hashish dens and rampant consumption by the poor, the use of the drug spread.

Box 12.1

Marijuana and the Cult of the Assassins

When Marco Polo returned from his travels to the east in A.D. 1297, he brought back a story that was destined to become a classic. He recounted that there was a terrorist, Hasan-ibn-Sabah, the "Old Man of the Mountain," who had a large band of followers willing to do anything, even commit murder or suicide, for their leader. This blind loyalty was attributed to a very clever recruitment strategy. Potential candidates were drugged and brought unconscious to an exquisite mountain garden. There, among the exotic flowers and waterworks, beautiful women were ready to minister to their every need. After experiencing the pleasures of this paradise, a recruit was drugged again and, upon awakening, was told that he could return to the garden in this life or after death only if he swore complete allegiance to the Old Man. The technique was apparently very successful, and the band of loyal followers became known as *assassins,* a name purportedly derived from hashish, the drug they were supposed to have been given.

The story was revived in the middle of the nineteenth century by a group of artists and intellectuals in Paris. A French psychiatrist, Dr. Jacques Joseph Moreau de Tours, read about the medicinal uses of hashish in combating plague and dysentery (now known to be fallacious) and decided to test the drug himself. Because he had read that hashish produces a profound change in perception, he thought that he might be able to administer it to sane individuals to reproduce the symptoms of psychosis and learn about the causes of mental illness. As part of his studies he asked an artist friend, Théophile Gautier, to try the drug and report his reaction. Gautier was impressed with the effect of hashish on his artistic senses and recommended it to a friend, the painter Boissard. Eventually the group expanded, meeting monthly in Boissard's hotel suite, where they consumed hashish while being observed by Dr. Moreau. The "Club des Hachichins," as they called their group, got its name from the story of the Old Man of the Mountain. The club eventually became a meeting place for the great French artists and writers of the day: Alexandre Dumas, Victor Hugo, Eugene Delacroix, and Charles Baudelaire. Not surprisingly, much of the art and literature of this period reflects a preoccupation with the newly discovered drug. Despite Moreau's eventual conclusion published in "On Hashish and Mental Alienation" (1845) that hashish use has deleterious effects on mental health, the use of the drug continued to spread. The fascination with hashish traveled across the English Channel to the British intellectual community, where W. B. Yeats, Oscar Wilde, and Ernest Dawson experimented with it to see if it could enhance creativity.

Almost 70 years later the story again made an appearance, this time in the United States as part of one of the most effective antimarijuana campaigns ever conducted. Harry Anslinger, who was struggling to justify the existence of the newly created Narcotics Bureau of the federal government, publicized stories about the horrors of marijuana "addiction" and depicted the pot smoker as a savage fiend whose aroused sexual desires and violent tendencies led to criminal activity and the use of stronger drugs. Anslinger revived the story of the "Old Man of the Mountain" but twisted it so that the drug was the cause of irrational violence. His campaign was so successful that states began to ban the use of marijuana in the 1920s and 1930s.

During Prohibition marijuana smoke floated up and down the Mississippi to the tunes of Dixieland. Jazz musicians found "moota" a pleasurable way to enhance their music without experiencing the stupefying effects of alcohol. They expressed their appreciation of the drug with songs such as Benny Goodman's "Sweet Marijuana Brown," Cab Calloway's "That Funny Reefer Man," "Texas Tea Party," and the "Mary Jane Polka."

FIGURE 12.6
The Dutch taught Europeans how to smoke, but the North Africans contributed the water pipe, or hookah, a device that cools smoke by drawing it through water. Lewis Carroll drew attention to the hookah in *Alice's Adventures in Wonderland* (1865) with his depiction of the Caterpillar sitting on a mushroom smoking a hookah and giving advice to Alice about parts of the mushroom to eat to make her grow tall again: "In a minute or two the Caterpillar took the hookah out of its mouth...and crawled away remarking...'One side will make you taller and the other side will make you shorter.'" Some authorities believe that Carroll purposely used a mushroom because he was familiar with reports that certain mushrooms cause hallucinations in which size and distance are greatly altered.

FIGURE 12.7

Two gentlemen smoking chiboques in the Turkish bazaar at the U.S. Centennial Exposition of 1876. It is thought that these pipes were filled with hashish since the custom of smoking hashish had recently become popular among the cultured and well traveled.

(Photo courtesy of the Fitz Hugh Ludlow Memorial Library.)

In the 1920s, moralistic outcries were raised and the ardent antimarijuana campaigning by the Narcotics Bureau led to increased condemnation of the drug. Marijuana was judged by the press and condemned by an emotional, ill-informed public. In the late 1920s and 1930s marijuana use was banned in Louisiana, Texas, and Illinois. Other states soon followed, and in 1937 the use of *Cannabis* came under federal jurisdiction with the passage of the Marijuana Tax Law, which heavily taxed but did not prohibit the drug.

Other countries had previously tried to reduce or prevent *Cannabis* use. South Africa passed the first anti-*Cannabis* law in 1870. By 1925 the League of Nations had agreed to an international statute against its use. These legal sanctions were not based on any substantial medical evidence of the detrimental effects of marijuana use and failed to halt its spread. Medical studies beginning in 1893 with the Indian Commission's thorough report failed to find any detrimental effects of moderate use. The explosive spread of pot smoking in the 1960s was a drastic demonstration of the failure of sanctions to halt marijuana use. Young people throughout the world, disillusioned with the ways in which the older generation was handling affairs, made marijuana smoking a symbol of their rebellion.

Ironically, at the same time that marijuana was being used to mock the status quo, medical researchers were learning about its virtues. Once THC (see Fig. 12.3) was isolated in 1965 and measured quantities could be used for testing, it was discovered that the active principal is a unique alcohol. Over the last 20 years research has shown that the compound is effective as a pain reliever, for combating hypertension, and in reducing pressure in the eyes of glaucoma patients. Use of the compound also reduces the nausea experienced by cancer patients undergoing radiation or chemotherapy. Since THC dilates the bronchial vessels, it provides relief for asthma sufferers, and it has proved useful in reducing the severity of epileptic seizures and symptoms of multiple sclerosis. Because of these useful medicinal qualities, marijuana cigarettes or concentrated THC in the form of pills can now be prescribed in some states by physicians.

The action of THC on the nervous system was uncovered in 1990 when a receptor to the compound was found in the brain by researchers at the National Institutes of Health who inadvertently discovered it while searching for new brain neuroreceptors. The finding of a receptor to which THC binds suggested that there is a natural compound in the body with a similar structure. In 1992 the ligand for the receptor, anandamide (from a Sanskrit word meaning "internal bliss"), was cloned (see Fig. 12.3). The discovery of this amide has raised hopes that synthetic drugs with the medicinal properties of THC but without the psychoactive effects can be designed.

Although study after study has been unable to find conclusive evidence of addiction or permanent deleterious medical effects with low or moderate use, heavy use has been correlated with several psychological and physiological problems. The drug is not physiologically addicting, but marijuana use easily becomes a habit, and heavy users are now known to have a reduced sex drive, lowered sperm counts, reduced motor coordination, and impairment of short-term memory. Since marijuana is usually smoked, heavy users can develop lung disorders similar to those incurred by cigarette smokers. Moreover, there is a high probability that heavy users will turn to "hard drugs." Nevertheless, statistics show that marijuana is less deleterious than its legal, more widespread counterparts alcohol and tobacco.

Opiates

Poppies are probably best known in the United States as garden ornamentals or for their edible seeds, but they rose to fame because of opium. As was discussed in Chapter 11, the capsules (fruits) from which the seeds are obtained are rich in an alkaloid-containing latex (Fig. 12.8). The latex has been extremely important in medicine, but codeine and morphine, the most abundant opium alkaloids, are today most often used as mind-altering drugs.

The collection of opium from fruits of the opium poppy, *Papaver somniferum* (Papaveraceae), extends back to at least 3000 B.C. Sumerian tablets from 2500 B.C. refer to opium as the "joy plant" and describe the ingestion of small balls of the latex to induce sleep and relieve pain. Opium is mentioned in every medical treatise written during ancient Greek and Roman times. The name opium comes from the Greek *opion* for poppy "juice." The early Greeks realized that the drinking of wine in which latex had been dissolved led to a trancelike state, and they therefore associated poppies with divinities such as Hypnos, the god of sleep, Morpheus, the god of dreams, and Thanatos, the god of death.

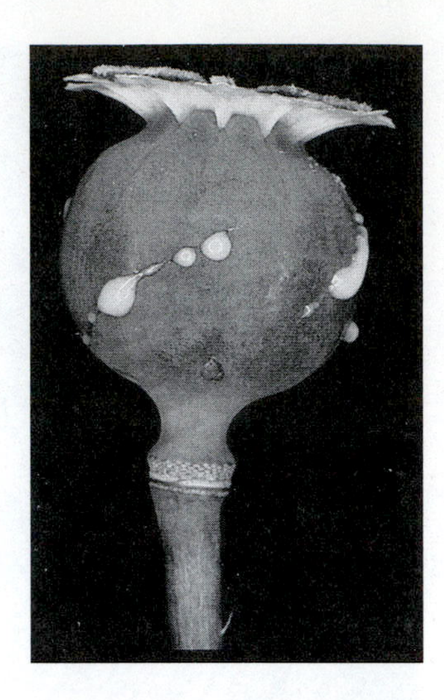

FIGURE 12.8
The pericarp of the developing opium poppy capsule is slashed so that the latex in the vesicles under the epidermis will exude.
(Photo courtesy of Wayne Ellisens.)

It is commonly thought that the Chinese introduced opium to the rest of the world, but China was ignorant of the drug until the seventh century A.D., when Arab traders first brought samples to the Orient. The original use in the East was to cure dysentery in much the same way that paregoric (tincture of opium) is used to cure diarrhea today. In the seventeenth century the Dutch introduced tobacco smoking in Taiwan and began to mix tobacco with opium in their pipes as a treatment for malaria. The practice of smoking opium spread to the mainland, where it was rapidly adopted throughout the country. Soon tobacco disappeared from the mixture, and smokers inhaled vapors from heated balls of latex that were dropped into the bowl of a pipe (Fig. 12.9).

Opium smoking became so popular that Chinese officials sought to ban the sale of the drug. Their antiopium edicts, issued as early as 1729, were ignored by both the Chinese populace and the Portuguese suppliers. By 1800 the British had gained a monopoly of trade rights with China. Since China had little use for Western goods, the English felt compelled to continue the Portuguese practice of opium dealing because it was one of the few commodities for which the Chinese would exchange coveted products such as silk and spices. Eventually England even established opium plantations in its Indian colonies to provide opium

FIGURE 12.9

A Chinese opium den showing users smoking the crude latex in small pipes.

for trade. The importation of opium into England was illegal at this time, but distance and the need to maintain international trade seem to have exonerated the practice elsewhere in the minds of the English. The Chinese lords tried repeatedly to halt this drug trafficking because they could see the debilitating effects on their people, but without the support of the British, their efforts failed.

The situation came to a head in 1839, when the Chinese confiscated and destroyed British opium supplies in Canton. England retaliated by invading China, touching off the first opium war, which ended with a treaty that ceded Hong Kong to the British. A second opium war broke out in 1865, and the Chinese were again defeated. This time they were forced to agree to British opium importation. Unfortunately, opium consumption among the Chinese continued to escalate. Trafficking in the drug was finally reduced in 1906, when the last emperor of China got the British to agree to a reduction in importation. Still, use did not drastically drop until the revolution and the establishment of the People's Republic of China in 1949. Opium trading and use are now strictly forbidden.

Opium consumption was not common in Europe until 1525, when Paracelsus (see Chapter 11) discovered, or rediscovered, a way to dissolve opium in alcohol. The tincture that was produced was known as *laudanum.* The ease with which laudanum could be consumed and its purported medical benefits made it the most popular drug of its time.

In 1803 morphine was isolated from opium (Fig. 12.10), producing a purified alkaloid that could be given in measured doses. On a per weight basis, morphine is 10 times as strong as opium, which contains at least 24 other alkaloids and numerous other compounds. Purified morphine can also be administered intravenously. When the hypodermic syringe was developed in 1853, it presented doctors with a rapid method of introducing this potent painkiller into the bloodstream, and morphine use rose.

Morphine's effectiveness was both a blessing and a curse. In many cases the speed with which a drug enters the bloodstream (and thus the intensity of the "rush") is correlated with the level of addiction it can produce. Only after 45,000 soldiers returned home from the Civil War addicted to the painkiller were the dangers of morphine use recognized.

In an attempt to develop a nonaddicting painkiller, scientists discovered in 1874 that morphine could be chemically altered by the addition of two acetyl groups. The end product is a semisynthetic compound known as **heroin.** Heroin is an even more powerful analgesic than morphine and, like morphine, was originally described as lacking addicting qualities. For a short time the virtues of this heroic drug (hence the name) were extolled by the medical community. The Bayer Company was the major marketer of the compound and in 1889 advertised a cough syrup that contained the wonder ingredient heroin. Although vestiges of the idea that heroin is nonaddictive persisted until 1905, it soon became apparent that it is even more dangerous than morphine because it crosses cell membranes more rapidly than does its natural counterpart. Because of this fast absorption (and hence cycling in the bloodstream), heroin must be administered more frequently than other opiates. In addition to the effects produced by heroin itself, users suffer

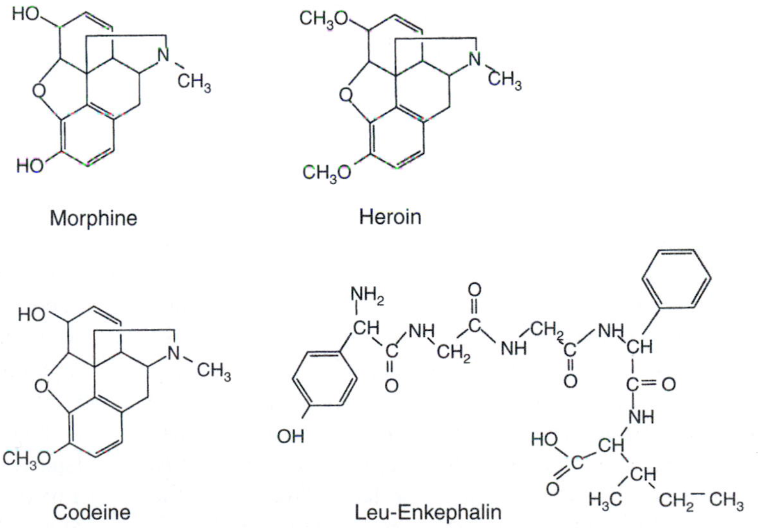

Morphine

Heroin

Codeine

Leu-Enkephalin

FIGURE 12.10

The major opium psychoactive alkaloids, morphine and codeine, are similar in structure. Heroin is a synthetic compound manufactured from morphine. All three alkaloids bind to specific brain receptors that are involved with feelings of well-being. Enkephalin, a peptide that is the naturally occurring binding compound, is shown for comparison.

from disruption of blood flow, infections, diseases such as hepatitis, and collapsed blood veins. Heroin is very physically addicting and produces pronounced withdrawal symptoms once the habit has become established. Two to three weeks of injecting 30 mg (0.01oz.) per day is considered enough to create a tenacious habit. The main cause of death for heroin addicts is overdose. The common use of adulterants and other drugs in combination with heroin causes doses to vary and makes it difficult to gauge "overdose" levels. The lifestyle of a heroin addict who no longer seeks the drug for pleasure but because he or she desperately needs it to avoid withdrawal tends to lead eventually to overall physical problems. The death rate for heroin users is more than twice the normal rate in any given age group.

In 1914 a law was passed prohibiting the possession of opiates for nonmedical purposes. In 1924 the manufacture of heroin was declared illegal in the United States, and in 1956 existing legal supplies of the drug were destroyed.

Because of fears of its addictive properties, morphine use has decreased in modern-day medicine. However, as was discussed in Chapter 11, morphine is extremely effective for treating severe pain and terminally ill patients. In contrast, the use of heroin has increased dramatically during the last 40 years. Figures from 1996 estimate that 2.4 million Americans have used heroin at some time and 216,000 continue to use it monthly. Since heroin is made from morphine, the total production of opium has increased correspondingly. Until 1972 most of the opium that eventually ended up in the United States was grown in Turkey. When Turkey and the U.S. government reached an agreement to limit opium poppy cultivation, the center of production shifted to Afghanistan, which produced 4600 tons in 1998, over three times as much as Myanmar, the second largest producer.

FIGURE 12.11
This Peruvian woman sells both coca leaves and oca, the lime (calcium oxide) in which Indians dip the leaves before masticating them. The lime is made by burning the leaves of other plants.

Cocaine and Crack

It is hard to believe that the modern American scourge cocaine originated as innocuous leaves chewed by Andean Indians for their mild stimulatory effects. The two most commonly used species, *Erythroxylum coca* and *E. truxillense* (Erythroxylaceae) (see Fig. 11.11), are native to the north-central Andes. Discoveries of bags of coca leaves and utensils in 3000-year-old Andean burial sites indicate that it was used by inhabitants of this region long before Europeans discovered South America. By the time Pizarro conquered Peru, coca was an integral part of Inca life and was considered a sacred plant. Peasants of the Inca empire depended on coca, as do an estimated 90 percent of their modern descendants, to mitigate the harsh environment of the high elevations of the Peruvian and Bolivian Andes.

The preparation of coca involves collecting the leaves, drying them, and sometimes allowing them to "sweat" (lightly ferment) to render them pliable rather than crisp. Indians normally dip the leaves into lime (calcium oxide) that they carry in a small bag before chewing them (Fig. 12.11). Lime in the form of ash raises the pH of the quid and aids in the extraction and absorption of alkaloids. The masticated residue is either expelled or swallowed. Chewing produces a feeling of well-being and lessens feelings of hunger and fatigue.

The Spanish conquerors tried to prohibit coca use until they realized that the Indians they enslaved would work harder if allowed to chew it. Coca was taken back to Europe by the Spanish, but in leaf form it never received much attention. In 1860 cocaine was isolated and, unlike the homely leaves from which it was extracted, soon became extraordinarily popular. Sigmund Freud publicized the drug in his treatise *Über Coca* (1884). Among other things, Freud recommended coca as a treatment for alcoholism and morphine addiction. He also lauded it as a local anesthetic and praised its use in psychotherapy to relieve depression. In both Europe and the United States, doctors began to prescribe its use (Fig. 12.12).

An enterprising Italian, Angelo Mariani, developed a coca wine beverage in 1860 that was soon the rage of Europe. Mariani's wine earned commendations from scores of celebrities, including Jules Verne, Ulysses S. Grant, President McKinley, Émile Zola, Henrick Ibsen, and

FIGURE 12.12
"What do you do, my dear, to look so beautiful?" "I drink Coca of the Incas every day," exclaims a French poster printed in 1876. Coca des Incas was a French tonic wine made from coca leaves, a forerunner of Coca-Cola.
(Courtesy of Tim Plowman.)

FIGURE 12.13
An advertisement from the era when Coca-Cola included extracts of coca leaves reads, "Tired? Then drink Coca-Cola. It relieves exhaustion. When the BRAIN is running under full pressure send down to the FOUNTAIN for a glass…you will be surprised how quickly it will ease the Tired Brain—soothe the Rattled Nerves and restore Wanted Energy to both Mind and Body. It enables the entire system to readily cope with any excessive demands made upon it."
(Photo from the Bettman Archives, Corbis.)

Thomas Edison. Coca-Cola, which originally included both caffeine-rich extracts from *Cola nitida* (see Chapter 13) and coca extracts, was first marketed in1886 as a headache remedy and a tonic (Fig. 12.13). Since 1904, however, federal law has prohibited the inclusion of cocaine in any beverage. Ironically, after the Coca-Cola Company complied with this law, it was sued for misleading advertising because the name implied that the beverage contained coca products. As a result, coca leaves, with the cocaine removed, are now used to flavor the syrup from which the cola is made.

As more and more reports of cocaine-induced violence and other effects of the drug circulated after the turn of the nineteenth century, the government took steps to curb its use. In 1914 the drug was formally declared illegal by the Harrison Narcotics Act. As a result, cocaine costs escalated. Despite illegality and high costs, its use continued to increase. Cocaine itself is a three-ringed alkaloid, but it is most commonly taken as a hydrochloride "salt." The salt, known as **coke,** is usually diluted with other substances (sugar, caffeine, lidocaine, etc.). The salt form is water-soluble and moves easily across the mucous membranes of the nose. Indulgers usually chop or grind crystals of cocaine hydrochloride before sniffing ("snorting") a line of fine powder. **Crack** is produced by treating the hydrochloride with boiling water and baking soda. If ether is used in the extraction process, freebase is produced. Crack is less dangerous than freebase because volatile solvents are not used to produce it. Both freebase and crack can be injected and, since they are unaffected by heat, can be smoked. They are also highly soluble in fatty compounds and thus pass quickly across cell membranes.

Just how dangerous is cocaine? Researchers have not been able to find any organic damage from prolonged coca chewing among Andean Indian users, but it should be remembered that Indians who chew coca leaves absorb only limited quantities of the drug because the alkaloids are present in the leaves in relatively small amounts. In the United States pure cocaine has become an extremely addictive and debilitating drug. An estimated 1.5 million Americans were chronic cocaine and crack users in 1998 according to a National Household Survey on Drug Abuse, down from 5.7 million users in 1985. Although the average native Andean user may chew 57 g (2 oz) of coca leaves a day, he or she extracts from them only about 0.002 g (0.00007 oz.) of cocaine. Heavy users can consume an average of 2 to 3 g (0.07 to 0.10 oz) a day of the drug plus adulterants. A lethal dose of cocaine varies from individual to individual, but it is now known that some people are extremely susceptible and can suffer heart failure with a single dose as small as 20 mg (0.007 oz). Unfortunately, it is impossible to determine in advance who will suffer such an effect or when.

Cocaine acts primarily by interfering with the reuptake of dopamine, an important brain neurotransmitter. Dopamine produces feelings of well-being when it is released from one transmitting brain neuron to another. After release, dopamine is normally pumped back into the original neuron. Cocaine prevents the reuptake of dopamine and therefore prolongs the feeling of well-being. In addition to blocking dopamine reuptake, there is now evidence that cocaine blocks the uptake of other neurotransmitters as well. The sustained use of cocaine produces physiological changes in the brain that lead to addiction and withdrawal. Unlike drugs such as morphine, however, the direct effects on adult users are primarily psychological. Nevertheless, cocaine addiction is associated with a number of physical effects because of the methods of ingesting the drug and the abandonment of sound health practices. If cocaine is inhaled for long periods, it can cause damage to the inner surface of the nose. Cocaine users tend to use increasing amounts to maintain the responses originally achieved. Since continual use eventually leads to a depletion of neuroreceptors, maintaining the original effect is never possible. Ultimately, chronic users take as many as 10 to 50 "hits" a day. Depression, feelings of paranoia, and often hallucinations accompany such doses. Lack of sleep and loss of appetite compound the mental problems addicts suffer. Recently, evidence of heart damage and increased risk of heart attacks has been uncovered. Researchers speculate that these cocaine-induced heart attacks occur when the heart is deprived of oxygen through the direct effects of increased or irregular heart rate or through the constriction of blood vessels or from the thickening of the blood, which increases the likelihood of developing blood clots.

One of the most tragic effects of the current cocaine epidemic has been the rising production of "crack babies" born to mothers who were users during pregnancy (Box 12.2).

Tobacco

Next to alcohol and caffeine, nicotine obtained from tobacco is the most widely used psychoactive drug in the world today. Common tobacco, *Nicotiana tabacum* (Solanaceae), and to a lesser extent its relative *N. rustica,* have become important commercial items in almost every country (Fig. 12.14). The genus *Nicotiana* is thought to be native to the New World, but at least one species, *N. suaveolens,* which occurs in Australia, was independently domesticated. At least 1000 years before Columbus landed in the West Indies, tobacco was smoked, eaten, and snuffed by native peoples throughout the New World (Fig. 12.15). Among Native Americans tobacco was used medicinally to ease the pain of childbirth and to stave off hunger on long hunts. The dried leaves were so valuable that they were used as money and incorporated into ritualistic and symbolic practices such as the smoking of a peace pipe.

Tobacco was considered sacred by many Native American tribes. Mayan priests thought that the smoke rising from their pipes carried messages to the gods (Fig. 12.16). Recent studies of Amazonian tribes have shown that the use of tobacco in divination rites continues today. Shamans of the Warao tribe of Venezuela, for example, fast almost to starvation during their initiation rites and then eat and smoke large quantities of tobacco. The resultant hallucinations produce the sensation of being transported to another world for meetings with the spirits that govern the lives of their people. Later in their careers shamans frequently turn to tobacco to help solve tribal problems.

Although Columbus brought "cigars" to Queen Isabella after his first voyage, it was not until Andre Thevet brought seeds of *N. tabacum* from Brazil to France in 1556 that tobacco cultivation started in Europe. Linnaeus named the genus after Jean Nicot, the French ambassador to Portugal who made a fortune importing and popularizing tobacco in Paris. Claims of tobacco's medicinal virtues as a remedy for female problems, a snakebite antidote, a lung strengthener, an ulcer remedy, a cure for the plague, and a potent aphrodisiac accelerated its sweep across Europe and Asia.

Tobacco use became so widespread that King James I was alarmed to find that purchases of tobacco were depleting England's silver supply. Neither the high cost of the leaves nor King James's blistering attack in his *Counterblaste to Tobacco* (1604) checked the spread of smoking or sniffing (Fig. 12.17). In other countries as well, leaders tried to discourage tobacco use. One Chinese emperor ordered smokers to be decapitated, and a Russian tsar ordered the nostrils of snuffers to be split so that they could no longer practice the habit. Still, tobacco continued to attract followers.

Eventually the British promoted tobacco cultivation in their colonies to ensure a national supply. Virginians started growing tobacco in 1612. Since an acre planted in tobacco yielded four times the revenue of an acre planted in corn,

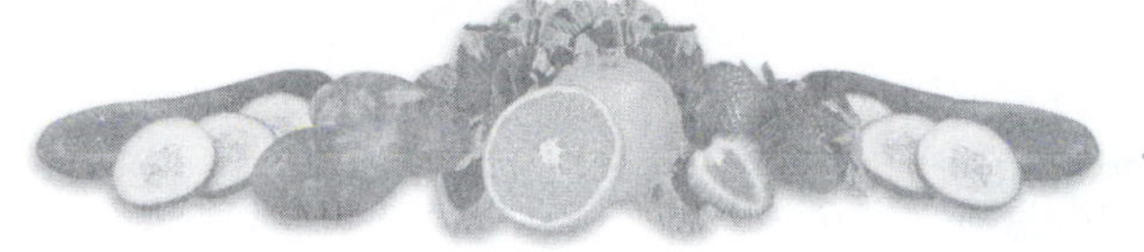

Box 12.2

Crack Babies: A Legacy of Abuse

The effects of a mother's abuse of drugs such as alcohol (Chapter 14) on her unborn baby have been known for some time, but the consequences of cocaine use by pregnant women were not suspected until the epidemic of "crack babies" began in the mid-1980s. Women often forgo many drugs, even heroin, when they are pregnant, but the craving for cocaine is so strong that addicts usually continue their habit during pregnancy. The deleterious effects of cocaine during this period are often twofold. First, habitual users neglect everything in seeking a cocaine high. As a result, women addicts often have no prenatal care and are in poor health due to faulty nutrition and lack of adequate rest. Second, there is now substantial evidence that using cocaine during pregnancy affects the fetus directly. Cocaine passes easily across the placenta, and it tends to remain longer in fetal tissues than in those of adults because the fetal liver is not sufficiently developed to break down the compound. A mother's crack binge can trigger spasms in a baby's blood vessels that can restrict oxygen and nutrient flow long enough to cause damage. Among the effects widely cited in the early 1990s were fetal stroke that led to the death of brain cells, malformed kidneys or genitals, and missing fingers or toes. In some instances, part of the intestinal tract died, necessitating immediate surgery to remove the necrotic section and rejoin the healthy parts. Spontaneous abortion rates are 5 to 10 times as high for cocaine users as for nonusers. For those women who do carry babies to term, babies are delivered prematurely and often with damaged organs. Crack babies stay in the hospital an average of 4 days longer than babies born to women who did not use the drug. The longer stay is necessitated by the babies' low birthweights (less than 5.28 lb) and their needs for additional treatment. They also need intensive care 1.5 times as often as those born to non-using mothers.

In 1991 it was estimated that additional hospital costs generated by cocaine babies exceeded $500 million per year. After birth, cocaine babies are often irritable and shun intimate contact, behavioral abnormalities that have been suggested to contribute later to their becoming child abusers. However, some studies have suggested that family environment is as important as drug use in causing low I.Q. and emotional problems.

Although the sensational reports of the number of crack babies and their problems published in the early 1990s now appear to have been exaggerated, subtle and costly effects were found in a 1998 analysis of 101 studies of babies born to known crack-using mothers. This study suggested that most crack babies are not hopelessly damaged and can become productive members of society with sufficient special educational help. The major long-term effect found was a low I.Q. associated with impaired language abilities. This effect translates to 3636–68,000 children (depending on the estimate) per year who could be diagnosed with special educational needs. The estimated cost of providing such special education could reach $352 million per year.

Obviously, the best solution to the problem of crack babies would be to eliminate the use of cocaine during pregnancy. Evidence indicates that if a woman stops using cocaine before the end of the first trimester, her baby will be almost as healthy as a child of a woman who abstained from the drug. Outpatient drug treatment programs have had some success, but too few are available and the treatment cost averages $6000 per person. The number of women needing such treatment is unknown, but the estimated costs of these programs are staggering. Even so, they are a sound investment when the costs of remediation are taken into account.

colonists turned more and more to tobacco farming. The Virginia monopoly on tobacco production granted by the queen was broken when settlers in Maryland began to cultivate the crop. Soon much of the eastern seaboard of the United States was engaged in tobacco production. For many years tobacco was the most important item of trade between England and North America.

Primitive cultivation of tobacco consisted of saving seeds from the previous year's crop and sowing them in a cleared area. Plants were harvested at maturity, and the leaves were dried in the sun. In contrast, modern tobacco farming is quite elaborate and employs various cultivation techniques for the many different cultivars and types of tobacco produced. On modern U.S. farms, tobacco is usually sown in seed beds and then the seedlings are transplanted into fields after 2 months of growth. The spacing of plants and the fertilizer regime vary with the final type of tobacco to be produced. Nitrogen-rich fertilizers are applied to encourage the development of large, pliable leaves suitable for cigar wrappers while the small, more brittle leaves favored for cigarettes are produced by limiting nitrogen and supplying phosphorus and potassium at critical stages.

While the crop is growing, the flower stalks and side shoots are usually removed to prevent the plants from diverting resources to seeds and low branch development. Plants producing leaves to be used for cigar wrappers are generally grown under cheesecloth, where the high humidity and protection from sunscald also promote the formation of large, thin, blemish-free leaves. Two to four months after planting, the crop is considered mature (Fig. 12.18). Good

FIGURE 12.14
Two species of tobacco, *Nicotiana tabacum* (right) and *N. rusticum* (left), were both domesticated by Native Americans and used for smoking.

FIGURE 12.15
Indians smoking a tabac (pipe). In the Amazon region of South America, tobacco is most commonly used as a snuff.
(Redrawn from Thevet, 1573.)

grades can be picked a leaf at a time, or whole stalks can be harvested. In either case the leaves or stalks are tied into bunches for drying and curing (Fig. 12.19).

During the curing process, the moisture content in the leaves is reduced from about 80 percent to 20 percent. Starches are converted to sugars, and some proteins are broken down enzymatically. Slow drying permits fermentation to take place but prevents the growth of molds and fungi. Curing can be done by circulating air or by smoking. Occasionally, bundles of leaves are sun-dried. After curing, the leaves are aged for a few months to a year. For cigarette and cigar fillings, the tobacco is remoistened before marketing and the veins and leaf bees (petioles) are removed. The softened leaves are then cut by machine into strips.

Depending on the final tobacco product, various substances can be added. Until 1994, it was generally thought that these substances—such as glycerin or diethyl glycol to retain moisture, cider concentrate, oil of hops, licorice, rum, or menthol for flavor—were safe. The American public was shocked in 1994 when the tobacco industry, under congressional pressure, released a list of 600 additives, several of which were considered harmful. More serious claims of targeting children in advertising or manipulating levels of nicotine were also

FIGURE 12.16
Mayan priests blew smoke to the four winds during religious ceremonies.
(Redrawn from a frieze at Palenque, Mexico.)

FIGURE 12.17
Despite James I's characterization of tobacco smoking as "A custome lothsome to the eye, hatefull to the nose, harmefull to the braine, dangerous to the lungs, and in the blacke stinking fume thereof, neerest resembling the horrible Stigian smoke of the pit that is bottomless" (*Counterblaste to Tobacco*), the British continued to smoke and went on to develop the genteel art of blowing smoke rings.

directed toward tobacco companies. Although the claims were denied by the industry, investigations showed that deliberate mixing of various tobacco cultivars and the addition of ammonia could increase absorption of nicotine. Nevertheless, cigarette manufacturers maintained that nicotine was non-habit-forming and that they did not "set" nicotine levels. Taking issue with these claims, four states (Florida, Minnesota, Mississippi, and Texas) brought lawsuits against the tobacco industry to recover Medicare costs for money spent treating diseases related to smoking. These states together won a settlement in 1997 and early 1998 of $40 million to be paid over 4 years. In November 1998, 46 additional states and five territories reached an agreement with cigarette producers by accepting a $206 billion plan (to be paid over 25 years) to settle their state lawsuits. The agreement also required cigarette makers to take down billboards, change their marketing practices, and finance campaigns to reduce or stop smoking by young people.

Tobacco smoking was widely enjoyed before 1880, but its popularity greatly increased after that year because of a change in the curing process. Before 1880 most tobacco was

FIGURE 12.18
A migrant worker harvesting tobacco on a farm in Virginia.
(Photo by K. Hammond, courtesy of USDA.)

FIGURE 12.19
Loading tobacco into a drying barn in Virginia as part of the curing process.
(Photo by K. Hammond, courtesy of USDA.)

cured by placing it directly over the hot smoke of a charcoal fire; after that year farmers began to use hot air brought to the drying rooms by flues. This indirect drying produced a milder form of tobacco than that produced by smoke-drying. Mild tobaccos were perfect for cigarettes, a relatively "genteel" way of smoking. However, such tobaccos produce an acid smoke when burned, whereas tobacco cured directly tends to produce an alkaline smoke. Acid tobacco smoke has little physiological effect on the body if it is simply puffed. Consequently, it must be inhaled to produce an exhilarating effect. The inhaled smoke is neutralized on the surface of the lungs, and the nicotine carried in the smoke is readily absorbed through the lung membranes. Chewers and smokers who inhale tend to become addicted to tobacco because they become physiologically dependent on the strong sensory reaction produced when nicotine is absorbed. Pipe and cigar smokers can become accustomed to smoking and dearly miss the habit if they quit, but in general there is little physical reaction to the withdrawal of nicotine since these smokers normally do not inhale.

Nicotine, the major alkaloid in tobacco is now known to be an extremely addictive substance that passes across the protective barrier of the brain faster than either heroin or caffeine. As are addictions to other substances, nicotine addiction is connected with the release of dopamine in the brain by acting on nicotinic acetylcholine receptors (Tables 12.2 and 12.3). It also stimulates the release of adrenaline and raises blood pressure. In its pure form, nicotine is a poison. Treated with sulfuric acid, nicotine yields a potent insecticide. In the doses usually consumed, nicotine is rarely toxic, but long-term effects include promotion of lung, bladder, breast, cervical, mouth, and throat cancers, heart disease, and stroke. Exposure to the tars and smoke cause yellowing of the teeth and fingernails and premature wrinkling of the skin. Smoking during pregnancy increases the risk of miscarriage and problems in early infancy. Children born to mothers who smoke tend to be underweight at birth and addicted to nicotine. In 1998, over 400,000 deaths in the United States were attributed to the effects of smoking tobacco. It is now clear that even secondhand smoke is carcinogenic and increases the risk of contracting pneumonia and bronchitis. A 1992 study by the U.S. Environmental Protection Agency (EPA) concluded that secondhand smoke causes 3000 lung cancer deaths a year among nonsmokers.

Belladonna, Henbane, and Mandrake

From Egyptian dynastic times through the Middle Ages one of the major sources of hallucinogenic compounds (and poisons) in Europe was a group of herbs belonging to the same family as tobacco. These species are rarely used today as sources of mind-altering drugs because of their toxicity. Historically, the most important were species of *Datura* (Fig. 12.20), *Hyoscyamus, Atropa belladonna,* and *Mandragora officinarum* (all Solanaceae) (Fig. 12.20). Many of these species were also used medicinally, and a few have become important modern sources of pharmaceutical compounds (see Chapter 11). All these plants have also been used as sources of murderous poisons from ancient Egyptian to modern times. The tropane alkaloids (see Fig. 11.17) (atropine, scopolamine, and hyoscyamine) these species contain are effective in slowing smooth muscle action. The alkaloids act by inhibiting acetylcholine receptors in both the central and peripheral nervous systems. Carefully measured amounts are thus useful in treating heart irregularities. In larger but sublethal doses the same alkaloids produce hallucinations.

Among the plants in this group, only datura is native to the New World (Fig. 12.21). Datura is particularly rich in scopolamine, which is considered the most hallucinogenic of the solanaceous alkaloids. In Central and South America seeds or roots of various species are eaten to produce visions, but even among people who use datura it is considered dangerous and too strong for general use. Usually datura was ingested only by boys during puberty rites or by trained shamans. In South America alcoholic beverages made from datura fruits were given to women and the slaves of dead warriors to stupefy them before they were buried alive with their deceased masters. Infusions of bark and leaves were also used by various tribes.

Species of datura native to Europe and Asia were less widely used than New World species in the same genus, but belladonna, mandrake, and henbane assumed great importance in Europe during the Middle Ages. The ancient use of solanaceous alkaloids in witchcraft and folk medicine led European scientists after the Renaissance to investigate their

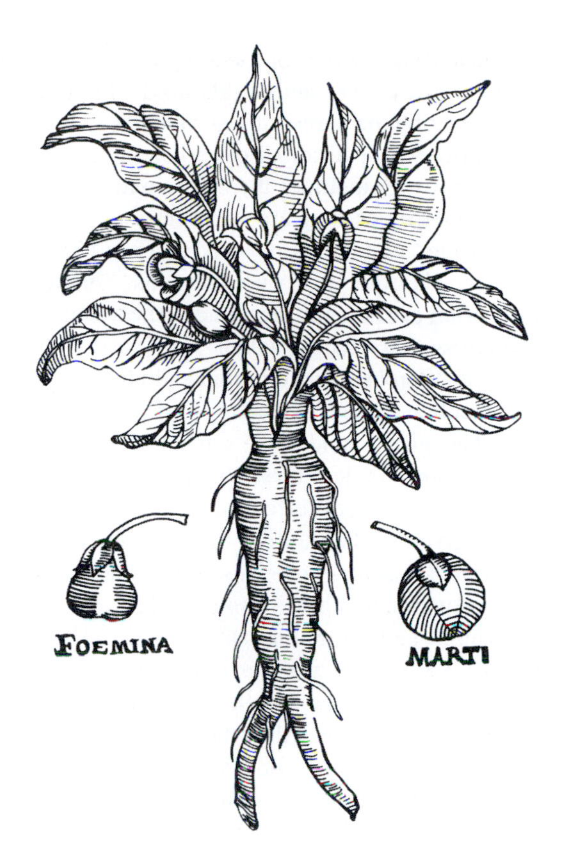

FIGURE 12.20
Mandrake, or *Mandragora* (from "man," and "dragon"), was long thought to be an aphrodisiac because of the human shape of its root. This belief is reflected in the biblical story of childless Rachel (Genesis 30), who exchanged an evening with her husband (Jacob) in return for her sister Leah's son's mandrakes. Mandrakes were considered to be male or female. The recognition of male and female mandrakes is odd in view of the fact that the plants are hermaphroditic. The practice presumably arose because of the belief that the plants represent entire human figures complete with sexual organs. Some roots would have part of their configuration resembling male genitalia; others, female. The smaller figures labeled *foemina* and *marti* are not depictions of the fruit, which is a capsule.
(From Matthias de L'Obel, 1581, Plantarum seu Stirpium icones, Antwerp.)

FIGURE 12.21
Datura acquired its common name, jimsonweed, in 1705 when British soldiers in Jamestown, Virginia, inadvertently ate the leaves in a salad and went mad for 11 days.

action and eventually to discover their medicinal properties. Similarly, their reputations as hallucinogens and poisons prompted authors and artists to include them in the writings and paintings of the time.

Atropine, obtained from either belladonna or henbane, can be absorbed directly through the skin, and during the Middle Ages witches rubbed ointments of fat or oil containing it on their bodies. The psychological sensations produced give one the feeling of flying. While under the influence of atropine, witches were supposed to be transported to rendezvous with spirits or demons. These meetings, called sabats, are often depicted in European art of the period. The modern portrayal of a witch riding a broom may stem from these ancient rites (Fig. 12.22). The association of witches with flying may come from the hallucinogenic sensation of being transported, and the broom purportedly represents the stick used to apply atropine-impregnated ointments to vaginal membranes for absorption.

Another figure commonly attributed to visions caused by solanaceous alkaloids is the werewolf. Like the jaguars and snakes seen when Native South Americans use caapi, wolves were associated with trances induced by henbane, belladonna, and related species.

Caapi

One of the most important groups of plants used by South American Amazonian tribes as a source of hallucinogenic compounds are members of the genus *Banisteriopsis* (Malpighiaceae), primarily *B. caapi* (Fig. 12.23). Depending on the tribe, this species is called caapi, ayahuasca, yaje, or cipo. Infusions of mashed bark or

FIGURE 12.22

Extracts of belladonna or henbane were mixed with fat or oil and applied to the skin. Atropine, the active compound in these plants, is absorbed through the skin or membranes and produces the sensation of flying, promoting the idea of witches riding on broomsticks.

stems are usually prepared, but stems can also simply be chewed to release the active alkaloid. People using ayahuasca describe the perceptions produced as feelings of having experienced death or of experiencing a separation of spirit and body. Sometimes the hallucinations are pleasant, and at other times they are apparently terrifying with vivid apparitions of jaguars or snakes (Fig. 12.24). Like many other alkaloid rich preparations, mixtures of ayahuasca often cause vomiting and diarrhea. Unlike many other psychoactive drugs taken by native peoples, ayahuassca is often used in communal rituals. It is thought that this practice enhances feelings of clairvoyance or telepathy. The group, under the supervision of a leader called a *yaquero,* is led through chants to experience shared hallucinations. This effect was reflected in the name *telepathine,* given to the mixture of active chemicals first isolated from *B. caapi.* Since this initial chemical work, the active compounds have been further purified and named harmine and harmaline (Fig. 12.25). These compounds, like LSD, have structures similar to that of the brain neurotransmitter serotonin (see Fig. 12.25) and mimic its effects. In some tribes the rituals commemorate an act of incest by a deity. In almost all cases the ceremonies are filled with sexual symbolism, but there is no sexual gratification during the experience.

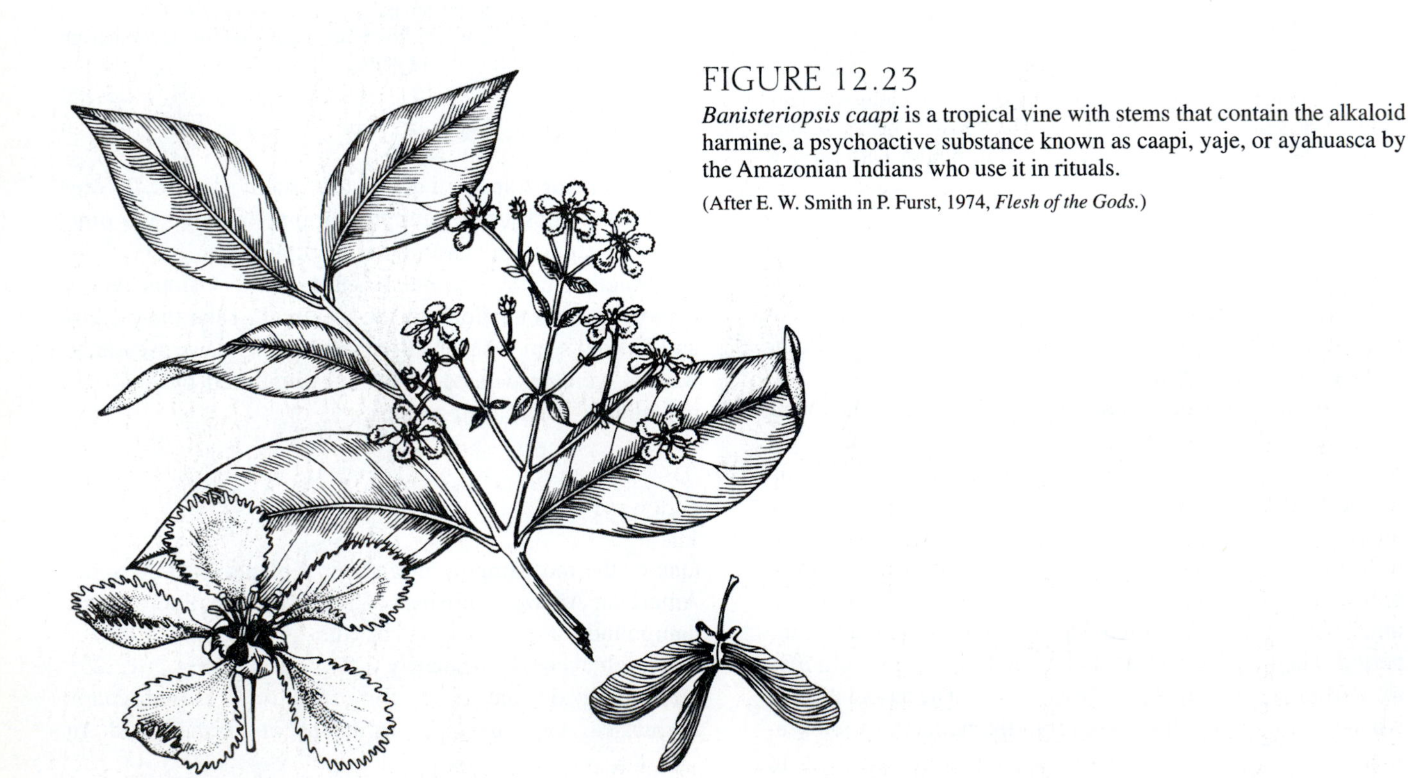

FIGURE 12.23

Banisteriopsis caapi is a tropical vine with stems that contain the alkaloid harmine, a psychoactive substance known as caapi, yaje, or ayahuasca by the Amazonian Indians who use it in rituals.

(After E. W. Smith in P. Furst, 1974, *Flesh of the Gods.*)

FIGURE 12.24
Many drugs produce hallucinations featuring animals. People under the influence of ayahuasca often report seeing jaguars or reptiles such as snakes. The solanaceous (tropane) alkaloids are associated with visions of wolves.
(Collage of animals from J. Harter, 1979, *Animals*. New York, Dover.)

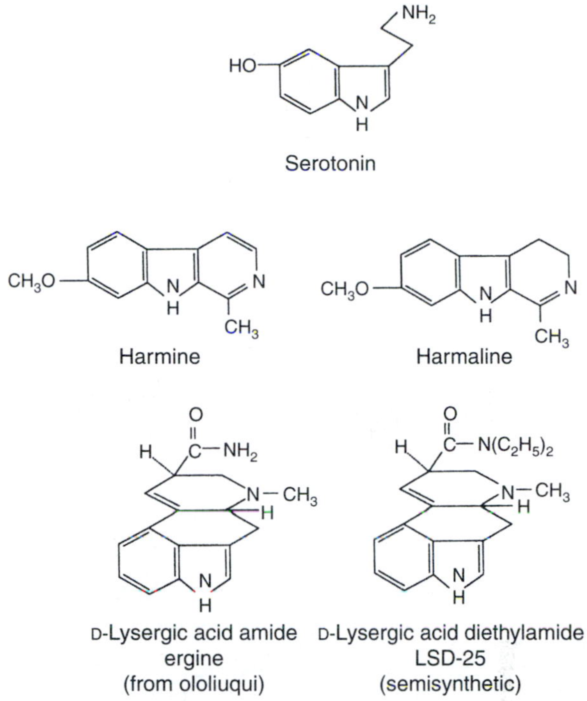

Serotonin

Harmine

Harmaline

D-Lysergic acid amide
ergine
(from ololiuqui)

D-Lysergic acid diethylamide
LSD-25
(semisynthetic)

FIGURE 12.25
Serotonin, a naturally produced neurotransmitter, and its mimetic alkaloids, harmine and harmaline (from *Banisteriopsis caapi,* Malpighiaceae), D-lysergic acid amide (from ololiuqui), and semisynthetic LSD-25 (D-lysergic acid diethylamide). Although D-lysergic acid amide structurally resembles its synthetic counterpart LSD-25, it is only one-tenth as powerful.

Kat

Americans are accustomed to reaching for a cup of coffee, tea, or another caffeine-containing beverage to get a "lift," but in other parts of the world similar effects are produced by consuming plants with different but equally effective stimulatory alkaloids. Among these is kat (also spelled chat, khat, or qat), *Catha edulis* (Celastraceae), a species native to Ethiopia that is now widely grown along the eastern edge of Africa. The alkaloids are ingested by chewing wads of freshly cut leaves usually mixed with small amounts of lime (calcium oxylate). Each lump, or quid, is masticated for about 10 minutes until all the juice has been expressed. The remaining cellulose mass is then swallowed. Kat produces effects similar to those of amphetamines, and its use eventually leads to addiction. Despite its stimulatory effects, the use of kat is accepted by Islamic people and its use is widespread among African Muslims.

Kava

In the South Pacific the official inebriating beverage is not alcohol but kava, made from a species of the pepper family, *Piper methysticum* (Piperaceae). Two methods are used to make the beverage. The traditional method is to chew fragments of the root and then spit out the masticated quid. The quid is then soaked in cold water or coconut milk, and the liquid is drunk. Chewing ruptures the cells of the root and appears to increase the potency of the drink. The second method, most commonly used today, is to grind the roots, soak them in water, and then filter the solution. Kava is used socially and served in kava bars. The beverage relaxes the body, reduces pain, and produces a mild euphoria. The effects are likened to those produced by diazepam (Valium). It does not dull the senses as alcohol does and appears not to be habit-forming. However, if used habitually, kava leads to skin lesions that will heal when use is curtailed. The compounds responsible for the effects are lactones that occur in various proportions in different plants. Some plants have a combination of compounds that causes nausea; others merely produce a pleasant sensation without a hangover.

Mescal Bean

Many species of the bean family (Fabaceae) are rich in alkaloids, and some of them have been used as hallucinogenic agents by various New World Indian tribes. Mescal, or red bean, *Sophora secundiflora* (Fig. 12.26), a common ornamental shrub or small tree of the American Southwest, was used by

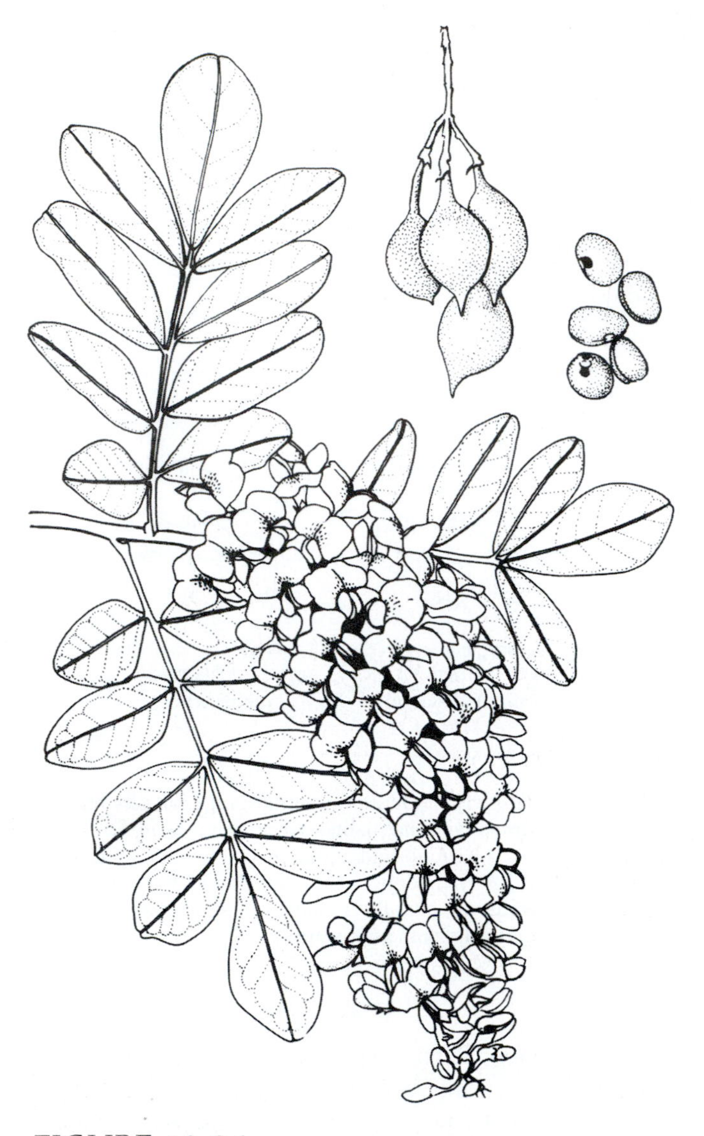

FIGURE 12.26

Sophora secundiflora, or Texas mountain laurel, is valued as an ornamental plant because of its fragrant purple flowers and glossy green foliage. The red and black seeds, sometimes called mescal beans, contain cytisine, an alkaloid that has been used to produce hallucinogenic trances but is extremely toxic and can produce respiratory failure and death.

the native peoples of that region to induce trances. The plant has received some recent publicity because mescal beans have been confused with mescaline (see Peyote on the next page) and unknowing youths have been tempted to try them. However, *Sophora* contains cytiscine, a much more dangerous alkaloid than mescaline and one that can easily cause death from respiratory failure if a slight overdose is taken.

Ololiuqui

Before 1937 no species of the morning glory family was known with certainty to be used for psychoactive purposes. In 1955 Humphrey Osmond described the use of *Rivea corymbosa* (Convolvulaceae) (Fig. 12.27) by Mexican natives, and in 1960 T. MacDougall detailed the use of another member of the family, *Ipomoea tricolor,* for altering perceptions. Still more startling was the discovery in 1960 that the seeds of these plants contained D-lysergic acid amide (LSD) (see Fig. 12.25). Before that time lysergic acid alkaloids were thought to be restricted in nature to a new fungi, including ergot (*Claviceps purpurea*), a fungus that infests grains. Eating rye infested with ergot causes ergotism, which can have mild to severe repercussions ranging from altered behavior to death (Fig. 12.28). Synthetically produced LSD had become part of the American counterculture in the 1960s and was regarded as something modern and exciting. Osmond and MacDougall's discoveries showed that the effects of these kinds of compounds had been recognized and utilized by native peoples for perhaps thousands of years.

Both of the convolvulaceous species now known to be used are called *ololiuqui* (pronounced *o-low-lee-oo-key*) (Fig. 12.27) by the peoples who use them. Very carefully measured quantities of seeds are ingested, usually only by experienced individuals during divinatory rituals. Because there is little difference between the quantities of seeds that produce vivid hallucinations and the quantities that produce death, there has been little lay experimentation with the seeds despite the fact that one of the species is a commonly cultivated ornamental plant.

FIGURE 12.27

Ololiuqui (*Rivea corymbosa*), a white-flowered vine of the morning glory family, contains the psychoactive compound, D-lysergic acid amide (LSD) (Fig. 12.25). True morning glories of the species *Ipomoea tricolor* are also called ololiuqui and have been used to induce hallucinations. Because of the extremely fine line between effective and lethal doses, the seeds of morning glories have rarely been used even by primitive peoples.

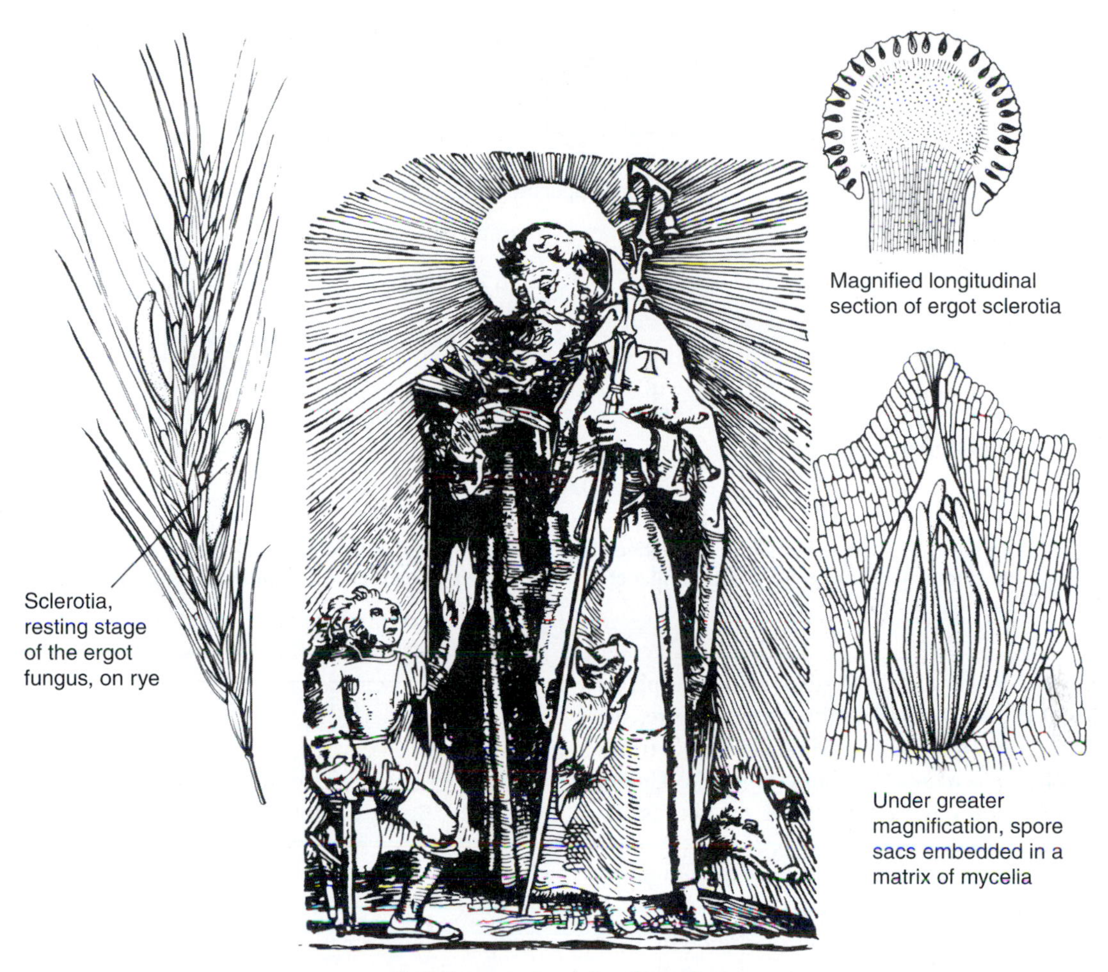

FIGURE 12.28
Ergotism is caused by the ingestion of compact masses of hardened tissue, called sclerotia, of the ergot fungus that parasitizes rye grain. Epidemics of ergotism were especially common among European peasants who unwittingly ate infested rye. The symptoms of intense burning pains, with the victim's limbs eventually becoming gangrenous, are depicted by the burning hand in this redrawn woodcut produced in Strassbourg in 1540. St. Anthony, also shown in the drawing, reflects the common name of a disorder, "St. Anthony's fire," sometimes applied to ergotism. In the late sixteenth century ergot was commonly employed by midwives to quicken labor and reduce the incidence of postpartum bleeding. The administration of ergot became a routine procedure after childbirth until medical practitioners realized its dangers and found other drugs that had the same beneficial effects.

Peyote

Several species of the cactus family are used as sources of hallucinogenic compounds. The best known is peyote, *Lophophora williamsii* (Cactaceae), a small, globose gray-green cactus native to the Rio Grande valley of Texas and northern Mexico (Fig. 12.29). No one is sure when peyote was first used, but sixteenth-century reports by European explorers describe its use by the Aztecs as a divinatory plant. These accounts refer to peyote as the "diabolic root,"

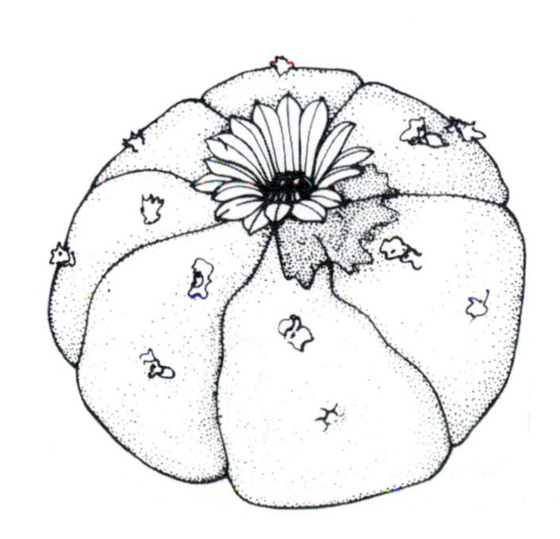

FIGURE 12.29
Mescaline is the psychoactive compound in the gray-green peyote buttons that are native to the southwestern United States and adjacent Mexico. Several other species of cactus also contain the alkaloid and cause hallucinations.

OH
HO — $CHCH_2NH_2$
HO

Norepinephrine

CH_3O — $CH_2CH_2NH_2$
CH_3O
OCH_3

Mescaline

CH_3O — $CH_2CH{=}CH_2$
O
H_2C — O

Myristicine

CH_3O — $CH_2CH{=}CH_2$
CH_3O
CH_3O

Elemicin

FIGURE 12.30
The chemical structure of mescaline, the dominant psychoactive alkaloid in peyote and related alkaloids which mimic the action of norepinephrine. Myristicine and elemicin are compounds with similar effects found in the spices mace and nutmeg (Chapter 8).

because the Spanish observed the Aztecs using the plants ritualistically. After the collapse of the Aztec empire, the use of peyote survived among a few Mexican Indian tribes such as the Huichol, Cora, and Tarahumara. In the United States the Plains Indians began to use it as late as the 1880s.

Indians harvested the plant by cutting off the top of the stem and leaving the sturdy taproot for regeneration. The stem tips, or buttons, were eaten fresh or dried for later consumption. Dried buttons require a period of softening and mastication before they can be swallowed. The initial reaction after swallowing is nausea. One or two hours later queasiness disappears and is replaced by kaleidoscopic illusions of vivid colors, sensual hallucinations, and distorted perceptions of time. During this period, which can last from 5 to 12 hours, the faithful report hearing the voices of their ancestors, who help them diagnose and cure their problems.

Thirty to 40 different alkaloids and numerous chemicals are ingested when peyote is consumed. The most active compound present is the alkaloid mescaline, which acts as a mimic of norepinephrine (Fig. 12.30). Isolated in 1896, mescaline was the basis for research into psychedelic compounds for the next 50 years.

There is no evidence that either peyote or pure mescaline is addicting, but both the plant and the compound are illegal to possess or sell in the United States. However, one religious sect, the Native American Church, uses peyote as an integral part of its services. In a case that went before the U.S. Supreme Court, the church received an exemption from the prohibition against peyote use. Originally founded by natives in northern Mexico, the sect holds Christian services modified by peyote and espouses as doctrine brotherly love, abstinence from alcohol, and high standards of moral conduct. Today there are over a quarter million members of the church in the United States (Fig. 12.31).

FIGURE 12.31
"A white man uses prayers out of a book, these are the first words on his lips. But with us, peyote teaches us to tell from the heart," says a Cheyenne peyote leader. Because of its importance in their worship services, peyote use is legal in the Native American Church. Peyote ceremony on a Kiowa reservation.

(Photo taken in 1892; courtesy of the Smithsonian Institution National Anthropological Archives.)

Plant Poisons

So many plants are poisonous to varying degrees that an enumeration here would be impossible. Regional agricultural stations usually publish lists of local plants that are dangerous to

FIGURE 12.32
To prepare strychnine poison, this man is scraping the bark of a *Strychnos* liana into a banana leaf, percolating water through it, and collecting the condensate.
(Photo courtesy of R. Wallace.)

humans or livestock, and many horticulture books indicate which ornamental plants can be toxic if eaten (see Kingsbury in Additional Readings). Some of these plants have been actively used by humans for their poisonous compounds. These are the ones in which we are interested here.

As was indicated earlier, numerous plants used for their psychological effects are poisonous in large quantities. It is difficult to know exactly how people discovered which plants were or were not toxic and the quantities that could be ingested safely. Ancient Greek and Roman medical records attest to an early interest in documenting the occurrence and usefulness of plant toxins. Administration of plant poisons as a form of capital punishment is well known from the story of the death of Socrates, who was sentenced to drink the juice of poison "hemlock," *Conium maculatum* (Apiaceae). In the Middle Ages "succession powders," some of which contained solanaceous alkaloids, were deviously employed to eliminate heirs to the throne. The employment of these powders became so common that the wealthy and powerful hired tasters as protection against sudden death from "food poisoning." Today scientists often seek out plants known by native people to be poisonous. Since many poisons are either alkaloids or steroids, such plants are potential sources of drugs or commercial poisons for use as insecticides, herbicides, or fungicides.

Some primitive tribes in the tropics use arrow or fish poisons such as curare and barbasco. Curare (Fig. 12.32) is made from extracts of the bark and stems of *Chondodendron tomentosum* (Menispermaceae) and from seeds of species of *Strychnos,* particularly *S. nux-vomica* (Loganiaceae). *Chondodendron,* a vine native to South America, contains tubocurarine. Species of *Strychnos* occur natively in South America, Southeast Asia, and Indonesia, and many yield the potent alkaloids curarine, strychnine (see Fig. 11.7), and toucine. Curare alkaloids, primarily tubocurarine chloride, have had limited applications in medicine. Because they block neuromuscular activity, they help relax muscles that remain contracted under anesthesia. Strychnine was formerly used medicinally as a central nervous system stimulant but has been replaced by safer synthetic compounds.

Barbasco and tuba are fish poisons used by Amazonian peoples. The first is prepared from various species of *Derris,* such as *Derris elliptica,* and the second from *Lonchocarpus nicou* (both Fabaceae). Recently, species of both genera have been shown to be effective insecticides because they contain rotenone. This isoflavanoid compound is fatal to insects but comparatively nontoxic to humans and it leaves no residue.

Other "natural" insecticides have been extracted from pyrethrum (*Chrysanthemum coccineum, C. cinerariifolium,* and *C. marschallii*) (Fig. 12.33) and from tobacco. Insecticidal compounds from species of *Chrysanthemum* are sold as Persian insect powder or Dalmation. Pyrethrins, the active compounds, are esters that break down quickly into acids and alcohols. Since they stun insects but do not instantly kill them, pyrethrins are usually mixed with other substances that ensure the death of the insect. In some

FIGURE 12.33
In Kenya women pick pyrethrum flowers that will be used as a natural source of a powerful insecticide.
(FAO photo.)

regions, notably Kenya, Tanzania, and parts of northeastern South America, pyrethrum is grown commercially as an insecticide.

A recent addition to the plant arsenal of insecticides comes from the neem tree, *Azadirachta indica* (Meliaceae), which is native to India and Myanmar. Farmers in those countries have traditionally placed leaves of the neem tree in bins of stored grain to kill weevils and have sprayed crops with a leachate of the leaves. Studies of the neem tree have found 20 compounds called limonoids that are structurally similar to insect hormones. When insects eat these compounds, they fail to molt or produce chitin for their exoskeletons. To date 200 insects tested are affected by neem's limonoids. Several companies in the United States are testing the compounds for eventual release in the American market.

It is estimated that there are at least 1200 plants potentially of use as sources of insecticides. However, the use of botanically-derived insecticides has decreased steadily in use since 1966. Pyrethrum is the only plant product that is today commercially important as an insecticide.

Undoubtedly, people will continue to search for additional plant sources of medicines, psychoactive drugs, and poisons. In such a search, Shakespeare's advice on plant poisons still applies:

> Virtue itself turns to vice, being misapplied,
> And vice sometimes by action dignified.
> Within the infant rind of this small flower,
> Poison hath residence, and medicine power.
>
> —*Friar Laurence in* Romeo and Juliet

13

Stimulating Beverages

The drinking of coffee, tea, and chocolate milk has become such an integral part of American life that people take it for granted. Many Americans profess that they could not face the morning without a cup of coffee, and a coffee break or tea time is enjoyed by most people in the Western world who need a lift during the day. People drink these beverages not only for the boost they provide but for their flavor as well. Other stimulating beverages, such as mate, kola, and guaraná, have become locally important for the same reasons, but coffee, tea, and cocoa reign supreme on a worldwide basis (Fig. 13.1).

All these beverages are stimulants because they contain the alkaloid caffeine and its relatives (Table 13.1 and Fig. 13.2), which cause particular physiological reactions in humans. Caffeine is a central nervous system stimulant and a mild diuretic. The caffeine in a cup of coffee reaches the bloodstream 5 minutes after the liquid is swallowed. As it circulates throughout the body, it stimulates the heart, increases stomach acidity and urine output, and causes a 10 percent rise in the metabolic rate. Caffeine makes a tired person more alert because it mimics the feelings produced when the body releases the hormone adrenaline. In excessive doses, however, caffeine can produce unpleasant symptoms. One gram of the compound (about 7 to 9 cups of coffee) causes anxiety, headache, dizziness, insomnia, heart palpitations, and even mild delirium. Heavy drinkers of tea or coffee can develop a tolerance to caffeine and can even suffer withdrawal symptoms such as headaches when they quit the habit. Caffeine ranks today as the most widely used psychoactive drug in the world not only because of the large number of coffee and tea drinkers but also because it is added to numerous soft drinks and various medications (Table 13.2). Unfortunately, children who have neither the body mass nor the tolerance for large amounts of caffeine consume appreciable quantities of soft drinks. There is some debate about the permanent effects of moderate caffeine intake on adults, but doctors now advise pregnant women to reduce or eliminate caffeine consumption because studies have shown that babies born to mothers who consume caffeinated beverages have a 4 percent lower birthweight than do those born to mothers who did not drink such beverages.

The way in which caffeine and related compounds work is something of a mystery, but biochemical studies have suggested that these compounds block adenosine, a chemical messenger that occurs naturally in the body. When adenosine attaches to

TABLE 13.1 Active Compounds in the World's Major Stimulating Beverages*

PLANT PART	CAFFEINE	THEOBROMINE	POLYPHENOLS
Coffee, unroasted, dried	1–1.5	—	—
Teas, dried leaves	2.5–4.5	—	25.0
Cacao			
Dried nibs	0.6	1.7	3.6
Fresh cotyledons	0.8	2.4	5.2
Kola, fresh seeds	2.0	—	—
Guaraná, dried fruit	3.0–4.5	—	—

*Figures given in percent weight. Amounts of the compounds in a particular beverage depend on how the beverage is made.

FIGURE 13.1
This drawing represents the world's three most important nonalcoholic beverages and the civilizations associated with them: coffee in an Arab's cup, tea being drunk by a Chinese, and cocoa in a Native American's goblet.

(From Dufour, *Traitez nouveaux et curieux du café, du thé, et du chocolat,* Lyons, 1671.)

TABLE 13.2 Amounts of Caffeine in Commonly Consumed Beverages and Medicines

ITEM	CAFFEINE, mg
Coffee	
5-oz cup, drip method	146
5-oz cup, percolator method	110
5-oz cup, instant	53
5-oz cup, decaffeinated	2
Tea	
5-oz cup, brewed 1 min	9–33
5-oz cup, brewed 3–5 min	20–50
12 oz, canned	22–36
Cocoa and chocolate	
6 oz, made with canned powder	10
1 oz milk chocolate	6
1 oz (1 square) baking chocolate	35
Soft drinks	
12 oz Mountain Dew	52
12 oz Dr. Pepper (regular or sugar-free)	37–38
12 oz Pepsi, regular	37
12 oz Coca-Cola	34
Nonprescription drugs	
Stimulants	
NoDoz (standard dose)	200
Pain relievers (standard dose)	
Excedrin	132
Midol	65
Anacin	64
Cold remedies	
Dristan	32
Diuretics (standard dose)	
Aqua-Ban	200
Weight-control aids (daily dose)	
Prolamine	280
Dietac	200

Source: Adapted from *Consumer Reports,* vol. 46. 1981, pp. 598, 599. Not all soft drinks and medications containing caffeine are included. Consumers should read the labels of such products to determine whether they contain caffeine. Several manufacturers have recently removed the caffeine from the form of their medications labeled "caffeine-free."

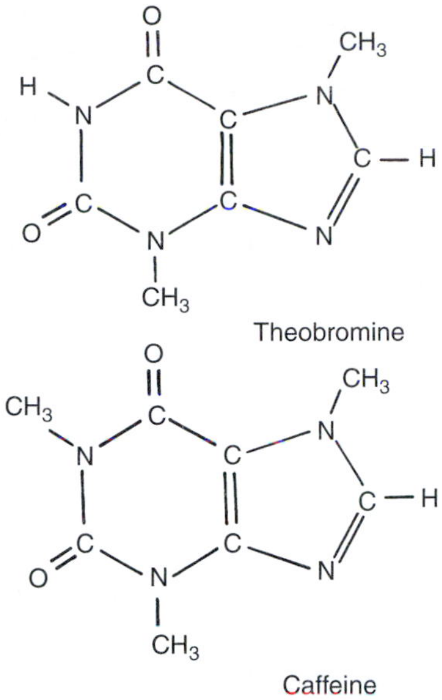

Theobromine

Caffeine

FIGURE 13.2
Caffeine and theobromine both have stimulating effects.

special receptors on the surfaces of brain cells, it dilates arteries, suppresses locomotor activity, and produces sedation. Caffeine and the related compounds theophylline and theobromine block the attachment of adenosine to the surfaces of these cells and thus maintain or cause a feeling of alertness and activity. These compounds, especially theophylline, also relax the smooth muscles of the bronchial system. As a result, theophylline is used to treat asthma.

In view of their worldwide importance, we discuss coffee, tea, and cacao in detail, looking at the places where humans first used them for beverages, their discovery by Europeans, their methods of cultivation, and the processing necessary before each reaches a cup or a glass. Because of their limited use, mate, kola, and guaraná are discussed only briefly. Table 13.3 lists the plants that yield the stimulating beverages and their adjuncts discussed in this chapter.

Coffee

Coffee is one of the world's most important commodities in terms of the value traded annually on the international market. Although the genus *Coffea* (Rubiaceae) is native to eastern Africa, comparatively few Africans drink coffee. The earliest records of coffee use come from Ethiopia, where natives chewed leaves and fruits gathered from wild trees growing in the understories of montane forests. A mixture of roasted and ground, or green, coffee fruits and fat was taken along on hunts as a survival staple similar to the pemmican (a mixture of dried buffalo meat and fat) used by Native Americans. Caffeine that dispelled fatigue and relieved hunger was leached out of the leaves or fruit during chewing. Beverage making seems to have been a later development, and some authors have suggested that an alcoholic beverage was made from the fruits before a nonalcoholic one was produced.

Coffee probably arrived in Yemen (Fig. 13.3) some time before the fourteenth century. Despite the name *Coffea arabica,* Ethiopians were the first to brew coffee (Fig. 13.4). Coffee drinking spread to Egypt by 1510 and to Italy by 1616. Alarmed at the rapidly spreading popularity of coffee drinking, the priests of Vienna urged the Pope to ban the drink as it had come from the land of infidels. After trying the brew, however, Pope Clement VIII promptly baptized it, making it a Christian beverage. By 1650 coffee had arrived in England, where it became an important part of the social and political environment. England alone had over 3000 coffeehouses by 1675, many of which functioned as forums for political and religious debates. The increasing attendance at

TABLE 13.3 Plants Discussed in Chapter 13

COMMON NAME	SCIENTIFIC NAME	FAMILY	NATIVE REGION
Achiote	*Bixa orellana*	Bixaceae	Mexico
Cacao	*Theobroma cacao*	Sterculiaceae	South America
Chicory	*Cichorium intybus*	Asteraceae	Europe
Coffee			Africa
Arabian	*Coffea arabica*	Rubiaceae	
Liberian	*C. liberica*		
Robusta	*C. canephora*		
Guaraná	*Paullinia cupana*	Sapindaceae	South America
Kola	*Cola nitida*	Sterculiaceae	West Africa
Mate	*Ilex paraguariensis*	Aquifoliaceae	South America
Tea	*Camellia sinensis*	Camelliaceae	China

FIGURE 13.3
The coffee plant on this ten rial note highlights coffee's importance in Yemeni history and culture.

FIGURE 13.4
A lithograph of an Arabian coffee service, redrawn from *A Thousand and One Arabian Nights.*

those houses and the spread of the ideas discussed within them so alarmed King Charles II that he labeled them "seminaries of sedition" and tried to have them closed. The uproar that resulted from his edict forced Charles to rescind his order, and soon the demand for coffee sent all of Europe looking for new sources of the seeds.

For years the Arabs monopolized the coffee trade and tried to prevent the cultivation of coffee by other countries. They shrewdly dipped the seeds in boiling water before marketing to kill the embryos and prevent germination. Eventually, the Dutch managed to secure live seeds from Mocha, a city in Yemen on the coast of the Red Sea, the traditional Arabian source of coffee. With these viable seeds, the Dutch started extensive plantations in the East Indies, breaking the Arabian monopoly. Trees from these plantations were sent to the botanical garden in Amsterdam in 1706, but only one survived the journey. Seeds from this single tree were later given to other botanical gardens in Europe, including the Jardin des Plantes in Paris.

The tree in the conservatory of the Paris garden was considered a curiosity by most people, but one man realized its potential value. This young Frenchman, Gabriel de Clieu, argued that if coffee grew well for the Dutch in the East Indies, it should grow equally well for the French in the West Indies. Stories vary as to how de Clieu obtained seeds, but they all agree that in about 1723 he arrived in Martinique with a single surviving offspring of the Parisian tree. From this small island, seeds were dispersed throughout the Antilles and to French Guiana. De Clieu was not the first to introduce coffee to the New World. Records show that the French had introduced coffee earlier into Haiti, and the Dutch had started plantings in Suriname in 1717. Eventually, growers in Brazil (Fig. 13.5), which now leads the world in coffee production (Table 13.4), obtained seeds from either Suriname or French Guiana. By 1729 Brazil was producing 200,000 bags of coffee annually.

Although *Coffea arabica* has figured most prominently in history and now accounts for more than 90 percent of the world's coffee production, it is not the only species of the genus that yields coffee-producing seeds. *Coffea canephora,* known as robusta coffee, is the source of 25 percent of the world's coffee, and *C. liberica,* or Liberian coffee, contributes about 1 percent. The predominant use of *C. arabica* can be attributed to several of its characteristics. First, it is self-fertilizing and self-compatible. The successful introduction of coffee into the New World was possible because the plant sent to Amsterdam was able to produce viable seed by self-fertilization, whereas the other two beverage-yielding species are self-incompatible. More important, however, coffee made from *C. arabica* has a better flavor than that made from the other two species. Robusta coffee is grown primarily for use in blended coffees or to make decaffeinated or instant coffee, in which the taste is disguised or altered. Liberian coffee, the most bitter of the three, is used primarily as a filler in mixtures with other coffees.

All cultivated species of *Coffea* are small trees with glossy leaves and fragrant, jasminelike white flowers produced in the leaf axils (Fig. 13.6). The fruits take 7 to 9 months to mature into what is called a cherry. Because it comes from an inferior ovary, the cherry is an accessory "berry" with a tough outer layer composed of the floral cup and exocarp, the fleshy mesocarp, and the thin, fibrous endocarp, or "parchment" (Fig. 13.6). Within the endocarp are two seeds pressed together in such a way that the inner side of each one is flattened. Each seed, commonly called a bean, is surrounded by a thin, silvery seed coat. The seed itself is composed mostly of endosperm surrounding a small, curved embryo.

Because coffee trees cannot tolerate sustained freezing, cultivation is restricted to tropical and subtropical latitudes. An average coffee plant produces its first crop after it is 3

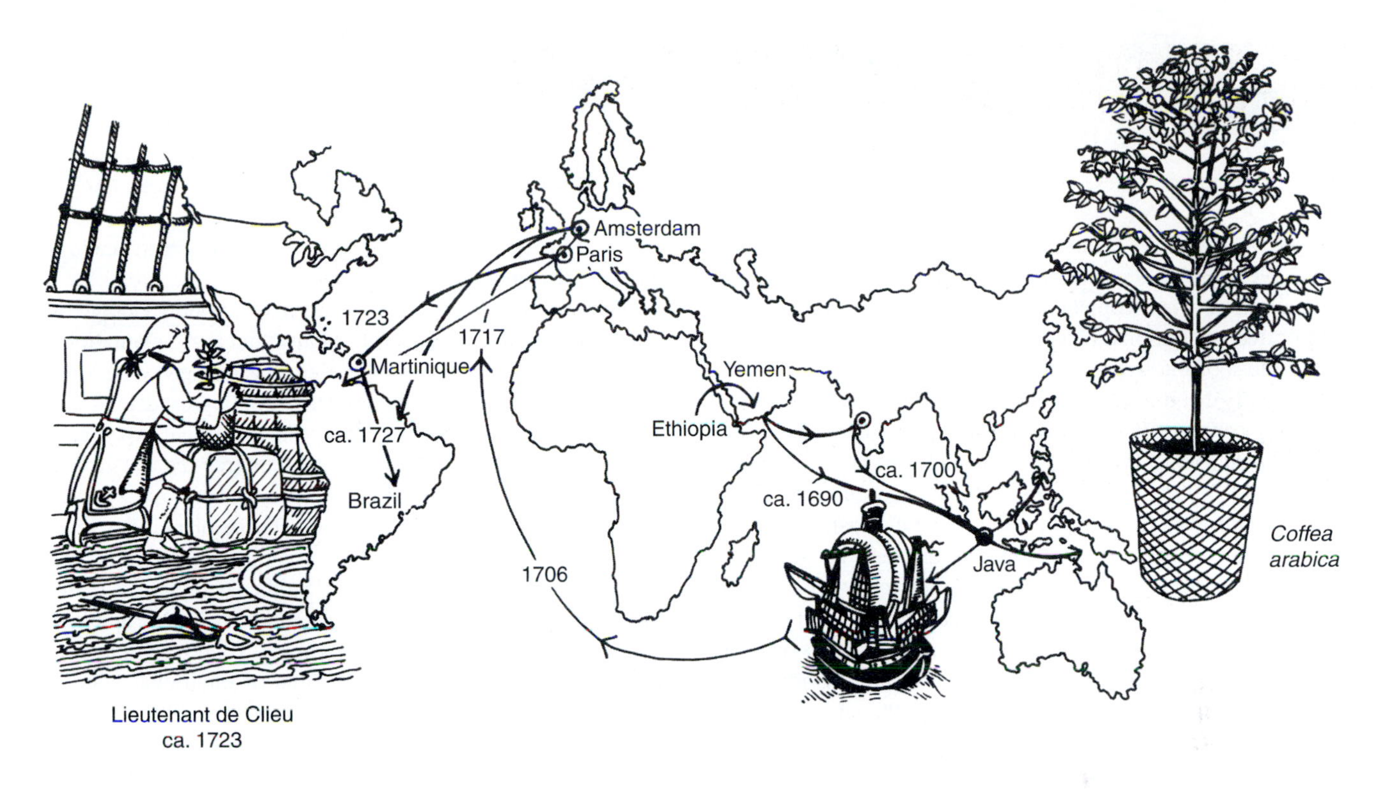

FIGURE 13.5

From its origins in Africa, virtually one clone of *Coffea arabica* spread around the world and established successful coffee production in the West Indies and South America. On the left is Lieutenant de Clieu tending the coffee plant he brought from France to Martinique.

(Adapted from N. W. Simmonds. 1976. *Evolution of Crop Plants.* New York, Longman.)

TABLE 13.4 1999 World Production of Coffee, Tea, and Cocoa

COMMON NAME	TOP 5 COUNTRIES	1000 METRIC TONS	TOP CONTINENTS	1000 METRIC TONS	WORLD, 1000 METRIC TONS
Coffee, green	Brazil	1,630	South America	2,576	6,476
	Colombia	648	Africa	1,232	
	Indonesia	455	North and Central America	1,136	
	Ivory Coast	365			
	Mexico	303	Asia	1,465	
			Oceania	66	
Tea	India	749	Asia	2,426	2,872
	China	722	Africa	371	
	Sri Lanka	280	South America	62	
	Kenya	220	Oceania	7	
	Indonesia	152			
Cacao beans	Ivory Coast	1,153	Africa	1,915	2,897
	Ghana	409	South America	384	
	Indonesia	350	Asia	468	
	Brazil	205	North and Central America	84	
	Malaysia	100			
			Oceania	46	

Source: FAOSTAT Database Gateway. Online at http://apps.fao.org/ (under "All databases" then Crops, Primary).

Box 13.1

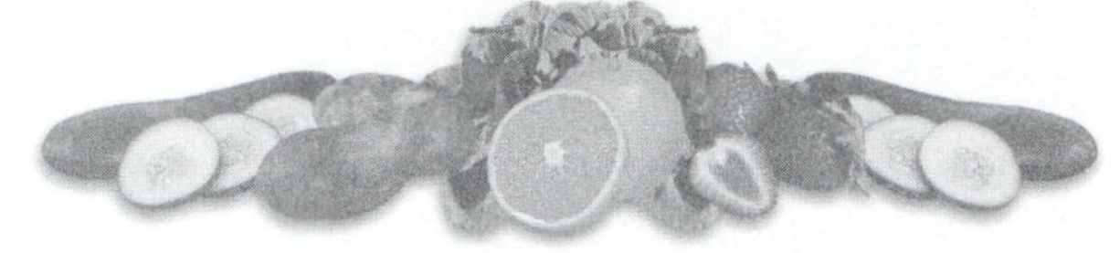

The Ecology of Cafetales

An Old World native, coffee is now most extensively grown in the New World and cultivated on mountain slopes naturally covered with tropical rain forest. Turning rain forest into coffee plantations, or cafetales, might seem like a sure road to loss of biodiversity, but traditional coffee cultivation has actually been a blessing in disguise. In Latin America, where 32 percent of the world's coffee is grown, coffee plantations have served for years as refuges for native plants and animals. Compared with many kinds of agriculture, traditional coffee plantations disturb the forest comparatively little. Arabian coffee, *Coffea arabica,* prefers to grow as understory shrubs, so mature, native rain forest trees were traditionally left in place when coffee trees were planted. Usually these native trees are a mixture of species with a variety of growth forms and biotic relationships. Traditional plantations retain a shade coverage of 60 to 90 percent. Other kinds of agriculture such as cattle raising or rice farming clear all vegetation, causing not only rapid depletion of nutrients from the soil and erosion but also the virtual elimination of the native fauna and flora. Cafetales managed under the traditional system are often productive for up to 30 years, use no fertilizers or pesticides, and have only about 1000–2000 trees per hectare. Studies have found that the diversity of insects was higher in coffee plantations than in native upland rain forest. Bird species diversities are also high, and the traditional plantations serve as wintering grounds for migrating birds. Studies also indicate that numerous bats and nonflying mammals inhabit the plantations as well. These diversities seem to be related to the structural complexity of the canopy trees, which provide shade, and to the tendency of owners to plant other fruit trees such as banana, tangerines, avocados, and mangoes among the coffee trees.

Recently, however, there has been an alarming change in coffee plantation management. Many "modern" plantations have switched to varieties of coffee that can grow without shade or only moderate shade. The varieties used in these "sun" plantations are smaller than the traditional types and thus easier to pick. In addition, 3000–4000 trees can be grown on one hectare, and times to first fruiting are 1 or 2 years faster than with traditional varieties. These plantations also employ fertilizers, herbicides, and insecticides that kill native species and pollute soils and rivers. Studies within these plantations show a drastically lower number of beetle and ant species relative to traditional plantations. Total bird diversity in sun plantations is less than half that of traditional plantations. Censuses conducted in North America showed a steady decline between 1980 and 1994 in the numbers of Baltimore Orioles, birds that winter in Latin America. Native American birds do not use coffee flowers as sources of nectar and generally ignore the fruits. In traditional plantations the canopy trees provide the nectar and fruit needed by these animals. Coffee is also well protected from insect herbivory because of the caffeine it contains. Consequently, pure stands of coffee support a very small insect community. All farmers want to maximize their income, but some balance should be sought by convincing plantation owners to farm in a more sustainable fashion. Actions might include monetary incentives or the labeling of coffee as "environmentally friendly" for traditionally grown coffee so consumers could sway the market in favor of the more ecologically favorable methods.

years old and has a productive life of about 40 years. Most coffee is grown today in open orchards, but some cultivators claim that superior coffee is produced by shaded trees (Box 13.1). Arabian coffee is generally grown on hillsides at altitudes of 1500 to 2000 m (4920 to 6560 ft), but robusta and Liberian coffee are generally grown at low elevations. Because of the size of the tree and the slope of the terrain on which it is usually grown, Arabian coffee is seldom harvested mechanically. Moreover, berries ripen sequentially on branches of the trees, so only a few can be picked from a branch at any time (Figs. 13.7 and 13.8).

Coffee Processing

To produce aromatic ground coffee, the processor must separate the seeds from the rest of the fruit and roast them. Either a wet or a dry process can be used to remove the outer fruit parts. In the dry process, the fresh fruits are dried in the sun and the pericarp is subsequently rasped away. In the wet process (Fig. 13.9), which produces a superior final flavor, the fresh fruits are depulped by a machine and the seeds are washed. The wet seeds are then allowed to ferment for 12 to 24 hours. This fermentation is not the same process involved in the production of alcohol (see Chapter 14). In the cases of coffee, cocoa, tea, and kola, fermentation refers to an enzymatic, chemical alteration of several compounds. Coffee fermentation produces substances that will eventually develop into the characteristic coffee aroma and taste. It also causes the seeds to take on a gray-green hue. After fermentation, the seeds are dried by being turned in the sun for at least 1 week. Any remaining endocarp and the seed coats are removed mechanically. The beans are usually graded before shipping (see Fig. 13.8). Commonly this grading is done by hand; individuals pluck out and discard cracked, diseased, or discolored beans as they pass in front of them on a conveyor belt. Subsequent processing typically occurs in the

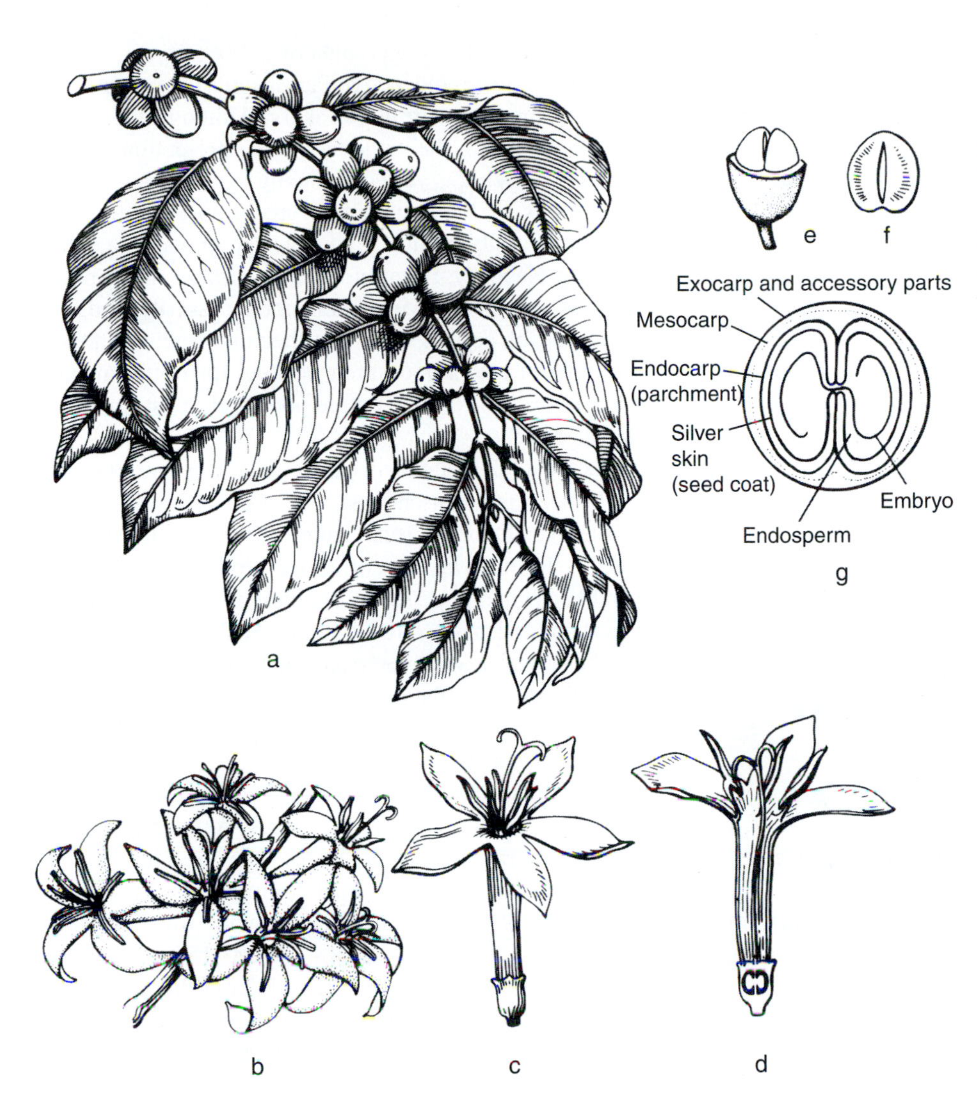

FIGURE 13.6
(a) A branch of coffee with clusters of berries in the leaf axils; (b) a cluster of jasminelike flowers; (c) an individual flower (d) in longitudinal section; (e) the position of the two seeds in the fleshy fruit; (f) the ventral (flattened) side of the seed; (g) a cross section of the whole berry showing the component parts.

([a] redrawn from Baillon.)

FIGURE 13.7
A coffee picker, such as this young girl from Malawi, can harvest 91 kg (200 lb) a day. About 2000 beans are needed to produce a pound (453 g) of roasted, ground coffee.

(Photo by A. Conti, FAO.)

country where the coffee is consumed. Roasting is done in cylinders that simultaneously tumble and heat the seeds. During the process water-soluble aromas develop and begin to diffuse. Beans are roasted at temperatures of 200° to 230°C (392° to 446°F) for 10–15 minutes, depending on the size of the roaster and the degree of roasting. As the beans roast, sugars are caramelized, starches are hydrolyzed, and the cellular matrix breaks down, allowing the beans to swell as gases are released. The kind of roast depends on how long the beans are held in the roasters and thus on the final temperature the beans are allowed to reach (200°C for light roast, 230°C for dark). The beans are quickly cooled with water or cold air to stop the process and slow the loss of aromatic oils.

Despite the statement that dark-roast coffees are "stronger" than light roasts, they are not stronger in the sense of containing more caffeine. Dark roasts seem stronger because they have more flavor than do light roasts. Dark-roast beans are larger than light-roast beans because they swell more during roasting, and they feel oily because

FIGURE 13.8
Grading coffee beans in Ethiopia.
(Photo by S. Pierbattista, FAO.)

their higher roasting temperatures cause some of the aromatic oils to come to the surface. Light- or medium-roast beans are dry on the outside. Almost all the roasting for American palates is of either the light or the medium type. Europeans and Latin Americans, by contrast, prefer dark roasts, one of which is commonly called espresso. It should be noted that **espresso** is a method of making coffee that involves forcing steam through the grounds. Espresso coffee can be made with light-roast coffee, but it is generally made with deeply roasted beans. Once roasting has been completed, the beans can be shipped whole or ground and packaged (often under a vacuum to preserve the flavor) before being shipped to market.

In 1994 researchers reported two compounds in coffee that raised blood cholesterol levels. Subsequent work showed that filters used in many coffee makers removed the compounds. However, coffee made with an espresso machine, a percolator, or a French press still contained the compounds.

There do seem to be some benefits to drinking coffee besides the lift it provides. Two reports in 2000, one of a study in Hawaii of over 8000 aging men and the other of a 10-year study in the Netherlands of 8000, 55-year-old people, found a lower incidence of Parkinson's disease in coffee drinkers compared with nondrinkers. Although the mechanism is unclear, the evidence for protection against Parkinson's disease is very convincing.

Methods of brewing coffee are almost limitless and range from simply boiling ground coffee in water to extraction using elaborate filtration systems. Many modern conveniences in coffee making, some not necessarily leading to a good cup of coffee, have also become popular. Among these devices are premeasured packets for coffee makers, throwaway filters for drip coffee pots, and, most important, instant coffee.

The first marketable instant coffee was introduced in 1909 by an American chemist who produced Washington's "soluble" Red E coffee. Today numerous brands and types are available. To make instant coffee, the producer puts coarsely ground beans into huge sealed stainless steel percolators and brews them under pressure for hours. Sealing is supposed to keep the aroma and flavor from escaping, but the damage done to the coffee by the high temperatures and constant recycling of the brew is irreparable. Aromas are often added later to make up for some of the lost ingredients. The coffee from the percolators is then sprayed under high pressure through nozzles into the top of a room several stories high. As the spray falls, it dries to a powder by the time it reaches the floor. Often, the instant powder is made to more closely resemble real ground coffee by being tumbled with water or steam, causing it to form small chunks. Freeze-drying is another method of producing instant coffee. During this process the brewed coffee is poured onto trays that are subjected to very low temperatures under a vacuum. Sharp-edged particles are produced when the sheet of freeze-dried coffee is broken into small fragments. In a few South American countries a coffee extract is made by slowly dripping cold water through ground coffee. A cup of coffee is made by pouring a little of the extract into a cup and adding hot water.

Another recent innovation is decaffeinated coffee. It was a German, Ludwig Roselius, who in 1906 developed the first successful process for removing caffeine from coffee. The young man was motivated to perfect the process because he believed that his father's death had been caused by drinking large amounts of coffee. Today two basic processes are used to remove caffeine from coffee beans: solvent extraction and water extraction. Most commercial decaffeinated coffees are produced by a solvent method in which beans that have been presoftened by steam are extracted with an organic solvent such as methylene chloride. The solvent is drawn off the beans, and any remaining traces left in the coffee are evaporated by steam or heat during the roasting process. The caffeine is removed from the solvent with water and then purified by crystallization. About 20 kg (44 lb) of caffeine is recovered from each ton of processed coffee. Because methylene chloride has been implicated in the destruction of the ozone layer, it has been

FIGURE 13.9
Steps in the processing of coffee. (a) Selective hand picking of only the red, ripe berries produces coffee with the finest flavor. (b) Depulping and washing the berries. (c) Fermenting the seeds. (d) Drying the beans. (e) Transporting the green beans to market. (f) An older method of roasting the beans. Modern roasters use automatic tumblers.

FIGURE 13.10
The same wild cornflower familiar as an introduced roadside weed provides the roots that are roasted and ground to make chicory.

banned as a solvent in Europe. Many companies have switched to ethyl acetate as a substitute. After solvent extraction, the solvents are recycled, the coffee beans are set aside for roasting, and the caffeine is sold.

In the water extraction process green beans are percolated with water that is saturated with all the water-soluble compounds in coffee except caffeine. Consequently, only caffeine is supposedly dissolved out of the coffee. Caffeine is removed from the water by filtration through charcoal treated to bind caffeine molecules. Many people prefer coffee decaffeinated in this way because no toxic organic solvents come into contact with the coffee. Nevertheless, it is more costly than the direct solvent extraction process because the caffeine cannot be recovered and sold as a by-product.

Over 900,000 kg (2 million lb) of caffeine is produced from the decaffeination of tea and coffee each year in the United States. Another 900,000 kg is imported each year. About three-fourths of this caffeine is used as an additive in soft drinks; the rest is added to headache and cold medicines (see Table 13.2). In 1990 it was estimated that 20 percent of the coffee drunk in the United States was decaffeinated. Because caffeine is removed from green coffee beans, not from ground coffee, it is possible to buy unroasted decaffeinated beans as well as decaffeinated roasted beans, ground coffee, and instant coffees.

A final treatment of coffee consists of the addition of other substances. In some cases such additions constitute adulterants or extenders; in others they serve as flavor enhancers. Adulterants include ground and roasted peas, beans, orris root, and grains such as wheat. A common coffee additive is chicory, *Cichorium intybus* (Asteraceae) (Fig. 13.10), which can be considered an adulterant or a flavor enhancer, depending on the reason for its addition to ground coffee. Chicory superimposes on the normal coffee flavor a distinctive, pronounced taste that many individuals enjoy (Fig. 13.11). Dried, powdered roots of this southern European native have been used alone to produce a beverage as well as mixed with coffee to "stretch" the beverage or substituted for coffee when the beans have been scarce. In the United States coffee with chicory is a regional specialty of Louisiana. Unlike coffee, chicory contains no caffeine.

A recent trend has been the production of designer, or flavored, coffees. Flavors (usually artificial) such as almond, chocolate, pecan, and vanilla are sprayed or poured over the beans after roasting. The volatile oils of the flavoring are leached into the beverage along with those of the coffee during brewing.

Cacao

Cocoa is the New World's contribution to the list of important nonalcoholic beverages, but the eating of chocolate surpasses the drinking of cocoa in popularity. The plant, *Theobroma cacao* (Sterculiaceae) (Fig. 13.12), from which chocolate products are obtained, is known as cacao. Despite the centuries-long history of cacao use, the procedures used to manufacture modern cocoa powder and chocolate have all been developed within the last 165 years. Cacao was collected by native peoples, and the seeds were roasted for use in a stimulating beverage long before Europeans came to America, but the beverage made by Central American Indians was quite different from modern sweet cocoa. The Mayans and Aztecs thought that cacao had a divine origin, a belief reflected in the name *Theobroma,* meaning "food of the gods."

Europeans first encountered cacao when Columbus landed in Nicaragua. Columbus reported the consumption of a strange beverage, but it was not until Hernán Cortés visited the royal Aztec court that its importance in the New World was appreciated. Like the Mayans, the Aztecs roasted cacao beans and used them in religious ceremonies. The Aztec god Quetzalcoatl (see Fig. 2.7) was credited with giving humans not only corn but also cacao. Montezuma, the last ruler of the Aztec nation, was reported to have drunk 50

FIGURE 13.11
"La chicorée, la plante qui fait du bien" (Chicory, the plant that makes you feel good). Steps in the processing of chicory root shown in the inset on the left include breaking and drying the roots, roasting, and grinding.

(Photo courtesy Le Roux et Companie.)

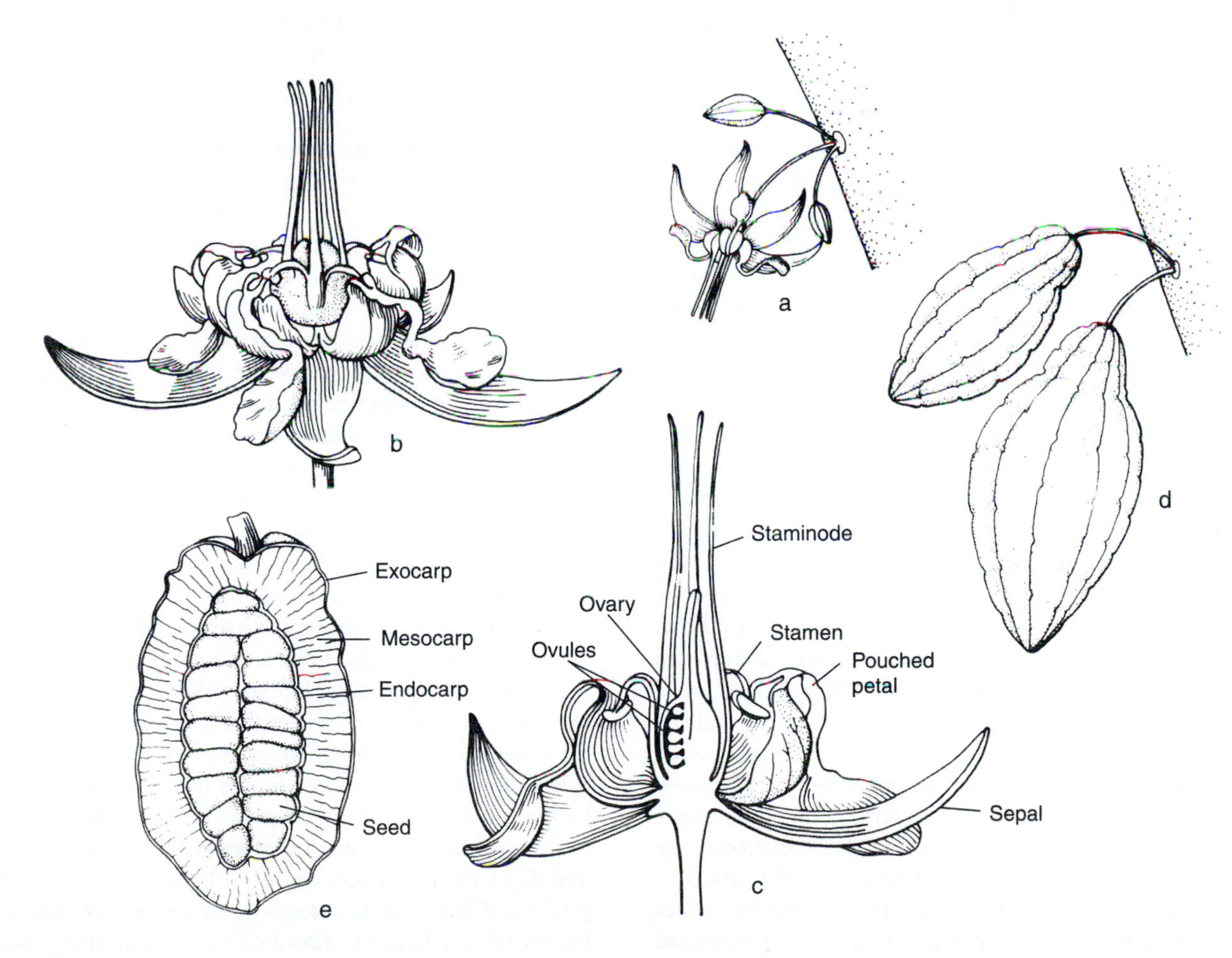

FIGURE 13.12
Flowers and fruits of a chocolate plant. (a) A cauliflorous flower; (b) flower greatly magnified; (c) longitudinal section of a flower; (d) fruits borne directly on the trunk; (e) longitudinal section of a fruit.

golden goblets of the bitter native beverage each day. This brew, offered to Cortés as a sign of respect, was concocted by mixing ground cacao seeds with ground achiote (annatto) seeds (*Bixa orellana,* Bixaceae) (see Figs. 15.30 and 15.31). Chili pepper and vanilla were added to the chocolate mixture, and the entire mass was heated until it turned into a homogeneous paste that could be molded into tablet-shaped pieces. A drink was made by dropping one of the tablets into hot water and stirring until the tablet dissolved. The Indians often thickened the drink by adding ground cornmeal.

When Cortés brought cacao back to Europe, it was not well received. The popularity of the beverage soared when Europeans later deleted the hot spices and added sugar and cinnamon. Sweet hot chocolate became the rage of Europe by the middle of the seventeenth century. Spaniards established cacao plantations on Trinidad and Hispaniola, commencing a long monopoly of the cacao trade. The Dutch were the first to break the Spanish monopoly by establishing plantations first in Curaçao and then in Southeast Asia in 1670. Germans later brought the plants to Samoa and New Guinea. African production began after 1879, when cacao was introduced into Ghana. As in the case of coffee, the production of cacao is now highest in areas far removed from its place of origin but within the same latitudes (within 20 degrees north or south of the equator). As shown in Table 13.4, Africa, rather than Central America, is today the world's largest producer.

Because cacao was first brought back to Europe from Mexico, it was initially assumed that the genus was native to Central America. However, studies have shown that the natural distribution of cacao extends from the Amazon region of South America to southern Mexico. However, cultivation was initiated 2000 years ago by the Mayans, who planted *Theobroma* in the Yucatán Peninsula. By the seventh century native peoples in Mexico were widely growing the species. The Aztecs later continued to increase its cultivation.

Cacao plants are small trees propagated either by seed or by cuttings. Because they naturally grow as understory trees, they have traditionally been planted in shaded orchards. Cacao belongs to a group of plants that bear their flowers directly on the trunks and branches (Fig. 13.12) rather than in leaf axils or on branch tips. The flowers are produced in flushes twice a year. After pollination by midges that live in the orchard debris, the flowers produce fruits that mature in about 3 months. Individual pods are 10 to 32 cm (4 to 13 in.) long and contain 20 to 60 seeds (Fig. 13.13). The pods are picked by hand and cracked open, and the seeds and pulp are scraped from the husks. The moist seeds are allowed to ferment in piles that are turned occasionally for 4 to 7 days (Fig. 13.14). During fermentation numerous structural and chemical changes occur. The pulp around the seeds ferments and forms alcohol and acetic acid. The acid eats through the seed coat, kills the embryo, activates enzymes that break down stored products, and generates volatile compounds responsible for the aroma and flavor of chocolate. When fermentation is complete, the seed coat has become purplish and the material inside the seed is an amorphous mixture of proteins, carbohydrates, and fat.

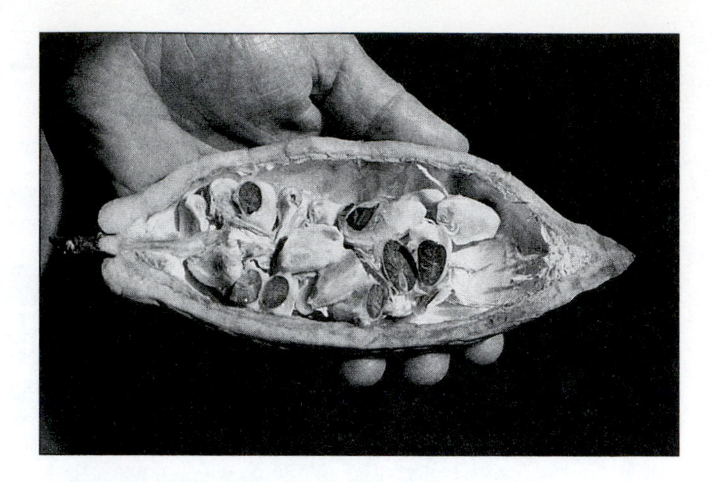

FIGURE 13.13
A dry cacao pod cut to show the seeds.
(Photo courtesy of M. Bonness.)

Once fermentation is complete, the seeds are dried and then "polished" by tumbling in a mechanical polisher to remove any remaining bits of pulp. At this point the cacao seeds are sorted, graded, inspected, and shipped to processing plants. The seeds smell good and have a slight chocolate flavor but are still very bitter and oily.

When the seeds arrive at the factory, they are cleaned to ready them for roasting. Roasting is done in batches or by moving the seeds on a conveyor belt through ovens. The beans are roasted at temperatures between 100° and 120°C (212° and 248°F) for 45 to 70 minutes. Roasting drives off water and acids and develops the final chocolate flavor. Once roasted, the seeds are cracked to release the large cotyledons. Fans blow away the debris consisting of the seed coats (commonly called shells or husks). The shells can be sacked for mulch, used for the extraction of theobromine, or pressed to yield cocoa butter. Cocoa butter has a milder flavor than regular chocolate and it can be made into white chocolate by the addition of milk solids or powdered milk. Extracted theobromine is converted to caffeine that is subsequently added to beverages and medicines.

The cotyledons left after roasting and separation from the seed coats are called **nibs.** It is from this part of the fruit that chocolate is actually obtained. The nibs are ground to a fine paste by a series of rollers that generate enough heat to melt the fat in the nibs and produce a thick, dark liquid called **chocolate liquor.** The fate of the liquor is determined by the eventual use of the chocolate. If the liquor is molded into small squares, it produces baking chocolate. Until the middle of the nineteenth century this was always the final product. Chocolate beverages were made by dissolving blocks of this kind of chocolate in hot water or milk (an English innovation) to which sugar had been added. The result was rich and very heavy because of the large amounts

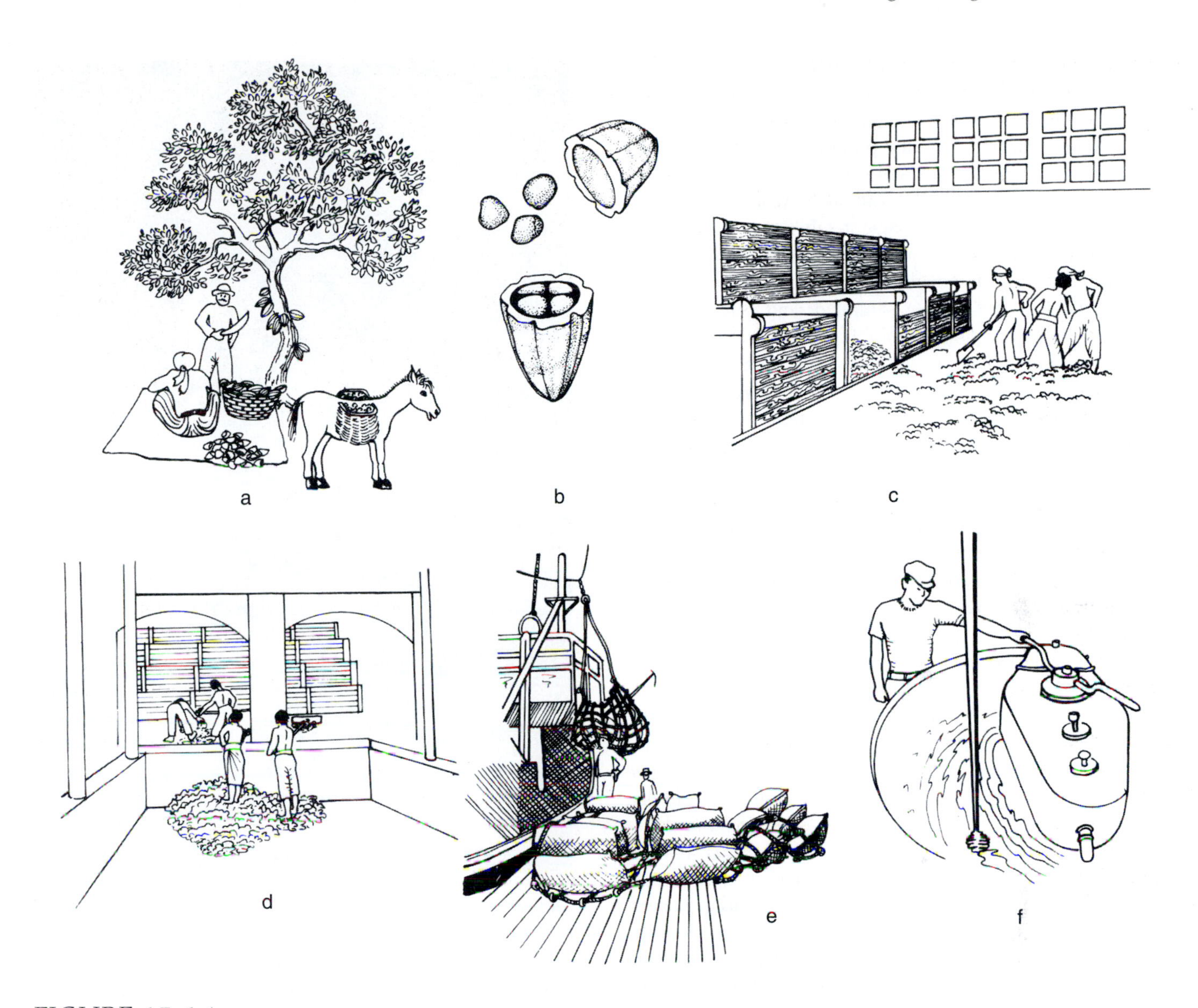

FIGURE 13.14

The steps in the processing of cacao. (a) Pods are picked and broken open (b) to obtain the seeds. (c) Once at the factory, the beans are placed in fermentation boxes and turned into the next bin each day. (d) The beans are washed in large roomlike bins and then (e) shipped to manufacturing plants in other countries. Cocoa nibs are produced by roasting the beans and removing their seed coats. At this point the nibs contain about 57 percent fat, 30 percent of which is later pressed out during the making of modern cocoa. (f) In the making of milk chocolate, cocoa nibs, sugar, cocoa butter, and condensed milk are stirred in revolving tubs to produce a smooth, homogeneous paste.

of fat in the chocolate. In 1828 in Holland, C. J. van Houten developed the first process for producing cocoa as it is known today. The process presses out much of the fat, yielding a relatively dry cocoa powder, with 30 percent less fat than that of the nibs. At about the same time the process of alkalization, or the adding of alkali to neutralize many of the organic acids, was also developed in Holland. **Dutching,** as this process is often called, darkens the cocoa, makes its flavor mild, and increases its solubility. About 90 percent of all cocoa is now dutched.

Another way in which the chocolate liquor can be used is to make chocolate candy. The first records of eating chocolate date to about 1847. The development of this form of chocolate could come about only after the discovery that defatting cocoa yields cocoa butter because eating chocolate is made by adding additional cocoa butter to the chocolate liquor. The additional fat allows the chocolate to be molded and shaped and adds smoothness.

Two additional European innovations led to the soft, creamy chocolate confections that have surpassed cocoa in popularity. In 1875 the Swiss company Nestlé started to add condensed milk to chocolate and produced the first solid milk chocolate. A short time later another Swiss company, Lindt, found that continual stirring of the chocolate mass (with or without condensed milk added) led to an extremely smooth, creamy texture (Fig. 13.14). Stirring causes the cacao particles to be ground into finer and finer pieces. When extra cocoa butter obtained from the production of cocoa powder is added, even creamier chocolate is produced. For chocoholics who think they must avoid such creamy confections because of the high fat content, there is good news. Laboratory studies have found that stearic acid, the most abundant fatty acid in cocoa butter, lowers total cholesterol despite the fact that it is a saturated acid (see Fig. 9.3). Chocolate contains polyphenols, multicyclical compounds, that act as antioxidants in the body. The fact that a single candy bar can contain the same quantity of polyphenols as all of the fruits and vegetables normally eaten in a day should provide some comfort to people who crave chocolate but feel guilty about eating it. In 1996, scientists reported that chocolate also contained three compounds that bind to the same brain receptor as marijuana. This report set off speculations that the chocolate "cravings" professed by many Americans was due to these compounds. However, subsequent work has indicated that "chocoholism" is culturally conditioned.

FIGURE 13.15
Training to conduct a Japanese tea ceremony includes mastering many of the Japanese arts. This deeply symbolic and highly ritualized ceremony takes place in a specially designed room located within a tea garden.

Tea

Tea is not as important an international commodity as coffee but is probably drunk by a greater number of people. Most of the tea produced is consumed locally, with the result that comparatively small quantities enter international trade.

Although the precise origin of tea, *Camellia sinensis* (Camelliaceae), is unknown, the species is undoubtedly native to China. Asians have many legends describing the first uses of tea. According to one story, an ancient emperor discovered tea when a leaf from a tea plant accidentally fell into his cup of boiling water. According to another legend, the first tea plant sprang from the eyelids of a pious monk who was a follower of Bodhidharma. The monk vowed to stay awake until Buddha's work on the earth was completed, but, although he tried valiantly, he fell asleep after many days. When he awoke, the monk was so mortified by his weakness that he tore off his eyelids. Buddha, pleased by the intentions of his follower, caused the eyelids to sprout into tea plants when they touched the ground. Tea was thus thought to have been a gift from Buddha to aid people in their vigils by keeping them awake and at the same time enhancing the peaceful, harmonic state necessary for contemplation.

Whatever the origin, elaborate methods of brewing tea were devised early in its history. One involved pouring boiling water over pieces of a crumbled cake of pressed tea leaves and rice. Sliced onions, ginger, and orange were added to the brew. The Chinese were drinking tea in the fifth century A.D., and by A.D. 780 the *First Tea Classic,* a Chinese compendium of information on the cultivation and preparation of tea, was written by Lo-Yu.

Brought to Japan in A.D. 801, the beverage also assumed a central role in Buddhist rituals there. Accordingly, each aspect of the Japanese tea ceremony (Fig. 13.15) reflects principles inherent in Zen Buddhist beliefs. The management of a tea garden with a portal so low that one must stoop to enter, a floral display of a precise form in the alcove, and a ritualized method of brewing tea leaves are all parts of the elaborate tea service. The tea is prepared by whipping powdered tea leaves into steaming water with a bamboo whisk. The green, foamy beverage is served in freshly rinsed raku bowls.

Europeans were first exposed to tea when the Portuguese began to explore the coast of China in the sixteenth century, but the Dutch introduced tea into England. Ironically, the first tea offered to the British public appeared in a London coffeehouse in 1657. By the eighteenth century tea had become an important trade item for Europeans. Both the British and the Dutch East Indian trading companies were involved in the purchase of tea in the Orient and its sale in Europe and the New World.

In colonial times many Americans of European descent were tea drinkers. Although tea was expensive, it was

FIGURE 13.16
A tea branch in flower.

FIGURE 13.17
Because the ends of the branches are constantly being picked, tea shrubs assume a rounded shape. Tea is ideally grown on hillsides where cool moist conditions produce leaves with the finest flavor.
(Photo courtesy of Japan National Tourist Organization.)

cheaper than coffee, and only small quantities were needed to produce a drinkable cup. The colonists were so accustomed to their tea that they became incensed when the British declared that they had to pay a tax on tea brought to the colonies. To show their resentment, the colonists staged the Boston Tea Party of December 16, 1773, during which they dumped large quantities of the British East India Tea Company's cargo into Boston Harbor.

The British initiated commercial plantings of tea in India between 1818 and 1834. India is now the world's largest producer of tea (Table 13.4). The planting of Chinese tea in India had an unexpected and dramatic effect on the tea industry. The variety of tea that had previously been grown in India was Assam tea, which has a heavier flavor than Chinese tea. The large-scale planting of Chinese tea led to crosses between the two varieties, which led to teas that were darker, more robust, and higher in caffeine than typical Chinese tea. These are now the most traded teas in the world. Sri Lanka, which is the third largest modern producer, began to grow tea only after 1800, when coffee rust, a kind of fungus that infects the plants, finally destroyed all of the British coffee industry there. In fact, it was only after the English coffee plantations on that island fell victim to the coffee rust and were replaced with tea orchards that tea supplanted coffee as the favorite beverage of the British and set the stage for the Victorian custom of having afternoon tea.

Tea plants (Fig. 13.16) are small evergreen trees. As indicated by the generic name, tea belongs to the same genus as the beautiful horticultural camellias. Plantations of tea were historically started with seed that had been carefully germinated in protected enclosures. When sufficiently robust, the young plants were set into fields. Today plants obtained from the rooting of cuttings are most commonly used because they ensure uniformity of a particular hybrid variety. Plantations are located primarily in areas with good rainfall and a constant cool temperature throughout the year. After about 3 years the trees are pruned, and they are repruned every 12 years or so. The type of pruning varies with the country in which the tea is grown, but the purpose is the same in all cases: to force the trees into a bushlike form (Fig. 13.17) that facilitates picking and encourages rapid shoot growth.

Picking can begin when trees are 4 years old. For fine teas, only the two or three youngest leaves and the terminal bud of each branch are picked (Fig. 13.18). It has been shown that these parts contain the highest quantities of caffeine and the other constituents that give tea its flavor. During the growing season, leaves can be picked from a shrub about once every 10 to 14 days.

Tea Processing

The processing of the leaves depends on the final kind of tea desired. For **green tea,** the leaves are usually steamed for 45 to 60 seconds and then dried and rolled at 90° to 110°C

FIGURE 13.18
Only the tips of the shoots of tea plants are picked, as these Japanese workers demonstrate.
(Photo courtesy of Japan National Tourist Organization.)

(219° to 256°F) for about 45 minutes. They are then rolled without heat for another 25 minutes and finally are dried and pressed at 50° to 60°C (148° to 166°F) for an additional half hour. Historically, processing was manual work done by rolling the leaves on a hard surface with the palm of the hand. Now rolling and twisting are done with mechanical rollers. Another method consists of firing the leaves in a pan at 250° to 300°C (507° to 598°F), followed by rolling and additional drying at a lower temperature. Green tea, which constitutes about one-fifth of all the tea produced, is usually consumed locally.

For the international market, most tea is processed into **black tea.** The processing of black tea is more complex than that of green tea and produces a final product that is more complex in flavor. Black tea is produced by first spreading the leaves out on a flat surface or placing them in drums or troughs to wither (Fig. 13.19); withering reduces the moisture content from 75 to 80 percent to 55 to 70 percent in 16 hours. Today withering is done in carefully controlled factory environments. The withered, flaccid leaves are then "disrupted" by machine rolling, crushing, and tearing. The purpose of this step is to crush

FIGURE 13.19

The steps in the processing of tea differ in different parts of the world and in many cases have changed from hand to mechanical methods. Various methods for harvesting and withering and some of the early machinery employed are shown. (a) Harvesting of tea by plucking the upper two or three leaves and the apical bud; (b) withering and fermenting the tea by hand; (c) drying leaves in bamboo baskets over charcoal in China; (d) mechanical cutting of tea by Reid's Original Teacutter (1870); (e) withering and fermenting tea on wicker trays, as is the custom in some parts of the Orient; (f) withering drums; (g) hand rolling in black tea production; (h) the Empire Roller, an early machine for the rolling of tea; (i) drying in a power tea dryer.

(Redrawn from photographs in Ukers, 1935. *All about Tea.*)

the cells of the leaves to release the enzymes in the cytoplasm. The rolled leaves are then allowed to sit for 40 minutes to 3 hours. During this time the temperature of the leaf mass rises as the released enzymes promote chemical reactions known as fermentation. In the case of tea, fermentation refers to the formation of brown-colored polyphenolic compounds. Caffeine is responsible for the stimulating effects of tea, and the tannins impart the characteristic bite and brown color, but the ratios of various polyphenols produced during fermentation determine the final flavor of the tea. After fermentation, the tea is dried with a hot air stream that stops all reactions and reduces the water content to about 3 percent. The dried tea is then tasted, graded, and packaged.

Oolong tea combines the taste properties of green and black teas because it is semifermented, or allowed to undergo a comparatively short enzymatic fermentation that occurs during the 6- to 8-hour withering and hand-rolling period. It is produced primarily in a small region of eastern China and northern Taiwan.

Tea is very rich in polyphenols that show antitumor and antimutagenic properties in animal studies. Polyphenols seem to work by preventing tumor cell growth and division. Green tea is particularly rich in these compounds, with about six times the quantity of that found in black tea.

Variations on a Theme

Drinking iced tea is a relatively recent phenomenon, as might be expected from the fact that ice was a scarce commodity until the advent of modern refrigerators. The introduction of iced tea to the American public, the only people who drink appreciable quantities of it, is credited to an enterprising Britisher trying to sell tea at the 1904 St. Louis World's Fair (the same fair at which ice cream cones made their debut). The gentleman, finding that no one wanted to drink hot tea in the middle of a St. Louis summer, added ice and immediately increased his sales.

Tea bags, now so commonly used in the United States to prepare tea, were initially made by a New York wholesaler who sent small samples of tea tied in silk bags to potential customers as a promotional scheme. The idea obviously went far beyond this advertising gimmick. Tea bags were made for many years from gauze. By 1934, 8 million yards of gauze was used annually in the manufacture of tea bags. Today, as discussed in Chapter 15, tea bags are made from a clothlike paper derived from plant fibers.

Instant tea dates from 1885, when an Englishman, John Brown, developed a dry paste made from tea extract, evaporated milk, and sugar. When hot water was added, the mixture produced a cup of tea suited to British tastes. However, instant tea did not become popular until the development of spray-drying in the 1940s. The United States is now the largest consumer of instant tea, with a variety of flavored and sweetened powdered teas available.

Less Important Stimulating Beverages

Mate

Mate is associated today with Argentina and to some extent Uruguay and Paraguay. The species from which mate comes, *Ilex paraguariensis* (Aquifoliaceae), is native to the mountains of northern Argentina, southeastern Brazil, and Paraguay. However, the name *mate* is derived from a Quechuan word, and historical and anthropological remains show that mate was widely used in pre-Columbian times by Andean Indians as well as by Indians of the Paraguayan lowlands. Other species of *Ilex* have been used to make caffeine-containing beverages, but today mate is the only species used on a large scale.

As in the case of tea, leaves are used to produce mate, but unlike tea, any of the leaves of the tall evergreen mate trees and even the twigs and bark of small branches can be used. The leaves are dried on a frame over a fire, crushed into rather coarse pieces, sieved, and shipped. Mate produced by this process is the commonly encountered green type. Some mate is toasted by exposing the leaves to more intense heat. In this case, the final product is brown, and the mate is sold as *mate cocido.*

Mate is traditionally drunk from a dry, hollowed gourd also called a **mate** (Fig.13.20). Crushed mate leaves are placed in the gourd (with or without sugar), and the gourd is filled with boiling water. After the mixture has steeped and settled, a metal straw (**bombilla**) with an expanded, flattened end is inserted into the ground. The flat end has holes large enough to allow movement of the hot tea but small enough to prevent pieces of the leaves from entering the straw.

FIGURE 13.20
These mate gourds are used for brewing the caffeine-rich mate leaves (foregound), and the infusion is then sipped up through the bombilla, which strains out the leaf particles.

Guaraná

Whereas the Andean Indians and cultures of southeastern South America used mate as a source of stimulating beverage, the Amazonian forest Indians used guaraná, *Paullinia cupana* (Sapindaceae). Guaraná plants are monoecious climbing vines (but sprawling shrubs under cultivation) native to central Brazil that produce caffeine-rich seeds. One to three seeds, each with a fleshy growth surrounding it, are borne in each capsule. Guaraná is prepared for beverage making by drying the seeds after the fleshy exteriors have been removed. The seeds are then roasted for 2 to 3 hours and shaken to remove the seed coats. The cooled, toasted seeds are then crushed with water and cassava flour into a paste that is formed into rods (**bastões**) about 2.5 cm (1 in.) wide and 20 cm (8 in.) long (Fig. 13.21). The final product looks like a segment of a rusted metal rod. A glass of guaraná is made by scraping the cylinder and mixing the resultant powder with water. Guaraná is second only to coffee as the most popular drink in Brazil. However, the term *guaraná* is now used, much like the term *cola* in the United States, to refer to a kind of soft drink.

Kola

A relative of cacao, kola or cola, *Cola nitida* (Sterculiaceae), is best known to Americans because of cola-flavored beverages. Kola is native to west Africa, where the seeds of various species have long been chewed or used to produce caffeine-rich beverages. In the original formulation of Coca-Cola, kola as well as coca (see Chapters 11 and 12) was used in the tonic. Now coca leaves from which the cocaine has been removed, artificial cola flavorings, and caffeine are added.

The flowers of the kola plant are borne in the axils of the leaves, not directly on the stems as they are in cacao. Each fruit contains about eight seeds that are scraped from the harvested pods. Kola fruits lack the pulp of cacao pods, and the fleshy seed coats are removed before the seeds are allowed to ferment, or "sweat," for a few days to develop their flavor. The seeds ("kola nuts") are simply dried and then chewed or pulverized into a powder that is mixed with boiling water to make a beverage. In addition to caffeine, kola contains kolanin, a cardiac stimulant.

FIGURE 13.21

Guaraná is a popular stimulating beverage in Brazil, where it is available in many forms. In the center are the native utensils used to prepare guaraná. The tongue of a large Amazonian fish, the *pirarucu*, is used to rasp a small amount of powder from the rods of ground guaraná seed paste. The powder is dissolved in water. On the right in the foreground is a figurine of the fish made of guarana paste. Next to the figurine are scales of the *pirarucu* (which are used as fingernail files). In the back are two tin jars of guaraná powder and a bottle of concentrated syrup.

14

Alcoholic Beverages

Alcoholic beverages are in many ways the opposites of stimulating beverages. Alcohol acts as a depressant rather than a stimulant in the human body, so people often use coffee to "sober up" after excessive drinking. One might think that humans would shun products that dull the senses, but the production of alcoholic beverages has been independently "discovered" or readily adopted by almost every society. Humans are also the only animals to drink alcohol to excess under natural conditions and are therefore the only animals that can become truly addicted to it. We concentrate in this chapter on the methods used to produce alcoholic beverages, the effects of alcohol as a drug, the various plants from which these beverages are made, and the history of the use of fermented beverages. Table 14.1 lists the plants discussed in our treatment of alcoholic beverages and their production.

The production of all alcoholic beverages depends initially on the process of fermentation by microorganisms, especially yeast. This process yields alcoholic beverages such as wines and beer with relatively low contents of alcohol. Distillation of these products results in a wide array of beverages with high alcohol contents.

Alcohol As a Drug

No informed person would deny that alcohol is a drug, yet it is used legally throughout most of the world, and approximately two-thirds of Americans over the legal drinking age consume it. Over half the money spent in the United States on beverages is used for those that contain alcohol, although by quantity, alcoholic beverages constitute only 34.5 percent of those purchased. Deaths related to alcohol consumption, including automobile accidents, rank fifth on the list of major killers of Americans. Perhaps it is people's long association with alcoholic beverages or the fact that until fairly recently most fermented beverages contained low levels of alcohol that explains the worldwide acceptance of alcoholic beverages. In low to moderate doses, little or no evidence has been found of persistent, harmful effects on the bodies of individuals consuming alcohol. In fact, epidemiological studies have shown that taking two drinks a day (equivalent to two bottles of beer, two glasses of wine, or 2 ounces of whiskey) lowers the risk of heart disease, probably because it raises the level of high-density lipoproteins in the blood. However, because alcohol

TABLE 14.1 Plants Discussed in Chapter 14

COMMON NAME	SCIENTIFIC NAME	FAMILY	NATIVE REGION
Agave	*Agave angustifolia* *A. palmeri* *A. tequilana*	Agavaceae	Mexico
Barley	*Hordeum vulgare*	Poaceae	Mediterranean
Corn (maize)	*Zea mays*	Poaceae	Mexico
Grape	*Vitis vinifera*	Vitaceae	Western Asia
Hop	*Humulus lupulus*	Cannabaceae	Eurasia
Juniper	*Juniperus communis*	Cupressaceae	North temperate regions
Marijuana	*Cannabis sativa*	Cannabaceae	Central Asia
Maguey	*Agave angustifolia*	Agavaceae	Mexico
Potato	*Solanum tuberosum*	Solanaceae	South America
Rice	*Oryza sativa*	Poaceae	China
Rye	*Secale cereale*	Poaceae	Eurasia
Wheat	*Triticum aestivum*	Poaceae	Mediterranean

is absorbed, metabolized, and excreted differently from most other drugs, it rapidly affects one's sense of reason and physical coordination. Social drinkers who drive cars or take tranquilizers while drinking often inadvertently harm themselves or others.

Beverage alcohol is **ethanol,** a two-carbon alcohol that is soluble in both water and fats (Fig. 14.1). Consequently, it moves rapidly across membranes, and almost all the alcohol consumed is completely absorbed in the stomach and the upper portion of the intestines and carried through the body to the liver, where it is broken down. The only way for alcohol to leave the system is by metabolic breakdown, a process that is linear over time. The breakdown products are carbon dioxide and water. However, ingestion rates usually greatly exceed the rate of metabolism. Once the quantities ingested exceed the amounts that can be metabolized, alcohol levels in the body rise. Women seem to be more strongly affected by alcohol than men, not only because of their average smaller body size but apparently also because alcohol is absorbed at a slower rate in women's stomachs than in men's. Alcohol's effect on humans is complex. Animal studies suggest that it affects normal neurotransmitter actions (Box 14.1).

In addition to its temporary effects, alcohol can lead to permanent physical damage. Drinking during pregnancy appears to be correlated with fetal abnormalities such as reduced brain size, small eyeballs, and malformations of the lips and jaw. This suite of characters, known as **fetal alcohol syndrome (FAS),** seems to be most pronounced if drinking is in "binges," is heavy throughout pregnancy, or occurs during critical times in fetal development. The number of newborns with FAS ranges as high as 5.9 per 1000 live births in some parts of the United States. In many cases the effects of seemingly minor FAS can persist throughout life.

Epidemiological studies show that individuals who both smoke and drink have higher risks of dying from cancer than do those who indulge in only one or neither of these activities. Alcohol can also be an addictive drug, with some individuals apparently more prone than others to addiction. Researchers have long sought a gene correlated with alcoholism. The tendency of the disease to run in families suggests such a genetic basis, but exposure to alcohol drinking could also explain the tendency of the children of alcoholics to drink. Within the last few years progress has been made in locating genes that have an influence on alcoholism. These genes appear to affect the functioning of dopamine, the chemical messenger that helps regulate pleasure-seeking behavior. Prolonged, excessive use often leads to severe malnutrition caused by lack of interest in eating properly, permanent brain damage, proliferation of fat in the liver (cirrhosis), and heart problems.

Attempts to make alcohol use illegal in 1920 with the passage of the Eighteenth Amendment (Prohibition) failed after its repeal in 1933 (Twenty-First Amendment). The recognition that alcohol is a dangerous drug did not change during the 13 years of Prohibition; people apparently decided that the numbers of deaths, proliferation of crime, and amounts spent for legal fees were higher under Prohibition than without it. For the past 67 years people have been able to buy and consume alcohol freely in the United States. This freedom is not without its consequences. Figures from 1996 estimated that alcohol contributed to over 100,000 deaths annually in the United States (Box 14.1). Twenty percent of the suicide victims were alcoholics. In addition alcohol is involved in 50 percent of the rapes, 66 percent of the homicides, and 70 percent of the assaults that occur annually. As alcohol becomes a proportionately more important mortality factor, there is pressure to institute legal sanctions against alcohol abuse while providing rational treatment for those who have become victims of alcoholism. Educational programs dealing with the use and abuse of alcohol have been initiated in many states, along with stiffer penalties for offenses such as driving while intoxicated

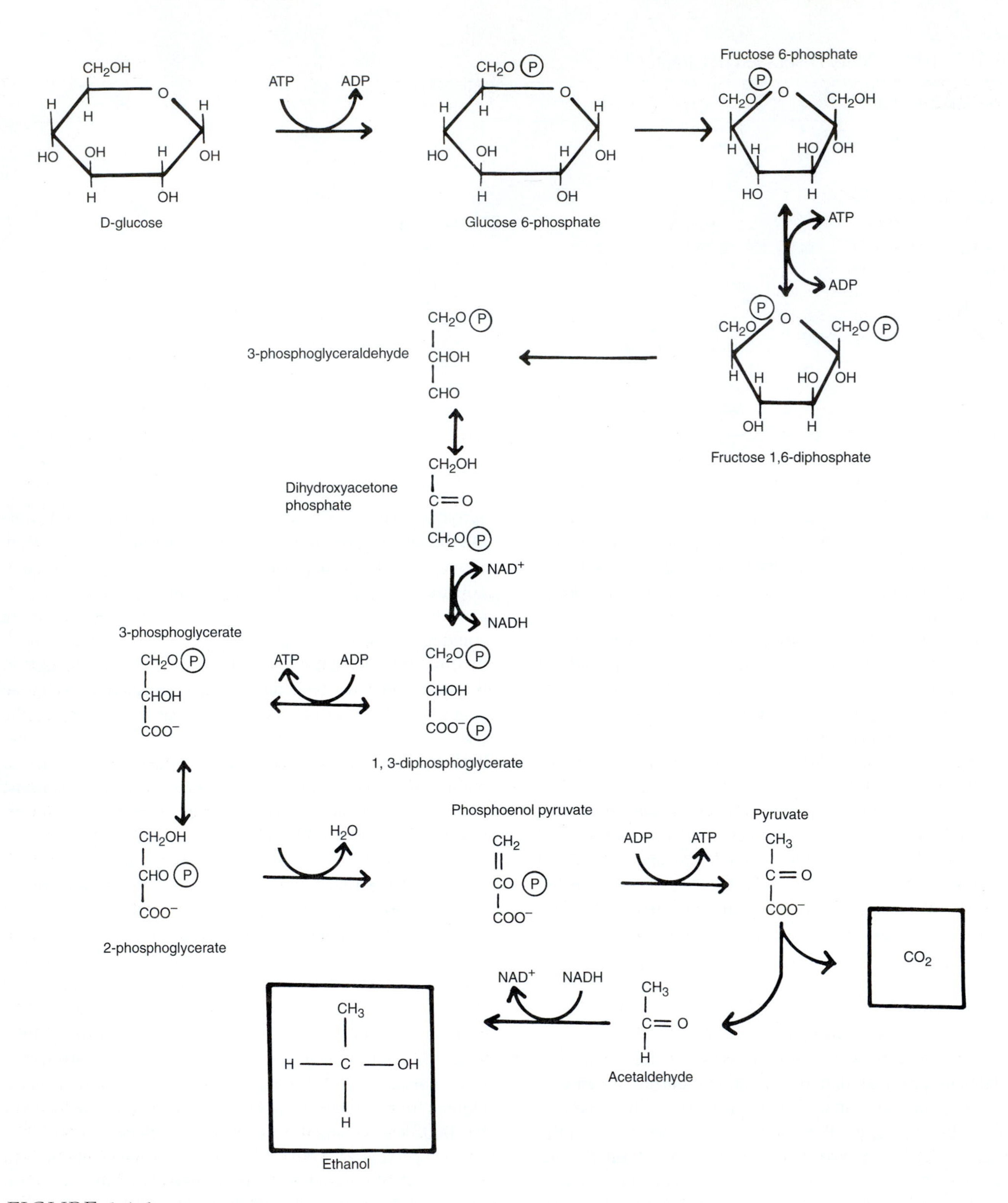

FIGURE 14.1
The process of fermentation, which involves the conversion of glucose or fructose to ethanol and carbon dioxide, is also known as the Embden-Meyerhof-Parnas pathway after the scientists who discovered it.

Box 14.1

Physiology of a Buzz

Most of us have experienced the series of sensations that follow from that first drink through several too many. First there is a feeling of excitement, then a feeling of relaxation or even euphoria. After awhile, although we may not be aware of it, we start to feel tired rather than excited, and people around us may notice that we are misspeaking or slurring words. The next day we do not remember the conversation or movie from the night before. Perhaps we even have a pounding headache and a queasy stomach. Only recently have scientists begun to understand the ways, often juxtaposed, in which alcohol works on the body and brain. Contrary to a commonly held notion, moderate drinking does not lead to the death of brain cells, merely to the temporary dysfunction of many of the brain's neurons by interfering with specific neuroreceptors. The receptors involved seem to be the gamma-aminobutyric acid (GABA), dopamine, endorphin, and glutamate receptors. The feelings of relaxation or reduction in stress can be traced to the effect ethanol has on the GABA receptors. Alcohol seems to increase the sensitivity of these receptors to GABA, a neurotransmitter that prevents neurons from firing. Since the firing of neurons is associated with excitation (and tenseness), the prevention or slowing of neural activity produces a calming effect. Feelings of well-being associated with alcohol appear to be linked to its ability to increase dopamine levels and stimulate endorphin production. Both of these neurotransmitters are part of the endogenous "reward" system, and both are stimulated (although differently) by psychoactive drugs such as cocaine and heroin.

Feelings of relaxation and the "glow" associated with alcohol consumption can also involve the glutamate receptors and usually occur within the first 10 minutes of consuming a glass of wine or beer. The glutamates are neurotransmitters that allow electrons to enter neurons, causing them to fire. Consequently, they are excitatory neurotransmitters. Molecules of ethanol can apparently bind to glutamate receptors and distort them so that they can no longer receive glutamate. After two drinks, 80 percent of the glutamate receptor function in the brain can be impaired; our cerebral neurons quit or slow down the sending of messages to neurons that control other parts of the body. If the neurons that fail to receive orders from the brain are those that ultimately control muscles, a feeling of relaxation, lethargy, and eventually difficulty in moving ensue. If the neurons to the muscles controlling speech are affected, speech becomes imprecise or slurred. Eventually, if involuntary muscle neurons do not receive adequate stimulation, the heart and breathing rates can slow, to the point of death in extreme situations. One particular type of glutamate receptor, the NMDA (N-Methyl-D-aspartate) receptor, seems to be involved in promoting the stimulation of the group of neurons involved in memory and learning. When alcohol prevents these receptors from receiving glutamate, memory is impaired. Drinking can result in the failure to remember events at a party or can even lead to complete blackouts. This process does not cause the loss of previous memories, it inhibits only the formation of memories that would have been formed during the time of alcohol influence.

For society as a whole, the effects of alcohol on the glutamate receptors is the most serious since it leads to impaired thought and coordination. It is estimated that each year in the United States, 20,000 fatal traffic accidents (38 percent) are alcohol-related. Similarly, half of the boating accidents, 47 to 65 percent of adult drownings, and almost 50 percent of industrial accidents involve alcohol consumption.

(Fig. 14.2). Perhaps this country will eventually find a balanced policy toward alcohol and other drugs that consistently takes into account their harmfulness to individuals and society.

Fermentation

Virtually all beverage alcohol is produced by various species of the genus *Saccharomyces.* This genus belongs to a group of fungi known as **yeasts.** Like all fungi, yeasts lack chlorophyll and cannot manufacture their own food. Yeasts differ from other fungi in that each cell lives independently and reproduces primarily by budding, or simple mitotic divisions (Fig. 14.3). Although many fungi can carry out fermentation, species of *Saccharomyces* are generally used because they are comparatively efficient at alcohol production and can tolerate higher levels of ethanol than can most other fungi. Also, during fermentation they produce compounds other than alcohol that are believed to influence the final flavor of the fermented liquid. Some strains of *Saccharomyces* also have the ability to clump into masses during the later stages of fermentation, facilitating their removal from the final beverage. The species of *Saccharomyces* that are used for alcohol production, primarily *S. cerevisiae* and *S. uuvarum,* are able to ferment sugar into ethanol under anaerobic (oxygen-free) conditions, usually in a solution.

Species of *Saccharomyces* live by ingesting sugar produced by other organisms. The only sugars that can be used by these yeasts are simple monosaccharides or disaccharides (Fig. 14.4). Monosaccharides are the most common type of simple sugar found in nature. Disaccharides include sucrose, table sugar, and maltose, a sugar important in the beer-brewing industry. In addition to sugars, yeasts need amino

Body weight (pounds)

Number of drinks	100	125	150	175	200	225	250
1	.03	.03	.02	.02	.01	.01	.01
2	.06	.05	.04	.04	.03	.03	.03
3	.10	.08	.06	.06	.05	.04	.04
4	.13	.10	.09	.07	.06	.06	.05
5	.16	.13	.11	.09	.08	.07	.06
6	.19	.16	.13	.11	.10	.09	.08
7	.22	.18	.15	.13	.11	.10	.09
8	.26	.21	.17	.15	.13	.11	.10
9	.29	.24	.19	.17	.14	.13	.12
10	.33	.26	.22	.18	.16	.14	.13
11	.36	.29	.24	.20	.18	.16	.14
12	.39	.31	.26	.22	.19	.17	.18

Sober (drinks 1–3); Driving impaired (drinks 4–7); Illegal (drinks 8–12)

FIGURE 14.2

The blood level concentration of alcohol in a person's body as related to the number of drinks consumed per unit of time, assuming that each drink is 1 ounce of 80 proof whiskey. In the last few years, many states and countries have lowered the legal level of alcohol permissible for driving from 0.10 to 0.08.

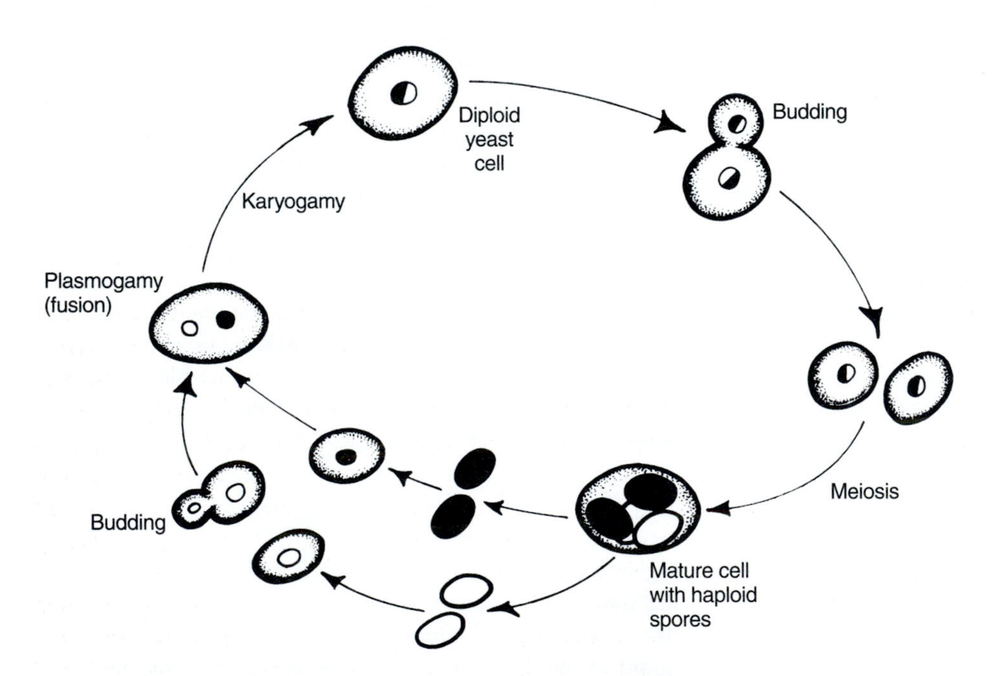

FIGURE 14.3

The life cycle of *Saccharomyces cerevisiae,* the yeast most commonly used in brewing processes, shows that yeasts are single-celled organisms that do not form mycelia or specialized spore-bearing structures. Although budding (mitotic divisions) of diploid cells is the most common form of reproduction, sexual reproduction involving meiosis can also occur.

acids to live and reproduce. The need for these two kinds of compounds is absolute: *Saccharomyces* will die in a medium that does not contain simple sugars and will grow poorly, if at all, in one that lacks nitrogen. Yeast cannot effectively metabolize starch even though starch is composed of units of sugar molecules. Consequently, in the production of all fermented beverages, the material that is to be fermented must contain natural mono- and disaccharides that have been produced by enzymatic degradation of starches. Essentially this means that enzymes in some form must be added to almost everything that is fermented except honey and some fruit juices.

During the process of fermentation simple sugars are broken down via a number of intermediate steps into ethanol and carbon dioxide. From the yeast's point of view, these are waste products. During fermentation the yeasts also produce small quantities of long-chain alcohols, acetaldehyde, acetic acid, and traces of sulfur-containing compounds. The chemical steps involved in fermentation are shown in Figure 14.1. Theoretically, taking into account only the energy required for the chemical reaction, 51.1 percent of the sugar could be converted into alcohol. In practice, because the yeast uses a portion of the sugar for growth and maintenance, about 47 percent is converted.

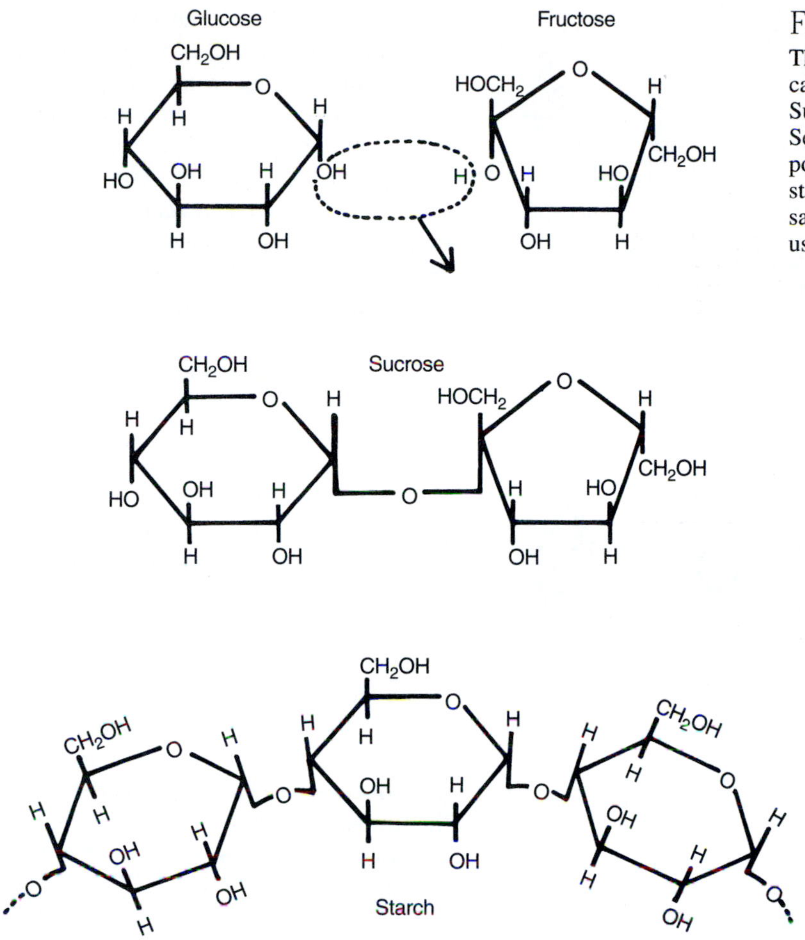

FIGURE 14.4
The monosaccharides glucose and fructose, both six-carbon sugars, can join to form sucrose, a disaccharide. Sucrose is the most common sugar transported in plants. Several monosaccharides can also link together to form polysaccharides or starches. Polysaccharides including starch must all be broken down to disaccharides or monosaccharides before they can be consumed by yeasts and used to produce alcohol.

As the yeast culture grows and metabolizes more and more of the sugar in the liquid, carbon dioxide builds up and causes the solution to become bubbly. This gas is often allowed to escape from the solution. With unlimited sugar, the alcohol level increases during fermentation until it reaches a concentration between 12 and 18 percent. Levels of alcohol above 18 or 19 percent are usually toxic to the yeast and lead to death of the cells. This tolerance limit places an upper value on the percentage of alcohol produced solely by fermentation. Achieving a higher concentration of alcohol requires that the solution be fortified by adding more concentrated alcohol or by distillation.

Wines, Mead, and Fortified Wines

By definition, **wine** is fermented fruit juice. In practice, however, the term is overwhelmingly used for the fermented juice of grapes, *Vitis vinifera* (Vitaceae). Other wines are generally specified by using the name of the fruit from which they are made, for example, peach wine and blackberry wine. Wine may be the oldest fermented beverage made by humans, although some authors contend that mead or even beer was produced earlier. **Mead** is a fermented solution of honey and water. The sugar in pure honey is so concentrated that *Saccharomyces* and other fungi and bacteria cannot live on it (hence its long shelf life). However, if honey is diluted with water, it provides an excellent medium for yeasts because it consists of simple sugars. Consequently, before humans could have consciously made mead, they would have had to gather wild honey and dilute it (by washing combs or sweetening water). Today mead is made by boiling a dilute solution of honey to which amino acids, and in some cases aromatic herbs, are added. Fermentation is completed in 6 to 8 weeks at temperatures between 15° and 25°C (59° and 77°F). Mead is classified according to the kind and dilution of the honey from which it was made.

In contrast to beer and mead, wine is produced in nature. The yeasts responsible for the fermentation of fruit sugars are usually present on fruit skins, and fermentation can occur naturally if the skin of a ripe, sweet fruit is punctured. Human production of wine would merely involve collecting fruits, bruising or crushing them, and letting them ferment. Still, no one knows when wine making began, and estimates varying from 8000 to 3000 B.C. have been proposed. The first concrete evidence of wine making comes from residues found in clay vessels from western Iran that have been dated to 5500 years ago.

FIGURE 14.5
This branch and flower of a wild mustang grape show the tendrils and the three- to five-lobed leaves characteristic of the grape family. In this species the greenish flowers have reduced sepals and five nectar-producing glands alternating with the stamens at the base of the two-celled ovary. The fruit is a two- to four-seeded berry.

Since grapes are the dominant fruit used for wine, we will concentrate on the production of grape wine. Botanical evidence suggests that the species of grape most widely used for wine was originally domesticated in western Asia in about 4000 B.C. Any wine making before that time would have used wild grapes (Fig. 14.5). Grapes, like most fleshy fruits, consist primarily of water, but they also contain an appreciable amount of fructose, a simple monosaccharide. Egyptians used wine primarily for religious ceremonies (Fig. 14.6), and it was only between 2000 and 1000 B.C. during the Grecian empire that wine became a popular beverage (Fig. 14.7). The Classical Greeks stored wine in vessels smeared with pine pitch to prevent leakage, a practice that may account for the Greeks' fondness for a resinous flavor in wine. Now this flavor is added to produce retsina wines. The Romans did not use pitch on wine vessels, which is one reason why Italian wines surpassed Greek wines in popularity under the Roman empire.

Wine grape cultivation spread from the eastern Mediterranean region (Fig. 14.8) in about 600 B.C. to France and later to Spain, Portugal, and Algeria. Until recently Europe and North Africa were the undisputed world leaders in quantity and quality of wine production. Now the United States, Argentina, and South Africa rank among the top 10 wine-producing countries (Table 14.2).

Wine making in the United States got off to a slow start, probably because the northeastern part of the country was too cold for growing European wine grapes. Although it could not be successfully grown in New England, the vinifera grape is hardy in many other areas of the New

FIGURE 14.6
Harvesting grapes and producing wine.
(Redrawn from an Egyptian wall frieze in the Tomb of Nakht.)

FIGURE 14.7

Bacchus as the god of wine.

(Engraved by Gabriel Muller [1771]. Courtesy of the Christian Brothers Collection.)

TABLE 14.2 1999 World Production of Wine

TOP COUNTRIES	1000 METRIC TONS	TOP CONTINENTS	1000 METRIC TONS	WORLD, 1000 METRIC TONS
France	6,264	Europe	20,592	27,943
Italy	5,806	North and Central America	2,217	
Spain	3,297	South America	2,077	
United States	2,045	Africa	1,004	
Germany	1,228	Oceania	802	
Argentina	1,254			
South Africa	877			
Australia	741			
Portugal	679			
Romania	650			
Chile	475			

Source: FAOSTAT Database Gateway. Online at http://apps.fao.org/ (under "All databases" then Crops, Processed).

FIGURE 14.8

According to Hebrew folklore, when Adam planted the first grapevine, the devil buried a pig, a lion, and a lamb at its roots so that those who drank too much of the juice of its fruits would become fat, ferocious, and feeble. Joshua and Caleb are shown in this woodcut carrying the "big grape."

(Woodcut by Jost Amman [A.D. 1531–1591]. Courtesy of the Christian Brothers Collection.)

World. Columbus introduced plants into the West Indies on his second voyage (1493), and the Spanish began its cultivation in California around 1769. By the middle of the nineteenth century California had a small but respectable wine industry.

It is lucky that Americans became involved in viticulture, because in 1860 European vines began to die from an infestation of *Phylloxera,* an insect commonly called a root aphid (it is not a true aphid). This insect is a native of North America and had been inadvertently carried to Europe between 1850 and 1860. Populations of the insect expanded rapidly in the fields of the susceptible *V. vinifera* and destroyed the vines. The French sent a commission to the United States, which determined that many North American species and hybrids were naturally resistant to the insect. As a result of the commission's findings, thousands of American cuttings and seeds were sent to Europe to replace the dying vines (Fig. 14.9). The plants produced from the cuttings and seeds, mostly from Texas, were used as rootstocks onto which the traditional varieties of the vinifera grape were grafted. Almost all European grapes are still grafted to American rootstocks, even though resistant strains of *V. vinifera* have been developed. In 1992, there was an outbreak of *Phylloxera* in the Napa Valley of California, where growers had been planting nongrafted grapes that were touted to be resistant to the insect. By 1999, the outbreak had run its course as growers succeeded in removing susceptible vines and replacing them with resistant plants.

Vines begin to produce several years after planting and can bear well for over 50 years. The deciduous plants are pruned between February and April in the Northern Hemisphere to remove the long branches produced during the previous year and to ensure symmetrical branch growth. The sprouts that emerge later in the spring are trained onto trellises. In many vineyards several varieties of grapes are grown together so that when the grapes are picked, they will automatically produce the desired blend of grape types. Blending can, however, be done at any number of points in the wine-making process, up to the final bottling.

Since the late 1990s there have been volumes printed about the potential benefits of drinking wine. The idea of beneficial effects were derived from the fact that there was a low incidence of arteriosclerosis in individuals living in southern Europe, where wine drinking is very common. This purported effect may be confounded by the eating habits of people in this region, which include a diet rich in olive oil and tomatoes. However, some studies suggest that flavonoids in wine slow the growth of cells in blood vessels

FIGURE 14.9

This allegorical statue commemorates the rescue of the French wine industry after the devastation by *Phylloxera* blight in the middle of the nineteenth century. The young woman represents the resistant American rootstocks that replaced the more susceptible European rootstocks. The statue is in Montpelier, France, the home of Gustave Foex, who was head of the Montpelier School of Viticulture at the time of the blight.

(Courtesy of the École Viticulture de Montpelier.)

that would eventually lead to narrowing of the arteries. Other studies indicate, however, that drinking more than one glass of wine, one beer, or one cocktail each day is correlated with an increased incidence of breast cancer in women.

Making Wine

Fine vintage wine is made today in much the same way that it was centuries ago, although the bulk wines that are now commonly produced are often treated slightly differently. Large wine-making companies in the United States monitor each step of the process with sophisticated chemical and computer analyses. Additional chemicals ranging from fructose to minute amounts of mineral salts are added whenever the analyses indicate a deficiency. In our discussion, we follow the traditional steps in wine making, pointing out some of the recent innovations.

Since grapes are naturally equipped with everything necessary for fermentation, the process of wine making is basically very simple. The grapes are crushed, and the juice is allowed to ferment. Juice can be expressed by stomping barefoot (Fig. 14.10) on the grapes (a method still used in some parts of Europe) or with hand-operated, electric, or fuel-powered presses. In areas where grapes are grown as part of a cooperative venture, they are often pressed at a central point and the liquid is transferred to trucks for transport to the winery. Sulfur dioxide is introduced into the closed container to kill bacteria. Grapes can even be shaken loose by mechanical pickers and pressed in the field. Again, the liquid is placed in an atmosphere of sulfur dioxide.

If the expressed juice is to be made into white wine, the free juice is run into fermentation tanks and the skins and stems are repressed (Fig. 14.11). The juice from the second pressing can be added to the original juice or used for lower-grade wines. For red wine, the skins go into the fermentation vat with the juice. The red color of red or rosé wine (from which the skins are removed after a short time) is due to pigments in the skins that dissolve in the juice. Consequently, white wine can be made from red grapes if the skins are removed right after pressing. Red wine cannot be made naturally from white grapes. For example, red wines made from white Thompson seedless grapes have the color added.

Once the juice is in the fermentation tank (Fig. 14.12), preferred strains of yeast are often added (although the process could proceed without additional yeast), and fermentation is allowed to continue for about 8 to 10 days, after which the incipient wine is drawn off the skins if they are still present. White wines are fermented at 10° to 15°C (50° to 59°F), and red wines at 25° to 30°C (77° to 86°F). At temperatures lower than 10° to 15°C white wines become musty flavored, and at higher temperatures they lose their fruitiness. For red wines, comparatively higher temperatures are needed to prevent an unpleasant, thin flavor, but temperatures that are too high can lead to an overcooked taste. Any additional liquid obtained from pressing the skins that remained throughout fermentation is considered inferior in quality to the initial juice and is used in the blending of poorer-grade wines or for vinegar.

After the initial fermentation the liquid is allowed to ferment for 20 days to about one month. During this second fermentation period particles and dead yeast cells settle to the bottom of the tanks. When the process is complete, the wine is drawn off this sediment and placed in aging tanks. The concentrations of acids in the grapes, principally malic and tartaric acids, change during the fermentation and throughout the aging process. The tartaric acid, which precipitates from the solution as "cream of tartar," was once one of the important commercial sources of this baking aid.

FIGURE 14.10
This European lithograph shows two old ways of expressing juice from grapes for wine making: using a wine press, shown in the background, or stomping on grapes.

FIGURE 14.11
Loading grapes into a vat for crushing.
(Photo courtesy of Champagne News and Information Bureau.)

Because particle sedimentation continues over time, the wine is often transferred across a series of tanks during aging in a process known as **racking,** each time leaving behind a sedimentary layer. Aging tanks are used over and over again but are cleaned and doused with sulfur to kill bacteria after each use. The large wooden tanks traditionally used in wine making can acquire individual characters, and some become known as producers of "superior" wine. The stainless steel tanks used today for fermentation and aging are efficient but lack the charm provided by these fragrant wooden vessels.

Fermentation is complete when there is no more fermentable sugar or the alcohol levels reach levels toxic to yeast. Therefore, fermentation must be stopped before all the sugar has been converted to ethanol or sugar must remain after the yeast have died for a wine to be sweet. Pasteurization can be used to kill any live yeast remaining in fermented wine, but this process profoundly affects the taste. Consequently, wine is usually filtered to remove the live (and dead) yeast after fermentation is deemed to be complete. Filtration does not remove bacteria, which are usually controlled by adding sulfur dioxide or sorbic acid.

FIGURE 14.12

The processing of vinifera and labrusca grapes is contrasted here. As the flowchart for red vinifera wines shows, natural yeasts are traditionally used, and filtration, if carried out, occurs at the end of the process. Sugar is rarely added. White vinifera wines undergo the same procedure, but the skins are removed before the juice is fermented. For labrusca grapes, the must, or unfermented juice, is sterilized before selected yeasts are introduced. Sugar is often added, resulting in sweeter wines. In both cases sulfur dioxide (a gas) is used to kill unwanted bacteria. To clear the wine of particles that are small enough to remain in suspension, "fining agents" such as diatomaceous earth are added. These agents attract and flocculate the particles so that they can settle out. This final step helps prevent the wines from developing off flavors during aging.

FIGURE 14.13
During the process of barrel aging, a wine's acidity is reduced and the aroma and final color develop. Workers periodically check a wine's development by removing samples for taste tests.
(Photo courtesy of Wine Institute.)

Time of aging in tanks varies. White wines are usually aged for from 1 year to 18 months, but red wines can be aged for as long as 5 years. At stages during aging the wine is sampled and judged by a wine master (Fig. 14.13). If the wine master deems the wine ready, it is bottled after the tank aging is complete or is used only for blending. Again, depending on the wine master's decision about the potential of the wine, it can be sold soon after bottling or aged further in the bottle. Some white wines benefit by aging in the bottle for up to 5 years, after which they tend to deteriorate. Red wines, in contrast, can continue to improve for 30 or perhaps 40 years. The quality of the wine that ultimately results depends on many factors, including the kinds of grapes from which it was made, the year in which the grapes were harvested, and the procedures followed during fermentation and subsequent aging. With proper aging, a truly fine wine can result.

The naming and labeling of wines vary considerably among countries. Within a given country, names and labeling practices can follow various schemes that give indications about the kind of grape used, the location of the vineyard, the slope of the area on which the vines were growing, the place where the wine was bottled, the level of sweetness, and of course the year in which the grapes were picked. There are hundreds of varieties of grapes, many of which are used to name a kind of wine, for example Cabernet Sauvignon, Chardonnay, Muscadet, Pinot Noir, and Syrah. Molecular studies of fragments and repetitive segments of DNA conducted in 1998 and 1999 have been able to reveal the origin of some of the popular wine grapes. For example, Cabernet Sauvignon is a hybrid between the Sauvignon Blanc and the Cabernet Franc grape. Chardonnay and the Gamay Noir grapes both resulted from selections of hybrids of Pinot and Gouais Blanc grapes. The year in which the grapes were picked and the wine was initiated is known as the **vintage** of a wine. The knowledge needed to read and assess a wine label properly takes time to acquire and is described in great detail in several of the references given in the Additional Readings.

Wines produced by simple fermentation of fruit juices are called **still wines** because they contain no gaseous carbon dioxide. The fermentation tanks are open or have outlets that allow the escape of gases as they are produced. If the fermentation ceases before all the sugars have been metabolized by the yeast, the wine will be sweet. If the yeast has converted all or almost all of the sugar before fermentation stops, the wine will be dry.

Champagne and Other Sparkling Wines

Quality sparkling wines, among them champagne—the effervescent wines from the region of La Champagne in France—are usually made by adding sugar and selected yeasts to a blend of still wine before bottling. The yeasts and sugar start a second fermentation in the bottle called *prise de mousse.* The carbon dioxide produced from the final spate of fermentation is trapped inside the bottle, causing the wine to effervesce or sparkle when opened. This late fermentation presents a problem, because the dead yeast cells and various particulate matter that settle out in the aging tanks of a still wine are sealed in the bottles. Ingenious methods have been devised for removing the sediment in the champagne without losing all the bubbles.

Removal of the sediment is facilitated by storing champagne bottles at an angle during the last part of fermentation and while the wine is aging. The bottoms of the bottles are kept higher than the necks (Fig. 14.14). The bottles are regularly turned a little to aid in the accumulation of the sediment in the bottle neck. Traditionally, the sediment was removed by opening the bottle and letting the champagne shoot outward. A little replacement wine was added, and the bottle was closed as rapidly as possible. Naturally, a great deal of champagne, as well as part of the effervescence, was lost. Today, the neck is plunged into a freezing solution which freezes the liquid in the neck containing the sediment (Fig. 14.15). The cork is removed, and the pressure from inside the bottle forces the frozen cylinder of champagne plus debris outward. The bottle is then immediately recorked. In another modern method, known as the transfer method, the second fermentation occurs in a closed tank. The sparkling wine produced is drawn off the sediment and bottled under pressure so that the carbon dioxide is not lost. For less expensive sparkling wines, carbon dioxide is simply pumped into a still wine at the time of bottling in the same way that it is pumped into carbonated beverages.

Fortified Wine

Fortified wines are fermented wines to which concentrated ethanol or a highly distilled beverage such as cognac is added. The practice of fortifying wines began in

FIGURE 14.14
Champagne bottles are stored on tilted racks and turned periodically to collect the sediment in the necks of the bottles.
(Courtesy of the Champagne News and Information Bureau.)

Europe in the seventeenth century. Specific regions such as Jerez in Spain, the Douro Valley in Portugal, and Madeira Island soon became centers for the production of these wines. These regions produce the classic fortified wines: sherry, port, and Madeira, respectively. They still produce the majority of, and to most connoisseurs the finest, fortified wines, although wines given the same names are now produced elsewhere. Differences in grape varieties and production procedures give each one a characteristic flavor. A second group of fortified wines includes aperitif wines such as Dubonnet and vermouth, which have, in addition to wine and spirits, flavorings from a variety of plants.

Beers, Ales, and Sake

As in the case of wine, no one knows when people first began to brew beer. Educated guesses usually converge on a date about 6000 years ago. Some anthropologists have correlated the brewing of beer with the establishment of

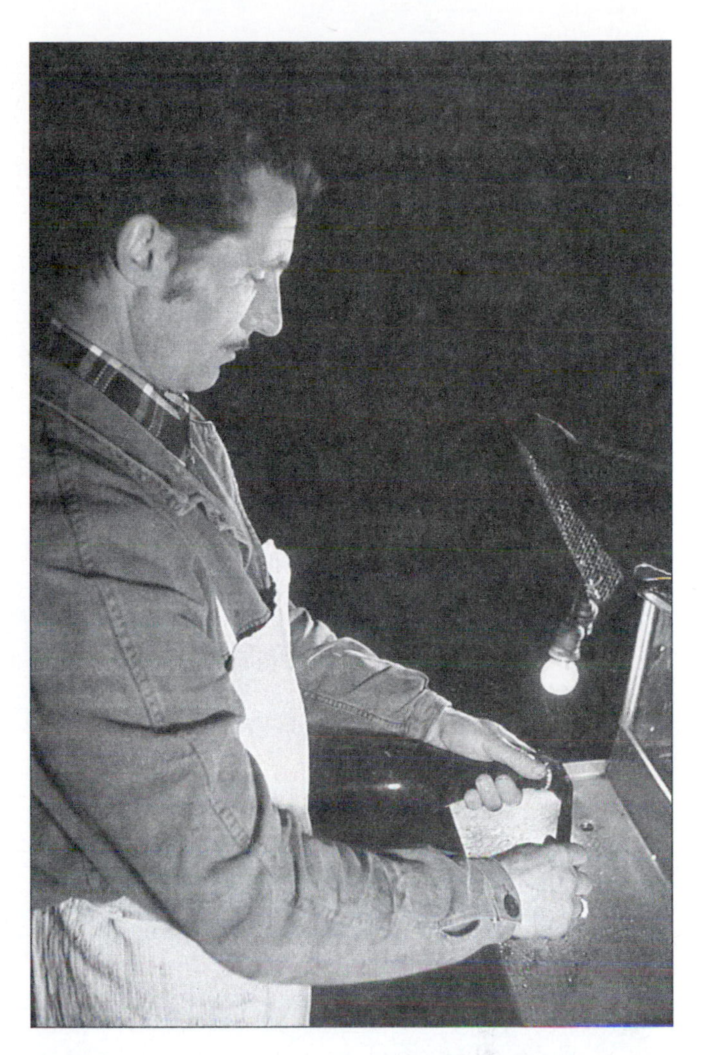

FIGURE 14.15
During "le degorgement" the cork of the champagne bottle is momentarily removed so that the sediment and frozen liquid in the neck of the bottle can extrude and be cut away. A fresh cork is expertly fastened to the bottle before more of the champagne can escape.
(Courtesy of the Champagne News and Information Service.)

permanent human settlements. Certainly the practice was well established by the beginning of recorded history. Written records indicate that much of the grain of the Sumerian civilization was used for making beer. A Greek myth claims that Demeter fled from Mesopotamia, disgusted at the beer drinking of its inhabitants. Early brewing is usually linked with bread making. As was described in Chapter 5, one of the early ways of making grains digestible was to sprout them. Barley breads were initially made from sprouted grain that had been dried and ground into flour. A soft dough of sprouted barley flour would be, as we shall see, a good place for yeasts to live. Egyptian beers in fact were made from a solution of water and pieces of dough made of sprouted barley flour that was subsequently allowed to ferment. After fermentation, the liquid was poured off the sludge that had settled on the bottom of the container. Although wine was never much used by Egyptians except for special occasions, beer was widely drunk, and many different kinds were made.

Early beers were relatively simple to make, but the liquid mixture to be fermented was initially left exposed to the air so wild yeasts and other microbes would fall into it. Sometimes bacteria and fungi other than species of *Saccharomyces* fell into the brew and would multiply and produce a foul-tasting or rotten-smelling batch of beer. Some quality control was ensured by using a small amount of the yeasty liquid from a previous good batch of beer. Uniform production of palatable beer is a relatively recent affair. In fact, the entire process of beer brewing, much more than that of wine making, has changed greatly in the last 200 years. Today brewing is a complicated process involving several ingredients never used in brewing in earlier times. Since most of the beer Americans drink is made in large breweries, we will follow the process as it now occurs in one of those operations (Fig. 14.16).

The three basic ingredients used in modern beer making are barley malt, hops, and water. In addition, most U.S. breweries also use adjuncts, or carbohydrates derived from plants other than barley. Before we go into the process of brewing itself, let us look into the origin and production of each important plant-derived ingredient.

FIGURE 14.16
Modern beer making is an exacting science, as the elaborate equipment and computerized control board of the Lone Star Brewing Company in San Antonio, Texas, show.

Malt

Strictly speaking, **malt** is any sprouted grain that has been subsequently dried, but in practice, the term refers to germinated barley grain. Barley, *Hordeum vulgare* (Poaceae), is preferred over the grains used in the past for several reasons. First, during malting, barley husks stay on the kernels, whereas other grains usually shed their husks once they begin to germinate. During the brewing process the husks add some flavor to the brew and later collect at the bottom of the mashing tank, where they form a bed through which the beer can be filtered before fermentation. Most important, however, is the fact that of all the possible malts, barley malt contains the largest amount of enzymes necessary for converting starches to fermentable sugars.

The first step in malting is to steep the grain in huge tanks where it is washed by a flow of water for 8 to 10 hours. The washing process causes the grains to absorb water and initiates the germination process. The barley then sits in still water for another 40 hours, after which the water is drained and the barley is conveyed to large, climate-controlled germination rooms where it is periodically turned over for a period of about 6 days.

The chemical processes set in motion by this artificially induced germination are the same as those that occur in nature. It was pointed out in Chapter 5 that the stored energy in a grain is starch. At the beginning of germination the embryo produces enzymes that break down the starch into sugars that can be readily absorbed by the developing seedling and used as a source of energy. From the point of view of the brewer, several important things happen during this brief period. The grain synthesizes an abundant supply of hydrolytic enzymes that convert some starch to sugar that can be used by the yeasts. During the process, the cell walls of the endosperm break down and many of the endosperm constituents other than starch are reduced to compounds with relatively low molecular weights. Among this last group of chemicals are many proteins that are degraded to shorter polypeptides and amino acids. Cutting proteins into small pieces helps eliminate the cloudiness that long-chain molecules can produce. In malting, the germination process is stopped when the emergent rootlets are about one-third as long as the grain itself (Fig. 14.17). Kilning, or heating, the sprouted grain to temperatures between 130° and 200°C (266° and 392°F) kills the emerging seedling and any microorganisms that may be present on the grain. Some large breweries malt their own barley, but others buy malt from huge malting corporations in Wisconsin, where most of the varieties of barley grown for malt are produced.

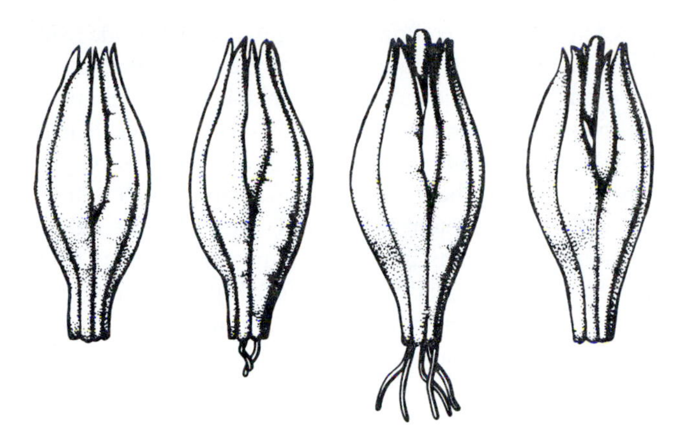

FIGURE 14.17
In the process of malting, barley is permitted to germinate until the emerging embryonic seedling is one-third the length of the fruit. At that point the fruits are kiln-dried (kiln-drying shrivels the emerging rootlet) and shipped to breweries.

Hops

The second important plant ingredient in beer is hops. Despite the modern assumption that hops have always been an integral part of beer, they were originally added to the beverage in A.D. 822, and the species was domesticated about 860. Wild hops grow natively from western Europe to Siberia and Japan. In classical Greek times they were added to bread and salad for flavor. The English began to use hops in brewing only after 1524. Before the use of hops became popular, other plants, such as bog myrtle, were often added to beer as flavoring agents. Until 1860 plant-derived flavoring agents other than hops were still added to beer in England. Hops probably rose to ascendancy because they not only imparted a pleasant taste and aroma to beer but also added enzymes that helped coagulate unwanted proteins. If too many proteins are present in beer, the brew tends to become cloudy. Hops, therefore, contribute to the production of a sparkling, clear beer. Compounds leached from the hops also appear to have antibacterial properties that help prevent spoilage of the beer.

Hops, *Humulus lupulus,* belong to one of only two genera in the Cannabaceae (the other is *Cannabis,* marijuana or hemp; see Chapters 12 and 15). Hops are dioecious vines that produce clusters of flowers (Fig. 14.18). Each female flower is subtended by a leafy bract. When mature, the female inflorescences resemble soft pinecones because all the bracts overlap one another. The bracts of the female flowers possess numerous glands that contain volatile oils.

Hops require a cool, dry climate for growth and are therefore grown in northern Europe and the northern parts of the United States. In southeastern Washington State, one of the largest hop-growing regions in the world, hops are grown in orchards consisting of rows of female vines trained onto trellises over 6 m (approximately 18 ft) tall

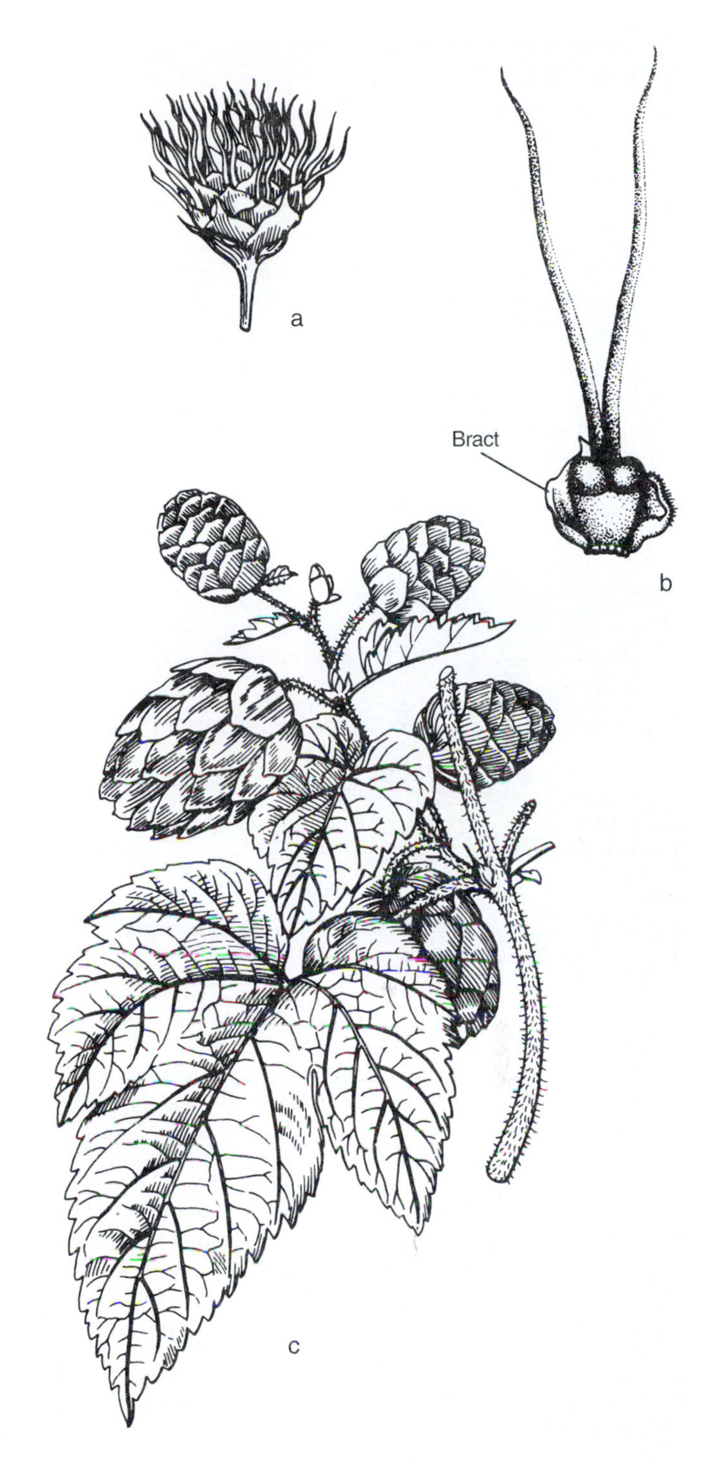

FIGURE 14.18
(a) The conelike inflorescence of the female hop plant is similar in structure to that of its relative, marijuana. (b) Both the flowers and the bracts of female hops have glands that produce the characteristic hop flavors, but since most hops that are grown are parthenocarpic, only the bracts provide the flavoring. (c) A branch with parthenocarpic fruit.
(After Baillon.)

FIGURE 14.19
Because hops are grown on a trellis system, harvesting consists simply of cutting the vines and laying them on a truck bed to haul to the processing plant.
(Photo by J. Neff.)

(Fig. 14.19). During harvesting, entire shoots are stripped from the trellises and the hops are removed. Brewers receive samples of hops and choose the lot with the qualities they desire. Before shipping, the hops can be dried and pressed into pellets. Pelletizing reduces shipping costs, prevents spoilage, and appears to have no effect on the properties of the hops.

Adjuncts

The last major ingredients now used in brewing are the adjuncts. **Adjuncts** are unmalted grains (barley, rice, and wheat), corn grits, corn syrup, or, more rarely, potatoes, that contain starches that can be converted into fermentable sugar. These carbohydrates can constitute today up to 30 percent by weight of the plant material used in brewing. Adjuncts are a recent brewing innovation and are much more commonly employed in the United States than in Europe. They are used because they are less expensive than barley malt. They also allow a brewer to use malt made from a lower grade of barley than would be the case if malt alone were used, because the flavor of the malt becomes relatively less important in the resultant light-flavored beers than in full-bodied European beers. A light flavor seems to be preferred by American beer drinkers. Alternative starch sources usually have less protein than barley does and thus reduce the number of clouding proteins that must be removed from the beer. Corn, rice, and potatoes are all precooked before use to convert their starches into a form the malt enzymes can degrade into sugars.

Brewing

When all the ingredients have been assembled, brewing can begin (Fig. 14.20). In large breweries, the malt and the adjuncts are mechanically conveyed to scales and the proper amounts are automatically weighed and added to produce the **mash.** Slightly acidified water is poured into the malt-adjunct mixture in a mashing tun, or vat. The water can be heated to a temperature between 68° and 73°C (154° and 163°F) before it is added, or it can be added cold and the entire mixture can be heated. In either case the mash is allowed to stand for 2 to 6 hours. During mashing, the enzymes in the barley malt diffuse into the solution and break down some of the starch of both the barley endosperm and the adjuncts. Proteins are also degraded into amino acids.

When mashing is complete, the liquid portion of the mixture contains simple sugars, some starches and other carbohydrates, proteins, amino acids, and various other compounds. This liquid, called the **wort,** is filtered, usually by draining it through the bed of barley husks that has formed on the floor of the mashing tun. The filtered wort is boiled to inactivate the enzymes and sterilize and concentrate the solution.

Hops are added a little at a time during the brewing process. The quantity of hops added varies from beer to beer and brewery to brewery, but it is always quite small compared with the amount of the other ingredients. Once brewed, the beer is cooled and the particles of hops are removed by filtering or centrifuging. Once these operations are complete, the wort is pumped to fermentation tanks and selected strains of yeast are added. It is essential that the cultures of yeast be continually checked in the laboratories maintained by the breweries to ensure that contamination has not occurred.

Fermentation is carried out today in large cylinders with conical bases and at cool or cold temperatures. If *Saccharomyces cerevisiae* is used and fermentation is allowed to proceed at room temperature, "top fermentation" occurs: the yeast rises to the top of the tank as the process continues, forming a frothy mass. At low fermentation temperatures, the mat of cells drifts to the bottom of the tanks, where it collects in the narrow part of the conical base. Ales are produced using this top-fermenting yeast. Bitters are ales that are well hopped and thus have a pronounced bitter flavor, and stouts are ales made from malts that have been so highly roasted that they produce a dark,

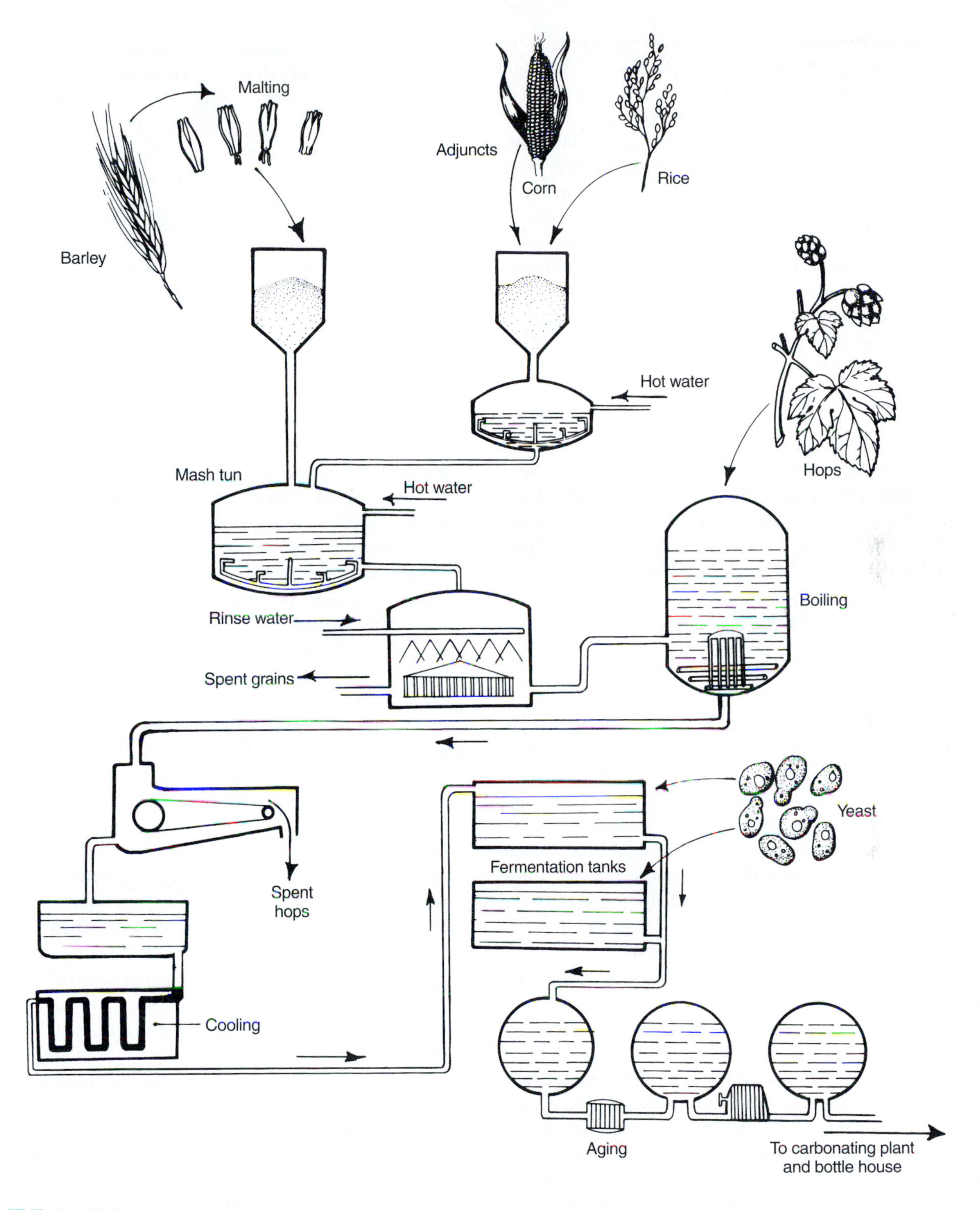

FIGURE 14.20
The major steps in the modern brewing of beer in the United States. The mash is made of a mixture of malted barley, other starches (adjuncts), and water. After filtering, the liquid, known as the wort, is boiled to kill any microorganisms it might contain. Hops are added at this stage. After filtering and cooling, yeast is added and the brew is fermented.

almost black, beverage. *Saccharomyces uuvarum* is by nature a bottom fermenter and is used to produce lager and Pilsner beers. Most U.S. beers are lagers. Lager beer can be classified into two types. One has a strong hop component and is known as a Pilsner beer; the other, brewed with comparatively few hops, yields a less bitter Munich beer. Much of the carbon dioxide produced during the fermentation is allowed to escape from the tops of the tanks. Fermentation continues for 7 to 12 days at cool temperatures. The resultant brew, called **green beer,** is run off from the fermentation tanks and allowed to age for about 2 to 3 weeks. During this time, insoluble proteins settle out of the liquid and some undesirable chemicals, such as polyphenols, are broken down. Most breweries in the United States pasteurize and then filter or centrifuge the beer. Since traditionally a low number of microorganisms (live or dead) per milliliter are desired in commercial beers, these processes ensure that no or very few yeast cells remain in the beer. However, pasteurizing and filtering remove the natural carbonation. Consequently, carbon dioxide is later readded before bottling. The few beers that are not pasteurized generally use Millipore™ or some other filtering system to remove the microorganisms.

TABLE 14.3 Calories per Serving of Alcoholic Beverages

BEVERAGES	AMOUNT	CALORIES
Beer, regular	1 can (12 oz)	151
Beer, light	1 can (12 oz)	70–96
Beer, dry	1 can (12 oz)	130
Gin, rum, whiskey (86 proof)	1 jigger (1.5 oz)	105
Wine, red	1 glass (3.5 oz)	87
Cola beverage (for comparison)	1 can (12 oz)	144

New Directions

A recent innovation in the production of beer is the brewing of "light" beer, or beer with fewer carbohydrates and consequently fewer calories than regular beer. Beer companies were capable of making light beers in the 1960s and attempted to introduce them to the American public. However, the first commercial successful light beer was not produced until 1975, when Miller began to market "Lite." Within 5 years more than 30 kinds of light beer were being sold in the United States. The success of light beers after 1975 can be attributed to the fact that beer companies had improved their taste coupled with a nationwide emphasis on good nutrition and the consumption of fewer calories. Table 14.3 shows the numbers of calories in regular, light, and dry beers and compares them with wine and sodas.

Light beer can be made in two ways. One way is to add fewer starches to the same amount of water used in normal brewing. In regular beer making, the yeast converts all the sugars in the wort into alcohol, but a large amount of unconverted (long-chain) carbohydrates is left in the liquid. In brewing light beer, the same amount of starch is converted to sugar by the malt enzymes and fermented, but fewer carbohydrates are left after the process is completed. In the second method, a different time and a different temperature cycle during the mashing process result in the production of a comparatively large amount of fermentable sugars. When this wort is fermented, a beer is produced that has a higher alcoholic content than regular beer but fewer carbohydrates. Water is added to make a final product that matches regular beer in alcoholic content. When marketed, light beers tend to have a slightly lower final alcoholic content than regular beers.

Four other trends involve the production of "dry beers," nonalcoholic beers, ice beers, and weisse beers. Dry beers first gained popularity in Japan but by the late 1980s were being promoted in the United States for their clean, crisp taste. Dry beers are made much like some light beers. The wort is brewed longer than for normal beer so that more of the carbohydrates are converted into fermentable sugars. The end product appears drier (less sweet) than regular beer because it has less sugar and fewer carbohydrates. In Japan dry beer often has an alcoholic content of 9 or 10 percent. In the United States the alcohol level is kept similar to that of regular beer at about 5.1 percent. Nonalcoholic beers, defined as malted beverages with less than 0.5 percent alcohol, represent another growing trend. For a nonalcoholic beverage to taste like its parent drink, it must be originally prepared in the standard way. However, in one method of producing low-alcohol beer the time and temperature are controlled so that the amount of fermentation is limited and the alcohol production is kept extremely low. In other systems the fermentation is allowed to proceed in the same way as for regular beer, and the alcohol is subsequently removed by one of two processes. It can be drawn off the liquid by heat, usually under a vacuum to keep the temperature low, or removed by dialysis or reverse osmosis using selectively permeable membranes. Ice beer is made by lightly freezing the beer and filtering out the ice crystals. Since water freezes at a higher temperature than does alcohol, the process removes only water, leaving a beer with a higher alcoholic content than regular beer. Weisse beers are made from wheat rather than barley and are occasionally flavored with herbs such as coriander and orange peel.

The success of modern beer brewing relies on understanding and controlling the complex series of chemical reactions that take place, beginning with the malting of the barley. The production of these barley enzymes, their action on starches and proteins, and the process of fermentation are all chemical processes. As in any chemical reaction, each one proceeds best at a certain temperature and in an environment with a specific acidity or alkalinity. Today these factors are increasingly regulated during the brewing process. Consequently, modern breweries, like some modern wineries, are like industrial plants, with huge tanks and pipes and processes controlled by centralized computers (Fig.14.16). Recent advances in beer making include using by-products such as the spent grain liquor and grain residues for making citric acid by fermenting them with *Aspergillus niger* and the genetic engineering of yeasts so that they will break down previously unfermentable sugars to glucose.

Sake

Sake, usually considered to be a rice wine, is the traditional beverage of Japan. Since it is made from fermented grain, sake is actually more properly considered a beer than a wine, but a beer made quite differently from those discussed above. The fermentation, like that of all other alcoholic beverages, is carried out by species of *Saccharomyces,* particularly *S. cerevisiae.* The distinction lies in the production of the fermentable sugars. Sake is made from rice, *Oryza sativa* (Poaceae). If the rice were simply malted or if enzymes were added to it, a form of regular beer would result. Instead, the conversion of rice starch to sugar is performed by another fungus, *Aspergillus.* Raw rice used to make sake is first polished by removing the pericarp–seed coat complex of the grains. The polished rice is washed, steeped, and then steamed for 30 to 60 minutes. Steaming changes the form of the starches present (the same reason for which it is cooked if it is used as an adjunct in beer) and sterilizes the mass. The rice is then cooled, spread in special rooms, and inoculated with cultures of *A. oryzae.* The rice is then mounded into piles, and the fungus is allowed to grow. By the end of 40 hours a mycelium, or layer of fungal hyphae, forms, and the rice begins to smell distinctly moldy. During this period the *Aspergillus* produces enzymes that break down some of the starch. The rice is then spread out, mixed with water, and heated. Lactic acid is added to prevent growth of other organisms. *Saccharomyces* cultures are added to the water-rice mixture over the next 25 days. During this period there is simultaneous breakdown of starch by enzymes released by the *Aspergillus* and fermentation of the sugars by *Saccharomyces.* By the end of the growth period of the two fungi the final alcoholic content of sake can reach 18 percent. The completely fermented thick slurry is placed in cloth filters and squeezed to force out all the liquid. In a sense, therefore, sake is a highly alcoholic beer. It is not aged as such but allowed to mature for about 40 days. Sake is filtered and pasteurized before marketing. Traditionally, sake is consumed within 1 year of its production. Unlike most other alcoholic beverages, it is typically served heated.

Chicha and Pulque

In Central and South America Indians commonly used corn to make a beer called chicha. Since barley malt was not available, they used a different source for amylases to break down the starch. Kernels of corn were chewed for a short period to mix them with salivary amylases, and the chewed mass was spit into a container of ground corn and water and allowed to ferment. Chicha is still produced in this way in remote regions of Peru and Bolivia.

In arid and semiarid regions of central Mexico the inhabitants tap the stems of various species of *Agave* and collect the exuded sweet sap, which is fermented into a beverage called pulque. When agave plants begin to flower at about the age of 10 to 12 years, the flowering stalk is cut at the base. A scar soon forms over the wound, which is punctured repeatedly to allow sap from the plant to collect in a bowl formed by the leaves. As much as 8 liters of sap (about 8.4 quarts) can collect in a day. This sap is collected using a large gourd called an **acocote.** The long gourd is hollow and has a hole at both ends, allowing the worker to suck the sap into the gourd while simultaneously avoiding the spines on the leaves. The sap is then allowed to ferment in large containers for about 14 days. In Mexico, pulque is more widely consumed than tequila, a distilled beverage more familiar to Americans.

Distillation

The process of distillation involves the separation of chemicals on the basis of their different boiling points (Fig. 14.21). A mixture of water and alcohol is heated until the alcohol begins to boil. The alcoholic steam is then condensed into pure alcohol. As long as the temperature is kept below 100°C (212°F), the water does not boil. The Arabs are usually credited with the discovery of this procedure about A.D. 700. During the Middle Ages the Arabs distilled scents (Chapter 8) and antimony salts, called *kuhl,* which were used by women to darken their eyelids. When the Arabs began to distill wine, they called it *al'kuhul,* from which our word *alcohol* is derived. In general, early uses were medicinal or the liquid obtained was used as a solvent. By the fifteenth century, however, the British and Scottish were distilling barley beer, and by 1688 brandy was being produced in Cognac, France. These early uses are examples of the two main classes of distilled beverages. The first includes the whiskeys, which are all distilled from solutions that would be classified as beers. The second group includes cognacs and brandies, which come from the distillation of wine.

Whiskeys

Whiskeys are all made in essentially the same way, by fermenting malted barley alone or malted barley mixed with other grains (Fig. 14.22) and distilling the product. The

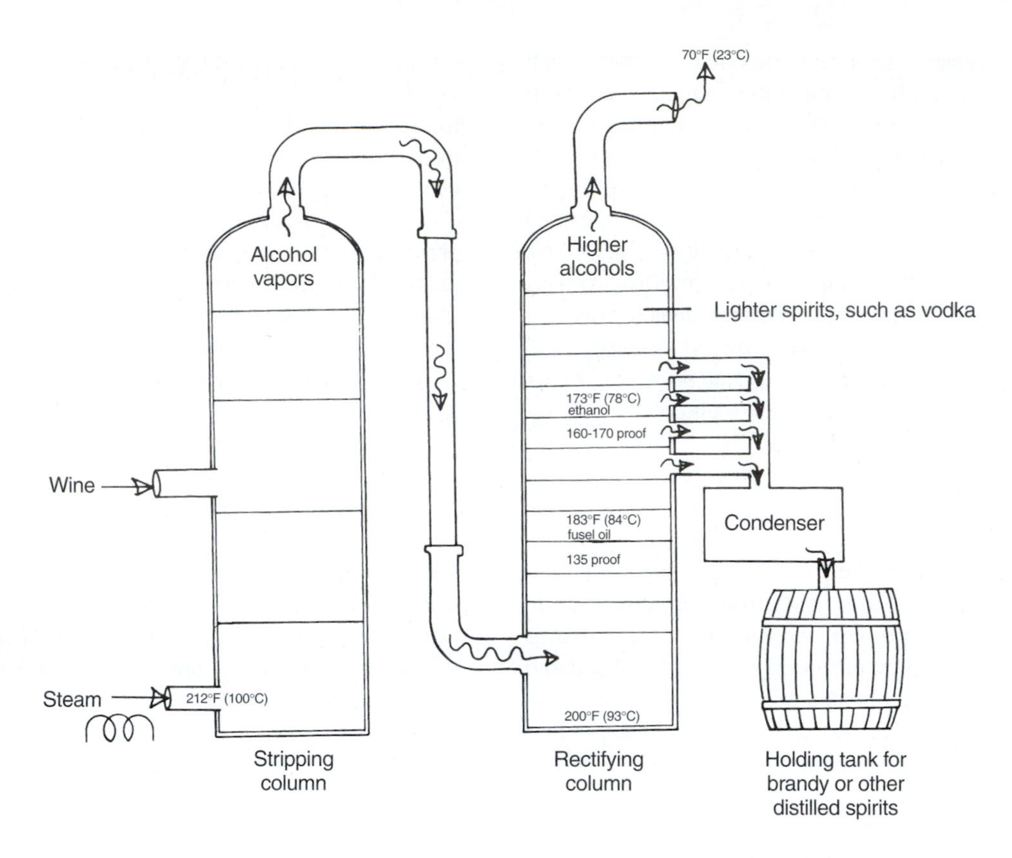

FIGURE 14.21

Distillation is a process that takes advantage of the fact that different compounds boil at different temperatures. Ethanol boils at 78.5°C (173.3°F), and water boils at 100°C (212°F). Other compounds present in the liquid to be distilled boil at different temperatures. In this diagram, wine, a beverage with the maximum alcohol content that can be produced by fermentation, is distilled into brandy. The wine is boiled, and the mixture of alcohol and water steam enters the rectifying column, which is hottest at the bottom and coolest at the top. As shown, in the area of the column where the temperature is about 78°C (173°F), an ethanol-water mixture of 80 to 85 percent alcohol (160 to 170 proof) condenses and is collected. Carried in this mixture are volatile oils that provide flavor and color to the distilled brandy. If purer, stronger ethanol, such as that found in vodka, is desired, the steam closer to the top will be collected and condensed.

(From J. Meyer, 1989. *Plain Talk about Fine Wine,* Santa Barbara, Capra Press.)

FIGURE 14.22

The mash in the steep room.

(Photo courtesy of Jack Daniels' Distillery.)

differences lie in the kinds of grains used, the places and length of time of aging, and the presence or absence of blending. Distillation of whiskey is not a simple operation; it involves a series of stills, each with several plates that allow the liquid to condense and revaporize several times. This type of multiple system produces a very pure solution. The concentration of ethanol that results can reach as high as 99.99 percent, but most distillations for beverages to be consumed are at values between 80 and 95 percent. The distillate is then diluted with water to the desired strength. The **proof** of a whiskey is equal to twice the concentration of ethanol. Thus a 90 proof bourbon contains 45 percent ethanol. Once distilled, and diluted, the raw whiskey is usually aged in some sort of barrel (always wood for good whiskeys) for at least 2 years (Fig. 14.23).

For many people scotch is the king of whiskeys. Scotch is supposed to be made from only barley malt. The characteristic taste is caused by the kilning of the malt over fires fueled by peat moss. Most scotch, however, is made not from pure malt but from 40 percent malt and 60

FIGURE 14.23
Cooperages have been kept in business by laws requiring that bourbon must be stored in new oak barrels during aging. This photo from 1913 shows where the old wooden casks were cleansed.
(Photo courtesy of Guinness Museum.)

FIGURE 14.24
Finely ground charcoal is tamped into this vat before newly distilled whiskey is dripped through to mellow the flavor before aging.
(Photo courtesy of Jack Daniels' Distillery.)

percent grain. The aging can take place in new or used casks that have been charred. The names of scotches are often derived from the hamlets or glens in which they are made.

Bourbon is an American invention, originally developed by Scotch-Irish immigrants in Kentucky and named after Bourbon County. By law, bourbon must be made from at least 51 percent corn, *Zea mays* (Poaceae), and distilled to produce a solution of less than 95 percent alcohol. It must also be aged for at least 2 years in new, charred oak barrels (Fig. 14.23). Since the barrels cannot be reused, they constitute a major part of the expense of producing bourbon. Tennessee sour mash whiskey is quite similar to bourbon but is brewed somewhat differently. For sour mash, some of the "spent beer" from the previous batch is added to the new mash. In addition, the fermented mash is filtered through maple charcoal (Fig. 14.24) before distillation.

Rye whiskey is made from at least 51 percent rye grain and distilled to no higher than 80 proof. It must also be aged for at least 2 years in new barrels. Irish whiskey is made from a mash of primarily barley malt, but wheat and rye are often added. It differs from scotch in that the malt from which it is made is not dried over peat.

Straight whiskey can be any of the distilled beverages discussed previously if it has been distilled to less than 80 proof and aged for at least 2 years in new charred barrels. Blended whiskeys are straight whiskey blended with neutral spirits (ethanol) or other spirits.

Gin, Vodka, and Rum

Gin and vodka differ from whiskeys in that they are distilled to a very high percentage of alcohol and therefore lack many of the flavoring agents that are carried across with the ethanol in other whiskeys. In addition, neither is aged. Gin has traditionally been made from a distillate of a fermented mash made from malt and other grains. Vodka can be made from malt and grains or from potatoes, whichever is cheaper. Gin is flavored with juniper, *Juniperus communis* (Cupressaceae), "berries" (actually fleshy cones) but can also have a number of other flavoring agents. Rum is the distillate of fermented molasses or sugar cane juice. Light and dark rums differ in the degree to which they were distilled and the amount of aging they receive.

Tequila and Mescal

Tequila and mescal, beverages developed in Mexico, have been produced only since the Spanish introduced the practice of distillation in 1521. Both are distillates of fermented agave "wine." For tequila, only *Agave tequilana* is used, but several species, primarily *A. angustifolia* (maguey) and *A. palmeri*, can be used for mescal. The production of both beverages starts with cooked "hearts" of the agave plant. The hearts are produced by cutting the plants at ground level, upending them, and cutting the leaves to produce a pineconelike core (Fig. 14.25). These hearts are cooked to

FIGURE 14.25
Trimmed stems of agave await processing at the Sausa Tequila factory in Mexico.
(Photo by J. L. Neff.)

FIGURE 14.26
In 1605 Carthusian monks living in the monastery of La Grande Chartreuse were given a recipe for an "elixir of long life," which included 130 medicinal and aromatic herbs. The brothers worked with the recipe to produce a therapeutic beverage and, later, Chartreuse liqueur, which they dispensed to the poor as medicine. Chartreuse is now sold worldwide as an aperitif, but it is still made by monks who have taken a vow of silence. Thus the recipe remains a guarded secret.
(Photo courtesy of the Chartreuse Distillery.)

carmelize the sugar. The slightly sweet-flavored hearts are then crushed with water, and the mash is fermented. This relatively mild "wine" is then distilled. Modern operations crush the cooked stems by machine and often ship the fermented liquid to Mexico City for distillation and aging (Fig. 14.25). Most tequila production, however, still centers around the small town of Tequila in west-central Mexico.

Brandies and Liqueurs

Brandies are distilled wines. The most famous are those from Cognac and Armagnac in France. The exquisite flavors of those brandies are due to the wines from which they are distilled and the fact that they are carefully matured for many years after distillation. Many distilled wines are labeled as brandies with a qualifying adjective indicating the kind of fruit wine from which they were distilled (e.g., peach brandy and raspberry brandy) or have special names such as calvados or applejack (apple brandies) or kirsch (cherry brandy). Fine brandies are always made by distilling a wine or flavored wine. Inexpensive brandies can be made by flavoring distilled wine or even relatively pure spirits.

Liqueurs and cordials differ from brandies in that they have sugar or syrup or both added to the distilled liquid. Liqueurs also have characteristic flavors. The flavors can be imparted to the liqueur before or after distillation, but fine liqueurs are all made by adding the flavoring agents before distillation and distilling the volatile oils along with the ethanol. The flavoring agents used include a broad spectrum of plants and plant products, herbs, fruits, and barks. The kinds of plants used and their proportions are often a carefully guarded secret. It is rumored that Chartreuse, a fine liqueur that has been made since 1605 by Carthusian monks in France and Spain, contains over 130 different flavoring agents (Fig. 14.26).

Fibers, Dyes, and Tannins

Plant and animal fibers in the broad sense have played a role in providing humans with shelter (Fig. 15.1), vessels in which to carry water and cook food, and thread for making fabrics. Most of the world has abandoned mud and waddle construction and baskets smeared with clay as water vessels or cooking utensils, but the importance of plant fibers as a source of weaving materials remains. Historically, fibers were dyed with plant-derived colors, but synthetic dyes have largely supplanted natural dyes except in the food, cosmetic, and pharmaceutical industries.

In prehistoric times humans probably obtained flexible plant fibers simply by pulling off strips of bark or cutting stems and leaves into thin, weavable ribbons. Although these materials can be lashed and interlaced into mats and baskets, they produce only coarse, stiff items. A major innovation was the discovery that individual fibers could be separated from surrounding cells and used to weave textiles. Undoubtedly animal skins predated woven material, but plant fibers were used long before animal fibers for weaving. In 1998, sandals over 8000 years old made of complexly woven plant fibers were found in a Missouri archaeological site. By 5400 years ago native people in Mexico were using cotton fiber. Linen was apparently woven 10,000 years ago in Turkey, at least 3000 years before the domestication of sheep and about 5000 years before the use of silk. Still, these dates are relatively recent considering the fact that the association of the genus *Homo* with plants and animals dates back hundreds of thousands of years. This recency is understandable as only a small number of animals (sheep, camels, vicuñas, guanacos, some goats and rabbits, and the silk moth) and relatively few plants produce fibers that can be twined or spun. Thus, humans had to appreciate the nature of fibers, learn which plants contained them, and learn how to extract them before spinning and weaving them.

Vegetable fibers share many features that distinguish them from animal or synthetic fibers. The primary difference between plant fibers and those obtained from other sources is that plant fibers are composed of cellulose (Fig. 15.2), long strings of glucose molecules attached to one another in a precise way. Animal fibers are made of protein (Fig. 15.2). This fundamental chemical difference determines how the two types react to heat, various chemicals, water, and predatory organisms. For example, cellulose molecules are not subject to denaturation by high temperatures, but heat cracks the protein backbone of animal fibers and makes them brittle. Because of this differential reaction to high heat, cotton sheets can be boiled when

FIGURE 15.1
Construction of a thatched roof in France is an ancient method of using plant fibers to create shelter.

they are washed but woolen garments are ruined by very hot water. Plant and animal fibers behave differently in dye baths as well. The complexity of the protein molecules of animal fibers promotes their acceptance of dyes, whereas plant fibers may require elaborate treatments to ensure successful color adherence. Animal fibers are particularly susceptible to attack by animal pests such as moths and silverfish. Plant-derived fabrics are essentially immune to these pests but, like paper, are readily attacked by fungi, mold, and even termites. Finally, plant fibers tend to be less elastic than animal fibers yet have a higher affinity for water and are thus more absorbent.

It is more difficult to compare plant fibers with synthetic fibers because the properties of synthetic fibers vary with the chemistry of different synthetic materials. Some synthetic fibers (discussed in Chapter 16) are made from cellulose, but most are manufactured from chemicals derived from petroleum. There has been some experimentation with using fatty acids cleaved from seed oils (see Chapter 9) as precursors of textile polymers. Synthetically produced fibers have challenged natural fibers for supremacy in terms of world production in recent years, but their cost advantage has dwindled as petroleum costs have risen. Even though they are often more expensive than their synthetic counterparts, natural fibers have never been totally replaced because people value their special properties: no synthetic material can match the comfort of cotton or the crispness of linen.

Vegetable fibers can be classified by their use (Table 15.1) or by the part of the plant from which they are obtained. A classification based on use divides fibers into those used for **textiles** (woven goods), brushes, plaiting or coarse weaving (Fig. 15.3), stuffing material, paper, and specialty products. Here we deal primarily with textile fibers, which are commonly grouped into seed and fruit fibers, soft, or bast fibers, and hard, or leaf fibers. **Bast fibers** come from the phloem tissues in the stems of various dicotyledons. **Hard fibers** come from the leaves of a few monocotyledonous plants.

Commercially important textile fibers are extracted from many unrelated species, but they can be functionally grouped into those that can be plaited, twined, woven, and spun into threads and those that cannot. The latter group includes fibers that are short, brittle, or slippery. Important plant fibers of this type include wood fibers used for items such as paper, fiberboard, cellophane, and rayon (see Chapter 16). Here we concentrate on fibers that can be interlaced, twined, or **spun** (aligned and twisted together) to form yarns and threads.

Spinning can occur only if the fibers have structural properties that cause the individual strands to clasp one another when twisted. The forces that keep spun fibers together in a strand are mechanical, but they prevent the strands from slipping free from one another and thus allow the production of a continuous yarn or thread. Most animal hairs (including human hair) and some plant fibers, such as those on milkweed seeds, cannot be spun because they are too slippery to stay together when twisted around one

TABLE 15.1 Classification of Vegetable Fibers by Use

USE	EXAMPLES
Textile fibers	
Soft, or bast, fibers	Hemp, jute, ramie, linen
Seed and fruit fibers	Cotton, coir
Hard, or leaf, fibers	Sisal, henequen, abacá, pineapple
Plaiting and weaving	Palms, stems of several grains, papyrus, pandanus, bamboos
Brush fibers	Palms, sorghum, broomroot
Filling fibers	Kapok, milkweed, cattails
Felting fibers	Paper, mulberry, lace bark
Papermaking fibers	See Chapter 16

a Celluolose Molecule

b Protein Molecule

Basic structure of amino acids

$H_2N-C(R)(H)-C(=O)-OH$

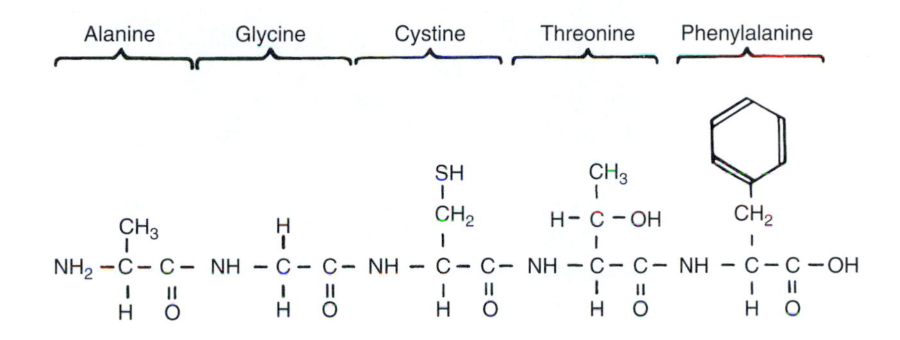

FIGURE 15.2

(a) Cellulose is a polysaccharide consisting of long chains of glucose molecules that are cross-linked to neighboring molecules, forming a rigid framework. (b) In contrast, the molecules that make up proteins are amino acids. Each amino acid has an amine (NH_2) group and a carboxyl (COOH) group attached to a central carbon atom. Amino acids differ from one another by having different side chains, indicated here by the letter R. When amino acids combine with one another, the amine end of one bonds with the carboxyl end of another, with the loss of a molecule of water. Protein molecules are usually composed of thousands of amino acids. The long unbranched chains often twist or coil into complicated configurations.

another. Plant fibers used for basketry, caning, and mat weaving are not separated from the other cells with which they occur, but they provide flexibility and strength to the strips cut from the plant material.

A fiber to be spun and woven into soft fabrics that can withstand repeated washings needs characteristics very different from those of one that is to be used in ropes that will be continuously wetted with seawater. The value of a fiber and the uses to which it lends itself are determined by a number of factors. The look and feel of a fiber are affected by its length and structure (Fig. 15.4). Fibers from different species can consist of single cells or groups of cells that remain together in long strands that vary in length from a fraction of a millimeter to over a meter. Round fibers feel silkier and have a smoother appearance than flat ones. The tensile strength of a fiber, the extent to which it can be stretched before breaking, varies with its cross-sectional area, the length of the individual cells of which it is composed, and the way in which those cells are held together. Elastic properties are a measure of the facility with which a fiber can regain its original shape after it has been stretched. The amount a fiber is twisted, the way in which the cells are held together, and the numbers of cells per fiber all contribute to its elasticity. Fibers also have different densities, or weights relative to an equal volume of water, which affect how fabrics made from them will drape. Finally, the chemistry of various fibers causes them to react differently with water, sunlight, heat, acids, alkalis, solvents, electricity, and microorganisms. All the various properties are taken into account in choosing a fiber for a particular purpose.

In the following discussions we point out the probable natural ecological roles of textile fibers produced in different plant organs as well as the ways in which humans have

FIGURE 15.3
In order to create willow baskets, artisans cultivate willow, strip off the bark, and then soak the branches to restore their pliancy for weaving.

artificially selected for specific characteristics. Table 15.2 lists the fiber plants discussed in this chapter, and Table 15.3 provides production figures for them. Table 15.2 also lists the dye plants discussed in this chapter. Tannin sources are described in Table 15.5.

Fiber Extraction

Despite the fact that various fibers come from different plant parts and from an array of plant species, the same basic procedures are used to separate them from the masses of other cells in which they are embedded. The primary processes are retting, scutching, decorticating, and ginning. **Retting** is an extraction process that rots away the soft plant parts and leaves the fibers intact. It is used mainly for bast fibers and takes advantage of the fact that fibers have thicker cell walls than do most other plant cells and are therefore comparatively resistant to breakdown by bacteria. Consequently, retting, or bacterial rotting, is often used to decompose soft plant tissues and dissolve the gums and pectins that hold plant cells together. Plant material can be retted by dumping it into tanks of stagnant water or simply by allowing it to remain on the ground, where it will be repeatedly covered with dew. During rettings, tissues absorb water and swell, causing the release of soluble compounds that help provide nourishment for the decomposing bacteria. The process takes several weeks, during which time the mass of rotting material must be tested continually to ascertain the point at which the soft tissues, but not the fibers, have disintegrated. If the plant matter is allowed to ret too long, even the fibers will fall apart.

Retting does not remove all the nonfibrous material, however. At the end of the process epidermal and thick-walled woody xylem cells can persist. These cells are

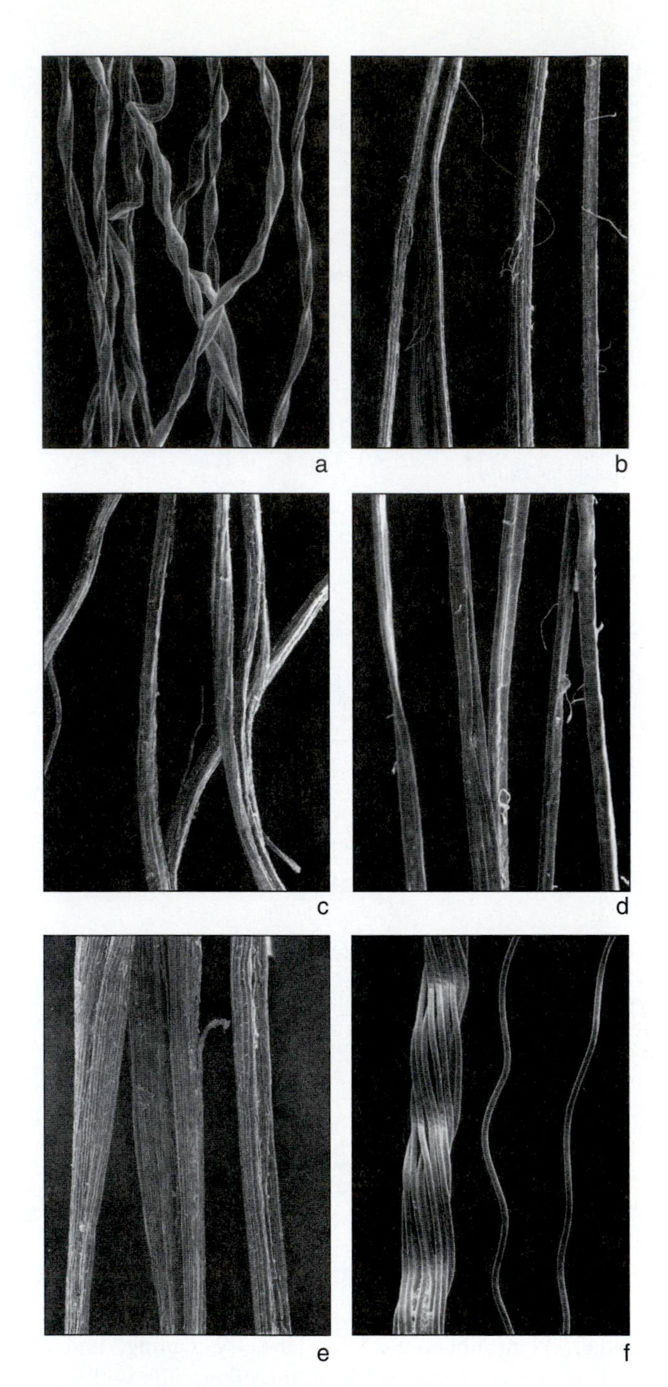

FIGURE 15.4
Scanning electron microscope photographs of selected plant fibers show that the semisynthetic rayon fibers (f) are smoother and more uniform than the natural plant fibers. The gradual twist of the cotton fiber causes the fibers to interlock and form threads when spun. (a) Cotton (× 110.5), (b) flax (× 23.2), (c) jute (× 40.5), (d) ramie (× 105.5), (e) sisal (× 23.2), (f) rayon (× 39).
(Photos by J. Mendenhall.)

TABLE 15.2 Fibers and Dyes Discussed in Chapter 15

COMMON NAME	SCIENTIFIC NAME	FAMILY	NATIVE REGION
	Fibers		
Abacá Manila hemp	*Musa textilis*	Musaceae	Philippines
Coir	*Cocos nucifera*	Arecaceae	Pantropical
Cotton	*Gossypium hirsutum*	Malvaceae	Central America
	G. barbadense		South America
	G. arboreum		Africa
	G. herbaceum		Africa
Flax	*Linum usitatissimum*	Linaceae	Near East
Hemp	*Cannabis sativa*	Cannabaceae	Eurasia
Henequen	*Agave fourcroydes*	Agavaceae	Central America
Jute	*Corchorus capsularis*	Tiliaceae	Eurasia
	C. olitorius		Eurasia
Kapok	*Ceiba pentandra*	Bombacaceae	Pantropical
Milkweed	*Asclepias syriaca*	Asclepiadaceae	Eastern North America
Ramie	*Boehmeria nivea*	Urticaceae	Tropical Asia
Sisal	*Agave sisalana*	Agavaceae	Central America
	Dyes		
Annatto	*Bixa orellana*	Bixaceae	Mexico
Bloodroot	*Sanguinaria isabellinus*	Papaveraceae	North America
	S. canadensis		
Butternut	*Juglans cinera*	Juglandaceae	North America
Henna	*Lawsonia inermis*	Lythraceae	Old World tropics
Indigo	*Indigofera tinctoria*	Fabaceae	Asia
Logwood	*Hymatoxylum campechianum*	Fabaceae	Mexico
Madder	*Rubia tinctorum*	Rubiaceae	Southern Europe, Asia
Safflower	*Carthamus tinctorum*	Asteraceae	Asia
Saffron	*Crocus sativus*	Liliaceae	Near East
Weld	*Reseda luteola*	Resedaceae	Mediterranean
Woad	*Isatis tinctoria*	Brassicaceae	Europe

Note: Most species listed in Table 15.5 are not included here.

removed from the fibers by washing and drying the material remaining after retting; it is then "broken" by being forced under fluted rollers. The breaking process crumbles the brittle material but not the more flexible bast or leaf fibers. The broken pieces of woody matter are then removed from the fibers by beating and scraping, a process known as **scutching** (Fig. 15.5). Finally, the fibers can be hackled to separate and align them. **Hackling** is accomplished by drawing a mass of fibers across a set of vertical pins resembling a comb (Fig. 15.5).

In some cases fibers can be most easily and inexpensively separated from other tissues by decorticating. **Decorticating** entails crushing the plant material and scraping the nonfibrous material from the fibers. In general, decorticating is used primarily for leaf fibers.

Ginning (see Fig. 15.11) is a process unique to seed fibers. During this process seeds are pulled free from the fibers covering them. In the case of cotton, the fibers undergo extensive further cleaning and combing.

Once extracted and cleaned, fibers can be further processed by bleaching or heating, or both, in an alkali solution. Fibers are usually bleached before dyeing so that their natural tan or brown pigments do not affect the final color. Later in this chapter we will explain how dyes adhere to fibers and which plant dyes have historically been used to color weavings and textiles.

Seed and Fruit Fibers

Although seed and fruit fibers come from different parts of the fruit, both aid in seed dispersal, but by different mechanisms. Long fibers on the surfaces of seeds promote dispersal by wind. Anyone familiar with the silky tufts of hairs on the tops of milkweed, *Asclepias* spp. (Asclepiadaceae), seeds knows how effectively they help the seeds float in the air. Fibers within the fruit wall generally protect the seed or provide buoyancy for water dispersal or do both.

In general, few plants have seed or fruit fibers that are long enough for purposes such as spinning. Cotton is a notable exception because it produces seed fibers that can be spun into thread. The seed fibers of milkweed and kapok, *Ceiba pentandra* (Bombacaceae) (Fig. 15.6), are too fine and slippery to spin. They are therefore used commercially

TABLE 15.3 1999 World Production of Major Textile Crops

COMMON NAME	TOP COUNTRIES	1000 METRIC TONS	TOP CONTINENTS	1000 METRIC TONS	WORLD, 1000 METRIC TONS
Cotton (lint)	China	3,830	Asia	10,587	18,239
	United States	3,691	North and Central America	3,847	
	India	2,073	Africa	1,656	
	Pakistan	1,495	South America	919	
	Uzbeckistan	1,100	Oceania	716	
Flax fiber	China	368	Asia	368	625
	France	66	Europe	239	
	Belarus	52	Africa	13	
	Russian Federation	24	South America	4	
	Netherlands	16			
	Belgium-Luxembourg	14			
Hemp fiber	China	28	Asia	125	70
	Romania	11	Europe	25	
	Korea, Democratic People's Republic of	11	South America	4	
	Spain	6			
	Russian Federation	4			
Jute and substitutes	India	1,850	Asia	2,847	2,856
	Bangladesh	812	Africa	5	
	China	100	South America	3	
	Myanmar	33			
	Nepal	15			
Ramie	China	165	Asia	167	171
	Brazil	4	South America	4	
	Philippines	2			
Sisal	Brazil	184	South America	199	383
	Kenya	26	Africa	75	
	Tanzania	24	North and Central America	58	
	Mexico	45			
	Madagascar	18	Asia	18	

Source: FAOSTAT Database Gateway. Online at http://apps.fao.org/.

primarily as stuffing material. Kapok was also formerly used for life preservers because it is light and water-resistant and has fibers that trap air.

Coconuts are also unique because they produce the only fruits from which fibers (coir) are extracted from the pericarp for commercial use. The entire coconut mesocarp consists of a fibrous pith that provides buoyancy for the water-dispersed fruits. Coconuts apparently can stay afloat in seawater without damage for months. The natural distribution of coconuts along tropical coasts around the world attests to the success of this method of dispersal.

Cotton is King

Cotton (Fig. 15.7) is by far the most important fiber today, rating globally as the most important nonfood plant commodity as well as the second most important oilseed crop (Chapter 9). The popularity of cotton can be ascribed to the large amount of fiber produced by each plant combined with the fact that the picking, processing, and manufacturing of textiles from cotton cost less than the processing of other plant fibers. The extensive use of modern machinery brought the cost of cotton down to its present level, while the processing of many other fibers is not as easily done by machine. Cotton is also a very versatile fiber that produces fine textiles that dye well and withstand vigorous washing.

Each cotton fiber is a single long epidermal seed coat cell (Fig. 15.8). These cells are so long that they resemble hairs. In many cultivated cottons there is a second layer of short fuzzy hairs underneath the long fibrous hairs. These short hairs, or **linters,** are removed and employed for papermaking when cottonseed is used as an oilseed crop (Chapter 9).

The taxonomy of the cotton genus, *Gossypium* (Malvaceae), is complicated, and there is still not complete agreement among botanists about the ancestors of the cultivated species. All agree, however, that certain species of cotton were independently domesticated in the Old and New

FIGURE 15.5
Processing flax includes breaking the stems to crumble the brittle woody matter, removing this unwanted material by scutching, and hackling the stems to free and align the long fibers.
(Corbis photo from the Scheufler Collection, ca. 1893, Postrekov, Czech Lands.)

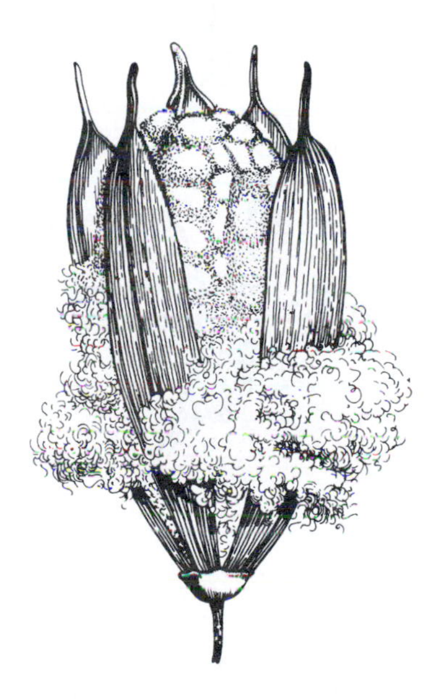

FIGURE 15.6
Kapok pods split at maturity, releasing seeds covered with long, silky hairs.
(After Baillon.)

Worlds. Some authors have suggested a preagricultural use of cotton around the eastern Mediterranean Sea, but archaeological evidence indicates that the first use in the Old World occurred in south-central Asia. The two fiber-bearing Old World species are *Gossypium arboreum,* known only as a cultigen, and *G. herbaceum.* Both are diploid. For many years it was thought that *G. arboreum* is a descendent of *G. herbaceum,* but recent studies using isozyme data have suggested that the two were independently domesticated from different ancestors. By the middle of the fifteenth century cultivation of the two species had spread throughout the Near East and into Europe, via Spain, by the Arabs. Both Old World cottons have short fibers, or **staples,** and both were eventually supplanted by New World cottons with longer fibers. Today Old World cottons are grown primarily in Iran and also in India and Pakistan.

Two species of cotton, *Gossypium hirsutum* and *G. barbadense,* both tetraploid, were domesticated in the New World. It has been known for many years that both species have two sets of chromosomes, one similar to those of the Old World cultivated cottons and one similar to many New World diploid species. Studies have shown that an Old World ancestor was the maternal parent that produced a hybrid that underwent polyploidy, whereas a New World diploid species was the paternal parent. It appears that there was an ancient long-distance colonization of the New World by a *Gossypium* belonging to the same group as *G. arboreum* and *G. herbaceum.* After this, colonization and speciation led to the development of a group of diploids to which the paternal parent belonged. More recently, a second colonization resulted in the hybrid that underwent polyploidy and eventually gave rise to the two New World domesticated species. One of the New World cultivated species, *G. hirsutum,* or upland cotton, has been found in archaeological excavations that date back to 3400 B.C. in Mexico, its presumed area of domestication. This species may have been originally domesticated on the Yucatán Peninsula. Columbus recorded being offered cotton thread by West Indian natives on October 12, 1492, upon his first landing in America. *Gossypium hirsutum,* known as cotton belt, upland, or West Indian cotton, accounts for about 95 percent of the world's crop. Early settlers chose this species for cultivation because it is more resistant to boll weevil attack than is *G. barbadense.*

The other important American cotton is *Gossypium barbadense,* commonly known as Sea Island, Egyptian, or pima cotton. None of these common names is particularly appropriate, but all reflect some aspect of its recent history. Despite its modern secondary status, *G. barbadense* appears to have been extremely important in the pre-Hispanic cultures of western South America. The earliest dates available from coastal Peru indicate that this species was semidomesticated by 2500 B.C. Looming of cotton textiles has been traced back to 1800 B.C. Over 2000 years before the height of the Inca empire in the thirteenth and fourteenth centuries, weaving had become an integral part of the culture of western South America. The enormous amounts of cloth used in Inca mortuary practices (Fig. 15.9) suggest that there was a large group of professional weavers who fulfilled their political obligations by becoming state weavers.

By the time of the arrival of Europeans, *Gossypium barbadense* had been spread by humans from its native region in western South America across the continent and into the West Indies. It is believed that it was given the name Sea Island cotton because it was initially grown in plantations

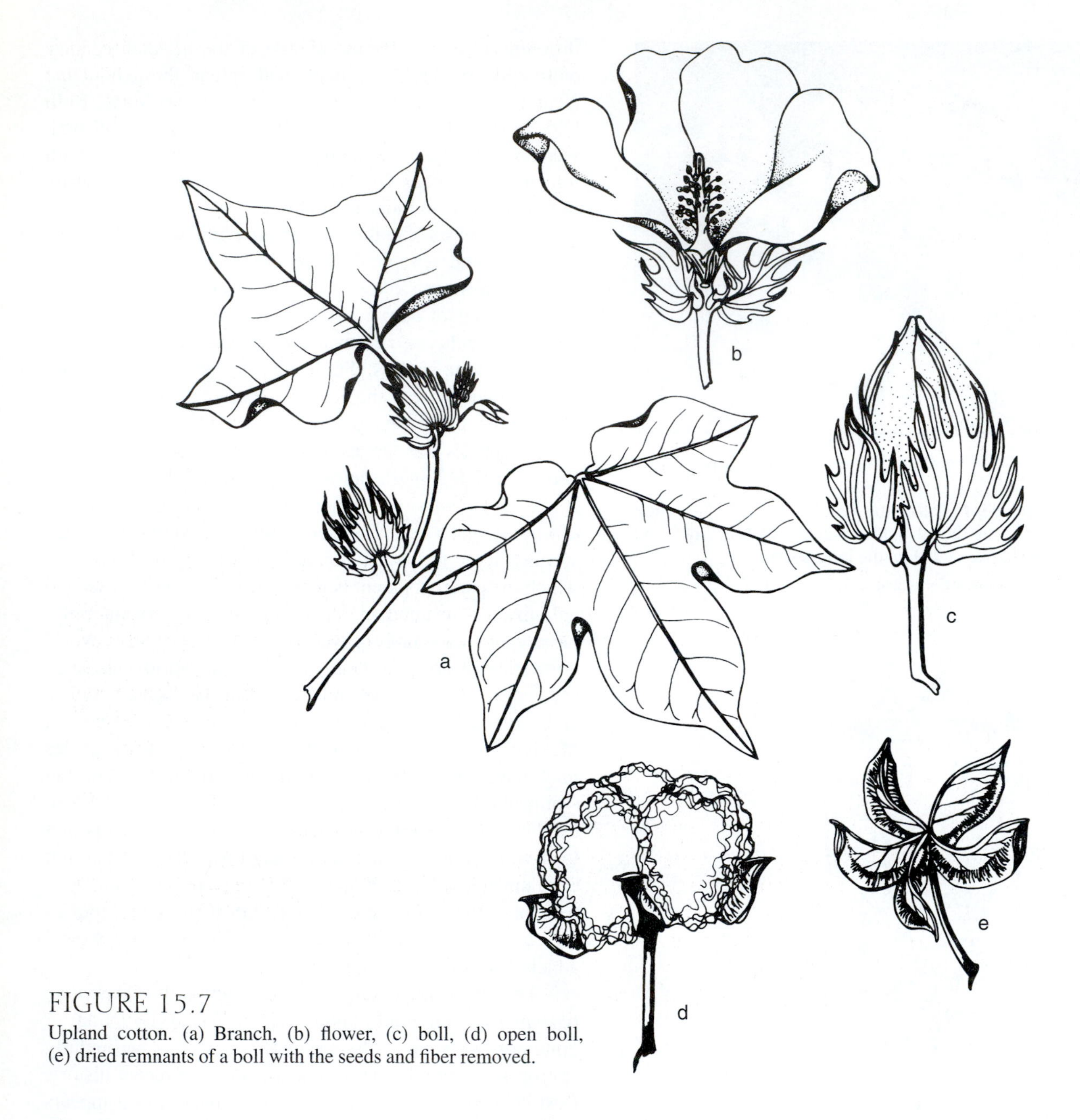

FIGURE 15.7
Upland cotton. (a) Branch, (b) flower, (c) boll, (d) open boll, (e) dried remnants of a boll with the seeds and fiber removed.

along the coast of the southern United States. Plants of this species were also successfully introduced to Egypt. The cultivars now included under the designation Egyptian cotton are all descended from a selection made between 1817 and 1819 by a Frenchman working in Cairo. The staple of these fine cottons can reach lengths up to 6 cm (2.4 in.) and produce fine- quality threads. Pima cotton, another cultivar of this species, was also selected from Egyptian plantings of *G. barbadense.* Its name is derived from its cultivation in Pima County, Arizona.

Most species of *Gossypium* are tropical perennials, but in cultivated species humans have selected for an annual habit and the ability to bloom and fruit in temperate latitudes. By growing cotton as an annual, farmers ensure short stature, uniformity of plant size (semicultivated perennial plants of both *G. hirsutum* and *G. barbadense* can be shrubs 4 m, 13 ft, tall), and synchronous fruiting. On modern farms the plants are often sprayed with defoliants when the cotton is mature so that the foliage will not get in the way of the harvesting machines that pluck the mature bolls from the plants (Fig. 15.10). Once picked, the seeds with their fibers have to be removed from the fruit, and then the fibers have to be separated from the seeds. Most Americans know that Eli Whitney invented the cotton gin, a machine that pulls cottonseeds from the fibers, in 1794 (Fig. 15.11), but the importance of this labor-saving device to the development of the cotton industry is not widely appreciated. In 1791, three years before Whitney's invention, the United States exported 400 bales of cotton. By 1800 exports had jumped to 30,000 bales (Fig. 15.12).

After ginning, cotton fibers undergo extensive processing. Different batches of cotton with staples of varying lengths are usually blended by feeding pieces of different

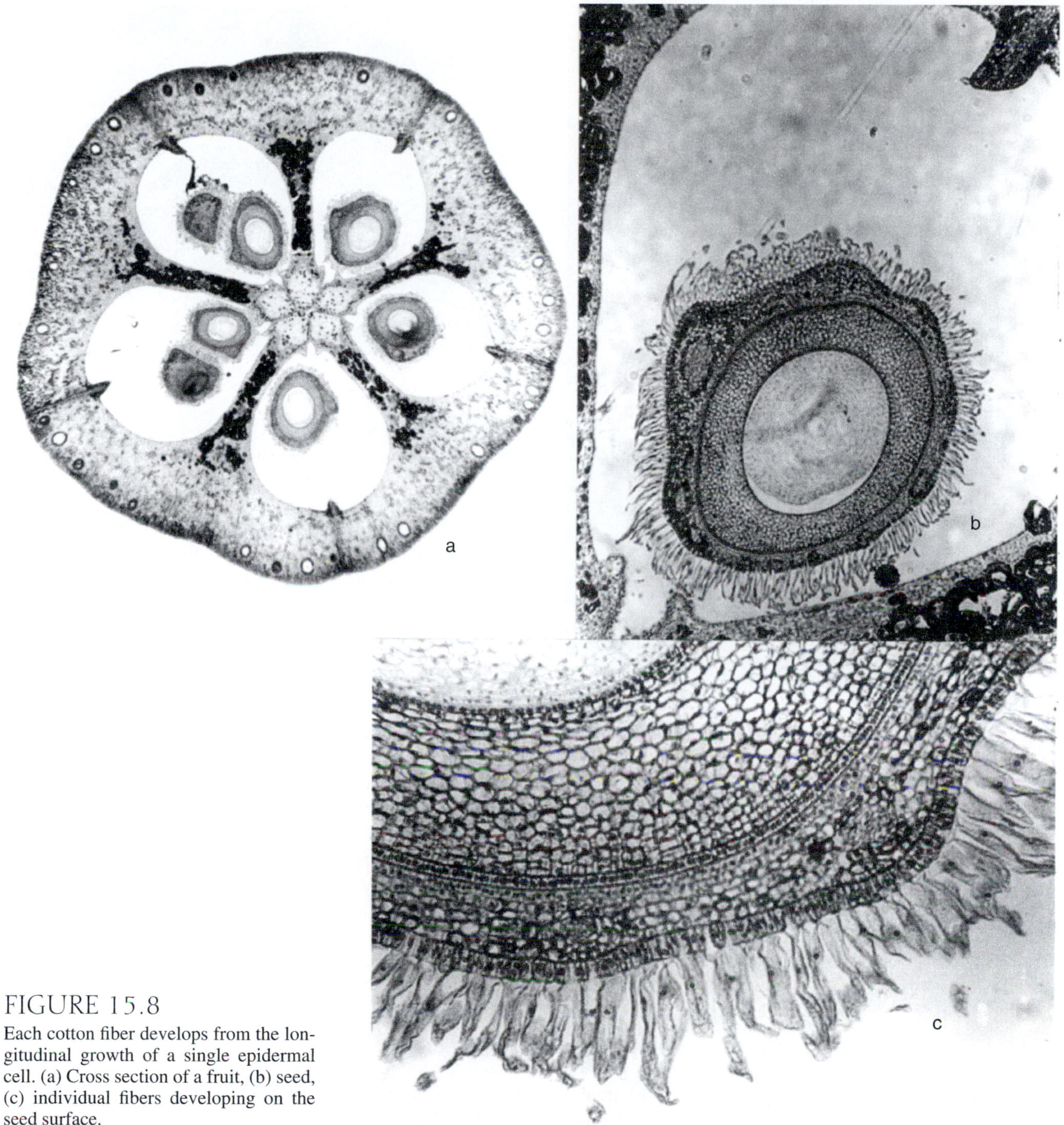

FIGURE 15.8
Each cotton fiber develops from the longitudinal growth of a single epidermal cell. (a) Cross section of a fruit, (b) seed, (c) individual fibers developing on the seed surface.

bales into bins. The cotton is then processed by picking machines that pluck and beat the fibers with steel spikes attached to rollers. During this procedure, unwanted matter is removed and the process of combing is begun. In large factories cotton bales are opened, blended, and picked in a single operation. After being picked, the fibers are **carded,** or combed parallel to one another. In earlier times cotton, like wool, was handcarded using two instruments resembling brushes that were repeatedly pulled across one another. The carded cotton was then spun and woven (Fig. 15.13). Today masses of fiber pass under huge rollers covered with wires of different lengths that tease the fibers apart. After being fed through a series of such rollers, the fibers are pulled rapidly downward and outward across a long comb, producing a thin web of relatively parallel fibers. The web is then gathered and twisted into a loose rope, called a **sliver,** about 2 cm (0.79 in.) in diameter.

Before spinning, slivers are mixed and drawn by being pulled with different amounts of pressure through a series of pairs of rollers to align the fibers parallel to one another and

FIGURE 15.9

Cotton weaving became a highly developed craft in the pre-Hispanic societies of the northern coast of Peru, and cotton textiles assumed an important place in mortuary practices from 900 to 200 B.C. up to the time of the Spanish conquest in A.D. 1532. This mummy of an 18- to 20-year-old woman interred in about A.D. 1200 was wrapped in 24 cotton cloths that had a combined area of 286 m^2 (3078 ft^2) and a weight over 150 kg (331 lb).

(Photo courtesy of James Vreeland.)

FIGURE 15.10

Cotton harvesting in Texas.

(Photo by D. Nance, courtesy of Agricultural Research Service, USDA.)

reduce the diameter of the original sliver. The fibers are then stretched and spun by being twisted tighter and tighter together as they are pulled onto rollers, resulting in a yarn or thread that is made up of many overlapping, parallel fibers held together by simple mechanical forces.

Once made, either the thread, or a textile woven from it, is cleaned and mercerized. Cleaning consists of boiling the thread or cloth with caustic soda for 8 hours and then bleaching it with a chemical such as hydrogen peroxide. These treatments remove pectins and waxes and make the fibers light in color so that they will readily accept dyes. Mercerization was invented in 1844 by John Mercer, an English textile maker, to increase the luster of cotton, promote its uptake of dyes, and heighten its durability. **Mercerizing** consists of placing a thread or woven textile, stretched under pressure, into a cold bath of caustic soda for several hours. The treatment causes the fibers to swell and the cellulose molecules to deform. The material is kept under pressure for some time and is not allowed to resume its former shape. Consequently, the fibers retain the swollen, circular form caused by the alkali treatment.

Before weaving, threads are sized to increase their strength. **Sizing** consists of the addition of a thick substance such as starch or a gel on the surface to stiffen it and fill surface irregularities. Sizing is particularly important for threads that will be used for the warp, or the major loom threads across which the weaving is done. The sizing material is washed out after the weaving or garment manufacture has been completed. Woven fabrics can also be singed, or passed rapidly over flames, to burn off any irregular, protruding pieces of fiber.

Cotton fabrics are often sanforized to reduce shrinkage when they are washed. **Sanforization** dates back only to 1970, when the Sanforizer Company introduced an ammonia process that swells cotton fibers and prevents their shrinking after they have been washed. Sanforizing eliminated the tradition of buying a cotton garment one size too large in anticipation of the inevitable shrinkage caused by the initial washing.

Another problem with cotton has been its tendency to wrinkle when washed. **Permanent press** cotton fabrics have been developed that have greatly reduced the need for ironing after laundering. Although several processes are used to produce a permanent press fabric, all involve the use of chemicals that cross-link the cellulose polymers of the cotton cloth and thus cause the fabric to retain the shape it had when the chemicals were applied. Not only does permanent press keep flat surfaces smooth; it also allows pleated or ruffled garments to retain their shape after washing.

There has been a recent effort to bring native cottons into commercial use. These cottons have staples that come

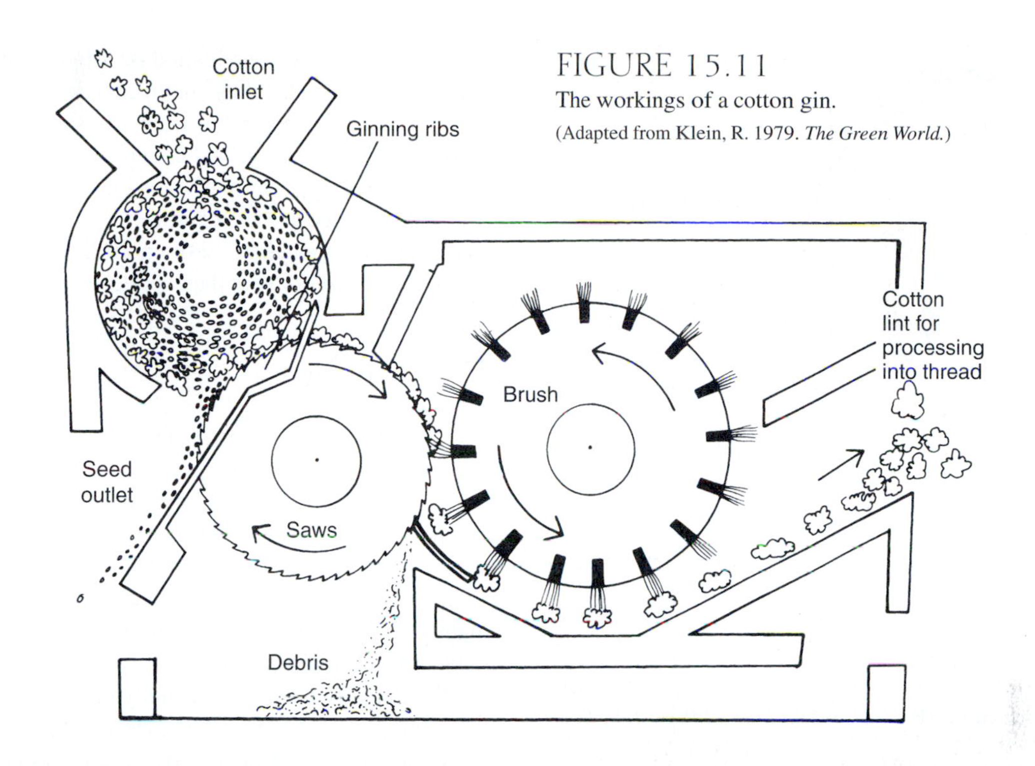

FIGURE 15.11
The workings of a cotton gin.
(Adapted from Klein, R. 1979. *The Green World.*)

FIGURE 15.12
Cotton bales ready for shipment in Navasota, Texas.
(Photo by J. L. Neff.)

FIGURE 15.13
Spinning cotton on the coast of Peru.
(Photo by J. Vreeland.)

in a variety of colors, usually tan, brown, cream, and mauve or green. There are native or semidomesticated forms of both *G. hirsutum* and *G. barbadense.* Levi-Strauss has begun to weave colored cottons derived from Mexican hirsutums into cloth for a line of jeans that it calls "naturals" since no bleaching, dyeing, or chemical processing is used in their manufacture.

Coir

The bulk of a mature coconut fruit consists of a thick, fibrous mesocarp that constitutes the source of a fiber known as coir. In American grocery stores, one usually

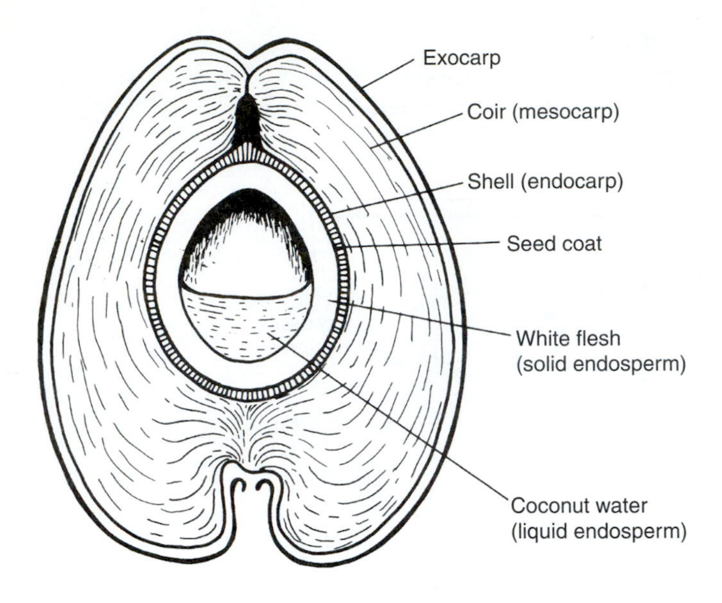

FIGURE 15.14
Coconut fiber, or coir, comes from the fibrous fruit coat that surrounds the seed, which is valued for its nutritious endosperm.

finds coconuts, *Cocos nucifera* (Arecaceae), that are about 15 cm (5.9 in.) in diameter and rough brown on the outside because all of the exocarp and the mesocarp have been stripped from the fruits before marketing. A whole, mature coconut is about twice the size of the supermarket items and is smooth green or greenish brown on the outside (Fig. 15.14). Coir fibers are made up of bundles of cells that are longer than cotton fibers but shorter than most bast or leaf fibers. High-grade coir is produced by harvesting and husking 10-month-old (immature) coconuts. The husks are retted for 8 to 10 months, usually in brackish water. When soft, the husks are thoroughly washed, beaten to remove all the pulpy remains, shaken, and washed again. The clean, pure fibers are spun into yarns that are used primarily for ropes and matting.

The fact that the most valuable coir fibers come from unripe coconuts presents an economic problem since the most valuable commodity obtained from coconuts is **copra** (dried coconut endosperm), which is eaten or used for oil (see Chapter 9). Copra can be obtained only from mature fruits. Consequently, although the mature fruits yield fibers that are tougher than those of young coconuts, most coir is extracted from the old husks that remain after copra has been removed from ripe fruits.

Fibers of mature husks are extracted in much the same way as they are from young fruits, but after retting the mass of fibers of each husk usually is decorticated using a machine with a drum of projecting nails that tear through the fibers. During this operation two types of fibers are produced. The first, **mattress fibers,** are relatively short fibers that fall to the ground under the drum and are collected later for use as stuffing materials. The combed longer fibers remaining in the mass are called **bristle fibers** and generally end up in brushes, brooms, and rough doormats.

Because of its natural resistance to salt water, coir is often used to produce netting for shellfish or seaweed harvesting. In general, however, coir fibers are inferior to those extracted from other plants, and the continued production of coir results largely from the availability of enormous quantities of husks, which are a by-product of the copra/coconut oil industry.

Bast Fibers

Bast, or "soft," fibers consist of the thick-walled phloem cells of various dicotyledonous species. In species such as flax and ramie, the fiber cells occur in groups just outside the conducting cells. In others, such as hemp and jute, there are clusters of fiber interspersed among the conducting cells (Fig. 15.15). The non-conducting fiber cells appear to function mainly to support the stems. Individual bast fibers can be made up of hundreds of cells and can be over 4.6 m (15 ft.) long. The fibers are usually separated from the stems by retting, followed, in some cases, by scutching. Depending on the quality of the fiber, most bast fibers can be bleached and dyed. Bast fibers are not mercerized or sanforized.

Jute

Jute, *Corchorus capsularis* or *C. olitorius* (Tiliaceae) (Fig. 15.16), is the world's foremost bast fiber and is second only to cotton in terms of production (see Table 15.3). The tall reedlike plants have been used since prehistoric times and probably yielded the fibers for some of the sackcloth referred to in the Bible. Large-scale production, however, dates only to the end of the eighteenth century, when Europeans began to search for an inexpensive substitute for flax. The primary use today is still for sacking and similar material, and this fiber is seldom worn, even in penance.

The species appears to be a native of the Mediterranean region, from which it was spread throughout the Middle East and Far East. India remains the primary producer (see Table 15.3). Jute plants are herbaceous annuals that can reach 5 m (16.4 ft) in height and yield fibers 1.8 to 3 m (6 to 10 ft) long. The fibers are also relatively inelastic and tend to disintegrate rapidly in water. They are separated from the stems by deep-water retting, which sometimes causes individual cells in a fiber strand to loosen, giving the fibers a rough feel. Their roughness, the brittleness of the fibers, and their inability to hold dyes promote the use of jute fibers for coarse goods such as carpet backing, canvas, twine, "gunny" sacks, wall coverings, and conveyor belts. Precarding with mineral oil and passing the fibers through rollers help soften the fibers. By-products of jute production are used for particleboard, paper, and rayon. The popularity of jute is due to its low cost, which results from its rapid growth and the ease with which its fibers can be extracted from the stems.

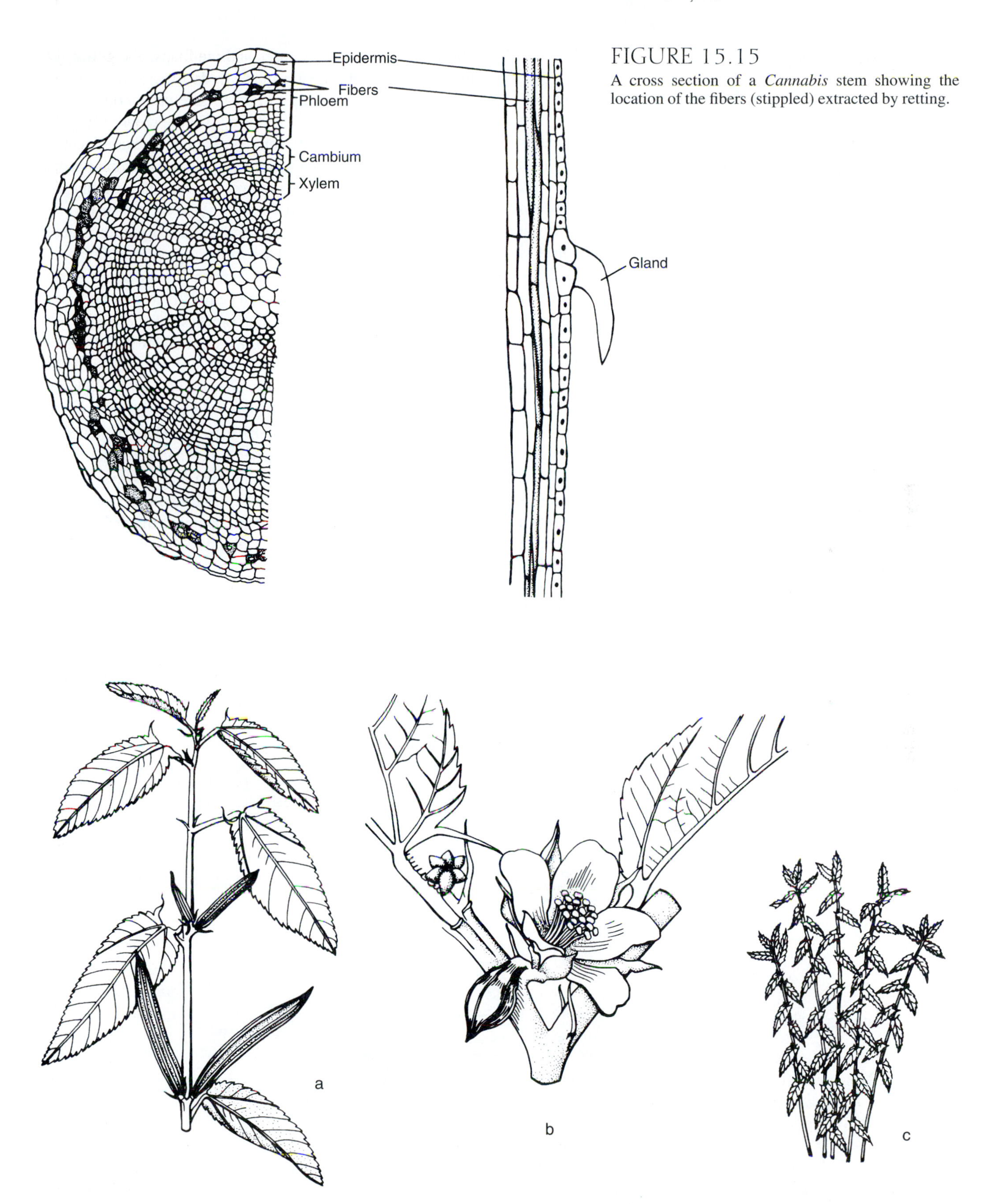

FIGURE 15.15
A cross section of a *Cannabis* stem showing the location of the fibers (stippled) extracted by retting.

FIGURE 15.16
Jute. (a) Fruiting branch, (b) flower, (c) plant.

Flax and Linen

Flax, *Linum usitatissimum* (Linaceae), is considered the oldest textile fiber used by humans. Archaeological digs have uncovered remains of a flax species in ancient settlements occupied by the Swiss Lake Dwellers about 10,000 years ago, and Egyptian mummies dated to be over 5000 years old were generally wrapped in linen cloth. Excavations in Turkey have yielded impressions of a fabric dated to be 10,000 years old that is thought to be linen. Although it is difficult to reconstruct with certainty the extent to which the Lake Dwellers or the inhabitants of Turkey used flax, carvings on Egyptian tombs document its cultivation along with wheat, figs, and olives (see Fig. 9.10). The Classical Greeks used linen, and the Romans spread the cultivation of flax all across Europe. The words *line* (as in a straight line) and *lingerie* and the generic name *Linum* are derived from the latin word for linen.

Flax is native to Europe and eastern Asia, but its geography has been so altered by humans that the original natural distribution is difficult to determine. Today there are no flax populations considered to be truly wild. The plants from which fibers are obtained are tall, little-branched annuals that yield linseed oil (see Chapter 9) as well as fiber (Fig. 15.17). Cultivars selected for either long fibers or high seed yield are planted today, although the same plants were used for both products in ancient times.

Flax fibers are naturally smooth, 0.3 to 0.9 m (1 to 3 ft) long, straight, and are two to three times as strong as cotton fibers. As a result, they are used for making buttonholes and button thread as well as hoses and mailbags. The beauty and luster of linen, the textile made from flax, have long been appreciated. The growing of flax for the production of linen was an important local industry in the Low Countries, England, and Ireland between the seventeenth and nineteenth centuries. Whereas wool was the principal textile fiber used in temperate areas during this period, flax was the leading plant textile fiber. Often the two kinds of fibers were used together. The linsey-woolsey of the Pilgrim settlers of North America was made by weaving wool across a warp of linen thread.

The modern ascendancy of cotton as a textile fiber is based largely on economics. Cotton can be readily and inexpensively processed by machines. Machine-processed cotton is far superior to that processed by hand. Flax generally is obtained by dew retting the cut stalks of mature plants, but since this process can take weeks, plants are now sometimes uprooted and chemically retted. Once the fibers have been lifted from the disintegrated tissues (Fig. 15.18), they are dried, scutched, and hackled (see Fig. 15.5). Because of the hand labor involved in traditional flax fiber extraction, linen has always been an expensive textile. Recently, retting has been made easier by the use before and after harvesting of chemicals that help dissolve the matrix between cells before retting begins.

Hemp

True hemp comes from the same species as marijuana, *Cannabis sativa* (Cannabaceae) (see Fig. 12.5). Despite the notoriety of the resinous constituents of hemp plants (see

FIGURE 15.17
Flax. (a) Flowering branch, (b) seed capsule, (c) seed in section.

FIGURE 15.18
Boy Scouts harvesting flax in England.
(Corbis photo, Hulton-Deutsch Collection, 1918.)

Chapter 12), *Cannabis* was initially spread around the world because of its fiber, not its chemicals. Like jute and flax, hemp has been cultivated since prehistoric times. A native of western Asia, hemp was used in the Chinese Neolithic Yang Shao culture, which flourished around 4000 B.C. The seeds were probably consumed along with millet, rice, barley, and soybeans in ancient China. The first indications of its use in the Mediterranean region come from the first century A.D.

Hemp plants are dioecious annual herbs (Fig. 15.19) that produce the best fibers when grown under temperate conditions with over 100 cm (39.4 in.) of rain a year. The fibers (Fig. 15.15) are like those of flax in appearance but tend to be stiffer and contain more dark-colored matter. Hemp fibers are the longest of any bast species, ranging from 1.5 to 4.6 m (5 to 15 ft) in length. They are extracted from the stems by mechanical "retting," hot-water retting, and dew retting, often followed by scutching and pounding. Well-processed hemp is creamy white and soft and has a silky sheen. Most of the fiber is extracted as quickly and inexpensively as possible, however, and tends to be dark and rough. As a result, hemp is typically used for cordage, rope, canvas, and sailcloth. There has been some discussion about the origin of the name jeans (Fig. 15.20) and of the cloth originally used for making Levi Strauss' first lot of working pants. Some say that hempen cloth was used, but this seems to be an unsubstantiated rumor. Whatever the original material used, the famous design of working pants with copper rivets at the seam joints was an instant success. Certainly the use of cotton fabric in all documented production helped make cotton the leading textile fiber.

Recently there has been renewed interest in the production of hemp, particularly in France, where new machines have been developed for scutching and fibers. The woody fragments separated from the fibers are collected and used in the manufacture of composition board.

FIGURE 15.19
A rendering of a hemp plant in Leonard Fuch's *De Historia Stirpium,* Basel, 1542.

Ramie

Boehmeria nivea (Urticaceae), or China grass, has never been one of the world's most important fibers because of the agronomic problems involved in growing the crop and the presence of undesirable chemicals in the fibers. Ramie is a monoecious perennial propagated by using pieces of rhizomes. Unfortunately, strands of ramie tend to mature unevenly, making harvesting difficult, and the fibers contain considerable amounts of gum and pectin that must be removed. However, once harvested and properly processed, ramie produces strong, durable fibers that are among the longest and silkiest plant fibers. Ramie is native to tropical Asia, where it has been gathered for fiber extraction for over 7000 years. Some Egyptian mummies (approximately 3300 B.C.) were bound with ramie strips rather than linen. A

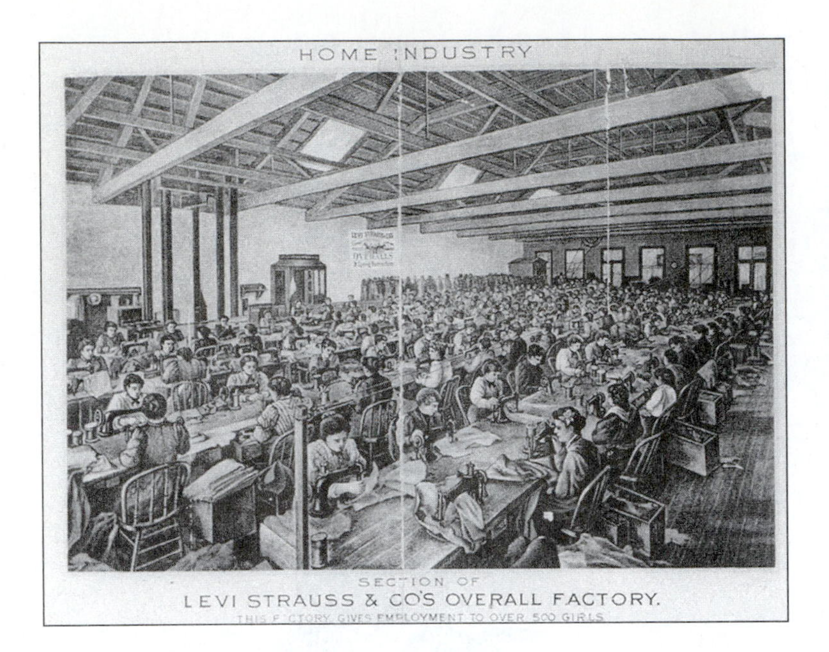

FIGURE 15.20
Debate still rages over the history of denim and jeans. The word denim is thought to be derived from a corruption of "serge de Nimes" for a cloth imported from Nimes, France. But this serge was made from a combination of silk and wool, and denim has always been made from cotton. The word jeans probably did arise from a fabric's origin. Jean was a cotton/linen and/or wool blend from Genoa, Italy, (Genes to the French). This lithograph shows the early jeans factory of Levi Strauss, a Bavarian immigrant who became famous for his pants which included metal rivets to increase their durability.
(Courtesy of Levi Strauss Company.)

related species, *B. cylindrica,* was used by New World Indians as a source of twine to attach spearheads and arrowheads to shafts.

The stalks of ramie are not retted to free the fibers. Instead, the bark and phloem tissues are peeled from the stems and the fibers are decorticated by beating and scraping. After the fibers are finally separated from the woody matter and soft tissues, they remain in ribbonlike strips because they are held together by gums. The gums can be dissolved by repeated washing and scraping or, in modern operations, by treating them with caustic soda. During the last 30 years ramie has been increasingly planted in the United States as a crop rotation species that is harvested, decorticated, and degummed by machine. Mechanized operations have reduced the price and partially eliminated the need for hand labor. As a result, ramie has become an increasingly important fiber for textiles and materials requiring exceptional strength, such as camping lantern mantles and backings in fire hoses.

FIGURE 15.21
Making rope from sisal in Africa.
(Photo by B. L. Turner.)

Leaf, or Hard, Fibers

The widespread use of leaf fibers is a comparatively recent phenomenon. Their use has escalated dramatically during the last 60 years as improved machinery and transportation systems have allowed greater exploitation of tropical crops. Most leaf fibers are obtained from rapidly growing tropical monocotyledons with fibers that can be easily and inexpensively extracted by decorticating machines. The fibrous strands occur with the phloem in the vascular bundles scattered in leaves and leaf bases. Within the leaves, the thick-walled fiber cells provide support. Once extracted, the fibers can be spun, but they are too stiff to be made into textiles suitable for modern clothing. The name *hard* applied to this type of fiber reflects this stiffness. Hard fibers, usually between 1.8 and 3.6 m (6 and 12 ft) long, tend to be shorter than most bast fibers. Nevertheless, leaf fibers make better ropes than do most bast fibers (Fig. 15.21). Although other plants are occasionally used, two genera, *Agave* and *Musa,* supply virtually all the commercially important leaf fibers.

Sisal and Henequen

Both sisal and henequen come from the leaves of species of *Agave* (Agavaceae): sisal from *A. sisalana,* and henequen from *A. fourcroydes.* Both species are native to Central America, where the Mayans and Aztecs are known to have extracted and woven sisal fibers into rough garments. In addition to fibers, *A. sisalana* has sharp spines on the ends of

FIGURE 15.22
Agave plants from which the lower leaves have been removed for sisal production.
(Photo by J. L. Neff.)

FIGURE 15.23.
Workers feed agave leaves under rollers that crush the leaves, squeezing water and soft matter from the fibers.
(Photo by J. L. Neff.)

FIGURE 15.24
Henequen, like sisal, is extracted from a species of Agave native to the New World. However, most henequen is now produced in Africa.
(Courtesy of B. L. Turner.)

its leaves that have been used by native peoples as needles. The provision of both the fiber and a sewing utensil gave rise to the common name "needle and thread plant." Today the fibers are used for sacking, mats, and tea bags and as reinforcements for materials such as rubber. The sap tapped from this agave can also be fermented into a mildly alcoholic wine known as pulque (see Chapter 14).

Fibers are removed from both *Agave* species in the same way. The outer, mature leaves are cut at the base (Fig. 15.22), carted to the factory, and fed between rollers that squeeze out most of the water and turn the soft tissues into an amorphous mush that is scraped away from the fibers (Fig. 15.23). The fibers are then washed and hung in the sun to dry (Fig. 15.24). They can be dyed or used directly since they are naturally a creamy white if properly washed and dried. Henequen is still grown in its native area of Mexico, but sisal has been planted in Brazil and Africa, where huge plantations produce most of the world supply (Table 15.3). In plantations, plantings are made using bulbils produced by older plants. One advantage sisal and henequen have over other fiber crops is that these plants are native to and grow best in arid regions. Consequently, they are excellent crops for regions little suited for other types of cultivation.

Abacá

Abacá, or Manila hemp, comes from *Musa textilis* (Musaceae), a relative of the banana. The fibers are extracted primarily from the outer parts of the leaf bases that make up the "stem" of these giant herbaceous plants. Within the last few decades abacá has been converted from a relatively obscure fiber plant known only in its native home in the Far East to one of the most widely used sources of plant fibers. Most people cannot recall seeing the fiber mentioned but have undoubtedly come into contact with tea bags, dollar bills, "Manila" envelopes, German or Italian salamis wrapped in clothlike casings, and filter-tipped cigarettes. All these products are made, or have been made, from abacá. These uses might seem to place abacá in a category similar to wood fibers, but abacá has also been used for textiles. Long before Europeans arrived in the Philippines, abacá had replaced bark cloth as the chief source of clothing materials. Ropes made from abacá are resistant to decay in water.

Like sisal and henequen, abacá fibers are often extracted by feeding the leaves into a decorticating machine. Harvesting can begin when plants are about 2 years old and can continue for 12 to 14 years or more. The expanded leaf blades are cut away, and the petioles are hauled to the processing plant. The fibers can also be obtained by peeling strips called **tuxies** from the outer edges of the petioles. The strips are placed on wooden blocks, and extraneous material is scraped from the fibers.

FIGURE 15.25
Roots of madder yield a red dye that has been used since ancient times. This illustration from Pomet, *Historie des Drogues,* was based on that originally published in Gerald's herbal of 1597.
(Courtesy of Rita Adrosko, Division of Textiles, Smithsonian Institution.)

Dyeing and Dye Plants

The ability to perceive color is a wonderful aspect of being human. Although animals such as butterflies are able to see various colors and some may even see a wider range of colors than people do, they cannot make use of color for enjoyment. Humans, in contrast, can pick, choose, and manipulate colors strictly for pleasure.

It is easy to see how humans learned about using **dyes** (water-soluble plant extracts). Children quickly see that colors are produced by mashing fruits or flowers, and parents can attest to the fastness of some of these extracts. Although many plant parts are pigmented, they do not yield good dyes because the chemicals responsible for the colors fade or turn muddy over time or do not adhere to objects that are to be dyed.

Despite the antiquity of dyeing, the chemistry of the processes by which dyes are bonded to fibers is not completely understood. For a dye to color a thread, a fabric, or another substance, it must be completely bound to the object it is to dye. Otherwise, the color will not be "fast" and will fade quickly or wash off when the object is immersed in water. It is believed that the direct bonding of charged parts of the fiber and dye molecules, hydrogen bonds, and hydrophobic interactions are all involved in the adsorption of classic dyes. It is also thought that in traditional plant dyeing processes the structural configuration of the cellulose polymers is altered so that the number of binding sites is increased, enhancing the affinity of the fiber for dye molecules.

By Egyptian times it was known that specific substances, called **mordants,** increase the adherence of various dyes to fabrics. The term *mordant* comes from the Latin *mordere,* meaning "to bite," because it was believed that the mordant literally bit into the fabric to make holdfasts for the dye. It is now known that mordanting agents, usually salts of metals such as tin, aluminum, iron, and chromium, form an insoluble compound with the dye. In ancient times these agents would have come from the metal vessels in which the dyeing was done or from additives such as dung and urine. In addition to producing color fastness, mordants can affect the final color of a dye. Iron is said to "sadden" colors, tin and chrome to "brighten" them, and copper to enhance bluish hues. A few dye plants, such as madder, *Rubia tinctorum* (Rubiaceae) (Fig. 15.25), produce particularly good dyes because they naturally contain substances that act as mordanting agents.

Every culture seems to have discovered its own repertoire of dyes. By 3000 B.C. Eurasians had already identified numerous plants as good dye sources. Henna, *Lawsonia inermis* (Lythraceae) was used by Greek and Egyptian women as a hair dye (Fig. 15.26). The leaves of this shrubby species can be ground with water into a paste that normally produces an orange dye with a great affinity for protein. Today chemically altered henna is used as the base for a wide array of hair colorants. Excavations of Pompeii revealed elaborate dye shops with evidences of a wide array of ancient dyestuffs. The use of safflower, *Carthamus tinctorius* (Asteraceae) (see Fig. 9.11) in India and other regions of the Old World dates from at least several centuries before Christ.

Historical associations with colors reflect the importance of former dye plants. Indigo, *Indigofera tinctoria* (Fabaceae) (Fig. 15.27), has been considered an opulent blue color since its first recorded use in China over 6000 years ago. The red stigmas of saffron, *Crocus sativus* (Iridaceae), dyed the royal robes of Irish kings. The green that people today associate with Robin Hood's men (approximately A.D. 1265) was produced by dipping their garments first in a dye bath of woad

FIGURE 15.26

Henna was used by women in ancient Greece to tint their hair a reddish brown. Today it is used as a hair rinse and leather dye. Classical hair care from a nineteenth-century lithograph. In Near Eastern cultures, it is also used to dye hands and feet for ceremonial purposes.

FIGURE 15.27

An indigo, or "anil," plant showing the compound leaves characteristic of the Fabaceae, or bean family. The pods are also visible.

(Courtesy of the Smithsonian Institution.)

FIGURE 15.28

Woad, a source of a blue dye, was one of the dyes used to make the green outfits worn by Robin Hood's men. The band of generous robbers would have known a lot about dyeing, which was actively practiced in the depths of Sherwood Forest. The dye process produced odors so foul that dyers were forced to carry out their trade away from towns. In fact, Queen Elizabeth I (1533–1603) decreed that no woad processing could take place within 5 miles of her residence so that she would not have to smell the fumes.

(Redrawn from Fuch's *De Historia Stirpium*, Basel, 1545.)

blue, *Isatis tinctoria* (Brassicaceae) (Fig. 15.28), and then in a bath of deep yellow weld, *Reseda luteola* (Resedaceae). The color of the jackets of the Tory redcoats in the American Revolution was produced by madder, and the gray of the southern forces in the Civil War was made with butternuts, *Juglans cinerea* (Juglandaceae).

Some dyeing processes are relatively straightforward, but others attest to human ingenuity. A case in point is indigo (Fig. 15.29), which comes from a legume with tiny reddish flowers. The plant that yields indigo and the process of indigo dye production originated in India. Because of the scarceness of good blue colors and the exceptional quality and colorfastness of the dye, indigo was an exceedingly important item of trade between India and other parts of the world for at least 2300 years. Although, as was mentioned earlier, Europeans had their own source of blue, woad never truly rivaled indigo for richness of color. In fact, woad continued to be used after indigo became readily available only because an international group known as the "woadites" lobbied for laws in Britain to prohibit the importation of indigo, which they labeled the "devil's weed."

When the New World was discovered, new dye plants such as red-orange annatto, *Bixa orellana* (Bixaceae) (Figs. 15.30 and 15.31); red bloodroot, *Sanguinaria canadensis* (Papaveraceae); and purple or black logwood, *Haematoxylum campechianum* (Fabaceae), were discovered. Logwood produces a dye that made a good black possible for the first time in Europe. It is used today to stain prepared microscope slides of animal (including human) tissue. Annatto (achiote), a red

FIGURE 15.29
An old illustration (1694) of indigo production in the West Indies shows the harvest of indigo (*negres coupant l'anil*), placing the stalks of the plants in vats of water to ret (*negres jetant l'anil dans l'eau*), agitating the stalks to enhance fermentation (*negres remnant l'anil dans l'eau*), and collecting the precipitate and hanging the dyed cloth to dry (*negres portant l'indigo dans les caisses pour faire secher*).
(Courtesy of Rita Adrosko, Division of Textiles, Smithsonian Institution.)

FIGURE 15.30
In this lithograph, annatto is being extracted from plants in the West Indies. The plant from which the dyes come is shown on the right, and an enlargement of the spiny fruit capsules is shown on the left. In the background is the vat in which the seeds are soaked and fermented. The seeds with their orange-red pulp were then ground to a paste in a mortar.
(Courtesy of Rita Adrosko, Division of Textiles, Smithsonian Institution.)

dye produced in the pulp surrounding seeds enclosed in a spiny pod, is still used in Latin American cooking and as a color agent for margarine, cheese, and cosmetics. Annatto is added seasonally to butter when cows are not getting fresh grass, and therefore enough beta-carotene, to produce cream that yields yellow-colored butter.

Many classic dyes are only of historical interest (Fig. 15.32) because of the development in the 1850s and 1860s of synthetic dyes made from derivatives of coal tar. These synthetic dyes, called **aniline dyes,** produce a wide array of vibrant colors that are much more colorfast than even the best natural dyes, such as indigo.

FIGURE 15.31
A branch of an annatto plant with flowers and fruit.
(After Baillon.)

There has been a resurgence of interest in the use of natural dyes. People who spin and weave natural fibers prefer to use natural dyes. In addition, there has been growing concern about the safety of synthetic dyes in products that people eat, ingest in medicines, or apply to sensitive areas such as the eyes in cosmetics. Consequently, increasingly large numbers of natural dyes are being incorporated into foodstuffs, pharmaceuticals, and cosmetics. Table 15.4 lists the important plant colorants approved for use in these kinds of products in the United States.

Tannins and Tanning

Unlike vegetable dyes, which have practically disappeared as fabric colorants since the advent of aniline dyes, natural tannins have remained important in the leather industry. Some large commercial operations use synthetic tanning agents. In the United States 15 percent of the processed leathers are tanned using vegetable tannins, and much of the other 85 percent are first tanned with chrome and then retanned with vegetable tannins. In underdeveloped and semideveloped countries most tanning is still done with vegetable tannins.

FIGURE 15.32
This lithograph of an American dye factory (approximately 1832) shows a worker lifting a bucket of dye from a furnace heater and other workers engaged in dyeing or hanging up dyed materials. Eighteen years later the discovery of aniline dyes revolutionized the dyeing industry.
(Courtesy of the Smithsonian Institution, Division of Textiles.)

Although tannins are present in almost all plants, their ecological roles are not fully understood. It is known that they are often concentrated in the heartwood and bark of trees and in structures, such as galls, produced in reaction to attack by insects. Perennial plants often have more tannins in their leaves than do annuals, and evergreen trees have higher tannin contents than do most deciduous trees. Consequently, many workers think that tannins deter feeding by various insects, especially since laboratory experiments have shown that some tannic acids interfere with the digestive processes of insects. Tannins may also inhibit microbial growth.

Like dyeing, **tanning,** the process of turning raw skins into leather, has traditionally been carried out using plant products. As a result of tanning, animal skins become resistant to the effects of water, heat, and various microorganisms. The chemicals responsible for this change are collectively called **tannins** and consist of a wide array of polyphenolic compounds. Animal skin contains fibers or proteins that are usually highly organized and resistant to degradation. However, certain regions of the fibers are unorganized and highly susceptible to microbial attack. It appears that tanning acts by impregnating the unorganized regions and chemically bonding with the protein molecules. In addition to tanning leathers, tannins have been used as mordants in dyeing operations. Because they have astringent properties, they have also been used medicinally. In recent decades tannins have been added to mud in oil drilling operations to increase their viscosity. Table 15.5 lists several important plant sources of tannins.

Early humans might have discovered which plants contained tannins by soaking hides in water into which pieces of particular plants had fallen or by trying to dye skins with various plants. No one knows for sure, but leather sandals 3300 years old found in Egyptian excavations clearly show that by that time the processes of tanning and dyeing leather were quite advanced.

Although there have been subtle improvements in tanning processes over the last several thousand years, the basic procedures are remarkably similar to those employed for centuries. The initial step in the tanning of leathers is to remove the hair on the outer surfaces. Originally the hides

TABLE 15.4 Plant Dyes Approved for Use in Foods, Medicines, and Cosmetics in the United States

ITEM	SCIENTIFIC NAME	FAMILY	LIMITATIONS
Food			
Annatto extract	*Bixa orellana*	Bixaceae	None
Beet powder	*Beta vulgaris*	Chenopodiaceae	None
Beta-carotene	Various yellow vegetables		None
Carrot oil	*Daucus carota*	Apiaceae	None
Corn endosperm oil	*Zea mays*	Poaceae	Chicken feed only
Fruit juices	Various fruits such as *Rubus* spp.		None
Grape color extract	*Vitis vinifera*	Vitaceae	Nonbeverage foods only
Grape skin extract	*Vitis vinifera*	Vitaceae	Still and carbonated alcoholic beverages
Marigold meal and extract	*Tagetes* spp.	Asteraceae	Chicken feed only
Paprika and paprika oleoresin	*Capsicum annuum*	Solanaceae	None
Saffron	*Crocus sativus*	Iridaceae	None
Toasted partially defatted cooked cottonseed flour	*Gossypium hirsutum*	Malvaceae	None
Turmeric and turmeric oleoresin	*Curcuma longa*	Zingiberaceae	None
Vegetable juice	Various vegetables		None
Drugs			
Annatto extract	*Bixa orellana*	Bixaceae	Can be used around eyes
Beta-carotene	Various species		Can be used around eyes
Logwood extract	*Haematoxylum campechianum*	Fabaceae	For nonabsorbable nylon sutures only
Cosmetics			
Annatto	*Bixa orellana*	Bixaceae	Can be used for eye cosmetics
Beta-carotene	Various species		Can be used for eye cosmetics
Henna	*Lawsonia inermis*	Lythraceae	Hair dyes only, not around eyes

Source: U.S. Food and Drug Administration, Center for Food Safety and Applied Nutrition, Office of Premarket Approval. Accessed online, November 1999, at http://vm.cfsan.fda.gov/~dms/opa-col2.html.

TABLE 15.5 Important Vegetable Sources of Tannins

COMMON NAME	PLANT PART	SCIENTIFIC NAME	FAMILY
Algarobilla	Pods	*Caesalpinia brevifolia*	Fabaceae
Catechu	Heartwood	*Acacia catechu*	Fabaceae
Chestnut	Bark, wood	*Castanea sativa*	Fagaceae
Chinese tannin	Galls	*Rhus semialata*	Anacardiaceae
Divi-divi	Pods	*Caesalpinia coriaria*	Fabaceae
Eucalyptus	Bark, wood	*Eucalyptus wandoo*	Myrtaceae
Hemlock	Bark	*Tsuga canadensis*	Pinaceae
Knoppern nuts	Fruits	*Quercus robur*	Fagaceae
Larch, European	Bark	*Larix decidua*	Pinaceae
Locust, black	Bark	*Robinia pseudoacacia*	Fabaceae
Mallee, brown	Bark	*Eucalyptus astringens*	Myrtaceae
Mangrove	Bark	*Rhizopora candelaria* *R. mangle*	Rhizophoraceae
Myrobalans	Fruits	*Terminalia chebula*	Combretaceae
Oak, cork	Bark	*Quercus suber*	Fagaceae
Pine, Scot's	Bark	*Pinus sylvestris*	Pinaceae
Pomegranate	Fruits, twigs	*Punica granatum*	Punicaceae
Quebracho	Heartwood	*Schinopsis lorentzii*	Anacardiaceae
Spruce	Bark	*Picea abies*	Pinaceae
Sumac, tanner's	Leaves	*Rhus coriaria* *R. typhina*	Anacardiaceae
Tara	Pods	*Caesalpinia spinosa*	Fabaceae
Turkish tannin	Galls	*Quercus infectoria*	Fagaceae
Wattle	Bark	*Acacia mearnsii*	Fabaceae

Source: From E. Haslam. 1966. *Chemistry of Vegetable Tannins.* New York, Academic Press.

were singed, but by Greek and Roman times the hair was removed by soaking the skins in a water solution of calcium oxide. Today enzymes are often used as depilatories. Originally tanniferous plant parts were included between skins to be tanned or thrown into water. It was soon discovered that concentrated solutions could be made by soaking the plant parts and then evaporating some of the water. Yet these potent solutions posed a problem with thick hides because the tannins quickly impregnated the outer skin layers and sealed off the inner layers. The inner layers consequently were improperly tanned. This problem is often avoided by pretreating thick hides with a "spent" (previously used) tannin solution. Another method of tanning used by American frontier settlers involved tying skins into a sack filled with a solution of tannins. Today tannins are prepared by soaking pieces of wood and bark in water. The tanniferous water extract is then spray-dried or concentrated. Powdered or solid tannins are redissolved in water during tanning operations. After tanning, leathers are washed and often treated with oil or grease for softness. The leather then can be finished by coating it with a layer of gum, resin, or wax. Different tanning procedures produce different types of leather.

In the Mediterranean region tannins were usually extracted from sumac, *Rhus* spp. (Anacardiaceae), or various oak, *Quercus* (Fagaceae), species. Later European sources of tannins included spruce, *Picea* (Pinaceae), and pomegranate, *Punica granatum* (Punicaceae). In North America the Indians treated hides with native plants. The colonists learned to use these indigenous sources of tannins and cut down hemlocks, *Tsuga canadensis* (Pinaceae), and stripped them of their bark. Originally hemlocks grew from Canada to Pennsylvania, with Boston as the center of the tanning industry. As the hemlock populations declined, hardwoods such as the chestnut, *Castanea dentata* (Fagaceae), began to be used. In the 1930s, as thousands of chestnut trees succumbed to the chestnut blight caused by the fungus *Endothia parasitica,* over 100,000 tons of tannin extract from dead trees alone became available. Pennsylvania, in the heart of the natural oak-chestnut forest region, became the center of the American tanning industry and remains so today.

During the last 100 years the heartwood of the quebracho tree, *Schinopsis lorentzii* (Anacardiaceae), of South America and species of wattle, *Acacia* (Fabaceae), from Africa and Australia have become important tannin sources. These two species account for 90 percent of the tannins imported into the United States. Mangroves, *Rhizophora* spp. (Rhizophoraceae), are also rich in tannins and would seem to provide an abundant tropical source of tannins because of their ubiquity in coastal tropical regions. In general, tannins have been extracted from wild trees, but some efforts have been made to cultivate species of wattle as tannin sources in Australia, South Africa, and Sri Lanka. There is also interest in cultivating New World species of sumac.

16

Wood, Cork, and Bamboo

People today may feel that they live in an age of petroleum, gas, concrete, and steel, but wood is still used to heat more homes and construct more family dwellings than any other material. The forests that yield this wood now cover about one-fifth of the earth's land surface. However, assuming that losses will follow the patterns of the past (Table 16.1), the land covered by forest will continue to shrink dramatically, creating hardships in both the less developed countries, which depend on wood for energy and construction, and the developed countries, which have become accustomed to consuming vast quantities of wood products such as paper and fiberboard.

In Chapter 1 we contrasted woody plants with herbs, stating that herbs do not contain wood, but we did not define wood. Although the word *wood* represents the same substance in a wide variety of plants, woods of different species have individual characteristics and appearances that make them useful for a wide variety of purposes. Wood has also become an important material in the manufacture of products as diverse as cellophane and cigarette filters. In this chapter we discuss what wood is, how it is formed, the differences between hardwoods and softwoods, and how the individual characters of woods influence their utilization. Table 16.2 lists the plants discussed in this chapter.

What is Wood?

As was outlined in Chapter 7, plants grow in length because of elongation of the cells produced by their terminal meristems (Fig. 16.1). Most cells formed by the meristems mature to form the epidermis, general matrix tissue (parenchyma), and primary xylem and phloem. In truly herbaceous dicotyledons and monocotyledons all or almost all of the vascular tissue is produced by terminal meristems. In plants that subsequently become woody a cambium layer forms between the primary xylem and phloem cells. The **cambium** is a lateral (expanding in width) meristem capable of dividing to produce additional xylem toward the inside of the stem and additional phloem toward the outside. The conducting tissues formed by the cambium are called the "secondary" xylem and phloem to differentiate them from the primary xylem and phloem derived directly from cell divisions of apical meristems.

TABLE 16.1 Global Changes in Forest Cover Compared with Grassland and Crop Regions Between 1700 and 1998

VEGETATION TYPE	Area (10^4km^2)						PERCENT CHANGE SINCE 1700	AREA CHANGE (10^6km^2) SINCE 1700
	1700	1850	1920	1950	1990	1998		
Forest and woodland	6,215	5,965	5,678	5,389	5,053	4,169	–33.0	–20.5
Grassland and pasture	6,860	6,837	6,748	6,780	6,788	6,967	1.5	10.7
Croplands	265	537	913	1,170	1,501	1,512	470.0	12.5

Sources: FAOSTAT at http://apps.fao.org; J. F. Richards. Land Transformations. Pages 163–178 in B. L. Turner II (ed.) 1990. *The Earth As Transformed by Human Action.* New York: Cambridge University Press.

TABLE 16.2 Plants Discussed in Chapter 16

COMMON NAME	SCIENTIFIC NAME	FAMILY	NATIVE REGION
Cork oak	*Quercus suber*	Fagaceae	Southern Europe, North America
Cucumber tree	*Magnolia acuminata*	Magnoliaceae	Eastern North America
Douglas-fir	*Pseudotsuga menziesii*	Pinaceae	North America
Eastern white pine	*Pinus strobus*	Pinaceae	North America
Eucalyptus	*Eucalyptus* spp.	Myrtaceae	Australia, New Zealand, South America
Hemlock	*Tsuga* spp.	Pinaceae	North temperate North America and Eurasia
Kenaf	*Hibiscus cannabinus*	Malvaceae	Africa
Leucaena	*Leucaena* spp.	Fabaceae	Africa
Mahogany	*Swietenia macrophylla*	Meliaceae	Tropical America
Mulberry, paper	*Broussonetia papyrifera*	Moraceae	Asia, Polynesia
Papyrus	*Cyperus papyrus*	Cyperaceae	Northern and tropical Africa
Red maple	*Acer rubrum*	Aceraceae	Eastern North America
Rice paper plant	*Fatsia papyrifera*	Araliaceae	China

The cells (Fig. 16.2) of secondary xylem are shorter than those of primary xylem. As they mature, the walls of these cells become hard and impregnated with lignin. The conducting xylem vessels are not fully functional until they are dead and their cytoplasm has disintegrated. **Wood** is the vernacular name for accumulations of secondary xylem.

As we saw when discussing stems in Chapter 7, isolated bundles of xylem and phloem cells occur in rings in some herbaceous dicotyledons. Strictly speaking, therefore, these herbs contain wood as we defined it. However, in everyday usage the term *wood* is applied only to the large quantities of secondary xylem that make up the bulk of a tree or a shrub's trunk and branches. This accumulation results from continued divisions by the ring of vascular cambium cells just inside the bark. Herbaceous plants are those plants that lack this extensive secondary growth even if a cambium is present.

An understanding of how cells are added through the lateral growth of woody stems shows why wood is composed of secondary xylem and not secondary phloem or a mixture of xylem and phloem cells. Think of a stem as a cylinder with the vascular cambium forming a circle near the circumference. During the year it adds xylem cells to the inside. As it does, the core of xylem increases girth. The cambium layer adds some cells to increase its own circumference to compensate for the expansion in diameter of the xylem. The old phloem on the outside of the vascular cambium cannot divide or stretch to accommodate the increase in the radius of the stem. Consequently it is torn and crushed as the stem grows.

The only part of the phloem that is composed of normal, functional cells is the part closest to the cambium. The xylem cells closest to the cambium are also the only functional (nonliving) conductive cells. The region of the xylem that actively conducts water is known as the **sapwood.** The older, no longer functional xylem cells remain intact and form the **heartwood.** Most of what people commonly think of as wood is the combination of these in large species that have accumulated appreciable amounts of secondary xylem (Fig. 16.3).

In addition to xylem, wood contains a number of other kinds of cells. Among these are living **ray parenchyma cells** in the functional part of the xylem that serve as a horizontal

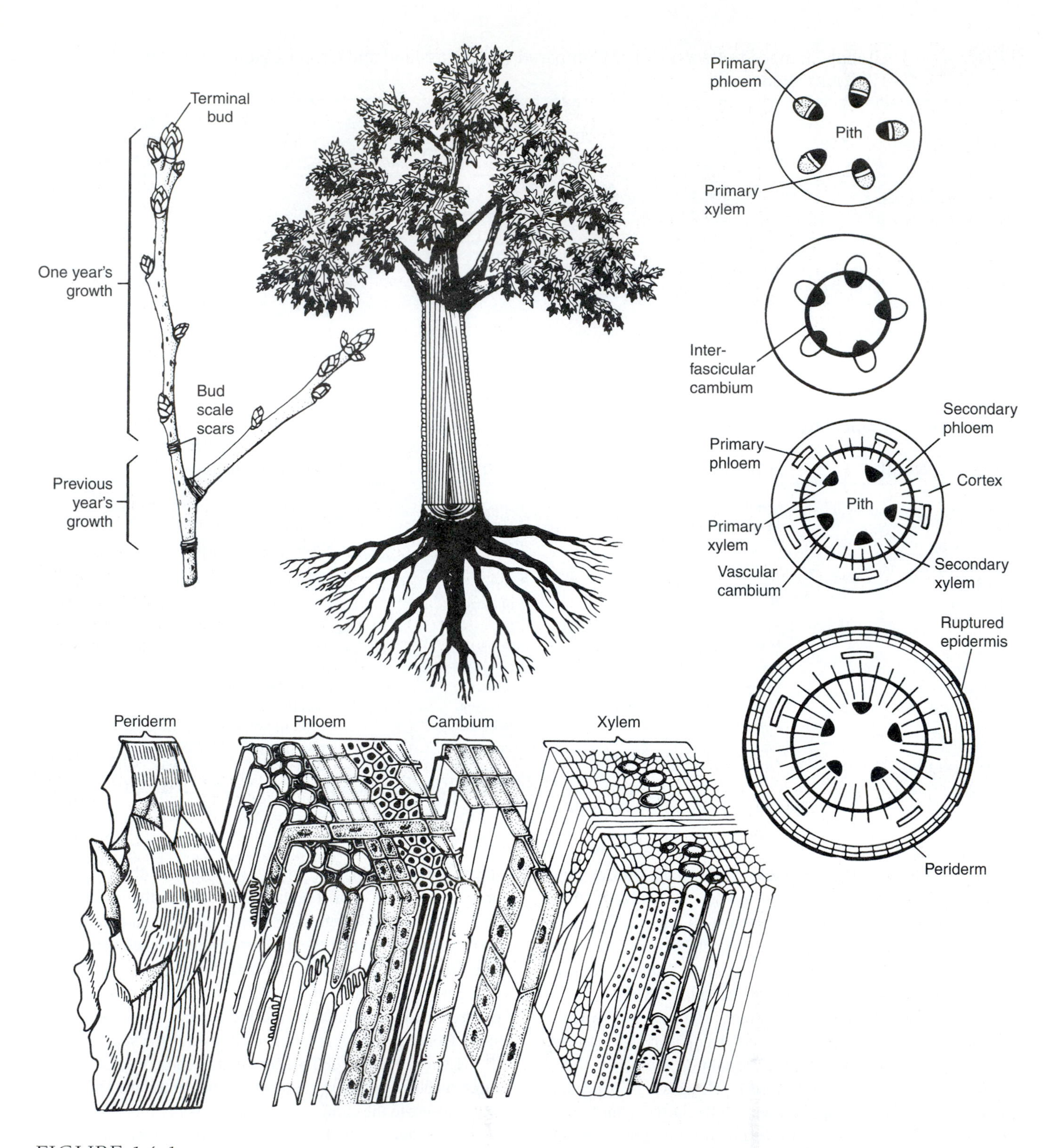

FIGURE 16.1

(Upper left) All increases in a tree's height occur at the very end of the branches, where new cells are produced by terminal meristems. Increases in girth result from cell divisions by the cambium, a lateral meristem. As the cambium cells divide, phloem cells are created to the outside, and xylem cells to the inside. As new phloem cells are produced, the old cells are pushed outward, crushed, and eventually sloughed off with the bark. Only the outermost cells of the xylem that form the sapwood actively conduct water and solutes, even though they, like the older xylem cells, are dead. The inner, nonfunctional xylem cells can eventually become impregnated with tannins, giving rise to a dense, dark-colored wood known as the heartwood. (Far right) Serial cross sections showing the development of the stem from the youngest portions of the plant. (Bottom) A magnification showing the different components of a woody stem.

FIGURE 16.2

(Left) Xylem elements from a dicotyledon such as this cucumber tree, *Magnolia acuminata* (Magnoliaceae), have pits along their walls through which water moves from cell to cell. (Right) The xylem of softwoods is composed of tracheids similar to those of the eastern pine, *Pinus strobus* (Pinaceae).

(Courtesy of W. A. Cote of the N. C. Brown Center for Ultrastructure Studies.)

conduction system (Fig. 16.3). The number of cells in the rays and their arrangement differ among species.

All the material external to the xylem is the **bark** (Fig. 16.3). Bark is therefore composed of living and dead phloem cells and the **periderm,** a tissue that replaces the epidermis as the protective layer. One part of the periderm is the **phellogen,** a secondary lateral meristem that produces **phellem,** or cork. The periderm of most species also contains **phelloderm,** a kind of parenchyma tissue. Because the periderm protects the vascular tissues and cambium, it must change continually to accommodate the growth of the trunk and branches. It changes by repeatedly making new phellogen and sloughing off the old, cracked layers of cork. Dead, crushed phloem cells are incorporated in the periderm and are sloughed off with the outer bark. The barks of different species of trees and shrubs are distinctive because of the variations in the ways in which the bark meristem produces new cells and sheds the old ones (Fig. 16.4).

Hardwoods and Softwoods

Wood is divided into two major categories, hardwoods and softwoods, by the lumber industry. These terms refer to the kinds of plants that produce the wood, not to the hardness or softness of the wood itself. All wood produced by gymnosperms (conifers) is considered softwood. Hardwood

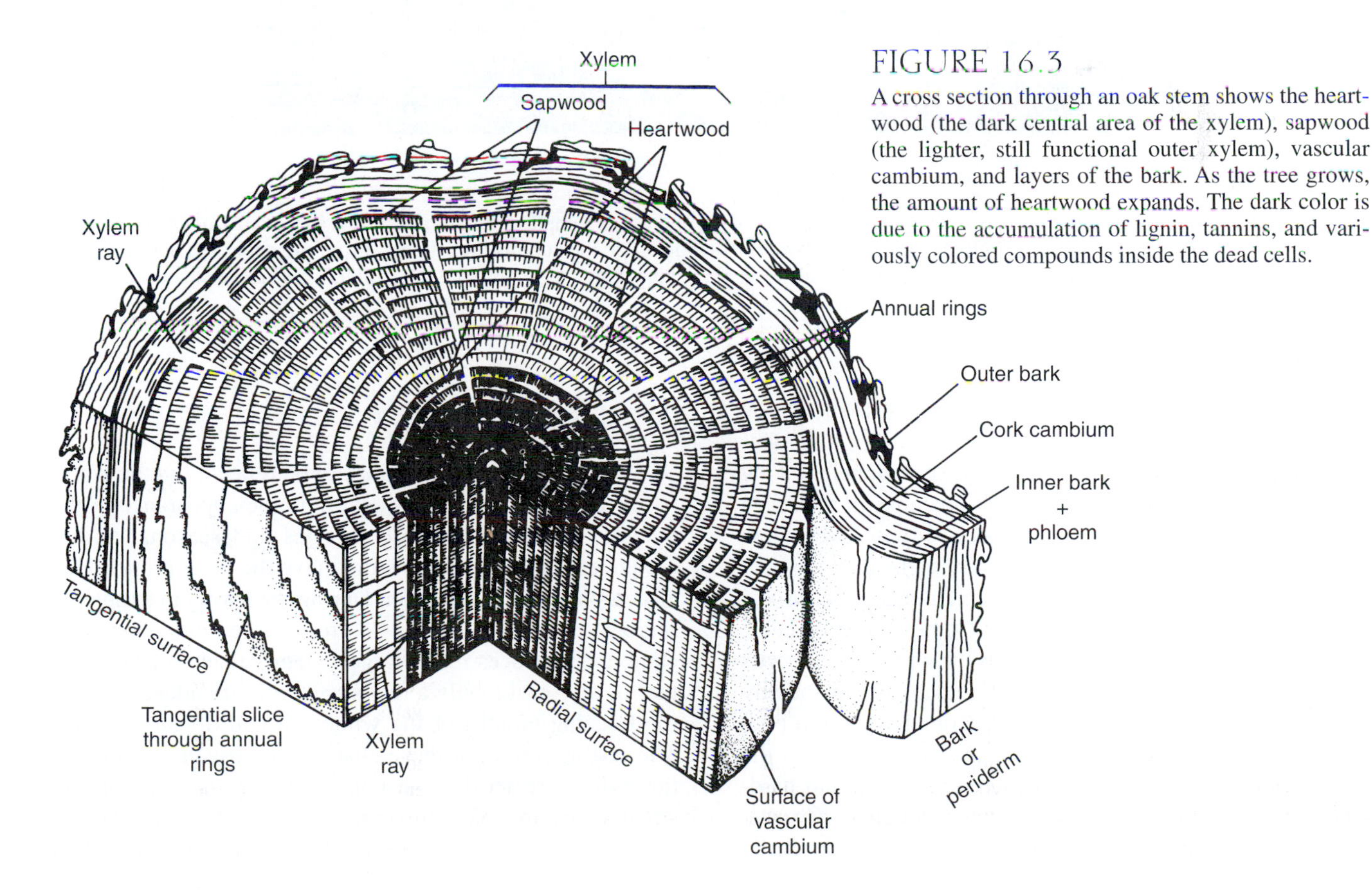

FIGURE 16.3

A cross section through an oak stem shows the heartwood (the dark central area of the xylem), sapwood (the lighter, still functional outer xylem), vascular cambium, and layers of the bark. As the tree grows, the amount of heartwood expands. The dark color is due to the accumulation of lignin, tannins, and variously colored compounds inside the dead cells.

FIGURE 16.4
With age, a tree's bark is shed in a way that is often characteristic of the species. A sample of bark types. Those indicated with an asterisk are turned 180 degrees from the way in which they are normally borne to emphasize the contrast in patterns. From the upper left across the top row: sycamore*, honey locust, pin oak*, Chinese elm; middle row: black gum, white ash, juniper*, white pine*; bottom row: shagbark hickory, tulip tree, white birch, and sugar maple*.
(Photos courtesy of Garry Fox.)

refers to any wood that comes from an angiosperm. These kinds of wood differ anatomically because the xylem of gymnosperms is composed primarily of **tracheids,** long cells that conduct water only through openings in their side walls (see Fig. 16.2). Almost all dicotyledons have xylem composed of **vessels,** relatively short cells that conduct water primarily through openings on their end walls. Mixed in with the vessels are tracheids and other kinds of cells. Because gymnosperm wood contains fewer kinds of cells than does angiosperm wood, it tends to be comparatively uniform (Fig. 16.5). Rays form a horizontal system in both types of wood. Resin canals (Chapter 10) can occur in the xylem, phloem, or even bark of both softwoods and hardwoods but are common in softwoods. Laticifers (Chapter 10) are found only in angiosperms, where they can occur in either the xylem or the phloem.

Tree Rings

Seasonal variations in climate can be reflected in the activity of the cambium and ultimately in the appearance of the xylem. Annual temperature changes and seasonal aridity are the climatic fluctuations that have the most pronounced effects on xylem production. During the spring in temperate regions or the wet season in tropical semiarid areas the cambium produces many large xylem vessels. In the summer or dry season only a few small cells are added. During the freezing months of the winter all cell division ceases. The visible results of these changes in growth are rings in the xylem created by the concentric layers of large and small cells. In cold temperate regions the rings can be counted to give an accurate idea of the age of a tree (Fig. 16.6). **Dendrochronologists,** who provide records of recent

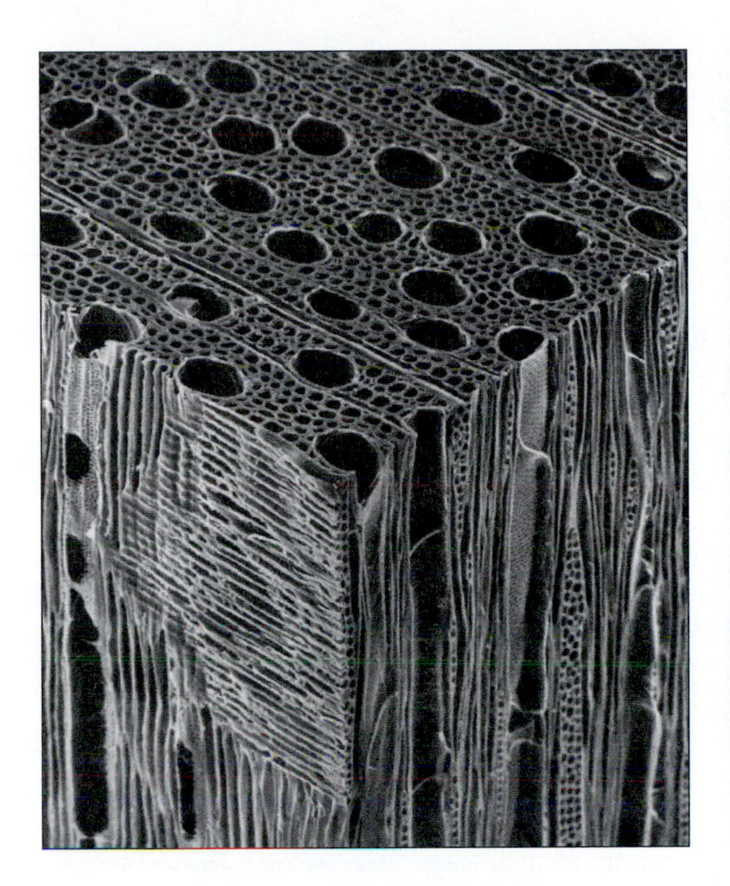

FIGURE 16.5

The scanning electronic micrograph on the left (× 123) is a piece of red maple, *Acer rubrum* (Aceraceae), a dicotyledon sliced to show the rays running perpendicular to the stem axis and an assortment of xylem elements with different patterns of secondary wall thickenings. The Douglas-fir, *Pseudotsuga menziesii* (Pinaceae), on the right (× 68) is a gymnosperm that has relatively uniform wood, in contrast to the maple. The differentiation between the small cells in the upper left corner and the large cells in front shows the appearance of annual rings.

(Courtesy of the N. C. Brown Center for Ultrastructure Studies.)

climatic events by studying the growth patterns of a tree's xylem, have helped date archaeological ruins and reconstruct climatic histories. Unfortunately, the rings produced by seasonal aridity in tropical habitats are poor age indicators because tropical arid seasons occur less predictably than do temperate winters.

Characteristics of Woods

Woods differ in color, porosity, grain, and figure, all of which contribute to their appearance (Fig. 16.7). The spectrum of wood colors, ranging from black to green, yellow, red, white, and even purple, is produced by the impregnation of the xylem

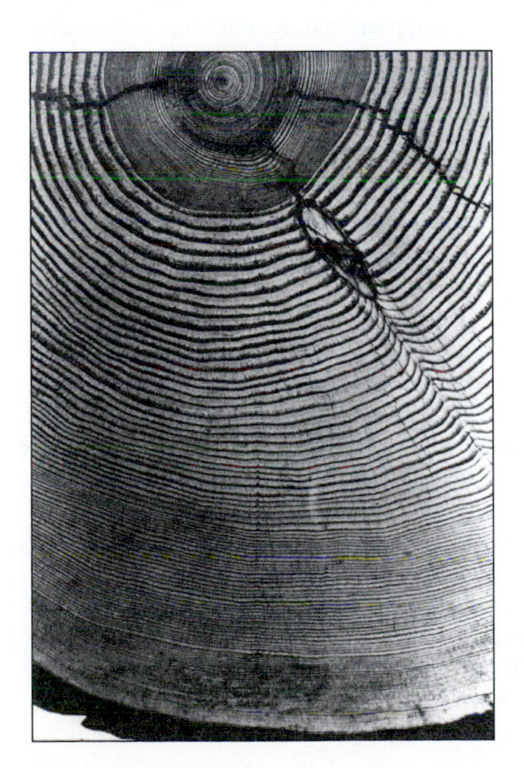

FIGURE 16.6

The age of trees growing in temperate regions and some of the history of their growth can often be read in their annual rings. In the case of this hemlock, *Tsuga* sp. (Pinaceae), the number and width of the rings indicate that the tree spent its first 75 years growing slowly in dense shade. In the seventy-sixth year the trees surrounding this tree were cut, allowing it to receive more sunlight. The effect of the increased sunlight is reflected in the increased wood production after age 76.

(From Brodatz, 1971. *Wood and Wood Grains.* New York, Dover.)

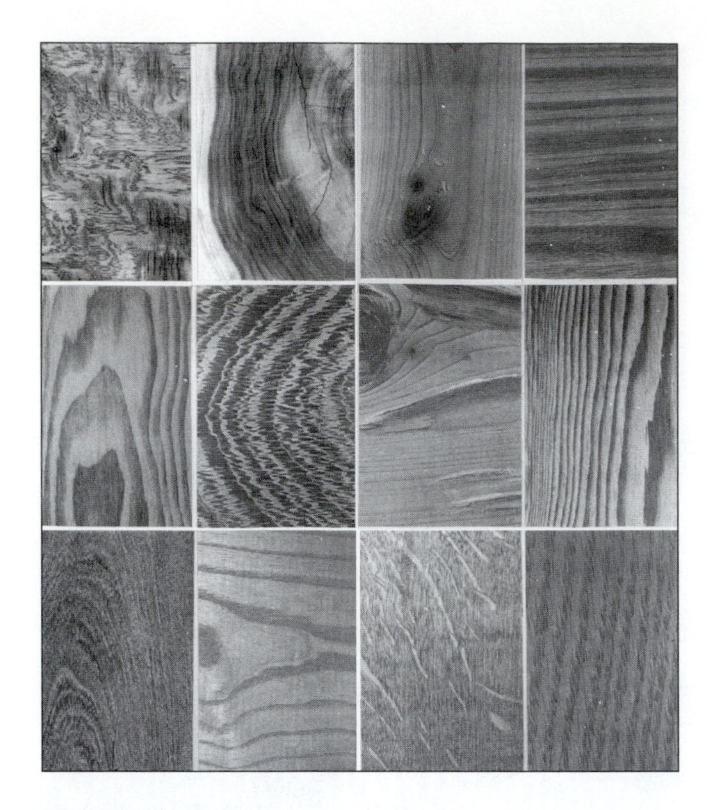

FIGURE 16.7
Figures of woods of different species. Top row, from the left: live oak, Texas ebony, cedar, zebra wood; middle row: yellow pine, wenge, pecky cypress, Douglas-fir; bottom row: mesquite, ash, crosscut oak, and plainsawn oak.

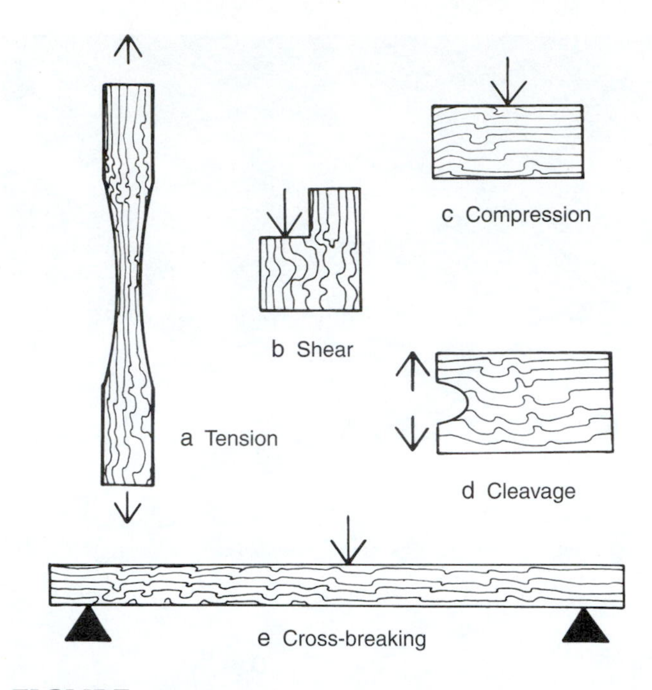

FIGURE 16.8
Tension, cross-breaking, shear, cleavage, and compression strength are mechanical properties of wood that are tested by using standardized pieces. Force is applied on each piece in the directions of the arrows in order to determine the minimum force needed to break the wood. Architects, engineers, and contractors use these measurements to help determine what kind, what size, and how many boards are needed for a particular job.

cells by different-colored compounds. **Porosity,** a feature of dicot wood, refers to the way in which large vessels are dispersed within a given part of the year's growth. Some hardwoods, such as maple (Fig. 16.5) and poplar, have a few large vessels scattered about within a ring. Others, such as oak and ash, have large vessels arranged parallel to the annual rings. **Grain** is a technical term referring to the alignment of the xylem vessels. All the cells can lie parallel to the vertical axis of the tree (Fig. 16.7) or can be variously tipped to produce an irregular pattern. In some cases the cells are arranged spirally or occur in bands oriented in alternate directions. The **figure** of a wood is determined by the number of rays, the porosity (in hardwoods), the grain, and the arrangement of the annual rings. The presence or absence of knots, which are formed by the inclusion of branches into the xylem, also contributes to the figure of a wood. Even within a given species of wood, the appearance can differ depending on the way in which it was cut (compare crosscut oak with plainsawn oak in Fig. 16.7).

Different species of tree produce wood that differs in density and mechanical properties. These structural qualities are important considerations for engineers, architects, carpenters, cabinetmakers, and even do-it-yourselfers. **Density** is defined as mass (in grams) divided by volume. An oven-dried piece of wood 1 cm^3 is usually used to calculate the density. This means that the mass is numerically equal to the weight. Moreover, since the mass of 1 cm^3 of water is 1 g, any wood with a value below 1 is "lighter" than water and will consequently float. One of the lightest woods is balsa, with a density of 0.13 g/cm^3. Lignum vitae has a density of about 1.23 g/cm^3 and sinks in water. Balsa is much too light to provide structural support, and lignum vitae is so dense that working with it is extremely difficult. Pine, the most common wood used in home construction, has a density of about 0.35 to 0.50, and a dense wood for fine furniture (such as oak), is usually around 0.60 g/cm^3. The mechanical properties of woods are determined using standardized pieces to ascertain characteristic features of tension, static bending, shear, and compression strength (Fig. 16.8).

When people buy wood, they balance its visual and structural properties against its prospective use (Fig. 16.9). If a wood is to be used in construction where it will not be seen, the structural properties and cost largely determine the choice. Color and figure are among the primary considerations for furniture and visible surfaces.

The structure and chemical composition of woods also determine their suitability as fuels (Fig. 16.10). Since softwoods and light hardwoods burn quickly with a flash of heat, they are best used as kindling to start a fire. Many conifer woods should be avoided because they are impregnated with resins that cause deposits of flammable material to build up in chimneys. Hardwoods with a medium to high density usually make the best firewood and charcoal. Charcoal is produced by burning wood slowly in an atmosphere with limited

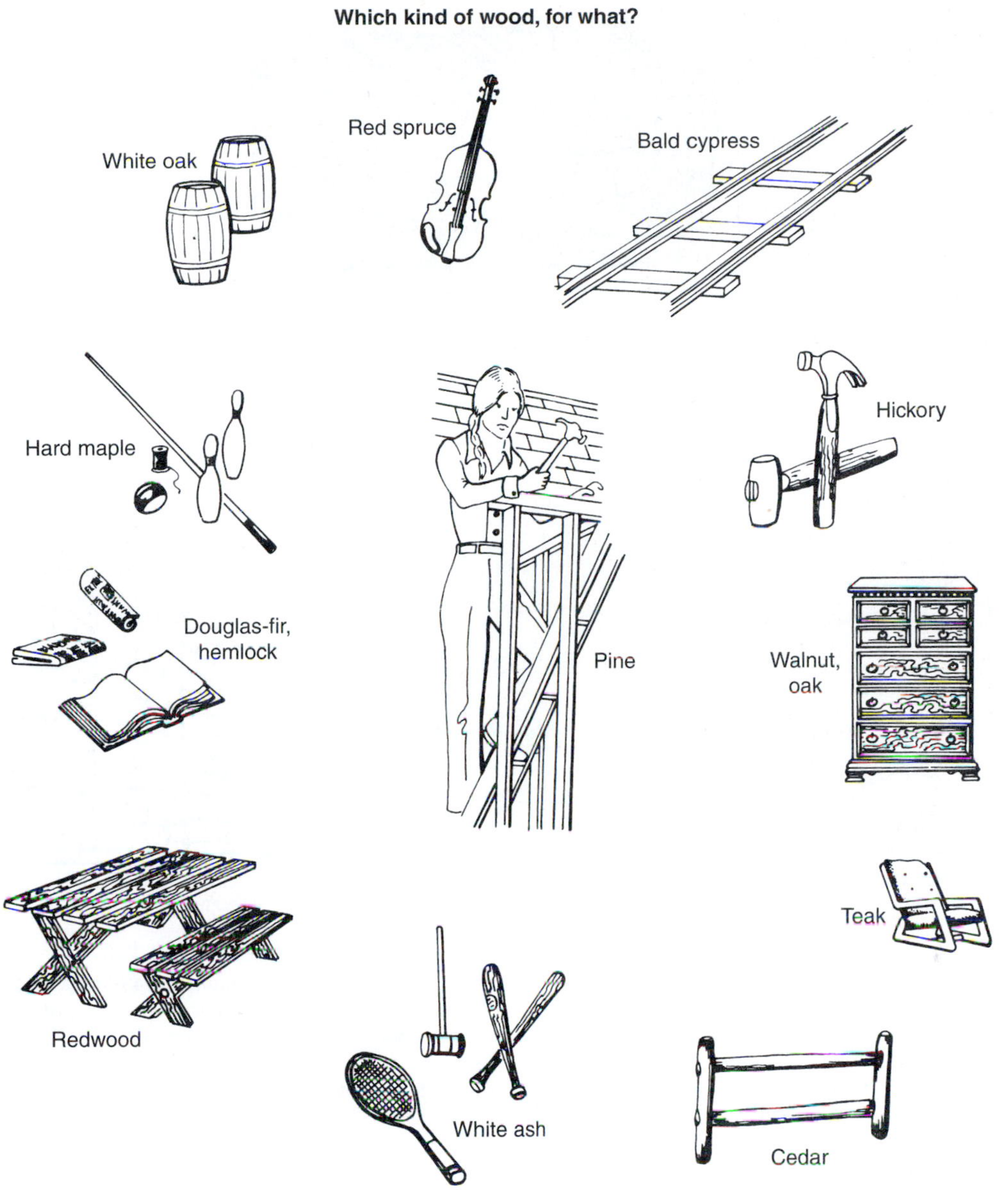

FIGURE 16.9
Because of differences in their characteristics, woods of different species lend themselves to different purposes.

oxygen (Fig. 16.11). During this initial burning, volatile hydrocarbons are driven off, leaving a material that burns at a much higher temperature than does wood. A charcoal fire can reach temperatures high enough to smelt metals.

Lumbering and Milling of Wood

In the building and furniture industry wood has to be harvested and milled (processed). The milling of lumber varies from one kind of wood to the next and from one commercial operation to another (Fig. 16.12). There are, however, several standard ways in which a large round trunk is reduced to boards of various dimensions (Fig. 16.13). Fresh (green) wood contains a large amount of water and cannot be used directly for construction because later drying results in warping. Consequently, lumber is allowed to season, or dry. Uncut or milled lumber can be air-dried in sheds. Such a passive system takes 20 to 300 days for every 2.5 cm (1 in.) of thickness of the wood. The great range in time reflects differences between woods and local humidity conditions. Air-drying can be hastened by using fans. A much more rapid method of drying wood involves the use of kilns, closed chambers with circulating hot air systems. When a kiln is used, drying time can be reduced to 2 to 50 days per 2.5 cm of thickness. Lumber can also be treated with preservatives (Fig. 16.14) or various insecticides or both.

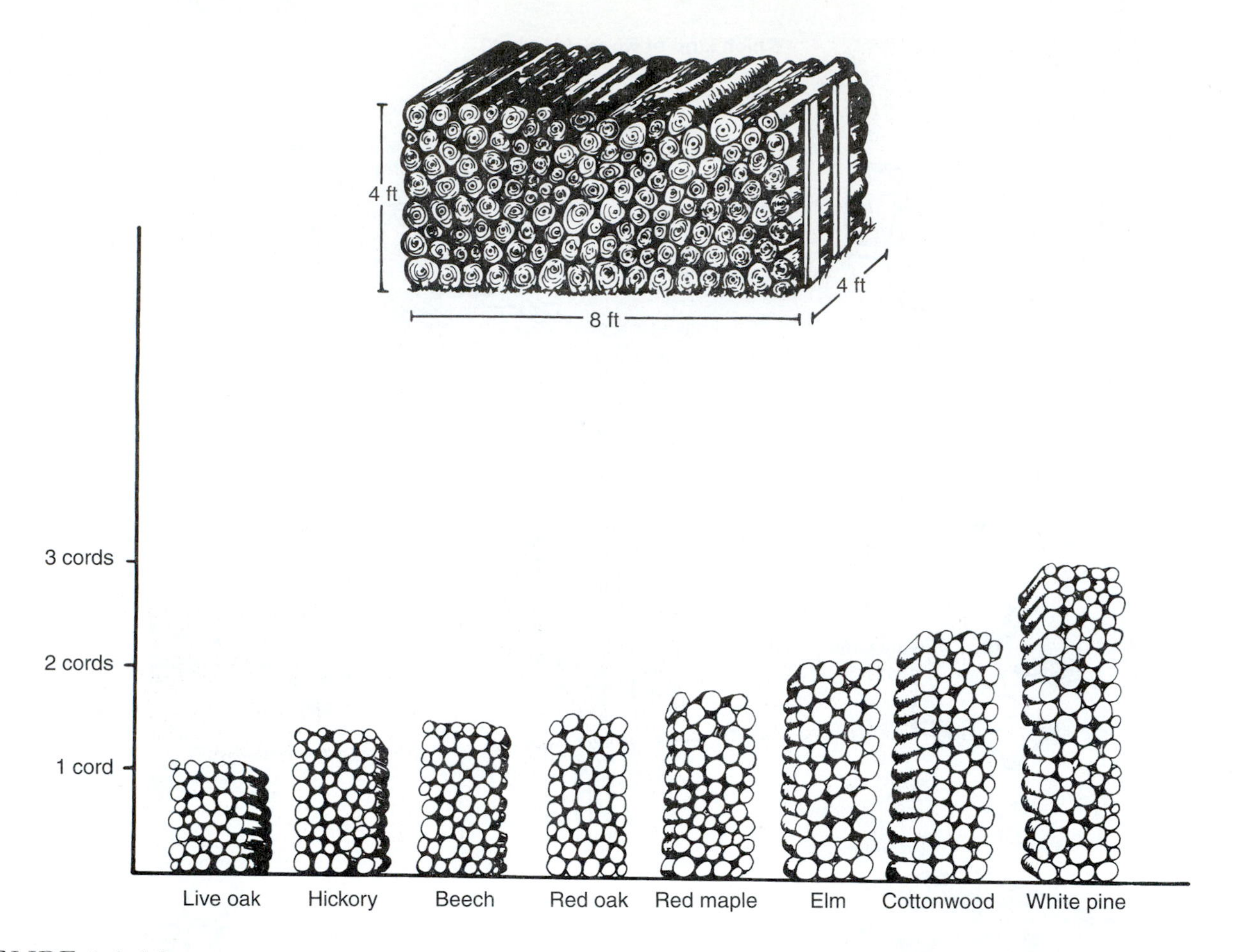

FIGURE 16.10

The standard unit of measurement for firewood is the cord, defined as a pile of wood 8 ft long, 4 ft high, and 4 ft deep as shown on the top left. A face cord has the same front dimensions but is only 2 ft deep. Thus, a face cord of 24-in. logs has half and a face cord of 16-in. logs has only one-third the volume of a full cord. When buying wood for home heating one should consider that the amount of heat given off by a particular kind of wood is a function of its density, chemical composition, and water content. The lower drawing shows the relative quantities of several kinds of wood needed to provide equivalent amounts of heat. Softwoods, such as white pine, are not only poor sources of heat but they can also cause problems with buildup of terpenes in home chimneys.

Wood that is not milled into lumber but is stripped of its bark, dried, and impregnated with chemicals (if necessary) is called **roundwood.** It is used for a number of purposes, such as fuel, posts, poles, mine timbers, and pulp. Logs can also be cut into relatively small, slightly wedge-shaped pieces called shingles that can be used for siding or roofing. Shakes, which are used for similar purposes, differ from shingles in that they are the same thickness at both ends. Both shingles and shakes are usually impregnated with chemicals that help them resist rotting.

Wood Products

Although the traditional uses of wood rely on its strength and workability, many modern uses depend on natural properties that allow wood to be used for semisynthetic products. Such manufactured wood products include veneers, composition boards, paper, and textiles.

Veneers are thin sheets of wood of uniform thickness that are often affixed to another surface, usually a different wood or composition board. The thin sheets can be made by shaving across a flat wood surface or by "peeling" a continuously revolving log (Fig. 16.15). Most veneers are now produced by the rotary (peeling) method because it yields larger, more uniform sheets. The thickness of a veneer sheet can vary from 0.25 mm (0.01 in.) to almost 1 cm (0.40 in.), but most are about 0.25 cm (0.10 in.) thick. After cutting, sheets of veneer are dried either in the air or in kilns. Rough pieces of veneer are used to produce crude objects such as orange crates and small berry boxes. Finer pieces are used for furniture, musical instruments (Fig. 16.16), and plywood surfaces. The cost of making furniture can be reduced by laminating a fine wood veneer onto a cheaper structural material. As tropical hardwoods have become increasingly scarce, this method of producing attractive surfaces has become increasingly common.

FIGURE 16.11
Charcoal is produced by stacking piles of wood so that they burn slowly in an oxygen-limited environment.
(Courtesy of Jack Daniels' Distillery.)

The most economically important use of veneer is in the manufacture of **plywood.** Plywood is made by gluing sheets of veneer together, with the grain of each sheet at right angles to the grains of the sheets above and below it. An odd number of sheets is always used, giving rise to the terms 3-ply, 5-ply, and so on. Some plywoods have a solid core consisting of small boards glued together. For solid-core plywoods, even numbers of sheets of veneer are used. The outer layers of veneer on plywood are often of a better quality and appearance than those of the inner layers.

In making a piece of plywood, care must be taken to balance the sheets correctly on either side of the center sheet or the core. An improper balance of sheets causes the plywood to warp and can lead to separation of the layers. Depending on the kind of wood, the number and thicknesses of the layers of veneer, and the kinds of glue used in their manufacture, plywoods can range from a material slightly stronger than a board of the same thickness to one that pound for pound is stronger than steel.

Although people probably think of plywood as a relatively modern product, pieces of boards with an analogous structure have been found in Egyptian tombs dated to be over 3500 years old. It is still not clear how the artisans of ancient civilizations were able to cut veneer sheets. The use of veneers as decorative objects dates from Roman times. During the reign of King Louis XIV of France the use of inlaid pieces of veneer in furniture to make designs, a process known as **marquetry,** was very popular and led to the production of beautiful works of functional art.

Plywood has limitations because its manufacture is dependent on the availability of large sheets of veneer. Much of the available wood is unattractive, irregularly shaped, or too small for making veneer, but these otherwise useless pieces of wood can be processed and made into particleboard, chipboard, or fiberboard. Consequently, processed boards have gradually supplanted plywood.

The manufacture of particleboard started in the 1940s after the development of synthetic resins and has continued uninterrupted ever since. For both particleboard and fiberboard, wood is first reduced to small chips by shaving or splintering. The particles are then sorted and graded. Once particles of the proper size have been produced, they can be mixed with resins, pesticides, and fire retardants and pressed into the desired shapes and sizes. There is less waste in the manufacture of particleboard than in the production of milled lumber. Often the two operations complement each other because the sawdust, bark chips, and small pieces of wood left from the milling process can be used for particleboard or fiberboard.

FIGURE 16.12
A board being cut with a large rotary blade.
(Courtesy of USDA.)

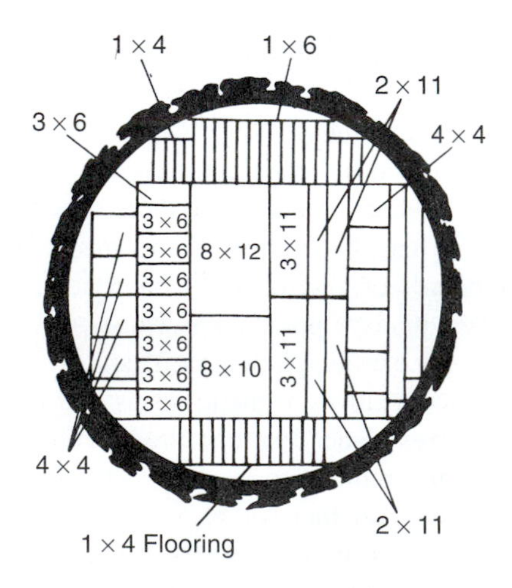

FIGURE 16.13
The mill operator decides what kind of cutting procedure will yield the most lumber from each log. Structural pieces are usually cut from near the center of a log, where the wood is dense and not likely to crack. The outer portions of logs are most suitable for smaller boards such as 1 by 4 in. flooring boards.

FIGURE 16.14
Timbers or boards are placed in cylinders, where they are treated with wood preservatives. Creosote (a coal tar product) and Wohlman salts (a poisonous mixture of fluorine, phenol, copper naphthenate, and arsenic salts) are often used to protect the wood from fungal decay and attack by insects. Since many of these compounds are carcinogenic, their use is now regulated.
(Photo courtesy of the Koppers Company.)

Fiberboard differs from particleboard in that wood fibers, not small pieces of wood, are used. The fibers are xylem elements that have been separated from one another by placing small wood particles in chemical solutions that dissolve the pectins that hold the fibers together. Fibers can also be produced mechanically by grinding wood chips. The processes used to reduce wood to fibers for fiberboard are the same as those used to make pulp for paper. Once the fibers have been separated, they can be mixed with various additives, such as synthetic resins, for different degrees of strength and with protective compounds for resistance to fire and pests. The pulp can also be pressed into sheets without the addition of resins to produce wet-felted fiberboard. The amount of pressure applied while the boards are drying and the amount of resin added determine the final strength and hardness of the fiberboard.

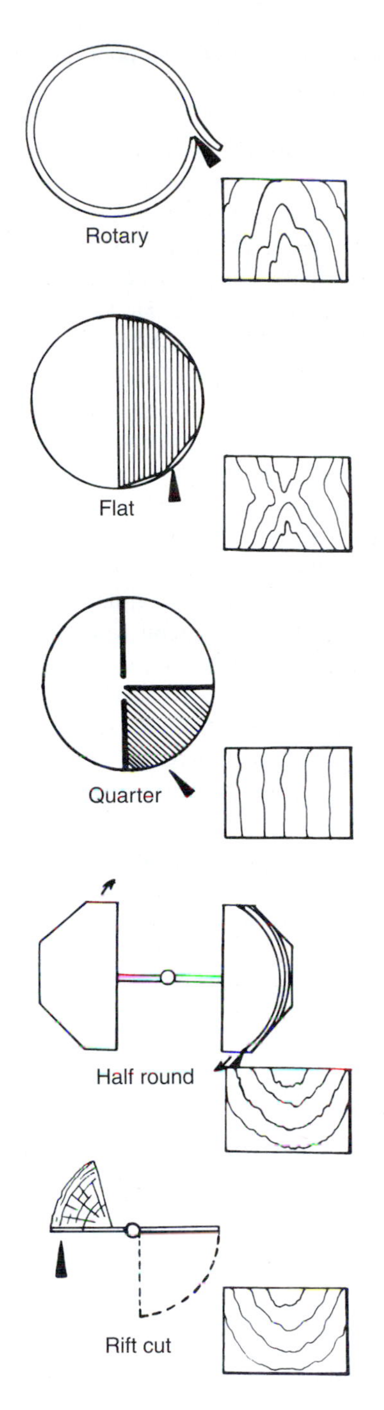

FIGURE 16.15
Methods of cutting veneer and the figures of the resulting products.
(Redrawn after Villiard, 1975. *A Manual of Veneering.* New York, Dover.)

FIGURE 16.16
Makers of fine guitars use different kinds of hardwood with varying hardness and density to produce instruments with beautiful tones.
(Photo taken at Collings Guitar, Dripping Springs, Texas.)

Paper

It is often said that the Egyptians invented paper because they wrote on sheets of papyrus. However, those sheets were formed by simply pressing together strips of the leaves of papyrus, *Cyperus papyrus* (Cyperaceae). Other paperlike substances have been made by various cultures. Among these are the famous rice "paper" of the Orient, which is made by pounding sheets spirally cut from the pith of the rice paper plant, *Fatsia papyrifera* (Araliaceae). Both the Mayans of Mexico and the Polynesians independently developed a method of producing paperlike sheets by pounding the bark of the paper mulberry, *Broussonetia papyrifera* (Moraceae). Inhabitants of the South Pacific, but apparently not the Mayans, used the bark cloth for clothing. None of these substances was true paper.

Paper, as it is known today, is made from plant fibers that have been separated from one another and then matted together in a thin sheet. The Chinese apparently first produced a paper made by this process in about A.D. 100. In its production they used the paper mulberry but separated the fibers and mixed them with other plant fibers such as flax and hemp and even fibers extracted from rags. The fibers, floating in liquid, were allowed to settle in a thin film on a screen. The screen was shaken as it was lifted to interlock the fibers. Upon drying, the resultant sheet of paper was peeled from the screen.

It took 1000 years for papermaking to spread from the Orient to Europe, reaching Spain when the Moors conquered it in the twelfth century. Some leaf and wood fibers were used to make early European papers, but most of these papers were made from fibers extracted from linen, cotton, or hempen rags. For several hundred years, paper was essentially made by hand, using variations of the process developed in China. Because of the hand labor involved, paper was expensive and not widely available. Hand production was finally supplanted with the invention of a papermaking machine in France after the French Revolution of 1789. After the British made improvements that sped up the process, the increased production created a shortage of rags. Rags were in such demand that even mummy wrappings were removed from Egyptian tombs and shipped to Britain. It soon became obvious, however, that new sources of fiber would have to be found. Wood was a practical source, but the initial methods of separating the wood fibers by rubbing logs on stones left much to be desired. The resultant fibers were short and weak and retained their pectins and lignins; their retention of pectins and lignins produced a paper that was short-lived and rapidly acquired a yellow tinge.

Chemical methods of separating wood fibers were developed about 1851. The first processes involved boiling wood chips in caustic alkali until the wood was reduced to a mass of fibers. Once washed, the fibers could be bleached. Improvements on this process involving different dissolving agents led to the sulfite process developed in Paris in 1857 and the sulfate process developed in 1884. Today almost all papers are made from wood using one of these two methods, but sheets labeled 100 percent rag bond are still made from fibers extracted from rags. Conifer woods are generally preferred over hardwoods for papermaking because their xylem tracheids are longer than the xylem vessels of hardwoods (about 2.0 to 4.0 mm versus 0.5 to 1.5 mm). Besides wood pulp, pulp from abacá (see Chapter 15), sugar cane pressings (bagasse), and bamboo is also used.

As in the case of fiberboard, the first step in making paper is the production of pulp. Either mechanical or chemical methods can be used. In either case, the logs are stripped of their bark and the wood is reduced to chips. Enormous machines are now driven into the forest to receive pulp trees as they are felled. The machines take entire logs, strip them, and reduce them to chips that are trucked to the paper plant. In other cases logs are floated to the mills, where the bark is stripped by raspers and is often subsequently used to fuel the mill. The logs are then chipped at the mill.

Mechanical pulping is used in the **ground wood process** and consists of grinding the chips to free the fibers, which can be bleached or used directly to make paper (Fig. 16.17). Several agents, including resins, gums, starches, or rosin, can be added to the pulp before the fibers are floated onto draining screens. They act as sizing agents to fill in surface irregularities and improve a paper's ability to accept ink. Most paper produced by a mechanical process has little sizing because the paper is of such poor quality that it does not justify the expense.

Once prepared, the slurry of fibers is flooded over a screen to produce uniformly thick sheets. The sheets drain as they move across a screen and onto rollers that press and mat the fibers before the sheets are rolled into wet pulp rolls. Paper produced from mechanically extracted fibers yellows rapidly and eventually crumbles because the pectins and lignins left in the pulp degrade. However, because of the demand for cheap newsprint, a large amount of paper is produced by this process.

Chemical processes for pulping (Fig. 16.17) differ from one another principally in the agents used to reduce the wood to pulp and dissolve the lignin. In the **sulfite process** the chips are cooked in a digester with bisulfites. Hot acid is then pumped in, and the cooking is completed. The softened fibers are then forcefully blown into a chamber to separate them.

The **sulfate process** is an alkaline rather than an acid process and uses as digestive agents sodium sulfate, sodium sulfide, and caustic soda (sodium hydroxide). This process is now the most widely used because unlike the sulfite process, it dissolves the resins out of the pulp and can therefore be used for gymnosperm woods such as Douglas-fir and pine. After digestion, the tenuously bound fibers are beaten to separate them.

Once separated from one another, the fibers produced by either process are washed and sometimes centrifuged to separate them from bits of bark and debris. The clean fibers are treated with a chlorine bleach that is later flushed out with sodium hydroxide. The fibers are then rinsed and mixed with enough water to make a slurry. Sizes, most commonly soaps made from rosin, are added to chemically produced pulp. Since sizing substances tend to bind poorly with the pulp, a binding agent such as alum (aluminum sulfate) is added. This is the same type of agent that is added to dyeing solutions to bind dyes to various fibers (Chapter 15), but in the textile industry they are called mordants. Binding agents also ensure the adherence of pigments to the fibers if the paper is to be colored.

Finally, the pulp is ready to be formed into paper. It is flooded into a large box with a slit in the bottom that allows an even, smooth flow of pulp across the screen (Fig. 16.18). The sheet drains as it moves with the screen until it reaches a first set of pressing rollers. It continues around a series of rollers through a drier. Modern paper plants can produce paper at the rate of 1219 m (4000 ft) per minute. If necessary, the paper is coated after it is produced so that it will be more suitable for the printing of halftones (photographs).

Paper produced by the sulfite process has a high acid content that causes brittleness and disintegration after about 100 years. Unfortunately, most books made after 1850 were printed on sulfite papers. In addition, sizings using aluminum were employed that formed acidic compounds over time. People are now facing the possible loss of millions of valuable books. The Library of Congress has embarked on an enormous

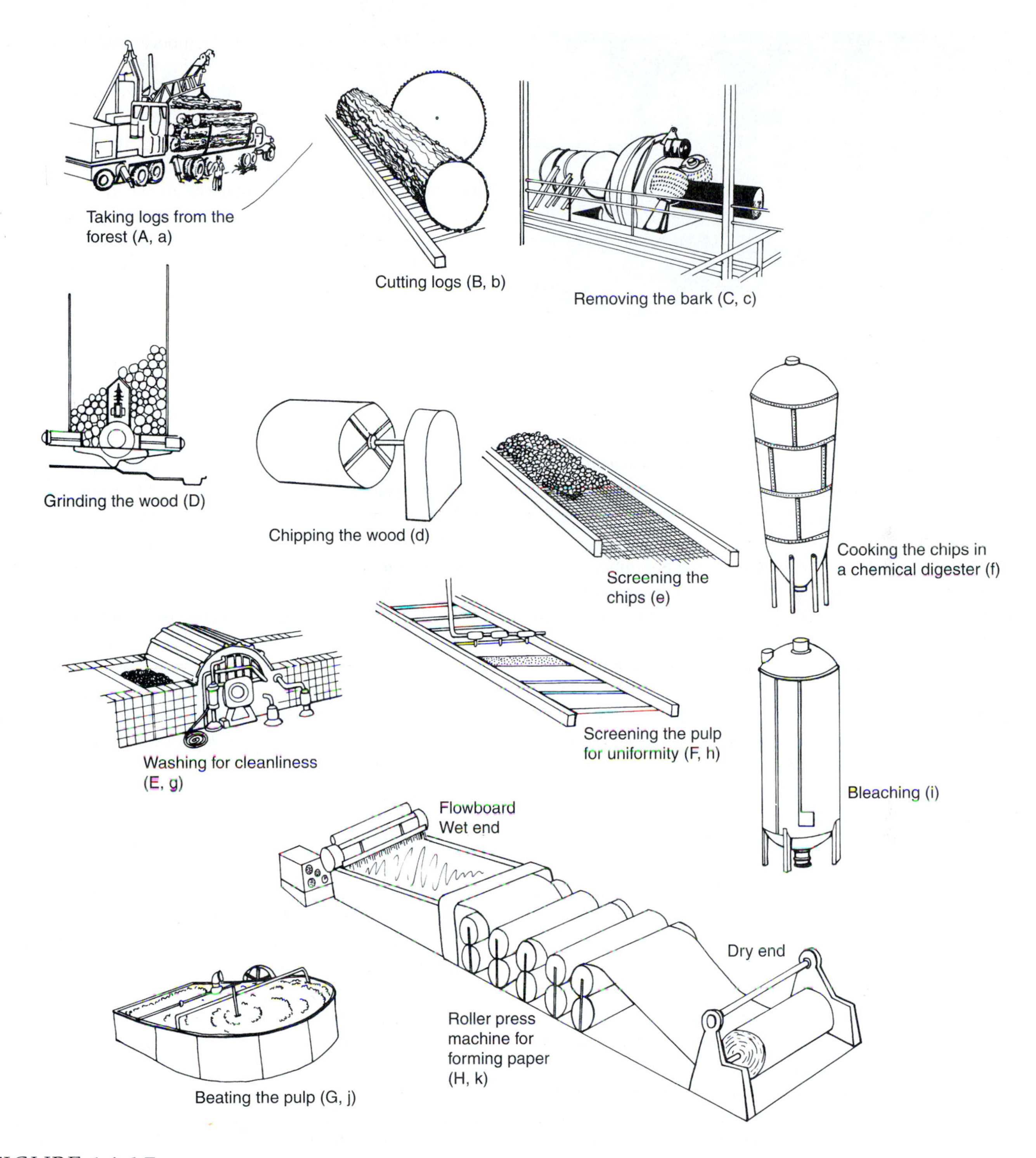

FIGURE 16.17

(A–C and a–c) Most paper is now made from wood (A,a) obtained from logs that are (B,b) cut, taken to the mill, (C,c) debarked, and (D,d,e) cut into small pieces or chipped. Fibers for making paper are then extracted by either mechanical or chemical processes. (A–H) In the mechanical ground wood process, wood is simply ground to separate the xylem fibers (D). The fibers are then (E) washed, (F) screened, (G) beaten, and (H) pressed into paper. Because this process leaves the pectins and lignins in the pulp, it produces a paper that turns yellow and brittle with age. (a–k) In the various chemical papermaking processes logs are (d) chipped, (e) screened, and (f) dumped into digesting tanks with chemicals that dissolve the pectins and lignins while freeing the fibers. Different chemicals are used to digest wood chips in the three processes used. The soda process utilizes a sodium hydroxide solution; the sulfite process uses a mixture of bisulfites and sulfuric acid to effect digestion; and the sulfate process uses a solution of sodium hydroxide and sodium sulfide. Of the three chemical processes, the soda process is the least used. The pulp produced from any of these chemical methods is (g) washed, (h) screened, (i) bleached for whiteness, (j) beaten, and floated onto screens where the water is drained off and the newly formed sheet is (k) passed through a series of rollers to press it before drying.

FIGURE 16.18
A headbox (rear, at the far end of the white screen) of a papermaking operation is used to regulate the flow of pulp onto the draining screen. (Courtesy of the Potlatch Information Service.)

project to treat its books with diethyl zinc, a compound that bonds with cellulose and neutralizes the acid. Many books and other documents are now labeled "acid-free" to indicate that they are made using an alkaline (sulfate) process and nonacid sizings and are not subject to premature disintegration.

The United States uses more paper and paper products than any other nation in the world, estimated in 1999 at 332 kg (733 lb) of paper per person per year. This translates to about 40 percent of all municipal garbage and an increase of 77 percent over the previous 10 years. Since 75 percent of U.S. solid waste is still disposed of in landfills, most of this paper ends up in a dump, where it is either burned or covered with earth. It might seem that allowing paper to rot is analogous to composting, but it turns out that most of the material in a landfill does not decompose because of the anaerobic conditions that develop when waste material is buried under layers of new debris and soil. An obvious solution to this problem is to reduce the amount of waste generated. Waste can be reduced by producing less in the first place or by recycling what people create.

The idea of recycling many waste materials is not new, but the public demand for such processes and the increased cost-effectiveness of recycling have only recently made the reuse of paper, glass, and various metals a viable option. Paper is particularly easy to recycle because the process consists of redissolving the wood fibers and flooding them onto a screen to form a new sheet of paper. The challenge has been to develop a process that can deal with the array of papers produced and the inks and different sizing substances that have been used on them. Recycling newsprint is rather straightforward, and for many years newspapers and plain white paper with standard black lettering were the only materials collected for recycling. In this process paper is simply shredded and dissolved with chemicals into a slurry. The black ink then floats from the paper as it is reduced to fibers. Globs of ink are removed from the slurry by centrifugation. Many glossy and colored papers are treated with sizing substances of various types to improve their surfaces or are impregnated with a color. Many of them are then printed with a range of chemical inks. These products could not be recycled with newspaper because many of the inks and print remained in the slurry after the centrifugation. Methods have now been developed that allow all papers, regardless of finish, color, or print type, to be collected and recycled together. This new process begins like the standard procedure with shredding, dissolving into a slurry, and removal of inks that float off easily but continues with a second dye and ink removal step that involves pumping air into the slurry. The foam that is developed traps the remaining ink particles and allows them to be removed with it. The pulp is then washed over a screen and finally rebeaten with water into a slurry that can be flooded onto the paper-forming screen.

Nevertheless, recycled paper is often inferior to paper made from virgin pulp. One way to increase paper quality without abandoning recycling would be to combine recycled and virgin pulp. However, because recycled pulp has weaker fibers than virgin pulp and requires different amounts of water and drying time, until recently it was impossible to work with combined pulps. New computers that can monitor precisely the slurry composition and continually adjust water quantities or drying time now allow mixtures to be used.

It has been estimated that recycling would save 12.5 million m^3 of "roundwood equivalent" per year if the current recycling rate of 29 percent of paper products were increased to 60 percent. Combining recycling with measures such as increasing lumbering and wood processing efficiency and reducing the use of disposable paper ingredients could halve the consumption of wood in the United States and greatly reduce this country's dependence on landfills.

Another way to reduce the environmental impact of papermaking would be to use fewer chemicals in the pulping of wood. Scientists have been using genetic engineering to produce trees that have less lignin and that therefore would need less processing to free the wood fibers and digest the noncellulosic compounds. Varieties of aspens have been developed that have significantly lower lignin levels and up to 15 percent higher levels of cellulose.

Rayon, Cellophane, Acetate, Arnel, and Lyocell

Rayon and cellophane are the same product in different forms. Both are made from pure cellulose derived from wood. Initial steps toward the production of rayon were taken in Europe in 1855, but suitable technology for mass production was not developed until after 1900. The first steps in the production of the two materials are the same as those employed in papermaking.

The cellulose pulp is first steeped in a 20 percent water solution of warm sodium hydroxide to swell the fibers. The swollen mass of cellulose is pressed and then shredded mechanically to yield tiny, fluffy particles. The alkali cellulose is then aged to depolymerize the cellulose before being placed in vats with carbon disulphide to form cellulose xanthate. The cellulose-xanthate particles are then dissolved in a very strong alkaline solution. These last two processes allow cellulose-xanthate particles to form a thick water suspension known as **viscose** that is "ripened" to make the cellulose even more soluble. Finally the viscose is filtered to remove any undissolved material and then is degassed. The final mixture is either extruded in thin sheets that are thoroughly washed to produce cellophane or forced through small openings to produce rayon threads. The threads are drawn through an acid bath that neutralizes the alkali and makes the fibers flexible. The strands are then washed, dried, and wound onto spools. High-performance rayons can be sanforized and mercerized (see Chapter 15), accept dyes well, and match cotton in general performance. The newest rayons (marketed as Avril) are like natural fibers in terms of the crimping or lobing of the individual strands.

Acetate and acetate fibers were once grouped with rayon and cellophane but now have come into their own. They differ from rayon and cellophane primarily in that they are made from purified cellulose to which one or more acetyl groups have been added. Despite this partially synthetic production, fibers made from acetates have many properties of completely natural plant fibers. Processes for the spinning of acetate fibers were developed in the 1920s, but it was only in about 1950, after the development of triacetates, which do not shrink, that they began to be used extensively.

The ingredients used to make acetates are wood pulp and sometimes the straggling fibers (linters) left on cottonseeds after ginning. The purified pulp, prepared by one of the chemical methods, is mixed with acetic acid, acetic anhydride, and a catalytic agent. The chemical reaction that results leads to the linkage of acetyl groups with the cellulose molecules of the fibers. The acetate is then chilled in water, forming flakes that are dissolved in acetone. Dyes can be added to the solution, resulting in deep, vibrant colors that do not fade when the fibers age. Triacetates are made by dissolving triacetate in alcohol and methylene chloride and then spinning it. These fibers, marketed under the name Arnel, and now used extensively in double knits, tricots, and various permanent press items. Acetate fibers dry rapidly, are soft, resist wrinkling, and are not attacked by moths or molds. Masses of acetate fibers that have not been spun are used to make the filters in some cigarettes.

Lyocell, commonly sold under the trade name Tencel™, is the first new textile fiber produced in 30 years. Like rayon and Arnel, lyocell is made from cellulose, usually cellulose extracted from plantation-grown trees, that is chemically treated. Strands of the fiber are produced by melting and spinning the cellulose using NMMO (N-methyl morpholine-N-oxide) as a solvent. Because the solvent is recovered and reused the manufacture of lyocell is more environmentally friendly than many other textile operations. Cloth made from lyocell is advertised as feeling like silk, hanging like wool, and able to be laundered like cotton, although many Tencel™ garments on the market develop a "hairy" surface if washed and thus should be dry cleaned. Truly machine-washable lyocell fabrics have to be treated with a special finish or modified when produced. Nevertheless, lyocell is expected to gain a large share of the textile market in the twenty-first century because it is made using cellulose from high-yielding tree plantations, does not produce industrial wastes when manufactured, and can be blended with a variety of other fibers to furnish an array of textures and finishes.

Cork

Strictly speaking, cork is not wood since it is part of the bark of a tree, not secondary xylem. Still, it is usually included in a discussion of wood because it is produced by trees and is often used for woodlike products. As shown in Figure 16.19, cork (phellem) is part of normal bark, but most trees produce only small quantities. One species, the cork oak, *Quercus suber* (Fagaceae), produces layers of cork several inches thick that can be stripped from trees without damaging them. The areas around the Mediterranean Sea (Fig. 16.20) where this species grows naturally are covered by an evergreen scrub vegetation similar in aspect to the scrub along the coast of southern California. As in California, this region is prone to frequent large-scale fires. The bark of the cork oak is an adaptation that provides protection against fire damage.

Cork acts as an insulating material because it is composed of numerous air-filled cells. These cells can withstand a compression of up to 10,000 lb/in^2 without rupturing. When the pressure is removed, the cells return to their former size. The air-filled cells also make cork lightweight and cause it to float. The Greeks and Romans appreciated the qualities of cork and used it for the soles of shoes and for fishing floats. People still use cork for shoe soles and fishing floats, but its modern uses have expanded because it has not been possible to produce a synthetic with all its qualities. One recent use has been for insulating spaceships.

Cork oaks are large evergreen trees that build up enough bark to be stripped when they are about 25 years old. This first stripping yields "virgin" bark that is considered inferior to the bark that replaces it. The virgin cork is mostly ground and used to make composition cork, flooring, and wall coverings., Successive crops of cork consisting of 2.5-cm- (1-in.-) thick layers from the lower 2 m (6 ft) of the trunk can be harvested about every 9 to 10 years after the first stripping (Fig. 16.21). The superior cork produced by the later layers is used for stoppers, barrel bungs, cork veneers, and specialty items.

Bamboo

Like cork, bamboo differs from wood because it does not contain any secondary xylem, but in the Orient bamboo is used as a structural material (Fig. 16.22) that fulfills many of

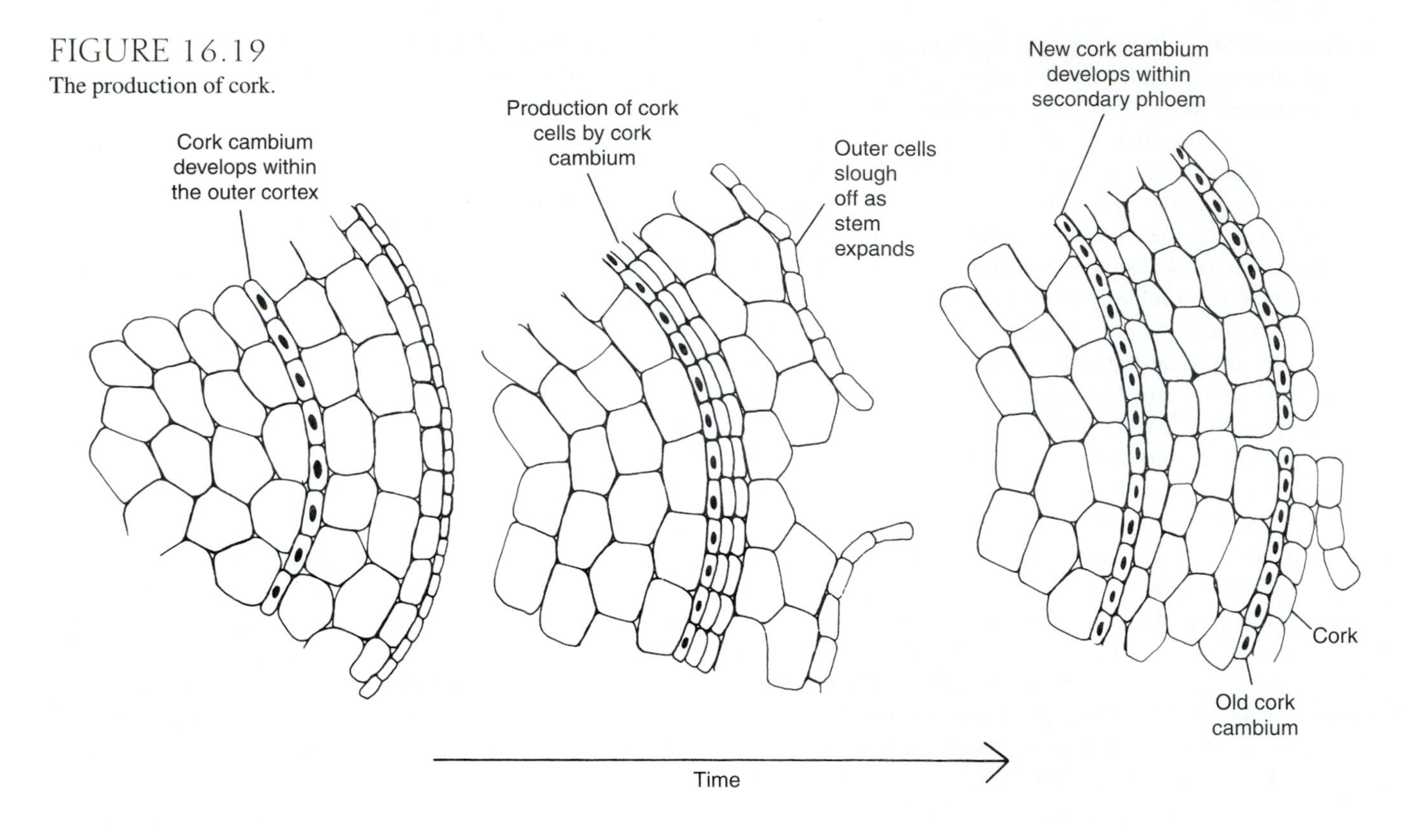

FIGURE 16.19
The production of cork.

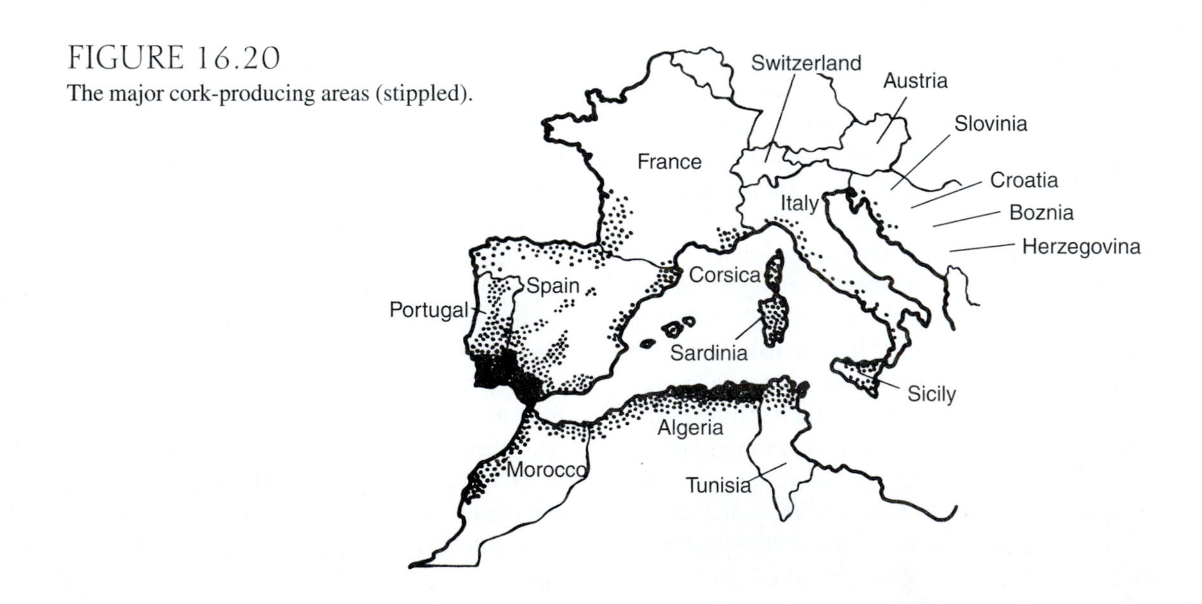

FIGURE 16.20
The major cork-producing areas (stippled).

the functions of wood. In China bamboo is used symbolically to represent resistance to hardship. The name *bamboo* can be applied to about 1000 species that belong to one subfamily of the grass family. New growth in bamboo is produced only by apical meristems (Fig. 16.23). Consequently, there is no secondary, or radial, growth. All growth is vertical and is caused by the production of new apical cells and cell elongation. Some bamboos can grow vertically 60 to 90 cm in a day. Stems of mature bamboo consist of segmented hollow tubes that give bamboo a unique combination of flexibility, light weight, and strength. Bamboo is used for a wide array of purposes, ranging from reinforcing rods in concrete to fishing poles and from lightbulb filaments to ornamental hedges. Bamboo has an appeal that stems not

FIGURE 16.21
Stripping bark from cork trees.
(Courtesy of the Armstrong Cork Company.)

FIGURE 16.22
Bamboo is being grown in Ghana as part of a project to provide building materials for home construction.
(Photo by P. Cenini, FAO.)

from its human use but from its mysterious reproductive system and association with the giant panda (Fig. 16.23 and Box 16.1).

Wood Supplies in the Future

There is growing concern about the ability of the earth's forests to meet the projected demands for wood and wood pulp that will come as populations increase. The need for wood and wood products is expected to rise gradually but steadily as countries such as China and India begin to use wood products in proportions similar to those of more developed countries. Half the original land covered by tropical forests has already disappeared, and there is some suggestion that logging has accelerated in the Amazon region since 1995 (Fig. 16.24). A 1992 estimate stated that 17 million hectares (6.9 million acres) or 4.4 percent of the land still covered by rain forest is cut each year. The amount of forest is actually increasing in the Northern Hemisphere due to reforestation. However, North America and Eurasia have already been largely deforested and most reforestation in these areas is in the form of tree plantations that lack the biological diversity of the original forests.

Large-scale deforestation has the potential for damage beyond the loss of natural forests. An increase in worldwide temperatures and the threat of changing global climatic patterns have been extensively discussed (see Chapter 19). The more local effects of logging operations are not as widely appreciated. The most common method of logging used in the past was **clear-cutting,** or the removal of virtually all the trees in a given area. Clear-cutting, more than any other type of harvesting, sets in motion a cycle of irreversible soil erosion and local population extinctions of other plants and animals. One devastating consequence is the loss of **mycorrhizae,** or the fungi associated with the roots of most woody plants. These fungi serve as a conduit for soil nutrients into the roots. Once an area has been clear-cut and erosion begins, many of these microorganisms die, making reforestation difficult or even impossible.

Although trees are a potentially renewable resource, the use of billions of cubic meters each year demands large-scale planning operations to ensure future supplies. Until 1950 virtually all wood was harvested by exploiting natural, often virgin (never-cut) forest lands. Now many American companies are trying to practice various methods of forest management. Among these methods are a number of practices designed to lessen the effects of deforestation. One such technique is "even-aged" cutting, which tries to ensure that a stand will regenerate into a forest similar to that which was logged in terms of the dominant species. The first step is to cut the larger understory layer of trees shorter than the desirable canopy. These trees are assumed to shade the seedlings of the dominant species and, if left standing, will become dominant when the canopy is cut. If these large

Box 16.1

A Good Strategy Turned Sour

The mystique generated by bamboo's unique shape and phenomenal growth is enhanced by its unusual reproductive cycle. Unlike other plants, most bamboos do not flower and fruit each year. Instead, individual plants flower and fruit once and then die, with the interval between germination and reproduction lasting from 15 to 30, 60, or even 120 years. Even more impressive is the fact that over huge geographic areas all the bamboo plants of a species flower and fruit on the same schedule, leading to mass blooms, immense fruit yields, and drastic diebacks. The evolutionary reasons for this unusual and lengthy periodicity and the physiological basis for the pattern are not completely known, but there have been attempts to explain both. One explanation for the large delay in bouts of bamboo reproduction is that it is an evolutionary "strategy" to protect seeds from predation. This explanation is based on observations that bamboos produce seeds that are avidly consumed by various vertebrates, including rodents such as wild rats. If a species of bamboo consisted of individuals that fruited each year, populations of seed predators would track the seed production and presumably destroy the bulk of the seed crop. By flowering massively only once every 15 or more years, bamboos avoid being tracked by vertebrates whose generation time is shorter than the flowering period. When bamboos do fruit in a region, the mass of seeds produced can be astounding. Seeds can cover the ground in a layer several centimeters deep. At the time the seeds are shed rodent populations are rather small. Consequently, the rodents are quickly satiated and there are ample seeds left for germination. Because of the abundant input of food from the bamboo seeds, the rodent populations explode the next year, only to find that there are comparatively few seeds. Eventually the rodent populations drift back to between-bloom levels while the bamboo populations are regenerated. Although this fruiting pattern has proved to be effective against seed predation, it has ironically contributed to the decline of panda populations in central and eastern China.

Pandas also feed on bamboo, but on the stems and leaves, not the seeds. In fact, bamboo is the only food giant pandas eat, and they require enormous amounts of it—up to 38 kg (85 lb) per individual per day—because they have a digestive system that is very inefficient in digesting plant matter. In the past, when a species of bamboo died back after flowering and fruiting, groups of pandas would migrate to an area with a different species of bamboo that was on a different flowering schedule. During the last century, however, the areas of China covered by wild bamboo have shrunk considerably as increased population levels have caused the conversion of natural areas to farming communities. Pandas have consequently been forced to live in reserves with a limited number of bamboo species. When the last mass bloom of one of the giant panda's major bamboo food sources occurred in 1983, many pandas apparently faced starvation. The Chinese then launched efforts to rescue the starving pandas, capturing approximately 10 percent of the pandas left in the world and placing them in zoos when they could be "cared for." However, the failure of giant pandas to reproduce well in these artificial situations has further imperiled the species and emphasized the need to maintain habitats for wild populations. This dilemma has become symbolic of the problems facing conservationists, as shown by the use of the panda as the symbol of the World Wildlife Organization.

Efforts to breed bamboos and discover the mechanism that triggers their periodic flowering have fared somewhat better. In 1988 Indian scientists reported that they were able to induce flowering in two species of bamboo artificially by germinating the seeds and then transferring pieces of the seedlings sequentially from one nutrient and hormone medium (a substance on which the tissue can grow) to another. The process resulted in flowering and fruiting and accomplished in a matter of weeks what would have taken decades to happen in nature. Eventually this research should help workers breed bamboos, learn about bamboo's strange reproductive behavior, and perhaps find a way to keep a supply of bamboo in the panda's natural habitats.

understory trees are cut, the seedlings of the desirable species receive increased sunlight and begin to grow. Once they reach a robust size, the canopy species can be cut. The young trees of the same species can then take advantage of the second new burst of sunlight and again become the dominant species. Another method is selective cutting of single large trees in an area. This is supposed to minimize the time and aerial extent of disturbance. Both kinds of logging, however, cause wider damage than would be indicated by the quantity of trees removed. Roads cut to allow crews and trucks into an area permit the invasion of weedy species and often human settlers. The destruction of vegetation by the logging operations is itself significant.

Many scientists, economists, and conservationists are concerned that forests are not regenerating as fast as they are being destroyed and that current logging practices lead to soil and nutrient loss and often to an unnatural dominance of one or two species of trees. Alternatives to logging virgin forests include the planting of tree farms (Fig. 16.25) and the successive reuse of areas that were logged in the past and now consist

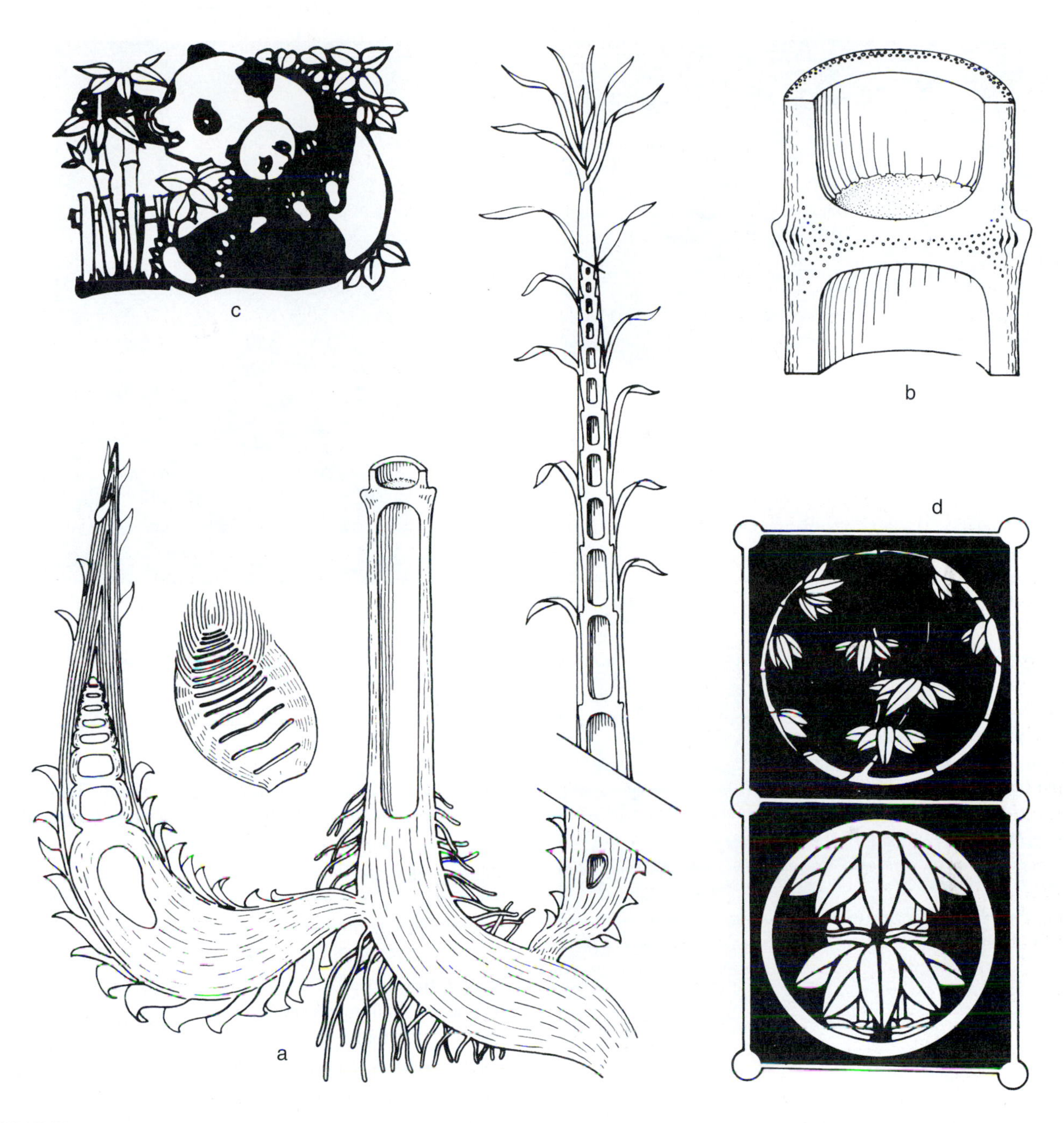

FIGURE 16.23

Bamboo is among the world's most curious and versatile plants. Like other monocots, bamboo has no vascular cambium, and its stems cannot increase in girth by adding lateral layers of cells. (a) The diameter of a stem is essentially determined at the time of sprouting. Bamboo spreads by sending up shoots from rhizomes. (a,b) Bamboo's structural strength and flexibility are due to the morphology of the stems, which are made up of a series of hollow segments separated by solid internodes. Vascular bundles and associated fibers are scattered in the outer walls of the segments. (c) In the Orient, where many species of bamboo are native, the plants are the sole food of the panda, and bamboos have come to symbolize resistance to hardship because they stay green all year and bend without breaking under the weight of snow. The long internodes are thought to represent virtue, while the hollow stems are reminders of the value of humility. (d) These associations and the importance of bamboo in construction and horticulture have led to the adoption of bamboo by many Oriental families as a symbol for a family crest.

FIGURE 16.24
Mahogany is a valuable hardwood that is fast disappearing from tropical forests. Here, mahogany logs are floated down a river to a harbor in the Ivory Coast.
(Photo by A. Defever, FAO.)

FIGURE 16.25
A 20- to 25-year-old red pine plantation.
(Photo courtesy of USDA.)

of **secondary forest,** a kind of forest that has regenerated in a region where the primary forest was cut. Tree farms are not a biological substitute for diverse natural forests since they are monocultures and often consist of species not native to the region in which they are planted. Nevertheless, their establishment is one way to lessen people's dependence on the few vir-

FIGURE 16.26
Leucaena species are among the trees being looked at as potential sources of wood in the future.
(Photo by S. Bauer, courtesy Agricultural Research Service, USDA.)

gin temperate and tropical forests that remain. Tree farms or plantations often consist of fast-growing trees, such as pines or species of *Eucalyptus* (Myrtaceae),which can be harvested in shorter intervals than natural forests. A previously unused genus, *Leucaena* (Fabaceae) (Fig. 16.26), is being grown as a potential new source of wood. In tropical areas trees of species of this genus, which has wood as dense as that of oak, can grow 8 m (over 24 ft) in height a year. Once cut, the stumps regenerate, making reseeding unnecessary. Similarly, a woody herb that yields the fiber kenaf, *Hibiscus cannabinus* (Malvaceae), is farmed as a possible source of paper pulp. One acre of this herb can supply five times the amount of pulp supplied by an acre of pine trees. Commercial operations of kenaf are now operative in southern Texas and Louisiana.

A final measure to help decrease the logging of forests would be to reduce the demand. Such reductions could come from recycling efforts, increasing the efficiency of milling operations, and eliminating the wasteful use of wood in many industries, especially construction. It has been estimated that revamping milling operations could reduce U.S. wood use by 25 percent.

17

Ornamental Plants

Not only do ornamental plants delight the senses, but in an increasing urban environment they often constitute people's only link with the natural, growing plant world. Although humans have planted gardens since the dawn of civilization, the expansion of modern cities has made horticulture a multi-billion-dollar business and gardening this country's number one pastime. In this chapter we examine the philosophies of gardening that have shaped American horticulture and assess the characteristics required of ornamental plants. We also look into the origins of common horticultural species and the changes that have occurred in them under the influence of artificial selection. Finally, we look at some future trends in the uses of ornamental plants.

Qualities of Ornamental Plants

In contrast to plants that yield utilitarian products, ornamental plants are appreciated for their aesthetic qualities or are used to beautify the appearance of another object (Fig. 17.1). Beauty is, of course, a subjective thing. In many cases practicality and utility (considered to be components of beauty by some but not by others) are also important factors in the choice of ornamental plants. The criteria for what is beautiful are in large part culturally determined and differ among countries, between parts of a country, among individuals, and from generation to generation. There are, however, a few basic elements of beauty common to art forms in general that can be applied to ornamental plants.

One of the primary elements of beauty is **color.** People perceive color as a combination of features, including the hue, or spectral wavelength, which is reflected by an object and received by the eyes as a combination of blue, green, or red; the lightness or darkness of the color; and the amount of color saturation. All these factors subtly influence the choice of ornamental plants.

Texture is a component of beauty that is often subconsciously perceived. From birth, people learn to associate certain visual patterns (Fig. 17.2) with tactile sensations. Eventually people develop visual impressions of textures that allow them to "feel" through their eyes. A good example is the visual contrast between a violet and a pansy. Both belong to the genus *Viola* (Violaceae) and can have flowers of the same color. However, the petals of the violet are smooth, whereas those of the pansy are

FIGURE 17.1
Plant motifs appear in many forms in art and architecture.

covered with short, upright hairs. Even without feeling a pansy, one knows that the petal surface is velvety because of the subtle play of color across the petal surfaces. People have learned to associate this shaded appearance with soft coverings of soft hairs (or threads in the case of true velvet). A similar process is involved when people see shaggy or smooth-barked trees or leaves with waxy or fuzzy surfaces (Fig. 17.3). People often infer the texture without actually feeling how rough or slick a surface is.

Line is another aspect of beauty that often is not consciously appreciated except by designers or architects. Yet one's choice of plants is frequently determined by their natural or inducible sizes or outlines. Vertical branching patterns of trees or shrubs guide the eye upward, whereas horizontal branches lead it toward the ground. People trim hedges to produce precise lines that guide the eye or to frame specific areas (Fig. 17.4).

Form is the last component of beauty involved in plants chosen for ornamental purposes. It is a three-dimensional quality that includes both shape and structure (Fig. 17.5). Differences in form can help determine how a plant is used as an ornamental (Fig. 17.6). Take, for example, the plants

FIGURE 17.2
Without touching them, one can perceive that these agave leaves and cactus stems have very different textures.

shown in Figure 17.6: the weeping willow, poplar, and elm. Weeping willows are used as landscape trees so that their unusual branching pattern can develop and be appreciated. Most cultivated poplars are fast-growing, narrow trees that often are planted to create hedges or screens. Elms have long been favorite street trees because of their vaselike shape and high canopies which allows free movement of pedestrians and traffic below.

Ornamentals are usually divided by the horticulture trade into three major groups: nursery plants, florist crops, and houseplants. Nursery plants are used for outdoor gardens, often to complement a structure or to beautify an area. Florist crops yield flowers or foliage for cut arrangements. Houseplants are growing plants that are confined to containers and used for interior decoration. Different qualities are sought in the plants that belong to each group because they serve different purposes. One quality—aesthetic appeal—is common to all. We will discuss each major group of ornamentals separately, but first we explain how ornamental plants are named.

The Naming of Ornamental Plants

Plants selected for use as ornamentals are often more difficult to assign to a biological species (see Chapter 1) than are other crops because many of them were created by humans. Some are produced by crossing (hybridizing) species. Such hybridizations can involve a series of elaborate steps if the plants to be crossed are quite different genetically. Many horticultural hybrids are sterile and must be propagated by cutting or grafts. Horticulturists also often seize upon bizarre mutants because of their novelty appeal and perpetuate them by asexual means such as cuttings. Since many ornamental plants cannot or do not sexually reproduce naturally and were created by artificial hybridization, they do not fit into concepts of natural species. Nevertheless, they are fitted as well as possible into the system of binomial nomenclature by using special provisions of the International Code of Botanical Nomenclature (see Chapter 1).

A horticultural crop can be given the Latin name of the species from which it was selected if a single parental species was involved. If the crop is an interspecific hybrid and one or both parents are known, a special designation called a **formula name** is used. This kind of name indicates that a species is a hybrid and can specify the two parents if they are known. For example, a hybrid species between *Rosa alba* and *R. nigra* would be written *Rosa alba* × *R. nigra*. Hybrids, particularly those of unknown parentage can be given a name with a multiplication sign before the species epithet to show that the plant is a hybrid (e.g., *Rosa* ×*wilsonii).* Intergeneric hybrids can use the same convention (×*Aster turneri).*

Plants selected for trade purposes often are given appealing names such as ‘Duchess of Windsor,’ ‘Crimson Star,’ and ‘Twinkle Toes.’ The single quotation marks, which are usually omitted, indicate an artificial (and often patented) form. When one buys an ornamental plant at a nursery this cultivar name is often the only one given. One might, therefore, buy ‘Andrew Jackson’ azaleas or ‘Purple Dawn’ African violets with no species name attached although, of course, each plant is referable to one (or a hybrid) species and has a proper Latin designation. In some cases **cultivars** or cultivated forms of a species or genus, are grouped into types that indicate a general kind of flower or growth form. Such groupings are commonly used for very popular plants such as roses and chrysanthemums. Roses, for example, are grouped into classes such as hybrid teas, climbers, and floribundas.

Nursery Crops

Nursery crops are plants sold to be grown outside houses, in parks and playgrounds, along highways and greenbelts, and as part of the landscaping for urban developments. Although often not thought of as such, all these places constitute kinds

FIGURE 17.3
The size, shape, and coloration of leaves all contribute to the perception of foliage texture.

of gardens, public or private plots of land where plants are cultivated. It has been said that the gardens of an age express an image of paradise for the people who create them. Consequently, the historical development of gardening styles often parallels that of the philosophical thinking of civilizations. Similarly, the roots of modern landscaping ideas have been shaped by those of previous cultures. For this reason, we outline the development of landscaping as an art form before discussing the major nursery crops in use today.

An appreciation for natural beauty seems to be common to all people. Flowers, leaves, and fruits have been used as ornaments for thousands of years. The development of gardens, however, had to wait until humans were settled. Only if people are settled in one place can they lay out, plant, and tend the flowers and trees of a garden. The first true Western gardens were planted in ancient Egypt. Eastern gardening started in China. Early records show that even at the inception of gardening in those areas, two very different concepts of gardens existed. Modern American gardens have elements of both styles but have been predominantly drawn from Western traditions. Most of our discussion of the development of gardens thus centers on those of Western civilizations.

The Development of Western Gardens

Egyptian interest in botany and gardens is well documented in wall paintings and hieroglyphics drawn as early as 2200 B.C. (Fig. 17.7). The Egyptians developed a concept of a garden as

FIGURE 17.4
A tree's branching pattern, the curve of a walk, and the edge of hedges, buildings, and plantings can all work as elements of line in landscape design.

an enclosed space, placing their houses within garden walls to keep out intruders and provide protection from desert winds (Fig. 17.8). Within the walls, date palms, figs, pomegranates, and grape arbors provided shade and sustenance, while T-shaped or rectangular pools were planted with lotus and papyrus and stocked with fish. It has been suggested that the pattern of irrigation ditches dictated the layout of these gardens, but the geometric, stylized forms of the paths and planting beds are consistent with the formal architectural style of other Egyptian art. In their search for plants to use in their gardens, the Egyptians organized the first plant-collecting expeditions (see Fig. 10.24). An early catalog of Egyptian botanical knowledge can still be seen on the walls of the temple at Karnak.

The formal Egyptian concept of gardens spread to Syria, Persia, and other parts of the Western world. In Persia, autocratic rulers ordered their subjects to plant groves of trees that became pleasure gardens and hunting preserves. These gardens were the forerunners of modern public parks. The word *park* arose when the Greeks mistranslated the Persian word *pardes,* meaning "paradise" or "garden."

The most famous gardens of ancient times were probably the Hanging Gardens of Babylon, about 47 km (60 miles) south of the modern city of Baghdad, Iraq. These gardens (Fig. 17.9) were created in 605 B.C. by King Nebuchadnezzar, who commissioned the building of a terraced garden to recreate the hillside scenery that his favorite, but homesick, Persian wife craved.

The Greeks considered the Babylonian gardens one of the seven wonders of the world but did little to advance gardening art. Their democratic system of government did, however, encourage the development of public parks, which became popular places for holding philosophical discussions. The Greeks also planted kitchen gardens in small, open courtyards, but their purpose was practical, not aesthetic.

FIGURE 17.5
With age, most trees and shrubs (top) assume a form characteristic of their species. By knowing what this shape will be or by pruning and training plants into desired shapes, a landscaper can use plants as architectural elements to create outdoor spaces. In the lower panel, from left to right, are elm, willow, and poplar trees. The elm and poplar guide the eye upward, whereas the willow brings the eye back toward the ground.

FIGURE 17.6
These sketches show how the arrangement of plants with different forms can create (top to bottom) open, static, linear, or intimate spaces.

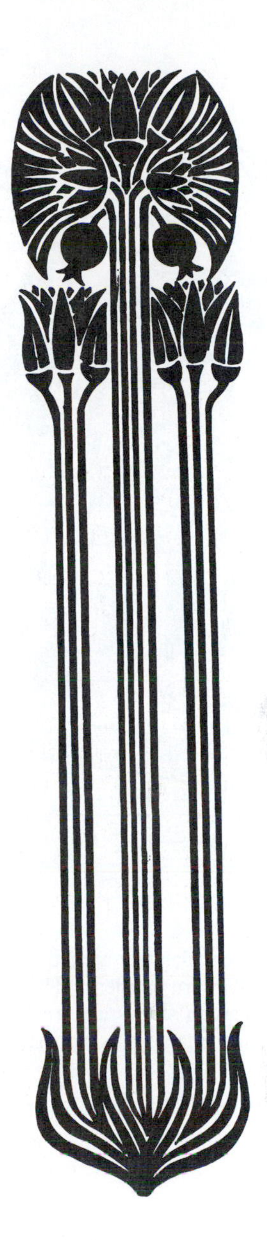

FIGURE 17.7
A stylized ornament incorporating the lotus and perhaps pomegranates, all highly esteemed in ancient Egypt.

Many of the artistic and cultural contributions of the Romans pale beside those of the Hellenistic Greeks, but the Romans excelled at gardening and became the most accomplished gardeners of the ancient world. This expertise can be attributed to several factors. The Roman empire lasted long enough for the development of a distinctive style of garden art. In addition, the Italian peninsula has a mild climate and good soils on which gardens thrive. The Romans had ample inspiration for their gardens from the tales brought back by soldiers from other parts of Europe, western Asia, and northern Africa.

Because the Roman empire was an urban civilization, most of its citizens lived in large city apartments. By planting window boxes and painting flowers on courtyard walls, people

FIGURE 17.8

Sun and limited water availability were the major influences in the design of the first Western gardens. This artist's concept of a wealthy Egyptian's garden shows a formal, symmetrical arrangement of pools, arbors, and partitions that creates a cool, shady oasis within the desert.

(Based on a drawing by D. Lancelot in *Histoire des Jardins,* Tours, 1883.)

brought nature into this new urban world. Wealthy Romans implemented garden designs that fit the grounds of villas outside the city, and for the first time, urban planners incorporated greenbelts into city designs. As Rome prospered, the cultivation of ornamentals thrived. The Romans built the first greenhouses, using mica or glass for windows. In these houses they learned to grow tender, exotic species and to force plants to bloom out of season. The art of topiary (Fig. 17.10) was a Roman innovation in which native species such as juniper, rosemary, and lavender were used. When Rome fell about A.D. 500, horticulture declined in Europe. For the next 1000 years gardening in Christian Europe was confined to monasteries where monks planted medicinal gardens and grew altar flowers (primarily lilies and roses).

In the Muslim world, gardening continued to flourish during the Dark Ages, and Muslims carried their visions of paradise and garden design to every land they conquered, including Spain, which was invaded by the Moors in the first century A.D. Spanish hillsides were terraced to create verdant displays, and villa gardens were laid out in a cruciform pattern to reflect the Muslim belief in the founding of civilization at the confluence of four rivers. Vine-covered arbors ("glorietas"), colonnades, fountains, and colorful tiles, which are still popular motifs in Spanish gardens, are legacies from this period of Islamic domination (Fig. 17.11).

During the crusades (1095–1291) interest in gardening was rekindled in other parts of Europe. The French developed their own form of pleasure gardens in the twelfth century consisting of small gardens enclosed by walls and planted with beds of flowers, clipped hedges, or even mazes of trimmed shrubbery often designed to enhance their function as places to entertain ladies.

As the Renaissance slowly spread across Europe, the revival of classical ideas extended from the fine arts to gardening and landscaping. Italians began to look at plants as architectural or sculptural objects that could add perspective to garden design. Plants were considered building materials to create outdoor corridors, vistas, and plazas on Italy's sloping hillsides (Fig. 17.12). The French readily adapted the Italian designs to their flatter terrain, producing elaborate formal gardens and a low colorful display called **parterre** that consisted of masses of bedding plants arranged so as to form patterns (Fig. 17.13). The most famous French gardens of this period are those at Versailles (Figs. 17.14 and 17.15).

In 1660 French gardening ideas crossed the English Channel and were emulated by Charles II and other nobles in England. By the end of the seventeenth century the stylized formal gardens of the Renaissance (Fig. 17.15) were being criticized as too costly and difficult to maintain. European philosophers began to lean toward naturalism. This swing toward natural expression was reflected in gardens that were reshaped with winding paths, water channels, and copses of trees and shrubs (Fig. 17.16).

Plants and ideas brought back to Europe by explorers of the New World, Asia, and Africa had a great impact on seventeenth-century gardens. These "exotics" piqued the curiosity of intellectuals, and people of the "upper classes" considered the study of horticulture an indispensable part of a well-rounded education. The wealthy began to maintain large private gardens in which to display new plants and animals. Public gardens such as the Royal Botanic Gardens at Kew (Fig. 17.17) are legacies of this era. Many species were first formally described from live specimens growing in these gardens, and in a few cases (e.g., coffee [see Chapter 13]) seeds from plants of exotic regions grown in European gardens were used to start plantations far from their native homes. For the most part, however, exotic plants in these gardens were like rare animals in zoos.

Not only the wealthy but also the common people began to engage in gardening as egalitarianism spread to all aspects of life. Everyone with a patch of land in England seemed to have a cottage garden. With the great variety of blooms and colors available, species were chosen that would provide a succession of colors and flowers. Since land was

FIGURE 17.9

An artistic reconstruction of the fabled Hanging Gardens of Babylon. Since the Babylonians used brick rather than stone for building, earthquakes and wars have left only legends about the most famous gardens of the ancient world. The fact that the Babylonians engineered lakes, canals, and reservoirs to try to control the erratic Euphrates River lends credence to the tales of a hydraulic device which drew water from the river and pumped it to the top of a pyramidal stack of hollow arches. These arches were constructed as huge planters filled with soil and planted with trees, shrubs, and flowers to create the illusion of a verdant hillside.

(Based on a drawing by D. Lancelot in *Histoire des Jardins,* Tours, 1883.)

FIGURE 17.10

Topiary—the creative pruning of plants into shapes such as animals—is shown at its most whimsical here at Green Animals, a garden maintained by the Newport Preservation Society, Rhode Island.

at a premium, careful planning went into the choice and arrangement of the flowering plants. The rules governing the design of English perennial borders today are an outgrowth of practices developed for these small cottage gardens.

In the eighteenth century cultural and economic exchanges between Europe and Asia resulted in an influx of Oriental ideas into Western art. Pagodas and ponds began to appear in gardens, but such additions remained in large part discordant elements in gardens originally laid out in a Western style.

American gardening did not come into its own until late in the nineteenth century, when the United States emerged as a major industrial power. Until that time most gardening in this country was pragmatic or copied from the homelands of the many immigrants who flooded into the United States. With industrialization, however, came a new lifestyle. Cities grew to sizes never before imagined. Some cities met the demand for natural areas within urban environments by incorporating public parks into city plans. Many famous large city parks, such as Fairmont Park in Philadelphia

FIGURE 17.11

The Moors, with their Koranic image of paradise, resuscitated Spanish gardens during the Middle Ages. Brightly colored tiles; deep, calm pools; and neatly arranged palms, cypress, citrus trees, and oleanders as well as sweet-smelling jasmines, narcissus, and spirea characterized these old Muslim gardens. This photo shows the lion's garden at the Alhambra as it looks today.

(Photo courtesy of the Spanish National Tourist Office.)

FIGURE 17.12

The Italians built terraces so that they could use their naturally hilly terrain. The main axes of these gardens can downslope so that shady evergreen corridors could be created for the display of sculpture and to orient strollers toward particular objects. Artificial waterfalls, fancy waterworks, pergolas, and grottoes were popular garden features of the day.

(Photo taken at the Tivoli Gardens; courtesy of Mae Ogorzaly.)

FIGURE 17.13

This floral clock from the Jardin des Acclimatation in Paris is the modern version of an eighteenth- and nineteenth-century garden feature. European gardeners tried to create accurate timepieces by planting flowers that opened and closed at specific times. As a guide for this pursuit, Linnaeus published the *Horologium florae* (*Sundial of the Plants*). It was soon discovered that climatic differences affect flowering times, so botanists from areas other than Sweden began to compile their own lists of plants useful for floral clocks. In the nineteenth century sundials were created from sculpted plants clipped to cast shadows on numerals on the face of the clock. Today buried electrical or mechanical clocks are used to tell time and are bordered by colorful displays of flowers.

FIGURE 17.14

Like other sixteenth- and seventeenth-century French gardens, Versailles is best viewed from above, where the forms of the ornate flower beds can best be seen. This *fleur de lis* "parterre" is executed in clipped boxwood and flowering annuals.

(1855), Central Park in New York (1858), Prospect Park in Brooklyn (1866), Golden Gate Park in San Francisco (ca.1870), the continuous park system of Boston (1880s), Huntington Gardens in San Marino (1903), and Forest Park in Saint-Louis (1904), date from this period. Attitudes toward botanical gardens began to change as they became not only public gardens but also sources of new material for potential horticultural use. Most of the major botanical gardens and arboretums (gardens featuring trees) in the United States (e.g., the Arnold Arboretum in Boston, the Morris Aboretum in Philadelphia, and the Morton Arboretum in Lisle, Illinois) are today actively concerned with the propagation of plants that have potential as beautiful and functional ornamentals.

Twentieth-century American landscaping was drawn from various components of traditional European gardens,

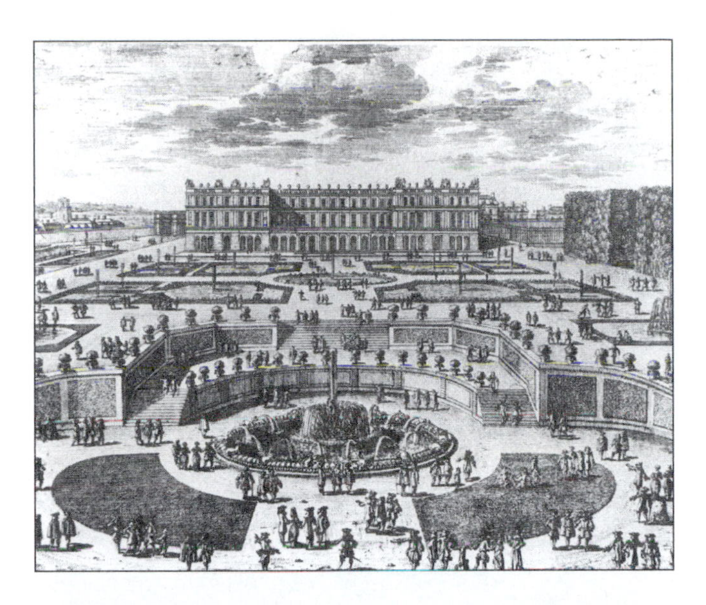

FIGURE 17.15
Gardens of the French romantic period reflect the influence of the Age of Reason. The classical symmetry of these "ordered universes" is evident in this engraving of the Versailles gardens from about 1684.

FIGURE 17.16
During the English naturalistic movement, gardeners tried to create the kind of "picture-perfect" landscapes popularized by painters of that time. For romantic effect, temples, ruins, hermitages, and even dead trees were placed within the gardens. This photo of the garden at Stourhead, Wiltshire, was provided by the British Tourist Authority.

but the modern constraints of urban life make many features of conventional European gardens inappropriate. Consequently, Americans have developed a renewed interest in Oriental landscape design. The reduced scale of Asian, primarily Japanese, gardens, their dependency upon a few well-chosen elements rather than mass plantings, and their use of evergreens make them particularly well suited to urban environments. In many cases, Europeans and Americans have incorporated elements of Oriental design, such as soft, naturalistic groupings of plants, into their gardens.

FIGURE 17.17
The conservatory at Kew Gardens.
(Photo courtesy of the British Tourist Authority.)

Oriental Gardening Philosophies

The Chinese cultivated food and medicinal plants in enclosed garden areas four centuries before their Egyptian counterparts. Although both Egyptian and Chinese gardens were precisely designed, the visual effect of Chinese gardens was different from that of Mediterranean plantings. The Eastern effect was one of naturalism, and throughout the history of Oriental gardens this emphasis, reflecting a deep reverence for nature, has remained apparent. The Chinese were the first people to create true pleasure gardens, and by 190 B.C. they were constructing extensive parks and public gardens. The Chinese considered landscaping a fine art interrelated with poetry and landscape painting and conceived of the plants they used as symbolic rather than architectural objects.

Buddhist missionaries carried their religion to Japan in the sixth century A.D. The Japanese adopted Buddhist philosophy and Chinese attitudes toward gardening but adapted them to their own climate, terrain, and indigenous flora. They also retained Shinto beliefs that all objects (water, stone, and plants) are inhabited by spirits, resulting in a unique style of gardening. After two centuries of Buddhism, religion had become so integral to gardening that the creation of gardens was entrusted only to individuals who understood its principles. In fact, the Japanese word for gardeners was *ishitatsen,* meaning "stone-setting priest," since monks handled garden design (Fig. 17.18) and setting stones was the first step in garden construction.

FIGURE 17.18
In this garden at the Imperial Palace in Kyoto, the stone beach in the foreground is symbolic of the existing world, the far shore represents the afterlife, and the curved bridge reminds visitors of the difficult path between this life and paradise.

FIGURE 17.19
The Zen garden at the Ryoanji temple in Kyoto is one of the most famous karesansui. Visitors sit on a veranda facing the garden to meditate on this composition of fifteen grouped stones to evoke thoughts of the perfect beauty of natural landscapes. Maintaining these austere gardens involves frequent meticulous raking performed by resident monks as part of their discipline.

FIGURE 17.20
In Japanese gardens one can discover a distillation of natural beauty that is part of the experience of a tea ceremony. In this private tea garden, created in the alleyway or "roji," on a tight urban lot, one has the sensation of being in a rustic, remote forest. Stepping stones are deliberately placed to guide one's path and spiritual transition. The water basin is placed low so that one stoops to be purified, and a lantern stands ready to illuminate those ablutions. In preparation for the tea ceremony, this garden has been ritually rinsed to create an aura of freshness and to reflect the dance of the light from the lanterns for evening services.

Buddhism brought with it a philosophical structure and a strict set of rules that were modified to apply to landscape design. The *Sakuteiki,* considered the world's oldest gardening manual, was written and published in the twelfth century and reflects this development. This reference provided advice for the orientation of the garden and explained how to move and place various elements within it. Like their Western counterparts, Buddhist gardens were recreations of paradise, replete with elements considered sacred to Buddhists, and were designed to enhance the strolling experience by sparking symbolic associations in the viewer's mind.

When Zen philosophy came to Japan, it had its own effect on gardening practices. In keeping with Zen asceticism, gardens were reduced to their essential form and created to foster meditation. The ultimate expression of this abstraction is the *karesansui* (Fig. 17.19), or dry garden, created with carefully placed rocks and raked gravel arranged to suggest natural landscapes and create a place for introspection.

From Zen also came the notion that everything that one does is a reflection of one's spiritual enlightenment, even the simple act of preparing a cup of tea. During the fifteenth century, Zen monks developed the tea ceremony, which was eventually moved into its own shelter within a specially designed garden (Fig. 17.20). The teahouse or garden was

FIGURE 17.21
Hand washing is a ritual act of purification that helps one prepare spiritually before the tea ceremony.
(Photo courtesy of the Japan National Tourist Organization.)

FIGURE 17.22
Water basins are provided so that visitors can rinse their hands before entering the sacred space within the garden at Heian-Jingu shrine, Kyoto.

built in the image of a rustic mountain retreat where one could contemplate quiet beauty. As participants pass through the garden, their path, pace, and views are guided by precisely placed stones. This guided movement is designed to be part of the spiritual preparation for the tea ceremony. A water basin to rinse one's hands and mouth (Fig. 17.21), stone lanterns, and a low portal so that the one has to stoop to enter the teahouse, bring to mind the Buddhist virtues of purification, enlightenment, and humility.

The Japanese prefer seasonal change over the displays of annuals so popular in Western gardens. The cherry blossoms in the spring, the warm hues of autumn, and even the sight of snow perched in pine boughs are highly prized. Except for these seasonal events and the *ikebana* arrangement placed in the teahouse, Japanese gardens tend to be composed of shades of green and brown, with evergreens pruned and trained into stylized shapes surrounded by carpets of moss.

The rustic nature of the tea gardens and the visual treats of the earlier paradise gardens were combined with the concept of movement through the garden to create the stroll gardens that are emulated today in city and country gardens of Japan. The features of these gardens (water basins for purification (Fig. 17.22), gates to symbolize the passage from the mundane world into the richly symbolic world of the garden, lanterns to light the path, streams that flow from the north, stepping stones to control one's path and pacing, "picture windows" where shaped trees or structures frame a view, and borrowed scenery to increase the garden's perspective) show the synthesis of centuries of Oriental garden art (Fig. 17.23). The "Japanese" gardens most frequently constructed as parts of American gardens have been based on stroll gardens designed by Oriental masters (Fig. 17.24). Americans ambling through one of these gardens are usually oblivious to the care that has gone into the placement of each item or to the symbolism of each element. However, the feeling of quiet beauty and tranquility is impossible to ignore. It is this feeling of a natural harmony with plants that has influenced American gardens and softened to some extent the linear architectural styles inherited from European cultures.

Uses and Types of Nursery Plants

The use of nursery plants today thus derives from a variety of aesthetic ideas and styles drawn from a wide array of cultures, but nursery plants often serve functional purposes as well. They enclose areas, provide privacy or form partitions, cover the ground, prevent erosion, and furnish shade. Recent studies have shown that landscape plants can substantially modify the climate of a localized area and consequently reduce energy costs. On a sunny day lawns have been shown to be 5° to 8°C (41° to 46°F) cooler than bare soil and 14° to 17°C (57° to 63°F) cooler than asphalt (Fig. 17.25). Turf also absorbs sunlight and reduces glare. Although the use of plants as windbreaks has long been appreciated, the effect of trees and shrubs as insulating factors for buildings has been emphasized only recently (Fig. 17.26). Proper use of shrubs or vines that provide shade in the summer and lose their leaves in cold winter months can save between 10 and 40 percent in heating and cooling costs (Fig. 17.27). Designs for solar heating of homes often include glassed areas for plants, which aid in the collection and circulation of heat from the sun.

For home use, ornamentals are usually chosen for the area to be landscaped. Often a yard is divided into several areas, including a public access area, which is for most homes the front yard. Plantings in such an area tend to be

FIGURE 17.23
This pattern of the path and bridge within the garden symbolizes the indirect path most of our lives take. In mythology, by crossing over this zigzag path a person can leave the devil (who can only travel in a straight line) behind and achieve spiritual growth.

FIGURE 17.24
The gardens at Ginkakuji, or the Silver Pavilion, in Kyoto demonstrate a combination of gardening styles, with the elements of the aristocratic hill and pond garden, and the raked gravel of Zen gardens, designed for leisurely, contemplative strolls that represent life's journey.

rather formal and showy. A second part of the yard includes the driveway and work areas, which are usually minimally planted or planted to screen trash or storage areas. Finally, the family living area, usually the backyard, is designed to reflect a family's lifestyle. In some cases the presence of a patio and swimming pool leaves little room for gardens. In other cases a family may decide to plant a series of flower beds that supply cut flowers throughout the growing season or allow as large an area as possible to serve as lawn for playing. All these decisions about land use require choosing plants with different characteristics.

For urban use, the primary emphasis is often on the ease of maintenance and durability. An abundantly flowering or fruiting tree may be rejected for city planting even though it is beautiful if it causes a serious cleanup problem with fallen branches, flowers, or fruits. Examples of formerly popular street trees that are no longer used include fruiting mulberry trees, which drop masses of blue-black fruits on sidewalks, streets, and cars, and silver maples, whose branches break in

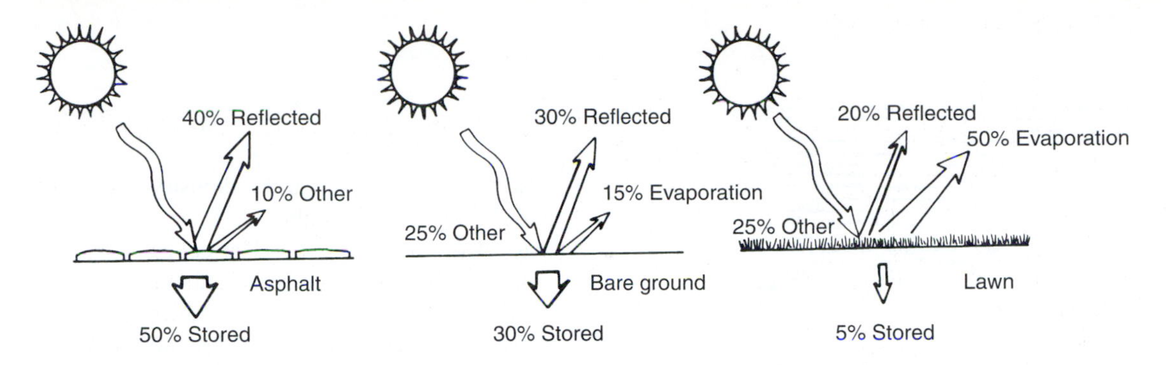

FIGURE 17.25
This diagram shows how heat is reflected from different surfaces. The grass-covered surface cools itself by evaporation, whereas asphalt and bare ground store heat and reradiate it to the air above. Heat loss due to convection and conduction accounts for the percentage labeled "other."

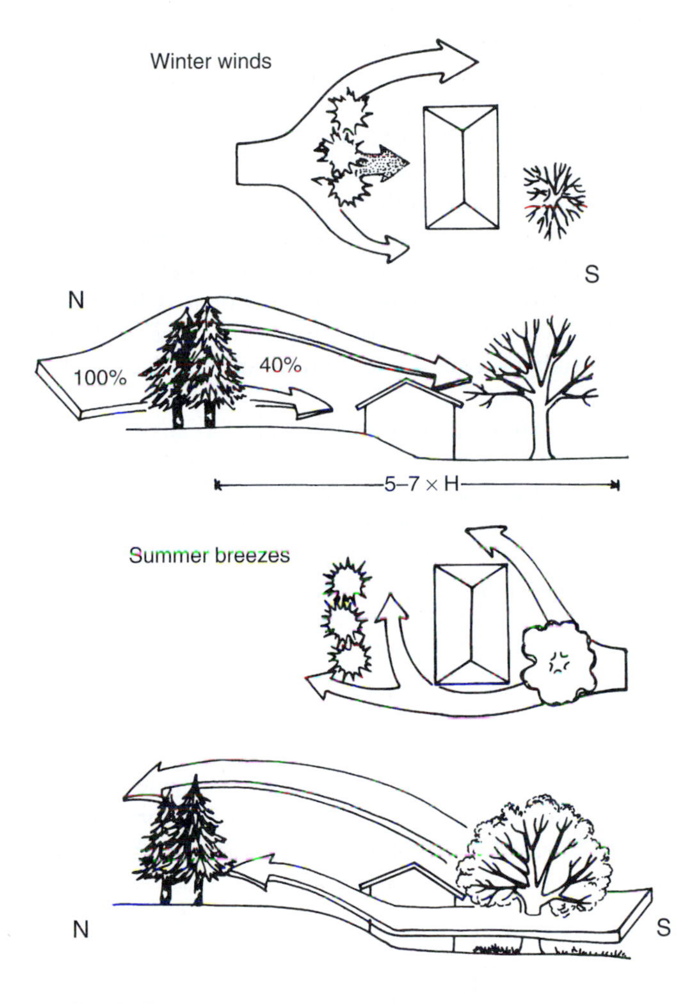

FIGURE 17.26
Windbreaks can be arranged to deflect cold winter winds (upper two figures) while allowing summer breezes (lower two figures) to cool a house. Both effects can be achieved because prevailing wind directions often differ between summer and winter. The height, density, and shape of the planting affect how much wind reduction is achieved. In general, a moderately dense (60 percent coverage) mass of evergreens performs better than does an impenetrable barrier which tends to form vacuums on its leeward side. As a rule of thumb, the greatest wind reduction occurs immediately in front of the barrier and beyond it for a distance of five to seven times the height of the barrier (H). Windbreaks have been shown to affect wind speeds for a distance of up to 30 times their height.

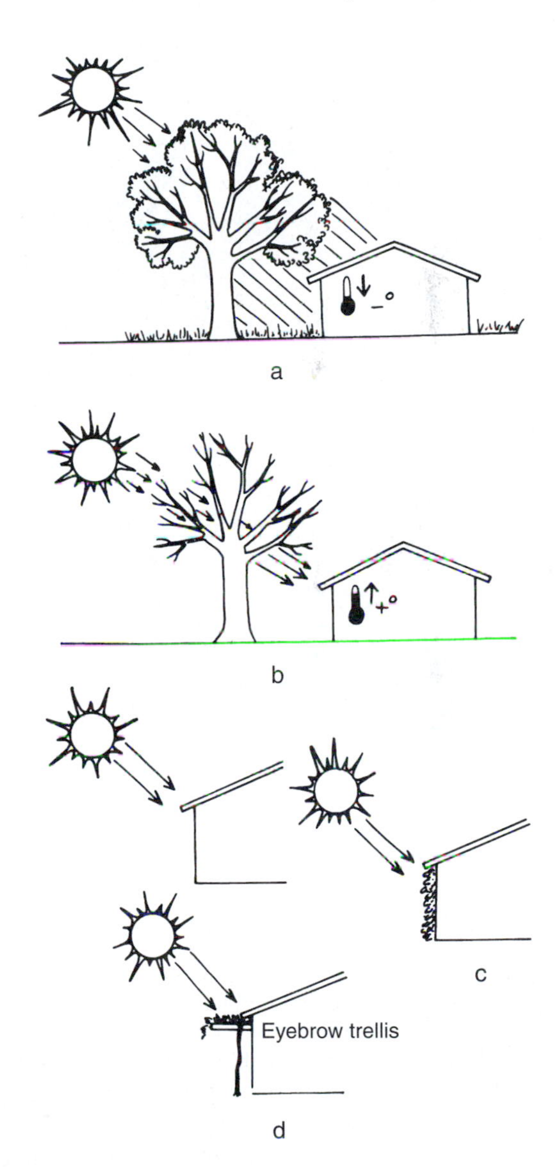

FIGURE 17.27
Judicious landscaping can modify the temperature around a home. (a) Deciduous trees on the south and west sides provide shade and reduce temperatures as much as 5° to 5.5°C (41°–42°F) inside. (b) When these trees lose their leaves in winter, the sun's rays are able to reach the house and provide heat. (c) Walls may be shaded by deciduous vines growing against a wall or trellis. (d) An eyebrow trellis keeps out the sun when it is high in the sky (summer) but allows the winter sun in to heat the house.

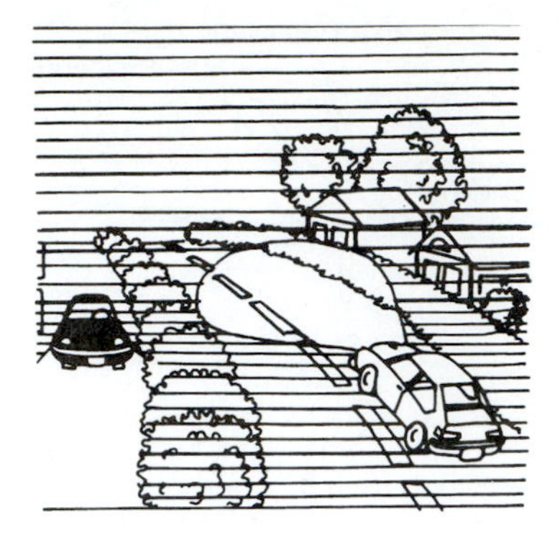

FIGURE 17.28
Highway plantings can help protect oncoming cars and nearby residents from automobile lights and traffic noise (top), keep drivers from being blinded by the sun's glare (middle) or lights from oncoming cars, and control erosion (bottom).

windstorms. In cities with high pollution levels from automobiles or industry and cities that use salt on icy roads, plants have to be chosen that can tolerate high levels of acid rain, carbon monoxide, smog (hydrocarbons plus nitrous oxide), and salt-impregnated soil. Plantings in public areas must also withstand the relatively high levels of physical abuse suffered when they are bumped, broken, carved upon, or visited by dogs. Highway plantings (Fig. 17.28) must require a minimum of maintenance and must also deter erosion on embankments.

With these ideas about the concepts of gardening and the functional necessities of ornamental plants in mind, let us look at the major kinds of nursery plants used: turf, ground covers, bedding plants, and deciduous and evergreen trees and shrubs.

Turf

The turf, or nonforage grass, industry in the United States is based on a limited number of species but constitutes a major part of the nursery trade. The concept of lawns may have arisen with herding nomads, although some people attributed the modern love of grass to the hypothesis that humans arose in the grasslands of eastern Africa. Mowed, aesthetic lawns as we now think of them originated in the late eighteenth century when André de Notre designed small lawn areas for the Palace of Versailles outside of Paris. The elite of England readily adopted the concept because of the ease and luxuriousness with which turf grasses grow in the mild moist British climate.

The turfgrass industry as it is known today had its inception with the invention of the lawn mower in 1830. This now-ubiquitous tool sprang from a very unlikely source. An enterprising inventor who had devised a machine to trim the threads of velvet to a uniform height saw the implications of an outdoor model that could clip blades of grass, and the idea of the lawn mower was born. Up to that time only the wealthy could afford the extensive hand labor involved in neatly trimming expanses of grass. With this tool, a neatly clipped carpet of green was possible for even small homeowners.

By breeding adaptable native and introduced grasses Americans have been able to create a reasonable facsimile of the English lawn. Unfortunately, in the absence of England's climate, maintenance of lawns in suburban yards and on golf courses is a costly endeavor. In 1996, North America had more than 32 million acres of lawn, more acreage than planted in any single food crop. Americans annually spend over $750 million on grass seed and over $45 billion on maintenance costs.

Nevertheless, the mania for lawns is understandable in terms of what they provide. They are easy to walk on, cooling because of the water they transpire, suitable for large open spaces, and effective in preventing the formation of dust and mud. Table 17.1 lists the major grasses used for turfs in the United States today.

Ground Covers

Ground covers—prostrate, dense-growing plants other than grasses—are often used in place of grass in areas where low maintenance is required, there is too much shade for grass to grow well, or slopes are too steep to mow. Ground covers are also used because they provide a texture different from that of grass and, in a few cases, a color contrast. All such covers, however, share the feature of low, creeping growth that provides a dense carpet over the surface of the ground. Unlike grass, ground covers are not suited to areas over which people walk because their stems are easily broken and their leaves are easily crushed. In addition, many of them provide an irregular cover, making walking difficult.

TABLE 17.1 Turf (Grass, Poaceae) Species Commonly Used in the United States

LOCATION, COMMON NAME	SCIENTIFIC NAME	NATIVE REGION
Northern states		
Bent grass	*Agrostis* spp.	Northern Hemisphere
Fine or red fescue	*Festuca rubra*	North America
Kentucky bluegrass	*Poa pratensis*	Eurasia
Perennial ryegrass	*Lolium perenne*	Europe
Wheatgrass	*Agropyron* spp.	Temperate regions
Southern states		
Bahia grass	*Paspalum notatum*	Mexico, South America
Bermuda grass	*Cynodon dactylon*	Old World
Carpet grass	*Axonopus* spp.	Americas
Centipede grass	*Eremochloa ophiuroides*	Southeast Asia
St. Augustine	*Stenotaphrum secundatum*	Tropical Americas
Zoysia	*Zoysia* spp.	Tropical Asia

TABLE 17.2 Ground Covers Commonly Used in the United States

COMMON NAME	SCIENTIFIC NAME	FAMILY	NATIVE REGION
Ice plant	*Carpobrotus edulis*	Aizoaceae	Southwestern Africa
Ivy	*Hedera helix*	Araliaceae	Europe, Africa
Japanese honeysuckle	*Lonicera japonica*	Caprifoliaceae	East Asia
Pachysandra	*Pachysandra* spp.	Buxaceae	Japan, North America
Periwinkle	*Catharanthus roseus*	Apocynaceae	Old World
Winter creeper	*Euonymus fortunei*	Celastraceae	China

FIGURE 17.29
The geometric keystone pattern of this parterre is a modern version of an ancient Egyptian design.

They are, however, well suited for planting along highway embankments and on other types of slopes. Several of the most common ground covers used in the United States are listed in Table 17.2.

Bedding Plants

Bedding plants are relatively small, usually annual plants grown in gardens, window boxes, or hanging baskets to provide displays of floral and foliage color. They are also used for carpet bedding, a special form of planting that produces designs (Fig. 17.29). Despite their small size, they constitute an important item in the nursery trade, in part because people tend to buy large quantities of them each year. As in the case of ornamental plants, relatively few species dominate the trade. One annual, *Petunia* ×*hybrida* (Solanaceae), accounts for 50 percent of commercial bedding plants (Fig. 17.30). Table 17.3 lists the common species used for bedding.

Trees and Shrubs

Trees and shrubs constitute the important structural features of ornamental plantings. Although both can be pruned to desired shapes, trees are usually allowed to assume their

TABLE 17.3 Some Bedding Plants Commonly Used in the United States

COMMON NAME	SCIENTIFIC NAME	FAMILY	NATIVE REGION
		Annuals	
Impatiens	*Impatiens balsamina*	Balsaminaceae	Asia
Marigolds	*Tagetes erecta*	Asteraceae	Central America
Pansy	*Viola ×wittrockiana*	Violaceae	Europe
Petunia	*Petunia ×hybrida*	Solanaceae	Europe
Phlox	*Phlox drummondii*	Polemoniaceae	North America
		Perennials	
Ageratum	*Ageratum* spp.	Asteraceae	New World tropics
Candytuft	*Iberis* spp.	Brassicaceae	Europe
Coleus	*Coleus ×hybrida*	Lamiaceae	Old World tropics
Columbine	*Aquilegia* spp.	Ranunculaceae	North temperate regions
Geranium	*Pelargonium* spp.	Geraniaceae	South Africa
Heliotrope	*Heliotropium* spp.	Boraginaceae	Tropics
Larkspur	*Delphinium* spp.	Ranunculaceae	North temperate regions
Lobelia	*Lobelia cardinalis*	Lobeliaceae	North temperate regions
Primrose	*Primula vulgaris*	Primulaceae	Europe
Sweet alyssum	*Alyssum* spp.	Brassicaceae	Eurasia
		Bulbs	
Crocus	*Crocus* spp.	Liliaceae	Mediterranean
Daffodil	*Narcissus* spp.	Amaryllidaceae	Europe, North Africa
Hyacinth	*Hyacinthus orientalis*	Liliaceae	Mediterranean
Iris	*Iris* spp.	Iridaceae	North temperate regions
Tulip	*Tulipa* spp.	Liliaceae	Asia

FIGURE 17.30
Despite the ever-increasing number and variety of plants available, petunias remain the most popular bedding plant sold.

normal growth form and shrubs are often trimmed. Trees and shrubs can be chosen for their attractive flowers, fruits, or foliage, but large trees are primarily used for shade. Because of their slow growth and the fact that they are less frequently grown for flowers than are shrubs, native species of trees are more widely used (or allowed to remain) than are native species of shrubs. Ornamental shrubs (Table 17.4) are usually exotics chosen from the variety of hardy types offered by local nurseries. Consequently, as in the case of cut flowers, bedding plants, and food crops, the same ornamental species are used repeatedly throughout large regions of the United States. The exact group of shrubs differs somewhat in different hardiness zones (Fig. 17.31) across the country, but compared with the number of shrub species available, only a handful are widely planted.

In contrast to ornamental shrubs, the species of trees planted or retained as parts of public or private landscape designs are so numerous that it is impractical to assemble a list. Several of the references in the Additional Readings provide descriptions of common tree species used in the various climatic regions of the United States. In addition, county extension agents supported by federal state, and local governments provide free advice and literature on trees, shrubs, and vegetable crops that are adapted for local conditions.

Florist Crops

Whereas the production of ornamental plants on a commercial basis is a product of urbanization, the cutting of flowers and foliage for personal and ceremonial use dates to prehistoric times. Excavations of Paleolithic burial sites have

TABLE 17.4 Some Ornamental Shrubs Commonly Used in the United States

COMMON NAME	SCIENTIFIC NAME	FAMILY	NATIVE REGION
Azalea	*Rhododendron* spp.	Ericaceae	Primarily Asia, also North America
Barberry	*Berberis verruculosa*	Berberidaceae	Old World
Bayberry	*Myrica pensylvanica*	Myricaceae	Eastern North America
Boxwood	*Buxus sempervirens*	Buxaceae	Old World
Bugle weed	*Ajuga reptans*	Lamiaceae	Europe
Butterfly bush	*Buddleja davidii*	Buddlejaceae	China
Cotoneaster	*Cotoneaster dammeri* *C. horizontalis*	Rosaceae	China
Deutzia	*Deutzia scabra*	Saxifragaceae	Japan
Fire thorn	*Pyracantha coccinea*	Rosaceae	Eurasia
Forsythia	*Forsythia ×intermedia*	Oleaceae	Old World
Holly	*Ilex* spp.	Aquifoliaceae	Old and New Worlds
Honeysuckle	*Lonicera* spp.	Caprifoliaceae	Eastern Asia
Hydrangea	*Hydrangea* spp.	Hydrangeaceae	North and South America
Juniper	*Juniperus* spp.	Cupressaceae	North temperate regions
Lilac	*Syringa* spp.	Oleaceae	Eurasia
Mock orange	*Philadelphus ×virginalis*	Saxifragaceae	North temperate regions
Ocotillo	*Fouquieria splendens*	Fouquieriaceae	New World
Oleander	*Nerium oleader*	Apocynaceae	Mediterranean
Oregon grape	*Mahonia aquifolium*	Berberidaceae	North America
Pieris	*Pieris japonica*	Ericaceae	Japan
Prickly pear	*Opuntia* spp.	Cactaceae	New World
Privet	*Ligustrum* spp.	Oleaceae	Old and New Worlds
Rhododendron	*Rhododendron* spp.	Ericaceae	Primarily Asia
Rose of Sharon	*Hibiscus syriacus*	Malvaceae	Eastern Asia
Viburnum	*Viburnum* spp.	Caprifoliaceae	Usually North American species

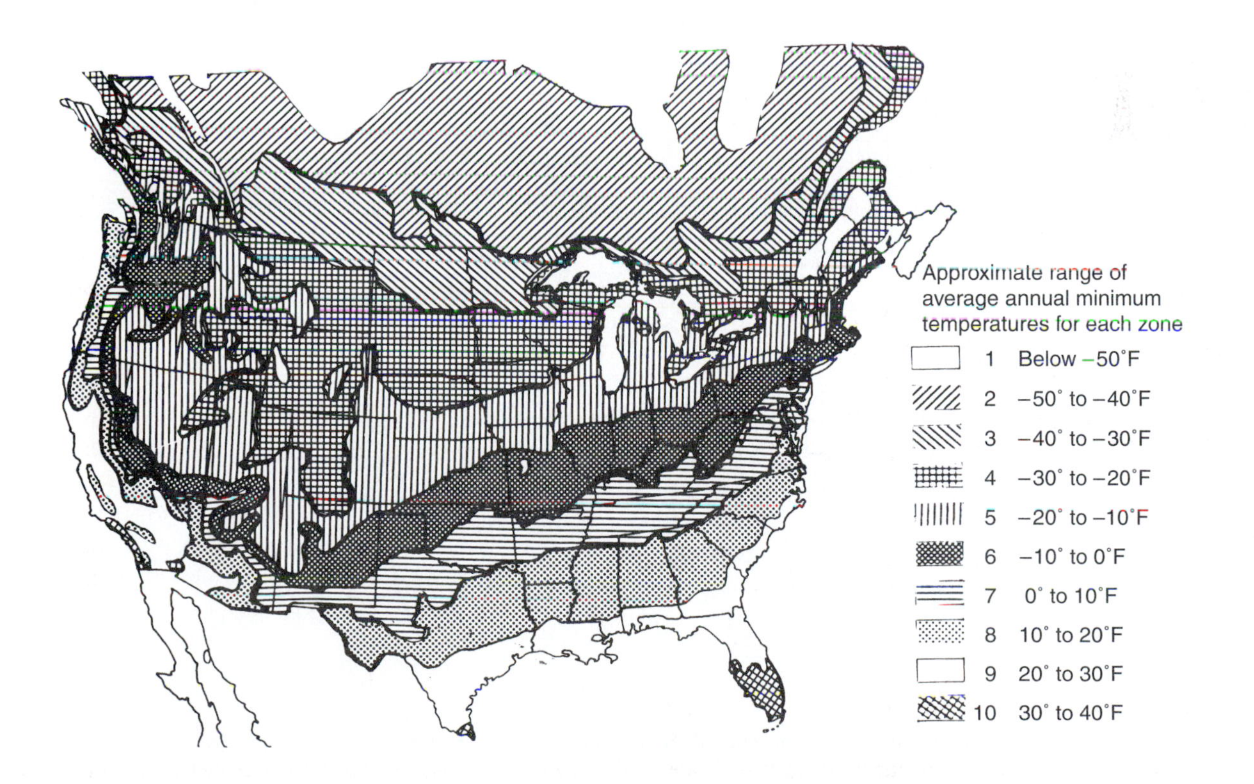

FIGURE 17.31

The most commonly used system for plant hardiness zones (numbers 1–10) as circumscribed by the U.S. Department of Agriculture helps gardeners and farmers determine which species and varieties to plant.

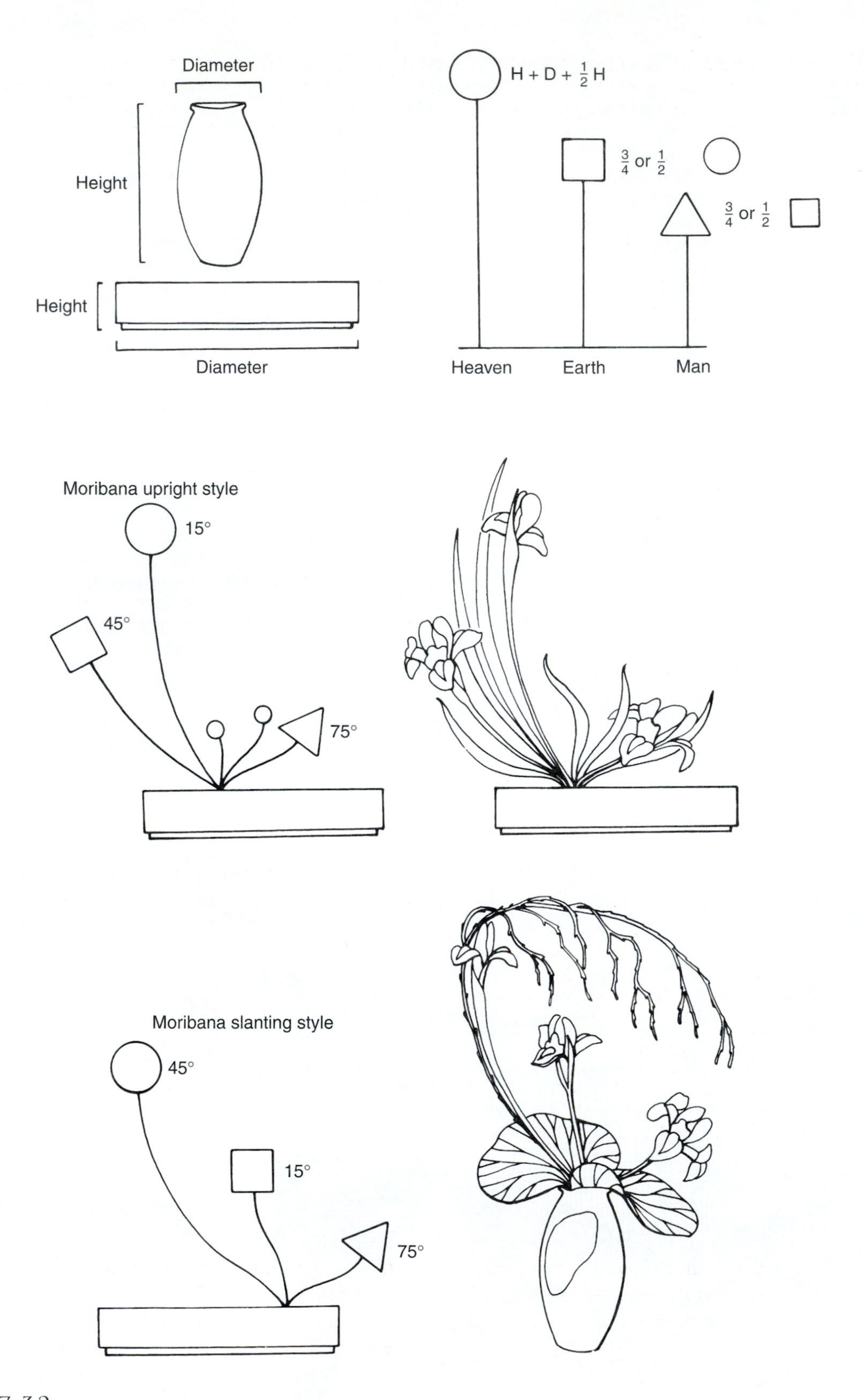

FIGURE 17.32

In Japanese flower arranging the three main elements symbolizing heaven, earth, and man are arranged at precise angles to one another and are cut to lengths of specified proportions. Shown are two variations of the *Moribana* style. (H = height; D = diameter.)

TABLE 17.5 Some Common Flowers and the Sentiments or Expressions They Represent

COMMON NAME	SENTIMENT/EXPRESSION	SCIENTIFIC NAME	FAMILY
Almond	Indiscretion	*Prunus amygdalus*	Rosaceae
Aloe	Sorrow, bitterness	*Aloe barbadensis*	Asphodelaceae
Apple	Temptation	*Malus ×domestica*	Rosaceae
Azalea	Temperance	*Rhododendron* spp.	Ericaceae
Bachelor's button	Hope in love	*Centaurea cyanus*	Asteraceae
Basil	Hatred	*Ocimum basilicum*	Lamiaceae
Belladonna	Imagination	*Atropa belladonna*	Solanaceae
Blue violet	Modesty	*Viola odorata*	Violaceae
Cabbage	Profit	*Brassica oleracea*	Brassicaceae
Carnation	Pride and beauty	*Dianthus caryophyllus*	Caryophyllaceae
Cashew	Perfume	*Anacardium occidentale*	Anacardiaceae
Chrysanthemum	Cheerfulness	*Chrysanthemum* spp.	Asteraceae
Cowslip	Pensiveness	*Caltha palustris*	Ranunculaceae
Daffodil	Deceitful hope	*Narcissus* spp.	Amaryllidaceae
Dandelion	Oracle	*Taraxacum officinale*	Asteraceae
Evening-primrose	Inconstancy	*Oenothera biennis*	Onagraceae
Flax	Domestic industry	*Linum usitatissimum*	Linaceae
Hellebore	Female inconstancy	*Veratrum viride*	Liliaceae
Hollyhock	Fruitfulness	*Althea rosea*	Malvaceae
Hop	Injustice	*Humulus lupulus*	Cannabaceae
Hoya	Sculpture	*Hoya carnosa*	Asclepiadaceae
Hyacinth	Play	*Hyacinthus orientalis*	Liliaceae
Iris	Message	*Iris* spp.	Iridaceae
Jasmine	Availability	*Jasminum grandiflorum*	Oleaceae
Lavender	Distrust	*Lavandula angustifolia*	Lamiaceae
Lily of the valley	Return of happiness	*Convallaria majalis*	Liliaceae
Madder	Calumny	*Rubia tinctoria*	Rubiaceae
Marigold	Mental anguish	*Targetes erecta*	Asteraceae
Mountain-laurel	Ambition	*Kalmia latifolia*	Ericaceae
Nightshade	Falsehood	*Solanum dulcamara*	Solanaceae
Orange blossom	Purity equal to loveliness	*Citrus sinensis*	Rutaceae
Pansy	Think of me	*Viola ×wittriockiana*	Violaceae
Periwinkle	Sweet remembrance	*Catharanthus roseus*	Apocynaceae
Poppy	Consolation of sleep	*Papaver somniferum*	Papaveraceae
Rhododendron	Danger—beware	*Rhododendon* spp.	Ericaceae
Rose	Love	*Rosa odorata*	Rosaceae
Saffron	Mirth	*Crocus sativus*	Iridaceae
Snapdragon	Presumption	*Antirrhinum majus*	Scropulariaceae
Sunflower	False riches	*Helianthus annuus*	Asteraceae
Sweet William	Gallantry	*Dianthus barbatus*	Caryophyllaceae
Tulip-tree	Fame	*Liriodendron tulipifera*	Magnoliaceae

Source: Adapted from L. Gordon. 1977. *Green Magic*. New York, Viking. During the Victorian era entire message were written using flowers. The reader had to decipher the meaning by using the sentiments represented by the flowers. Gordon describes many other flowers and their sentiments as well as examples of some messages in his book. We include here only flowers common in the United States, in particular those mentioned in this text. The scientific names are those that appear to belong to the common names given by Gordon.

shown that sprigs of flowers placed around bodies were important in interment rites. Since those very early times, differing cultures have adopted diverse species as favorite ornamental plants and have developed personal styles in the use of cut flowers. In Japan, flower arranging developed into an art form that is an integral part of many aspects of the culture, such as the tea ceremony (Fig. 17.32). In Victorian England, flowers were given symbolic connotations and were used to send messages (Table 17.5).

Traditionally, cut flowers were available only in the summer or at very high cost from local greenhouses at other times of the year. Today, because of the efficiency of modern transportation, people are no longer limited to locally produced flowers (Fig. 17.33).

The cut flower market is increasing globally and is growing at the rate of 6 to 9 percent per year. In 1985, world consumption of cut flowers was US $12.5 billion; in 1998 it was US $35 billion. About 75 percent of the international importation of cut flowers and potted plants is within Europe, with Germany alone accounting for 30 percent of the world's imports. The Netherlands leads the world in exports, accounting for about 65 percent of the global trade,

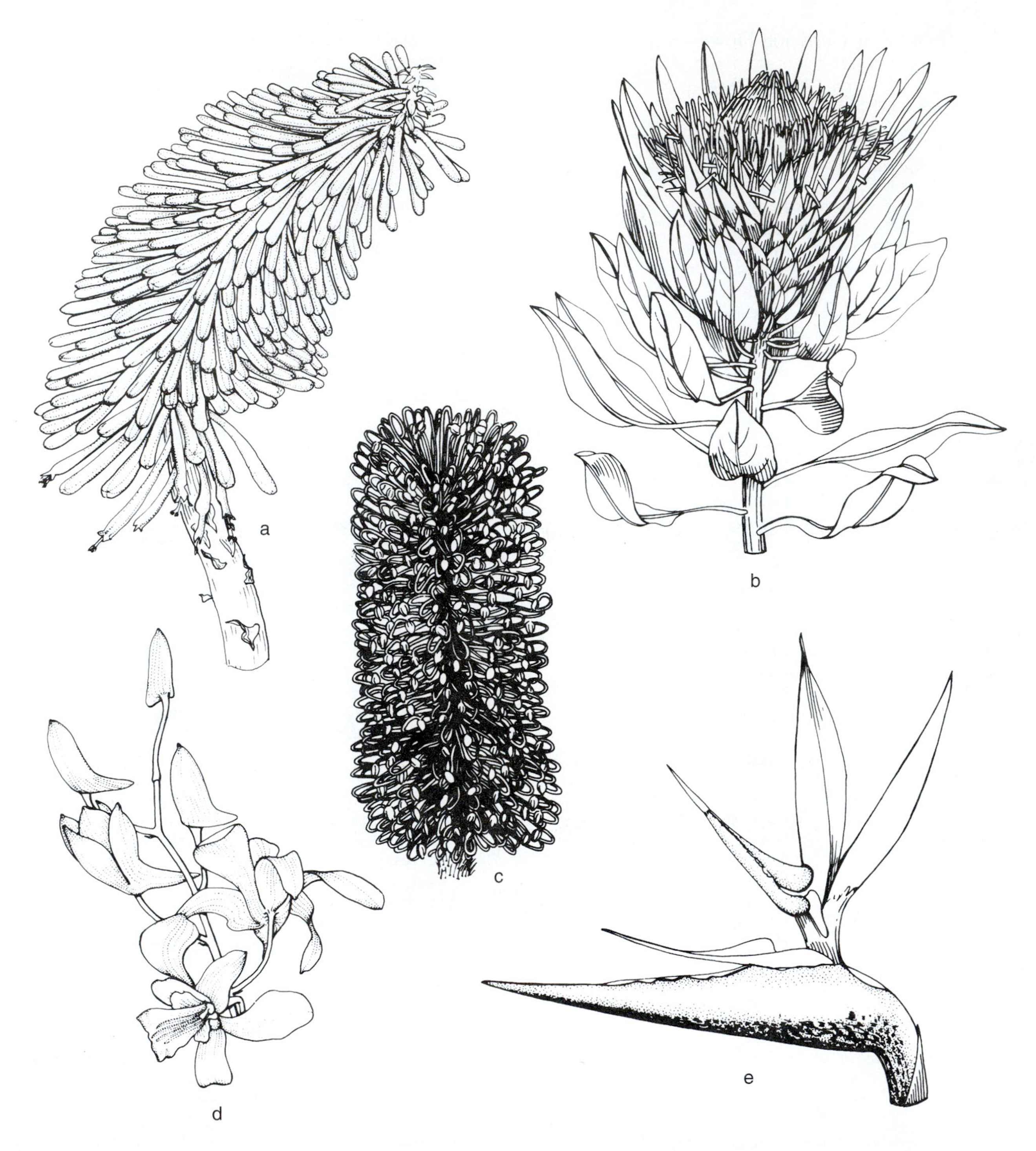

FIGURE 17.33
Some of the exotic blooms used in the cut flower industry include (a) *Kniphofia* (Asphodelaceae), (b) *Protea cynaroides* (Proteaceae), (c) *Banksia* spp. (Proteaceae), (d) *Dendrobium* spp. (Orchidaceae), and (e) *Strelitzia reginae* (Musaceae). (b and c after Baillon, a, d, and e from nature.)

but it is facing increased competition from Africa (particularly Kenya), Asia (Thailand and Malaysia), and Latin America (Mexico and Colombia). North America supplies about 90 percent of its own domestic market, with California the major supplier. The main flowers produced for the cut flower industry are chrysanthemums (Fig. 17.34), roses, carnations (Fig. 17.35), and lilies, but exotic flowers (Fig. 17.36) and plants are increasingly available.

Chrysanthemums, *Chrysanthemum* spp. (Asteraceae), have been prized in the Orient as ornamentals and symbols

FIGURE 17.34
Chrysanthemums, native to the Old World, have been modified from the original form most similar to that on the center right to produce a whole array of showy flowers.

of long life since at least 500 B.C. Over the period of their use chrysanthemums have changed greatly from the original forms, which had few ray flowers and a central button of numerous disk flowers. Most of the modifications have involved the selection of forms with an increased number of ray flowers (usually sterile) and fewer or no disk flowers (Fig. 17.34). The petaloid part of the ray flowers has also been greatly modified under the action of selection, yielding large, showy heads. Within the modern horticultural cultivars, the major groups are singles (which retain the button, daisy look), semidoubles, "anemones," pompoms, commercial decoratives, spiders, and 'Shasta' daisies. The genus is native to the Old World.

To many people the rose, *Rosa* spp. (Rosaceae), is the queen of cut flowers (Box 17.1). Few flowers other than the rose can boast of entire gardens devoted to their culture. Roses account for over $500 million of the world's cut flower market each year. During the history of selection for ornamental roses, there has been a pattern parallel to that found with chrysanthemums. In this case, however, enhancement

FIGURE 17.35
(a) Chrysanthemums, (b) roses, (c) carnations, (d) orchids (*Cattleya*), (e) snapdragons, and (f) gladiolus are among the most popular flowers in the floral industry today.

of the flower was due not to the petal-like conversion of individual flowers within a head but to the modification of stamens into petals in an individual flower. The wild ancestors of cultivated roses have a single whorl of five petals, numerous stamens, and many separate pistils, as can be seen in native North American species (Fig. 17.37). Mutants in which the stamens became petaloid were chosen and perpetuated until, in many cases, the blossoms consisted of an almost infinite series of petals and no stamens.

Roses today are placed into three main classes, depending on their blooming behavior. Plants that bloom once a year or season belong to the "old rose" group. Many

FIGURE 17.36
Orchids of many species are now found in florist's shops. (a) *Cymbidium,* (b) *Vanda,* (c) *Phalaenopsis,* (d) *Paphiopedilum,* (e) *Oncidium.*

climbing roses and the damask roses are in this class. The second major class, the "perpetual roses," includes species and varieties that bloom twice a year, usually in the summer and fall. Hybrid perpetuals, cultivars that resulted from a complicated series of crosses in the early part of the nineteenth century, are members of this class, as is the original American Beauty developed in 1875. The last group is the "overblooming hybrids," which, as the name implies, flower almost continuously throughout the growing season. This includes the polyanthas, the floribundas, and the hybrid tea roses. The classic tea roses are descended from tea-scented Chinese forms that reached Europe and North America in about 1810.

Carnations, also called "pinks" because the edges of the petals look as if they have been cut by pinking shears, rival roses as the world's most loved flower. Included within the carnation genus *Dianthus* (Caryophyllaceae) are the grass-leafed, tall, clove-scented florist flowers and the humble Sweet William used as bedding plants (Fig. 17.38). Greenhouse cultivars are all derived from *D. caryophyllus,* a

Box 17.1

A Rose is a Many-Splendored Thing

In 1986 a century of controversy ended when *the* rose was chosen as the national flower of the United States. Four states (Georgia, Iowa, New York, and North Dakota) and the District of Columbia had previously claimed various roses as their representative flowers. Iowa and North Dakota chose wild roses, but Georgia, New York, and Washington, D.C., named varieties descended from Asian ancestors. The rose of the national floral emblem does not, unfortunately, have any particular designation to indicate that it is a native species.

Fossils that have been identified as roses and dated to be over 60 million years old indicate the age of the rose group. Botanists believe that the genus *Rosa,* which today contains about 250 species, originated in Asia and subsequently spread across the temperate areas of Europe and North America. Different species and selected rose cultivars have been prized for centuries and have been linked with human history for over 5000 years. Perhaps it is their variety of flower colors and sizes, delicate fragrances, and ability to last several days once cut that have inspired poets, clergymen, physicians, scientists, and suitors. The ancient Greeks credited Aphrodite, the goddess of love, with the origination of roses.

In Persia, where roses were grown for their fragrant oil, sultans slept on mattresses stuffed with rose petals. Cleopatra banked on their romantic powers when she received Marc Antony in a room filled knee-deep with rose blossoms. The Egyptians reportedly maintained a cut flower industry, growing and exporting roses to Rome, where the flower achieved an unprecedented popularity. Roses later came to represent the decadence of the Roman empire as farmers grew them instead of grain. Nero arranged to have his dinner guests showered by rose petals released from bags suspended from the ceiling. The rationale was that roses protect one from drunkenness. The rose also came to symbolize secrecy. A rose hung in a meeting chamber indicated that the conversation was meant to be kept private, or "*sub rosa.*"

Christians initially shunned roses because of their association with pagan Rome, but eventually they too incorporated red roses into their symbolism. They considered each of the five petals to represent one of Christ's wounds, and the red to symbolize his blood. The shape and form of rose windows in many famous cathedrals in Europe also reveal the influence of these flowers. During the struggle between the houses of Lancaster and York for rule of England in the fifteenth century, each of the feuding sides chose a rose for its emblem (red for Lancaster, white for York). When the War of the Roses, as it has since been called, finally ended, England chose as its national flower a variegated red and white rose that combined the symbols of the two warring houses.

Perhaps nowhere have science and art been more entwined than in the breeding of roses. The surge in Western breeding began toward the end of the eighteenth century with the arrival of four roses (the China "studs") in Europe. Since then, more than 25,000 cultivars have been developed by crossing Asiatic, European, and New World species. Despite all this diversity, breeders have struggled for centuries to find or reproduce the legendary blue rose first mentioned in a thirteenth-century Arabian text. Flowers of other families can be a rich blue color, and species of roses may once have had a form of the pigment gene that produced a blue shade. However, if it was present, it seems to have been lost or is no longer expressed. The advent of genetic engineering has rekindled interest in the blue rose. Once biochemists learned that the same precursor compounds can be converted to cyanadin, which produces red, orange, and pink flowers, or delphinidin, the blue pigment of delphiniums and violets, the stage was set for isolating the enzyme that shunts the pathway toward delphinidin. Calgene, an American biotech company, isolated the gene for delphinidin from a blue petunia. After the gene is incorporated into a rose cell, plants must be grown using tissue culture and must be tested for hardiness and blooming ability. Despite these efforts, it remains to be seen if there is a market for blue roses. True blue is so foreign to the concept of a rose that blue blossoms might seem artificial or dyed.

The rose probably will retain its position as the queen of flowers with yearly cutthroat competition between different cultivars for the coveted designation "Rose of the Year" by the American Rose Society. Red, yellow, white, or even blue, the rose will continue to be the ultimate extravagance and the most emphatic declaration of admiration that a gentleman or lady can tender his or her beloved.

FIGURE 17.37

Through the selection of mutants with petaloid stamens, the tea rose with its many whorls of petals was developed from a wild rose, similar to this one, which had one whorl of petals and many separate stamens and pistils.

FIGURE 17.38
Sweet William is in the same genus as the spicy-scented florist's carnation.

Eurasian species. Sweet William has been selected from *D. barbatus,* a native of western Europe. A comparison of the two again shows Americans' propensity for showiness in cut flowers. The florist's crop has exaggerated flowers with scores of petals, whereas the bedding plant resembles the wild state with five free petals.

The term *lily* is used for a variety of species with lily-like flowers. These include day lilies (*Hemerocallis, Liliaceae*), alstroemerias (*Alstromeria, Alstroemeriaceae*), calla lilies (*Calla, Araceae*), peace lilies (*Spathiphyllum, Aracaeae*), and freesias (*Freesia, Iridaceae*). However, true lilies belong to the genus *Lilium,* which contains about 80 species native to the north temperate regions of the world but with most cultivated species originally from Eurasia. They are stately plants with a variety of showy flowers. Among the most common are the Madonna lily (*L. candidum*), Mountain lily (*L. auratum*), scarlet Turk's cap (*L. chalcedonicum*), Easter lily (*L. longiflorum*), and showy Japanese lily (*L. speciosum*).

Lily flowers have been associated with nearly every major female religious figure in the Western world, including the Roman goddess Juno, the Greek goddess Hera, and the Middle Eastern goddess Astarte. Likewise, in the Judeo-Christian world lilies have held a place of honor. One legend says that when Eve left the Garden of Eden she shed real tears of repentance and that those remorseful tears gave rise to lilies. The spiritual principle suggested here is that true repentance is the beginning of beauty. A mark of purity and grace throughout the ages, the regal white lily became a symbol of the greater meaning of Easter. Its flowers, which embody joy, hope, and life, now grace millions of homes and churches each spring.

Houseplants

Houseplants are often sold in florists' shops but differ from traditional florists' crops because they are sold in containers rather than as cut flowers. In addition, they are chosen not simply for their flowers but also for their ability to survive under the growing conditions imposed by human dwellings. Most foliage plants are native to tropical areas, where they generally occur in understory habitats. Because they are adapted to growing in the shade of forest trees, they can tolerate the low light conditions found in most buildings. Such plants must also be able to withstand small soil volumes (resulting from the constraints of the pots) and the variable humidity caused by central heating and air-conditioning. Since houseplants receive continuous muted light and are restricted in size by their containers, they rarely flower. There are exceptions, such as African violets, but the majority of species of houseplants never flower inside a house, apartment, or office. Still, houseplants have become tremendously popular. The fact that the wholesale value of potted plants increased by 61 percent between 1990 and 2000 reflects the importance they have in interior design, where they soften both home and office environments. Table 17.6 lists a few of the most common houseplants and their places of origin.

In contrast to most houseplants, some potted plants are sold strictly for their flowers. These plants last only for a limited time because they are forced to bloom for a particular holiday, season, or market period. Examples of such potted plants are poinsettias, red tulips, and Easter lilies. All these flowers are raised by nursery owners, who time their blooming to coincide with Christmas (Fig. 17.39), Valentine's Day, and Easter. Although the flowers of these plants last much longer than those of cut flowers, the plants usually die when they finish blooming unless they are subsequently carefully tended.

Future Trends

Several philosophical attitudes toward the use of ornamental landscape and garden plants have emerged in the last few decades. At first they appear to be diametrically opposed, but they need not be. Recent increases in the cost of fertilizer, pesticides, and water, along with concern about the pollution caused by runoff, have prompted people to reevaluate their use of high-maintenance, exotic ornamental species. In response, Americans are experimenting with artificial plants and native species and are even landscaping with plants that can be eaten.

The logical extreme of the tendency to choose low-maintenance or no-maintenance plants is the employment of artificial plants such as silk or plastic plants. Although such a trend is understandable in terms of resolving the conflict

TABLE 17.6 Principal Potted Plants Used in American Homes and Offices

COMMON NAME	SCIENTIFIC NAME	FAMILY	NATIVE REGION
	Flowering Plants		
African violet	*Saintpaulia inonantha*	Gesneriaceae	Tanzania
Begonia	*Begonia* spp.	Begoniaceae	New World tropics
Coleus	*Coleus* spp.	Lamiaceae	Old World tropics
Crown of thorns	*Euphorbia milii*	Euphorbiaceae	Madagascar
Geranium	*Pelargonium* spp.	Geraniaceae	South Africa
Jade plant	*Crassula argentea* *C. arborescens*	Crassulaceae	Africa
Kalanchoe	*Kalanchoe integra*	Crassulaceae	Africa
Lantana	*Lantana camara*	Verbenaceae	New World tropics
Wandering jew, spiderwort	*Tradescantia* spp. or *Zebrina pendula*	Commelinaceae	New World tropics
Wax plant	*Hoya carnosa*	Asclepiadaceae	China and Australia
	Foliage Plants		
Air plant	Bromeliads: e.g., *Cryptanthus* spp. *Quesnelia marmorata,* *Vriesea carinata*	Bromeliaceae	Brazil
Aralia	*Schefflera* spp.	Araliaceae	Old World tropics
Asparagus fern	*Asparagus densiflorus*	Liliaceae	South Africa
Corn plant	*Dracaena fragrans*	Agavaceae	New Guinea
Dumb cane	*Deiffenbachia* hybrids	Araceae	Tropical America
Ivy	*Hedera helix*	Araliaceae	Eurasia, Northern Africa
Jade plant	*Crassula argentea*	Crassulaceae	Africa
Philodendron	*Philodendron* spp.	Araceae	Tropical America
Rubber plant	*Ficus,* primarily *F. elastica*	Moraceae	Asia
Screw pine	*Pandanus veitchii*	Pandanaceae	Polynesia
Snake plant	*Sansevieria trifasciata*	Agavaceae	Asia, America
Split-leaved philodendron	*Monstera deliciosa*	Araceae	Central America
Spider plant	*Chlorophytum comosum*	Liliaceae	South Africa
Wandering jew, spider plant	*Tradescantia* spp.	Commelinaceae	Americas
Watermelon plant	*Peperomia argyreia*	Piperaceae	Tropical South America
Weeping fig	*Ficus benjamina*	Moraceae	Southeast Asia, Australia
	Forced-Flowering Plants		
Azalea	*Rhododendron* spp.	Ericaceae	Asia, North America
Camellia	*Camellia japonica*	Camilliaceae	Japan, Korea
Christmas cactus	*Schlumbergera bridgesii*	Cactaceae	Brazil
Chrysanthemum	*Chrysanthemum fructescens* *C. ×morifolium*	Asteraceae	China, Canary Islands
Crocus	*Crocus* spp.	Liliaceae	Eurasia
Daffodil	*Narcissus* spp.	Amaryllidaceae	Eurasia
Easter lily	*Lilium longiflorum*	Liliaceae	Japan
Gardenia	*Gardenia jasminoides*	Rubiaceae	China
Hyacinth	*Hyacinthus orientalis*	Liliaceae	Eurasia, Africa
Poinsettia	*Euphorbia pulcherrima*	Euphorbiaceae	Mexico
Primrose	*Primula vulgaris*	Primulaceae	China
Tulip	*Tulipa* spp.	Liliaceae	Asia

between the human desire for elements of nature and the practical constraints of space and maintenance, it seems contradictory to use a manufactured object to help people forget their alienation from nature.

A second trend is toward the incorporation of more native plants into the repertoire of ornamental species. In many instances the use of ornamental plants has been dictated mostly by custom, not by an intrinsic superiority of the species. Particularly if selections and hybrids were made with local, native species, many beautiful plants could be added to the list of bedding plants and shrubs that are now used. Since native trees, shrubs, and herbs are usually adapted to the local pests and climatic conditions of an area, the use of fertilizers, pesticides, and irrigation water is much lower than that necessary to maintain introduced ornamentals. A subset of this movement has led to the

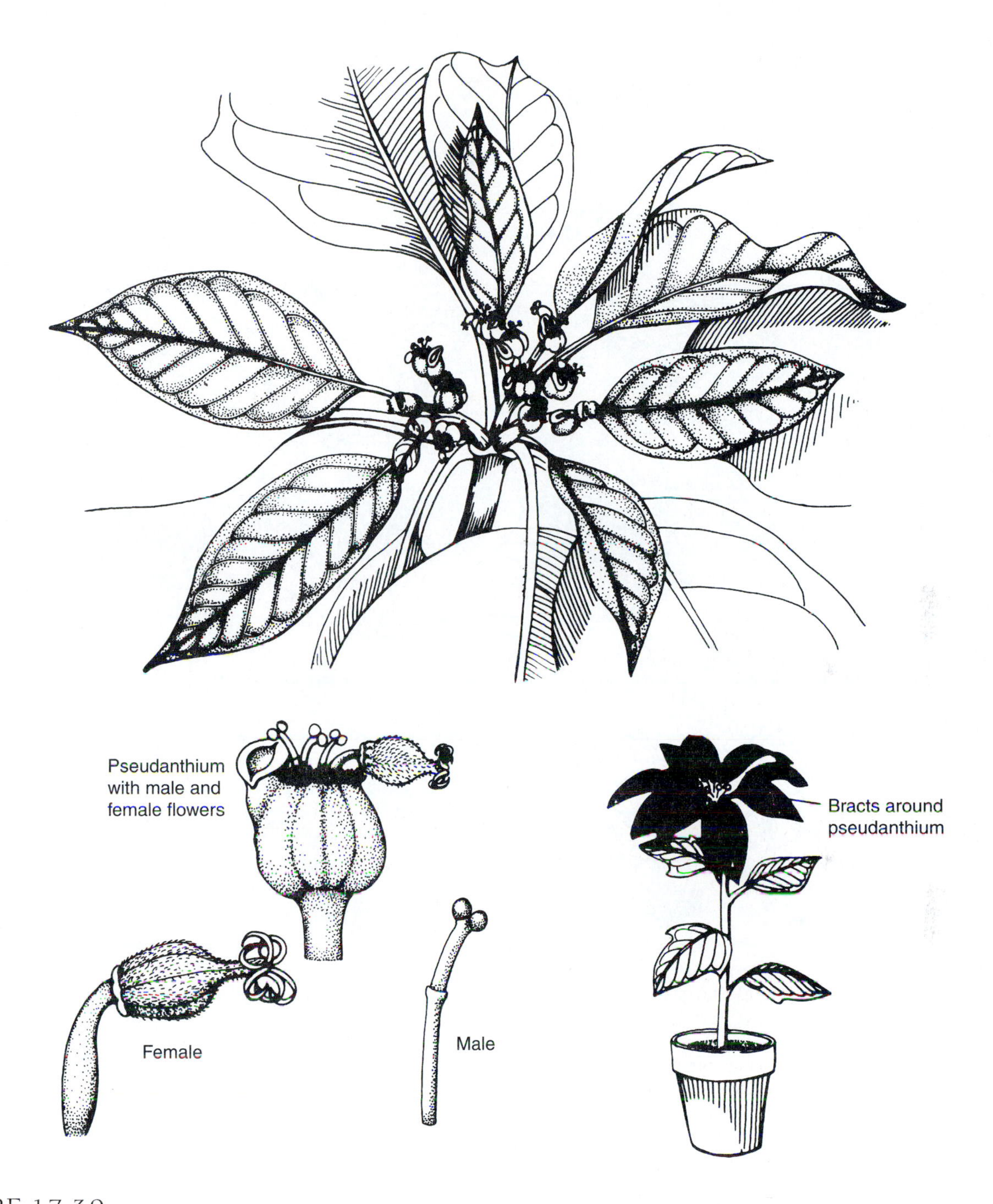

FIGURE 17.39
What looks like a flower of a poinsettia plant is actually a cluster of small flowers in inflorescences called pseudanthia, surrounded by red leafy bracts. The plants naturally bloom when days become less than 11 hours long. Nursery gardens can easily alter light levels to cause flowering in time for Christmas.

development of xeriscape gardening. These gardens usually lack covers of grass and consist of plants native to arid and semiarid regions. Mulching materials are generally applied between plants to reduce evaporation, discourage weed growth, and cool the soil surface. Encouraging the use of indigenous species not only increases the variety of plants from which to choose and avoids the costs associated with trying to grow exotic species, it also helps provide a means of preserving many species that might otherwise become extinct.

A final direction in garden design is that of edible landscape. A small garden of edible plants may seem little different from earlier kitchen gardens, but there is a distinction. In edible landscapes the plants are chosen and arranged for their aesthetic value, with flavor and nutrition being ancillary qualities. Herbs and leafy vegetables and fruit-bearing

FIGURE 17.40

One can apply design criteria to this edible landscape. The geometric form of the stone paths, the rows of onions bordering the paths, and upright corn stalks function as line in the design. Textural interest is provided by the contrast between the delicate foliage of the thyme, the fernlike parsley, the broad leaves of leeks and sorrel, and the coarse foliage of the artichokes. Color is contributed by red chard, okra, and beet leaves; the interplanting of marigolds; the edible flowers of squash, violets, and nasturiums; and the varying shades of the ripening tomatoes, peppers, and eggplants. Grapevines growing on a trellis mask the carport and work area behind.

annuals, shrubs, and trees are evaluated for their color, texture, line, and form just as ornamentals would be (Fig. 17.40). Many landscapers welcome the extension to the original definition of ornamentals as plants valued solely for their artistic merit. They emphasize that it is time to challenge the unwritten taboo against mixing flowers and vegetables and keeping food plants out of yard areas as an unnecessary waste of resources. Rosalind Creasy dubs this taboo the "edible complex" and argues that devoting resources that would otherwise be put into purely decorative landscapes to the production of food plants is a way to make landscapes more environmentally sound. A parallel trend involves the planting of gardens that encourage visits by animals such as butterflies and birds by providing nectar-filled flowers and a diversity of berries and grains for food (Box 17.2).

Box 17.2

Wildscaping

"If you build it they will come" is the motto for bird and butterfly gardeners, for, with habitat shrinking, wildlife are eager to accept invitations to move into our yards. Wildscaping, as this growing trend in landscaping is called, involves knowing the basic needs for shelter, food, and water of the fauna you would like to host and providing them in your landscape. It is a way to help balance the human need for both yards and contact with nature.

Providing shelter for wildlife can be as easy as letting a portion of your property revert to its unmanicured, wild state. Rock piles, wood stacks, and piles of brush serve as homes to many animals. Wildscapes to support a diverse population will have layers of vegetation, with canopy trees, understory shrubs, and a mix of annuals, perennials, and grasses along the edges. Birds and squirrels will use the tall trees for nesting, while butterflies will hide out in cracks in old wood. Birds will appreciate a source of nesting materials (e.g., twigs, strips of bark from particular trees). Butterflies appreciate sunny patches for basking and areas that provide wind protection. Both birds and bats are willing to move into man-made homes, if they are built according to specifications that the National Wildlife Federation and other similar agencies can provide.

Ponds, waterfalls, and birdbaths can provide water for wildlife as well as aesthetically pleasing garden features. Some butterfly gardeners install leaky pipes to create muddy areas that are especially appealing to male butterflies, who like to engage in puddling behavior, or other insects that use mud for nest construction.

Native trees and shrubs with edible seeds or fruits, such as juniper, viburnum, or cherries are natural food sources for birds. Homeowners have long augmented native food sources with feeders filled with sunflower seed mixes and specialty seeds to attract particular birds, such as thistle seed for goldfinches. Butterfly and hummingbird feeders filled with a simple mixture of sugar and water or commercial nectar spiked with amino acids can also supplement natural food sources. Red admiral, hackberry, and question mark butterflies have reputations as dipsomaniacs for their propensity to feed on tree sap and fermenting fruits. Feeders designed to keep these butterflies around are filled with overripe fruits and enough water to promote fermentation. Chances are, if a plant is a good nectar source, a wide selection of butterflies and hummingbirds will be able to use it. Some butterflies, such as the queen butterflies devoted to *Eupatorium,* have preferences for particular plants.

The challenge to cultivating butterflies and getting them to reside involves considering the other feeding stage in the butterfly life cycle. Butterfly gardeners realize that most kinds of caterpillar can eat only a certain plant or plants. These so-called host plants include dill, fennel, and parsley for black swallowtails (Fig. 17.41), passionflower vine eaten by gulf fritillaries and zebra longwing larvae, and *Aristolochia* spp., which are devoured by pipevine swallowtail caterpillars. The famous monarch migration every spring and fall is paced by the availability of its obligate food, milkweed species, along the east and west flyways. As butterfly gardeners develop

continued

FIGURE 17.41

Butterfly gardeners need to provide both nectar plants to attract butterflies and the host plants on which their larvae feed. The black swallowtail caterpillar (left) can eat dill, fennel, parsley, or rue. If it is threatened, it will stick out its scent horns and squirt the offender with a stinky fluid. This caterpillar will develop into a stunning adult (right) with wings of black, blue, yellow, white, and red.

(Caterpillar photo courtesy of D. Millard.)

Box 17.2 *cont*

their gardens they can consult butterfly references for more extensive lists of caterpillar host plants.

Butterfly gardeners must tolerate periodic decimation of plants provided as caterpillar hosts in order to sustain a crop of caterpillars. Pesticide use is discouraged since it can kill caterpillars and butterflies and harm nontarget organisms. Even biological controls can cause problems. *Bacillus thuringiensis* (Bt) is a natural pathogen of caterpillars, and thus incompatible with butterfly gardening. For wildscapers, release of beneficial predators such as ladybugs and lacewings combined with other organic methods, such as stiff water sprays to combat spider mites, are the best options for keeping pest insect populations in check.

The aesthetics of wildscapes have been perceived as controversial in communities that have adhered to conventional standards of tightly trimmed lawns and hedges. But wildscapes are not necessarily untidy. Landscapers have been able to create habitats that can function for dual purposes. These softer, more natural yard treatments are enhanced by the sounds of water bubbling, insects buzzing, and toads calling. The sight of butterflies floating past, purple martins swooping down to snatch mosquitoes, and the erratic flight patterns of hummingbirds help create lively environments and allow a close-up view of animal-plant interactions. Wildscaping is receiving support from many gardening sectors since it is compatible with other trends in landscaping, such as increased use of native plants, xeriscaping, water gardening, and organic gardening. As acceptance of this more natural aesthetic grows, we are seeing increasing numbers of homeowners letting their yards go to the birds, bats, and butterflies.

18

Algae

Algae may be one of the most maligned and misunderstood groups of organisms in the world. Many professional botanists find them fascinating but disagree on their classification, and laypersons are more likely to know about their association with polluted waters and toxic shellfish than about their contributions to society. Looking beyond their murky reputation, one discovers their vital importance in the evolutionary history of plants, their amazing variety, and their potential as a source of food and industrial products. Aquatic algae occur throughout the world in both freshwater and water of every level of salinity, floating as **plankton** (microorganisms drifting in the water) or reaching down to the depths of effective light penetration. Surprisingly, algae can be found in such extreme situations as snow surfaces and Antarctic ice to transient ponds in deserts and can even thrive in hot springs and within rocks. They are not just confined to aquatic habitats; many can be found in terrestrial and subterranean ecosystems. In addition, algae exist in symbiotic associations with fungi to form lichens and with animals such as corals and sea anemones (Fig. 18.1). People living close to the sea have employed algae directly for food, as a gelling agent, for medicine and animal food, as fertilizer, and in industrial processes (Fig. 18.2).

Scientists have long struggled to classify the algae in a meaningful way. Linnaeus recognized 14 algal genera, only 4 of which (*Conferva, Ulva, Fucus,* and *Chara*) are considered algae by modern definitions. Making sense of the seaweeds was hampered before microscopes were available to observe them in detail and because of the belief that they were asexual. The sporadic fossil record of this diverse group has offered little help in unraveling its evolutionary history. Fortunately, tools such as the electron microscope and advances in biochemistry and molecular biology have shed more light on the algae by providing taxonomic criteria on which to base classification decisions.

Defining algae is not easy, since the group encompasses an array of heterogeneous organisms ranging in size from single cells 0.2 to 2.0 μm (bacterial size) to multicellular giant kelps up to 60 m (196 ft) long (Fig. 18.3). Although all algae have chlorophyll and build carbohydrates and proteins comparable to those found in vascular plants, their definition hinges on a biological technicality. The algae are distinguished from vascular plants by their sexual apparatus: all (with the exception of the Charophyta) lack a multicellular enclosure around their spore- and gamete-bearing

FIGURE 18.1
Algae live interesting lives in symbiotic association with sea animals. Within corals they help in reef building by absorbing calcium from seawater and leaving deposits of calcium carbonate (limestone) when they die. Living endozoically within sea anemones, they are thought to contribute to the phototactic abilities that have been observed in some of these animals.
(Andrew J. Martinez/ Photo Researchers, Inc.)

FIGURE 18.2
An assortment of the growing number of seaweed products available in supermarkets in Ireland.
(Photo courtesy of M. Guiry.)

structures. Even the multicellular algae produce gametes in unicellular containers or in multicellular containers in which every cell is fertile (Fig. 18.4). Asexual reproduction is accomplished by simple division, by fragmentation, by the creation of specialized vegetative propagules (e.g. *Sphacelaria*), or by means of flagellated or nonmotile spores.

Modern treatments group most algae together in the kingdom Protista, which includes the protozoa. Within this kingdom, algae are currently divided among anywhere from 4 to 13 divisions (or phyla). The blue-green algae (Cyanobacteria), which historically were placed with other algae, are now considered to belong to the Monera along with nonphotosynthetic bacteria. These organisms are considered **prokaryotes,** organisms that lack a nucleus and have their genes in a circular piece of DNA in the cytoplasm. Early **phycologists** (people who study algae) based their taxonomic systems on the pigmentation of an organism (blue-green, green, red, brown). This system holds true today, although the determination of the division to which a group belongs is also based on the structure of organelles such as flagella, nuclei, chloroplasts, eyespots, and **pyrenoids,** which usually function as centers of starch formation. Nevertheless, the fact that algae do not constitute a natural evolutionary group becomes obvious when one considers that one group of algae (currently placed within the green algae) apparently gave rise to all green plants. Table 18.1 shows some of the currently accepted orders of algae included in this chapter, with their characteristics. Specific genera and species are also shown.

The history of organisms traditionally considered algae stretches back nearly 3.5 billion years into the Precambrian, when bacterial forms similar to modern blue-green algae appeared in rocklike formations in Australia (Fig. 18.5). These organisms known as cyanobacteria (or Cyanophyta, if they are placed with the algae) are important not only for their critical role as nitrogen fixers but also because they are photosynthetic. It is believed that forms similar to blue-green algae were the first organisms to release oxygen that subsequently diffused into the atmosphere. This release of oxygen helped create the environment required for aerobic respiration and led to the formation of the protective ozone layer that shields the earth. Both of these events set the stage for the eventual colonization of land. The development of **eukaryotes,** organisms that possess a nucleus containing chromosomes and surrounded by a double membrane, is placed at around 1.5 billion years ago, with the appearance of green algae in fossils also dated to this time. Other groups of algae appear in the fossil record by 900 million years ago.

People are only beginning to appreciate the present and potential value of algae within the biosphere and as worldwide commodities. In aquatic environments algae are vital as primary producers, providing food for marine life and excreting organic compounds into the surrounding medium. Their photosynthetic activities also release oxygen into the water and are of critical importance in maintaining the global CO_2 balance. Despite their ubiquity, algae constitute an underutilized source of food and raw materials for industrial purposes. Our failure to exploit this versatile resource has resulted from many factors. Early uses were restricted because algae had to be harvested from the wild, making industry dependent on the vagaries of climate and tidal action. In many cases the life histories of the organisms were not understood. Research into the biology of *Porphyra,* for example, led the way to its cultivation and to the adoption of breeding techniques that resulted in a multimillion-dollar industry. Now that studies are allowing people to cultivate algae and technological advances are permitting efficient harvesting, economic uses of algae may burgeon. Businesses are exploring the potential of algae in food

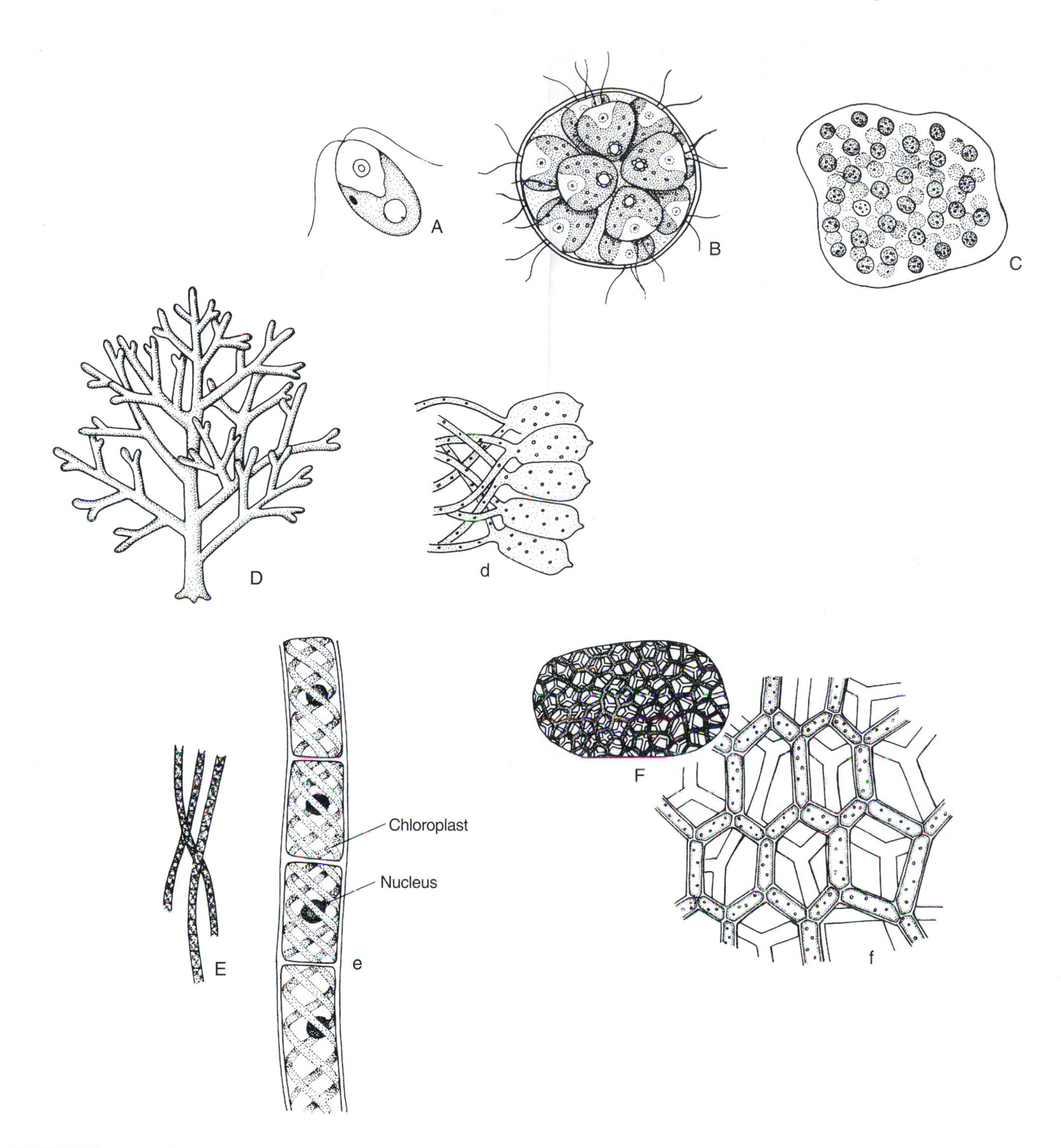

FIGURE 18.3A

The morphological diversity of the algae, the different complements of chlorophylls in modern forms, and the range of sexual variations they exhibit all contribute to the belief that the group is derived from more than one ancestral type. Here the range in size and structure is illustrated with: (A) *Dunaliella,* a motile single-celled organism; (B) *Pandorina,* which produces motile colonies of determinate size; (C) *Microcystis,* which is composed of colonies of indeterminate size held together in a gelatinous matrix; (D) *Codium,* a macroscopic alga composed of interwoven siphons that create a dense surface layer; (d) a portion of a branch in cross section; (E) and (e) *Spirogyra* with its characteristic spiral chloroplast within filaments of cylindrical cells created by division in one plane; (F) and (f) *Hydrodictyon* with its cylindrical cells arranged to create netlike colonies that can grow up to 1 m in size; (G) and (g) *Enteromorpha,* whose unique structure is a result of a

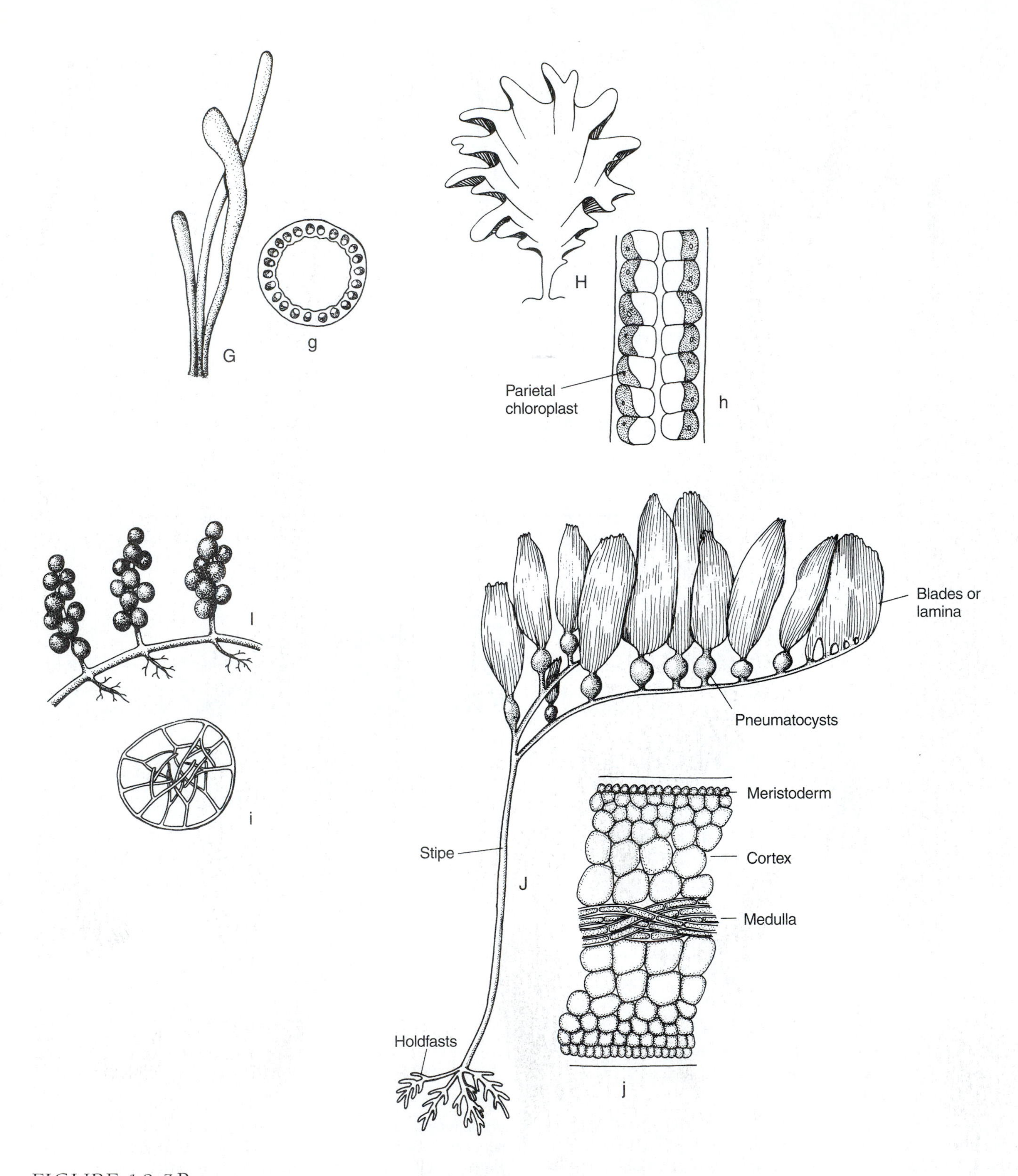

FIGURE 18.3B

series of cell divisions that create a single-layered, then a multilayered, and finally a tubular plant body; (H) *Ulva,* or sea lettuce, whose leaflike fronds are just two cells thick (h) as a result of collapse of a tubular thallus during development; (I) *Caulerpa,* or sea grapes, whose thallus is composed of a wall-layer made rigid with a matrix of ingrowths, shown in cross section (i); (J) *Macrocystis,* whose fronds exhibit a complex branching pattern and an internal structure that approaches that of "higher" plants. The thallus consists of a holdfast at the base that resembles a root but is not specialized for mineral uptake and from 1 to 100 long tubular stipes from which a number of leaflike blades interspersed with flotation bulbs (pneumatocysts) emerge; and (j) a cross section showing the meristoderm (a meristematic epidermis), the cortex, and the medulla (a series of interwoven cells).

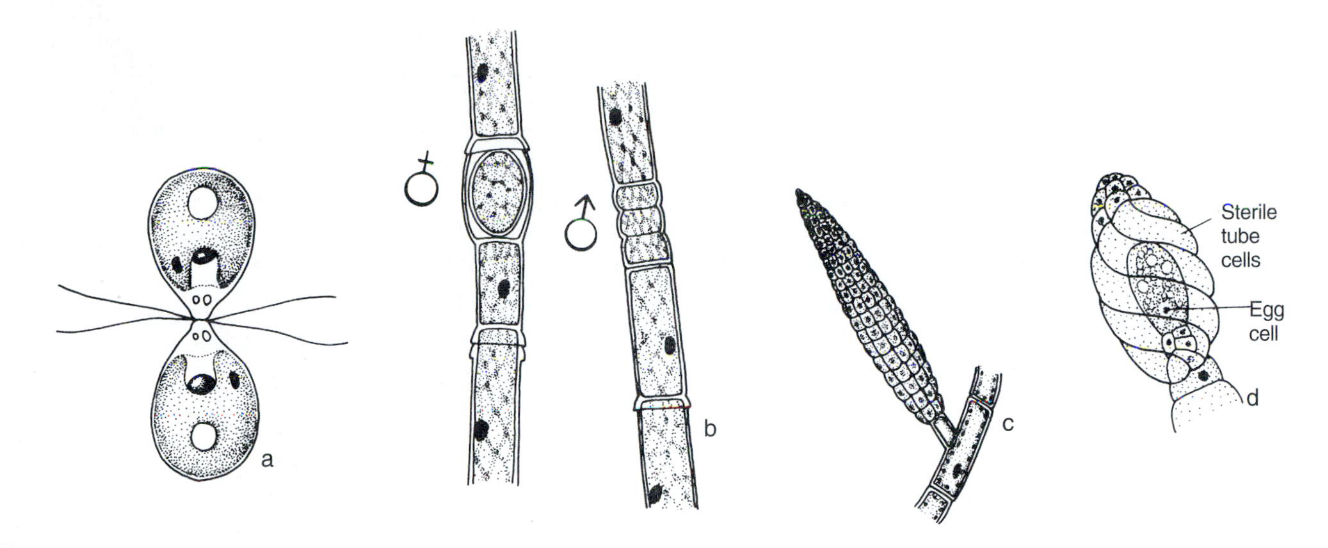

FIGURE 18.4
(a) In unicellular algae such as *Chlamydomonas,* entire cells function as gametes and unite in sexual reproduction. In most multicellular algae, every cell of the gamete-bearing structures is fertile. (b) In *Oedegonium,* particular cells in the strings of cells differentiate and yield either an egg or several sperm. (c) In *Ectocarpus,* multicellular gametangia can be found where every cell is fertile. (d) The exception to the rule that all sex cells are fertile is provided by the charophytes, such as *Nitella,* whose gametes are enclosed in a layer of sterile cells.

production, in industry, and as a source of medicines. Scientists are also trying to identify and avoid blooms of toxic algae and to control weedy species that become a nuisance when the balance of water systems is upset. This chapter deals with these topics in turn.

Algae As Food

Nearly 7 million metric tons of dried seaweed entered the world market in 1997. Chapman and Chapman (see Additional Readings) conjecture that 160 species of algae are used directly for food, but the bulk of the edible algae comes from three genera: *Porphyra, Laminaria,* and *Undaria.* These three seaweeds have been eaten throughout recorded history not only for their food value but also for their flavor, color, and texture. Although they are composed primarily of structural carbohydrates that are largely indigestible, up to 20 to 25 percent by dry weight are soluble carbohydrates and edible protein. Their vitamin C content can equal that of citrus fruits, and they are rich sources of vitamins A, D, B, B_{12}, E, riboflavin, niacin, pantothenic and folic acids and trace elements such as iodine (Table 18.2).

Species of *Porphyra* have been eaten wherever people have found them growing. Different species of this red alga (Rhodophyta) have been called everything from *laver* in England and the United States to *luche* in Chile, *karengo* among the Maoris in New Zealand, *nori* in Japan, *slack* in Scotland, and *sloke* in Ireland. In England the seaweed is boiled, and the resulting *mush* is fried or served with butter. Rice, fish, and nori constitute the basic components of the Japanese diet. This seaweed is eaten for its taste as well as its nutritional benefits. Nori wrappers are essential to the creation of sushi, an item of Japanese cuisine now considered trendy in the United States and Europe (Fig. 18.6).

Nori has been collected from the wild in Japan since at least A.D. 533 to 544. Cultivation (Fig. 18.7) dated from A.D. 1623 to 1649, when Japanese fishermen discovered floating pieces of bamboo on which *Porphyra* was growing. They started cultivating the algae by "planting" bamboo twigs in the marine **intertidal zones** (the inshore areas between high tide and low tide) where the plants were found growing attached to rocks and other substrates. A few weeks after "planting," tiny *Porphyra* plants could be seen attached to the bamboo, and when the plants grew large enough, they were harvested. This practice continued relatively unchanged until an English phycologist made a remarkable discovery about the mysterious "purple laver."

Until the early 1950s fishermen placed branches in the water in the early fall, only to harvest the seaweed a few months later. For the rest of the year the plant disappeared, and no one knew where the spores went. Farmers were forced to plant the poles in places where *Porphyra* was known to grow. In 1949, Dr. Katherine Drew took mature fronds of *Porphyra umbilicalis* back to her laboratory, and the spores from them germinated, growing onto a piece of shell she had placed in the culture dish. Thanks to her observations, it was realized that the plant spent half the year masquerading as a branched filamentous organism that looked to the naked eye like a small rosy patch, a form that had once been considered the separate

TABLE 18.1 Characteristics of the Algae Discussed in Chapter 18

DIVISION	EXAMPLES	PIGMENTS	STORAGE PRODUCT	CELL WALL COMPOSITION	FLAGELLA AND INSERTIONS
Cyanophyta, or Cyanobacteria (blue-green algae)	*Anabaena, Lyngbya, Microcystis, Nostoc, Oscillatoria, Phormidium, Schizothrix*	Chlorophyll a, phycobilins, carotenoids	Glycogen-like granules	Various amino acids	Absent
Chlorophyta (green algae)	*Caulerpa, Chlamydomonas, Chlorella, Codium, Dunaliella, Enteromorpha, Hydrodictyon, Pithophora, Scenedesmus, Spirogyra, Tetraselmis, Prototheca, Ulva*	Chlorophyll a and b, cartenoids, xanthophylls	Starch (sometimes oil)	Cellulose and other polymers	1, 2–8 usually equal subapical
Charophyta (stoneworts)	*Chara, Nitella*	Chlorophyll a and b, carotenoids, xanthophylls	Starch	Cellulose, some calcified	2, equal, subapical
Phaeophyta (brown algae)	*Ascophyllum, Durvillaea, Ecklonia, Ectocarpus, Fucus, Hizikia, Laminaria, Macrocystis, Sargassum, Undaria*	Chlorophyll a and c, beta-carotene, xanthophylls (including fucoxanthin)	Laminaran, mannitol	Cellulose, alginic acid	2, unequal, lateral
Chrysophyta (golden and yellow-green algae, including diatoms)	*Chaetoceros, Prymnesium, Pseudonitzschia, Skeletonema*	Chlorophyll a and c, carotenoids, xanthophylls (notably fucoxanthin)	Laminaran, oil	Cellulose, silica, calcium carbonate, some with chitin	1 or 2, unequal or equal, apical
Pyrrophyta (dinoflagellates)	*Gambierdiscus, Protogonyaulax, Ptychodiscus Pfiesteria*	Chlorophyll a and c, beta-carotene, several xanthophylls	Starch (some with oil)	Absent or made of cellulose in plates	2, unequal (one girdling, one trailing)
Rhodophyta (red algae)	*Acanthopeltis, Ahnfeltia, Chondrus, Dilsea, Eucheuma, Gelidium, Gracilaria, Porphyra, Digenia Pterocladia, Chondria*	Chlorophyll a, some with d, phycobilins	Floridean starch (most with plastids of other shapes: ribbons, discoidal, etc.)	Cellulose and sulfated galactans	Absent

Source: Adapted and modified from H. Bold and M. Wynne. 1978. *Introduction to the Algae.* Englewood Cliffs, N.J., Prentice Hall.

species *Conchocelis rosea.* This filamentous growth was diploid and in nature produced another set of spores. These spores germinated and subsequently underwent melosis, leading to the familiar **macrophytic** ("large-leaved") haploid *Porphyra* generation that is harvested (Fig. 18.8). The discovery of this alternation of morphologically distinct generations in *Porphyra* and the recognition that the diploid, branched filamentous generation could be grown on the surface of any calcareous material revolutionized the nori industry.

Within a few years the life cycle of the commercial species of nori, *P. tenera,* had been worked out, and by 1960 the use of diploid cultures had produced a surge in nori production. Other improvements, such as floating net culture and the selection of improved varieties, have also contributed to making nori the billion-dollar item of retail commerce it is today.

Today artificial cultivation of the diploid phase is effected by placing oyster shells on the bottom of a tank

TABLE 18.2 Food Values of Commonly Eaten Algae, Based on 100-g Edible Portion

SEAWEED, RAW	WATER, %	FAT, g	CARBO-HYDRATE (FIBER), g	ASH, g	CALCIUM, mg	PHOSPHORUS, mg	IRON, mg	SODIUM, mg	POTASSIUM, mg
Agar	16.3	0.3	0.7	3.7	567	22	6.3	—	—
Irish moss	19.2	1.8	2.1	17.6	885	157	8.9	2,892	2,844
Kelp	21.7	1.1	6.8	22.8	1,093	240	—	3,007	5,273
Laver	17.0	0.6	3.5	11.0	—	—	—	—	—

Source: Adapted from Composition of Foods, USDA Handbook No. 8. 1975 printing.

FIGURE 18.5

These large limestone formations known as stromatolites are still being formed by secretion of calcium carbonate by cyanobacteria in coastal waters such as this in Shark Bay, Australia. Microscopic examination of fossilized stromatolites provides evidence that this group has remained relatively unchanged throughout geological time. The activity of blue-green algae released oxygen into the atmosphere forcing anaerobic organisms to adopt strategies to deal with oxygen or remain waterborne.

(Photo by John Reader, Science Photo Library/Photo Researcher.)

filled with seawater (Fig. 18.9). Mature, fertile spore-bearing fronds of *Porphyra* are placed in the tanks and are induced to release their spores by dropping the temperature, increasing the light levels, or artificially lengthening the daylight hours or by using some combination of those treatments. Dilute seawater containing released spores can also be used as a spore source. The spores from either source germinate and bore into the shells, forming small branching filaments that soon cover the inside half of the shell. The nets can be suspended in the water from poles with bags of the shell fragments that serve as sources of colonizing spores. After spore attachment, the poles are placed horizontally at a carefully determined angle to ensure exposure to tidal action, which allows a drying out period each day to reduce damage by fungi (Fig. 18.10). Alternatively, huge spools of netting can be rotated for 1 to 5 days in large tanks in which the **conchospores** (diploid spores produced by microscopic branched filaments) are suspended. These nets can then be suspended in the ocean from buoys. Nets seeded by these or other methods are submerged until individual plants reach 2 to 3 cm (0.78 to 1.2 in.) in size. Once the blades reach this size, portions of the netting can be rolled up and frozen for cultivation later in the season. The "plantlets" that are left to grow can be harvested by boat 40 to 50 days later.

After harvesting, the fronds are washed in seawater and brought ashore where they are chopped very finely, put in large barrels, and covered with water. This slurry is poured onto frames 18 by 21 cm (7 by 8.3 in.) and shaken to homogenize and interlock the fibers in much the same process that is used in making felt or paper (Chapter 16). After drying in the sun for a few days or processing in a mechanical dryer, the sheets are peeled off the frames and packaged for sale. In the past eight people could produce up to 1500 sheets of nori a day. Modern machines allow one or two people to produce that many sheets in an hour.

Traditionally, early harvests of nori were considered the best because they had better flavor, a softer texture, and coincidentally, a nutritional value greater than that of later collections. The discovery that baby plantlets could be frozen and then placed out for cultivation up to a year later has extended the harvest (through April rather than March) while providing a higher-quality product throughout the season.

Other macroalgae cultivated in the Orient for human consumption include two brown seaweeds (Phaeophyta): *Laminaria* (Fig. 18.11) and *Undaria. Laminaria,* or kelp, is known as *haidai* in China and *kombu* ("delight") in Japan. It has been cultivated since 1730 in Japan with techniques that

FIGURE 18.6

Asians take nori seriously, making judgments about the quality and use of the product by using elaborate criteria and rigorous standards comparable to those Westerners reserve for evaluating the quality of wines. The finest nori is lustrous black with a greenish tinge. The persistence of a green color after toasting is correlated with a higher ratio of chlorophyll compared with the phycobilin pigments. The luster is a quality correlated with rapid growth of the nori, as is the texture. The most tender nori comes from early season, first harvests of the seaweed. The flavor is influenced by growing conditions and is coincidentally associated with the amino acid balance and thus the nutritional value of the nori. (a) In making sushi, the dried nori is placed on a bamboo mat and a thin layer of rice is pressed onto its surface, leaving a strip a few inches wide on the far edge. (b) The fillings (cooked, raw, or pickled vegetables, fish, meat, or eggs) are laid in a row along the edge and then rolled up, using the mat as an aid and the moistened edge of the seaweed to seal the cylinder. (c) Cutting the roll produces the neat finger foods that have become popular in Western cuisine.

have changed little since then. Stones are usually set out in September to provide a substrate for the germination of released spores (Fig. 18.12). Other techniques for encouraging seaweed growth include letting the spores settle onto a floating raft, providing clean concrete blocks, and creating a fresh substrate for spore germination by blasting the reefs with dynamite. When the plants grown from spores have reached a sufficient size, they are harvested from open boats, typically between July and August of the following year (Fig. 18.13).

China has also perfected a method of artificial cultivation that allows plants to be grown off its southern coast even though the water temperature there is too high for spore formation. Temperatures are kept cool in a laboratory, where the spores are produced and allowed to germinate and grow into plantlets. Plantlets that can grow well in warm water are then set out along the coast.

Once harvested, the seaweed is laid high on the beach to dry (Fig. 18.14) and then prepared in any of several ways. One of the most common methods includes boiling the seaweed in iron or copper vats with the aniline dye malachite green to impart a uniform green color to the fronds. The resulting plant material is dried, stacked, and pressed together before being shredded with a plane. These thin strips of dried seaweed are sold as an ingredient for soups and vegetable dishes. *Kombu* can also be processed to produce a brewed beverage or even coated with pink or white icing and sold as a confection. Kombu is held in such high

FIGURE 18.7

The traditional method of cultivating and processing *Porphyra* to make nori. (a) Placing the bundles of brush to create holdfasts on which the seaweed can grow. (b) Washing the laver after harvest. (c) Sorting and chopping the weed. (d) Making the sheets by pulling a screen up through a vat of the suspended laver fragments and setting them out to dry.

(Redrawn from lithographs in the *Bulletin of the Bureau of Fisheries,* 1904.)

esteem by the Japanese that it is widely employed as part of New Year's festival decorations.

Another important edible genus of brown algae is *Undaria* (also a kelp, it is called *qundai-cai* in China and *wakame* in Japan (Fig. 18.15). It is usually found attached to rocks from 1 to 8 m below the man low tide level. *Undaira* has been introduced to or invaded waters outside its native range and now occurs in Europe, New Zealand, Australia, and Tasmania. It is sometimes cultivated along with *Laminaria* as their growing seasons and cultural requirements are complementary. Wakame is typically harvested from February through June (before the *Laminaria* harvest season) by simply drying and baling it. The resulting product can be cut up and sold for consumption as is ("wakame chips"), toasted and eaten, coated with sugar ("ito-wakame"), or mixed with boiled rice. Chapman and Chapman report that extracts of wakame are reputed to increase intestinal absorption of calcium, promoting the formation of bones.

Besides the macroalgae, a few microscopic algae have been directly consumed by humans (Fig. 18.16). *Chlorella* is the only microscopic green alga (Clorophyta) that is of current commercial importance. One blue-green algae, *Spirulina,* has provided food for people in both the Old World and the New World. The Spanish conquistadores discovered the Aztecs consuming *Spirulina* (Fig. 18.16) as a major protein source. Women in the Lake Chad area of north Africa still harvest the algae by pouring the salty, alkaline

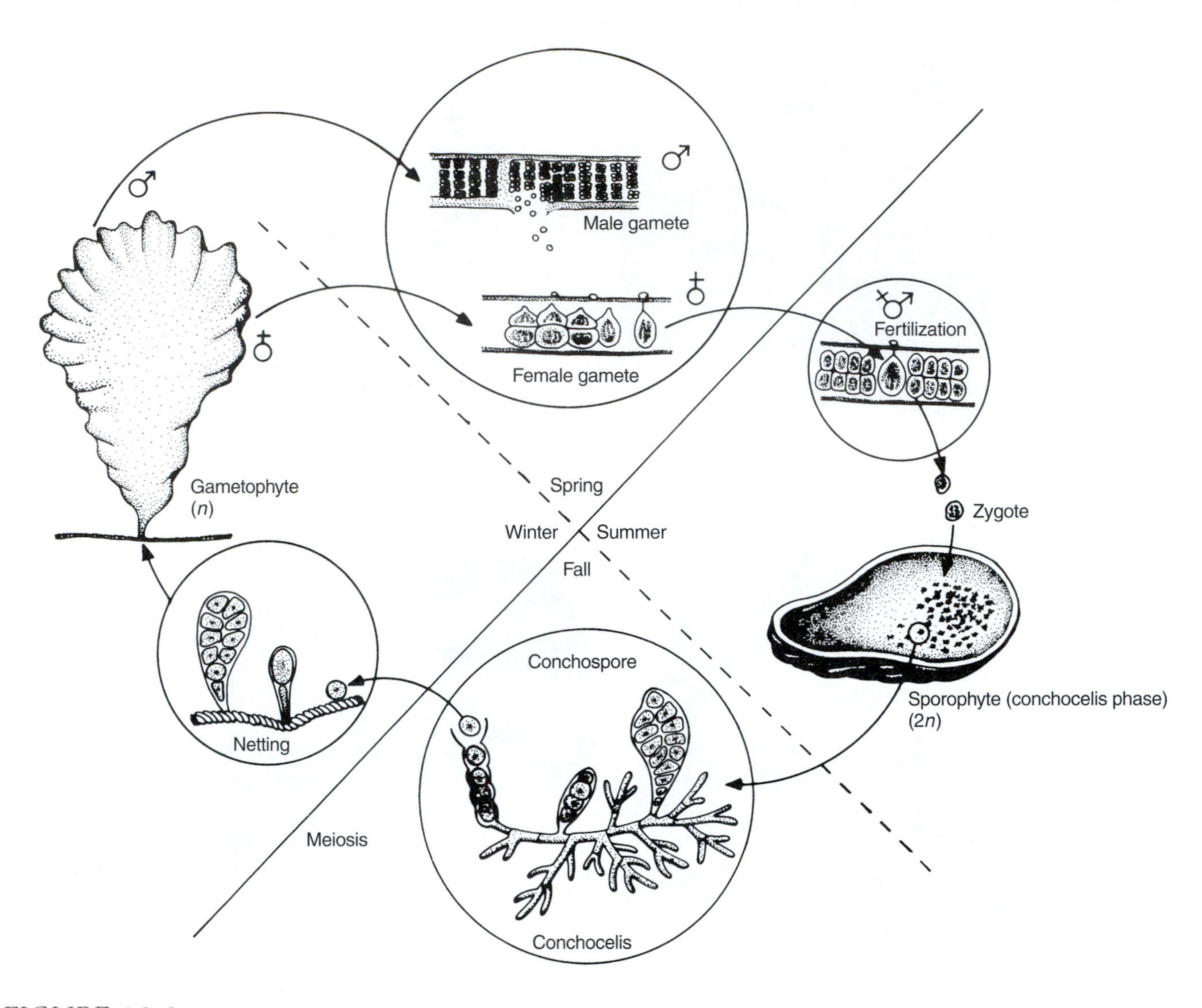

FIGURE 18.8

The life history of *Porphyra,* showing the haploid macroscopic plant with fertile regions on the frond edges, the release of male gametes and the fertilization of female gametes, the release and subsequent germination of the zygote to produce the microscopic diploid phase on the shell (or other calcareous) surface, the release of conchospores in the conchocelis phase and their germination (with meiosis) to complete the life cycle. Note that the largest, most conspicuous phase of the life cycle is the haploid gametophyte.

FIGURE 18.9

In large-scale cultivation of *Porphyra,* the conchocelis phase is grown for approximately five months on oyster shells suspended on ropes. After the shells have developed greyish-purple spots, they are placed on the bottom of tanks into which nets wound around drums are rotated. Microscopic examination of the fibers of the nets reveals whether enough spores have settled on them. These nets are then strung on poles or floated in the water for development of the macrophytic stage.

(Photo courtesy of M. Ohno.)

FIGURE 18.10

In the pole system of nori cultivation, seeded nets are suspended between fixed poles and exposed to air during low tide. The floating system, which allows expanded cultivation of nori, is effected by suspending nets from floating buoys in the open water, where they do not experience a drying period. This photo shows one of the first efforts to cultivate nori in N. America, utilizing the floating system in Cobscook Bay, Maine.

(Photo courtesy of D. Cheney.)

FIGURE 18.11

Laminaria harvest in Japan.

(Photo courtesy of S. Kawashima.)

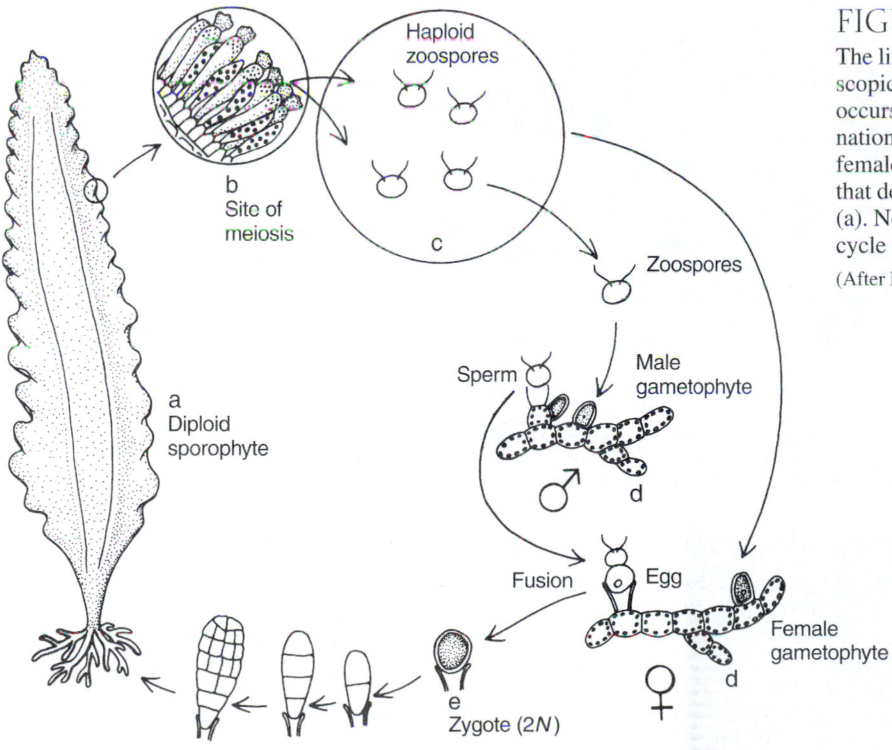

FIGURE 18.12

The life cycle of *Laminaria,* showing (a) the macroscopic plant, (b) a section of a frond where meiosis occurs to produce (c) haploid zoospores, (d) germination of these spores into microscopic male and female gametophytes, and fusion to create a zygote that develops into the conspicuous sporphytic phase (a). Note that the most conspicuous phase of the life cycle is the diploid sporophyte.

(After Bold 1973.)

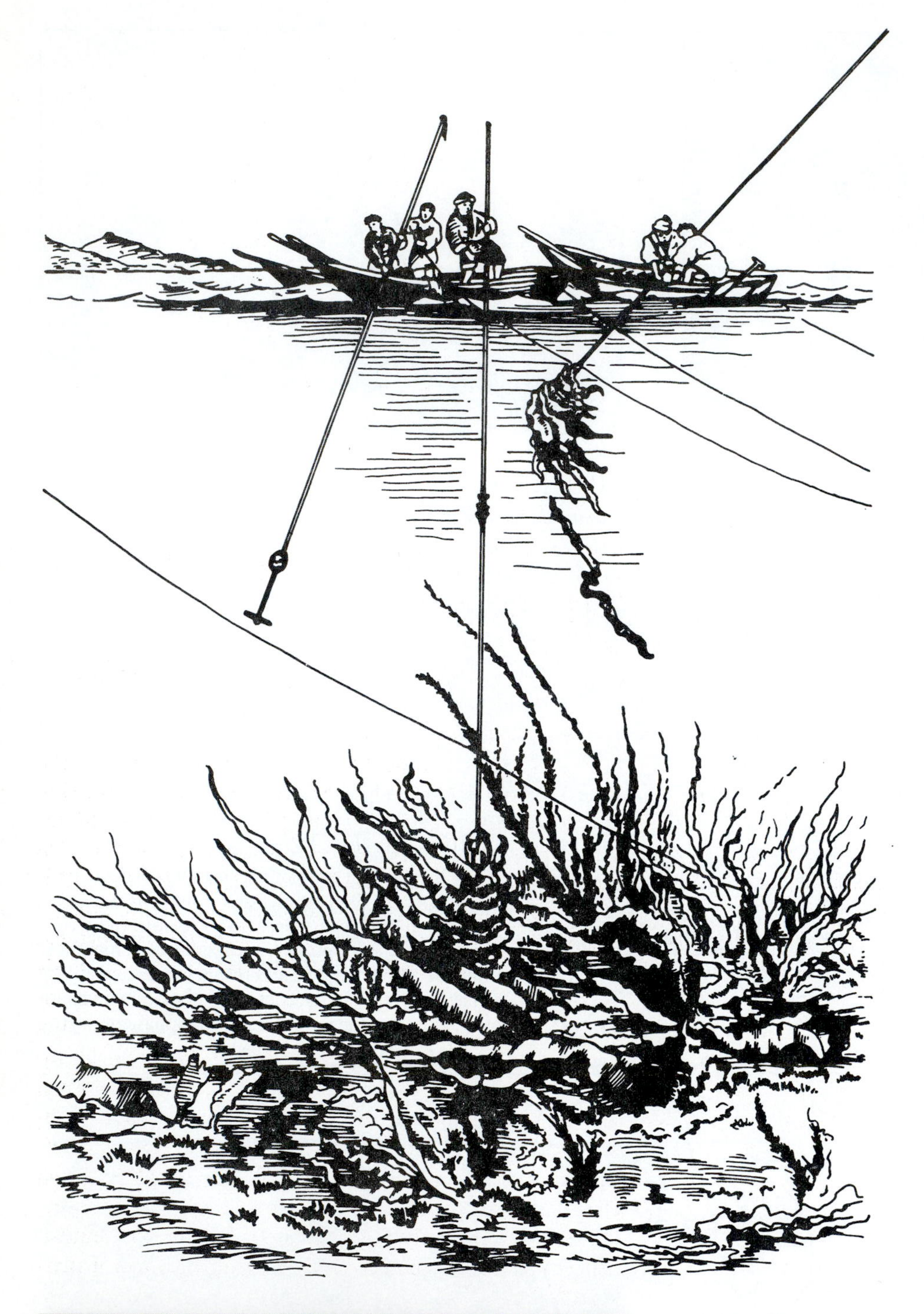

FIGURE 18.13
Gathering kelp with poles in Hokkaido, Japan.
(Redrawn from a lithograph in a 1904 Bureau of Fisheries Publication.)

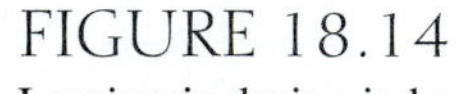

FIGURE 18.14
Laminaria drying in bundles on the beach.
(Photo courtesy of S. Kawashima.)

lake water through muslin cloth and sun-drying the resulting filtrate before cutting it into cakes. *Spirulina* has a high protein content (up to 72 percent dry weight) and a good amino acid balance, being only mildly deficient in the sulfur-containing amino acids. Because it has a protein content superior to that of all other vegetable products and equal to that of meat and dairy products, it has been investigated as a

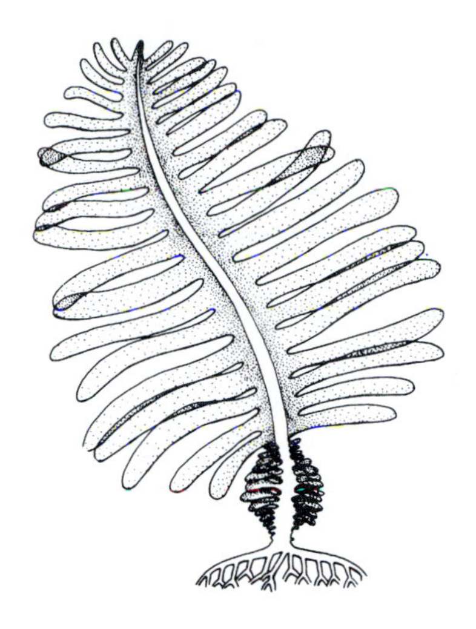

FIGURE 18.15
Undaria, the softest of the brown algae, is washed, dried, and chopped to produce wakame.

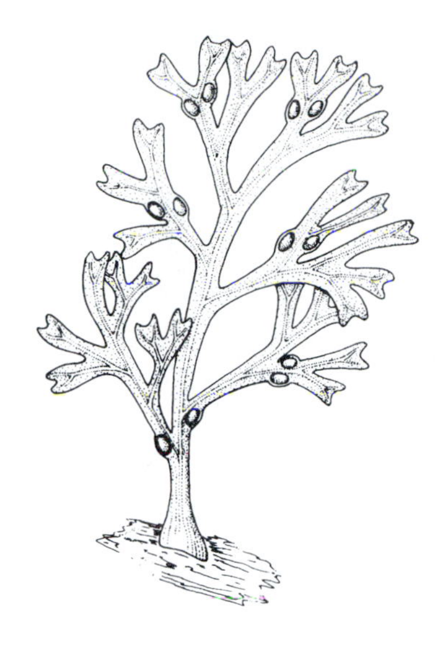

FIGURE 18.17
Fucus, also known as rockweed, is abundant in the waters of the North Atlantic, where it grows attached to rocks in the intertidal zone.

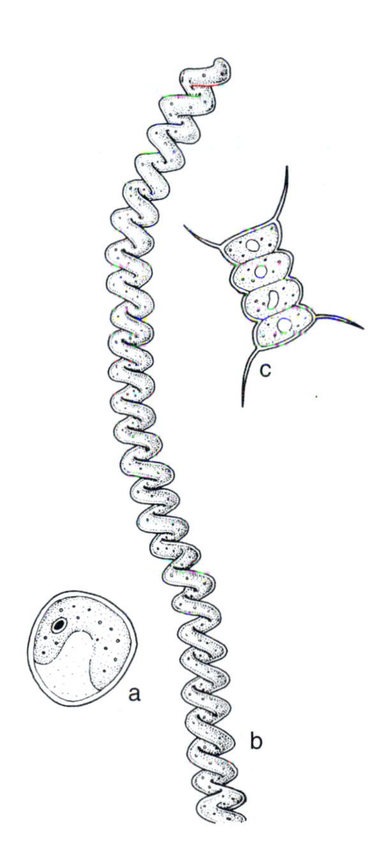

FIGURE 18.16
(a) *Chlorella,* (b) *Spirulina,* and (c) *Scenedesmus* are a few of the microscopic algae consumed by humans.

potential food source by researchers in southern France. In those investigations it was found to be significantly more productive per acre than grain or cattle. *Spirulina* was found to yield 10 tons of protein per acre compared with 0.16 ton for wheat and 0.016 ton for beef. In Mexico an enterprising British businessman, Christopher Hills, noted the proliferation of *Spirulina* in Lake Texcoco near Mexico City. He began to drag the algae from the lake, dry it, and sell it around the world. When quality control became an issue, a plant was built in which the *Spirulina* could be separated from debris. The operation has grown into a successful business that markets the dried algae as a high-protein, high-carotene additive for chicken feed. *Spirulna* is now commercially grown in an increasing number of sites worldwide, including the Imperial Valley of southern California, where artificial ponds have been created in the saline compacted soil of the Colorado Desert, and on the Kona coast of Hawaii. Production of *Spirulina* as food, animal feed, and chemicals has grown to over 1000 metric tons per year (1998) and will surely rise if early data claims on the medical virtues of this potent blue-green alga lead to a marketable drug.

Wild animals have been known to eat seaweeds, and humans have used algae as fodder for domesticated animals. Today genera such as *Laminaria, Macrocystis, Ascophyllum,* and *Fucus* (Fig. 18.17) are commercially harvested, dried, and ground into meal for incorporation into animal feed. The carotenoids naturally present in these algae are credited with producing especially golden yolks and yellow body fat when consumed by poultry. A single-celled planktonic alga, *Dunaliella bardawil* (Fig. 18.3a), produces such large amounts of beta-carotene and glycerol that a process has been patented for raising the algae in order to extract those compounds.

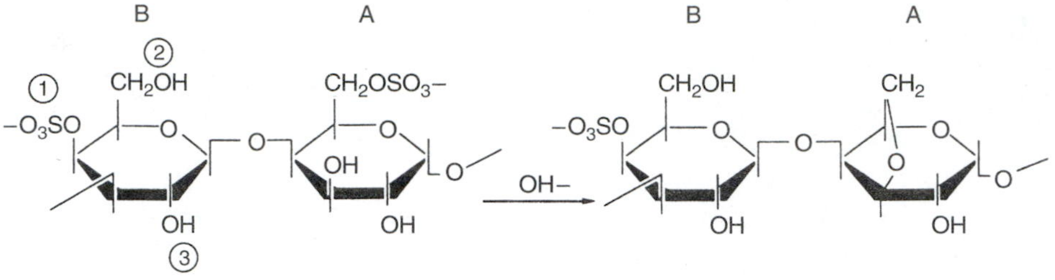

FIGURE 18.18

Alginates, carrageenans, and agars all consist of long chains of sugar molecules. They differ in the kinds and numbers of the groups attached at positions 1, 2, and 3.

Industrial Uses

Traditionally, algae have been used industrially to produce iodine for photographic and medicinal purposes and soda ash for the manufacture of glass and gunpowder. Today the most important chemicals extracted from algae are the **phycocolloids,** hydrogels that are used in the food, cosmetic, and pharmaceutical industries as emulsifiers and gelling agents (see Chapter 10 for a description of how these work). The name *phycocolloids* is derived from *phykos,* meaning "seaweed" in Greek, and *colloids* refers to their behavior as colloids (substances that are suspended but not dissolved in water). The following sections will discuss the three most important groups of these compounds: the alginates, the carrageenans, and the agars (Fig. 18.18).

Alginates

Alginates, which are long molecules composed of alginic acid and its salts, were discovered by the British chemist E. C. C. Stanford, who mistakenly described them as nitrogenous compounds because he was working with impure substances. In 1896 an extract was correctly prepared and called *tangsäure,* meaning "seaweed acid." Like pectins in the higher plants, alginate compounds are components of the cell walls and the mucilaginous matrix between cells. Unlike land plants, however, aquatic plants must have cells that are flexible but have enough tensile strength to withstand wave and tidal action. As polymers with sugar acids such as mannuronic and guluronic acids, alginates form gelatinous substances that can absorb up to 200 to 300 times their own weight in water and provide the needed flexibility.

The alginic acid industry started in Japan in 1923, when a professor noticed that a leftover solution of sodium sulfide spread on a pile of discarded seaweed caused the seaweed to change consistency. This discovery, which demonstrated the hydrophilic nature of extracts from the alga, led to the commercial production of alginates in the United States in 1929. Most of the seaweed is harvested from the wild, though the Chinese have devised a system of artificial support for the cultivation of *Laminaria* for phycocolloid production.

FIGURE 18.19

Marcocystis harvesting in California, by law, involves cutting only to 4 feet beneath the water surface. This removes the upper portion of the canopy tissues, including the apical meristems. New growth comes from the apical meristems, which lie below the depth of the cutting blade. Regeneration takes place in approximately 2 to 4 months, permitting two to three harvests per year.

(Photograph by D. Cheney.)

Alginates are extracted from various brown algae (Phaeophyta), primarily *Macrocystis, Laminaria, Ascophyllum, Ecklonia,* and *Durvillaea.* The last four are typically harvested by hand from boats, although *Laminaria* is harvested mechanically in Norway. *Macrocystis* (Fig. 18.3j) harvesting differs because it is a large deep-water species with bulky, floating fronds. Consequently, it is usually harvested mechanically with a large reciprocating cutter that crops the surface growth while allowing future regeneration (Fig. 18.19). Commercial production (Fig. 18.20) usually includes leaching with hydrochloric acid to extract the soluble mineral salts, followed by alkaline treatment (sodium carbonate or ammonia) and filtering to yield an initial algin complex. Further treatment with acid precipitates out the algin, which is then refiltered and washed. This crude extract of calcium alginate is chemically treated to produce the free acid form or other salts for different purposes. New methods of mass culture are being developed that should substantially increase the harvest (Fig. 18.21).

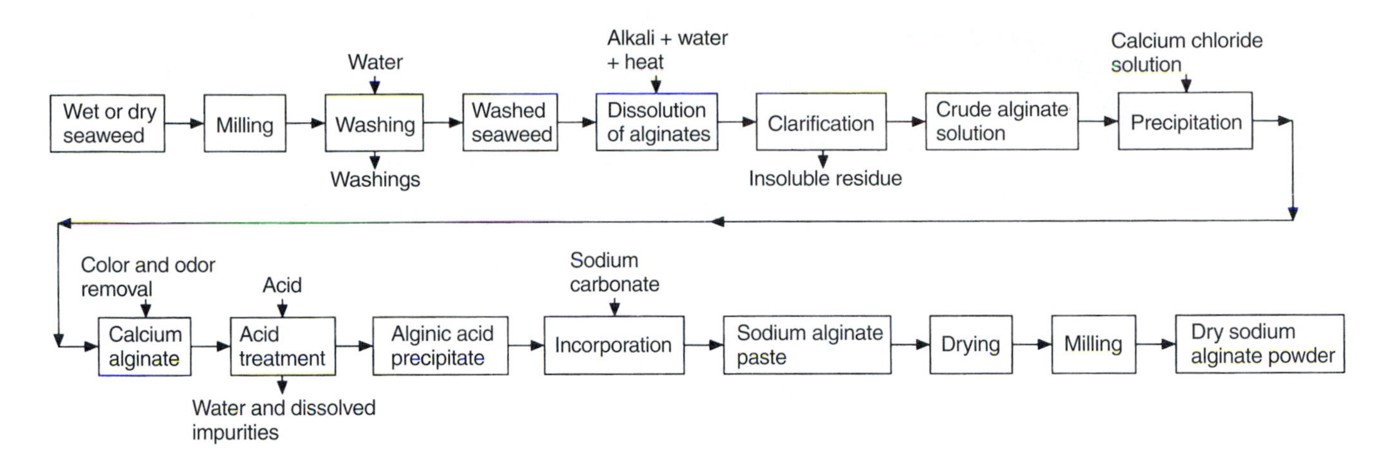

FIGURE 18.20

Flowchart showing the commercial production of alginates.

(Reproduced by permission of Merck & Co. [Rahway, N.J.] Kelco Division, U.S.A., and appearing in *Algae and Human Affairs,* Cambridge, U.K., Cambridge University Press 1988.)

Sodium alginate and potassium alginate salts have good water solubility and form thick fluids in solution. They are commonly used in the paper industry for sizing and polishing, in paints to keep pigments in suspension over a range of temperatures, as adhesives in the manufacture of products such as charcoal briquettes, and as suspending agents in foods, cosmetics, and medicines. People consume alginates every day in a wide range of foods. They are added to prevent ice crystal formation in ice cream, to keep icings from separating, as clarifiers and foam stabilizers in beer, as a filler in candy bars, and as emulsifiers in salad dressings.

Alginate salts formed with metal ions have poor water solubility and tend to form gels. These compounds find wide use in the production of cellophane-like materials, the canning of fruits, and the manufacture of artificial "foods" such as fake cherries, imitation caviar, and reconstituted onion rings. Because the alginates exhibit "pseudoplasticity"—that is, the solutions in which they are incorporated become less thick as more shear is applied—they are ideal in paint manufacture. While the paint is being applied, alginates help minimize brush drag and cause the brush marks to disappear quickly. After application, the paint recovers its viscosity and sagging and runoff are prevented.

The reaction of alginic acids with heavy metals such as mercury, beryllium, copper, and cobalt results in salts that are water-insoluble. They are used in the manufacture of plastic because they can be molded when wet but dry to a hard state when set. Specially treated alginate salts are used to produce fibers that can be spun, woven, and turned into cloth. This use was temporarily abandoned when it was discovered that the resulting thread would dissolve in soap solutions. However, improvements in the process have yielded more durable products that indicate that alginates could provide a nonflammable artificial silk. Alginate gels are used for dental impressions, and alginate fibers have been processed to produce high-quality audio speakers.

Carrageenans

Whole plants of Irish moss, *Chondrus crispus* (Fig. 18.22b), have been used for centuries in the making of jellies and milk puddings (blancmanges). This red alga (Rhodophyta) was imported into the United States until the early nineteenth century when the mayor of Boston, Dr. J. V. C. Smith, recognized that the same plants were growing off the coast of Massachusetts and initiated local harvesting. In 1862 E. C. C. Stanford, who earlier had discovered alginates, isolated the compound responsible for the thickening ability of the moss and called it carrageen (from the Irish *carrageen,* meaning "rock moss"). The name was later changed to carrageenan.

Carrageenans are phycocolloids that have an even wider range of applications than the alginates. These polymers are extracted principally from *Chondrus crispus,* in the North Atlantic and *Eucheuma* (Fig. 18.22a) species from the Philippines. The former are collected from the wild by hand either from small boats or along beaches. The latter are cultivated on networks of lines attached to poles or mangrove stumps that stand below the level of low tide (Fig. 18.23). Since 1971 improvements in techniques for cultivating *Eucheuma* have resulted in increases in production from 500 metric tons per year to over 25,000 metric tons per year.

The processing of seaweeds to produce carrageenans consists simply of washing the weeds in seawater to remove sand and then drying and bleaching them in the sun. The weeds can be used as is or can be further processed by digesting them with a hot solution of calcium or sodium hydroxide that promotes a swelling of the tissues and forces the components of the cell wall into solution. This treatment also removes sulfate groups from the polysaccharide chains, enhancing gel strength and reactivity to protein. The carrageenan is then separated by filtration or centrifugation

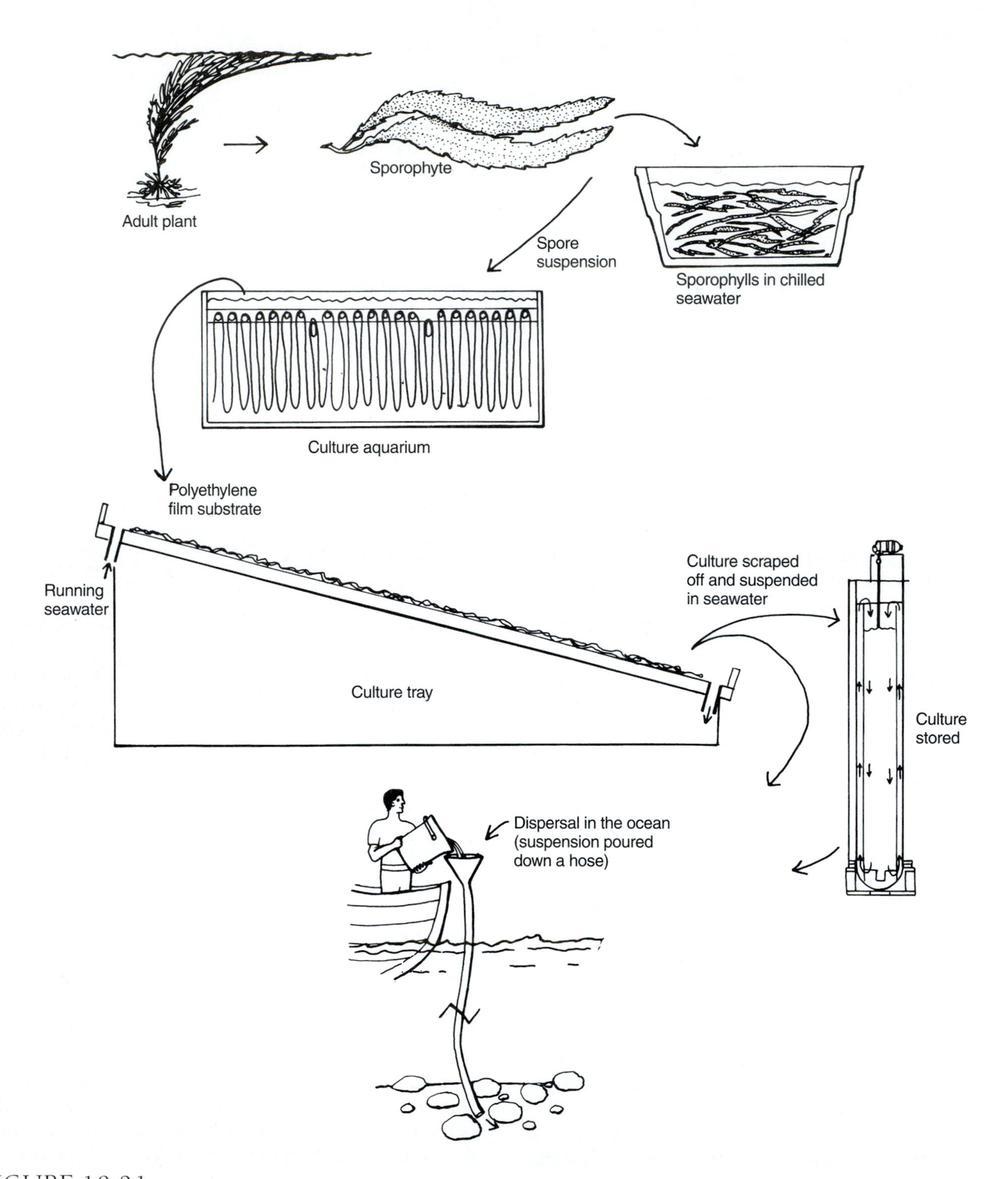

FIGURE 18.21

New techniques for mass culturing kelp include culturing sporophytes, scraping the spores from the substrate, and dispersing the embryos in the sea. This method has shown great promise for restoring the giant kelp beds of coastal California.

(From W. North, 1976. Aquacultural techniques for creating and restoring beds of giant kelp, Macrocystis spp. *Journal of the Fisheries Research Board of Canada* 33:1015–1023.)

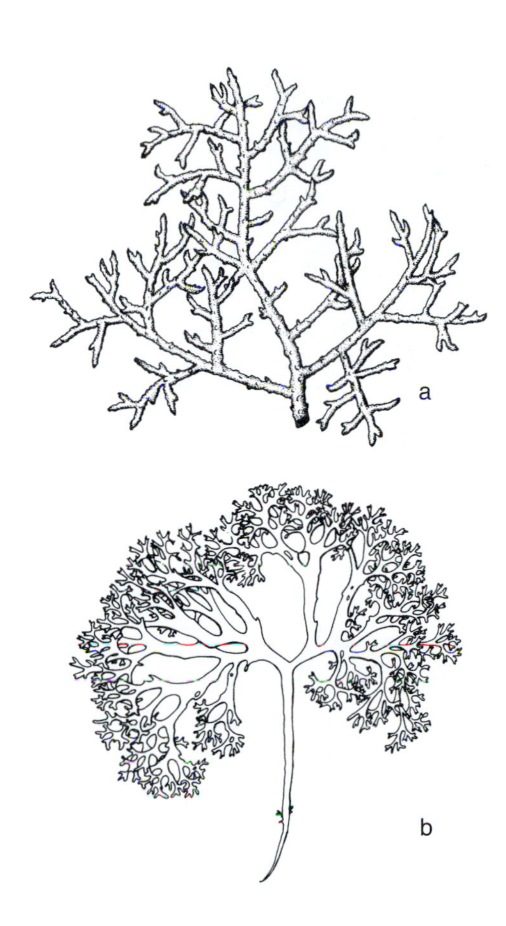

FIGURE 18.22
(a) *Eucheuma* sp. and (b) *Chondrus crispus* are important commercial sources of carrageenans.

followed by filtration. Other recovery techniques include drying, precipitation with alcohols, and a freeze-thaw process similar to that used in agar production.

Carrageenan is valued because its gelling, thickening, and suspending properties are thermally reversible, or capable of reversing states when the solution is heated to a liquid and then cooled back to a gel. Carrageenan is especially effective when in solution with proteinaceous materials. When it is added to hot milk that is subsequently cooled, bonds between the carrageenan and the casein protein molecules of the milk are formed, creating a creamy thick texture. Because of this unique property, carrageenan has replaced alginates in the manufacture of chocolate milk and packaged milk products such as yogurts, eggnog mixes, and ice creams. Because of its high melting point, carrageenan is especially useful for making dessert gels in hotter climates. Since it is immune to degradation by many common enzymes (unlike cellulose gum), it is used in toothpastes to bind the components and impart a sheen to the teeth. Besides its broad use in the food and toiletry industries, carrageenan is important as a sizing agent in the textile and leather industries. It is also used in an aqueous solution to suspend oil-based inks so that they can be combed or swirled to create marbleized designs on paper and fabric.

FIGURE 18.23
Chondrus (Irish moss) harvest off the coast of Prince Edward Island, Canada.
(Photo courtesy of D. Cheney.)

Agars

Perhaps the most indispensable phycocolloid is agar. The first reference to this substance occurred in Japan in 1658, where it was called *kanten,* meaning "cold sky," because it was traditionally produced during cold weather (Fig. 18.24). A physician's wife, Frau Franny Eilshemius, first discovered its microbiological applications. Her husband showed it to the renowned bacteriologist Robert Koch, who has received credit for the discovery. Agar quickly became an indispensable substrate for growing cells of all types. Agar is so important in scientific studies that during World War II, when Japanese supplies were cut off, the Allies were forced to find other sources. It turned out that sources lay close to home because many red algae contain agar and red algae as a group are distributed worldwide. Today agar is extracted predominantly from the red alga *Gelidium* (Fig. 18.24a) and to a lesser extent, from *Gracilaria, Pterocladia, Acanthopeltis,*

FIGURE 18.24

Polysaccharides from red algae had been used for centuries in the Orient, but the purification process was supposedly not discovered until 1658, when an innkeeper discarded some leftover jellies by throwing them outside on a cold evening. The freezing, thawing, and drying yielded flakes that, when rehydrated, produced a clearer, better textured jelly. Whereas the historical method of purifying agar entailed a series of freeze and thaw cycles, modern processing typically entails squeezing the agar in a hydraulic press. This illustration shows (a) *Gelidium,* one of the red algae commercially exploited for agar production, (b) a traditional furnace for boiling the seaweed, (c) a press used for purifying the gel, (d) pouring liquid agar into cooling trays, and (e) tools for cutting the gel into cakes and bars.

(Redrawn from a lithograph in a 1904 *Bulletin of the Bureau of Fisheries.*)

and *Ahnfeltia.* Different species yield agars with varying properties. The seaweeds are cleaned mechanically or chemically to remove sand, salt, pigments, and epiphytes; heated in hot water to extract the water-soluble agar; and centrifuged or filtered, or both, to remove the insoluble residues and concentrate the compounds. The agar is purified by submitting the concentrate alternately to freezing and thawing, using a hydraulic press to squeeze out the water and impurities. After drying, the resulting agar is sold in the form of sheets, flakes, or powders.

Chemically, agar is similar to carrageenan. However, the mixture of compounds in agar behaves differently, producing stiff gels in even smaller concentrations (as little as 1 to 2 percent aqueous solution) and demonstrating a remarkable capacity to reverse states when the solution is heated to a liquid and then cooled back to a gel. Because of this property, agar has been employed as an unsurpassed microbiological and tissue culture medium. When purified, agarose (the neutral galactose fraction) is used to produce the high-quality gels necessary for chromatographic and electrophoretic work. Because of its hydrophilic nature, agar is added to bakery products to keep them moist. Because it so readily complexes with proteins, agar is used to clarify wines, juices, and vinegars. After agar is added to these solutions, the offending proteins (which tend to "cloud" the product) aggregate with the agar to form a gel that can be removed by filtration or centrifugation. Agar has been used as a suspending agent for radiological solutions, a binder for medicinal tablets and capsules, and a bulk laxative.

Cyanobacteria As Fertilizers and Algae As Soil Conditioners

Seaweeds have been employed as soil amendments in coastal areas for hundreds of years. Records of the use of algae as a substitute for animal manure date to the fifth century A.D. In 1665, King Charles II of England granted permission to his subjects to collect seaweeds in tidal areas for agricultural purposes. Easements the width of one oxcart were granted by King Charles as a right-of-way for the collecting of seaweeds by farmers. These easements are still part of the land records in Rhode Island.

Because of their ability to secrete mucilage, soil algae such as *Chlamydomonas mexicana* are effective in loosening highly compacted soil. The algae cause the soil to aggregate enough for oxygen and water to move freely through it and thus allow cultivation. After the eruption of Mount Saint Helens in Washington in 1980, the ash layer became so compacted that farming was impossible. After *Chlamydomonas* (Fig. 18.4a) was introduced into the ash, the soil became loose enough for farming.

Seaweeds make good green manure because they are rich in potassium and nitrogen but low in phosphate. This profile makes seaweed a perfect complement to manure, which is high in phosphate (three times the phosphate content of seaweed), equivalent in nitrogen, but only one-third as rich in potassium. Seaweed has the advantage of being free of terrestrial weeds and fungi pathogenic to crops but has the disadvantage of needing to be hauled to farming areas. In Europe and north Africa calcareous algae have been used to reduce soil acidity. Liquid fertilizers made from algae have also been produced. Despite these achievements, the most interesting prospect for the use of algae as fertilizers is represented by the employment of living nitrogen-fixing blue-greens to enrich rice fields.

Although many nitrogen-fixing cyanobacteria could be employed to furnish nitrogen to agricultural crops, the use of the blue-green alga, *Anabaena azollae,* which lives in symbiotic association with the tiny fern *Azolla,* has the longest history. *Azolla* has been used to fertilize rice fields since the eleventh century in China (Chapter 5) and is still used there today and in Vietnam, India, and west Africa. Since *Anabaena* grows within cavities on the undersurface of fern leaves, its growth parallels that of the fern, and the size of the fern populations provides a simple indication of the size of the *Anabaena* population (Fig. 18.25). This feature has permitted small farmers in rural areas to adopt its use. In contrast with free-living *Anabaena,* the symbiotic species produce two to six times the number of **heterocysts,** structures in which nitrogen fixation occurs. The use of *Azolla* and *Anabaena* has been credited in Chinese studies with an 18 percent increase in rice yields. In a study by the International Network on Soil Fertility and Fertilizer Evaluation for Rice, the inoculation of fields with 0.2 kg of *Azolla* per hectare (0.4 lb/acre) was found to provide the equivalent of 30 kg per hectare (66 lb/acre) of commercial nitrogen-containing fertilizer.

If the use of *Azolla* has obvious appeal, why is it not more widespread? The limiting factor in the cultivation of *Azolla* is the availability of phosphorus. Since phosphorus is a nonrenewable resource, it does not make sense to grow *Azolla* and *Anabaena* for nitrogen fixation if a cheap alternative source of nitrogen is available. This is in fact why the use of *Azolla* has decreased in recent times. There is also the problem of producing inocula of *Azolla* and *Anabaena* adapted to local conditions. Still, if strains could be selected that could tolerate a range of conditions and have limited phosphorus requirements, this partnership between a blue-green alga and a fern could be a technologically appropriate way to improve rice yields and could even lead to other uses for heterocystous species.

Harvesting Fossil Algae

The long history of algae on the earth has left a rich fossil legacy that provides not only clues to evolution but useful "mineral" products as well. It would be hard to overestimate the importance of the raw materials provided by these ancient relics of early life on the earth. Accumulations of calcareous algae have created reefs that are vital marine habitats and have left limestone deposits that lend themselves to use as building materials and in the manufacture of cement, plastics, glass, and chemicals. Ground limestone has long been added to soils to neutralize acidity. It has more recently been employed to reduce the acidity produced in the atmosphere and water bodies by pollution from the burning of sulfur-laden coal. Finally, the silica-containing remnants of diatomaceous algae provide an indispensable raw material with a host of applications.

The diatoms (Bacillariophyceae) are placed in the Chrysophyta, along with other golden-brown and yellow-green algae. They are unique in design; they sequester silicon dioxide in their cell walls, forming a hard case (**frustule**) that persists after the protoplast dies (Fig. 18.26). The structure and ornamentation of the frustule surface not only are beautiful (Fig. 18.27) but also present a large surface area that is highly porous. This unique combination of characteristics lends itself to uses in products ranging from filtering agent to flea, cricket, and cockroach killers, enamel polishes, and embalming compounds.

Deposits of diatomaceous earth formed from the slow settling and compaction of marine and freshwater diatoms over millions of years have been found in pure layers up to 300 m (982 ft) in depth. These sediments have been used in the manufacture of lightweight brick since at least A.D. 532, when the Emperor Justinian specified that the dome of the

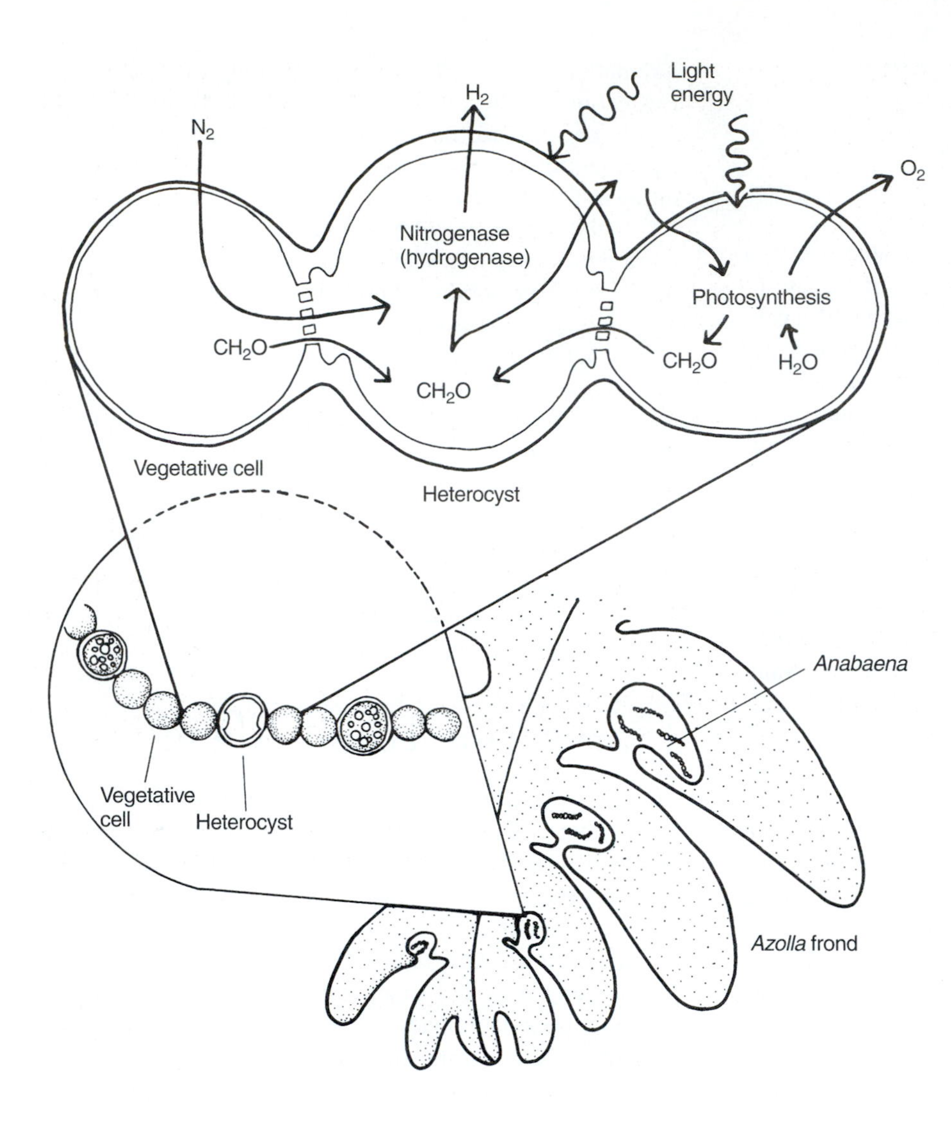

FIGURE 18.25

A schematic drawing showing *Anabaena* growing within dorsal lobe leaf cavities within a frond of *Azolla* and the flow of nitrogen and photosynthate from vegetative cells into thick-walled cells called heterocysts. These cells are impermeable to oxygen since the process of nitrogen fixation is inhibited by its presence. Interestingly, blue-greens that live in symbiotic associations fix both nitrogen and carbon at the same time and have two to six times as many heterocysts as do their free-living counterparts.

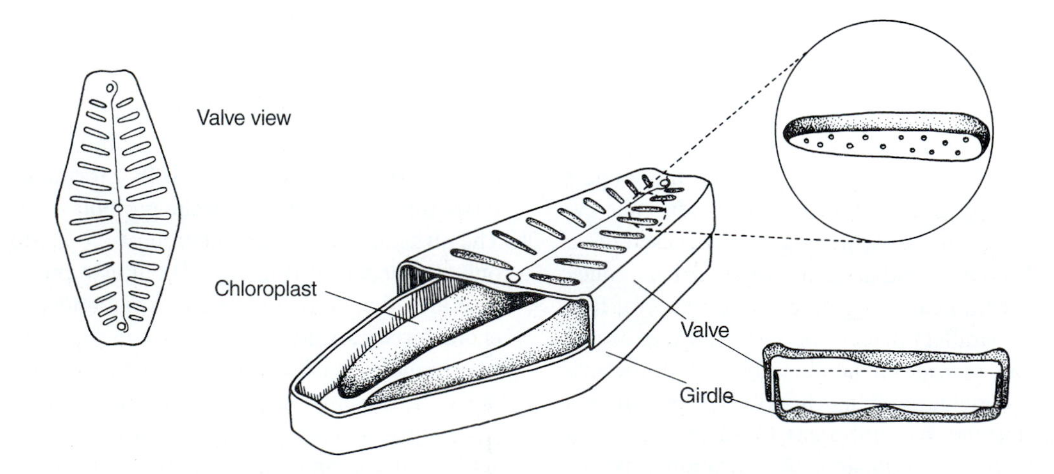

FIGURE 18.26

The cell walls of diatoms fit together like the halves of a petri dish, with the valve like the top and the girdle like the bottom. The patterns of sculpturing and perforations on the frustule (hard case) surfaces are characteristic and are used in identifying species.

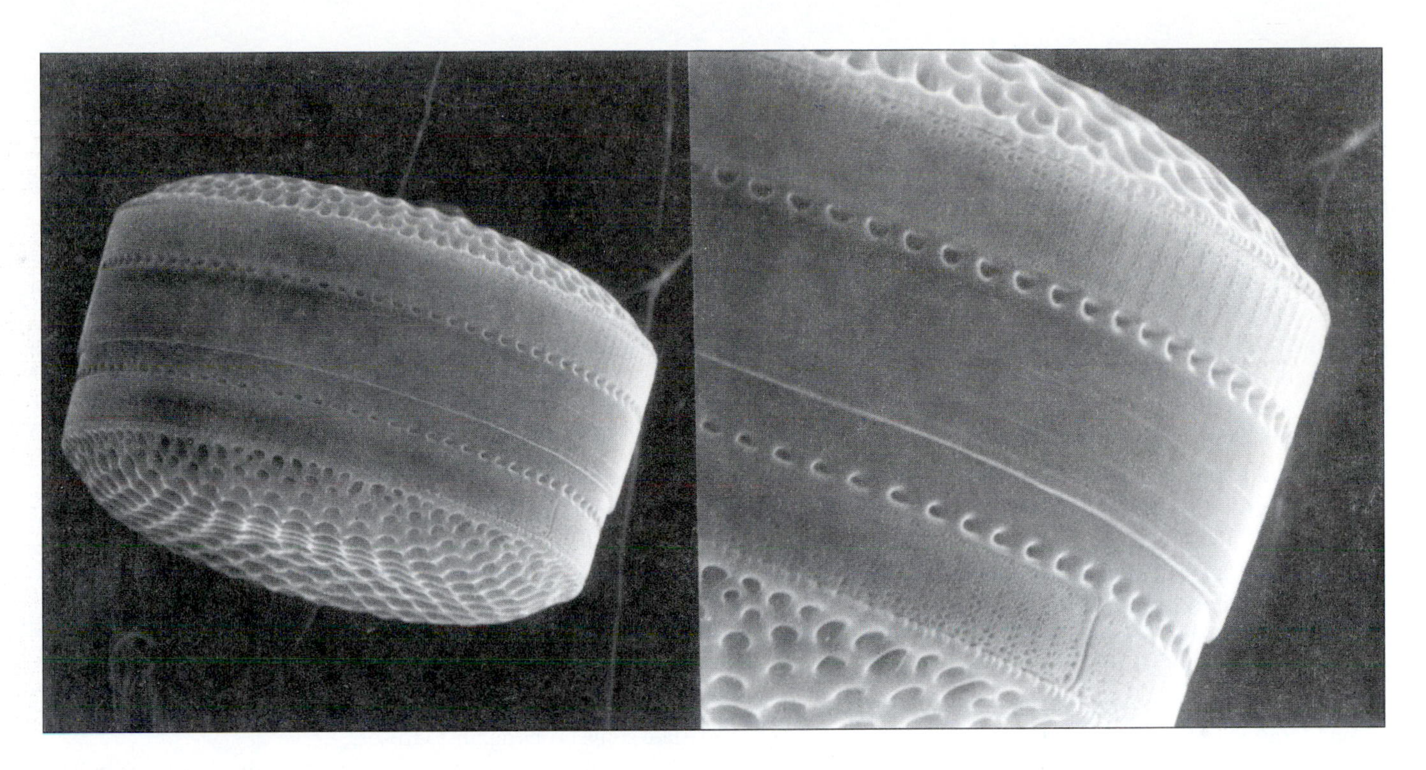

FIGURE 18.27
The elaborate patterns on the surface of diatom frustules led to their use in testing the resolution of lenses. This photo taken with a scanning electron microscope shows a whole cell and a side, or girdle view, of *Thalassiosira oestruppi*, a small common open ocean diatom from a warm ring of the Gulf Stream in the North Atlantic.
(Photos courtesy of G. Fryxell.)

church of Hagia Sophia in Istanbul be constructed from diatomaceous bricks. Commercial mining of diatomaceous deposits or diatomite started in the 1850s in Germany. The first uses were as polishing agents, as the silica-containing frustules act as fine abrasives that leave few scratches, since they collapse under pressure. Their insulation properties and utility as filtering agents were recognized later. In 1876 diatomite was used to filter sewage in the United Kingdom, and in 1914 sugar cane manufacturers began to use it to filter sugar liquors. Alfred Nobel discovered that diatomite stabilizes nitroglycerin in the production of dynamite, and undertakers figured out that the cell walls of the dead algae constitute a cheap, inert filler material. Other uses include the incorporation of diatomite into match heads and cigars to control burning, into paints to reduce glossiness (especially important in camouflage), into fertilizers and dried foods as anticaking agents, as ingredients in pet litter because of their absorptive capacities, and for use on pets to render their fur inhospitable to fleas.

The bulk of the world's diatomaceous silica is still produced by the United States, with Russia, France, and Denmark also being important suppliers. Most of the diatomite is mined in open pits, particularly in the United States, although some underground mining is practiced in Europe. After being extracted from deposits, the diatoms are milled, dried from their original 60 percent to 5 percent moisture, and sorted according to size. Because of concerns about waste disposal of diatomite cake, the development of new filtration technologies, such as membrane filters, and the diminished use of diatomite as fillers in paints, production has been slowing. Meanwhile, new applications, such as diatomite filters for biotechnology may create renewed demand for these fossil filters.

Algae in Wastewater Treatment

The employment of the skeletal remains of diatoms as effluent filters is not the only role algae play in sewage treatment. Several pilot plants are proving the viability of capitalizing on the natural role of algae as water purifiers to create integrated microbial algae-bacterial systems to handle waste.

Traditional sewage treatment begins with the physical settling out of the solid fraction of waste, a process known as **primary treatment. Secondary treatment** is usually effected with a mixture of fungi and bacteria that act on the waste to convert the soluble organic matter into an insoluble form that precipites out as sludge. Secondary treatment significantly reduces the **biological oxygen demand** (BOD; the amount of oxygen required for oxidative decomposition of materials present in the water) of the final liquid fraction that is discharged into local lakes, rivers, or the ocean. A low BOD means that comparatively little oxygen is removed from ("demanded of") the water into which this "treated sewage" is discharged. Sewage discharged with a high BOD would deplete the oxygen of the water into which it was released.

When a suspension of an activated algal-bacterial culture is maintained in sewage and exposed to natural light, a process called bio-oxidation takes place. The microbes convert the soluble organic matter to insoluble forms as described above, but in the process the algae release oxygen into the media that helps maintain the aerobic bacteria. The algae and bacteria used in this process also absorb nitrogen and phosphorus from the sewage suspension and use them as nutrients for their own growth. Under photosynthetic oxidation, the BOD is reduced 90 percent, and 80 percent of the nitrogen and phosphorus are removed, thus effecting a degree of **tertiary treatment.** In practice, treatment plants using this technique (such as the one in Hollister, California) have proved to be cost-effective without even considering the reduced environmental impact of released effluent that has undergone tertiary treatment.

Algae in Medicine

In the Western Hemisphere the role of algae in medicine has generally been restricted to the use of phycocolloids as binders and emulsifiers for pharmaceutical preparation and wound dressings. By contrast, Oriental medicine has relied on algae both for their physical properties and as sources of bioactive compounds. As Western medicine broadens its scope, the uses of algal hydrocolloids are being expanded and the therapeutic reputations of drugs derived from algae are receiving more attention from scientists.

Although the employment of phycocolloids began with industrial uses, their medicinal role is expanding. Recent uses include polymerization of alginate fibers to create calcium alginate "wools" that are used as absorbent wound dressings. Because of their ability to complex with metals, alginates and carrageenan have been suggested as **chelating agents** (compounds that complex with metal ions) for the treatment of heavy metal and radioactive nucleotide poisoning. Dried, sterilized *Laminaria* stipes are used during obstetric and gynecological procedures because of their colloidal properties. After insertion into the cervical canal, the "*Laminaria* tent" takes up fluids from the surrounding tissues, slowly swelling to three to five times its original circumference; resulting in gradual dilation without the trauma produced by mechanical devices.

As a drug source, algae have been used in Oriental folk medicine for the treatment of everything from parasites to coughs, gout, goiter, hypertension, venereal disease, and even tumor growth. In modern Japan two of these applications are medically significant. *Laminaria,* a rich source of iodine, is prescribed for the treatment of **goiter** (an enlargement of the thyroid gland accompanied by abnormal metabolism resulting from a deficiency of iodine), and an extract of *Digenia simplex* is administered for internal parasites. The use of *Digenia* as a **vermifuge** (a substance that expels parasitic worms) was purportedly learned by observing and emulating the eating habits of the long-lived dugong, a marine mammal that inspired the myth of mermaids. In addition to sea grasses, this mammal ingests large amounts of *Digenia,* which is now known to contain kainic acid, a potent vermifuge.

The potential for extracting medicinal drugs from algae is multifaceted. Fatty acids, terpenes, tannins, and bromophenols obtained from algae have all shown antibiotic activity in experimental situations. Polysaccharides obtained from red algae have been shown to control the herpes virus that produces cold sores and genital warts and are the active compound in the drug Zovirax. The action is due to the ability of these polysaccharides to interfere with the virus's attachment to human cell membranes. Alginic acid, an algal extract marketed under the name Nomozan is used to inoculate some crops against viruses.

In anticancer studies extracts from *Sargassum* sp., *Codium pugniformis* (Fig. 18.3d), and *Laminaria japonica* have been shown under laboratory conditions to inhibit tumor growth. In addition, *Spirulina* extracts inhibit the growth of oral cancer cells, and algal pigments known as phycocyanins, which are present in both red and blue-green algae, show some antitumor effects and boost the immune system by increasing the number of spleen lymphocytes. Since algae are a rich source of beta-carotene, which has been linked with low rates of cancer (see Chapter 7), they may eventually constitute an important source of this compound.

Recently, a number of cyanobacteria have shown particular promise as sources of antiviral compounds. *Lyngbya majuscula* produces a compound known as curacin-A, which has been shown to inhibit cell division, suggesting its use in the fight against cancer. Water extracts of *Spirulina* contain a compound known as calcium-spirulan, which has been shown, in vitro, to inhibit replication of HIV-1, herpes simplex virus, human cytomegalovirus, and the viruses that cause influenza-A, mumps, and measles. Calcium-spirulan is a polymerized sugar molecule that has also been shown to enhance the activity of the cells and organs of the immune system. In Russia, where the children of Chernobyl were suffering from radiation poisoning, extracts of *Spirulina* were administered to improve their immune system and reduce the allergenic sensitivities they were experiencing. A freshwater blue-green algae, *Nostoc ellipsosporum* has yielded a different compound that is active against HIV. Cyanovirin-N (CVN) is a protein that forms an irreversible bond with HIV that prevents it from infecting cells. Unfortunately, CVN cannot be used as a vaccine because the antibodies produced in response to vaccination inactivate the protein's continued activity, and the resulting complexes of antibodies could prove dangerous in their own right. Still, it is thought that it might be used on a one-time basis to help stop the transmission of the virus. For instance, it might be used in cases such as accidental needlesticks or to protect infants from maternal transmission. Researchers are also looking into the possibility of vaginal microbiocides containing CVN, in hopes that the buildup of antibodies might be avoided with vaginal administration.

Seaweeds that contain sterols and related compounds have been shown to reduce cholesterol levels in the blood and lower blood pressure. *Laminaria japonica* is just one seaweed administered for hypertension. Its effectiveness may be related to its iodine content. Many algae have been found to possess compounds that are thought to regulate blood cholesterol in the same way that naturally occurring thyroid hormones operate. *Spirulina, Chlorella,* and *Scenedesmus* (Fig. 18.16c) have also demonstrated the ability to lower cholesterol. *Spirulina* provides the added benefit of behaving as an appetite suppressant, which may help patients who are at higher risk of hypertension because of obesity. *Spirulina* is often eaten by vegetarians because of its high content of the B vitamins. However, the ingestion of over 50 g (1.8 oz) a day causes nausea unless a tolerance to the alga has been acquired. There is also experimental evidence of potent anticoagulant effects of polysaccharides extracted from the brown alga *L. digitata* and the red alga *Dilsea edulis.* Carrageenan extracted from a variety of sources has proved to have a weak anticoagulant activity.

Other important applications for algae are as research tools and in diagnostic work. The vermifuge mentioned previously, kainic acid, breaks down nerve dendrites at higher doses, producing symptoms similar to those seen in patients with Huntington's disease. Modern scientists have employed kainic acid in research on epilepsy. Phycobiliprotein pigments have proved to be good sources of sensitive fluorescent dyes. In diagnostic applications, these dyes are being used to track antibodies.

Toxins from Algae

Once one realizes that algae, like higher plants, possess bioactive compounds, it is not surprising to learn that they can produce deadly toxins. As with vascular plants, many of these compounds are thought to be valuable to the plant primarily as a way to deter herbivory, but because of the differences in predators (grazing invertebrates and fish) and habitats, the distributions and modes of action of algal toxins are unique. Because they manufacture their own food, algae were thought to rarely act as pathogens or parasites. Only *Chlorella* (Fig. 18.16a), the achlorophyllous *Prototheca,* and more recently *Pfiesteria,* have been implicated in directly infecting people and animals.

Toxic seaweeds are found predominantly in the tropics, where there is a high level of grazing and predation. In the

FIGURE 18.28
A close-up of *Caulerpa racemosa,* one of the edible species of *Caulerpa.* Other, toxic species are responsible for wreaking havoc in coastal ecosystems where they have escaped and taken over. (Photo courtesy of J. Huisman.)

United States and Europe people are able to avoid consuming these species, and so, with the exception of *Hizikia,* which contains arsenic, and *Caulerpa* (Figs. 18.3i and 18.28), whose less toxic species are collected for salads in the Pacific, toxic seaweeds are not much of a direct threat to humans. The economic importance of toxic algae may result from our ability to utilize them as sources of pesticides, as in the case of *Chondria armata.* The pesticidal properties of this species were discovered when someone noticed that flies landing on the dried fronds died.

The species of microalgae that pose a health threat to humans are for the most part a problem only when they "bloom." That is, as a result of ecological conditions that favor the growth of a particular species, that species proliferates and dominates the ecosystem. Different factors have been identified as being associated with blooms, including upwellings of marine water or freshwater that bring nutrients to the surface and heavy rains that wash phosphates into the sea and lower its salinity. Low salinity promotes the growth of dinoflagellates (a group of primarily marine, plantlike organisms with flagella). Temperature increases also promote the germination of **hypnocysts** (the resting stage of some of the culprits) and the growth of the population of these cells. The adverse impacts of these algae are threefold: they can act directly on humans; they can lead to massive fish kills; or their toxic products can be sequestered in filter-feeding organisms that are consumed by animals that then become sick.

Direct effects on people include dermatitis caused by a range of algae. "Swimmer's itch" results from contact with water containing large quantities of *Oscillatoria nigroviridis.* Contact dermatitis has also been produced by blooms of the freshwater blue-green algae *Anabaena, Lyngbya majuscula,* and *Schizothrix calcicola.* Two of the active compounds in

Lyngbya, an alkaloid and a phenol, are responsible for the irritation and have produced tumors in experimental animals. Various red algae have also been shown to produce respiratory distress when people breathe air containing them. **Silicosis** (chronic lung infection resulting from breathing silica dust) has been reported to result from the inhalation of diatoms; this fact might make one think twice about using diatoms on dogs and cats to repel fleas.

Fish kills are most often produced by algae either directly as a result of the movement and accumulation of toxins within the food chain or indirectly through oxygen depletion caused by the respiratory activity of the algae or that of the bacteria during decomposition of dead biomass. The alga most often associated with fish kills is *Prymnesium parvum.* This organism of brackish or marine water produces a toxin that alters the permeability of gill membranes to oxygen and does not need to be abundant to affect fish.

The most sinister algal toxins are those involved in the massive fish kills and shellfish poisonings associated with "red tides." The first reported case of this deadly phenomenon occurred in 1530 in the Virgin Islands. Since then there have been over 500 documented deaths of humans from red tide, with thousands affected by eating infected shellfish. The organisms involved are dinoflagellates (Pyrrhophyta), some of which are not red and many of which exhibit bioluminescence. Filter-feeding shellfish that consume planktonic organisms are unaffected, but humans who eat the tainted shellfish are susceptible. Savvy natives learned to recognize luminescent tides as a sign of danger. Today federal monitoring of shellfish has kept casualties to a minimum. Although red tides do not constitute a major health threat, their insidious nature has caught the imagination of seafood fanciers, who tend to avoid all ocean fish (even nontarget organisms) when an alert has been called.

In 1991, another dinoflagellate with a fascinating life history was linked to fish kills and human symptoms not associated with the consumption of shellfish. *Pfiesteria* is an ambush predator with a complex life cycle consisting of two dozen life stages, a few of which produce toxins. This toxic algae whose blooms have been directly linked to water pollution, has been shown to proliferate and develop into poisonous forms in the presence of fish secretions or excrement. Found in brackish waters and estuaries on the U.S. East Coast, it stuns fish, which first become lethargic and later develop characteristic lesions. The fish are not killed directly by the alga but most often by secondary infections of these lesions. The effect on humans has only recently been established. Unlike other dinoglagellate poisonings, this toxic syndrome comes not from the consumption of shellfish but from contact with infected fish or the water. Symptoms include memory loss, confusion, respiratory and gastrointestinal distress, and skin problems.

Another type of poisoning resulting from dinoflagellates is known as paralytic shellfish poisoning (PSP). Paralysis is caused by an array of a dozen different poisons that block the movement of sodium in the nerves. PSP is

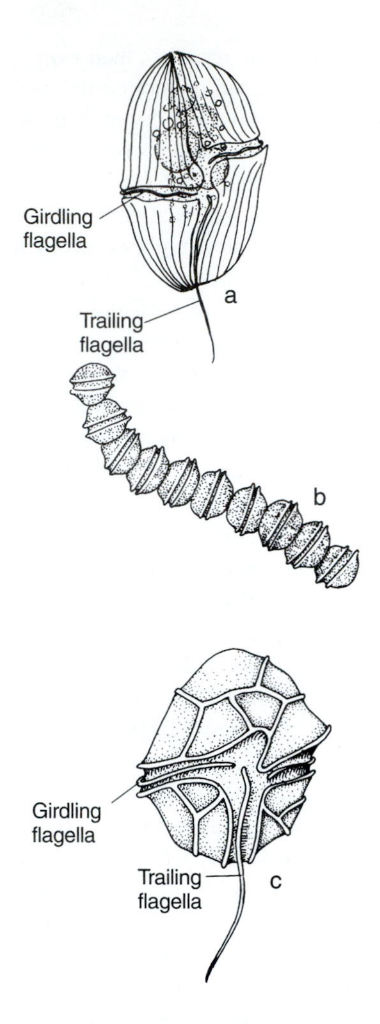

FIGURE 18.29
Dinoflagellates are characterized as being (a) naked or (c) armored, depending on the visibility and thickness of the cellulose plates covering their surface. (c) They are also typically biflagellate, with one flagellum girdling their midsection and another trailing. Shown are (a) a species of *Gymnodinium* and (b, c) two species of *Gonyaulax* one of which is unicellular and one which is composed of chains of cells linked together. These species have been implicated in causing toxic red tides.

associated with blooms of *Protogonyaulaux* sp. (*Gonyaulax*) (Fig. 18.29 b,c) and produces nausea, vomiting, diarrhea, tingling of the extremities, disorientation, and paralysis that can lead to death. This same organism can cause fish kills when its toxins become concentrated in the filter-feeding zooplankton. Neurotoxic shellfish poisoning (NSP) is known from red tides in Florida, where *Ptychodiscus* (*Gymnodinium brevis*) (Fig. 18.29a) has produced numbness and symptoms resembling food poisoning but no known fatalities. The toxin in this case is lipid-soluble, with a different mode of action. The bottom-dwelling dinoflagellate *Gambierdiscus toxicus* (Fig. 18.30) produces a syndrome known as ciguatera poisoning. The toxins are found concentrated in the organs of coral reef fish, and people who eat them suffer nausea, abdominal cramps, and muscle weakness.

Box 18.1

The Sinister Side of Algal Toxins

What does Alfred Hitchcock have to do with harmful algae? Scientists now think that the inspiration for his classic psychothriller *The Birds* was the result of an incident involving a toxin from an unlikely algal source. This newly discovered phytotoxin has now been traced up the food chain and implicated in other dramatic ecocatastrophies along the U.S. West Coast.

In August of 1961 hundreds of seabirds started acting erratically, flying into buildings and streetlights, making noises like babies crying, and attacking citizens in the Santa Cruz, California, area. It wasn't until 30 years later when cormorants and brown pelicans exhibited similar bizarre behaviors that biologists were able to speculate about the cause of the frightening bird behavior on which Daphne DuMaurier based a short story on and Hitchcock brought to the big screen.

The detective work in the 1991 mystery included examining the content of the cormorants' and brown pelicans' stomachs and discovering that the anchovies they had ingested contained domoic acid, a toxin produced by several algae including a pennate diatom *Pseudonitzschia australis* found in those waters. Domoic acid was first discovered and identified in 1987, when 107 people had become ill and 4 died after consuming cultured blue mussels on Prince Edward Island. Because their symptoms included both gastrointestinal distress and a host of neurological symptoms such as dizziness, seizures, short-term memory loss, respiratory distress, and even coma, this syndrome was dubbed amnesiac shellfish poisoning (ASP).

Scientists were fascinated with ASP for many reasons. First, it was produced by a diatom, an alga in a whole different taxonomic order from the dinoflagellates that were responsible for previous algal poisonings. Another interesting aspect is how the toxin accumulates in the food chain, not only in shellfish but also in crabs and schooling fish (anchovies and sardines). Domoic acid's effect on mammalian physiology is also curious, as it is a potent neurotoxin with a mode of action similar to (but obviously stronger than) the food additive monosodium glutamate (MSG).

Like MSG, domoic acid enters the bloodstream, passes across the blood-brain barrier and binds to nerve cells. The resulting excitation by MSG causes headaches in some people. With domoic acid, the effects are more dramatic and long-lasting, since it is an amino acid that can reside in the brain a long time with the potential to kill or incapacitate the mammalian nervous system.

Though ASP outbreaks have been rare, the sensational nature of the symptoms never fail to capture the public's imagination and cause concern. Current monitoring programs along the West Coast are not thought to be adequate to anticipate occurrences, and the conditions leading to *Pseudonitzschia* blooms need more research. Still scientists suspected that there might be a link between elevated algae levels in water samples and the 1998 death of 400 sea lions along the coast of California. Using a DNA probe to track the toxin, they have been able to follow domoic acid through the food web, establish it as the culprit, and add to our knowledge of this insidious poison.

FIGURE 18.30
An S.E.M. of *Gambierdiscus,* one of the first dinoflagellates from coral reef habitats found to produce ciguatoxins, responsible for ciguatera poisoning.
(Photo courtesy of D. Tindall.)

Whether toxic algae blooms and incidences of shellfish poisoning are on the increase or simply our detection and awareness have been heightened is the subject of continued debate. Toxic algae catastrophies in 1961, 1991, and 1998 involved both (Box 18.1).

Algae As Nuisance Organisms

Compared with the specter of algae as precursors of red tide epidemics, slimy layers on ship hulls may seem a trivial concern. Therefore, it may come as a surprise to learn that marine biofouling perhaps has the most significant economic consequences of anything discussed in this chapter.

Proliferations of algal weed populations can be nuisances that humans have yet to control.

Researchers estimate that a slime layer only 1 mm (0.039 in.) thick can result in a 15 percent loss in ship speed and an 80 percent increase in skin friction compared with a clean hull. The process of biofouling occurs on ship hulls, submarines, offshore oil and gas installations, and oceanographic equipment. As soon as such structures are submerged, the process of ecological succession begins, with bacteria, diatoms, and blue-green algae colonizing the surfaces and setting the stage for macrofouling by diatoms, filamentous algae, and then larger animals and plants. The filamentous algae involved are usually species of *Enteromorpha* (Chlorophyta) (Fig. 18.3g) and *Ectocarpus* (Phaeophyta) (Fig. 18.4c), which can produce spores that disperse in the water and tolerate the wide fluctuations in salinity and temperature encountered on ocean voyages. The effects of friction from biofouling and the added burden of the biomass itself were estimated by the Center for Biofouling and Bioinnovation to cost $6.5 billion in 2000.

For at least 2500 years humans have tried to control algal fouling. Early methods included coating or impregnating wooden hulls with animal grease, tar mixtures, and toxic compounds such as arsenic or sulfur mixed with oil; charring the surface of the vessel; and applying copper or lead sheets. Copper sheeting gained favor in England in the mid-eighteenth century but became inappropriate when iron-hulled ships were developed because of interactions between the metals. Today control treatments include mechanical removal, electrochemical chlorination, and the application of antifouling coatings (paints to which copper compounds have been added). The compounds leach out at a constant rate and discourage algal growth. With the discovery of several copper-tolerant algae in the 1950s, organic compounds containing tin were developed. Since the environmental safety of these compounds is being questioned (several countries have outlawed their use), the search for biological controls continues. One branch of research is investigating substances naturally produced by corals to ward off microbial attack.

Algae are considered noxious weeds when they proliferate to nuisance levels in bodies of water, usually as the result of runoff from streets, lawns, golf courses, sewage treatment systems and farms. *Legionella* (a bacterium) in association with blue-green algae form mats in air-conditioning systems that can cause musty odors and foul air. As a result of nitrogen and phosphorus enrichment, algae can grow out of control, clogging irrigation and drainage channels and providing shelter for mosquitoes and other pests. Pernicious algal growth can deplete the water system of oxygen when respiration exceeds photosynthesis (on cloudy or cold days, for example) and after death, when the algae are consumed by bacteria. When large algae overtake aquaculture systems, fish production can drop by half. Epiphytic algae can become so numerous that they kill off sea grasses by coating the leaves and blocking the sun. Filamentous mat-forming algae, such as *Pithophora, Spirogyra,* and *Hydrodictyon* (Fig. 18.3f) thrive in nutrient-enriched water, reducing the appeal of waterfront property for homeowners and swimmers.

In some cases, the damage goes beyond aesthetics, as voracious seaweed growth threatens the existence of once-productive ecosystems. *Caulerpa verticillata,* a species native to the Florida coast, has responded so well to heightened nutrient levels and lower salinity brought on by fertilization and water diversion that its growth is threatening the reef community. Since this species generates its own toxins, fish do not eat it, so it eventually replaces other vegetation, forming a vast monoculture that is inhospitable to organisms further up the food chain.

Another species of *Caulerpa* native to the Caribbean has created a full-scale ecological disaster. In this case, the disaster is not due to nutrient enrichment but to the alga's introduction into a foreign ecosystem where it has no natural predators. *Caulerpa taxifolia,* a common species in tropical aquariums, escaped into the Mediterranean off the coast of France and has now become an unmanageable invader. From a square-yard patch discovered in 1988 populations have grown more than 50 percent per year and now cover an area equal to 6000 football fields. Traditional methods of controlling alga growth with algicides containing copper sulfate (which breaks down easily in water) are not feasible on this scale, and the alga is quickly spreading to other areas of the Mediterranean Sea with the same loss of diversity that Florida experienced. Research is now being conducted on importing sea slugs to try and control the spread of the "alga that swallowed the Mediterranean," and a scandal over the government inaction that allowed the situation to get out of hand has erupted.

In light of the potential to shut down marine ecosystems, the economic consequences of weedy algae are hard to estimate. Although algicides are practical in small scale situations, the world is now facing much more dire consequences of our tipping the ecological scales. Part of the solution is to cut back the source of the problem by advocating moderate fertilization practices, curtailing the use of detergents containing phosphates, building drainage ponds for erosion control, and reducing the diversion of water from shallow lakes, but more needs to be done. In the face of current ecocatastrophies, biological control, even if at a snail's pace, may be the last chance we have to restore the balance in these already devastated areas.

Use of Algae in the Future

As humans have learned more about algae, it has become apparent that they represent a vast, largely unexplored resource with a potential for manipulation and creative uses beyond what has been mentioned in this chapter.

Aquaculture is a relatively efficient means of producing protein, comparing favorably in terms of the ratio of weight gain to feed intake attained in poultry and cattle production.

Since fish are cold-blooded, they expend less energy on respiration. Filter-feeding shellfish are even more efficient than fish, since they lead such sedentary lives. Algae are a major component of aquaculture and mariculture either as a crop (as discussed earlier) or as vital primary producers in the system. In shrimp cultivation, phytoplankton cultures are used for shrimp larvae until they are large enough to consume zooplankton and eventually larger food such as deep-sea organisms, chopped mussels or clams, or artificially constituted diets. In shrimp farming many species are used, including diatoms such as *Chaetoceros, Skeletonema* (Chrysophyta), and the green flagellate *Tetraselmis* (Chlorophyta). If diatoms are used, they are cultured in a medium enriched with silica and added daily to the stock tanks.

In the Orient, where industrial uses of algae have always been prominent, aquaculture practices are being employed to grow everything from eels to oysters. With more information becoming available from research into the needs of both the plants and the fish, the methods should spread and improve the nutrition and income of rural populations as well as providing epicurean delights for gourmets.

If people could safely and economically combine the technology of domestic waste treatment with algae and aquaculture, it might be possible to use sewage effluent as a source of nutrients for the algae. This is being done on a small scale but needs to be pursued more widely.

Algae are being looked at as sources of pulp for paper, and if they prove a good source, deforestation might be slowed. Since they are natural indicators of pollution, algae could be used more widely as bioindicators to alert people to potential problems. One of the major uses of algae is for the detection of pollutants in water systems. Conversely, algae can be employed to remove or detoxify water polluted by heavy metals and other toxins. In fact, recent research has shown that genetically altered *Chlamydomonas reinhardtii* cells are able to behave like "heavy metal sponges," selectively removing copper, zinc, lead, cadmium, mercury, and nickel from contaminated water. Algae may also be increasingly used in microbial mining operations to recover and concentrate gold ions from the ocean or from sludges made from mining residues.

One of the most intriguing prospects for algae is to prolong human space travel (Fig. 18.31). Algae are thought to be the ideal organisms for such semiclosed systems since they exhibit rapid growth and have relatively controllable metabolisms. On long missions it is hoped that algae could remove CO_2 from the air and replace it with O_2, help in the recycling of human wastes and water, and provide a proteinaceous food source. Research has already been conducted on the common genera *Spirulina, Chlorella,* and *Scenedesmus* (Fig. 18.16).

Despite the promise of algae in space, there are many challenges to overcome. There is some fear that the accumulation of toxic contaminants (including carbon monoxide) could pose problems. There are also questions about the feasibility of algae as a food source. The presence of indigestible components (cell walls, etc.) and unpleasant flavors, textures, and odors would have to be overcome before space voyagers could be satisfied with algae on their plates. In the meantime, research aimed at tailoring systems in space is expanding the frontiers of knowledge about the potential uses of this vast array of organisms.

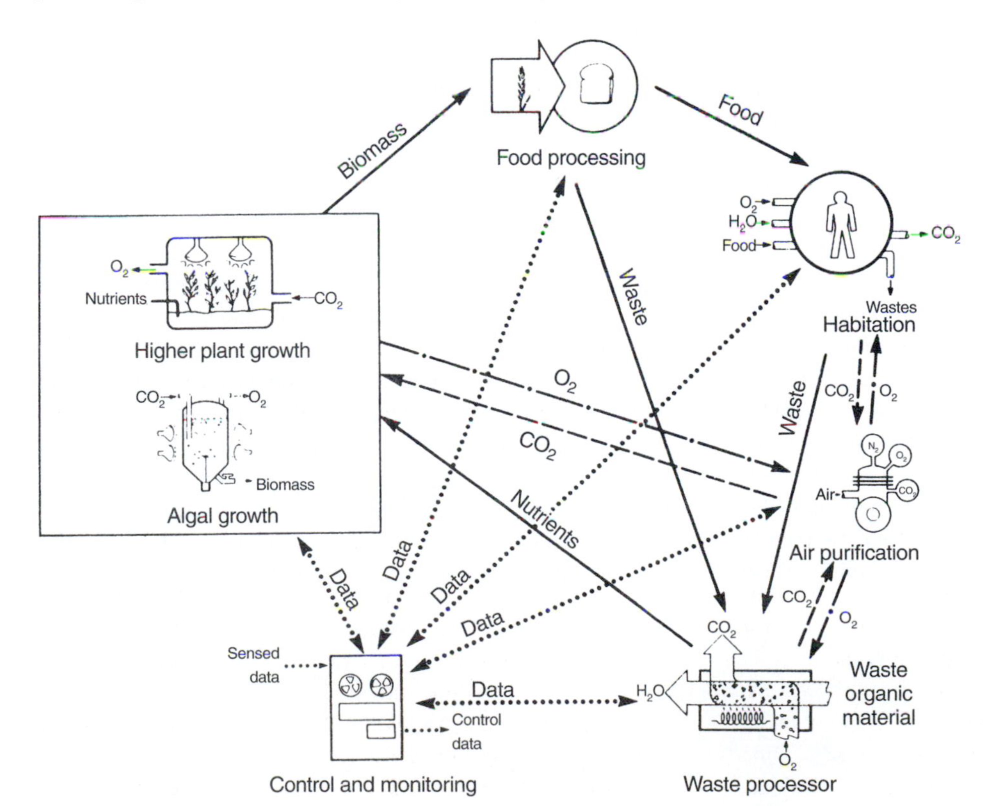

FIGURE 18.31

This diagram shows how algae could theoretically function in a life-support system. The arrows indicate the interaction between algae and higher plants, food and waste processing, and air purification. A bioregenerative system might enable people to carry out prolonged space missions.

(Diagram provided courtesy of NASA.)

19

Uses of Plants in the Future

Throughout this book we have discussed the past and present uses of plants as foods, fibers, sources of bioactive compounds, fuels, and construction materials. In this chapter we look ahead to examine how plants will serve people in the twenty-first century (Fig. 19.1). Included in this discussion will be the challenges humans will face in meeting future demands for plants and their products. Certainly, a major challenge will be improving agricultural efficiency to meet the needs of the world's growing population. However, those efforts must be balanced with an exploitation of the earth's resources in a sustainable fashion—that is, in a manner that prevents the degradation or collapse of natural ecosystems. Therefore, as we discuss the potential for innovations, we will try to point out the inherent problems and limitations.

Plants As Food

Human populations, unlike those of all other organisms, have grown at an unchecked rate for the last 2000 years (Fig. 19.2). In 1999, the world's population passed the 6 billion mark and is estimated to increase to 8.9 billion by 2050. The figure 8.9 billion is lower than the 9.5 billion projected just a few years ago because estimates now include the probability of huge death tolls in Africa due to AIDS and the fact that the birth rate in Nigeria has declined. Most of the growth will be in undeveloped countries, but even the U.S. population is expected to grow by over 23 million as a result of a combination of births and immigration.

Many people are seriously concerned about whether we can produce enough food to feed the projected future human population. In 1999, the World Health Organization estimated that 1.2 billion people, roughly half of the world's population, already suffer from some form of malnutrition (Fig. 19.3). Lamentably, it is the countries that are growing the fastest that will have the most difficulty expanding their agricultural productivity because of poor soils and insufficient rainfall. To make matters worse, these countries do not have an economic base that will allow them to import needed food or a political infrastructure that will allow the efficient use of donated food. In light of these problems, some demographers believe that the **carrying capacity** (Fig. 19.4) of the earth, or the stable number of individuals the world can support, has nearly been reached. Others, looking back to

FIGURE 19.1

Speculation about the future is a favorite human pastime. These headlines from 1996–2000 encapsulate issues that recently captured the interest of readers.

past successes in agricultural technology, believe that there is still room for improvement of crop yields to support more people. This chapter will discuss the basis for both attitudes and the potential contribution of various methods for increasing food production. These strategies include the development of new varieties of traditional crops that are more productive or have superior nutritional value relative to current varieties, the introduction of new natural or engineered crops, the expansion of agriculture into new areas, and the implementation of innovative agricultural practices such as hydroponics. All these strategies will be discussed in light of their economic feasibility and environmental costs.

In the first edition of this book (1986) we talked of the optimism inspired by the "Green Revolution," a phenomenon that took place in the 1960s. After a drought in India in 1965 and 1966, there were widespread crop failures. When normal rainfall returned in 1967 and 1968, India experienced record-breaking yields and the press began to talk of a Green Revolution in agriculture. This increase in yield was due not only to the renewal of adequate rainfall but also to the introduction of new varieties of traditional crops. The spectacular increase in yields led to the belief that human ingenuity could overcome any agricultural problem.

For example, wheat was traditionally difficult to grow in tropical areas because its tall, slender stalks were often knocked over by the heavy downpours. Once it was matted to the ground, the wheat rotted easily in the humid air. In 1953 Dr. Norman Borlaug began to cross and select wheat from varieties sent to him in Mexico from around the world. By 1963 he had developed a dwarf wheat that had a high yield, resistance to rust infection, and no dormancy requirement so that it could be planted in the spring and harvested that summer. These wheats prospered in tropical countries and provided breeding stocks for new temperate varieties as well. For his work, Dr. Borlaug received a Nobel Peace Prize in 1970.

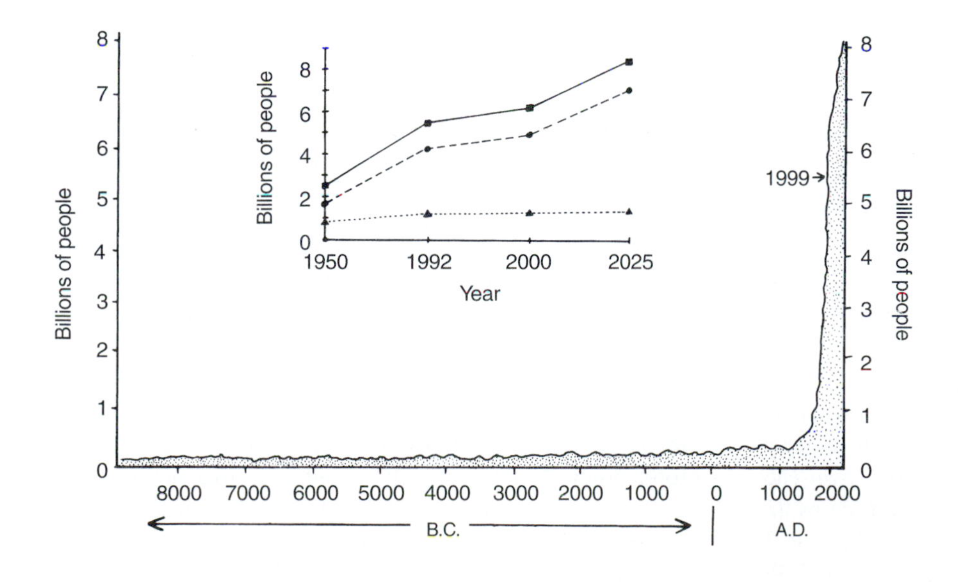

FIGURE 19.2

The human population has exploded over the last 2000 years and is predicted to reach over 8 billion by 2020 (lower figure). The insert shows the total population growth and projected growth since 1950 (solid line), the portion of this growth that has and will occur in developing countries (dashed line), and the growth of developed countries (dotted line). It is obvious that the major contribution to the earth's population will be in the developing countries.

(Insert modified from: International Union for Conservation of Nature, 1990. *Conserving the World's Biological Diversity.* Gland, Switzerland, IUCN.)

FIGURE 19.3
Half of all child deaths worldwide are associated with malnutrition, with an estimated 190 million children considered underweight and 230 million stunted. Children in Bhutan being fed as part of the World Food Programme.
(Photo by F. Mattioli, FAO.)

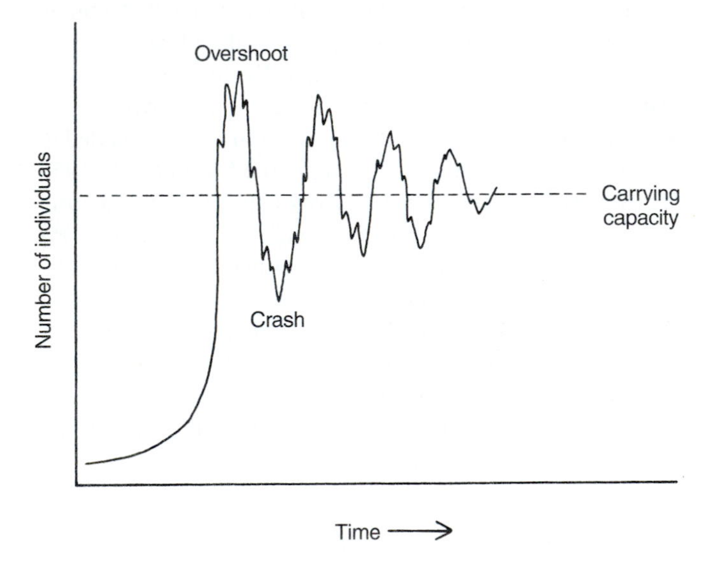

FIGURE 19.4
Natural populations show a characteristic growth pattern, with a slow start followed by a period with a high rate of growth. Such populations often reach high levels that overshoot the carrying capacity, or the number that can be sustained in that environment. When carrying capacity is exceeded, the population crashes and then rises again. After a number of years the number of individuals in the population tends to fluctuate around the carrying capacity. Figure 19.2 shows that the human population is still in a rapid-growth phase and could easily overshoot the earth's carrying capacity. If it does, there will be mass starvation.

Rice was another crop that naturally grew too tall for successful planting in humid tropical areas. It is also generally slow-growing, so only one crop could be grown each year. Research on rice began at about the same time as that on wheat, in this case in the Philippines. By 1973, IR-24 (International Rice variety 24) had been produced and distributed. This variety had high-yielding short stalks and was fast-maturing. The grains had soft kernels, a characteristic desired by many people who eat a diet high in rice. These rice varieties were hailed as miracle plants because their combined characters led to yields that were double those of previous cultivars. Since that time continued breeding and selecting have continued with the development of pest-resistant IR-36 in 1976 and, more recently, IR-62 and IR-64, which have higher resistance to an increasing array of pests. Semidwarf varieties, such as IR667-98, that are photoperiod-insensitive and nitrogen-responsive have allowed rice yields to rise dramatically.

Since the 1920s breeding and selection have also led to an improvement in quality that has been as valuable as or more valuable than high yields (Fig. 19.5). Cases in point here are high-lysine corn and sorghum. It was pointed out in Chapter 5 that lysine and the sulfur amino acids are often in short supply when grains are used to complement legumes. High-lysine corn was produced in American research stations in 1963, and since that time strides have been made toward breeding greater palatability into these nutritionally superior varieties. Similarly, breeding with sorghum, a crop that can be grown on relatively arid land, has led to varieties that have much higher lysine levels than do varieties that were used previously. Wheat lysine contents have also been increased in experimental strains from 12 to 20 percent of the total grain protein.

Many scientists still believe that the yields of most tropical crops could be doubled if these crops were subjected to intensive breeding and selection programs. Most tropical crops, even widely cultivated ones such as cassava, the millets, and bananas, have not been subjected to intense selection regimes. Even if one takes into account the potential gains from these underexploited resources, there is still a finite limit to the amount of primary plant production that can occur on this planet. Moreover, with any selection regime, a limitation to crop improvement is imposed by the availability of genetic variability. One way to maintain such variation is by placing seeds of wild relatives and **land races** (comparatively unselected local forms of crops generally grown by indigenous peoples) of crop species in gene banks. In these storage facilities the seeds are periodically retrieved, germinated, and used or returned as fresh seeds to the bank. Contrary to some people's opinion, the newly discovered techniques of genetic engineering do not diminish or negate the need for preserving genetic diversity in this fashion. In fact, the genetic information stored in seed banks assumes new importance as a source of raw materials for use in biotechnology.

In 1986, we explored how far increases in food production resulting from traditional crop improvement methods might go. Between 1940 and 1998, people managed to keep food production apace with population size increases on every continent except Africa. On a worldwide basis, 80 percent of this increase in food production resulted from higher-yielding varieties of traditional crops. Although the

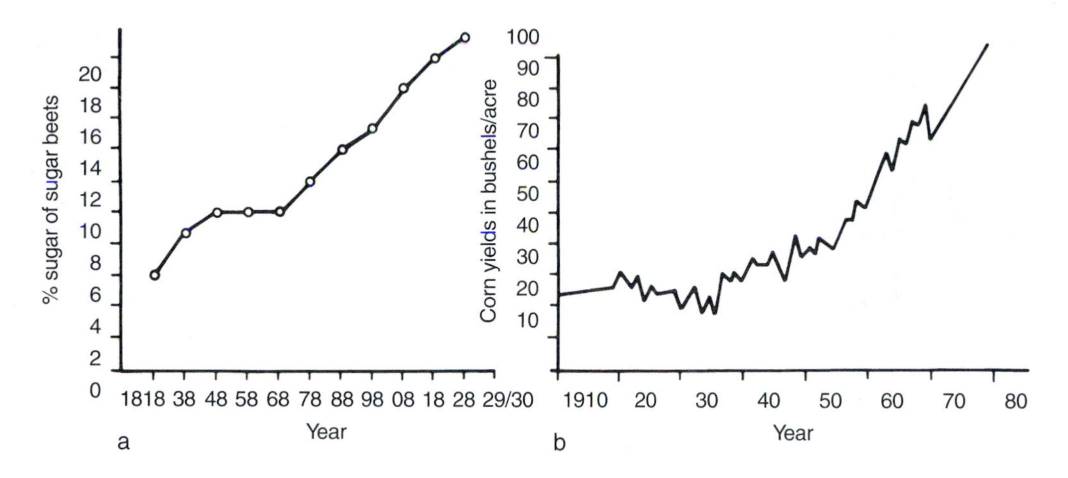

FIGURE 19.5

Increases in yields of crops brought about by artificial selection. (a) Changes in sugar concentration in the sugar beet over time caused by selection. (b) Increases in corn yields over time caused by selection.

(From Pimentel et al. 1973. Food production and the energy crises. *Science* 182:443–449.)

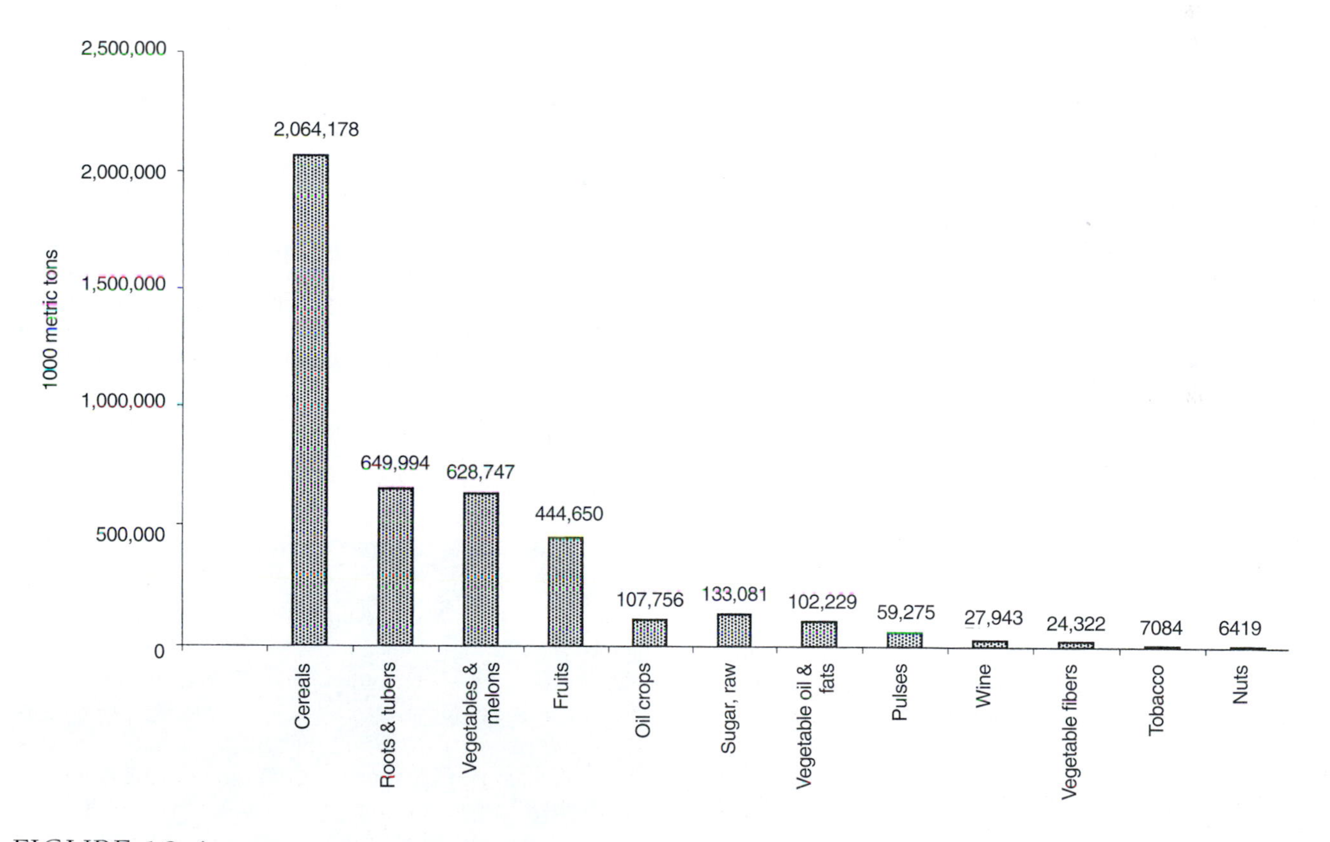

FIGURE 19.6

Worldwide production of the major kinds of plant products used by people, excluding lumber and fuel. These figures show the overwhelming dominance of grains (cereals) in terms of production per year. Pulses = dry seed legumes; veg. fibers = plant fibers. Numbers above the bars are the actual figures for 1999.

(From FAOSTAT database available at http://aps.fao.org.)

land devoted to agriculture in the United States increased only 11 percent from 1940 to 1985, harvests doubled because of plants developed during the Green Revolution. On a global scale, grain production tripled between 1950 and 1985.

There is no denying the importance of the gains that were made during the Green Revolution, but they were all made with the same crops that were the foundation of the first agricultural civilizations (Figs. 19.6 and 19.7). As new varieties of these crops are adopted, the diversity within

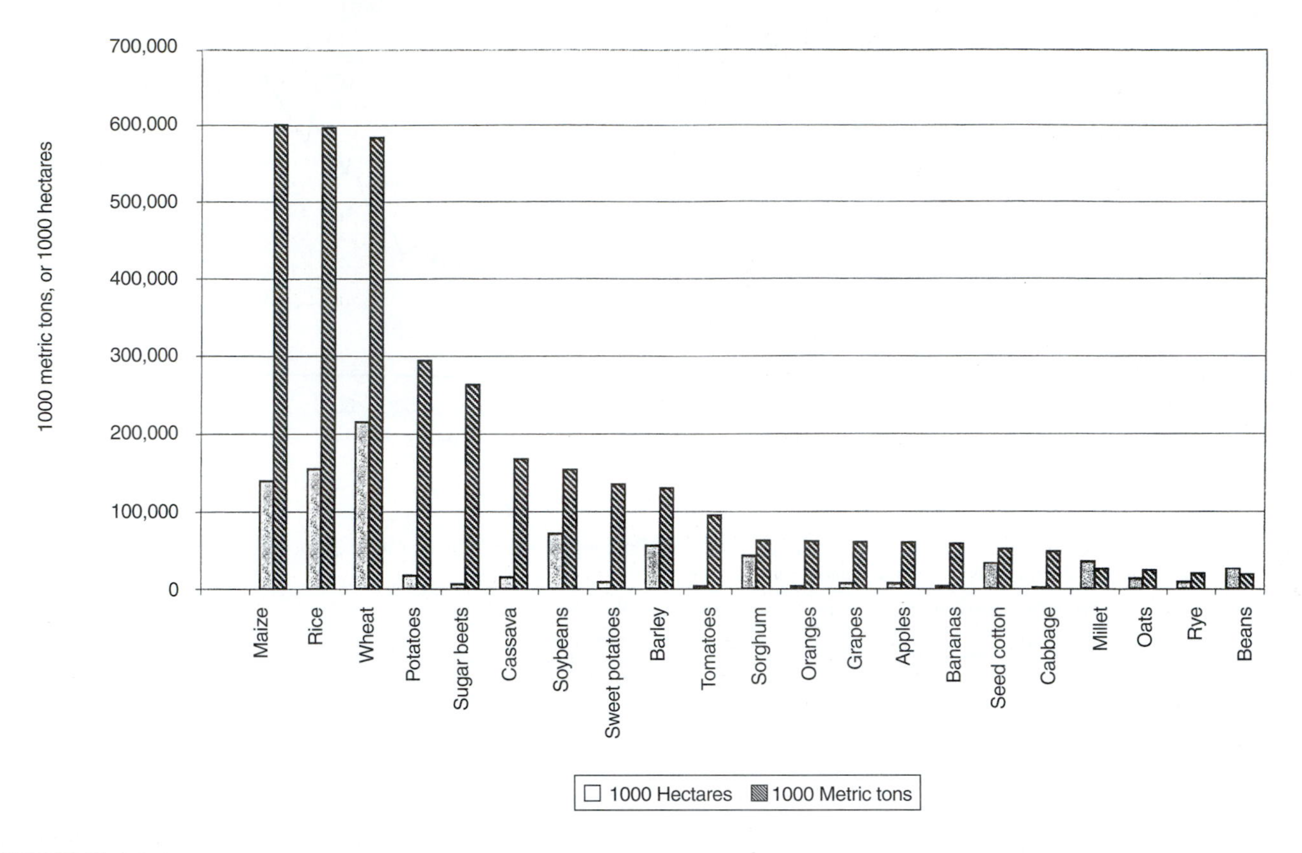

FIGURE 19.7

The 20 major vegetable crops of the world in terms of metric tons produced each year. Stippled bars show hectares devoted worldwide to the crop. Hatched bars show metric tons produced. It is obvious from the figure that there are greater yields per hectare of root crops, such as sugar beets, potatoes, and cassava relative to grain crops such as wheat, rice, and corn. Note also that oats, rye, and especially millet have low yields relative to the acreage on which they are planted. These low yields reflect the fact that these grains are often grown on poor soils. In terms of raw tonnage, sugarcane (not shown) leads all crops, with an estimated 1,274,697,080 metric tons produced each year. Since most of the weight of sugar cane is waste material, it is excluded here.

(From FAOSTAT database available at http://aps.fao.org.)

crops dwindles. In 1993, 96 percent of the peas grown commercially came from two varieties; 71 percent of the corn, from six varieties; and 65 percent of the rice, from only four varieties. Perhaps even more disturbing is the fact that since 1997, total production of grain has decreased and since 1983 per capita production has been falling.

Potential New Crops

Many of the trends we predicted more than a decade ago have come to pass, with one notable exception: no major new food has been incorporated into the human repertoire. Although advances in food production have been made through traditional plant breeding and better land use, one must ask why "promising" crops such as quinoa, amaranth (Fig. 19.8), and winged bean have not become staples. Their cultivation has increased during the past 10 years, but agriculture on a global scale is still dominated by wheat cartels and companies that deal in corn, sorghum, and rice.

FIGURE 19.8

Whatever happened to winged beans, buffalo gourd, and other plants that showed promise as crops? Perhaps ingrained eating habits thwarted their acceptance. Genetically engineered plants today face a different sort of opposition.

(Photo by N. Vietmeyer.)

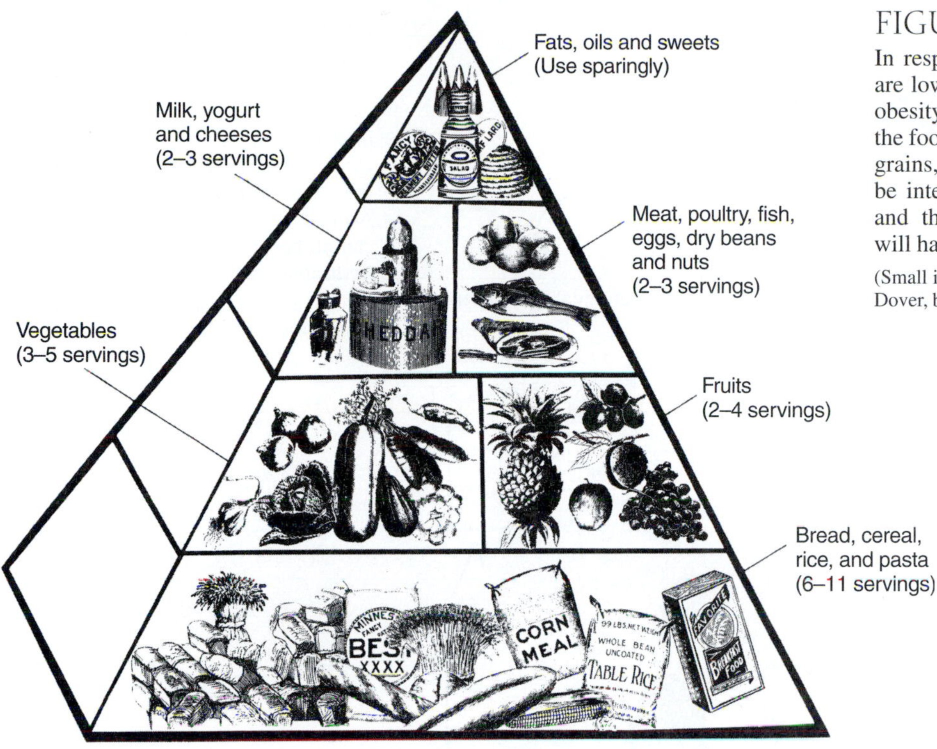

FIGURE 19.9
In response to evidence that high-fat diets that are low in fiber may contribute to heart disease, obesity, and some cancers, the FDA has redrawn the food pyramid to emphasize the importance of grains, fruits, and vegetables in the diet. It will be interesting to see whether the new pyramid and the new "consumer-friendly" food labels will have an impact on national eating habits.

(Small illustrations from J. Harter, 1979. *Food and Drink*, Dover, by permission.)

Probably the most important reason why these crops have not fulfilled our expectations is the difficulty of changing entrenched eating habits.

Another anticipated trend that did not materialize was for people in the developed countries to follow the recommendations of the U.S. Food and Drug Administration (FDA) (Fig. 19.9) and begin eating "lower" on the food chain. Not only would this kind of diet be healthier, but because of the calories invested in livestock and the energy loss from respiration at each stage, it is energetically more efficient for people to eat vegetable protein rather than animal protein. Although this recommendation seems sensible, statistics show that in every country that has experienced an increase in net income level, there has been a tendency to increase the amount of animal protein in the diet. There is a clear progression as wealth increases from an all-starch diet, to one with starch and legumes, to one with scraps of meat, and finally to a diet rich in animal protein. Because the wealthy countries (which are the countries developing new plant crop varieties) do not have a protein deficiency, there has been little impetus to select for high-protein varieties of crops, even though such an increase might be produced using either conventional methods or genetic engineering.

One trend that we remarked on in our second edition (1995), the rise in production of "yuppie" or novel foods with gourmet appeal, has continued in the Western world, with brocco-flower, broccolini, plucots, and doughnut peaches now common fare. Variations in color of standard plant foods such as yellow kiwis, orange beets, maroon carrots, and yellow watermelon are marketed for eager buyers. Exotic foods, primarily from the Orient, have proliferated. These include nori (Chapter 18), which has accompanied the craze for sushi, and spices such as lemon grass and kaffir lime leaves. One of the fastest growing sectors of the fresh food industry is the production of "ready to eat" packaged vegetables. These include washed, bite-sized pieces of lettuce, field greens, spinach salads, baby carrots, and broccoli florets. Luckily, this trend makes it easier for Americans to follow the advice to eat these health-enhancing foods (see Box 7.1).

A trend that was alluded to in earlier editions of this book that has now become a major thrust in agriculture is the production of genetically engineered or genetically modified (GM) food (Fig. 19.10). However, as highlighted in the next section, genetic engineering has not produced any completely new crops; instead it has focused on improving qualities and yields of existing crops.

Genetic Engineering

The use of genetic engineering in agriculture was widely touted in the early 1900s, but few pundits expected the rapidity with which it swept across American agriculture in the last part of that decade (Fig. 19.11). Fewer still anticipated the violent reaction that has sprung up in many parts of the world. Chapter 1 explained the principles of genetic engineering and Chapters 4 and 5 showed how tinkering led to new qualities in tomatoes and rice. Table 19.1 provides a review of the components of traditional artificial selection and compares them with those of tissue culture and genetic engineering. In may 1994, the Flavor savr™ tomato, a tomato that had been engineered to remain in the "vine-ripened" state, was released (see Box 4.2). Unfortunately,

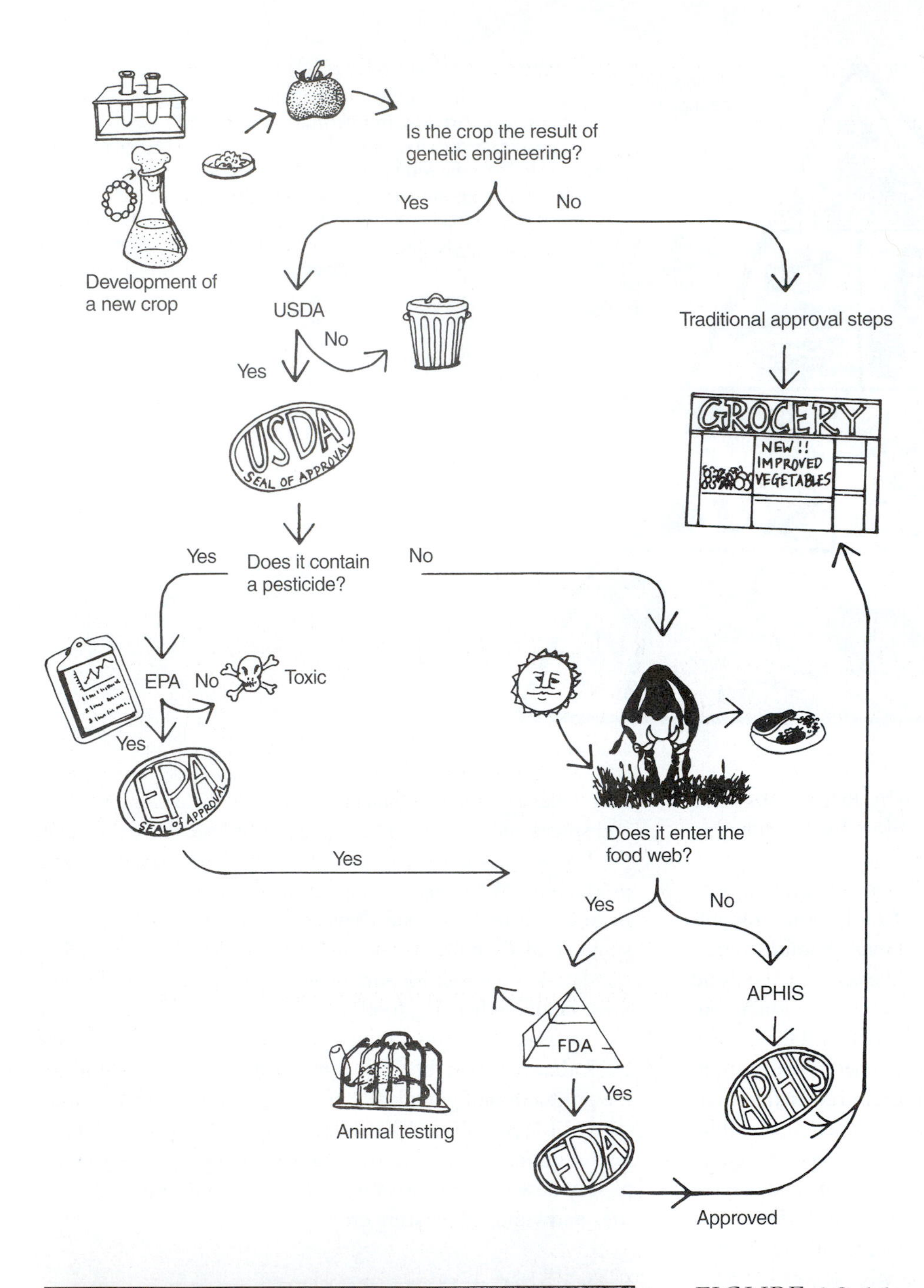

FIGURE 19.10
This flow diagram shows the regulatory steps for genetically engineered crops. Depending on the nature of the gene inserted into the engineered crop, the U.S. Department of Agriculture, the Food and Drug Administration, and the Animal, Plant, Health Inspection Service can all become involved in its approval or disapproval.
(Design by Dr. Charles Arntzen; figure by M. Simpson and M. C. Ogorzaly.)

FIGURE 19.11
A geneticist examines wheat cells that have had new genes implanted by bombardment in the gene gun enclosure.
(Photo by B. Prechtel, Agricultural Research Service, USDA.)

because of problems with shipping, this initial offering of a genetically modified (GM) food was not a success, but soon after other modified crops did succeed.

In 1999, about 50 percent of the soybean, 33 percent of the corn, and 35 percent of the cotton crops in the United States and 62 percent of the canola crop in Canada were

TABLE 19.1 Differences between Whole Organism Selection, Tissue Culture (Cellular), and Genetic Engineering (Molecular) of Crops

	WHOLE ORGANISM SELECTION	TISSUE CULTURE	GENETIC ENGINEERING
Kind of process used	Natural or induced mutation, breeding selection	Tissue culture (protoplast, anther, embryo, callus)	DNA manipulation
Level of understanding of why changes occur	Unknown	Partially understood	Known, and used to direct manipulation
Level of control over changes	Random	Partially random	Nonrandom, controlled
Number of variants	Large	Intermediate	Small
Degree of restriction	Primarily within a species	Within and between species	Within and between species
Level of acceptance in population	Very high	Intermediate	In flux
Ability to answer risk questions	Low in terms of understanding fundamental processes, but high because of long-term experience	Intermediate	Potentially high but still untested

TABLE 19.2 Kinds of Characters Being Introduced into Commercial Crops

TRAIT	EXAMPLES OF CROPS WITH TRAIT INSERTED
Resistance to:	
Herbicides	
Bromoxynil	Cotton, tomato
Glyphosate	Corn, cotton, rapeseed, soybean, tomato
Glufosinate/bialophos	Alfalfa, corn, canola, soybean, tomato
Sulfonylurea	Cotton, tomato
2, 4D	Potato
Fungi (*Rhizoctonia*)	Potato
Insects	
European corn borer	Corn
Leptodopteran insects via Bt-endotoxins	Apple, corn, cotton, potato, canola, rice, tobacco, tomato, walnut
Bacteria	Potato
Viruses	Alfalfa, cantaloupe, cucumber, corn, potato, squash, tomato
Tolerance to adverse growing conditions	
Cold, using fish "antifreeze"	Tomato
Heavy-metal tolerance	Tobacco
Stress	Potato
Qualities of taste or nutrition	
Altered ripening	Tomato
Altered fatty acid structure/content	Canola
Altered amino acid content	Corn, rice, soybean, sunflower
Increased solids or dry matter	Potato, tomato
Increased lysine production	Tobacco
Increased simple sugar	Potato
Increased starch content	Potato
Production of vitamins	Rice

Source: Modified from D. A. Kessler, M. R. Taylor, J. H. Maryanski, E. L. Flamm, and L. S. Kahl. 1992. The safety of foods developed by biotechnology. *Science* 256:1747, and R. P. Wrubel, S. Krimsky, and R. E. Wetzler. 1992. Field testing of transgenic plants. *BioScience* 42:282.

planted with genetically modified seed. Since 1996, when genetically engineering corn and soybeans were first grown commercially on 1.7 million hectares (4.2 million acres), the land planted in these crops has swelled to 39.9 million hectares (98.8 million acres). Ninety-nine percent of these crops were grown in the United States (72 percent), Argentina (17 percent), and Canada (10 percent).

What sort of characteristics are engineered into crop plants? Technology is being employed to improve crops in three major ways (Table 19.2). First, genetic engineers are trying to improve the quality of a food for the consumer by altering such characters as taste, fatty acid profile, levels of protein, sugar composition, and resistance to spoilage. The second and third sets of traits affect production levels by introducing into crops genes that confer disease, herbicide, insect, and stress resistance and tolerance to adverse conditions. It is this last suite of engineered traits that has led to the explosion of genetically modified crops in the United States and Canada because they reduce costs and losses to predators and weeds. Soybeans and canola have been engineered to resist herbicides such as Roundup™ so that a farmer can spray weedkillers over an extended period of time. Corn and cotton have been modified with Bt, a gene from *Bacillus thuringiensis,* that enables plants to manufacture their own pesticides. One other goal is to turn plants into biofactories by introducing genes that allow them to synthesize plastics, lubricants, enzymes, hormones, and vaccines.

Despite the production gains genetic engineering has allowed and the ready adoption of GM crops by American farmers, consumer resistance to GM products, primarily food items, has grown steadily. The antipathy toward genetic modification comes from a variety of perspectives. Some individuals hold the religious conviction that humans should not play God by modifying creatures of the earth. However, since people have altered plants and animals for thousands of years, this argument seems to be one of degree. Introducing new genes into a crop by crossing with a related species seems to be acceptable, whereas placing a gene isolated from a bacterium into a plant is not. Another reason for decrying the use of genetic engineering is simply the fear of genetically altered or "unnatural" food. The media have fostered this fear by calling modified organisms "Frankensteinfood" after the monster (mistakenly) called Frankenstein. In part because of two recent food scares (mad cow disease and dioxin-tainted food), Europe has become particularly adamant in its refusal to accept genetically altered foodstuffs and has called for labeling that clearly indicates when an altered product is present in a food. American companies have fought labeling requirements, claiming that such labeling would be too expensive. The expense comes from tracking ingredients. For example, genetically altered soybean oil could be sold to a company that roasts peanuts in the oil. The peanuts could be sold to a company that manufactures candy bars. Under labeling laws, the candy bar would have to state that it contains genetically altered ingredients. Tracking such products is probably possible in the United States and Canada, but it would currently be impossible in developing countries that lack sophisticated computer technology.

A final suite of criticisms of the rapid adoption of genetically altered crops come from biologists who warn that adequate testing has not been done either on the effects of such foods on humans or on the potential ecological consequences. Health risks range from the possibility of anaphylactic shock from a compound (e.g., peanut protein) moved from one species into another. The modified food would probably not be suspected of harboring a compound that can produce an allergenic reaction. Another concern is the potential increase in "superbugs" that are resistant to the antibiotics produced by genetically engineered organisms. It has been shown that genetically altered crops can hybridize with wild relatives and thus it would be possible for pest or herbicide resistance to move into wild species. A highly publicized but later refuted study suggested that pollen of genetically altered corn could blow onto host plants of benign butterflies such as the monarch and kill the caterpillars.

The refusal of Europeans to accept genetically altered foods and the criticisms within the United States have led to new rules for biotechnology companies that require publishing scientific information about the safety of a modified animal or human food. Some manufacturers have boasted that they have removed all genetically engineered products from baby food or taco chips, but the same companies use oils and syrup from altered corn in other products, and soybean oil, much of it altered, is ubiquitous in processed foods.

Despite these fears and claims, genetic engineering holds great promise of bringing about a "Double Green Revolution," in terms of both quantity and quality. Scientists have engineered rice that produces beta-carotene, making the grains yellow and a possible solution to vitamin deficiency diseases. At this time, an estimated 100 million children suffer from vitamin A deficiency; half a million of them go blind and 2 million die each year. The potential of moving dwarfing traits into other crops could lead to substantial increases in yield. When weighing the pros and cons of using genetically engineered crops, the British Nuffield Council on Bioethics stated in its May 1999 report:

> The probable costs of the (mostly remote) environmental risks from GM crops to developing countries, even with no controls, do not approach the probable gains of GM crops concentrated on the local labor-intensive production of food staples. Are lower safety standards justified because, by producing more or better food and more jobs for the undernourished, or by reducing agrochemical use, GM crops save many more lives than they cost and improve more lives than they worsen?

These concerns and issues of the control of genetically altered plants (and animals) are some of the major issues that will be facing the world in the next decade.

Expanding Agriculture into New Areas

Efforts to increase food production have not been limited to manipulating major crops in highly advanced temperate countries. Both wet, tropical lowland areas and arid areas are potential regions into which agriculture could expand. Research is being directed toward utilizing crops that are better adapted to these areas than are the traditional crops developed in temperate latitudes. Similarly, new farming methods are being sought to overcome the problems of poor soils, high leaching, high predation levels, and aridity or salt buildup that have thwarted the use of those regions in the past.

Most of the uncultivated land across the world that is not covered by ice or tropical rain forest is not used for farming because of aridity. Conventional methods of agriculture, even those involving drought-tolerant crops, often cannot be employed in truly dry regions because of saline soils. Even if an area is not initially saline, salt can accumulate when an arid or semiarid region is irrigated because of the unnatural, repeated applications of large amounts of water. Irrigation water flooded onto a field percolates rapidly through the soil, dissolving inorganic salts as it moves downward. However, because the air is so dry, there is intense evaporation at the soil surface, which soon begins to pull the water and dissolved salts upward again. As the water vaporizes into the air, the salts it contained are left behind, forming a deposit on the surface of the soil.

In several areas, notably California and Israel, new methods of irrigation that combat salt accumulation by using a drip technique have been employed successfully for many years. Drip irrigation is achieved by applying water drop by drop through emitters spaced along hoses or leaky pipe hoses that are laid in fields near plant bases. The water is absorbed quickly by the soil and from there by the plants. Although it requires large amounts of labor, drip irrigation greatly reduces the amount of water needed and drastically lowers salt accumulation. Such small amounts of water are administered per unit time that large-scale percolation and subsequent evaporation are minimized and, as a consequence, salt deposition does not occur. An additional benefit is that fertilizers can be added to the irrigation water when needed. Using drip irrigation combined with other methods, Israel reduced the amount of water applied to each hectare by 36 percent between 1951 and 1990. As a result, the amount of land irrigated tripled but the amount of water used only doubled over the same period of time.

Another scheme for conserving water and increasing yields is **hydroponics** (Fig. 19.12), or the growing of plants in nutrient solutions. Because there is no loss of soil and minerals from leaching and runoff and because evaporation from leaf surfaces is reduced, hydroponic growers use only 10 percent of the water necessary for soil agriculture. In addition, they reap harvests that would normally be produced on 30 times as much land. In most hydroponics operations the plants are "rooted" in an inert substance (e.g., marble, gravel, or sand) to give them support while a solution of water, minerals, and air is pumped over them. Since hydroponic operations are enclosed, the environment can be controlled and pest problems are reduced. The drawback of hydroponic cultivation is that it is expensive, requiring large amounts of energy and hand labor. Hydroponics also does not lend itself to the production of crops such as wheat, corn, and rice, which are usually grown on huge acreages. Nevertheless, in regions with little available arable soil, adequate sunlight throughout the year, and a premium on fresh vegetables, such systems have proved to be profitable.

Putting new areas, such as the lowland tropics and deserts, into cultivation carries a price: it will necessarily lead to a further reduction in the number of species from which people can find new sources of food, medicines, and genetic material. As previously unused land is cleared for cultivation or timber, vast areas of natural vegetation, and with them many plant and animal species will be eliminated. Humans use over 37 percent (4.9×10^9 hectares, 1.21×10^{10} acres) of the land surface of the earth; 30 percent of that is planted to crops, and 70 percent is used as pasture (Fig. 19.13). Although some additional land can be used for agriculture, the ability to expand into new, profitable agricultural areas had largely disappeared by 1950, leading to the push for increasing crop yields.

FIGURE 19.12

In this hydroponic greenhouse, thousands of heads of lettuce mature every 28 to 35 days. Because the water is recycled, 90 to 97 percent less water is used than in equivalent field operations.

(Photo courtesy of Hydroculture, Inc.)

FIGURE 19.13

Overgrazing has contributed to desertification of more than a third the total of degraded dry land. Goats are shown feeding on a lone *Acacia* shrub in the Sudan.

(Photo by R. Faidutti, FAO.)

The Energy Costs of Agriculture

Another challenge to producing greater supplies of food in temperate agricultural areas or even in previously unused arid or tropical regions is that it must be accomplished in the face of dwindling supplies of fossil fuels and the products obtained from them. Almost all estimates of possible increases in food production assume a technology equivalent to that now used on highly productive U.S. farms. On these farms large machines fueled primarily by diesel oil are essential for plowing, cultivating, and harvesting. In some areas huge irrigation circles that are run by electricity (generally derived from fossil fuels) keep production levels high. Pesticides and fertilizers synthetically produced in factories that use large amounts of energy are applied routinely. By the time American food crops have been planted, harvested, shipped, processed, and put on the table, six times more energy has been expended in their production than will be derived from their consumption. With increasing shortages of fossil fuels, the costs of this energy imbalance will continue to rise. Costs of food production have risen to the point where even affluent countries have turned to lower-energy methods of agriculture.

In the search for such methods, attention has been focused on primitive techniques such as the **chinampa** agriculture of the Mayans. This system was similar to what is today hailed as the "modern" French intensive, or raised-bed, method of crop production. Similarly, increasing interest is being expressed in growing rice. Where possible, the Chinese introduce *Azolla,* an aquatic floating fern, into rice paddies. *Azolla* can not only coexist with rice but also has a symbiotic association with a nitrogen-fixing (see Chapter 6) blue-green alga (see Chapter 18). When the rice is harvested, the fields are drained, and the ferns, with a rich supply of nitrogen in their nodules, sink to the ground and decompose, becoming in essence a green manure crop.

Agriculturists are also interested in ways of reducing the consumption of fossil fuels on land that is now heavily used. One way to lower fuel use is to reduce the number of times tractors or cultivators cross the fields and limit the use of synthetically produced fertilizers and pesticides. Among the most successful methods now employed for reducing the use of heavy machinery is the practice of no-till, or low-till, agriculture. **Tillage** refers to the process of plowing fields to remove weeds that are competing with crops. Tractors usually plow just below the soil surface between the crop rows to sever the roots of the weeds. In low- or no-till agriculture, all cultivating is eliminated. Plowing, seeding, and the application of herbicides are done in a single initial operation. Since 1980 there has been a dramatic shift to no- or low-till agriculture in the United States.

Herbicides, or chemicals that inhibit plant growth, are used in grain fields to prevent the growth of broad-leafed weeds and in fields of dicotyledonous crops to prevent infestation by weedy grasses. The most commonly used broad-leaved "weed" killers are glyphosate (Roundup™) and bromoxynil. A common herbicide used for grasses is dalpon. In addition to lowering the fuel consumption of farm machinery, low-till agriculture reduces runoff, soil erosion, and consequently the leaching of fertilizers. Less fertilizer is ultimately needed, and the pollution of water systems is reduced. Herbicides are, of course, manufactured products, and fuel is used during their synthesis and application. On the whole, the total energy consumption and costs of a farm are substantially reduced by the use of low-till methods, but the costs of such heavy reliance on pesticides are not only monetary.

The use of modern pesticides (the collective term for fungicides, herbicides, and insecticides) has been credited with increasing yields about 33 percent simply by reducing insect damage. In areas where huge acreages are devoted to a single crop (monocultures), pesticides have become mandatory since pest populations can build up to large sizes and spread very rapidly (Fig. 19.14). When DDT, the first modern insecticide, was developed, it was hailed worldwide not only because it prevented crop destruction but also for its role in eliminating the malaria mosquito. Millions of lives were saved, and there was optimism that humans would continue to develop chemicals that could solve all pest problems. When Rachael Carson published *Silent Spring* in 1962, the world was shocked. She convincingly argued that pesticides were not the panacea they appeared to be and that insidious dangers were involved in the haphazard use of agricultural chemicals. Despite her warnings, 2.5 million tons of pesticides were applied in the United States in 1992. However, an estimated 37 percent of all potential food crops were still lost to plant pests and pathogens. Ironically, the amounts of pesticides used in the United States increased 10-fold between 1945 and 1989 yet losses to insect pests alone rose from 7 to 13 percent.

We are now aware of the numerous drawbacks of extended pesticide use. Populations of all kinds of pests can become resistant to pesticides because natural selection favors mutations that confer resistance. DDT has been ineffective against the common house fly since 1946, and group after group of other insects have become resistant to one insecticide after another (Fig. 19.15). Sometimes different kinds of pesticides act antagonistically toward one another. Similarly, the loss of natural weed predators through the use of a general insecticide can increase pressure from weed competition, and the use of herbicides can force more plant-feeding insects onto crop species. Beneficial insects such as pollinators and insect pest predators can also be affected detrimentally by insecticides.

Humans are not immune to the effects of pesticides. Some pesticides have been shown to cause cancer in laboratory animals and presumably could cause cancer in humans as well. Short-term data on the modern pesticides that are now GRAS (generally recommended as safe in FDA language) indicate that most are broken down quickly and thus pose little threat to humans, but no one knows what the long-term effects will be. The annual costs of health care and lost jobs caused by pesticide poisoning were estimated to be $787

FIGURE 19.14
The timeless competition between people and plant pests is dramatically illustrated in this engraving of a locust plague in Europe.

million in the United States in 1992. In 1991 the World Health Organization estimated there were 1.1 million cases of pesticide poisoning, of which 20,000 were fatal. Although economic benefits to agriculture from the use of pesticides seem to outweigh their disadvantages at the moment, the balance may turn as energy costs force the price of pesticides and synthetic fertilizers higher and the ecological and human health care costs become insupportable.

Making Agriculture Safer

Research is now being directed toward developing agricultural methods that will reduce the need for manufactured pesticides. The group of practices that reduce pest pressure without the need for heavy pesticide applications is known as **integrated pest management** (IPM). Individual methods involved in IPM have their own particular goals and efficacy in given situations, but the strength of the management system lies in the simultaneous use of as many of these methods as possible (Fig. 19.16). The major kinds of control that are currently integrated into this system are listed in Table 19.3.

Classic biological control, one of the methods of IPM, involves the introduction of a pest's predator to keep populations of that pest at a relatively low level. This system has worked admirably in several situations. One example was the control of scale insects on citrus crops in California. Ladybugs (ladybird beetles), a natural predator of cottony scale, were introduced into California and managed to control the scale insects. Saint John's wort, or klamath weed, *Hypericum perforatum* (Hypericaceae), an introduced weed of U.S. rangeland, was finally brought under control by the

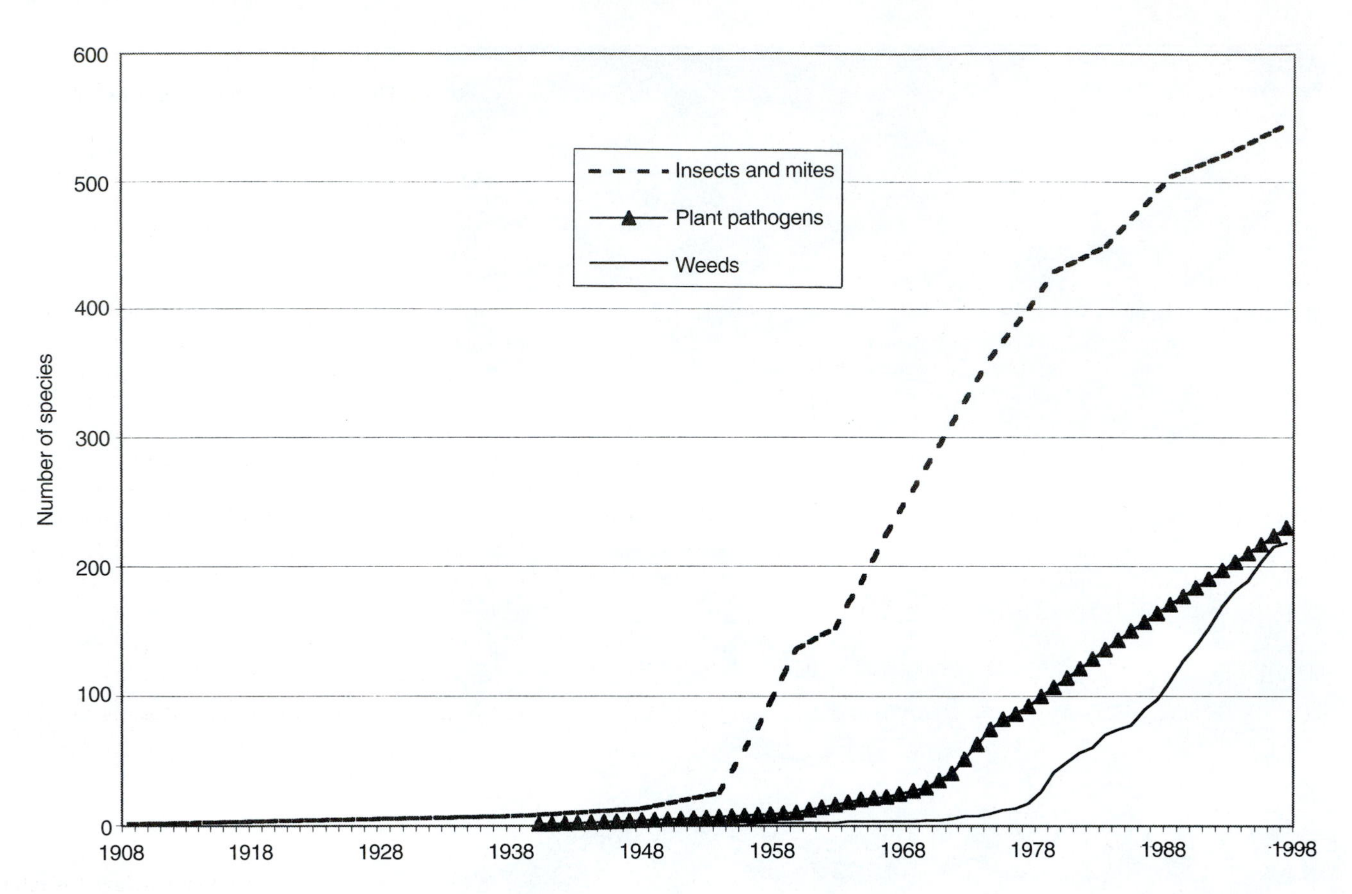

FIGURE 19.15
Number of pesticide-resistant species, 1908 to 1998.
(From Vital Signs. 1999. WorldWatch Institute. Accessible at http://www.worldwatch.org.)

purposeful introduction of its predators, European beetles of the genus *Chrysolina.* The rampant, aggressive spread of *Opuntia* across Australia after its introduction from North America was checked by the importation of its natural predator, the moth *Cactoblastus cactorum.*

Unfortunately, such simple solutions are not always possible. Biological control of this type is generally most effective for pests that have been introduced into an area. By contrast, native pests have already reached an equilibrium with their native predators. For classic biological control to work effectively, a pest must have a specific predator that can successfully be introduced into the area where that pest is a problem. The predator should be specific, because it must concentrate its efforts on the pest organisms while not posing a threat to desirable organisms. The establishment of such a predator, even if one is found, is often impossible because the proposed area of introduction lacks the environmental conditions it needs for establishment and reproduction.

Monocultures of crops can lead to rapid population increases of the native pests, sometimes with and sometimes without subsequent increases in predator populations. One of the methods employed in IPM is the maintenance of native predators through the preservation of natural areas and the restrained use of broad-spectrum pesticides that could eliminate them as well as the pests.

Other methods that can be employed as parts of an IPM scheme include the release of massive quantities of laboratory-reared predators. Large-scale releases can sometimes knock back outbreaks of pests, but the rearing programs are costly. Similar to this approach is the release of laboratory-produced male-sterile individuals of an insect pest species. The theory behind this procedure is that the females of the pest species (preferably a species in which the females mate only once) will mate with the sterile males and not produce any viable offspring. Male sterilization techniques have been successful in some situations as a control measure, but as in the case of pesticides, natural selection eventually leads to circumvention of the technique. Females that avoid mating with the male-steriles are selected, and researchers eventually have to screen the strains that are released in particular localities very carefully.

Another measure that can be employed to keep pest populations at a low level is **intercropping,** or the growing of unrelated species intermixed in fields. By avoiding large expanses of a single crop, this procedure tends to thwart large buildups of insect pest populations. For pest species that feed preferentially on a given crop, intercropping makes

TABLE 19.3 Controls Employed in Integrated Pest Management

CONTROL	ACTION
Classic biological	Introduces a pest predator that will reproduce and survive in the area in which the pest is active. Usually this method is effective only if the pest has also been introduced into the area.
Inundative biological	There is a periodic release of mass-reared predators of the pest that will cause a massive reduction in the pest population. These predators usually cannot maintain high natural population densities in areas where the pest is active.
Conservative biological	Practices the conservation of natural pest predators in the area where the pest is a problem. Plants that help maintain high parasitoid densities are encouraged. These aims are accomplished partly through multiple cropping and the retention of natural areas.
Competitive	Uses innocuous organisms that compete with pests for crop plants. Included in this category are the use of male-sterile insects and that of so-called trap plants, which divert pests from crops.
Biorational	Uses chemicals that modify the behavior of pests. These include pheromones, juvenile hormones, attractants, and antifeedants.
Chemical	Uses natural or synthetic compounds such as traditional pesticides as well as hormones, chemical sterilants, growth regulators, and microbial toxins.
Bioengineering	Like chemical, except that the insecticides are produced by the plant itself as a result of the introduction of a gene for the production of the compound.
Cultural	Manages agricultural systems through the physical techniques of proper tillage, irrigation, light traps, crop isolation, pruning, timing, and quarantine.

Source: Modified from G. F. v. Emden and D. B. Peakall. 1996. *Beyond Silent Spring*. New York, Chapman and Hall; and S. Batra. 1982. Biological control in agroecosystems. *Science* 215:134–139.

FIGURE 19.16
Entomologists examine a papaya fruit trap that contains Oriental fruit fly eggs that have been parasitized by wasps as part of research into biological control of the pests.
(Photo by S. Bauer, courtesy of Agricultural Research Service, USDA.)

the preferred species difficult to find. In other cases intercropping consists of interplanting species repellent to pests with preferred host species. A report of the National Academy of Sciences proposed intercropping as one of several approaches toward sustainable agriculture in tropical regions where it is necessary to preserve natural biodiversity while providing for the people living there (Table 19.4).

During the last decade dramatic strides have been made in using alternative pesticides and engineering pesticide resistance into crops. Dried cultures of *Bacillus thuringiensis* (Bt) are now commonly sprayed onto crops and even ornamental trees to prevent insect damage. This bacterium produces a crystalline protein (endotoxin) that interferes with the digestive systems of moth and butterfly larvae and some other kinds of insects. Genetic engineers have isolated the gene for this substance and inserted it into cotton, tobacco, potatoes, and several other crops (see Table 19.2). Again, problems are created by this and other kinds of tinkering with "natural" interactions. For example, as a result of the common use of the Bt endotoxin, insects are becoming increasingly immune to the compound. By 1998, 300 insect and mite species were known to be resistant to Bt endotoxins.

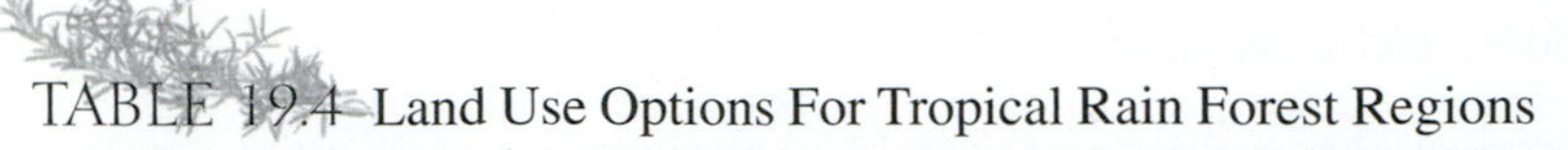

TABLE 19.4 Land Use Options For Tropical Rain Forest Regions

1. Intensive cropping similar to the chinampa system of intricate ditches and irrigation channels for water control.
2. Shifting cultivation (temporary agriculture in clearings planted for a few years with annual or perennial crops and then allowed to remain fallow in order to recover)
3. Practicing agropastoral techniques in which livestock are raised and crops are grown in a synergistic manner
4. Cattle ranching using appropriate levels of grazing and applications of fertilizer to avoid the damage incurred by overgrazing
5. Utilizing agroforestry, which integrates the cultivation of herbs and perennial tree species with annual crops and livestock to ensure ecological stability and sustainable production
6. Cultivating a mixture of trees so that a combination of firewood, food, medicine, and construction materials are grown together
7. Planting mixtures of perennial crops such as pineapple, sugar cane, bananas, and rubber to provide constancy over long periods while allowing high production per year
8. Establishment of plantation forests on damaged lands for the production of wood for fuel or industrial uses
9. Regenerating secondary forest when land becomes deforested and abandoned to facilitate the change to a more primary forest
10. Logging forests so that damage to the forest structure is reduced
11. Permitting a low level of habitation of some forests to demonstrate how people can live with forests without having a severe impact on them
12. Setting aside forest reserves that are protected from cutting but can be used for scientific study, recreation, and tourism

Source: Modified from *News Reporter*. 1993. 43(3):22.

Presumably, agronomists will be able to alter the molecular configuration of the Bt toxin molecule in a number of ways, thus producing "variants" of the inhibitory substance so that selection for resistance to any one type can be averted. Selection of resistant species can also be slowed by integrating genetic engineering with other aspects of IPM, such as the use of other insecticides, biological control, and rotation of crops.

Agriculture and Global Change

The wild card in speculation about the potential for increasing food production through agricultural technology is global change (Figs. 19.17 and 19.18). It is now obvious that several trace gases in the atmosphere are increasing worldwide. What is not so easy to assess is the future effect of these greenhouse gases. The term **greenhouse gas** refers to gases that produce an atmospheric warming similar to that created within greenhouses covered with plastic or glass.

Under normal circumstances, when sunlight reaches the earth, much of it is absorbed, converted into heat energy, and reradiated back into the atmosphere as infrared rays that are eventually lost to space. Since 1800, humans have been generating abnormally large quantities of carbon dioxide, methane, and nitrous oxide that reduce the reradiation into space. Unlike the major gases in the atmosphere (nitrogen, oxygen, and argon), these gases absorb infrared energy and send it back to the earth's surface. The result is containment of the heat and a warming of the earth's atmosphere.

Using cores from Arctic and Antarctic glaciers, scientists have been able to document the changes in atmospheric carbon dioxide over the last 160,000 years. These studies have shown that there have been normal fluctuations in carbon dioxide levels, with low levels corresponding to glacial periods. The levels reached over the last 200 years, since the beginning of the industrial revolution, however, are unprecedented. These high levels, which are caused primarily by the burning of fossil fuel (yielding carbon dioxide and nitrous oxides), increases in the land surface devoted to rice cultivation (yielding methane), and increases in the number of cattle (generating methane), have led to a significant rise in the earth's atmospheric temperature (Fig. 19.17).

Using tree ring and lake sediment data to estimate temperatures over the last 600 years, scientists have concluded that the decade of the 1990s was the warmest in the entire six-century period. Since warming causes water to expand, a warmer global climate is expected to produce more frequent and more violent storms than in cooler times. Increased temperatures have been associated with droughts in several parts of the world, including the south-central United States and Africa. Another effect is the melting of the ice caps, which is expected to raise ocean levels between 20 cm (7.86 in.) and 1.4 m (4.6 ft). Current estimates are that within the next 60 years, one-fourth of the land within 152.7 (500 feet) of the U.S. Atlantic coast will disappear because of erosion by high sea levels, storms, and wave action.

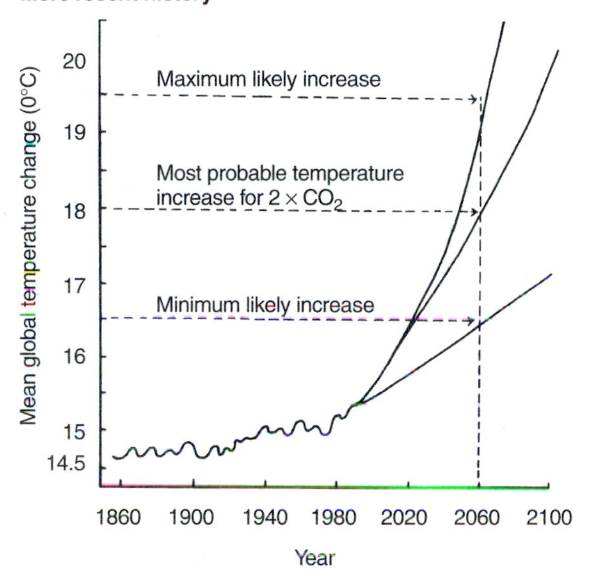

FIGURE 19.17

Changes in the earth's temperature over (top) geological time and (bottom) recent and future times. As shown at the top, the average global temperature for the last 800,000 years has been colder than it is at present. Only at a few times did temperatures exceed the current mean temperature of 15°C. Because of the rise in the amount of greenhouse gasses in the atmosphere, average global temperatures are expected to rise between 1.5 and 4.5 celsius degrees between now and the year 2060. The graph at the bottom shows various estimates, with a rise of 3°C considered to be the most likely.

(Adapted from Gates, D. M. 1993. *Climate Change and Its Biological Consequences.* Sinauer.)

The Search for New Medicines

Some of the most exciting developments in future plant use probably will come from pharmaceutical companies. There has been a resurgence of interest in screening plants for bioactive compounds thanks to a new appreciation of their

FIGURE 19.18

As part of research on the effects of global warming, a scientist checks the level of solar radiation within various growth tunnels with a light-sensing bar. Carbon dioxide levels within the different tunnels range from the 350 parts per million we have now to the 200 ppm present during the last ice age.

(Photo by P. Reich, courtesy of Agricultural Research Service, USDA.)

potential and the realization that many species from old forests will become extinct before people analyze them chemically. To spur this screening process, pharmaceutical companies have formed partnerships with botanists and have launched programs for the collection and screening of plants in both temperate North America and the tropical areas of South America, Africa, and Southeast Asia.

Such efforts have already resulted in the production of taxol (Chapter 11), an extract from relatives of the Pacific yew tree (*Taxus brevifolius*), which has been touted as the most promising treatment for ovarian and breast cancer yet developed. These enterprises have not been without problems, however. The development of taxol was delayed for many years by bureaucratic bungling, and the search for cures in the tropics has met with some resistance in less developed countries.

Historically, pharmaceutical companies have followed clues provided by native healers (Fig. 19.19) to guide them to medicinal plants. If the compounds proved to be useful, the companies would pay local people low wages for gathering plants for extraction or would develop synthetic compounds based on the structures of the chemicals identified in the plants. In neither case would the native people benefit significantly from the income generated by the commercial drug. Countries are now demanding the receipt of some of the profits derived from their indigenous knowledge and natural resources (Box 19.1). Exactly how they can be compensated is not clear, however. One touted agreement involved Merck and InBio (Instituto Nacional de Biodiversidad) in Costa Rica. In 1991 Merck, a pharmaceutical company, gave Costa Rica $1 million to train Costa

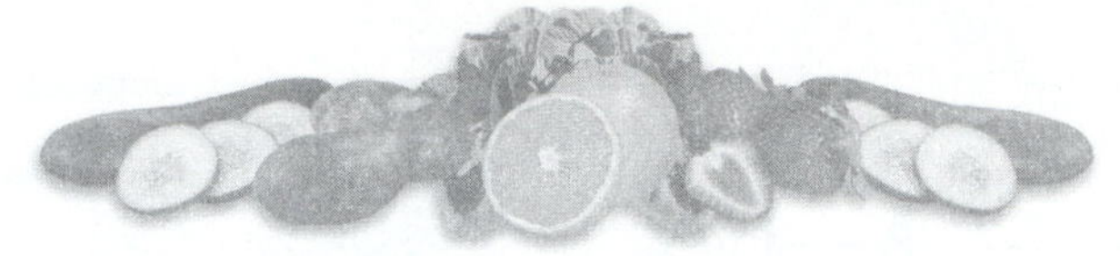

Box 19.1

The Ethics of Ethnobotany

Since 1990 there has been growing concern over the rights of indigenous people to intellectual and biological property. As investors from developed countries scramble to claim rights to the world's flora and fauna and capitalize on the knowledge of native people, the world community struggles to come up with sensible regulations. The need for such agreements stems from exploitive practices by individuals and pharmaceutical and seed companies from the developed world. Flagrant examples of the kinds of activities that have caused concern include the attempt by the USDA and W. R. Grace (a chemical and health care company) to patent the neem tree, a plant indigenous to India, where it has been used medicinally for centuries. The European Patent Office eventually revoked the patent. Similarly, the patents of one American businessman for turmeric, a commonly used spice native to India, and of another American businessman for caapi, a hallucinogenic plant employed ritually by native people of Brazil and Venezuela, were nullified.

Biopiracy is also being practiced in a much more subtle manner. For example, seed companies routinely take a crop native to one part of the world and through breeding and engineering produce a high-yielding variety that the native country must subsequently purchase from them. Farmers can be threatened with lawsuits if they buy seeds one year and then save seeds from that crop to plant the following year. The practice of saving seeds for subsequent planting has been around for a long time and is particularly cost-effective for poor farmers. In order to discourage this practice, Monsanto even engineered seeds with a "suicide" gene that caused infertility (but not production) of seeds from engineering plants to ensure that people would have to buy new seeds each year. After much public outcry over the burden this practice would place on poor farmers and the risk that this "terminator" gene could invade wild populations, Monsanto was forced to withdraw grains with this engineered characteristic from the market.

Under the International Biodiversity Agreement signed by almost all countries except the United States, citizens have rights to plants and animals native to their countries and to the knowledge of indigenous people. In the next few years, governments and courts will be struggling over how to define those rights, determine what should be patented, what is fair and reasonable compensation, and who has jurisdiction over these new laws. This is a daunting task as the stampede continues to claim the world's genetic and intellectual resources.

Ricans to collect and identify potentially useful plants and to develop methods to conserve the tropical rain forest. It also signed an agreement to provide royalties to InBio for any drug that resulted from collaborative efforts in Costa Rica. This sounds like a good solution and is certainly more enlightened than the previous policy, but it has been criticized because the government, not the local *curandero* or the landowner, will receive a share of the profits. Table 19.5 shows steps that are followed in discovering a new medicinal compound from indigenous tropical plants.

Pharamaceutical companies and biotechnology firms are also actively experimenting with plants as "biofactories" for "pharming" bioactive compounds. Such compounds include human hormones such as insulin, sex hormones, vaccines, and traditional medicinal compounds such as digoxin. Hormones and animal enzymes can be produced by plants if the genes for their production are transferred to the plants. This technology usually involves the use of a vector such as a virus or a plasmid. Bacteria are already used in the production of insulin, but the growing of these bacteria requires an energy input such as sugars in the medium. Green plants make their own food, so production is much cheaper. The gene for interferon has been transferred successfully into turnips, and both potatoes and tobacco can be engineered to produce harvestable quantities of human serum albumin.

Psychoactive Drugs in the Future

Chapter 12 discussed the use and abuse of psychoactive drugs. It is possible that there will be a shift in the use of psychoactive drugs from plants in the future. Already sentiments about marijuana have softened in light of evidence of its medicinal properties and relatively nonaddictive properties. By contrast, tobacco has come under increasing scrutiny. Smoking has been shown to be unequivocally harmful to both smokers and those around them. Nicotine has proved to be very addictive (more so than heroin!) despite denials by the tobacco lobby. Recent rulings that cigarette companies lied about the harmful effects of nicotine and the successful lawsuits by numerous states in the United States have paid off in a slow but steady worldwide decline in cigarette production. This is the first decline since the first U.S. Surgeon General's report in 1964 on smoking and health. In 1999, the World Health Organization announced that it would seek a global ban on cigarette smoking.

FIGURE 19.19
Native peoples such as this Peruvian curandera have used natural products as medicines for thousands of years. Pharmaceutical companies need to compensate indigenous people if they exploit their knowledge.

Fibers in the Future

As nonrenewable resources such as petroleum become scarcer and prices climb, there will be an increasing emphasis on plant-derived fibers and semisynthetic fibers made from plants. Cotton has become an expensive crop to produce because of the need for large applications of pesticides. During the 1980s in Texas, for example, pesticides were routinely applied as many as 50 times per season. With insecticidal compounds such as the digestive inhibitors from *Bacillus thuringiensis* inserted into cotton plants, applications of pesticides were expected to decrease. The first field tests of Bt-engineered cotton were in 1989, and by 1999 successful pest-resistant plants were being grown in Texas. However, in 1996, the Texas cotton crop failed because of large populations of boll worms that resulted from the higher than average temperatures and increases in the acreage planted to corn, another host of the boll worm. A drought that year seemed to weaken the engineering plants, making them susceptible to insect attack. Despite this failure, 55 percent of the cotton planted during 1999 in the United States was engineering for insect resistance. Although concerns such as those discussed in the section on genetic engineering remain because of the use of cottonseed for edible oil, farmers readily plant GM (genetically modified) cotton because it requires substantially fewer pesticides than regular cotton.

Frighteningly, tests in 1998 showed that 300 anthropod species were already resistant to Bt, and farmers may be able to stay ahead of the pests by using a combination of slight variations in the molecular structure of the Bt toxin and by using a planting scheme that intersperses nonresistant cotton with resistant cotton. This kind of planting scheme, which lowers the selection for resistance, was urged in 1999 for all Bt-protected crops. Some scientists even suggested that 50 percent of the total acreage be used as "refuges" (areas planted to nonengineered corn) with farmers being allowed to spray with pesticides if needed.

In Chapter 15 we mentioned the appearance of Tencel™, a new fiber made from cellulose. This fiber might replace cotton in some garments, thereby reducing the acreage devoted to this very insect-prone crop. Other sources of cellulose, such as blue-green algae, can be used for fiber, and substitution of naturally colored cotton could

TABLE 19.5 Steps in the Discovery of Plant-Derived Drugs from Tropical Regions

1. Determining indigenous uses by interviewing traditional healers. Specimens are collected, and their use is confirmed by other healers.
2. Identifying the plant species precisely. This identification must be done by a skilled taxonomist, since many species look alike but contain different concentrations of compounds.
3. Screening material of the plant species for bioactive compounds. Ideally, the field researcher dries or prepares the material in the same way as the traditional healer before it is sent to a pharmacologist for screening.
4. Isolating and purifying potentially active compounds.
5. Testing the presumed active compounds and identifying the effective compound.
6. Conducting a classical pharmacological evaluation to determine whether the compounds have unique properties.
7. Determining the active compound's structure and potential for synthesis.
8. Applying for patent rights and clinical trials.
9. Signing a contract with the indigenous people for royalties from marketing of drugs that use the active compound.

lessen the need for bleaching and dyeing commonly practiced in regular cotton processing.

Future Trends in Paper, Fuel, and Building Material

Most of the world still depends on wood for heating and cooking, and as population sizes have increased the demand for roundwood for heating, lumber, and pulp rose continually until 1990; production in that year (3447 million m^3, or 123092 ft^3) was more than double the 1950 level. During the 1990s, however, production leveled off owing in large part to the decline of the industry in the economically troubled former Soviet Union. Of the wood that is harvested, about 55 percent is used directly for fuel or firewood and 45 percent for industrial lumber, including pulp wood for paper. The consumption of these forestry resources is extremely disproportionate, however, with the industrial countries using over half of that produced. Per capita use in developed countries is 12 times that in developing countries except for fuelwood. Paper consumption continues to rise, with the United States, China, and Japan the leading producers and consumers. Predictions are that paper consumption will triple between 2000 and 2010, an increase that will necessitate greater and greater forest destruction. Much of the world's roundwood production still comes from the cutting of natural forests (8.9 billion acres). Of the estimated original 6 billion hectares (15 billion acres) of natural forest, only about 3.6 billion hectares remain. Fourteen million hectares (35 million acres) are cut every year, 90 percent of which is in the tropics. Our challenge is to meet the world's need for wood and paper through prudent management of tree farms and to make protection of the earth's remaining forests an economically viable option.

At the rate at which people are currently harvesting timber, there are not enough plantations or selectively cut forests to meet current demand. Japan, for example, is completely dependent on the wood it obtains from forests in Southeast Asia to supply its appetite for wood and wood pulp. In this case, external demand for wood has contributed significantly to deforestation. Chapter 16 briefly mentioned ways to harvest trees selectively from forests without destroying the entire forest ecosystem. By using this type of management, people can ensure a supply of some important hardwoods. Wood for many construction purposes and for pulp for paper can also come from tree "farms." As long as such farms are planted on reclaimed farming land, no new forest is destroyed. Such plantations using fast-growing woods can contribute significantly to the supply of wood and pulp. Another source of pulp is fast-growing herbaceous plants such as jute (see Chapter 15), kenaf (*Hibiscus cannabinus*), and bamboo (see Chapter 16). Kenaf can produce two to four times the yields of pulp per hectare produced from southern pine.

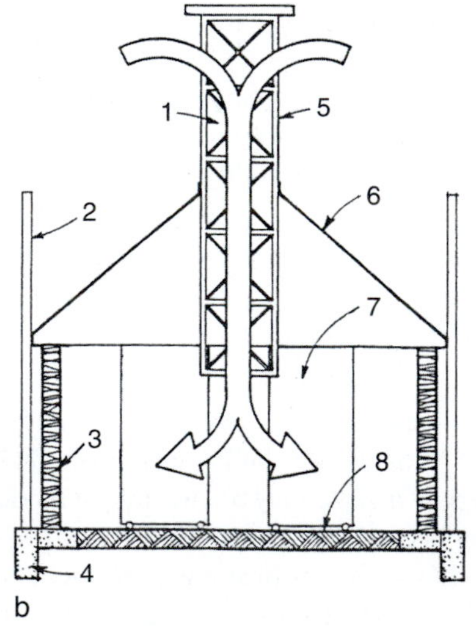

FIGURE 19.20

This alternative building system utilizes low-cost indigenous construction methods and is a model of a sustainable climate-conscious design that utilizes available and renewable resources. (a) The external features of the building. (b) The construction. In this design, the walls (3) provide insulation and are composed of fly ash material that is a waste product of the coal industry. Steel trusses (5) help create cooling towers (1) that circulate air throughout the building and help support the sheet-metal roof (6). Recycled oil-drilling poles (2) are set into a concrete foundation (4), and the floor (8) is made of compacted earth. The only wood used in the building is in the mesquite doors (7).

(Courtesy Pliny Fisk III; design by Max's Pot.)

FIGURE 19.21
In tropical slash-and-burn agriculture, new land is cleared with fire and sometimes is cultivated for only a single year before the fertility is depleted and it is abandoned to weeds and erosion. A newly cleared site in Central Sumatra.
(Photo by H. Null, FAO.)

Another alternative is to reduce the demand for new harvests every year. This reduction can be done in part by recycling wood, paper, and wood products. An enormous amount of landfill is taken up by paper. Discarding paper not only necessitates the use of excessive amounts of land for solid waste disposal; it also requires a continuous supply of pulp. More and more towns, cities, and even countries in the developed world are now recycling paper and paper products. Less commonly practiced but an obvious next step is the recycling of wood and wood products. Alternative building methods (Fig. 19.20) that use low amounts of wood can also reduce forest clearing.

In Brazil the pattern of deforestation is somewhat different. There has been destructive harvesting of hardwoods for agricultural use rather than for wood and wood pulp. In this region, as in Africa and parts of southeast Asia, population pressure has been the major force leading to forest destruction (Fig. 19.21).

For a variety of ecological and historical reasons, the lowland wet tropics contain the highest diversity of life (measured as the number of species per unit area) (Fig. 19.22) of any habitat on the earth. Much of this diversity still remains essentially unknown. Perhaps as many as 500,000 of the estimated 3 million plant and animal species in the tropics have never been adequately described (have no scientific name or proper documentation in a written text), and many of these undescribed species might produce substances of importance to humans. The current rate of destruction of tropical areas will ensure the extinction of most of these species before they can be collected and studied.

FIGURE 19.22
This engraving of a tropical rain forest gives some indication of the denseness and diversity of the natural vegetation in lowland humid forests.
(From Von Marilaun, K. 1896. *The Natural History of Plants*. London, Blackie and Sons.)

How can one convince people who need land for raising food to preserve the land? Is it possible to use rain forest (or native grasslands) while at the same time conserving most of its biodiversity? Several approaches have been taken to meet these challenges. First, botanists have tried to estimate the value of a piece of rain forest and thus demonstrate that it is economically viable to preserve the forest. A study by Peters and colleagues (see Additional Readings) of land near Iquitos, Peru, estimated that the sustainable use of a hectare of land for fruit and rubber latex would ultimately yield $6330. If a low level of selective logging were part of the system, the value would increase to $6820. In contrast, if the hectare were simply completely logged, the total value of the timber and pulp would be $3184. Subsequent use of the land for cattle pasture would add an additional $2956, yielding a total of $6140. Not only is the total net present value more under sustained use; the rain forest is left intact under a sustainable harvesting land use. Nevertheless, such estimates rarely jibe with the lifestyles of slash-and-burn farmers or the needs of large corporations that want to establish huge ranches or plantations. Table 19.4 lists several strategies for reducing the impact of logging in tropical forest areas while managing the forest in a sustainable fashion.

In this chapter we have put forth many suggestions about future plant use. A copy of the collage of headlines from the first edition (1986) (Fig. 19.23); compare with Fig. 19.1) shows that some of the predictions made by the media 16 years ago have materialized, some have come to naught, and the same problems still plague us today. Advances in agricultural technology and in the ability to model global climate and economics have created new opportunities to face the challenges ahead. However, unless humans formulate solutions with a global outlook, the rich will get richer and the poor will get poorer, leading inevitably to decline of the earth's habitability. It will be interesting to see if people of the world can come together and make progress toward a sustainable global ecosystem.

FIGURE 19.23

A composite of headlines from 1982 to 1984 shows that the major issues of boosting agriculture production, dwindling energy sources, mounting population, and genetic engineering have not changed substantially over the last two decades. Compare Figure 19.1.

Additional Readings

Chapter 1

Capon, B. 1990. *Botany for Gardeners: An Introduction and Guide.* Portland, OR, Timber Press.

Cronquist, A. 1988. *The Evolution and Classification of Flowering Plants,* 2d ed. Bronx, New York Botanical Garden.

Hawkes, J. G. 1983. *The Diversity of Crop Plants.* Cambridge, MA, Harvard University Press.

Heiser, C. B., Jr. 1990. *Seed to Civilization. The Story of Food,* new ed. Cambridge, MA, Harvard University Press.

Huxley, A. 1987. *Green Inheritance.* Garden City, NY, Anchor Press.

Janick, J., R. W. Schery, F. W. Woods, and V. W. Rutlan. 1981. *Plant Science: An Introduction to World Crops,* 3d ed. San Francisco, N. H. Freeman.

Johns, T. 1990. *With Bitter Herbs They Shall Eat It: Chemical Ecology and the Origins of Human Diet and Medicine.* Tucson, AZ, University of Arizona Press.

Klein, R. M. 1986. *The Green World: An Introduction to Plants and People,* 2d ed. New York, Harper & Row.

Levetin, E., and K. McMahon. 1999. *Plants and Society.* Dubuque, IA, WCB-McGraw-Hill.

Lewington, A. 1990. *Plants for People.* New York, Oxford University Press.

Mauseth, J. D. 1995. *Botany: An Introduction to Plant Biology.* Philadelphia, Saunders College Publishing.

Qui, Y-L., J. Lee, F. Bernasconi-Quadroni, D. E. Soltis, P. S. Soltis, M. Zanis, E. A. Zimmer, Z. Chen, V. Savolainen, and M. W. Chase. 1999. The earliest angiosperms: Evidence from mitochondrial, plastid, and nuclear genomes. *Nature* 402:404–407.

Raven, P. H., R. F. Evert, and S. E. Eichhorn. 1999. *Biology of Plants,* 6th ed. New York, W.H. Freeman.

Solits, P. S., D. E. Soltis, and M. W. Chase. 1999. Angiosperm phylogeny inferred from multiple genes as a tool for comparative biology. *Nature* 402:402–404.

Starr, C., and R. Taggart. 1998. *Biology: The Unity and Diversity of Life.* New York, Wadsworth.

Chapter 2

Anderson, E. 1971. *Plants, Man, and Life.* Berkeley, CA, University of California Press.

Balter, M. 1998. Why settle down? The mystery of communities. *Science* 282:1442–1445.

Candolle, A. de. 1886. *Origin of Cultivated Plants.* New York, Hafner.

Cohen, M. N. 1977. *The Food Crisis in Prehistory.* New Haven, CT, Yale University Press.

Cowan, C. W., and P. J. Watson. 1992. *The Origins of Agriculture.* Washington, DC, Smithsonian Institution.

Harlan, J. R. 1992. *Crops and Man,* 2d ed. Madison, WI, American Society of Agronomy.

Harris, D. R., and G. C. H. Hillman (eds.). 1989. *Foraging and Farming.* London, Unwin Hyman.

MacNeish, R. S. 1991. *The Origins of Agriculture and Settled Life.* Norman, OK, University of Oklahoma Press.

McCorriston, J., and F. Hole. 1991. The ecology of seasonal stress and the origins of agriculture in the Near East. *American Anthropologist* 94: 905–925.

Piperno, D. R., and D. M. Pearsall. 1998. *The Origins of Agriculture in the Lowland Neotropics.* New York, Academic Press.

Pringle, H. 1998. The slow birth of agriculture. *Science* 282:1446–1450.

Rindos, D. 1984. *The Origin of Agriculture: An Evolutionary Perspective.* Orlando, FL, Academic Press.

Sauer, C. O. 1952. *Agricultural Origins and Dispersals.* New York, American Geographical Society.

Sauer, J. D. 1993. *Historical Geography of Crop Plants: A Select Roster.* Boca Raton, FL, CRC Press.

Smith, B. D. 1997. The initial domestication of *Cucurbita pepo* in the Americas 10,000 years ago. *Science* 276:932–934.

Smith, B. D. 1998. Between foraging and farming. *Science* 279:1651–1652.

Vasey, D. 1992. *An Ecological History of Agriculture 100,000 B.C.–A.D. 10,000.* Ames: Iowa State University Press.

Vavilov, N. I. 1951. *The Origin, Variation, Immunity, and Breeding of Cultivated Plants.* Chronica Botanica 13. Waltham, MA, Ronald Press.

Chapter 3

Hora, B. (ed.). 1986. *The Oxford Encyclopedia of Trees of the World,* 2d ed. London, Oxford University Press.

Janick, J., and J. N. Moore (eds.). 1996. *Fruit Breeding.* Vol. 2: *Vine and Small Fruit Crops.* New York, Wiley & Sons.

Janick, J., and J. N. Moore (eds.). 1996. *Fruit Breeding.* Vol. 3: *Nuts.* New York, Wiley & Sons.

Jennings, D. L. 1995. Raspberries and blackberries. Pages 429–434 in J. Smartt and N. W. Simmonds (eds.), *Evolution of Crop Plants,* 2d ed. London, Longman Scientific and Technical.

Monselise, S. P. 1986. *CRC Handbook of Fruit Set and Development.* Boca Raton, FL, CRC Press.

Price, R. 1954. *Johnny Appleseed: Man and Myth.* Bloomington, IN, Indiana University Press.

Watkins, R. 1995. Apple and pear. Pages 418–422 in J. Smartt and N. W. Simmonds (eds.), *Evolution of Crop Plants,* 2d ed. London, Longman Scientific and Technical.

Watkins, R. 1995. Cherry, plum, peach, apricot and almond. Pages 423–428 in J. Smartt and N. W. Simmonds (eds.), *Evolution of Crop Plants,* 2d ed. London, Longman Scientific and Technical.

Wilhelm, S. 1974. The garden strawberry: A study of its origin. *American Scientist* 62:264–271.

Zohary, D., and M. Hopf. 1993. *Domestication of Plants in the Old World,* 2d ed. Oxford, Clarendon Press.

Chapter 4

Bates, D. M., R. W. Robinson, and C. Jeffrey (eds.). 1990. *Biology and Utilization of the Cucurbitaceae,* Ithaca, NY, Comstock.

Bates, D. M., and R. W. Robinson. 1995. Cucumbers, melons and water-melons. Pages 89–96 in J. Smartt and N. W. Simmonds (eds.), *Evolution of Crop Plants,* 2d ed. London, Longman Scientific and Technical.

Bompard, J. M., and R. J. Schnell. 1997. Taxonomy and systematics. Pages 21–47 in R. E. Litz (ed.), *The Mango: Botany, Production, and Uses.* New York, CAB International.

Heiser, C. B., Jr. 1979. *The Gourd Book.* Norman, University of Oklahoma Press.

Heiser, C. B., Jr. 1995. Peppers. Pages 449–451 in J. Smartt and N. W. Simmonds (eds.), *Evolution of Crop Plants,* 2d ed. London, Longman Scientific and Technical.

Litz, R. E. (ed.). 1997. *The Mango: Botany, Production, and Uses.* New York, CAB International.

Merrick, L. C. 1995. Squashes, pumpkins and gourds. Pages 97–105 in J. Smartt and N. W. Simmonds (eds.), *Evolution of Crop Plants,* 2d ed. London, Longman Scientific and Technical.

Morton, J. F. 1987. *Fruits of Warm Climates.* Miami, J. F. Morton.

Rick, C. M. 1995. Tomato. Pages 452–457 in J. Smartt and N. W. Simmonds (eds.), *Evolution of Crop Plants.* 2d ed. London, Longman Scientific and Technical.

Roose, M. L., R. K. Soost, and J. W. Cameron. 1995. Citrus. Pages 443–448 in J. Smartt and N. W. Simmonds (eds.), *Evolution of Crop Plants,* 2d ed. London, Longman Scientific and Technical.

Smith, B. D. 1997. The initial domestication of *Cucurbita pepo* in the Americas 10,000 years ago. *Science* 276:932–934.

Spiegel-Roy, P., and E. E. Goldschmidt. 1996. *Biology of Citrus.* Cambridge, Cambridge University Press.

Yamaguchi, M. 1983. *World Vegetables: Principles, Production and Nutritive Values.* Westport, CT, AVI Publishing.

Yang, S.-L., and T. W. Walters. 1992. Ethnobotany and the economic role of the Cucurbitaceae of China. *Economic Botany* 46:349–367.

Chapter 5

Beadle, J. 1980. The ancestry of corn. *Scientific American* 242:112–119.

Carson, G. 1976. *Cornflake Crusade.* New York, Arno Press.

Chang, T. T. 1995. Rice. Pages 147–155 in J. Smartt and N. W. Simmonds (eds.), *Evolution of Crop Plants,* 2d ed. London, Longman Scientific and Technical.

Davies, M. S., and G. C. Hillman. 1992. Domestication of cereals. Pages 199–224 in G. P. Chapman (ed.), *Grass Evolution and Domestication.* Cambridge, Cambridge University Press.

de Wet, J. M. J. 1992. The three phases of cereal domestication. Pages 176–198 in G. P. Chapman (ed.), *Grass Evolution and Domestication.* Cambridge, Cambridge University Press.

Doebley, J. 1992. Mapping the genes that made maize. *Trends in Genetics* 8:302–307.

Evans, G., M. 1995. Rye. Pages 166–170 in J. Smartt and N. W. Simmonds (eds.), *Evolution of Crop Plants,* 2d ed. London, Longman Scientific and Technical.

Feldman, M., F. G. H. Lupton, and T. E. Miller. 1995. Wheats. Pages 184–192 in J. Smartt and N. W. Simmonds (eds.), *Evolution of Crop Plants,* 2d ed. London, Longman Scientific and Technical.

Fussell, B. 1992. *The Story of Corn.* New York, Knopf.

Harlan, J. R. 1992. Origins and processes of domestication. Pages 159–175 in G. P. Chapman (ed.), *Grass Evolution and Domestication.* Cambridge, Cambridge University Press.

Harlan, J. R. 1995. Barley. Pages 140–147 in J. Smartt and N. W. Simmonds (eds.), *Evolution of Crop Plants,* 2d ed. London, Longman Scientific and Technical.

Iltis, H. H. 1983. From teosinte to maize: The catastrophic sexual transmutation theory. *Science* 222:886–894.

Iltis, H. H. 2000. Homoeotic sexual translocations and the origin of maize (*Zea mays,* Poaceae): A new look at an old problem. *Economic Botany* 54:7–42.

MacNeish, R. S., and M. W. Eubanks. 2000. Comparative analysis of the Río Balsas and Tehuacán models for the origin of maize. *Latin American Antiquity* 11:3–20.

Mangelsdorf, P. C., and R. G. Reeves. 1938. The origin of maize. *Proceedings of the National Academy of Sciences (U.S.A.)* 24:303–312.

Oka, H. I. 1988. *Origin of Cultivated Rice.* Amsterdam, Elsevier.

Pringle, H. 1998. The slow birth of agriculture. *Science* 282:1446–1450.

Rasmusson, D. C. 1985. *Barley.* Madison, WI, American Society of Agronomy.

Smith, P. M. 1995. Temperate forage grasses. Pages 208–218 in J. Smartt and N. W. Simmonds (eds.), *Evolution of Crop Plants,* 2d ed. London, Longman Scientific and Technical.

Thomas, H. 1995. Oats. Pages 132–136 in J. Smartt and N. W. Simmonds (eds.), *Evolution of Crop Plants,* 2d ed. London, Longman Scientific and Technical.

Zohary, D., and M. Hopf. 1993. *Domestication of Plants in the Old World.* Oxford, Clarendon Press.

Chapter 6

Caradus, J. R., and W. M. Williams. 1995. Other temperate forage legumes. Pages 332–343 in J. Smartt and N. W. Simmonds (eds.), *Evolution of Crop Plants,* 2d ed. London, Longman Scientific and Technical.

Debouck, D. G., and J. Smartt. 1995. Beans. Pages 287–294 in J. Smartt and N. W. Simmonds (eds.), *Evolution of Crop Plants,* 2d ed. London, Longman Scientific and Technical.

Delgado-Salinas, A., T. Turley, A. Richman, and M. Lavin. 1999. Phylogenetic analysis of the cultivated and wild species of *Phaseolus* (Fabaceae). *Systematic Botany* 24:438–460.

Duke, J. A. 1981. *Handbook of Legumes of World Economic Importance.* New York, Plenum.

Gepts, P. 1998. Origin and evolution of the common bean: Past events and recent trends. *HortScience* 33:1124–1130.

Hymowitz, T. 1995. Soybean. Pages 261–266 in J. Smartt and N. W. Simmonds (eds.), *Evolution of Crop Plants,* 2d ed. London, Longman Scientific and Technical.

Kaplan, L., and T. F. Lynch. 1999. *Phaseolus* (Fabaceae) in archaeology: AMS radiocarbon dates and their significance for pre-Columbian agriculture. *Economic Botany* 53:261–272.

Ng, N. Q. 1995. Cowpea. Pages 326–332 in J. Smartt and N. W. Simmonds (eds.), *Evolution of Crop Plants,* 2d ed. London, Longman Scientific and Technical.

Nwokolo, E. 1996a. The need to increase consumption of pulses in the developing world. Pages 3–11 in E. Nwokolo and J.

Smartt (eds.), *Food and Feed from Legumes and Oilseeds.* New York, Chapman & Hall.

Nwokolo, E. 1996b. Pigeon pea (*Cajanus cajan* (L.) Millsp.). Pages 64–73 in E. Nwokolo and J. Smartt (eds.), *Food and Feed from Legumes and Oilseeds.* New York, Chapman & Hall.

Purseglove, J. W. 1974. Leguminosae. Pages 199–332 in J. W. Purseglove (ed.), *Tropical Crops: Dicotyledons.* New York, Wiley & Sons.

Singh, A. 1995. Groundnut. Pages 246–250 in J. Smartt and N. W. Simmonds (eds.), *Evolution of Crop Plants,* 2d ed. London, Longman Scientific and Technical.

Zohary, D. 1995. Lentil. Pages 271–274 in J. Smartt and N. W. Simmonds (eds.), *Evolution of Crop Plants,* 2d ed. London, Longman Scientific and Technical.

Zohary, D., and M. Hopf. 1993. *Domestication of Plants in the Old World.* Oxford, Clarendon Press.

Chapter 7

Ayensu, E. S., and D. G. Coursey. 1972. Guinea yams: The botany, ethnobotany, use and possible future of yams in West Africa. *Economic Botany* 26:301–318.

Cobley, L. S. 1976. *An Introduction to the Botany of Tropical Crops,* 2d ed. Revised by W. M. Steele. New York, Longman.

Correll, D. S. 1962. *The Potato and Its Wild Relatives.* Renner, TX, Texas Research Foundation.

Cribb, P. J., and J. G. Hawkes. 1986. Experimental evidence for the origin of *Solanum tuberosum* subspecies *andigena.* Pages 383–404 in W. G. D'Arcy (ed.), *Solanaceae: Biology and Systematics.* New York, Columbia.

Grun, P. 1990. The evolution of cultivated potatoes. *Economic Botany* 44:39–55.

Hawkes. J. G. 1990. *The Potato: Evolution, Biodiversity and Genetic Resources.* Washington, DC, Smithsonian Institution Press.

Heinz, D. J. 1987. *Sugarcane Improvement through Breeding.* Amsterdam, Elsevier.

Hobhouse, H. 1985. *Seeds of Change.* New York, Harper & Row.

Hosaka, K., and R. E. Hanneman, Jr. 1988. The origin of the cultivated tetraploid potato based on chloroplast DNA. *Theoretical and Applied Genetics* 76:172–176.

Lancaster, P. A., J. S. Ingram, M. Y. Lim, and D. G. Coursey. 1982. Traditional cassava-based foods: Survey of processing techniques. *Economic Botany* 36:12–45.

Lathrap, D. W. 1973. The antiquity and importance of long-distance trade relationships in the moist tropics of pre-Columbian South America. *World Archaeology* 5:170–186.

Liebig, Baron J. von. 1863. *Natural Laws of Husbandry.* New York, Appleton.

Nye, M. 1991. The mis-measure of manioc (*Manihot esculenta,* Euphorbiaceae). *Economic Botany* 45:47–57.

Olsen, K. M., and B. A. Schaal. 1999. Evidence on the origin of cassava: Phylogeography of *Manihot esculenta. Proceedings of the National Academy of Sciences (U.S.A.)* 96:5586–5591.

Seigler, D. S., and J. F. Pereira. 1981. Modernized preparation of cassava in the Llanos Orientales of Venezuela. *Economic Botany* 35:356–362.

Smartt, J., and N. W. Simmonds (eds.). 1995. *Evolution of Crop Plants,* 2d ed. London, Longman Scientific and Technical.

Ugent, D. 1970. The potato. *Science* 170:1161–1166.

Vaughan, J. G. 1977. A multidisciplinary study of the taxonomy and origin of *Brassica* crops. *BioScience* 27:35–40.

Wilson, L. A. 1970. The process of tuberization in sweet potato [*Ipomoea batatas* (L.) Lam.]. Pages 2–26 in D. L. Plucknett (ed.), *Proceedings of the Second International Symposium on Tropical Root and Tuber Crops.* Honolulu, University of Hawaii.

Yamaguchi, M. 1983. *World Vegetables.* Westport, CT, AVI Publishing.

Yen, D. E. 1974. The sweet potato and Oceania. *Bulletin of the Bernice P. Bishop Museum* 236:1–389.

Zimmer, K. S. 1991. The regional biogeography of native potato cultivars in highland Peru. *Journal of Biogeography* 18:165–178.

Chapter 8

Ackerman, D. 1990. *A Natural History of the Senses.* New York, Random House.

Andrews, J. 1984. *The Domesticated Capsicums.* Austin, University of Texas Press.

Apt, C. M. (ed.). 1978. *Flavor: Its Chemical, Behavioral, and Commercial Aspects.* Boulder, CO, Westview Press.

Billing, J., and P. W. Sherman. 1998. Antimicrobial functions of spices: Why some like it hot. *Quarterly Review of Biology* 73:3–49.

Caterina, M. J., M. A. Schumacher, M. Tominaga, T. A. Rosen, J. D. Levin, and D. Jullus. 1997. The capsaicin receptor: A heat-activated ion channel in the pain pathway. *Nature* 389:816–824.

Hirasa, K., and M. Takemasa. 1998. *Spice Science and Technology.* New York, Marcel Dekker.

McGee, H. 1998. In victu veritas. *Nature* 392:649–650.

Mori, K., H. Nagao, and Y. Yoshihara. 1999. The olfactory bulb: Coding and processing of odor molecule information. *Science* 286:711–715.

Muller, P. M., and D. Lamparsky (eds.). 1991. *Perfumes: Art, Science, and Technology.* London, Elsevier Applied Science.

Nabhan, G. P., and G. Braasch. 1997. Why chilies are hot. *Natural History* 106:24–29.

Purseglove, J. W., E. G. Brown, C. L. Green, and S. R. J. Robbins (eds.). 1981. *Spices.* Vols. 1 and 2. New York, Longman Scientific and Technical.

Sherman, P. W., and J. Billing. 1999. Darwinian gastronomy: Why we use spices. *BioScience* 49:453–463.

Swahn, J. O. 1991. *The Lore of Spices.* New York, Crescent Books.

Chapter 9

Corner, E. J. H. 1966. *The Natural History of Palms.* Berkeley, University of California Press.

Heiser, C. B., Jr. 1976. *The Sunflower.* Norman, University of Oklahoma Press.

Lehninger, A. L., D. L. Nelson, and M. M. Cox. 1993. *Principles of Biochemistry.* New York, Worth.

Nwokolo, E., and J. Smartt. 1996. *Food and Feed from Legumes and Oilseeds.* New York, Chapman & Hall.

Patterson, H. B. W. 1989. *Handling and Storage of Oils, Fats, and Meal.* New York, Elsevier Applied Science.

Penfield, M. P., and A. M. Campbell. 1990. *Experimental Food Science.* New York, Academic Press.

Röbbelen, G., R. K. Downey, and A. Ashri. 1989. *Oil Crops of the World: Their Breeding and Utilization.* New York, McGraw-Hill.

Salunkhe, D. K., R. N. Chavan, R. N. Adsule, and S. S. Kadam. 1992. *World Oilseeds: Chemistry, Technology, and Utilization.* New York, AVI.

Weiss, E. A. 1971. *Castor, Sesame, and Safflower.* Boston, Barnes and Noble.

Chapter 10

Davidson, R. L. (ed.). 1980. *Handbook of Water-Soluble Gums and Resins.* New York, McGraw-Hill.

Furia, T. E. (ed.). 1972. *Chemical Rubber Company Handbook of Food Additives,* 2d ed. Cleveland, Chemical Rubber Co.

Howes, F. N. 1949. *Vegetable Gums and Resins.* Waltham, MA, Chronica Botanica.

Howes, F. N. 1950. Age-old resins of the Mediterranean region and their uses. *Economic Botany* 4:307–316.

Hymowitz, T. 1972. The trans-domestication concept as applied to guar. *Economic Botany* 26:49–60.

Langenheim, J. 1969. Amber: A botanical inquiry. *Science* 163:1157–1169.

Mantell, C. L. 1947. *The Water Soluble Gums.* New York, Reinhold.

Mantell, C. L. 1950. The natural hard resins: Their botany, sources and utilization. *Economic Botany* 4:203–242.

Morell, V. 1993. Window to the world. *Discover* (August): 45–51.

National Academy of Sciences. 1977. *Guayule: An Alternative Source of Natural Rubber.* Washington, DC, National Academy of Science.

Poinar, G. 1999. *The Amber Forest.* Princeton, NJ, Princeton University Press.

Schultes, R. E. 1976. The taming of wild rubber. *Horticulture* 54(11):10–21.

Schultes, R. E. 1977. The odyssey of the cultivated rubber tree. *Endeavour* 1 (new series):133–138.

Sivak, M. N., and J. Press (eds.). 1998. *Starch: Basic Science to Biotechnology.* New York, Academic Press.

Webster, C. C., and W. J. Baulkwill. 1989. *Rubber.* New York, Longman Scientific.

Whistler, R. L., and T. Hymowitz. 1979. *Guar: Agronomy, Production, Industrial Use and Nutrition.* West Lafayette, IN, Purdue University Press.

Chapter 11

Balandrin, M. F., J. A. Klocke, E. S. Wurtele, and W. H. Bollinger. 1985. Natural plant chemicals: Sources of industrial and medicinal materials. *Science* 228:1154–1160.

Blumenthal, M. 1997. *Popular Herbs in the U.S. Market.* Austin, TX, American Botanical Council.

Blumenthal, M. 1998. *The Complete German Commission E Monographs: Therapeutic Guide to Herbal Medicines.* Boston, Integrative Medicine Communications.

Braun, S. 1996. *Buzz: The Science and Lore of Alcohol and Caffeine.* New York, Penguin Putnam.

Brevoort, P. 1998. The booming U.S. botanical market: A new overview. *Herbalgram* 44:33.

Cowan, R. 1990. Medicine on the wild side. *Science News* 138:280–282.

Duke, J. A., C. R. Gunn, E. E. Leppik, C. F. Reed, M. L. Solt, and E. E. Terrell. 1973. *Annotated Bibliography on Opium and Oriental Poppies and Related Species.* ARS-NE-28. Beltsville, MD, USDA.

Evans, W. C. 1996. *Trease and Evans' Pharmacognosy.* Philadelphia, W.B. Saunders.

Farnsworth, N. R. 1984. How can the well be dry when it is filled with water? *Economic Botany* 38:4–13.

Gilbert, L. E. 1980. Ecological consequences of a coevolved mutualism between butterflies and plants. Pages 210–240 in L. E. Gilbert and P. H. Raven (eds.), *Coevolution of Animals and Plants,* 2d ed. Austin, University of Texas Press.

Harborne, J. B. 1982. *Introduction to Ecological Biochemistry,* 2d ed. New York, Academic Press.

Kingston, D. G. 1994. Taxol: The chemistry and structure-activity relationships of a novel anticancer agent. *TIBTECH* 12:222–227.

Klayman, D. L. 1985. *Quinghoasu* (artemisinin), an antimalarial drug from China. *Science* 228:1049–1055.

Kreig, M. B. 1964. *Green Medicine: The Search for Plants That Heal.* Chicago, Rand McNally.

Lawrence, G. H. M. 1965. Herbals: Their history and significance. Pages 1–35 in *History of Botany.* Pittsburgh, Hunt Botanical Library.

Lewis, W. H., and M. P. F. Elvin-Lewis. 1977. *Medical Botany.* New York, Wiley.

Mann, J. 1992. *Murder, Magic, and Medicine.* Oxford, Oxford University Press.

Morton, J. F. 1977. *Major Medicinal Plants.* Springfield, IL, Charles C. Thomas.

Raffauf, R. F. 1996. *Plant Alkaloids: A Guide to Their Discovery and Distribution.* New York, Food Products Press.

Readers' Digest. 1986. *Magic and Medicine of Plants.* Pleasantville, NY, Readers' Digest Association.

Reports, C. 1999. Herbal Rx: The promises and pitfalls. *Consumer Reports* (March):44–48.

Rock, J. F. 1922. Hunting the chaulmoogra tree. *National Geographic* 41:242–276.

Singh, Y. N., and M. Blumenthal. 1997. Kava: An overview. *Herbalgram* 39:33–55.

Stern, W. L. 1974. The bond between botany and medicine. *The Bulletin (Pacific Tropical Botanical Garden)* 4:41–60.

Stone, T., and G. Darlington. 2000. *Pills, Potions and Poisons.* New York, Oxford University Press.

Swain, T. (ed.). 1972. *Plants in the Development of Modern Medicine.* Cambridge, MA, Harvard University Press.

Taylor, N. 1965. *Plant Drugs That Changed the World.* New York, Dodd, Mead.

Weissmann, G. 1991. Aspirin. *Scientific American* 266:84–90.

Withering, W. 1785. *An Account of the Foxglove and Some of its Medical Uses, with Practical Remarks on Dropsy and Other Diseases.* London, C. G. J. and J. Robinson.

Woodham, A., and D. Peters. 1997. *Encyclopedia of Healing Therapies.* New York, Dorling.

Wrangham, R. W., and J. Goodall. 1989. Chimpanzee use of medicinal leaves. Pages 22–37 in P. G. Heltne and L. A. Marquardt (eds.), *Understanding Chimpanzees.* Cambridge, MA, Harvard University Press.

Chapter 12

Anderson, E. F. 1980. *Peyote, the Divine Cactus.* Tucson, University of Arizona Press.

Austin, G. A. 1979. *Perspectives on the History of Psychoactive Substance Use.* National Institute on Drug Abuse. Washington, DC, U.S. Government Printing Office.

Barth, I. 1997. *The Smoking Life.* Columbus, MS, Genesis Press.

Blackwell, W. 1990. *Poisonous and Medicinal Plants.* Englewood Cliffs, NJ, Prentice Hall.

Booth, M. 1996. *Opium: A History.* New York, St. Martin's Press.

Castaneda, C. 1973. *The Teachings of Don Juan.* New York, Ballantine.

Evans, W. C. 1996. *Trease and Evans' Pharmacognosy.* Philadelphia, W.B. Saunders.

Furst, P. T. (ed.). 1972. *Flesh of the Gods.* New York, Praeger.

Heiser, C. B. 1969. *Nightshades, the Paradoxical Plants.* San Francisco, W.H. Freeman.

Iversen, L. L. 2000. *The Science of Marijuana.* Oxford, Oxford University Press.

Kingsbury, J. M. 1972. *Deadly Harvest: A Guide to Common Poisonous Plants.* New York, Holt, Rinehart and Winston.

Lingeman, R. R. 1974. *Drugs from A to Z: A Dictionary,* 2d ed. New York, McGraw-Hill.

Merlin, M. 1992. *Kava: The Pacific Drug.* New Haven, CT, Yale University Press.

Phillips, J. E., and R. D. Wynne. 1980. *Cocaine: The Mystique and the Reality.* New York, Avon Books.

Schivelbusch, W. 1992. *Tastes of Paradise.* New York, Pantheon.
Schleiffer, H. (ed.). 1973. *Sacred Narcotic Plants of the New World Indians.* New York, Hafner.
Schleiffer, H. (ed.). 1979. *Narcotic Plants of the Old World.* Monticello, NY, Lubrecht and Cramer.
Schmutterer, H. E. 1995. *The Neem Tree* Azadirachta indica *A. Juss. and Other Meliaceous Plants.* New York, Weinheim.
Schultes, R. E. 1969. Hallucinogens of plant origins. *Science* 163:245–254.
Schultes, R. E., and A. Hofmann. 1973. *The Botany and Chemistry of Hallucinogens.* Springfield, IL, Charles C. Thomas.
Schultes, R. E., and A. Hofmann. 1992. *Plants of the Gods: Their Sacred, Healing and Hallucinogenic Powers.* Rochester, VT, Healing Arts Press.
Schultes, R. E., and R. F. Raffauf. 1990. *The Healing Forest.* Portland, OR, Dioscorides Press.
Schultes, R. E., and R. F. Raffauf. 1992. *Vine of the Soul: Medicine Men, Their Plants and Rituals in the Colombian Amazonia.* Oracle, Synergistic Press.
Stone, T., and G. Darlington. 2000. *Pills, Potions and Poisons.* New York, Oxford University Press.
Taylor, N. 1966. *Narcotics: Nature's Dangerous Gifts.* New York, Dell.

Chapter 13

Braun, S. 1996. *Buzz: The Science and Lore of Alcohol and Caffeine.* New York, Penguin Putnam.
Coe, S. D., and M. D. Coe. 1996. *The True History of Chocolate.* New York, Thames and Hudson.
Eden, T. 1976. *Tea,* 3d ed. London, Longman.
Erickson, H. T., M. P. F. Correa, and J. R. Escobar. 1984. Guarana (*Paullinia cupana*) as a commercial crop in Brazilian Amazonia. *Economic Botany* 38:27–286.
Gutman, R. L., and B.-H. Ryu. 1995. Rediscovering tea. *Herbalgram* 37:33–48.
Harler, C. R. 1964. *The Culture and Marketing of Tea,* 3d ed. New York, Oxford University Press.
Johnson, M. D. 2000. Effects of shade-tree species and crop structure on the winter arthropod and bird communities in a Jamaican shade coffee plantation. *Biotropica* 32:133–145.
Kolpas, N. 1977. *The Chocolate Lovers' Companion.* New York, Quick Fox.
Pendergrast, M. 1999. *Uncommon Grounds: The History of Coffee and How It Transformed Our World.* New York, Basic Books.
Perfecto, I., R. A. Rice, R. Greenberg, and M. E. Van der Voort. 1996. Shade coffee: A disappearing refuge for biodiversity. *BioScience* 46:598–608.
Porter, R. H. 1950. Mate: South American or Paraguay tea. *Economic Botany* 4:37–51.
Purseglove, J. W. 1974. Pages 599–612 (tea) and 571–598 (cacao) in *Tropical Crops: Dicotyledons.* New York, Wiley.
Schapira, J., D. Schapira, and K. Schapira. 1975. *The Book of Coffee and Tea.* New York, St. Martin's Press.
Schivelbusch, W. 1992. *Tastes of Paradise.* New York, Pantheon.
Simmonds, J. (ed.). 1976. *Cocoa Production.* New York, Praeger.
Sivetz, M., and N. W. Desrosier. 1979. *Coffee Technology.* Westport, CT, AVI Publishing.
Ukers, W. H. 1935. *All about Tea,* 2 vols. New York, The Tea and Coffee Trade Journal Co.
Ukers, W. H. 1935. *All about Coffee,* 2d ed. New York, The Tea and Coffee Trade Journal Co.
Willson, K. C., and M. N. Clifford. (eds.). 1992. *Tea: Cultivation to Consumption.* London, Chapman and Hall.
Wood, G. A. R., and R. A. Lass. 1985. *Cacao,* 4th ed. New York, Longman Scientific and Technical.
Wrigley, G. 1988. *Coffee.* New York, Longman Scientific and Technical.
Young, A. M. 1994. *The Chocolate Tree: A Natural History of Cacao.* Washington, DC, Smithsonian Institution Press.

Chapter 14

Bahre, C. J., and D. E. Bradbury. 1980. Manufacture of mescal in Sonora, Mexico. *Economic Botany* 34:391–400.
Braidwood, R. J., J. D. Sauer, H. Helbaek, P. C. Manglesdorf, H. C. Cutler, C. Coon, R. Linton, J. Stewart, and J. Oppenheim. 1953. Did man once live by bread alone? *American Anthropologist* 55:515–526.
Braun, S. 1996. *Buzz: The Science and Lore of Alcohol and Caffeine.* New York, Penguin Putnam.
Corran, H. S. 1975. *A History of Brewing.* Newton, MA, David and Charles Abbot.
Dudley, R. 2000. Evolutionary origins of human alcoholism in primate frugivory. *Quarterly Review of Biology* 75:3–15.
Einset, J., and C. Pratt. 1975. Grapes. Pages 130–153 in J. Janick and J. N. Moore (eds.), *Advances in Fruit Breeding.* West Lafayette, IN, Purdue University Press.
Gibbons, B. 1992. Alcohol, the legal drug. *National Geographic* 181(2):2–35.
Johnson, H. 1978. *The World Atlas of Wine: A Complete Guide to Wines and Spirits of the World.* New York, Simon and Schuster.
Johnson, H. 1989. *The Story of Wine.* New York, Simon and Schuster.
Lichine, A. 1982. *Alexis Lichine's New Encyclopedia of Wines and Spirits,* 3d ed. New York, Knopf.
Motluk, A. 1999. Jane behaving badly. *New Scientist* (27 November):28–33.
Neve, R. A. Hops. Pages 33–35 in J. Smartt and N. W. Simmonds (eds.), *Evolution of Crop Plants,* 2d ed. New York, Longman Scientific and Technical.
Olmo, H. P. 1995. Grapes. Pages 485–490 in J. Smartt and N. W. Simmonds (eds.), *Evolution of Crop Plants,* 2d ed. New York, Longman Scientific and Technical.
Reports, C. 1999. Alcohol: The whole truth. *Consumer Reports* (December):60–61.
Rose, A. H. (ed.). 1977. *Alcoholic Beverages.* Economic Microbiology I. New York, Academic Press. [Chapters of special interest: R. W. Goswell, and R. E. Kunkee, Fortified wines, pp. 477–535; K. Kodama, and K. Yoshizawa, Sake, pp. 423–475; A. Jarczyk, and W. Wzorek, Fruit and honey wines, pp. 387–421; A.H. Rose, History and scientific basis of alcoholic beverage production, pp. 1–41; A. C. Simpson, Gin and vodka, pp. 537–593.]
Schivelbusch, W. 1992. *Tastes of Paradise.* New York, Pantheon.
Steinmetz, G. 1992. Fetal alcohol syndrome. *National Geographic* 181(2):36–39.
Stone, T., and G. Darlington. 2000. *Pills, Potions and Poisons.* New York, Oxford University Press.
Tomlan, M. A. 1992. *Tinged with Gold: Hop Culture in the United States.* Athens, University of Georgia Press.
Webb, A. D. 1984. The science of making wine. *American Scientist* 72:360–367.
Wile, J. 1978. *Frank Schoonmaker's Encyclopedia of Wine.* New York, Hastings House.

Chapter 15

American Fabrics and Fashion Magazine (eds.). 1980. *Encyclopedia of Textiles,* 3d ed. Englewood Cliffs, NJ, Prentice Hall.
Barber, E. J. W. 1991. *Prehistoric Textiles: The Development of Cloth in the Neolithic and Bronze Ages.* Princeton, NJ, Princeton University Press.

Brooklyn Botanic Garden. 1973. *Natural Plant Dyeing: A Handbook. Plants and Gardens* 29:1–64.

Cook, J. G. 1968. *Handbook of Textile Fibres,* 4th ed. *Vol. I: Natural Fibres.* Watford, England, Merrow Publishing.

Dempsey, J. M. 1975. *Fiber Crops.* Gainesville, University of Florida Press.

Haslam, E. 1979. Vegetable tannins. Pages 475–523 in T. Swain, J. B. Harborne, and C. F. van Sumere (eds.), *Recent Advances in Phytochemistry*, vol. 12. New York, Plenum.

Monro, J. M. 1987. *Cotton,* 2d ed. New York, Longman Scientific.

Phillips, L. L. 1963. The cytogenetics of *Gossypium* and the origin of New World cottons. *Evolution* 17:460–469.

Saltzman, M. 1992. Identifying dyes in textiles. *American Scientist* 80:474–481.

Schetky, E. M., C. H. Woodward, and E. Scholtz. (eds.). 1980. *Dye Plants and Dyeing—A Handbook.* Brooklyn, NY, Brooklyn Botanic Garden.

Spencer, J. E. 1953. The abaca plant and its fiber, Manila hemp. *Economic Botany* 7:195–213.

Vreeland, J. 1977. Ancient Andean textiles. Clothes for the dead. *Archaeology* 30:167–178.

Wendel, J. F., and V. A. Albert. 1992. Phylogenetics of the cotton genus (*Gossypium*): Character-state weighted parsimony analysis of chloroplast-DNA restriction site data and its systematic and biogeographic implications. *Systematic Botany* 17:115–143.

Wendel, J. F., C. L. Brubaker, and A. E. Percival. 1992. Genetic diversity in *Gossypium hirsutum* and the origin of upland cotton. *American Journal of Botany* 79:1291–1310.

Chapter 16

Abramovitz, J. N., and A. T. Matton. 1999. Protecting the forest products economy. Pages 60–77 in L. R. Brown (ed.), *State of the World 1999.* New York, W. W. Norton.

Britt, K. W. (ed.). 1970. *Handbook of Pulp and Paper Technology* 2d rev. ed. New York, Van Nostrand Reinhold.

Clark, T. F. 1965. Plant fibers in the paper industry. *Economic Botany* 19:394–405.

Constantine, A., Jr. 1975. *Know Your Woods.* Revised by H. J. Hobbs. New York, Scribner's.

Crane, J. C. 1947. Kenaf: Fiber plant rival of jute. *Economic Botany* 1:334–350.

Department of Energy. 1979. *Project Retrotech: Home Weatherization Manual.* Department of Energy Conservation Paper No. 28C (DOE/CS-0106). Washington, DC.

Farrelly, D. 1984. *The Book of Bamboo.* San Francisco, CA, Sierra Club.

Forest Products Laboratory. 1989. *Handbook of Wood and Wood-Based Materials for Engineers.* New York, Hemisphere Publishing.

Goombridge, B. 1992. (ed.). *Global Biodiversity: Status of the Earth's Living Resources.* London, Chapman and Hall.

Harrar, E. S. 1947. Veneers and plywood: Their manufacture and use. *Economic Botany* 1:290–305.

Hunter, D. 1947. *Paper Making: The History and Technique of an Ancient Craft,* 2d ed. New York, Alfred Knopf.

Jane, F. W. 1970. *The Structure of Wood,* 2d ed. Revised by K. Wilson and D. J. B. White. London, Adam and Charles Black.

Marden, L. 1980. Bamboo, the giant grass. *National Geographic* 158(4):502–529.

McClure, F. A. 1966. *The Bamboos: A Forest Perspective.* Cambridge, MA, Harvard University Press.

Parker, H. 1994. *Simplified Design of Structural Wood,* 5th ed. New York, Wiley.

Wilson, E. O. 1992. *The Diversity of Life.* Cambridge, MA, Belnap Press.

Zuckerman, S. 1991. *Saving Our Ancient Forests.* Los Angeles, CA, The Wilderness Society, Living Planet Press.

Chapter 17

Bailey, L. H., and E. Z. Bailey. 1976. *Hortus Third, Revised and Expanded.* New York, Macmillan.

Bush-Brown, J., and L. Bush-Brown. 1980. *America's Garden Book.* Revised by the New York Botanical Garden. New York, Scribner's Sons.

Creasy, R. 1982. *The Complete Book of Edible Landscaping.* San Francisco, Sierra Club Books.

Ellefson, C. L., T. L. Stephens, and D. Welsh. 1992. *Xeriscape Gardening: Water Conservation for the American Landscape.* New York, Macmillan.

Foster, K. P. 1999. The earliest zoos and gardens. *Scientific American* (July):64–71.

Foster R. S. 1978. *Homeowner's Guide to Landscaping That Saves Energy Dollars.* New York, David McKay.

Halfacre, R. C., and J. Barden. 1979. *Horticulture.* New York, McGraw-Hill.

Heriteau, J. 1990. *National Arboretum Book of Outstanding Garden Plants.* New York, Simon and Schuster.

Hobhouse, P. 1985. *Color in Your Garden.* Boston, Little Brown and Company.

Janick, J. 1979. *Horticultural Science,* 3d ed. San Francisco, CA, Freeman.

Johnson, H. 1973. *The International Book of Trees.* New York, Simon and Schuster.

Thomas, G. S. 1990. *Perennial Garden Plants.* Portland, OR, Timber Press.

Thomas, G. S. 1992. *Ornamental Shrubs, Climbers, and Bamboos.* Portland, OR, Timber Press.

Waddington, D. V., R. N. Carrow, and R. C. Shearman (eds.). 1992. *Turfgrass.* Madison, WI, American Society of Agronomy.

Wyman, D. 1990. *Shrubs and Vines for American Gardens,* enlarged and revised ed. New York, Macmillan.

Wyman, D. 1990. *Trees for American Gardens.* Revised and enlarged ed. New York, Macmillan.

Chapter 18

Arasaki, S., and T. Arasaki. 1983. *Vegetables from the Sea.* Tokyo, Japan Publications.

Ayehunie, S., A. Belay, Y. Hu, T. Baba, R. Ruprecht. 1996. Inhibition of HIV-1 replication by an aqueous extract of *Spirulina platensis (Arthrospora platensis).* 7th IAAA Conference, Knysna, South Africa, April 17.

Blakeslee, D. 1998. Blocking HIV transmission the natural way: 2 new ideas. *Journal of the American Medical Association HIV/AIDS Report*, available at http://amaassn.org/special/hiv/newsline/special/jamadb/cvn/htm.

Bold, H., and M. Wynne. 1978. *Introduction to the Algae.* Englewood Cliffs, NJ, Prentice Hall.

Chapman, V. J., and D. J. Chapman. 1980. *Seaweeds and Their Uses.* New York, Chapman and Hall.

Drew, K. 1949. Conchocelis-phase in the life history of *Porphyra umbilicalis. Nature* 164:748–749.

Environmental Protection Agency. 1998. *What you should know about* Pfiesteria piscicida. EPA document No. 842-F-98-011, Washington, DC., available at http://www.epa.gov/owow/estuaries/pfiesteria/fact.html

Graham, L. E., and L. W. Wilcox. 2000. *Algae.* New York, Prentice Hall.

Higgins, M. 2000. Toxic algae blamed for sea lion toll. *ENN News,* January 13, available at http://www.enn.com/news/ennstories/2000/01/01132000/sealion_8965.asp

Jackson, D. F. 1964. *Algae and Man.* New York, Plenum Press.

Kozlenko, R., and Henson, R. 1998. Latest scientific research on *Spirulina. Spirulina* Health Library, available at http://www.spirulina.com/SPLNews96.html

Lee, R. E. 1999. *Phycology,* 3d ed. New York, Cambridge University Press.

Lembi, C., and J. Waaland. (eds.). 1988. *Algae and Human Affairs.* Cambridge, Cambridge University Press.

Lobban, C. S., and M. J. Wynne. (eds.). 1981. *The Biology of Seaweeds.* Boston, Blackwell Scientific.

Mayell, H. 2000. Seaweed destroying Florida reefs. *ENN News,* June 13, available at http://www.msnbc.com/news/419949.asp

Meinesz, A. 1999. *Killer Algae: The True Tale of a Biological Invasion.* Translated by D. Simberloff. Chicago, IL, University of Chicago Press.

National Science Foundation/National Oceanographic and Atmospheric Administration. Harmful Algae Page. Distribution of HABs in the US, maintained website at http://redtide.whoi.edu/hab/HABdistribution/HABmap.html

Ohno, M., and A. Critchley. 1993. *Seaweed Cultivation and Marine Ranching.* Yokosuha, Japan, International Cooperation Agency.

Pillay, T. V. R. 1990. *Aquaculture Principles and Practices.* Cambridge, Cambridge University Press.

Sea Grant Media Center. 1999. *The Threat of Domoic Acid-Potent Neurotoxin in Humans.* http://www.seagrantnews.org/news/19990120cpbrief_health/silver.html

South, G. R., and A. Whittick. 1987. *Introduction to Phycology.* Oxford, Blackwell Scientific.

Stein, J. R., and C. A. Borden. 1984. Causative and beneficial algae in human disease conditions: A review. *Phycologia* 23:485–501.

Chapter 19

Abramoviyz, J. N. 1998. Sustaining the world's forests. Pages 21–40 in L. R. Brown (ed.), *State of the World 1998.* New York, W. W. Norton.

Balick, M. J., and P. A. Cox. 1996. *Plants, People, and Culture: The Science of Ethnobotany.* New York, Scientific American Library.

Barton, J. H. 1991. Patenting life. *Scientific American* 264(3):40–46

Brown, L. R. 1998. Struggling to raise cropland productivity. Pages 79–95 in L. Brown (ed.), *State of the World 1998.* New York, W. W. Norton.

Conway, G. 2000. Food for all in the 21st century. *Environment* 42:9–18.

Cotton, C. 1996. *Ethnobotany: Principles and Applications.* New York, Wiley & Sons.

Emden, G. F. v., and D. B. Peakall. 1996. *Beyond Silent Spring.* New York, Chapman and Hall.

Fowler, C., and P. Mooney. 1990. *Shattering: Food, Politics, and the Loss of Genetic Diversity.* Tucson, The University of Arizona Press.

Gendel, S. M., A. D. Kline, D. M. Warren, and F. Yates. (eds.). 1990. *Agricultural Bioethics: Implications of Agricultural Biotechnology.* Ames, IA, Iowa State University Press.

Hails, R. S. 2000. Genetically modified plants—the debate continues. *TREE* 15:14–18.

Langridge, W. H. 2000. Edible vaccines. *Scientific American* 283:66–71

Mann, C. C. 1999. Crop scientists seek a new revolution. *Science* 283:310–314.

Martin, G. J. 1995. *Ethnobotany: A methods manual.* New York, Chapman and Hall.

Nuffield Council. 1999. *Genetically Modified Crops: The Ethical and Social Issues.* London, U.K. Nuffield Council on Bioethics.

Paarlberg, R. 2000. Genetically modified crops in developing countries: Promise or peril? *Environment* 42:19–27.

Peters, C. M., A. H. Gentry, and R. O. Mendelsohn. 1989. Valuation of an Amazonian rainforest. *Nature* 339:655–656.

Pierce, J. T. 1990. *The Food Resource.* New York, Longman Scientific and Technical.

Pimentel, D., M. S. Hunter, J. A. La Gro, R. A. Efroymson, J. C. Landers, F. T. Mervis, C. A. McCarthy, and A. E. Boyd. 1989. Benefits and risks of genetic engineering in agriculture. *BioScience* 39:606–614.

Plotkin, M. J. 1993. *Tales of a Shaman's Apprentice: An Ethnobotanist Searches for New Medicines in the Amazon Rain Forest.* New York, Viking.

Plotkin, M. J., and L. M. Famolare. (eds.). 1992. *Sustainable Harvest and Marketing of Rain Forest Products.* Washington, DC, Island Press.

Plucknett, D. L., N. J. H. Smith, J. T. Williams, and N. Murthis Anishetty. 1987. *Gene Banks and the World's Food.* Princeton, NJ, Princeton University Press.

Tuxill, J. 1999. Appreciating the benefits of plant biodiversity. Pages 96–114 in L. R. Brown (ed.), *State of the World 1999.* New York, W. W. Norton.

Glossary

absolute Volatile oils obtained by washing a concrete (which see) with alcohol and evaporating the alcohol.

acocote A long, narrow gourd hollow in the middle and punctured at both ends so as to serve as a straw to suck out sweet liquid accumulated in the center of an agave plant after the inflorescence has been cut.

adjunct An addition; in brewing, the addition of starches or sugar to malt for subsequent fermentation.

adventitious Produced in an unusual location. Adventitious roots are those arising from locations other than from the primary root and its branches.

agamospermy The production of seeds without fertilization.

aggregate fruit A fruit composed of the fused, ripened ovaries within a single flower, which can include additional structures (e.g., strawberry).

agriculture The science of raising crops and livestock.

ale A kind of beer made using the top-fermenting yeast *Saccharomyces uuvarum.*

aleurone The outermost layer of the endosperm in grasses that secretes substances to promote germination.

alginates Long-chain molecules composed of alginic acid and its salts that occur as components of algal cell walls and the matrices between them.

alkaloids A heterogeneous group of compounds that contain nitrogen and that usually have a basic reaction. Their names generally end in *ine* (e.g., caffeine, nicotine).

allelopathy The inhibition of plant growth by compounds produced by another plant or soil microorganism.

allopolyploid A polyploid (which see) derived from the hybridization of individuals with different chromosome complements.

amber Fossilized plant resin.

AMS Accelerator mass spectrometry; a method used to date samples by taking advantage of the differences in masses of isotopes of an element and hence their acceleration rates.

amylopectin A branched form of starch.

amylose A linear form of starch.

analgesic A substance that relieves pain.

angiosperms Vascular plants characterized by having a closed carpel and double fertilization (which see).

aniline dyes Synthetic dyes derived from coal tar.

anther The portion of a stamen that produces pollen.

antisense technology A technique used in genetic engineering that takes advantage of the fact that DNA is a double helix with complementary strands of nucleotides.

aril An outgrowth, usually fleshy, of the funiculus or hilum (which see) that completely or partially surrounds a seed (e.g., the fleshy covering of a pomegranate seed).

autopolyploid An individual with more than two sets of chromosomes from the same individual or species.

awn An elongate, needle-like projection, such as that extending from the top of the glumes in a grass spikelet.

---

bark A nontechnical term for all of the tissues outside of the vascular cambium in woody plants.

bast fiber A fiber obtained from the phloem tissues of herbaceous dicotyledons.

bastões Rods formed from mashed *Paullinia cupana* seeds mixed with cassava powder.

bedding plants Ornamental plants planted in beds or other defined spaces.

beer An alcoholic beverage made by fermenting a solution containing malt and (usually) hops.

beriberi A deficiency disease caused by the lack of thiamine, or the inability to absorb it, that results in inflammation and degeneration of the heart, intestinal tract, and nervous system.

binomial A two-part name consisting of a generic name and a specific name.

biological oxygen demand (BOD) The amount of oxygen required for oxidative decomposition of materials present in water.

bleaching The removal of coloring agents; in the vegetable oil industry, diatomaceous earth or other biological filters are used. In the textile and paper industries, bleaching agents such as sodium hydroxide are used.

bolt Rapid elongation of the flowering stalk of an herbaceous plant, usually associated with the senescence of other plant parts.

bombilla A straw with an end that serves as a strainer for leaf particles when drinking mate (which see).

bran The pericarp and fused seed coat of a grass fruit (grain).

bulb A short underground stem surrounded by swollen storage leaf bases.

---

callus tissue A group of many undifferentiated plant cells that can be induced to give rise to a new plant.

calyx A collective term for the sepals of a flower.
cambium A meristematic vascular plant tissue that produces secondary phloem and xylem cells.
card To comb fibers parallel to one another.
carpel A single unit of an ovary believed to have resulted from the folding of a leaflike structure to enclose the ovules borne on its surface. A single ovary in a flower can have one or several carpels.
carrying capacity The maximum stable number of individuals an ecosystem can support.
caryopsis A fruit of members of the Poaceae (grass family) that consists of a single seed with the seed coat fused with the layers of the pericarp. Also known as a true grain.
Casparian strip A band or strip containing lignin and suberin on the radial and transverse walls of endodermal cells of roots.
cellulose A polymer consisting of glucose units linked by a beta linkage.
chaff The bracts surrounding the florets and spikelets in cereal grains.
chelate A compound that complexes with metal ions.
chinampa A farming system of the Mayans that used raised beds for intensive agriculture.
chocolate liquor The initial fluid product obtained by crushing and heating cacao nibs (which see).
clear-cutting The removal of virtually all standing trees during logging or land-clearing operations.
climacteric Referring to a fruit that has an abrupt transition from immaturity to ripeness.
clone A group of genetically identical individuals produced by some form of asexual reproduction.
coke Cocaine that has been altered to form a hydrochloride salt.
color The aspect of objects and light sources that may be described in terms of hue (spectral wavelength), lightness (brightness), and saturation.
combine A machine that combines the operations of harvesting and threshing grain.
conchospores Diploid spores produced by the microscopic branched filaments of *Porphyra.* After germination these spores undergo meiosis and grow into a haploid macrophytic plant.
concrete A class of odorants (which see) obtained by dissolving volatile oils and waxes from plant matter using hydrocarbon solvents.
conditioning In the vegetable oil industry, referring to the preparation of seeds for extraction by breaking, rolling, flaking, and subjecting to heat.
copra Dried coconut endosperm.
cord A standard unit of measurement for firewood; a pile of cut wood 8 ft long, 4 ft high, and 4 ft deep.
corm A compressed underground stem enclosed by dry, scale-like leaves.
corolla A collective term for the petals of a flower.
cortex The ground tissue of plants between the vascular tissues and the epidermis.
cotyledon An embryonic leaf of a seed plant.
crack The drug produced by treating cocaine hydrochloride with boiling water and baking soda.
culm A hollow stem of a grass plant.
cultivar A word derived from cultivated variety. It refers to a selected form of a species that has particular characteristics.
cultivation Preparation for the raising of crops; the caring for crops.
cupule A cup-shaped depression; in corn (*Zea mays*), the cup-shaped portion of the axis in which spikelets are borne.
cyanobacteria A group of photosynthetic blue-green algae.

---

decoction A solution prepared by soaking a substance in alcohol in order to extract compounds from it.
decorticate The removal of cortical and outer tissues from fibers using pressure.
degumming The removal of plant gums from vegetable oils by mixing the crude oil with water and separating, by centrifugation, the oil from the water layer containing the gums.
dehiscent Referring to a structure that opens to release its contents upon maturity.
dendrochronology The study of time using data provided by the annual rings of trees.
density Mass (in grams) divided by volume (in cubic centimeters).
deodorize To remove odoriferous compounds. In the vegetable oil industry, deodorizing is accomplished by blowing steam through the oil.
dicotyledons A plant group defined by the presence of two embryonic leaves, or cotyledons, in the seed.
dioecious Meaning "two houses"; plants within a species that are unisexual, with some plants bearing only female flowers and others only male flowers.
distillation The process of heating solutions to a vapor or gaseous stage and then recondensing to a liquid stage. Because compounds differ in their boiling points, the process can be used to obtain pure compounds or to separate compounds in a liquid mixture.
domestication The genetic alteration of plants and animals caused by human actions.
double fertilization A reproductive mode unique to angiosperms that involves the fusion of one sperm with the egg cell in the embryo sac and the fusion of a second sperm with one or more polar nuclei. The first fusion forms the zygote; the second, the endosperm.
drupe A fleshy fruit resulting from the maturation of a single carpel in which the innermost layer of the pericarp (endocarp) is hard.
dutch The term used for the addition of alkali to chocolate to neutralize bitter acids.
dye A coloring agent.

---

egg The female gamete.
endocarp The innermost layer of the pericarp, or fruit, wall.
endodermis The innermost layer of the cortex forming a sheath around the vascular tissues; in roots, these cells have a suberized band called a Casparian strip on the radial and transverse walls.
endosperm A nutritive tissue of angiosperms formed by the fusion of a male gamete and the polar nucleus or nuclei in the embryo sac. The tissue is usually polyploid (which see).
enfleurage A process of extracting volatile oils from plant matter by placing pieces of plant tissue on animal lard.
ensile To prepare and store silage.
essential For oils, the term refers to those compounds that produce an essence, or perfume.
ethanol An alcohol with the formula CH_3CH_2OH.

eukaryote An organism that possesses a true nucleus bound by a double membrane.
eutrophication Nutrient enrichment, usually of waterways, that often leads to unnaturally rapid increases in algal population sizes followed by their decline and that promotes an increase in oxygen-using bacterial populations. The result is a depletion of oxygen in the water and death of plants and animals.
exocarp The outermost layer of the pericarp or fruit wall.
expeller A screw press used to press oils from seeds or fruit pulps.

---

farming The practice of agriculture.
FAS Fetal alcohol syndrome; a suite of characters including facial changes, low birthweight, and small brain size associated with heavy drinking by a mother during pregnancy.
fatty acid A water-insoluble hydrocarbon consisting of a straight or branched carbon chain with a methyl group (CH_3) on one end and a carboxyl (COOH) group on the other.
favism A genetically based disease stemming from the lack of glucose-6-phosphate dehydrogenase that produces hemolytic anemia. This condition is aggravated by eating broad beans (*Vicia faba*).
figure The appearance of a wood's surface due to a combination of grain, porosity, the number of rays, the number of annual rings, and the way in which the wood was cut.
filament The stalk that bears the anther in a flower.
fixative In perfumery, a substance that retards the diffusion of perfume oils, thereby "fixing" the scent.
flagellates Single-celled plants and animals that use flagella for movement.
flagellum (pl. flagella) A whiplike organelle used for movement.
floret A small flower; one of the flowers in a head of the sunflower family or in a spikelet of the grass family.
Floridean starch A polysaccharide stored by red algae.
form A three-dimensional character consisting of shape and structure.
formula name A name of a hybrid plant, derived from a combination of the specific epithets of the two parental species.
frustule The siliceous case of diatoms.
funiculus The stalk connecting the ovule to the ovary wall.

---

gamete A haploid cell formed by meiosis and cytoplasmic division of a germ cell.
gametophyte A haploid structure that produces the gametes, or sex cells. In flowering plants, pollen grains contain the microgametophytes (male), and the megagametophytes (female) are situated in the ovules.
generative cell The cell in the male gametophyte (pollen grain) that divides to form the two sperm nuclei.
germ The female reproductive cell; in grasses, referring to the embryo.
gin To separate the seeds of cotton from their epidermal fibers.
glochid A minute barbed bristle or hair such as those commonly surrounding reduced leaves of cacti.
glucosinolate A sulfur-containing cyclical compound found in members of the Brassicaceae.
glume Bract at the base of a grass spikelet (which see).
gluten A mixture of water-soluble proteins (gliaden and glutenin) found in bread.
glycoside A compound that yields glucose on hydrolysis.
goiter An enlargement of the thyroid gland accompanied by abnormal metabolism resulting from a deficiency of iodine.
graft The fusion of a scion (the aboveground stem or bud of a woody plant) with a rootstock of another individual by aligning the cambial layers of the two components.
grain (1) A fruit of members of the Poaceae (grass family) that consists of a seed with the seed coat fused with the layers of the pericarp. (2) The appearance of a wood due to the alignment of the xylem elements.
grain legume The name given to dry, edible legume seeds.
green beer Newly fermented beer, or beer that has not been aged for 2–3 weeks.
greenhouse gases Gases in the atmosphere that lead to warming.
groat A hulled (pearled) grain that has been broken into pieces.
ground cover A prostrate or low-growing ornamental plant that is used instead of grass to cover the ground.
ground tissue Tissues of a plant other than the vascular tissues and the epidermal tissues.
ground wood process A process that yields wood fibers by mechanically grinding wood chips.
gum The term for a number of different polysaccharides that swell when they are dissolved in water or that absorb water.
gymnosperms Vascular plants with exposed ovules and single fertilization.
gynoecium The designated female part of a flower usually consisting of an ovary containing ovules, a style, and a stigma.

---

haploid A set of chromosomes with only one member of each homologous pair and usually written as *n*.
hard fibers Fibers extracted from the parenchyma of some monocotyledons.
haustorium (pl. haustoria) The specialized root of a parasitic plant that can penetrate the vascular tissues and absorb water, minerals, or nutrients from the host plant.
head The terminal inflorescence of a member of the Asteraceae or sunflower family consisting of tightly packed flowers surrounded by bracts; the form of some crop plants that consists of a greatly reduced stem surrounded by leaves as in iceberg lettuce or cabbage.
heartwood The portion of the secondary xylem of trees that is no longer conducting water and nutrients.
hesperidium Fruit of *Citrus* and its allies derived from a superior ovary and consisting of a rind with a leathery exocarp containing cavities filled with volatile oils, a spongy mesocarp, and an endocarp that develops elongate juice-filled sacs that fill the carpels and surround the seeds.
heterocysts Specialized cells in cyanobacteria where nitrogen fixation takes place.
heterosis Hybrid vigor; superior size or other qualities in plants produced by hybridization.
hilum The scar on a seed that marks the place of attachment of the funiculus (which see).
hormone A signaling compound produced in one part of an organism that has a profound effect on cells and organs in other parts of the body.
horticulture The small-scale growing of crops; the early stages of agriculture; the growing of ornamental crops.
hydrogel A substance that modifies water by slowing the movement of the water molecules and thereby "thickening" it.
hydrogenation The addition of hydrogen to molecules such as unsaturated fatty acids.

hydrolysis The splitting of a chemical bond by the addition of water.
hypnocysts Resting spores of certain algae.
hypocotyl The region where the stem and root merge.

---

imperfect flower A flower that contains either functional male or functional female organs but not both.
inbreeding Reproduction within an individual (selfing) or between genetically very similar individuals.
indehiscent Referring to a structure (fruit, anther, etc.) that does not split open to release its contents at maturity.
inferior ovary An ovary that is located below the place of divergence of the sepals and petals and to which the bases of the sepals and petals are fused.
inflorescence A cluster of flowers borne on a flowering stalk.
infusion A liquid prepared by steeping or soaking a substance in water in order to extract compounds from it.
integument The outer layer or layers of cells that surround the nucellus (which see) in a flower.
intercropping The growing of unrelated species in a mosaic in the same field.
internode The parts of a stem between nodes (i.e., areas from which leaves do not emerge).
intertidal zone The shore region between the limits of high and low tide.
involucre One or more whorls of bracts that occur at the base of a flower or inflorescence.
iodine value A number reflecting the amount of iodine reacting with the double bonds of a fatty acid. The higher the number, the larger the number of double bonds and the higher the unsaturation.

---

lager A kind of beer made using the botton-fermenting yeast, *Saccharomyces cerevisiae*.
land race A local variety of a crop developed by indigenous people.
latex A mixture of organic and inorganic compounds contained in canals or pockets known as laticifers.
laticifer A single cell, clusters of cells, or canals or fused cells that contain latex.
legumes The fruits of members of the Fabaceae (bean or legume family) resulting from the maturation of a single carpel that splits along two opposite sides at maturity; members of the Fabaceae.
lemma The second of the bracts below a grass flower; *see* palea.
ligand A molecule that binds to another.
lignin Complex organic polymers associated with cellulose in plants.
lodge In agriculture, referring to the tendency for tall thin plants such as many cereal crops to fall and mat, leading to rotting of the stems.

---

macerate To soften or separate a material by steeping in a solution; in perfumery, fragmented pieces of plants are steeped in hot oil or fat.
malt Grain that has been steeped and allowed to germinate and then killed; to treat something with, or use, the previously steeped grain in an operation.
marquetry A decorative process in which elaborate patterns are formed by inserting pieces of wood, shell, or ivory into a wood veneer that is applied to a piece of furniture.
mash The mixture of grains, water, and sometimes adjuncts (which see) that are the basis of fermentation for beer and whiskeys subsequently distilled from them.
mate The dried leaves of *Ilex paraguariensis*; the beverage made by steeping the leaves of *Ilex paraguariensis*; the gourd from which mate is drunk.
mead Fermented diluted honey to which various flavoring agents can be added.
megagametophyte The female gamete-bearing part of a flower.
Mendelian trait A character determined solely by the alleles of a single gene that segregates simply during meiosis in the manner described by Gregor Mendel.
mercerize The treatment of cotton fibers or cloth with caustic soda.
meristem A cluster of cells that have the ability to divide and form additional cells. Primary meristems are present in the embryo and continue to divide throughout the life of the plant. Secondary meristems develop from primary meristems, usually from more differentiated cells.
mesocarp The middle layer of the pericarp (fruit wall) of a fleshy fruit.
mesophyll The middle layer of a leaf, generally consisting of cells containing abundant chloroplasts, where photosynthesis occurs.
microgametophyte The male gamete-bearing structures.
micropyle The opening between the integuments (which see) in an ovule.
monocotyledons A plant group defined by the presence of a single embryonic leaf (cotyledon) in the seed.
monoecious Meaning "one house"; plants within a species that have unisexual flowers but bear both male and female flowers on the same plant.
mordant Chemicals that fix a dye to the object to be colored by combining with the dye to form an insoluble compound.
multiple fruit A fruit formed by the fusion of the fruits from separate flowers and their inflorescence axis (e.g., a pineapple).
mutagen Any substance that causes mutations.
mycelium The diffuse mass of long narrow filaments that constitute the vegetative body of a fungus.
mycorrhizae Nonparasitic fungi that form symbiotic associations with the roots of many kinds of plants in which the fungi facilitates the uptake of nutrients and the plant provides food in the form of carbohydrates for the fungi.

Neolithic The period in human cultural development defined by the production of stone tools that are smoothed and shaped and by the adoption of agriculture.
neuron A nerve cell.
neurotransmitters Chemical substances released by nerve cells that affect other nerve cells or an organ.
nib The amorphous embryo of cacao that is pressed to form chocolate.
nitrogen fixation The conversion of inert nitrogen (N_2) into an ionic form such as ammonium, nitrate, or nitrite ions.
node The part of a stem where one or more leaves are (or were) attached.
nodule A spherical mass; used in the case of legumes to refer to the spherical growths on the roots that are inhabited by bacterial rhizoids that fix nitrogen.

nonvolatile oil An oil that has a sufficiently high mass that it does not diffuse into the air.
note In perfumery, a group of compounds that convey a characteristic type of fragrance.
nucellus The megasporangium, or portion of an ovule that gives rise to the female gametophyte.
nursery crops Ornamental plants sold to be grown outside the house, in parks, or in other public places.

---

odorants Basic components of a perfumer's trade that are combined to form a finished perfume.
oolong A semifermented tea produced in eastern China and Taiwan.
outcrossing The fusion of gametes from genetically different individuals.
ovary The basal, often swollen part of a flower that contains the ovules; the base of the gynoecium (which see).
ovule The egg-bearing parts of the ovary and their surrounding structures that will develop into seeds after fertilization.

---

paddy Flooded land on which rice is grown; threshed, unmilled rice.
palea The first of the two bracts beneath a grass flower; *see* lemma.
palisade A layer of the mesophyll that consists of chloroplast-rich elongate cells perpendicular to the epidermis.
parenchyma cell A live, nucleated nonvascular plant cell with various functions.
parterre A method of planting contrasting bedding plants so as to form a design.
parthenocarpic Producing a fruit without fertilization. Parthenocarpic fruits can be seedless or can contain seeds produced parthenogenetically.
parthenocarpy The production of a fruit without fertilization.
pearl To remove, usually by polishing, the outer pericarp or seed covering, or both, especially in grains.
pectin The name given to any of several plant polysaccharide derivatives of polygalacturonic acid that form gels.
pellagra A deficiency disease caused by lack of niacin and characterized by diarrhea, dementia, and dermatitis.
pepo The fruit of members of the Cucurbitaceae (squash family) consisting of a hard rind derived from the basal parts of the calyx and corolla fused to the ovary, with a fleshy layer composed of mesocarp tissue in which the seeds are usually embedded.
perfect flower A flower that contains functional male and female organs.
perianth The collective name for the calyx and corolla of a flower.
pericarp The part of a fruit derived from the ovary wall.
pericycle A layer of tissue between the vascular cylinder and the endodermis. In flowering plants it is present in roots but generally lacking in stems.
periderm The outer, protective layer of woody plants that consists of the phellem, the phellogen, and the phelloderm.
pharmacognosy The part of pharmacology that deals with simple drugs.
pharmacology The science of drugs, including herbs, therapeutics, and toxicology.
phellem Cork; the secondary tissue produced to the outside of the cork cambium.
phelloderm The inner part of the periderm, consisting of the tissue produced on the inner side of the cork cambium.
phellogen Cork cambium; a lateral layer of meristematic cells in the stems and roots of gymnosperms and many dicotyledons that produces cork (phellem) to the outside and phelloderm to the inside.
phloem The principle food-conducting tissue of a vascular plant.
photosynthesis The capture of energy from the sun and the use of the energy to synthesize water and carbon dioxide into sugar that can be converted into cellulose, starch, and other organic substances.
phycobilins Accessory pigments in cyanobacteria and red algae.
phycocolloids Hydrogels derived from algae.
phylogeny The evolutionary pattern of ancestry and descent of organisms.
phytolith A siliceous particle formed in cells of plants that take up silica from the soil. The size and shape of these particles are often distinctive for a species and can even differ between wild and domesticated forms of the same species.
pistil The "female" part of a flower in which eggs are produced and generally consisting of an ovary, a style, and a stigma.
plankton A collective term for organisms that move passively with currents or swim weakly in bodies of water.
Pleistocene The geological epoch from about 2.5 million to 10,000 years before present and roughly equivalent to the Ice Ages.
plywood A kind of "wood" made by gluing layers of veneer to one another, with the grain of each layer perpendicular to the grain of the layer to which it is glued.
polar nucleus A nucleus in an angiosperm embryo sac that will fuse with one of the sperm nuclei to form the endosperm.
pollen The male gametophyte of an angiosperm or a gymnosperm.
polyploid A plant with more than two sets of chromosomes.
pome Fruit of members of the Maloideae (the apple subfamily of the Rosaceae, or rose family) characterized by a fleshy floral cup surrounding the multicarpellate ovary.
porosity The arrangement of large-diameter vessels in the annual growth ring of a woody dicotyledon.
primary compounds Plant compounds that are the primary products of photosynthesis, used in plant metabolism, or part of a plant's hormonal system.
prokaryote An organism that does not possess a true nucleus or organelles enclosed in double membranes.
proof In the alcoholic beverages industry, a measure of the alcohol content equal to double the percent ethanol on a volume basis.
protoplast A plant cell with the cellulose wall removed and thus bounded only by the plasma membrane.
pulse The seed or plant of an edible legume.
pyrenoids Areas in the protoplast that usually function as sites of starch formation.

---

racking The operation of moving fermenting wine across a series of barrels to decant it from the precipitated sediment.
radicle An embryonic root.
ray A horizontal conducting tissue in the stems of woody plants.
refining A process in the vegetable oil industry that neutralizes an oil by using alkali to remove the free fatty acids.

resin In plants, naturally occurring polymerized terpenes that are synthesized in, or secreted into, specialized ducts.
resinoids A class of odorants derived from semisolid plant secretions, primarily resins.
respiration The process of taking in oxygen and breaking down sugars to yield energy and carbon dioxide.
restriction endonucleases Enzymes extracted from bacteria that are able to cut DNA of other organisms at locations with precise sequences of nucleotides.
ret To rot; in textiles, the freeing of fibers by rotting away the softer tissues.
rhizome An underground, often elongate, horizontal stem that superficially resembles a root.
root hair An extension of a root epidermal cell that functions in the absorption of water and solutes.
rootstock The root and basal stem onto which a scion is fused during grafting.
round wood Wood that is produced by simply stripping tree trunks or stems of bark.
rust Fungi that infest plants and produce reddish brown lesions.

---

sanforize To treat cotton products with ammonia in order to prevent shrinking.
saponin A steroid (or triterpenoid alcohol) that contains a glucoside moiety and produces foam when mixed with water.
sapwood The portion of the secondary xylem in trees through which water and nutrients move.
saturated Referring to a fatty acid that has no double bonds between carbon atoms in the chain.
scion The stem or bud that is transferred to a rootstock or stem during grafting.
scurvy A disease marked by spongy gums and loose teeth caused by the lack of vitamin C.
scutch To beat and scrape retted plant matter in order to separate the fibers from the brittle tissues.
scutellum The single modified embryonic leaf (cotyledon) in a monocot seed.
secondary compounds Plant compounds that are not part of a plant's photosynthetic or metabolic system. They are synthesized for color, protection, or attraction.
secondary forest A forest that has grown up and replaced the original forest.
self-compatible The ability of pollen from a plant to germinate on the stigmas of flowers and fertilize ovules on the same plant.
self-fertilization The fusion of male and female gametes produced by the same individual.
self-incompatible The inability of male and female gametes of a given individual to form viable zygotes.
shattering The character of wild grains that leads to the breaking apart of the inflorescence as a means of fruit dispersal.
silage The product of anaerobic fermentation of animal forage or the stalks of grasses such as corn or sorghum.
silicosis A lung disease resulting from inhaling siliceous particles.
sinsemilla Literally, "without seeds"; used to refer to marijuana inflorescences that do not contain seeds.
size or **sizing** Substances spread on cloth or paper to fill in irregularities and provide stiffening or a smooth surface.
soap A metal salt of a fatty acid.
spikelet A small or secondary spike; the basic unit of a grass inflorescence usually consisting of three florets, each subtended by two bracts (palea and lemma) with the entire unit subtended by two bracts called glumes.
spin In textiles, referring to the aligning and twisting together of fibers to make a thread.
spongy parenchyma A layer of the mesophyll that consists of chloroplast-rich, irregularly shaped, loosely packed cells.
spore A haploid regenerative cell produced by meiosis.
sporophyte A diploid organism that produces spores by meiosis of special cells.
stamen The "male" part of a flower that produces pollen and generally includes both an anther and a filament.
staple Fibers on cotton seeds.
starch A polysaccharide formed by repetitive units of glucose linked by a simple, or alpha, linkage.
steroid A tetracyclic triterpenoid with a characteristic structure and methyl groups at positions 10 and 13.
stolon An elongate horizontal stem growing along the surface of the soil and rooting periodically at internodes.
stomate (pl. stomata) An opening in the epidermis of the aerial parts of a plant through which gas exchange occurs; the opening is flanked by two guard cells that regulate the size of the opening.
stone cell A square to round cell with a highly lignified cell wall found in some fruits (pears, quinces) of the Maloideae subfamily of the Rosaceae.
stone fruit Fruits from the Prunoideae (plum subfamily) of the Rosaceae (rose family), characterized by a stony or bony endocarp.
stout An ale made from strongly roasted malt and consequently of much darker color and richer flavor than regular ale.
stromatolite A large, stonelike structure composed of accumulated growth of cyanobacteria over billions of years.
suberin A waxlike compound that is impermeable to water.
sulfate process A chemical method of producing wood pulp that uses sodium sulfate, sodium sulfide, and sodium hydroxide.
sulfite process A chemical method of producing wood pulp that uses bisulfites and acid.
superior ovary An ovary that is located above the region from which the sepals and petals diverge from the floral axis.
sustainable Referring to the use of resources so that none is completely depleted and the ecosystem remains healthy.
swidden An impermanent agricultural plot formed by cutting and burning the vegetation cover.
syconium A hollow, multiple, urn-shaped fruit of the genus *Ficus*, which is formed by the folding of a broad, flattened inflorescence on which numerous flowers are borne.
symbiosis A mutualistic interaction between two organisms that is beneficial to both.
synapse The space across which neurotransmitters travel.

---

tan To convert raw animal hide into leather.
tannin Either of two major classes of complex polycyclic compounds that can turn animal skins into leather.
textile A fiber used to make cloth; a woven cloth.
texture Overall structure, usually of a surface.
tracheids Elongate, thick-walled, nonliving, conducting and supporting cells with tapering ends that are present in secondary xylem.
thresh To separate grains from the enclosing bracts.
tillage The plowing of fields to remove weeds.
tiller A sucker or branch arising from the bottom of a stem.

tinctures Alcoholic solutions of compounds extracted from plants.
tropane alkaloid A two-ringed nitrogen-containing compound derived from proline.
tuxies Strips pulled from the outer petiole edges of abacá and used as a source of fiber.

---

unsaturated Referring to a fatty acid that has one or more double bonds between carbon atoms in the chain.

---

variety A taxonomic category below the species and subspecies level; a form of an agricultural or horticultural crop.
vegetative cell The haploid cell in the pollen grain that is responsible for the growth of the pollen tube down the style.
veneer Thin sheets of wood of uniform thickness.
vermifuge An agent that rids the body of parasitic worms.
vessel Cells of intermediate length with variable arrangements of pits on their walls and plates at their ends that conduct water and nutrients in most angiosperms.
vintage The designation of a wine according to a particular type, region, or year.
volatile Used to refer to oils of low molecular weight that can diffuse into the air.
vulcanization The addition of sulfur and lead oxide to isoprene rubbers in order to cross-link the isoprene chains and stabilize the rubber.

---

wax A plant product formed by the linkage of an alcohol and a fatty acid.
wine Fermented fruit juice.
winnow To separate chaff from the fruits of grains by tossing the grain-chaff mixture in the air and allowing the wind to blow away the light chaff.
winterizing In the vegetable oil industry, a process that removes compounds that precipitate at moderately low temperatures and cause the oil to cloud. The process consists of lowering the temperature of the crude oil until precipitation occurs and then filtering the oil.
wood Secondary xylem; in the vernacular, the large accumulations of secondary xylem that make up the bulk of the trunks and stems of trees.
wort The liquid substrate of yeast fermentation in beer brewing; the liquid portion of mash (which see).

---

xylem The vascular plant tissue that conducts water and solutes.

---

yeast One of the groups of fungi that are unicellular and lack a mycelium.

---

zoospore A swimming spore.
zygote The initial cell formed from the fusion of a male and a female gamete.

Index